Metal Dihydrogen and σ-Bond Complexes

Dear MyCopy Customer,

This Springer book is a monochrome print version of the eBook to which your library gives you access via SpringerLink. It is available to you at a subsidized price since your library subscribes to at least one Springer eBook subject collection.

Please note that MyCopy books are only offered to library patrons with access to at least one Springer eBook subject collection. MyCopy books are strictly for individual use only.

You may cite this book by referencing the bibliographic data and/or the DOI (Digital Object Identifier) found in the front matter. This book is an exact but monochrome copy of the print version of the eBook on SpringerLink.

MODERN INORGANIC CHEMISTRY

Series Editor: John P. Fackler, Jr., *Texas A&M University*

Recent volumes in the series:

CARBON-FUNCTIONAL ORGANOSILICON COMPOUNDS
Edited by Václav Chvalovský and Jon M. Bellama

COMPUTATIONAL METHODS FOR THE DETERMINATION OF FORMATION CONSTANTS
Edited by David J. Leggett

COOPERATIVE PHENOMENA IN JAHN–TELLER CRYSTALS
Michael D. Kaplan and Benjamin G. Vekhter

GAS PHASE INORGANIC CHEMISTRY
Edited by David H. Russell

HOMOGENEOUS CATALYSIS WITH METAL PHOSPHINE COMPLEXES
Edited by Louis H. Pignolet

INORGANOMETALLIC CHEMISTRY
Edited by Thomas P. Fehlner

THE JAHN–TELLER EFFECT AND VIBRONIC INTERACTIONS IN MODERN CHEMISTRY
I. B. Bersuker

METAL COMPLEXES IN AQUEOUS SOLUTIONS
Arthur E. Martell and Robert D. Hancock

METAL DIHYDROGEN AND σ-BOND COMPLEXES
Gregory J. Kubas

MÖSSBAUER SPECTROSCOPY APPLIED TO INORGANIC CHEMISTRY
Volumes 1 and 2 • Edited by Gary J. Long
Volume 3 • Edited by Gary J. Long and Fernande Grandjean

MÖSSBAUER SPECTROSCOPY APPLIED TO MAGNETISM AND MATERIALS SCIENCE
Volumes 1 and 2 • Edited by Gary J. Long and Fernande Grandjean

OPTOELECTRONIC PROPERTIES OF INORGANIC COMPOUNDS
Edited by D. Max Roundhill and John P. Fackler, Jr.

ORGANOMETALLIC CHEMISTRY OF THE TRANSITION ELEMENTS
Florian P. Pruchnik
Translated from Polish by Stan A. Duraj

PHOTOCHEMISTRY AND PHOTOPHYSICS OF METAL COMPLEXES
D. M. Roundhill

A Continuation Order Plan is available for this series. A continuation order will bring delivery of each new volume immediately upon publication. Volumes are billed only upon actual shipment. For further information please contact the publisher.

Metal Dihydrogen and σ-Bond Complexes

Gregory J. Kubas
Los Alamos National Laboratory
Los Alamos, New Mexico

Springer Science+Business Media, LLC

Library of Congress Cataloging-in-Publication Data

Kubas, Gregory J., 1945–
    Metal dihydrogen and [sigma symbol]-bond complexes: structure, theory, and reactivity/Gregory J. Kubas.
    p. cm.
    Includes bibliographical references and index.

    DOI 10.1007/978-0-306-47597-9
    1. Metal complexes. 2. Hydrogen. 3. Chemical bonds. I. Title.

QD474 .K82 2000
541.2′242—dc21

00-059283

©2001 Springer Science+Business Media New York
Originally published by Kluwer Academic/Plenum Publishers, New York in 2001
MyCopy version of the original edition 2001

http://www.wkap.nl/

10 9 8 7 6 5 4 3 2 1

A C.I.P. record for this book is available from the Library of Congress

All rights reserved

No part of this book may be reproduced, stored in a retrieval system, or transmitted in any form or by any means, electronic, mechanical, photocopying, microfilming, recording, or otherwise, without written permission from the Publisher

www.springer.com/mycopy

To my wife Chrystal, who encouraged me to take on this challenge;

To my family—Kelly, Sherry, Joseph, and Esther

To my main mentors and colleagues—John, Du, Bob, and Phil

Foreword

The discovery of the Kubas complex was a defining event in the historical development of coordination chemistry. By 1920, the Lewis ideas on the role of electron pairs in bonding had already associated the coordinate bond with the donation of a lone pair to a metal. For example, donation of the ammonia lone pair to Co^{III} was implicated in the classical Werner cobalt–ammonia complexes. Subsequent developments extended the coordination concept beyond lone-pair donors. Around 1950, a series of discoveries by Wilkinson, Chatt, Fischer, and others showed how the π electrons of unsaturated ligands such as cyclopentadienide ion and ethylene can also bind to metal ions. These π complexes stimulated the modern development of organometallic chemistry and homogeneous catalysis.

The 1984 Kubas report of the molecular hydrogen complex provided the archetypical example of a σ complex — one in which a σ bonding electron pair binds the ligand to a metal. This extended the coordination concept to the third and last category of electron pair. Because molecular hydrogen is unique in containing just one pair of σ-bonding electrons, the nature of the bonding was unambiguously established. Although scattered examples of σ complexes had been found from 1960, the field had remained something of a backwater. With the intense activity in the field after 1984, many new and unexpected phenomena came to light. Quantum exchange coupling, dihydrogen bonding, stretched dihydrogen complexes, and nonclassical fluxionality pathways are just some of these. Although hydrogen is the simplest element, having but one proton and one electron, it is capable of an astonishingly wide range of behavior.

In this book we have Kubas' thoughts, not only on the dihydrogen complex problem with the perspective of the 17 years that have passed since the initial report, but also on the other main areas of σ complexation. For example, σ binding of C–H bonds is a critical first step in alkane activation, the goal of which is the selective functionalization of alkanes. With the expected rise of methane as a starting material for the chemical industry, this area is likely to gain increased prominence in the future. σ binding of Si–H bonds is of importance in a number of industrially important silation reactions using transition metals as catalysts.

The book can be considered the definitive account of 20th century work in the area of σ complexation. I have no doubt, however, that the field will continue to surprise us in the future with new and exciting phenomena. Already there are strong indications that enzymes that use molecular hydrogen, such as the hydrogenases, involve σ complexes as intermediates along the reaction pathway. Similarly, their probable role in heterogeneous catalysis remains to be probed.

Some new discoveries have implications that are worked out within a few years. Others, like that of the Kubas complex, lead to rich veins of intellectual ore that continue to produce for years to come and forge links among disparate areas of science. Those of us who are interested in metal–hydrogen chemistry therefore live in a fortunate time.

R.H. Crabtree

Bethany, Connecticut

Preface

The synthesis and characterization of fascinating new molecules that could not have been imagined to be stable lie at the heart of chemistry. The Nobel Prize-winning discovery of buckminsterfullerene is a perfect example of the charm and beauty of chemistry, which even nonscientists can appreciate. However, there are relatively few such paradigm-shifting discoveries or new bonding concepts that we all remember from our early science courses. A technological example from when I was a freshman at Case Western Reserve University in the early 1960s was the laser, for which the potential uses were initially vastly underestimated. In 1965 the first complex of "inert" dinitrogen was reported by Allen and Senoff, and there is a remarkable parallel with our even more unexpected discovery of dihydrogen complexes nearly 20 years later. I also well remember learning about the "banana-bond" representing the 3c-2e bonding in electron-deficient boron hydrides such as diborane. This multicenter "nonclassical" chemical bonding first described by Longuet-Higgins and later by Lipscomb had a very important role in the development of valence theory and initially had only a few prominent members: boranes and hydride- and alkyl-bridged metal complexes.

Little did I imagine at this time that I and my colleagues at Los Alamos would discover in 1983 the stable coordination of the σ bond in H_2 to a transition metal center in $W(CO)_3(PR_3)_2(H_2)$, thereby adding to nonclassical bonding. This is another marvelous example of serendipity in science but required both keen perception and determination. My scientific background, mentors, and colleagues obviously played major roles, as did the pure joy that chemical synthesis has provided me for nearly 40 years. One of my pastimes as a student was synthesizing Werner-type coordination complexes such as cobalt and nickel amine complexes and marveling at their luminous colors and crystallinity. The eye is the most convenient and cheapest spectrometer, and color changes were key to finding H_2 complexes. Perhaps no other science arouses the senses as much as synthetic transition metal chemistry.

Beginning my undergraduate transition metal research in John Fackler's group at Case, I was inspired by his elegant metal–sulfur-ligand systems as well as by the superb synthetic skills of mentors such as Dimitri Coucouvanis, a graduate student at the time. At Northwestern, my own skills were honed by my thesis preceptor, Duward Shriver, who literally wrote the book on manipulating air-sensitive compounds and emphasized Lewis-acid-base interactions, a crucial feature in $M-H_2$

bonding and in all chemistry. Prophetically perhaps, my thesis work concerned hydride-bridged zinc systems. Coincidentally, one of my fellow graduate students in Shriver's group, Basil Swanson, is a colleague at Los Alamos and one of the coauthors of our seminal 1984 paper on H_2 coordination.

Postdoctoral studies with Tom Spiro at Princeton and Lew Jones at Los Alamos emphasized vibrational spectroscopic studies, which were crucial in the discovery that I seemed guided toward. At Los Alamos, research on $M-SO_2$ complexes ultimately led to the unraveling of metal-H_2 complexes in the early 1980s. My closest colleague, Bob Ryan, directed my attention toward the structure and bonding principles of SO_2 complexes, and his mentoring and collaboration have been invaluable throughout my career. Phil Vergamini was a wonderful coworker in never doubting that I was onto something in my quest to unravel the structure of the H_2 complex, even when most dismissed $M-H_2$ coordination as preposterous. His neutron studies of $W(CO)_3(P^iPr_3)_2(H_2)$ and subsequent X-ray work by a postdoc, Harvey Wasserman, provided key evidence for the side-on bonded structure of molecular H_2 coordination. In the subsequent development of the field, I have been fortunate to have excellent collaborators, particularly Juergen Eckert, Jeff Hay, Cliff Unkefer, and Carol Burns at Los Alamos, and Carl Hoff at Miami, Ken Caulton at Indiana, and Kurt Zilm at Yale.

The far-reaching research field emanating from $M-H_2$ bonding now defines the structure, dynamics, and bonding principles of species referred to as σ complexes, wherein the two electrons in any $X-Y$ bond interact with a vacant metal site. The structural and bonding principles developed add to the general understanding of electron-deficient bonding. Nonclassical carbonium ions (carbocations) were initially studied by Nobel Laureate George Olah at Case Western Reserve University while I was a undergraduate there, and 35 years later their bonding concepts still continue to be developed. Over 800 publications have appeared concerning $M-H_2$ complexes alone and many more regarding $C-H$ and other σ-bond activation, making σ complexes one of the hottest areas in chemistry. Often the crucial step in catalysis or enzymatic processes is activation of a σ bond, and hydrogenations such as hydrodesulfurization of petroleum are among the world's largest man-made reactions. Activation of $C-H$ bonds in, e.g., methane, is one of the "Holy Grails" in chemistry, and M−alkane complexes are now observable by low-temperature NMR. The theoretical aspects of the structure, bonding, and dynamics of σ complexes are heavily stressed throughout this book because of their synergism with experiment. Modern density functional quantum calculations of structures and bond distances now accurately mimic experimental results and offer true predictions confirmed experimentally. What the future holds is anyone's guess, and indeed the first stable Xe complex, $[AuXe_4]^{2+}$, was reported by Seidel and Seppelt in *Science* while this book was being edited!

I would like to acknowledge several of the researchers in this field for looking over drafts of selected chapters of this book and/or providing preprints, illustrations, or other helpful information: Bob Morris, Bob Crabtree, Mike Heinekey, John Bercaw, Juergen Eckert, Bruno Chaudret, Odile Eisenstein, Agusti Lledos, Carl Hoff, John Hartwig, John Peters, Jack Norton, Ken Caulton, Mike Hall, Rick Ernst, Al Sattelberger and many others.

Gregory Kubas

Los Alamos

Contents

1. Introduction	1
1.1. Nonclassical Bonding in σ Complexes	1
1.2. Small-Molecule Activation on Transition Metal Complexes	5
1.3. σ Complexes As "Arrested Intermediates" along the Reaction Coordinate for σ-Bond Cleavage	7
1.4. The Importance of Backdonation in σ-Bond Coordination and Cleavage	10
1.5. Reactivity of σ-Complexes: σ-Bond Metathesis, Acidity, and Heterolytic Cleavage of X–H Bonds	12
References	16
2. Background and Discovery of Dihydrogen Coordination	17
2.1. Homogeneous Activation of Dihydrogen on Metal Complexes	17
2.2. Discovery of Dihydrogen Complexes	20
2.2.1. Background	20
2.2.2. Discovery of Dihydrogen Coordination	24
2.2.3. The Development of Dihydrogen Coordination Chemistry	29
References	30
3. Synthesis and General Properties of Dihydrogen Complexes	33
3.1. Stable Dihydrogen Complexes	33
3.1.1. Introduction	33
3.1.2. General Properties	39
3.1.3. Complexes Synthesized by Addition of H_2 Gas to an Unsaturated Precursor	42

3.1.4. Complexes Prepared from H_2 Gas by Ligand Displacement or Reduction	45
3.1.5. Protonation of a Hydride Complex	46
3.1.6. Other Methods of Preparation	47
3.2. Dihydrogen Complexes Unstable at Room Temperature	48
3.2.1. Organometallic Complexes Observed at Low Temperature in Rare Gas or Other Media	48
3.2.2. Binding of H_2 to Bare Metal Atoms, Ions, and Surfaces	50
References	51

4. Bonding and Activation of Dihydrogen and σ Ligands: Theory versus Experiment ... 59

4.1. Introduction	59
4.2. Development of the Theoretical Aspects of Metal–Dihydrogen Bonding and Calculational Methodologies	61
4.3. Theoretical Studies of Oxidative Addition of Dihydrogen	63
4.4. Palladium (Dihydrogen) — The First Prediction of a Stable Dihydrogen Complex	67
4.5. Theoretical Studies of $M(CO)_3(PR_3)_2(H_2)$ and Bonding Model for Dihydrogen Coordination	69
4.6. Relative Strengths of Dihydrogen as a π Acceptor and as a Donor Ligand Compared to Other Ligands	75
4.7. Dihydrogen versus Dihydride Coordination: Activation of Dihydrogen toward Oxidative Addition	79
4.7.1. The Crucial Influence of the Ligand trans to H_2	81
4.7.2. Other Ligand/Metal/Charge Effects on H_2 Binding and Activation: Why $FeH_2(CO)_4$ Is a Dihydride	84
4.7.3. The Reaction Coordinate for OA: The Point at Which the H–H Bond Is Broken	88
4.7.4. Elongated Dihydrogen Complexes: Extraordinarily Delocalized Dynamic Systems	97
4.7.5. Further Aspects of Electronic and Steric Control of Dihydrogen versus Dihydride Bonding	104
4.8. Polyhydride–Dihydrogen Complexes	107
4.9. Cp and Tp Complexes, Including d^0 Systems	111
4.10. Polyhydride Complexes with CO versus Halide Ligands	114
4.11. Interaction of a Coordinated σ Bond with a cis Ligand: The cis Effect in Hydride–Dihydrogen Complexes	115
4.12. Intramolecular Hydrogen Exchange, Polyhydrogen Complexes, and σ-Bond Metathesis	121
4.13. Dihydrogen and Methane Binding to Naked Metal Ions	128
4.14. Interaction of Dihydrogen with Metal Surfaces, Metal Oxides and Hydrides, and Nontransition Metal Systems	131
References	135

5. Structural and NMR Studies of Dihydrogen Complexes ... 143

5.1. Introduction ... 143
5.2. Diffraction Methods ... 145
5.3. Solid State NMR ... 148
5.4. Solution NMR: J_{HD} Coupling and Isotope Effects ... 150
5.5. NMR Relaxation Time T_1, Deuterium Quadrupole Coupling, and Effects of Dihydrogen Rotation ... 158
5.6. H–H Distance and Other Crystallographic and Structural Aspects of Dihydrogen Coordination ... 164
References ... 167

6. Intramolecular Dynamics of Dihydrogen–Hydride Ligand Systems: Hydrogen Rotation, Exchange, and Quantum-Mechanical Effects ... 171

6.1. Introduction ... 171
6.2. Dihydrogen Rotation in $M(\eta^2\text{-}H_2)$ ... 173
    6.2.1. Determination of the Barrier to Rotation of H_2 ... 174
    6.2.2. Inelastic Neutron Scattering ... 176
    6.2.3. Origin of the Barrier to Rotation of $\eta^2\text{-}H_2$ ... 178
        6.2.3.1. Metal-Hydrogen Binding ... 178
        6.2.3.2. Effect of the Metal Center ... 179
        6.2.3.3. Ligand Effects ... 182
        6.2.3.4. Intramolecular Interactions and Crystal Packing Effects .. 184
    6.2.4. Two-Dimensional Parameterized Model for H_2 Rotation ... 186
6.3. Intramolecular Hydrogen Rearrangement and Exchange ... 187
    6.3.1. Extremely Facile Hydrogen Exchange in $IrXH_2(H_2)(PR_3)_2$ and Related Systems ... 190
    6.3.2. Other Systems that Exchange Hydrogens in H_2 and Hydride Ligands ... 192
    6.3.3. Quasi-Elastic Neutron Scattering Studies of H_2 Exchange with cis Hydrides ... 194
    6.3.4. Quantum-Mechanical Exchange Coupling and Hindered Rotational Phenomena ... 198
References ... 203

7. Thermodynamics, Kinetics, and Isotope Effects of the Binding and Cleavage of σ Ligands versus Classical Ligands ... 207

7.1. Themodynamics of Dihydrogen and σ-Bond Coordination to Metal Centers ... 207
    7.1.1. Binding Energies of H_2 and σ Ligands to Stable Complexes ... 207
    7.1.2. Binding Energies of H_2 and σ Ligands to Thermally Unstable Metal Complexes ... 210

7.2. Binding Strength of Dihydrogen and σ Ligands versus
    Classical Ligands; Importance of Entropy 213
7.3. Kinetics of Formation and Substitution of Dihydrogen, Alkane, and
    Related σ Ligands 222
7.4. Reaction Profiles and Kinetics for Coordination and Oxidative
    Addition of Dihydrogen and Other σ Ligands 226
7.5. Isotope Effects in σ-Ligand Coordination and Oxidative Addition 233
    7.5.1. Equilibrium Isotope Effects for H_2 versus D_2 Binding 234
    7.5.2. Origin of the Inverse EIE for H_2 and Other σ-Ligand Binding
        and EIE for OA 235
    7.5.3. Kinetic Isotope Effects for H_2 and Alkane OA and
        Reductive Elimination 238
References 241

8. Vibrational Studies of Coordinated Dihydrogen 245

8.1. Vibrational Modes for η^2-H_2 245
8.2. Normal Coordinate Analysis of $W(H_2)(CO)_3(PCy_3)_2$ 249
8.3. Bonding Character of Dihydrogen Complexes Suggested by
    Vibrational Analysis 253
8.4. Highly Mixed H–H and M–H_2 Modes in Elongated Dihydrogen
    Complexes: New Normal-Mode Definitions 255
References 257

9. Reactions and Acidity of Dihydrogen Complexes 259

9.1. Introduction 259
9.2. Homolytic Splitting of Coordinated Dihydrogen 261
9.3. Isotopic Exchange and Other Intramolecular Hydrogen
    Exchange Reactions 265
9.4. Heterolytic Cleavage and Acidity of Coordinated Dihydrogen 270
    9.4.1. Introduction 270
    9.4.2. Thermodynamic and Kinetic Acidity of H_2 Ligands 273
    9.4.3. Intermolecular Heterolytic Cleavage of Coordinated H_2 278
    9.4.4. Intramolecular Heterolytic Cleavage 280
        9.4.4.1. Proton Transfer to Ancillary Ligands and H_2
            Bonding 280
        9.4.4.2. Proton Transfer to Anions 283
9.5. Catalytic Hydrogenation and Related Reactions 284
    9.5.1. Direct Transfer of Hydrogen from H_2 Ligands 284
    9.5.2. H_2 Complexes as Precursors for Catalytic and Other Reactions .. 288
    9.5.3. Ionic Hydrogenation and Zeolite-Catalyzed Hydrogenation 290
References 291

10. Activation of Hydrogen and Related Small Molecules by Metalloenzymes and Sulfur Ligand Systems ... 297

10.1. Hydrogenases: Biorganometallic Formation and Splitting of Dihydrogen 297
    10.1.1 Introduction ... 297
    10.1.2. [NiFe] H-ases ... 300
    10.1.3. A Metal-Free Hydrogenase ... 306
    10.1.4. [Fe] Hydrogenases: Highly Organometallic Active Sites ... 307
10.2. Nitrogenases and Nitrogen Fixation ... 312
10.3. Hydrocarbon Activation by Oxygenase Enzymes ... 316
10.4. Models for Biological and Industrial Activation of Hydrogen on Sulfur Ligands and Sulfides ... 319
References ... 322

11. Coordination and Activation of Si–H, Ge–H, and Sn–H Bonds ... 327

11.1. Introduction ... 327
11.2. Synthesis and Characterization of $M(\eta^2\text{-Si–H})$ Complexes ... 330
11.3. Si–H Bonding to Metals Compared to M–H_2 Bonding ... 342
    11.3.1. Comparisons of H–H versus Si–H σ-Bond Activation and OA Processes ... 344
    11.3.2. Theoretical Studies of Si–H versus H–H σ-Bond Activation ... 349
11.4. Reactions and Dynamics of σ-Silane Complexes ... 352
    11.4.1. Heterolytic Cleavage of $M(\eta^2\text{-Si–H})$ and Related Reactions ... 353
    11.4.2. Dynamics of Silane Ligands and Exchange with cis Hydrides ... 355
11.5. Germane and Stannane σ Complexes ... 357
References ... 362

12. C-H Bond Coordination and Activation ... 365

12.1. Introduction ... 365
12.2. Agostic C–H Coordination and Cyclometallation ... 367
    12.2.1. Structures and Strengths of M $\cdots$ H–C Interaction ... 367
    12.2.2. Cyclometallation and Oxidative Addition of Agostic C–H Bonds ... 374
    12.2.3. Agostic Interactions in Phosphine Complexes ... 378
12.3. Alkane Coordination ... 383
    12.3.1. Experimental Evidence ... 383
    12.3.2. Theoretical Studies of Alkane Activation and C–H versus C–F Bond Coordination ... 388
12.4. Evidence for σ Complexes in the Oxidative Addition and Reductive Elimination of C–H Bonds ... 395

12.4.1. Isotopic Exchange in Intermediates in RE and OA
of Alkanes and Other Transformations 395
12.4.2. Direct Spectroscopic Evidence for Alkane Complexes as
Intermediates in OA of Alkanes 399
12.4.3. Shilov Chemistry and Related Electrophilic Systems for
Homogeneous Alkane Oxidation 403
References 411

13. Coordination and Activation of B–H and Other X–H and X–Y Bonds ... 417

13.1. B–H Bonds 417
    13.1.1. Metal Borohydride and Agostic-Like Systems 417
    13.1.2. σ Complexes of Neutral Boranes 421
13.2. X–H σ Interactions Where X Is Also a Lone-Pair Donor (N, P, S) 429
13.3. Interactions of σ Bonds X–Y Where X and Y Are Not Hydrogen 431
    13.3.1. C–Halogen Bonds 431
    13.3.2. C–C, C–Si, C–N, and C–B Bond Interactions with
Metal Centers 432
    13.3.3. Activation of C–P bonds in Phosphines 435
13.4. Summary and a Glance to the Future 436
References 437

Abbreviations Commonly Used in the Text 441

Index 445

1

Introduction

1.1. NONCLASSICAL BONDING IN σ COMPLEXES

There is an air of magic to certain discovery processes that even the term serendipity fails to capture. The revelation in 1983 of a new type of chemical bonding, coordination of an intact dihydrogen molecule (H_2) to a transition metal complex, certainly falls into this category, not only because it was totally unexpected but also because of how it came about.[1] As will be described in Chapter 2, such a complex, $W(CO)_3(P^iPr_3)_2(H_2)$ (Figure 1.1), was not even being sought, and several years elapsed between the synthesis and proof of its structure. Molecules containing only strong "inert" σ bonds such as H–H in H_2 and C–H in alkanes had always been believed to be incapable of stable binding to a metal. Yet transition metals and their compounds had long been known to catalyze reactions of H_2 with organic molecules, which is the basis of one of the world's most important chemical reactions — catalytic hydrogenation. Metal hydride complexes formed by oxidative addition of the H–H bond were known to be a part of the catalytic cycle for some time and

η^2-H_2 complex   dihydride complex

were well-characterized species. However, the dihydrogen complexes (referred to as η^2-H_2 or H_2 complexes), which were assumed to be unobservable intermediates in dihydride formation, were completely unrecognized until 1983. It is remarkable that even the theoretical basis for interaction of a σ bond with M was still in its infancy this late in the history of inorganic chemistry.

Thus H_2 coordination surely ranks as one of the most well-hidden "secrets" in all of chemistry. Part of the reason for this was the stubborn notion that such a complex could not be stable, as was clearly illustrated by the controversy over the initial report of its existence. However hundreds of H_2 complexes have now been identified and are part of a rapidly growing class of compounds called *σ complexes*, a most appropriate term referring to the three-center interaction of the *bonding*

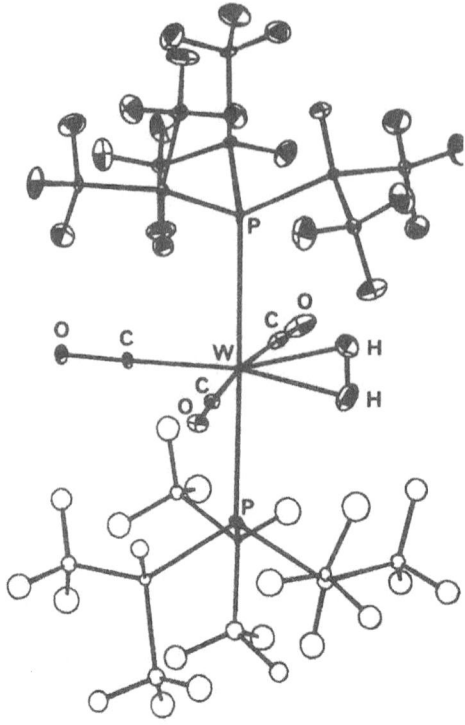

Figure 1.1. ORTEP drawing of the neutron structure of W(CO)$_3$(P^{i}Pr$_3$)$_2$(H$_2$) at 30 K, showing intact H–H bond elongated to 0.82(1) Å. Lower phosphine is disordered.

electron pair in H–H or other X–Y bonds with M. These complexes perfectly complement classical Werner-type compounds, where a ligand donates electron density through its *nonbonding* electron pair(s) and π complexes such as metal–olefin complexes in which electrons are donated from bonding π electrons.

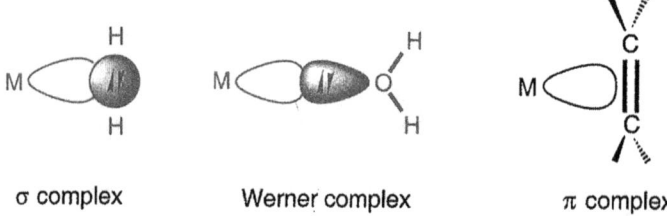

σ complex   Werner complex   π complex

The σ ligand is side-on (η^2) bonded to M, and the bonding in M–η^2-H$_2$ and other σ complexes has been termed *nonclassical*, analogous to the three-center, two-electron (3c-2e) bonding in nonclassical carbonium ions (now called carbocations) and boranes (Figure 1.2). Indeed transition metal fragments, CH$_3^+$, and H$^+$ are regarded as *isolobal* species by Hoffmann,[2] as they possess similar chemical

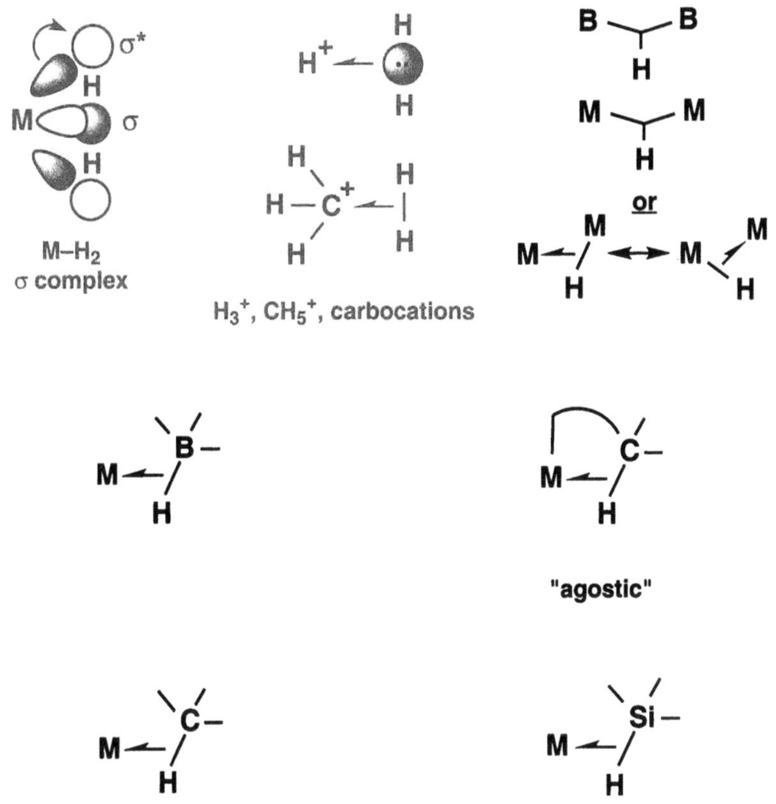

Introduction

Figure 1.2. Examples of nonclassical 3c-2e bonding.

bonding properties, especially toward nonclassical coordination of H_2. Positively charged fragments such as $[ML_n]^+$, CH_3^+, and H^+ are all strong electrophiles

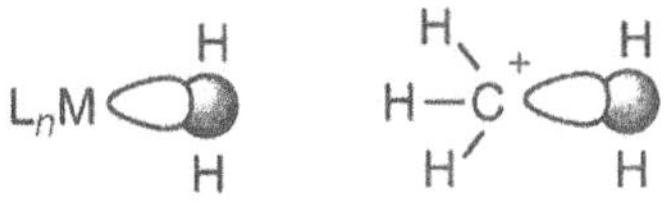

("superelectrophiles" in the extreme sense[3]) toward the Lewis basic H_2, but, as will be shown, transition metals can uniquely stabilize H_2 and other σ-bond coordination by *backdonation* or *backbonding (BD)* from d orbitals, which main-group analogues cannot do. This bonding is then remarkably similar to the Dewar–Chatt–Duncanson model[4] for π complexes where M → L BD is crucial, and indeed H_2 is a good π-acceptor ligand like CO and ethylene. It is important to note that the linkage in LM_n–H_2 systems is identifiable as *the* bond between two species, each of which is capable of independent existence — $LM_n + H_2$. This principle should be remembered whenever a question arises concerning the validity of "true" σ complex, i.e., one not stabilized by a primary linkage such as in intramolecular σ-bond interactions, commonly known as agostic interactions (Figure 1.2), or an ionic bonding component. In this regard, Parry and Kodama point out that the entire field of coordination chemistry began through attempts to explain the fact that molecules such as NH_3 and $CoCl_2$, which are stable in their own right, combine to give stable adducts.[5] Nonetheless, coordination of σ bonds even in an agostic sense is also important from the standpoint of relieving electronic unsaturation in coordinatively unsaturated complexes. Complexes that might not otherwise be stable can be isolated, and indeed the 16e precursor to the first H_2 complex, $W(CO)_3(PR_3)_2$, is stabilized by an intramolecular $W \cdots H–C$ interaction (see Chapters 2 and 12).

The number and variety of σ bonds already found to interact inter- or intramolecularly with metal centers is impressive. It should be noted that a *σ com-*

σ complex	agostic (β, γ, ...)	α-agostic
H–H, B–H	C–H, Si–H,	C–H, Si–H,
C–H, Si–H,	B–H, N–H	S–H, P–H,
Ge–H, Sn–H	C–C, Si–C	C–P, C–B, C–N

plex should not be referred to as an agostic complex, which implies intramolecular interaction rather than external ligand binding. In principle *any X–Y bond can coordinate to a metal center M* providing steric and electronic factors are favorable, e.g., substituents at X and Y do not block the metal's access. Even more generally speaking a vacant metal orbital can interact with any nearby electron density. Hence complexes of metals with a rare gas containing a closed-shell octet of electrons should be possible, and $M(CO)_5(Xe)$ has been observed in liquid Xe. A stable complex of this type or an alkane σ complex would have been unthinkable before 1983 but now would not be surprising and indeed $[AuXe_4][Sb_2F_{11}]_2$ has recently been isolated.[6] It is astonishing but alkane σ complexes are believed to be intermediates in Pt^{II}-catalyzed methane conversions in reaction media as harsh as sulfuric acid at 200°C (see Chapter 12), despite a small $M–CH_4$ binding energy of only about 10 kcal/mol!

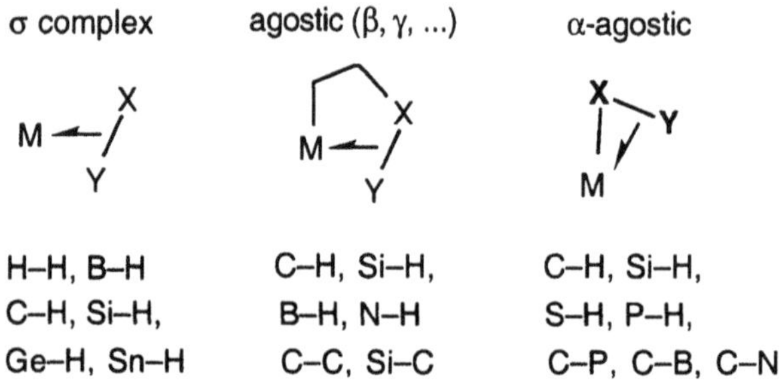

Another exciting aspect is the ability to "see" breaking of a chemical bond by effectively taking snapshots along the entire reaction coordinate. The lengthening and eventual rupture of a σ bond (or any bond) by binding it to M is referred to as "activation."

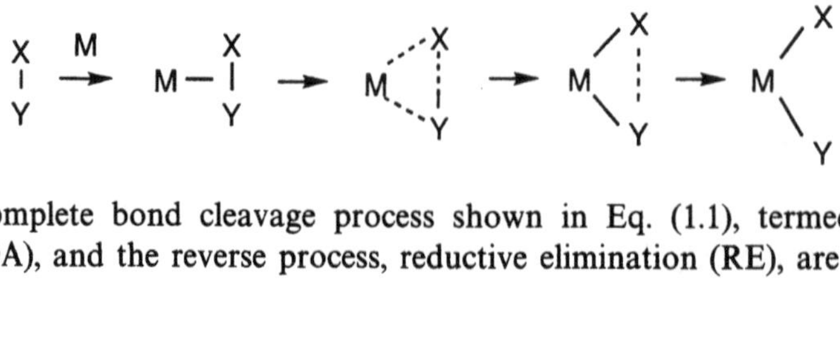

$$(1.1)$$

The complete bond cleavage process shown in Eq. (1.1), termed oxidative addition (OA), and the reverse process, reductive elimination (RE), are fundamentally important to catalytic processes such as hydrogenation. Depending on the electronic character of M, which is influenced by other ligands L on M, determines whether Eq. (1.1) can proceed to completion without any sign of the intermediate structures. However, in certain cases cleavage of XY can be "arrested" along the reaction coordinate, thus giving a σ complex where the X$\cdots$Y distance (d_{XY}) can vary greatly. Thus the coordination and activation of H_2 and other molecules containing simple two-electron σ bonds on metal complexes is immensely important in terms of fundamental science. Furthermore, the dynamic processes in Eq. (1.1) and in the interactions and exchange of σ ligands with ligands bound cis to them are perhaps the most widely encompassing and complex in all of chemistry, as will be shown in Chapter 6.

The subject of this book is coordination and activation of H–H and other σ bonds, both from a theoretical and an experimental point of view. Perhaps no other field of chemistry has seen such effective interplay between experiment and theory as M–η^2-H_2 and M–hydride systems. Obviously the innate simplicity of the H_2 molecule is the primary factor here, along with its importance in applied chemistry. The interrelationships between activation of H_2 and other σ-bonded molecules such as alkanes and silanes is also highly significant because conversion of methane and other alkanes is being pursued intensively. The CH_3–H and H–H bond energies are practically identical (104 kcal/mol), and the C–H and H–H bonds are not too dissimilar in polarity. Both molecules have σ and σ^* orbitals of reasonably similar shape, energy, and extent, although the differences in overlap with metal d orbitals can be critical. Thus the H_2 molecule (as well as silanes and germanes) can be coordinated to M in a stable fashion at ambient temperature whereas M-bound alkanes cannot, although an alkane complex has just recently been observed at low temperature. Among the greatest challenges in chemistry are the quest for a stable alkane complex and the attendant conversion of light alkanes such as CH_4 in large reserves of natural gas to liquid fuels, e.g. methanol.[7]

1.2. SMALL-MOLECULE ACTIVATION ON TRANSITION METAL COMPLEXES

Transition metal complexes containing coordinated small molecules such as CO, O_2, and N_2 have been known for many decades and are of critical importance in industrial and biological processes and coordination chemistry as a whole (Table 1.1). Of all the industrially relevant molecules, H_2 is arguably the most important. Catalytic hydrogenation reactions represent the most massive man-made chemical reactions in the world: All crude oil is treated with hydrogen to remove sulfur and nitrogen in hydrodesulfurization/hydrodenitrogenation processes. Ten billion tons of ammonia is produced worldwide by the Haber process:

$$N_2 + H_2 \xrightarrow[\text{Fe catalyst}]{400-550^\circ C} NH_3 \qquad (1.2)$$

Many trillions of dollars worth of products are derived directly or indirectly from hydrogen reactions. Although most of these are well-established processes, even a

Table 1.1. Transition Metal Small-Molecule Binding and Conversions in Industry and Biology

Oxygen: O_2	Respiration	$Fe-O_2 \underset{\text{(hemoglobin)}}{\rightleftharpoons} Fe + O_2$
Nitrogen: N_2	Produce ammonia (mimic nitrogenase)	$N_2 + H_2 \xrightarrow[700°C]{Fe} NH_3$
Carbon monoxide: CO	To gasoline	$M-C\equiv O \xrightarrow{H_2} C_6H_{14}$
Methane: CH_4	To liquid fuels	$CH_4 \xrightarrow{M}$ Gasoline, methanol
Water:	Splitting to hydrogen	$H_2O \xrightarrow{energy} H_2 + O_2$
Hydrogen: H_2	Catalytic hydrogenation Mimic hydrogenase	$H_2 + H_2C=CH_2 \rightarrow H_3C-CH_3$ $H_2 \rightleftharpoons 2H^+ + 2e^-$

slight improvement in efficiency could mean substantial savings. For example, ammonia production is very energy-intensive, and for several decades chemists have been seeking better routes by studying the structure and function of biological enzymes such as nitrogenases and hydrogenases. The latter reversibly produce hydrogen from protons and have dinuclear active sites with Fe–Fe or Fe–Ni bonds and CO ligands that are remarkably organometallic in nature (Chapter 10):

$$2H^+ + 2e^- \leftrightarrow H_2 \tag{1.3}$$

The structure and bonding principles in σ coordination and activation of H_2 appear to have direct application in biology, and it is now evident that Nature scooped us over a billion years ago in this regard! Understanding the binding site in hydrogenases may lead to efficent biomimetic methods for converting seawater to hydrogen. Iron is now recognized to be the key metal in the active site of hydrogenase, nitrogenase, and oxygenase enzymes, including methane monooxygenase (MMO)[8] that converts CH_4 to methanol.

Hydrogen clearly holds promise as the fuel of the future as it is expected to replace or supplement natural gas and other hydrocarbons. Combustion of H_2 and its use in fuel cells to generate electricity both produce only water, which could prove crucial if CO_2 from hydrocarbon fuels continues to be a threat to global climate.[9] Production of hydrogen using metal compounds as catalysts will involve H_2 complexes at least as intermediates, and H_2 complexes have, e.g., been implicated in solar energy conversion schemes based on the photoreduction of water.[10] Industrially important water–gas shift and related H_2-producing reactions undoubtedly proceed via transient H_2 complexes because their converses—hydrogenation reactions—clearly can and do involve direct transfer from H_2 ligands (see Chapter 9). Understanding how H_2 binds to other substances may also be critical for developing safe and efficient methods for hydrogen storage.

Small-molecule binding to metal complexes is thus an exciting field in coordination chemistry as well as being important for understanding chemical bonding to M. Reversible binding of dioxygen[11] and dinitrogen[12] to simple, nonheme transition metal complexes was among the most significant and exciting chemical discoveries of the 1960s and has invigorated synthetic inorganic chemistry for decades to come. The characterization and report of the first N_2 complex, $[Ru(NH_3)_5(N_2)]^{2+}$, by

Introduction

Allen and Senoff actually involved a considerable amount of serendipity and initial disbelief. Their seminal paper was so astounding at the time that it was initially rejected by four referees![12] Looking back to this landmark finding, there is a remarkable parallel to our discovery almost 20 years later of the first stable molecular H_2 complex, $W(CO)_3(P^iPr_3)_2(H_2)$, which forms the basis for this book (see Section 2.2.2).

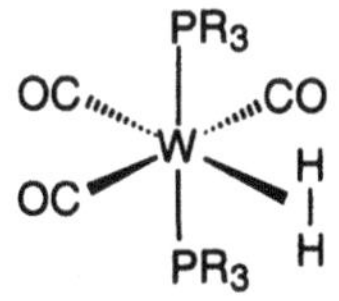

$$\begin{array}{c} PR_3 \\ OC_{\prime\prime\prime\prime\prime}|_{\prime\prime\prime\prime\prime}CO \\ OC\!-\!W\!-\!H \\ |\quad\;|\\ PR_3\;\;H \end{array}$$

H_2 coordination was even more startling because there is a monumental difference in the bonding of N_2 and H_2 to metal d orbitals that required yet another "leap of faith" in chemical bonding concepts analogous to that needed for metal π bonding. From fundamental bonding concepts, it would seem improbable that an H_2 or alkane molecule, with no lone pair or π electrons, could be a strong enough electron donor or acceptor to give a stable metal complex. How could a bonding electron pair interact with metal d orbitals? There were two possibilities, shown in Eq. (1.4), whereby H_2 could act either as a base or as an acid. However such a species

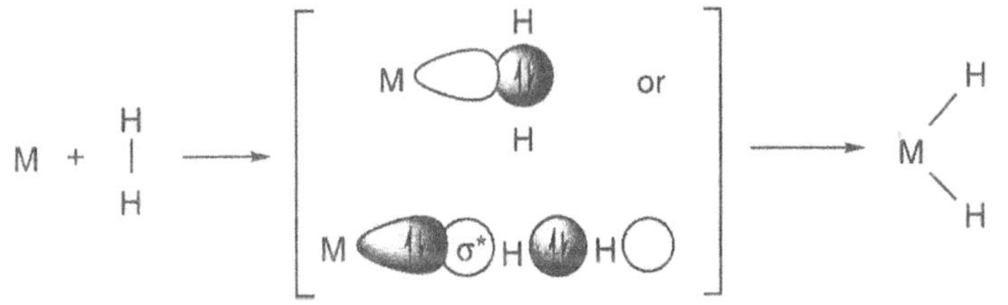

$$M + \begin{array}{c}H\\|\\H\end{array} \longrightarrow \left[\begin{array}{c} M\!\!-\!\!H \\ \quad\;\;H \\ \text{or} \\ M\!\!-\!\!\sigma^*\!H\;H \end{array} \right] \longrightarrow M\!\begin{array}{c}\diagup H\\ \diagdown H\end{array} \qquad (1.4)$$

would seem to be only an unstable intermediate on the reaction path to a metal dihydride complex. Complete splitting of the strong H–H bond (104 kcal/mole) on M has long been thought to be dogmatic in the mechanism of catalytic hydrogenation because H_2 itself is virtually unreactive except for free-radical processes, e.g., combustion. Theoretical calculations of OA initially did not suggest an H_2 complex as a stable intermediate along the reaction coordinate for H–H bond-breaking. For these reasons, virtually no one believed that a M–H_2 complex could be "isolated in a bottle," as had been done for M–N_2 complexes. As we shall see, not only was this accomplished, but hundreds of H_2 complexes have now been synthesized and studied experimentally and theoretically by over 80 research groups worldwide.

1.3. σ COMPLEXES AS "ARRESTED INTERMEDIATES" ALONG THE REACTION COORDINATE FOR σ-BOND CLEAVAGE

In 1983, the first H_2 complex, $W(CO)_3(P^iPr_3)_2(\eta^2\text{-}H_2)$ (**1**), not only was isolated and characterized but later shown to exist in dynamic equilibrium with a dihydride form (**2**) in solution. The tautomeric relationship shown in Eq. (1.5) is extremely significant because it proves that the activation and splitting of H_2 on M occurs via side-on (η^2) bonding of H_2 to the metal. Binding and cleavage of a σ bond on M

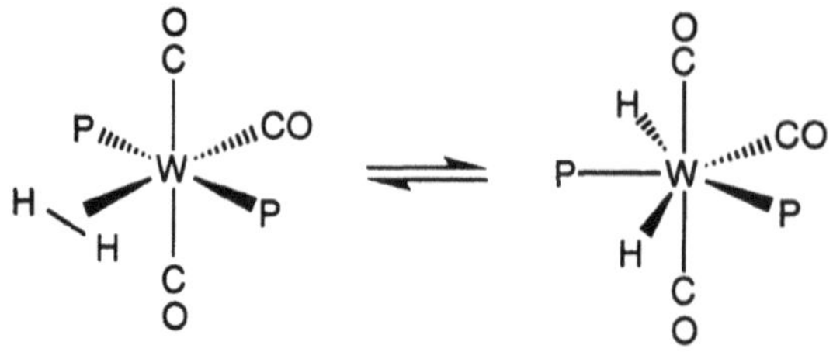

$$(1.5)$$

(1)   (2)

can now be studied in exquisite detail, along with its reverse, i.e., σ-bond formation and elimination. It is astonishing to realize that one of the strongest chemical bonds is weakened and is breaking and reforming dozens of times a second without shining a laser on it or adding energy of any kind. Equation (1.5) provides an excellent example of the powerful synergy between experiment and theory that can be applied hand in hand to these systems. The trigonal bipyramidal structure for **2** is more confidently proposed on the basis of calculations rather than experimental findings. Computations based on modern density functional theory (DFT) mimic experimental systems exceptionally well now and in some cases even predict experimental findings or challenge uncertain data.

To add further excitement to the field, *elongated* H–H bonds of over 1 Å were found around 1990. Until then only d_{HH} in the range 0.85–0.90 Å had been observed by solid state NMR and diffraction methods (after correction for H_2 librational motion). "Elongated" is of course a relative term since the H–H bond is always stretched to some degree on coordination. Thus not only can a chemical bond be snapped like rope, it can effectively be stretched like a rubber band until the bond is virtually gone, i.e., a dihydride forms as shown in Eq. (1.6). A large number of

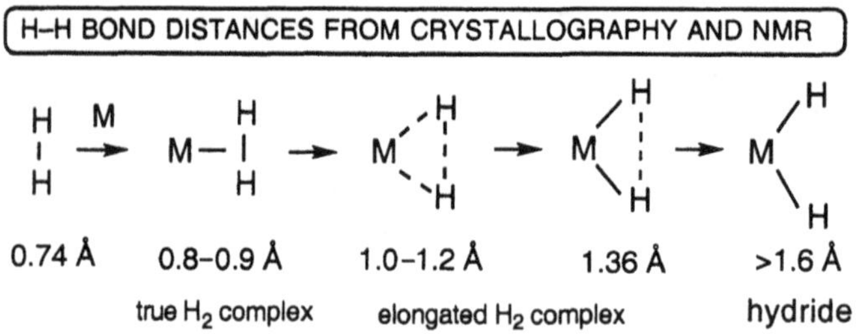

H–H BOND DISTANCES FROM CRYSTALLOGRAPHY AND NMR

$$(1.6)$$

0.74 Å   0.8–0.9 Å   1.0–1.2 Å   1.36 Å   >1.6 Å

true H_2 complex   elongated H_2 complex   hydride

complexes displaying a near *continuum* of d_{HH} have now been structurally characterized throughout the entire transition metal series. The OA process can be arrested at various points along the reaction coordinate merely by varying the M–L sets and changing the electronics at M. Although not the first example of an elongated H_2 ligand, a type of complex originally synthesized in 1971 in Taube's group,[13] $[Os(H_2)(en)_2(acetate)]^+$ (**3**), shows a very long d_{HH} and is on the verge of becoming a dihydride, which it was originally believed to be! Structure **3** also created another paradigm shift because of the simplicity of its Werner-like ligand set: No bulky phosphines, strong π acceptors such as CO, or exotic "organometallic" ligand sets are needed to stabilize nonclassical interactions.

$$\underset{(3)}{\left[\begin{array}{c} \overset{1.34}{H\text{---}H} \\ \overbrace{N_{\cdots}\overset{|}{\underset{|}{Os}}\cdots N}^{} \\ N\quad\ \ N \\ | \\ OAc \end{array}\right]^{+}}$$

Little H–H bonding interaction remains when d_{HH} becomes greater than 1.1 Å, so intriguing questions arise, such as "at what point is the bond broken?" Since there can be no lone-pair repulsions for hydrogen ligands and little steric interference in general, H atoms can get as close together or as far apart as electronics at M dictate. Theoretical analyses of H_2 complexes (see Chapter 4) currently show that the energy barrier for stretching the H–H bond from 0.85 all the way to 1.6 Å is surprisingly low, on the order of 1 kcal/mol! Thus the "elongated" H_2 molecule is extremely delocalized about M, and the H atoms in H_2 essentially undergo large-amplitude vibrational motion along the reaction coordinate for OA and also librate/rotate about the $M-H_2$ axis at a much slower rate. This is but one aspect of the amazing dynamic properties of η^2-H_2, which can also readily enter the realm of quantum-mechanical behavior (see Chapter 6).

It is now clear that η^2-H_2 binding followed by OA serves as a prototype for *all* σ-bond activation processes, including C–H and Si–H. As will be seen in Chapter 11, silanes have similarly been demonstrated to bind in η^2-Si–H fashion, and, as for η^2-H_2, the η^2-SiH_4 structure in **4** can also exist in equilibrium with its OA tautomer **5** with virtually identical thermodynamic parameters. Delocalized bonding analogous to that in 3 undoubtedly also occurs since coordinated Si–H distances varies

$$\underset{(4)}{Mo-\underset{SiH_3}{\overset{H}{\diagdown}}} \rightleftharpoons \underset{(5)}{Mo\underset{SiH_3}{\overset{H}{\diagup}}} \tag{1.7}$$

widely in a way that is similar to that in H–H activation. An invaluable spectroscopic yardstick for measuring activation in $M(\eta^2\text{-}X\text{-}H)$ bonds is *the value of the NMR coupling constant J_{XH} compared to that in the free ligand.* There is typically a 50–80% reduction in J_{HD} in unstretched HD complexes, a 74% reduction in $J(^{13}CH)$ for cyclopentane coordination in $CpRe(CO)_2(C_5H_{10})$, and 65% in $J(^{11}BH)$ in complexes of neutral borane ligands. There is, however, a contrast here to J_{SiH} in silane complexes, which are always closer to those in OA products and more analogous to J_{HD} in elongated H–H bonds. Thus despite the obvious similarities, there are subtle differences in the properties and bonding in the various types of σ complexes.

SiH_4 binding and Si–H cleavage are exciting because they directly model the binding of CH_4 to M and subsequent C–H bond-breaking. By analogy with H–H, presumably any σ bond may be stretched on a M until it breaks. However, the situation becomes more complex because substituents at C or Si can change both electronics and sterics and alter the activation. Nonetheless, all of the remarkable phenomena above can be explained in a relatively simple, coherent bonding model that applies to all coordinated σ bonds.

1.4. THE IMPORTANCE OF BACKDONATION IN σ-BOND COORDINATION AND CLEAVAGE

As will be shown in Chapter 7, the binding energy of H_2 to $W(CO)_3(P^iPr_3)_2$ is nearly identical to that for the aquo ligand and other weak to moderate σ bases.

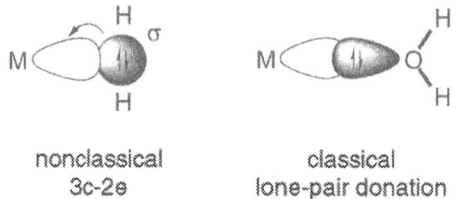

nonclassical
3c-2e

classical
lone-pair donation

Remarkably, one of the simplest conceivable H_2 complexes, $[Ru(H_2O)_5(H_2)]^{2+}$, can be formed by displacement of an aquo ligand from the hexaquo complex by pressurized H_2.[14] It seems astonishing that interaction of a *bonding* electron pair could be on a par with that for a lone pair. What is unique about the three-center bonding in M–H_2 and other σ-bond complexes that stabilizes them and sets them apart from species such as carbocations is BD, i.e., donation of electrons from a filled metal d orbital to the σ* orbital of the H–H bond, similar to metal donation to π* orbitals in the Dewar–Chatt–Duncanson model for olefin coordination.

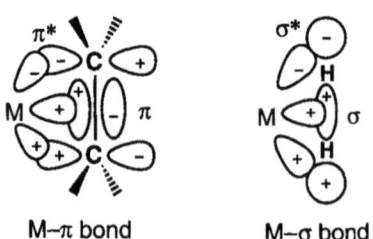

M–π bond   M–σ bond

BD is the crucial component in both aiding the binding of H_2 to metals and activating the H–H bond toward cleavage. If this backbond becomes too strong, e.g., if more electron-donating coligands are put on M, the σ bond breaks to form a dihydride because of overpopulation of its antibonding orbital. This is analogous to a high degree of olefin activation as represented by the metallocyclobutane structure **7**, except that the C=C double bond in **6** can only be weakened to a near single bond and is never completely broken. σ-bond complexes can simplistically be looked upon as "arrested" OA, as originally suggested for Si–H bond coordination by Kaesz,[15] but the arrest can be *anywhere* along the reaction coordinate. This is dramatically demonstrated by the "stretching" of d_{HH} [Eq. (1.6) and structure **3**], which is controlled primarily by the ability of the metal to backdonate electrons. In fact,

Introduction

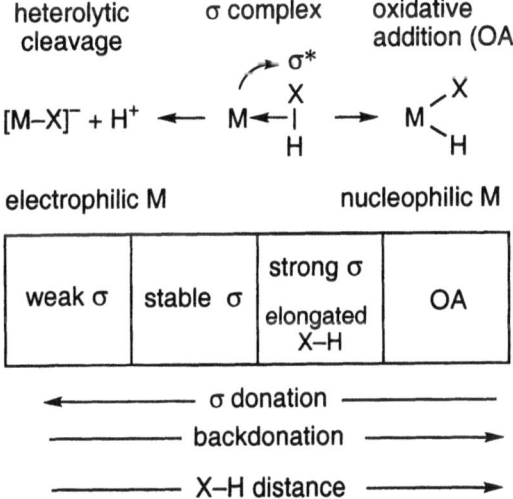

(6) (7)

whether H_2 binds molecularly or adds oxidatively to give a dihydride to a particular metal fragment can often be predicted by examining the NN, CO, or SO stretching frequencies of the corresponding N_2, CO, or SO_2 complexes as a gauge of a metal's backbonding ability. If there is one principle to keep in mind while reading this book and reaching an understanding of σ-bond activation, it is that *BD controls σ-bond activation toward cleavage and that a σ bond cannot be broken solely by sharing its two electrons with a vacant metal d orbital.* Although the latter interaction is generally the predominant bonding component, a σ-bond complex is unlikely to be stable at room temperature without at least a small amount of BD. Thus stable H_2 complexes of main group elements (e.g., pure Lewis acids such as BX_3) are unknown.

However, H_2 can bind in stable fashion to very electron-deficient metals, which are weak backbonders, nearly as well as to more electron-rich metals. Calculations show that for highly electrophilic metals *the reduction in BD is almost completely offset by increased electron donation from H_2 to the electron-poor metals.* It is important to note there are two completely different pathways for cleavage of H–H and X–H bonds: homolytic cleavage (OA) and heterolytic cleavage. Both pathways

have been identified in catalytic hydrogenation (see Chapter 9) and may also be available for other σ-bond activations such as C–H cleavage. Heterolytic cleavage of X–H bonds via proton transfer to a basic site on a cis ligand or to an external base is a crucial step in both biological and industrial processes, as will be highlighted below and detailed in Chapters 9–12.

As a reflection of the dual activation pathways, H_2 is in essence the perfect ligand because it is *amphoteric*, i.e. it is essentially both a Lewis acid and a Lewis base. H_2 is perhaps the most versatile "weak" ligand just as amphoteric SO_2 is the most versatile "strong" coordinating agent toward both transition metal and main-group compounds.[16] Virtually every known unsaturated transition metal fragment either molecularly binds or oxidatively adds H_2. The primary difference between H_2 (and other nonclassical ligands) and classical ligands such as SO_2 is that electron donation originates from a *bonding* electron pair to give a three-center interaction.

One can learn much about a metal center's electronic properties by binding H_2 to it and examining the d_{HH} or NMR parameters such as J_{HD}. The relative binding strength and stereochemical position of other σ-bond ligands coordinated to the same menu of metal fragments can then be compared to give relative acceptor abilities of the σ ligands. For example, is BD to Si–H bonds in silanes stronger than that to an H–H bond? There is a great deal of synergism in comparing the binding of a variety of nonclassical σ ligands and also classically bound small molecules such as N_2 on a variety of metal centers.

1.5. REACTIVITY OF σ COMPLEXES: σ-BOND METATHESIS, ACIDITY, AND HETEROLYTIC CLEAVAGE OF X–H BONDS

Not only are σ-bonded ligands stable, it is possible to detect their interaction with cis ligands on M (cis effect; see Chapter 4). Extremely facile exchange occurs between H_2 and hydride ligands on metal complexes, which is believed to take place

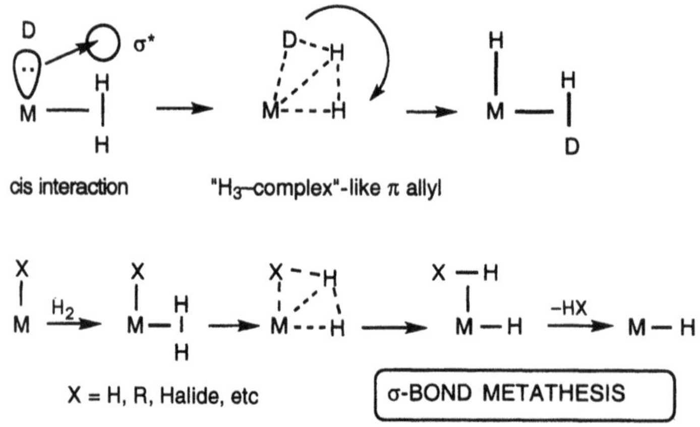

via a transient H_3 complex similar to the four-center interactions in heterogeneous activation of hydrogen and σ-bond metathesis. σ-bond metathesis has been invoked to explain reactions on d^0 metal centers where OA is not possible. The theoretically predicted open H_3 ligand, which is analogous to a π-allyl ligand, can then easily rotate about the M–H_3 axis to exchange atom positions, as observed by isotopic scrambling using deuterium labels. The energetics for hydrogen exchange in H$_2$–hydride complexes has been measured as low as 0.5 kcal/mol even in the solid state, i.e., virtually a barrierless process. Thus this can be considered a form of "bond–no-bond resonance" as proposed in nonclassical carbonium ion systems and is a concept with far-reaching implications. Conceivably *any* type of σ bond can be metathesized with *any* cis ligand by this mechanism, as in olefin metathesis, with the only limitations being thermodynamic and steric factors.

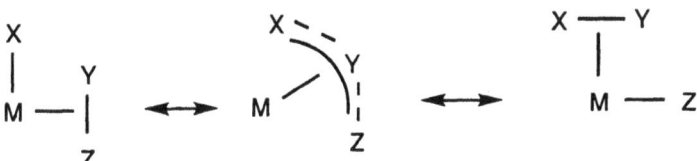

One of the most important questions is whether *direct transfer* of hydrogens from an η^2-H$_2$ ligand rather than the dihydride with a fully broken H–H bond takes place in catalytic hydrogenation. This is difficult to prove conclusively, but there is evidence that this can occur in some reactions, e.g., styrene hydrogenation (see Section 9.2). In some cases, H$_2$ can become a stronger acid than sulfuric acid upon binding to cationic electrophilic metal centers, attaining a pK_a as low as −6, a decrease of over 40 units from the pK_a of free H$_2$! The complex can then undergo heterolytic cleavage, e.g., protonate bases B such as ethers and form a monohydride [Eq. (1.8)].

Intramolecular cleavage will occur if a cis ligand is more basic than an external base, e.g., if L has a lone pair to attract the partially positively charged hydrogen. The kinetic acidity of an H$_2$ complex is greater than that of the corresponding dihydride (see Section 9.4.2). Equation (1.8) shows that a metal hydride can be protonated to form a cationic H$_2$ complex that is either stable (a common synthetic route) or may readily eliminate H$_2$. Here the kinetic site of protonation is normally the M–H bond, even though the thermodynamic site of proton transfer can be M (an H$_2$ complex forms initially and rearranges to a dihydride). Other σ bonds can be cleaved heterolytically as in Eq. (1.8), particularly on electrophilic *cationic* metals. For coordinated Si–H bonds, the coordinated bond becomes polarized in the opposite sense Si$^{\delta+}$–H$^{\delta-}$, i.e., the Si becomes positively charged [Eq. (1.9)].

(1.8)

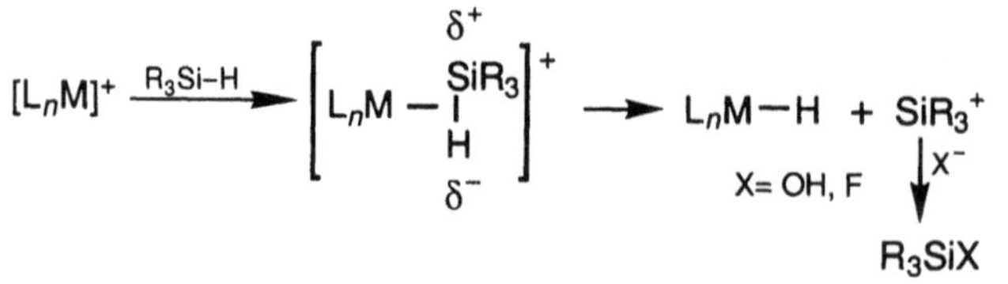

$$[L_nM]^+ \xrightarrow{R_3Si-H} \left[L_nM - \overset{\overset{\delta^+}{SiR_3}}{\underset{\underset{\delta^-}{H}}{|}}\right]^+ \longrightarrow L_nM-H + SiR_3^+ \quad \downarrow X^- \atop R_3SiX \quad X=OH, F$$ (1.9)

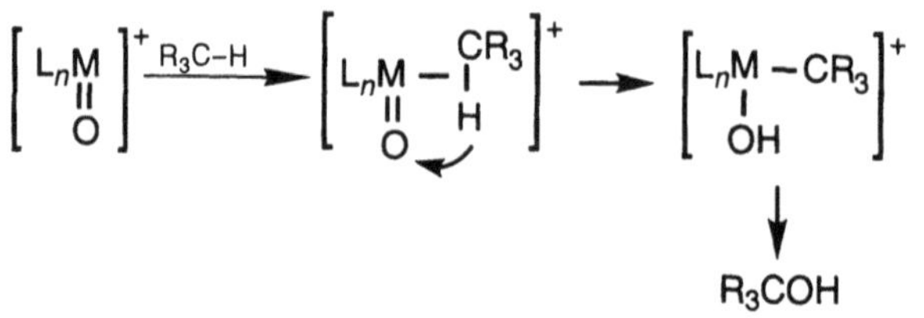

$$\left[L_nM\atop{\|\atop O}\right]^+ \xrightarrow{R_3C-H} \left[L_nM - \overset{CR_3}{\underset{H}{|}}\atop{\|\atop O}\right]^+ \longrightarrow \left[L_nM - CR_3 \atop OH\right]^+ \downarrow R_3COH$$ (1.10)

Very reactive silylium ions that scavenge nucleophiles such as water or abstract fluoride from normally unreactive anions such as $B(C_6F_5)_4^-$ are eliminated. An important question is whether C–H bonds in alkanes, particularly CH_4, can be split in this manner and whether elimination of protons or carbocations will occur on "superelectrophilic" metal centers. Intramolecular heterolytic cleavage of C–H in a fleeting σ-alkane complex may occur in alkane conversions such as alkane oxidation by oxygenase enzymes, where the proton becomes acidic and transfers to a basic ligand [Eq. (1.10); see also Section 10.3] and Shilov-like conversions of CH_4 (see Chapter 12). In such systems transfer of protons is very facile because of the extremely high mobility of H^+, and even short-lived, very weak σ-omplexes can be the crucial intermediates. The C–H bond is thus more likely to be polarized toward $C^{\delta^-}\cdots H^{\delta^+}$ on an electrophilic M, and H^+ immediately leaps to the basic oxo ligand as soon as the alkane contacts M (Sleeping Beauty effect; see Section 10.3). The M center may be considered to be a superelectrophile isolobal with H^+, mimicking superacid-induced carbocation chemistry, i.e., a σ complex such as M^+-CH_4 is equivalent to CH_5^+.

Metal–H_2 coordination and activation toward OA convey a rare elegance in chemistry, and most of the structure, bonding, and reactivity principles derived from H–H bond activation should apply to all types of metal σ-bond interactions. The major points discussed above and in the rest of this book are summarized here in terms of H_2 activation:

1. σ-bond coordination to M is a form of electron-deficient, 3c-2e bonding, and depending on the electronics at M, σ ligands can bind more strongly than H_2O or N_2 or as weakly as rare gases. Intramolecular agostic coordination of σ bonds is important in relieving electronic unsaturation in coordinatively unsaturated complexes that might not otherwise exist.

2. A complete analogy can be drawn between M σ-bond and M π-bond coordination, e.g., σ ligands are good donors to strong electrophiles but can also be good π acceptors on more electron-rich metals. This dual interaction allows H_2 to bind to or undergo OA with virtually all metal fragments. Computational chemistry is a very powerful synergistic tool for characterizing, understanding, and even predicting σ-bond interactions.

3. BD controls σ-bond cleavage, which can effectively be arrested anywhere along the reaction coordinate as manifested by a continuum of d_{HH}. The existence of BD is proven experimentally by measuring a small barrier (1–3 kcal/mol) to rotation of H_2 about the $M-H_2$ axis using inelastic neutron-scattering methods.

4. H_2 binding and cleavage on M are prototypes for all σ-bond X–Y coordination and activation processes, as illustrated by the existence of two different X–Y (H–H and Si–H) coordinated to the same M. The nature of the ligand trans to the σ ligand is crucial in σ-binding versus activation toward OA.

5. σ-bond interactions with metals are arguably the simplest yet paradoxically the most dynamic, complex, and enigmatic chemical topologies known. The H_2 molecule can simultaneously bind to and dissociate from M, undergo equilibrium OA, rapidly rotate about the $M-H_2$ axis, and exchange with cis ligands, all on the same metal fragment. Exchange of $MH(H_2)$ can proceed via an $M-H_3$ intermediate,

$$M + \begin{array}{c}H\\|\\H\end{array} \rightleftharpoons M-\begin{array}{c}H\\|\\H\end{array} \rightleftharpoons M\begin{array}{c}{}^{\nearrow H}\\{}_{\searrow H}\end{array}\!\!H \quad \text{fluxional} \tag{1.11}$$

$$\begin{array}{c}H^*\\|\\M-\!\!\!\begin{array}{c}H\\|\\H\end{array}\end{array} \longrightarrow \begin{array}{c}H\\|\\M-\!\!\!\begin{array}{c}H^*\\|\\H\end{array}\end{array} \circlearrowright \quad \text{rotation, exchange} \tag{1.12}$$

and pairwise exchange of two adjacent hydride ligands can go through an $M-H_2$-like transition state. In most cases these dynamics cannot be frozen out on the NMR timescale even at the lowest accessible temperature because the proton hops between ligands nearly unimpeded. It is remarkable that the hydrogens in elongated H_2 complexes "vibrate" almost freely along the reaction coordinate for cleavage over a distance of nearly an angstrom in the solid state. For σ ligands other than H_2, such as silanes, substituents at Si can affect bonding electronics, sterics, and dynamics.

6. Coordinated H_2 and other X–H can become strong acids and undergo extremely facile proton transfer [heterolytic cleavage of X–H, Eqs. (1.8) and (1.10)] and are thus capable of direct reaction in catalysis. *The mechanism of many homogeneous and heterogeneous catalysts undoubtedly involve heterolytic cleavage of σ bonds on unsaturated metal sites (even on metal oxides[17]) generated by the reaction conditions.* This is particularly true for electrophilic systems, e.g., cationic complexes where the H in the X–H bond becomes highly protonic and the σ complex can be viewed as a protonated M–L bond whether formed this way or by the addition of H_2 to a precursor. This is similar to viewing bridging hydrides as protonated M–M bonds and nonclassical B–H–B bonds as protonated B–B bonds. The exceptionally mobile H^+ can migrate extremely quickly and reside at the most basic site on a molecule. Protonating a hydride ligand is also faster than protonating the metal center because it involves less geometric and electronic rearrangement.

7. Biological activation of hydrogen in hydrogenases mimics organometallic chemistry (and vice versa) in that acceptor ligands such as CO are used to control

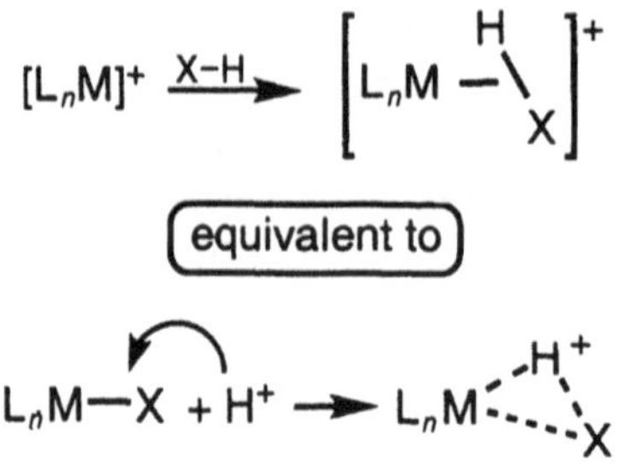

$$[L_nM]^+ \xrightarrow{X-H} \left[L_nM - \overset{H}{\underset{X}{\diagdown}}\right]^+$$

equivalent to

$$L_nM-X + H^+ \longrightarrow L_nM \cdots \overset{H^+}{\underset{X}{\cdots}}$$

the electronics of the active metal site. Reversible H_2 σ coordination and heterolytic cleavage is favored on such electrophilic sites over OA. Nitrogenases and oxidases may also involve σ complexes.

REFERENCES

1. Kubas, G. J. *Acc. Chem. Res.* **1988**, *21*, 120.
2. Elian, M.; Chen, M. M. L.; Mingos, D. M. P.; Hoffmann, R. *Inorg. Chem.* **1976**, *15*, 1148.
3. Olah, G. A. *Angew. Chem. Int. Ed. Engl.* **1993**, *32*, 767.
4. Dewar, M. J. S. *Bull. Soc. Chim. Fr.* **1951**, *18*, C79; Chatt, J.; Duncanson, L. A. *J. Chem. Soc.* **1953**, 2929.
5. Parry, R. W.; Kodama, G. *Coord. Chem. Rev.* **1993**, *128*, 245.
6. Seidel, S.; Seppelt, K. *Science* **2000**, *290*, 117.
7. Arndtsen, B. A.; Bergman, R. G.; Mobley, T. A.; Peterson, T. H. *Acc. Chem. Res.* **1995**, *28*, 154; Crabtree, R. H. *Chem. Rev.* **1995**, *95*, 987; Hall, C.; Perutz, R. N. *Chem. Rev.* **1996**, *96*, 3125; Shilov, A. E.; Shul'pin, G. B. *Chem. Rev.* **1997**, *97*, 2879.
8. Wallar, B. J.; Lipscomb, J. D. *Chem. Rev.* **1996**, *96*, 2625.
9. Marks, T. J. et al. *Chem. Rev.*, in press.
10. Sutin, N.; Creutz, C.; Fujita, E. *Comments Inorg. Chem.* **1997**, *19*, 67.
11. Vaska, L. *Science* **1963**, *140*, 809; Valentine, J. S. *Chem. Rev.* **1973**, *73*, 235.
12. Allen, A. D.; Senoff, C. V. *J. Chem. Soc., Chem. Commun.* **1965**, 621; Reviews: Allen, A. D.; Harris, R. O.; Loescher, B. R.; Stevens, J. R.; Whiteley, R. N. *Chem. Rev.* **1973**, *95*, 11; Hidai, M.; Mizobe, Y. *Chem. Rev.* **1995**, *95*, 1115.
13. Hasegawa, T.; Li, Z.; Parkin, S.; Hope, H.; McMullan, R. K.; Koetzle, T. F.; Taube, H. *J. Am. Chem. Soc.* **1994**, *116*, 4352; Malin, J.; Taube, H. *Inorg. Chem.* **1971**, *10*, 2403.
14. Aebischer, N. Frey, U.; Merbach, A. E. *Chem. Comm.* **1998**, 2303.
15. Andrews, M. A.; Kirtley, S. W.; Kaesz, H. D. *Adv. Chem. Ser.* **1978**, *167*, 229.
16. Kubas, G. J. *Acc. Chem. Res.* **1994**, *27*, 183, and references therein.
17. Over, H.; Kim, Y. D.; Seitsonen, A. P.; Wendt, S.; Lundgren, E.; Schmid, M.; Varga, P.; Morgante, A.; Ertl, G. *Science* **2000**, *287*, 1474.

2

Background and Discovery of Dihydrogen Coordination

2.1. HOMOGENEOUS ACTIVATION OF DIHYDROGEN ON METAL COMPLEXES

Several decades prior to the discovery of H_2 complexes, reactions of H_2 with transition metal complexes had been studied intensely because of the widespread importance of catalytic hydrogenation and oxidative addition processes in chemistry. Many reviews have been published on this subject, and a retrospective account of homogeneous catalytic hydrogenation was published in 1980 by Jack Halpern,[1] a pioneer in the field. He points out that Brian James's 1973 monograph, *Homogeneous Hydrogenation*,[2] cites nearly 2000 references from the preceding 20 years, with that number probably doubling during the following 7 years. It is remarkable that only two documented examples of homogeneous catalytic activation of H_2 by metal complexes were reported prior to 1953 when Halpern began his studies and heterogeneous hydrogenation predominated. The first dates back to 1938 when Calvin[3] reported that Cu^I salts such as Cu^I acetate catalyzed reduction by H_2 of substrates such as Cu^{II} and quinone under mild homogeneous conditions. It is surprising that this discovery was not appreciated or followed up for 15 years, despite the fact that it was only heterogeneous activation of H_2 that had been known up to that time. Halpern also comments that activation of H_2 had long been believed to be attributable to the distinctive properties of solid surfaces, which implies that H_2 had originally been thought to be unreactive with metal complexes in solution. This was understandable considering that the first transition metal hydride complex, $H_2Fe(CO)_4$, was discovered by Hieber only in 1931, and his claim of a covalent M–H bond remained controversial for many years afterwards![4] It is also surprising that it was only around 1955 that M–H bonds became better characterized and generally accepted.

The second homogeneous system involved hydrogenations catalyzed by cobalt carbonyl complexes, particularly hydroformylation of olefins (the oxo process):

$$RCH=CH + H_2 + CO \xrightarrow{[Co_2(CO)_8]} RCH_2CH_2C(=O)H \qquad (2.1)$$

This reaction was also discovered in 1938 by Roelen[5] and has been studied extensively, as will be discussed later.

There are three known mechanisms for homogeneous hydrogen activation:

1. *Homolytic Cleavage*: Both atoms of H_2 incorporate onto the metal as hydride ligands [oxidative addition (OA)] via a transient with an initially unknown struc-

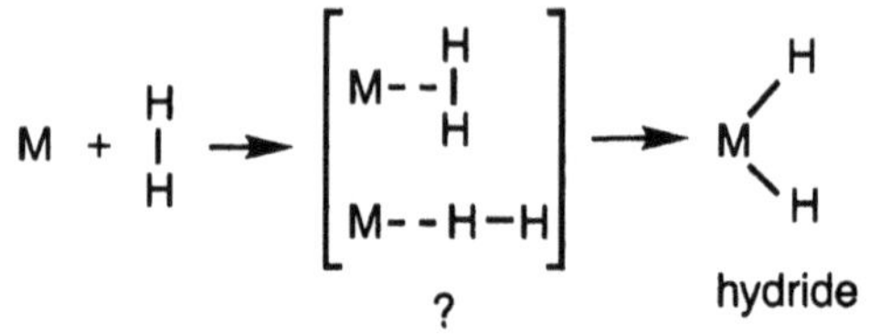

$$M + \begin{matrix}H\\|\\H\end{matrix} \rightarrow \begin{bmatrix} M\text{-}\text{-}\overset{H}{\underset{H}{|}} \\ M\text{-}\text{-}H\text{-}H \end{bmatrix} \rightarrow M\begin{matrix}\diagup H\\ \diagdown H\end{matrix}$$
$$\qquad\qquad\qquad ? \qquad\qquad\text{hydride}$$

ture. It is now clear that the transient is a σ complex of H_2, most likely the η^2-H_2 complex.

2. *Heterolytic Cleavage*: The H–H bond is split heterolytically with the hydride moiety (H^-) ligating to the metal and the proton (H^+) migrating to either an external Lewis base (intermolecular heterolytic cleavage) or to an ancillary ligand or anion (intramolecular heterolytic cleavage).[6] In intramolecular cleavage, the base is the ligand L (or anion) and a four-center intermediate can be involved whereby the

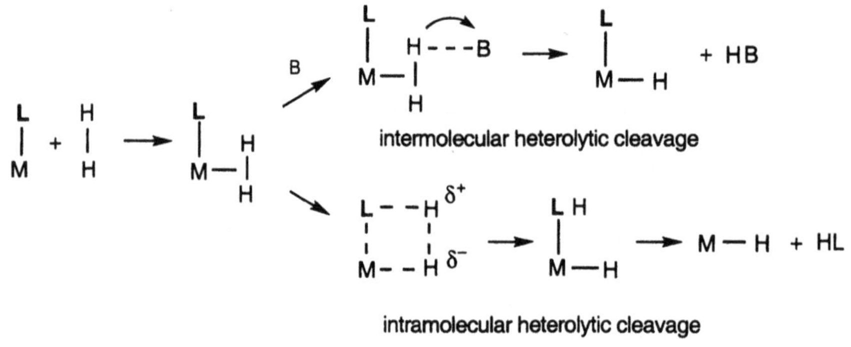

$$\underset{M}{\overset{L}{|}} + \underset{H}{\overset{H}{|}} \rightarrow \underset{M-|}{\overset{L}{\underset{H}{|}}}H \begin{array}{c} \overset{B}{\nearrow} \quad \underset{M-|}{\overset{L}{\underset{H}{|}}}\overset{H\text{-}\text{-}\text{-}B}{\rightarrow} \underset{M-H}{\overset{L}{|}} + HB \\ \text{intermolecular heterolytic cleavage} \\ \searrow \quad \underset{M\text{-}\text{-}H}{\overset{L\text{-}\text{-}H^{\delta+}}{\underset{\delta-}{|}}} \rightarrow \underset{M-H}{\overset{L\;H}{|}} \rightarrow M-H + HL \\ \text{intramolecular heterolytic cleavage} \end{array}$$

HL moiety remains coordinated or dissociates from M. For salts such as copper acetate, the acetate anion is protonated to form acetic acid. When coordinated to an electrophilic M, H_2 becomes very acidic and protonates the nearest Lewis base (see Chapter 9).

3. *Electron Transfer*: These are rarely observed reactions whereby H_2 is a reducing agent:

$$Hg^{2+} + H_2 \rightarrow Hg^0 + 2H^+ \qquad (2.2)$$

The activation on copper acetate involves heterolytic cleavage of hydrogen to form CuH (the hydrogenating agent) and acetic acid while that on $Co_2(CO)_8$ is homolytic.[1,2]

One of the breakthrough discoveries was that of Vaska and DiLuzio in 1962,[7] who found that the Ir^I complex that bears Vaskas name reacted reversibly with H_2 to give an Ir^{III} dihydride. Equation (2.3) and homolytic H_2 cleavage in the related

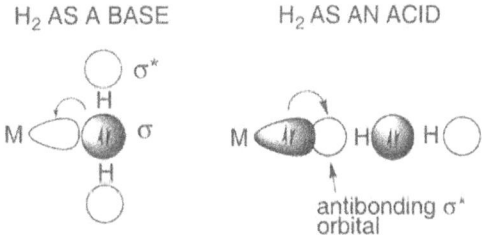

Wilkinson catalyst, $RhCl(PPh_3)_3$,[8] are prototypes of the OA reaction critically important to hydrogenation and homogeneous catalysis in general. It was now possible to directly observe the electronic and stereochemical properties of binding of a gaseous diatomic molecule on a metal complex and understand the factors that determine its activation toward cleavage.

Prior to our experimental finding of side-on bound H_2 there was considerable debate about how H_2 interacted with a metal center as the initial step in H–H cleavage. Some researchers believed that H_2 would act as a Lewis base, i.e., σ-donation to a vacant metal orbital, while others thought that it would behave as a Lewis acid, accepting electrons from a filled metal orbital into the σ* orbital (end-on bonded H_2). In 1959 both Halpern[9] and Syrkin[10] speculated that activation of H_2 involved attack of the bonding electrons of H_2 on a vacant metal d orbital via a three-center transition state as above (η^2-H_2). Other researchers believed the opposite: transfer of electrons from an occupied d orbital to the antibonding (σ*) orbital of η^1-H_2.[8,11] As will be shown in Chapter 4, the answer turned out to be essentially a combination of both interactions, where $M \rightarrow H_2$ σ* backbonding fulfilled the function of H_2 as a Lewis acid (electron acceptor). Regardless of structure, the initial M–H_2 interaction was regarded as only transient, and quantum-mechanical calculations were not even attempted for this problem until two decades later.

After our discovery of H_2 complexes in late 1982 detailed below, it was surprising to come across a review article by Orchin[12] published in 1972 on the mechanism of the oxo (hydroformylation) reaction that contained the precise

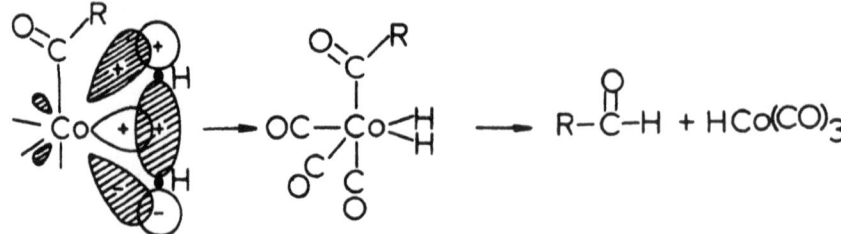

Figure 2.1. The hydrogenolysis of acylcobalt carbonyls by molecular hydrogen. Reprinted with authors' permission from Orchin and Rupilius.[12]

bonding picture (Figure 2.1) for η^2-H_2 coordination that we now know to be true (see Chapter 4). Orchin even stated that the bonding involves backdonation to σ^* that is entirely analogous to that in M–olefin π-bonding (although no attempt was made to justify this). The possibility of actually isolating an η^2-H_2 species was not mentioned, but Orchin did note that "the incipient dihydro species is then capable of hydrogenolyzing the Co–COR bond," as shown above. Theoretical calculations done many years later do indeed support $Co(\eta^2$-$H_2)$ intermediates in the oxo reaction.[13]

One of the first questions asked about η^2-H_2 ligands was whether they could directly react with substrates in catalytic processes without first cleaving to hydride ligands. As will be discussed in Chapter 9, there is now evidence for this. Also, η^2-H_2 when bound to cationic electrophilic metal centers can be highly acidic and readily undergo direct heterolytic cleavage. It can be deprotonated by external bases as well as transfer a proton internally to an ancillary ligand, two extremely important reactions similar to the earliest homogeneous cleavage reactions of H_2.

2.2. DISCOVERY OF DIHYDROGEN COMPLEXES

2.2.1. Background

Discoveries in modern synthetic inorganic chemistry have rarely been chronicled since most are either straightforward or not truly paradigm-shifting, but the clandestine nature of molecular H_2 coordination and the importance of this unexpected finding make this story worthy of telling.* As in most discoveries, serendipity played a key role, but what must be noted emphatically here was our adherence to three classic scientific principles: keen observation, recognizing and exploring anomalies, and maintaining an open mind. As Pasteur appropriately stated, "Chance favors only the prepared mind."

As I noted in the Preface, my particular scientific background and the pure joy of chemical synthesis played major roles in the discovery. My thesis work concerned metal hydrides, in particular Lewis acid–base complexes of dialkylzinc species with hydride ion, which in some cases were in equilibrium with hydride-bridged com-

---

*For similar accounts see also Kubas.[14]

plexes.[15] As with M–H$_2$ systems, M–H–M bonding is also nonclassical and can be viewed as a three-center interaction of M with a M–H bond. This research was

$$\underset{R'}{\overset{R}{>}}Zn \xrightarrow[\text{thf}]{\text{NaH}} \left[\underset{R'}{\overset{R}{>}}Zn-H \right]^- Na^+ \xrightarrow{R_2Zn} \left[\underset{R'}{\overset{R}{>}}Zn\underset{H}{\overset{\diagdown}{\diagup}} Zn\underset{R'}{\overset{R}{\diagdown}} \right]^- Na^+$$

$$>Zn \diagup + \diagup Zn \diagdown^{\ominus}_{H} \rightleftharpoons >Zn \leftarrow \diagup Zn \diagdown^{\ominus}_{H}$$

$$M + \overset{H}{\underset{H}{|}} \rightleftharpoons M \leftarrow \overset{H}{\underset{H}{|}}$$

in some sense prophetic of reversible H$_2$ binding. Many steps actually led up to the discovery of H$_2$ complexes, which followed an inverse Murphy's law: "Everything that could go right, did." Failure to unearth M–H$_2$ coordination loomed everywhere along the chain of events and observations detailed below. It was important that I performed all laboratory work myself rather than leaving it to students or technicians, as direct observation rather than second-hand reporting was critical throughout the 4 year period, 1979–1983, during which the W–H$_2$ structure was ultimately proven, and I often wonder if this discovery could have been made otherwise. There were actually two major findings: the serendipitous synthesis of the coordinatively unsaturated Mo and W precursors that reversibly bound H$_2$, and the recognition and proof that the complex coordinated H$_2$ molecularly.

In the mid to late 1970s we had been investigating the binding of SO$_2$ to metal complexes that causes acid rain. SO$_2$ is a strong ligand much like CO but more versatile, behaving as a σ base (planar M–SO$_2$), σ acid (pyramidal M–SO$_2$), or π-acid (η^2-S,O) toward M.[16] Its amphoteric bonding to virtually any type of unsaturated M has a remarkable parallel with that for the H$_2$ ligand. A series of octahedral, zero-valent, d^6 complexes, cis,trans-Mo(CO)$_2$(PPh$_3$)$_2$(SO$_2$)L, was synthesized in which the M–SO$_2$ geometries were highly sensitive to varying just one cis ligand, L.[17] If L was a strong π acceptor such as CO, SO$_2$ acted as a base, but if L was an electron donor, SO$_2$ switched to its π-acceptor mode to take advantage of the increased electron density at M available for $d\pi$–π^* backbonding. This sensitivity toward electronics at M is almost exactly mirrored by H$_2$, and a metal's backbonding ability is critical to σ-bond activation. Conversely, the electronic properties of specific sites on complexes can be precisely defined by studying SO$_2$ (and H$_2$) coordination geometry at the site. These studies led to uncovering H$_2$ coordination.

Crucial to any unexpected scientific finding is the first identifiable event that diverts one from the ordinary research path. This happened in 1979 and strangely enough turned out to be a mundane occurrence for a synthetic chemist: a low reaction yield! However, this particular case was intriguing because there was no obvious reason for only a 20% yield for a two-step synthesis of the Mo–SO$_2$

complex shown in Scheme 1. In order to explore steric effects, a complex with bulky phosphines, P^iPr_3 was synthesized wherein the latter coordinated in trans fashion,

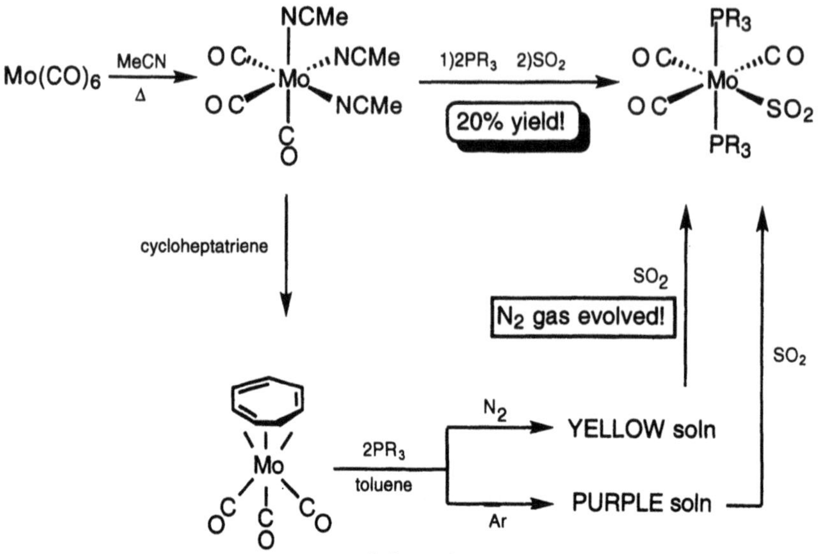

Scheme 1

apparently forcing the carbonyls to reorient to the less electronically favored meridonal geometry. The η^1-planar SO_2 indicated that the metal fragment is relatively electron-deficient, a crucial feature that would later favor H_2 coordination over OA to a dihydride. The synthesis involved stepwise replacement of labile MeCN ligands in fac-$Mo(CO)_3(NCMe)_3$ by two equivalents of phosphine in benzene followed by treatment with excess SO_2. Painstaking (and seemingly superfluous) efforts to increase the yield were made utilizing a different starting material, $Mo(CO)_3(cht)$ (cht=cycloheptatriene), and the PR_3 and SO_2 displaced cht to give the desired complex, but still in low yield. However, key observations were made while carrying out many reactions—notably evolution of a gas when SO_2 was added to the reaction mixture generated by the addition of P^iPr_3 to the Mo–cht complex under a N_2 atmosphere in a Schlenk apparatus. Serendipitously, when a N_2 tank emptied, Ar was substituted as the inert atmosphere, whereupon it was noted that gas evolution did not occur and that the normally orange $Mo(CO)_3(cht)$–phosphine reaction mixture attained an unusual deep-purple color just before SO_2 was added. It was surmised that the evolved gas could be N_2 displaced by SO_2 from a yellow reversibly bound N_2 complex and that the purple color was due to a coordinatively unsaturated species, $Mo(CO)_3(P^iPr_3)_2$ (Scheme 2).

Solutions of the purple species immediately turned yellow on exposure to N_2 and back to purple *in vacuo*. I had recalled that coordinatively unsaturated $RhCl(P^iPr_3)_2$ was also purple-lilac in color and reversibly bound N_2.[17] The possibility of a five-coordinate group 6 complex that could bind small molecules such as N_2 much like Vaska's complex was quite exciting because such a species was unknown. 16e $Cr(CO)_5$ fragments were unstable and were generally created by photolysis of $Cr(CO)_6$ in rare-gas matrices at very low temperatures.[18] These species had shown extraordinary binding capabilities toward weak ligands, coordinating CH$_4$ and even rare gases in the vacant site:

$$Cr(CO)_6 \xrightarrow{-CO} Cr(CO)_5 \to Cr(CO)_5L \qquad L = N_2, CH_4, Ar \qquad (2.4)$$

These species could only be studied spectroscopically, but we now had *stable* analogues that could be studied crystallographically. A series of 16e M(CO)$_3$(PR$_3$)$_2$

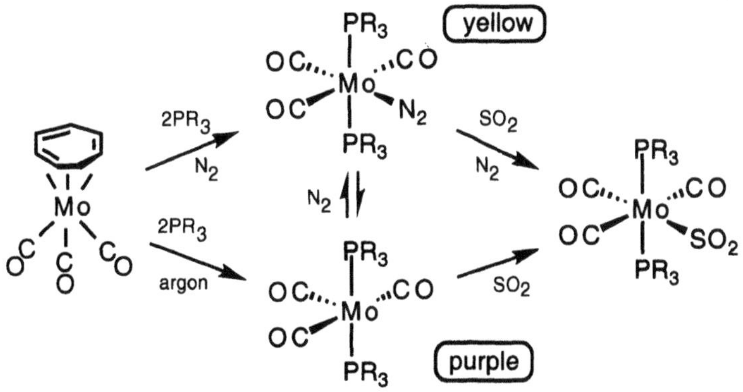

Scheme 2

species for M = Mo and W and R = ^{i}Pr and Cy were characterized[19] and stereoselectively coordinated, often reversibly, virtually any small molecule that could fit inside the congested coordination sphere. Only later did a crystal structure reveal an agostic interaction with a C–H bond from the phosphine alkyl groups that effectively "reserved" a site for incoming ligand binding and stabilized the electronic unsaturation (Figure 2.2).[20] Such agostic interactions (see Chapter 12) were among the first σ complexes, although they were internal in nature and could compete with weak external ligand binding mainly for entropic reasons (see Chapter 7). M(CO)$_3$(PR$_3$)$_2$ is structurally very dynamic and undergoes a large variety of reactions, including OA.[21]

The syntheses of M(CO)$_3$(PR$_3$)$_2$ succeed only for bulky phosphines and, curiously, only for P^{i}Pr$_3$ and PCy$_3$ whereas similar-sized phosphines such as P^{i}Bu$_3$ do not work. By good fortune I had chosen the only reaction and the only reactants

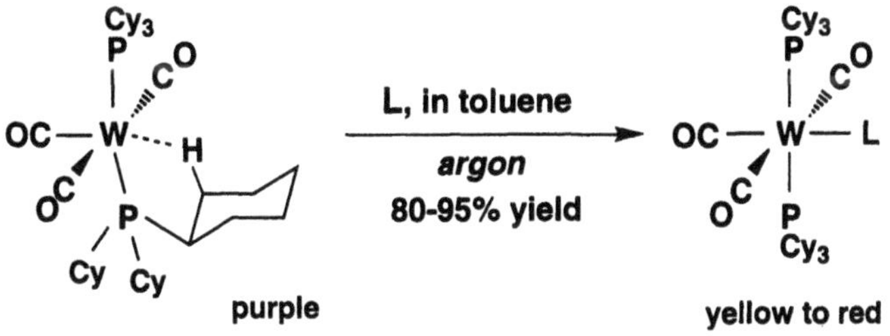

Figure 2.2. Types of coordination and reactivity with W(CO)$_3$(PR$_3$)$_2$.

that would give $M(CO)_3(PR_3)_2$, and alternate routes to these complexes could not be found despite numerous attempts. We later became aware that Moers and Reuvers had carried out a similar reaction of $Mo(CO)_3(cht)$ with PCy_3, but in benzene at 60°C under an unspecified atmosphere and had obtained only light-yellow $Mo(CO)_4(PCy_3)_2$.[22] Thus prior discovery of $Mo(CO)_3(PCy_3)_2$ was missed probably because of exposure to N_2 and/or atmospheric oxygen or extended heating, which leads to disproportionation to $Mo(CO)_4(PCy_3)_2$. Another opportunity to isolate $M(CO)_3(PR_3)_2$ was missed because of reaction conditions: photolysis of $W(CO)_4(P^iPr_3)_2$ in a hydrocarbon–glass matrix at 77 K was postulated to form $W(CO)_3(P^iPr_3)_2$,[23] but no further studies were done. Such are the vagaries of science: Allen and Senoff's discovery of N_2 complexes might not have materialized if their reaction had been carried out under an inert atmosphere other than N_2.

2.2.2. Discovery of Dihydrogen Coordination

The above findings were quite satisfying but merely a stepping-stone to the real breakthrough. Soon after reversible N_2 binding to $Mo(CO)_3(PR_3)_2$ was observed, H_2 was also noted to add reversibly to give yellow complexes with properties similar to the N_2 adducts. The catalytic activity of $M(CO)_3(PR_3)_2$ toward hydrogenation of ethylene under mild conditions was tested, but no reaction occurred because H_2 and C_2H_4 only displace each other on the one available coordination site. However, the complexes from H_2 addition to $W(CO)_3(PR_3)_2$ were intriguing because of their unusually facile H_2 loss *in vacuo* or under Ar and their anomalous IR frequencies — *the principal clue to the discovery of molecular coordination*. I had made it a standard practice to record the Nujol mull spectrum of every solid reaction product. Ironically, simple techniques often lead to important discoveries, e.g., mass spectroscopy in the case of buckminsterfullerene. The recognition of anomalies is usually the critical factor, and my experience in vibrational spectroscopy paid off here. Instead of the expected v_{WH} at 1700–2300 cm^{-1} and δ_{WH} at 700–900 cm^{-1} that would be characteristic of a seven-coordinate dihydride complex, three widely spaced bands near 1570, 950, and 465 cm^{-1} were observed (Figure 2.3). These bands underwent large shifts to 1140, 700, and 320 cm^{-1} when deuterium was substituted for hydrogen, a valuable technique for locating M–H modes that I learned in graduate school. In another stroke of luck, CsBr salt plates were routinely used along with an IR spectrometer capable of recording bands down to 250 cm^{-1}. Otherwise the telltale weak band at 465 cm^{-1} shifting to 320 cm^{-1} would have escaped notice because most spectrometers then in use did not record below 400 cm^{-1}. This low-frequency instrument was available because of a strong research program in vibrational spectroscopy at Los Alamos. Indeed I was fortunate to have first-rate colleagues engaged in IR and NMR spectroscopy and X-ray and neutron diffraction.

The anomalous low-frequency mode along with the high lability of the H_2 suggested that "the bonding of the hydrogen to these metal complexes may be novel," an exquisite understatement in the initial 1980 report.[19] However, it must be remembered that at the time, the mere possibility that molecular H_2 binding could be present was astounding to everyone, including myself. A testament to the general

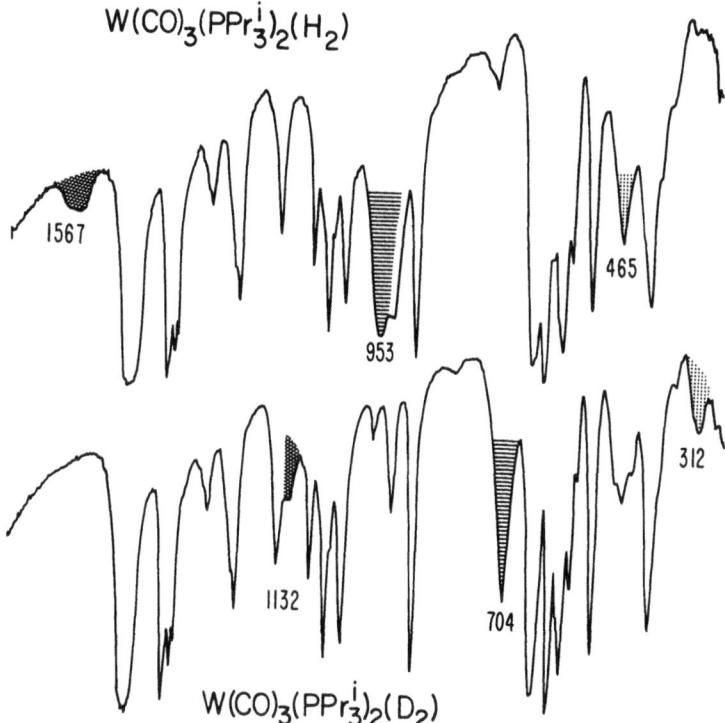

Figure 2.3. Nujol mull IR spectra (band positions in cm^{-1}) of H$_2$ and analogous D$_2$ complex. Reprinted with permission from Kubas et al.[25] Copyright 1986 American Chemical Society.

disbelief in H$_2$ coordination was the fact that virtually no one else pursued this research even after our provocative remarks and similar earlier speculation by Singleton on a ruthenium–hydrogen system (see below) were published. Presentations of these results at the Spring American Chemical Society and Inorganic Gordon Conferences in 1980 also elicited little fervor, but at least I had no worry about being scooped. However, it was like working in a vacuum. I was not even aware of Singleton's work and of most of the other contemporary theoretical and experimental studies relating to H–H and other σ-bond coordination.

X-ray diffraction studies of M(CO)$_3$(PCy$_3$)$_2$(H$_2$) produced only frustration because crystallographic disorder prevented location of hydrogen positions. This problem was destined to plague virtually all of our subsequent structural studies of H$_2$ complexes, the only bad luck encountered. What was encouraging was the fact that the CO and PR$_3$ were in octahedral arrangement with a vacancy in the sixth site hopefully containing the H$_2$ molecule. Efforts turned to neutron diffraction to locate the hydrogens since they have a much larger cross section toward neutrons than X-rays. However, neutron beam time was (and can still be) difficult to obtain. Fortuitously, a long-time colleague of mine, Phillip Vergamini, was setting up a single-crystal neutron diffractometer at the Los Alamos Neutron Scattering Center and supported my belief in H$_2$ coordination. Considerable time and effort were spent obtaining the large single crystals (ca 10 mm^3) needed for neutron diffraction. The very soluble W(CO)$_3$(P^{i}Pr$_3$)$_2$(H$_2$) complex was synthesized, and large crystals were obtained from hexane solutions. Phil was finally able to collect data in late 1982, but it had to be taken on a prototype instrument at room temperature, using a pulsed neutron beam from an accelerator source. Although a complete structure could not be obtained, partly because of disorder in one of the phosphines, a difference-Fourier map phased on the nonhydrogen-atom coordinates from room-temperature X-ray studies tenuously indicated the presence of a side-on bonded H$_2$ ligand. It was a thrilling moment, to at last find what we were hoping to find. A subsequent low-temperature X-ray study by Harvey Wasserman confirmed η^2-H$_2$ (Figure 2.4), showing d_{HH} of 0.75 (16) Å (X-ray) and 0.84 Å (neutron, ΔF). Since the distance in free H$_2$ is 0.74 Å, it was reasonably clear that the H$_2$ molecule was essentially intact and had not dissociated into well-separated (>1.6 Å) hydrides.

This was an exciting time, but we knew that the usual uncertainty in X-ray hydrogen locations and the marginal neutron evidence was destined to cause skepticism concerning our claim of H$_2$ coordination. Further proof was sought, and a consultant, the late Russ Drago, suggested an experiment that led to incontrovertible spectroscopic evidence for the presence of a weakened H–H bond. The idea was elegant in its simplicity: Synthesize the HD complex and look for a large HD coupling constant in the proton NMR that would show that the H–D bond was still mostly intact. The ^{1}H NMR of W(CO)$_3$(P^{i}Pr$_3$)$_2$(H$_2$) had already been observed to give a broad single resonance (-4.2 ppm) for the η^2-H$_2$ ligand, showing no coupling to P or W. This in itself was unusual in comparison to classical hydrides, but the HD substitution proved to be diagnostic: a clear 1:1:1 triplet (deuterium spin = 1) with $J_{HD} = 33.5$ Hz resulted (see Figure 5.5). This is almost as high as the value for HD gas, 43.2 Hz, and the coupling between hydride and deuteride ligands on a metal is usually unobservably small. Thus, a weakened H–D bond had to be present in W(CO)$_3$(P^{i}Pr$_3$)$_2$(HD) as also confirmed by vibrational spectroscopy. IR spectra of the HD complexes showed bands at positions intermediate to those for the H$_2$ and D$_2$ complexes and not the superimpositions of MH and MD modes previously observed for hydride–deuteride species. Finally, to cap it all off, a broad, weak absorption was located in the IR at 2360 cm^{-1} assignable to ν_{HD}. Then ν_{HH} was located at 2690 cm^{-1} in W(CO)$_3$[P(C$_6$D$_{11}$)$_3$]$_2$(H$_2$), which required deuterated phosphines to avoid interference from C–H modes (Figure 2.5). Analysis of the expected vibrational modes for η^2-H$_2$ and Raman studies by Basil Swanson added further evidence for side-on bonded H$_2$. Four of the six expected modes (see Chapter 8) were now observed. The Raman data showed shifts in several low-frequency bands at 400–600 cm^{-1} on isotopic substitution by D$_2$ indicative of mixing of unobserved deformational M–H$_2$ modes with M–CO bands. Such multiple low-frequency modes had never been seen in hydrogen-containing complexes.

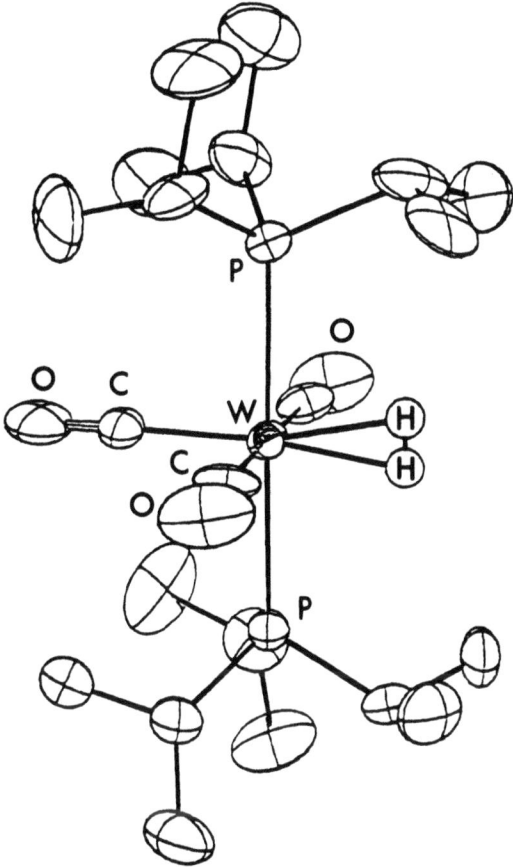

Figure 2.4. ORTEP drawing of X-ray structure of $W(CO)_3(P^iPr_3)_2(H_2)$ at $-100°C$. Reprinted with permission from Kubas et al.[24] Copyright 1984 American Chemical Society.

Bolstered by the new evidence, we presented the findings at the Spring 1983 American Crystallographic Association and American Chemical Society meetings, fully 4 years after the H_2 complexes were first synthesized. However, as we feared, the frail neutron data combined with the general prejudice against accurate location of hydrogen by diffraction methods and the improbable nature of the findings still fueled much skepticism. Because there have been notable debacles in chemistry (polywater and later cold fusion), such controversy over a new type of compound or chemical bonding was expected. In interesting parallels two decades earlier, Allen and Senoff's paper on N_2 complexes was summarily rejected initially and both Saul Winstein and H. C. Brown told Nobel Laureate George Olah that he most probably was wrong when he claimed at a conference to have obtained long-lived nonclassical carbonium ions.

Speculation over our provocative findings continued for months, especially after they were prematurely reported in a C&E News article in April 1983, highlighting new findings at the Seattle ACS meeting. We had not yet submitted a paper on our

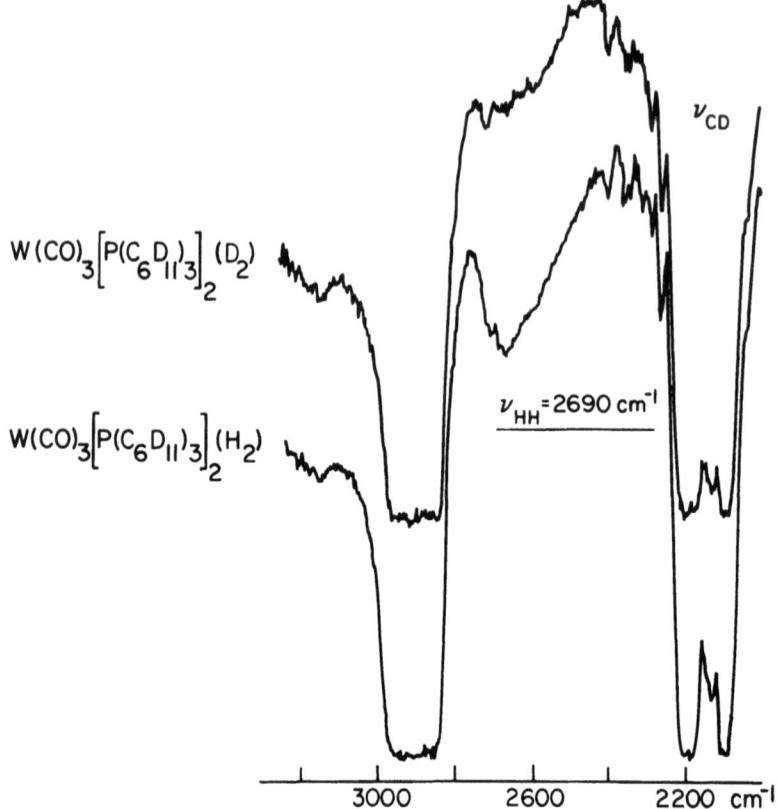

Figure 2.5. Nujol mull IR spectrum of $W(CO)_3[P(C_6D_{11})_3]_2(H_2)$ in ν_{HH} region. Reprinted with permission from Kubas et al.[25] Copyright 1986 American Chemical Society.

work to a refereed journal, and this C&E story added to the controversy until our seminal communication was finally published in *JACS* in January 1984.[24] The NMR and IR data proved to be indisputable evidence, and we later did obtain improved low-temperature neutron data from Brookhaven National Laboratory (see Figure 1.1). Although a large degree of disorder was still present in the C–H positions of one of the P^iPr_3 ligands (which hindered publication of the structure), the side-on bonded H_2 ligand was seen clearly with a d_{HH} of 0.82 Å. As mentioned in Chapter 1 and as will be discussed further on, tautomeric equilibrium of $W(CO)_3(PR_3)_2(H_2)$ with its dihydride form was soon found and elegantly demonstrated that H_2 complexes are true intermediates in the OA of H_2.[25]

During the period of our discovery and for considerable time afterward, H_2 binding seemed unique to our complexes because the bulky phosphines sterically inhibited OA to a seven-coordinate dihydride. Herb Kaesz, who also was a consultant to our group at the time, viewed this as "arrested oxidative addition," a term he had used to describe the bonding in $CpMn(CO)_2(\eta^2\text{-HSiPh}_3)$ as noted in Chapter 1. The Mn–η^2-Si–H system was indeed one of the first examples of a true σ-bond complex but was unrecognized as such because it lacked the superb clarity of the η^2-H_2 ligand, which does not contain any electrons outside the H–H bond. Steric factors such as interligand repulsions had also been initially invoked to help explain the arrest in this Si–H system. However, *electronic* rather than steric factors determine whether H_2 coordinates molecularly or scissions to hydrides. The hundreds of H_2 complexes that would eventually be synthesized after our discovery were totally unimaginable to us at the time. We even had difficulty knowing where to search for new examples, and it would take over a year before some were found by other researchers. Several complexes containing H_2 ligands had been prepared years before but were thought to be hydrides. In retrospect this was understandable because of the near invisibility of hydrogen atoms in X-ray crystallography and the rare use of neutron diffraction methods. The identification of O_2 and N_2 coordination by crystallography suffered no such problem. Also, virtually no one believed that an H_2 molecule, which lacks lone-pair or π electrons, could ligate to a metal in stable fashion.

2.2.3. The Development of Dihydrogen Coordination Chemistry

Molecular hydrogen binding represents one of the greatest hidden phenomena in chemistry. This was brought to light in 1986 when, using NMR methods, Crabtree examined several polyhydride complexes prepared decades earlier and found that they contained H_2 ligands.[26] A year earlier he had characterized a complex of his own, [IrH(H_2)(PPh$_3$)$_2$(bq)]$^+$, and demonstrated by J_{HD} that it possessed bound H_2, the first confirmation of our findings.[27] However, his revelation that complexes prepared as early as the late 1960s contained η^2-H_2 was an eye-opener. A list of these complexes is given in Table 2.1. The most interesting case is RuH$_2$(H_2)(PPh$_3$)$_3$, which was originally reported in 1968 by Knoth[28] and is perhaps the earliest unrecognized stable H_2 complex. The unusual nature of this complex elicited comments in the literature by Singleton in 1976 and at scientific meetings about the "dihydrogen-like nature" of the binding (similar to those in our 1980 paper).[29] Ironically, obtaining definitive proof for η^2-H_2 in this particular complex turned out to be a difficult challenge even long after H_2 binding was established.[30] Suitable

Table 1.1. Polyhydrides Later Recognized to Contain H_2 Ligands

Complex first synthesized	Reference	Identified as H_2 complex	Reference
RuH$_2$(H_2)(PPh$_3$)$_3$ (1968)	28	M–H_2 proposed (1976)	29
		NMR T_1 data (1986)	26
		Solid state NMR (1993)	31
FeH$_2$(H_2)(PEtPh$_2$)$_3$ (1971)	36	NMR T_1 data (1986)	26
		Neutron structure (1990)	37
[FeH(H_2)(Ph$_2$PC$_2$H$_4$PPh$_2$)$_2$]$^+$ (1977)	38	X-ray structure (1985)	35
ReH$_5$(H_2)(PPh$_3$)$_2$ (1982)	39	NMR T_1 data (1988)	26
		Neutron structure (1991)	40
RuH$_2$(H_2)$_2$(PCy$_3$)$_2$	41	NMR T_1 data (1988)	42
		X-ray structure (2000)	43

crystals could not be obtained for diffraction studies, the hydrogens exchanged too rapidly on the NMR timescale to determine J_{HD}, and the rotational motion of η^2-H_2 could not be observed by inelastic neutron scattering (Chapter 6). NMR T_1 and solid state ^{1}H NMR measurements (see Chapter 5) finally provided evidence for a short d_{HH}(0.93 Å).

The first spectroscopic evidence for a M–H_2 interaction was obtained by Sweany just prior to our initial neutron diffraction study in 1982. Photolysis of $Cr(CO)_6$ in the presence of H_2 in a rare-gas matrix was claimed to give $Cr(CO)_5(H_2)$ based on v_{CO}.[32] However, these results were admittedly difficult to publish until after our seminal paper appeared. At about the same time related papers also showed that $Cr(CO)_5(H_2)$ could be formed in liquid Xe or cyclohexane but is stable for only seconds at room temperature.[33] It is important to note that these findings proved that sterically demanding coligands are unnecessary for H_2 coordination per se but can impart thermal stability electronically and/or by stabilizing the 16e precursor by an agostic interaction.

Crabtree's report of NMR T_1 evidence for $[IrH_2(H_2)_2(PCy_3)_2]^+$ established an alternate NMR criterion for H_2 cooordination that is useful in highly fluxional polyhydride complexes.[34] This T_1 method would also prove to yield good estimates of d_{HH} in the solution state (see Chapter 5), and several complexes thought to be classical hydrides were reassigned as H_2 complexes (Table 2.1). The Ir complex was also the first bis(H_2) complex and the first synthesized by protonation of a metal hydride, a now common technique for generating H_2 ligands (see Chapter 3). Morris[35] determined the second crystallographic structure of an η^2-H_2 complex, trans-$[FeH(H_2)(dppe)_2]^+$, which had actually been prepared years earlier by several other groups but was unrecognized as containing H_2. The situation wherein the true identity or unique feature of a compound was revealed (or popularized) by a subsequent researcher is common in chemistry. However, our original Mo and W H_2 complexes had not been previously prepared, and we were unaware of any outside efforts concerning H_2 coordination.

The number of well-established new H_2 complexes has been rapidly increasing, and there are presently over 350 including ligand variations. The number of research papers in the field exceeds 800, including over 100 theoretical papers. There has been great deal of synergism between experiment and theory because of the simplicity of the H_2 ligand. Metal–H_2 systems and σ ligands in general are much more sophisticated and varied in bonding and reactivity than could have been imagined, as will be seen throughout this book.

REFERENCES

1. Halpern, J. *J.Organometal. Chem.* **1980**, *200*, 133.
2. James, B. R., *Homogeneous Hydrogenation*, Wiley: New York, 1973. See also: Harmon, R. E.; Gupta, S. K.; Brown, D. *J. Chem. Rev.* **1973**, *73*, 21.
3. Calvin, M. *Trans. Faraday Soc.* **1938**, *34*, 1181; Calvin, M. *J. Am. Chem. Soc.* **1939**, *61*, 2230.
4. Crabtree, R. H., *The Organometallic Chemistry of the Transition Metals*, Wiley: New York, 1988.
5. Roelen, O., U.S. Patent 2,327,066, **1943**; Roelen, O. *Angew. Chem.* **1948**, *60*, 62.
6. Brothers, P. J. *Prog. Inorg. Chem.* **1981**, *28*, 1.
7. Vaska, L.; DiLuzio, J. W. *J. Am. Chem. Soc.* **1962**, *84*, 679.

8. Osborn, J. A.; Jardine, F. H.; Wilkinson, G. *J. Chem. Soc. A* **1966**, 1711.
9. Halpern, J. *Adv. Catal.* **1959**, *11*, 301.
10. Syrkin, Ya. K., *Usp. Khim.* **1959**, *28*, 903.
11. Nyholm, R. S. *Proc. Int. Congr. Catal., 3rd*, **1964**, 25; Vaska, L.; Werneke, M. F. *Trans. N.Y. Acad. Sci.* **1971**, *31*, 70.
12. Orchin, M.; Rupilius, W. *Catal. Rev.* **1972**, *6*, 85.
13. Versluis, L.; Ziegler, T. *Organometallics* **1990**, *9*, 2985; Sola, M; Ziegler, T. *Organometallics* **1996**, *15*, 2611; Torrent, M.; Solà, M.; Frenking, G. *Chem. Rev.* **2000**, *100*, 439.
14. Kubas, G. J. *Acc. Chem. Res.* **1988**, *21*, 120; Kubas, G. J. *Comments Inorg. Chem.* **1988**, *7*, 17.
15. Kubas, G. J.; Shriver, D. F. *J. Am. Chem. Soc.* **1970**, *92*, 1949; Kubas, G. J.; Shriver, D. F. *Inorg. Chem.* **1970**, *9*, 1951.
16. Ryan, R. R.; Kubas, G. J.; Moody, D. C.; Eller, P. G. *Struct. Bonding (Berlin)* **1981**, *46*, 47.
17. Van Gaal, H. L. M.; Moers, F. G.; Steggerda, J. J. *J. Organometal. Chem.* **1974**, *65*, C43; Van Gaal, H. L. M.; van den Bekerom, F. L. A. *J. Organometal. Chem.* **1977**, *134*, 237.
18. Perutz, R. N.; Turner, J. J. *Inorg. Chem.* **1975**, *14*, 262; Perutz, R. N.; Turner, J. J. *J. Am. Chem. Soc.* **1975**, *97*, 4791; Turner, J. J.; Burdett, J. K.; Perutz, R. N.; Poliakoff, M. *Pure Appl. Chem.* **1977**, *49*, 271; Andrews, L.; Moskovits, M. *The Chemistry and Physics of Matrix Isolated Species*, Elsevier: Amsterdam, 1989.
19. Kubas, G. J. *J. Chem. Soc., Chem. Commun.* **1980**, 61.
20. Wasserman, H. J.; Kubas, G. J.; Ryan, R. R. *J. Am. Chem. Soc.* **1986**, *108*, 2294.
21. Butts, M. D.; Kubas, G. J.; Luo, X.-L.; Bryan J. C. *Inorg. Chem.* **1997**, *36*, 3341.
22. Moers, F. G.; Reuvers, J. G. A. *Rec. Trav. Chim. Pays-Bas* **1974**, *93*, 246.
23. Black, J. D.; Boylan, M. J.; Braterman, P. S.; Wallace, W. J. *J. Organometal. Chem.* **1973**, *63*, C21.
24. Kubas, G. J.; Ryan, R. R.; Swanson, B. I.; Vergamini, P. J.; Wasserman, H. J. *J. Am. Chem. Soc.* **1984**, *106*, 451.
25. Kubas, G. J.; Ryan, R. R.; Wroblewski, D. *J. Am. Chem. Soc.* **1986**, *108*, 1339; Kubas, G. J.; Unkefer, C. J; Swanson, B. I.; Fukushima, E. *J. Am. Chem. Soc.* **1986**, *108*, 7000.
26. Crabtree, R. H.; Hamilton, D. G. *J. Am. Chem. Soc.* **1986**, *108*, 3124; Hamilton, D. G.; Crabtree, R. H. *J. Am. Chem. Soc.* **1988**, *110*, 4126.
27. Crabtree, R. H.; Lavin, M. *J. Chem. Soc., Chem. Commun.* **1985**, 794.
28. Knoth, W. H. *J. Am. Chem. Soc.* **1968**, *90*, 7172.
29. Ashworth, T. V.; Singleton, E. *J. Chem. Soc., Chem. Commun.* **1976**, 705.
30. Gusev, D. G.; Vymenits, A. B.; Bakhmutov, V. I. *Inorg. Chim. Acta* **1991**, *179*, 195.
31. Wishiewski, L., Zilm, K.; Kubas, G. J.; Van Der Sluys, L. S., unpublished results **1993**.
32. Sweany, R. L. *J. Am. Chem. Soc.* **1985**, *107*, 2374.
33. Upmacis, R. K.; Gadd, G. E.; Poliakoff, M.; Simpson, M. B.; Turner, J. J.; Whyman, R.; Simpson, A. F. *J. Chem. Soc., Chem. Commun.* **1985**, 27; Church, S. P.; Grevels, F.-W.; Hermann, H.; Shaffner, K. J. *Chem. Soc., Chem. Commun.* **1985**, 30.
34. Crabtree, R. H.; Lavin, M. *J. Chem. Soc., Chem. Commun.* **1985**, 1661.
35. Morris, R. H.; Sawyer, J. F.; Shiralian, M.; Zubkowski, J. D. *J. Am. Chem. Soc.* **1985**, *107*, 5581.
36. Aresta, M.; Giannoccaro, P.; Rossi, M.; Sacco, A. *Inorg. Chim. Acta* **1971**, *5*, 115.
37. Van Der Sluys, L. S.; Eckert, J.; Eisenstein, O.; Hall, J. H.; Huffman, J. C.; Jackson, S. A.; Koetzle, T. F.; Kubas, G. J.; Vergamini, P. J.; Caulton, K. G. *J. Am. Chem. Soc.* **1990**, *112*, 4831.
38. Giannoccaro, P.; Sacco, A.; Ittel, S. D.; Cushing, M. A. *Inorg. Synth.* **1977**, *17*, 69.
39. Baudry, D.; Ephritikine, M.; Felkin, H. *J. Organomet. Chem.* **1982**, *224*, 363.
40. Brammer, L.; Howard, J. A.; Johnson, O.; Koetzle, T. F.; Spencer, J. L.; Stringer, A. M. *J. Chem. Soc., Chem. Commun.* **1982**, 1388.
41. Chaudret, B.; Commenges, G.; Poilblanc, R. *J. Chem. Soc., Chem. Commun.* **1982**, 1388.
42. Arliguie, T.; Chaudret, B.; Morris, R. H.; Sella, A. *Inorg. Chem.* **1988**, *27*, 598.
43. Borowski, A. F.; Donnadieu, B.; Daran, J.-C.; Sabo-Etienne, S.; Chaudret, B. *Chem. Comm.* **2000**, 543.

3

Synthesis and General Properties of Dihydrogen Complexes

3.1. STABLE DIHYDROGEN COMPLEXES

3.1.1. Introduction

Over 350 stable H_2 complexes have now been synthesized and characterized (Table 3.1), and about 100 more are either thermally unstable, transient species or are proposed to contain H_2 ligands. Every metal from V to Pt is represented, and one lanthanide complex is known. Only the very early transition metals and actinides have not been observed to form stable H_2 complexes. The great majority of complexes contain octahedrally coordinated d^6 metals, which are relatively low-valent, primarily because of the favorable electronic situation. Virtually all are coordinatively saturated, and the few that are not normally contain π-donating halide or pseuodohalide ligands, e.g., $RuHX(H_2)(PR_3)_2$ (X = Cl, I, SR). The exception is $[IrH_2(H_2)(P^tBu_2Ph)_2]^+$, which contains η^2-H_2 with one of the shortest observed d_{HH} (0.85 Å). Paramagnetic H_2 and σ complexes are virtually unknown, but unstable 19e H_2 complexes may be intermediates in H_2 cleavage on metal radicals.[1]

Most H_2 complexes are cationic because the increased electrophilicity of the M reduces BD that leads to OA, e.g., many varieties of H_2 complexes containing a Cp-type ligand have been isolated but all except $CpMn(CO)_2(H_2)$ are cationic. Neutral complexes normally contain a mixture of donor ligands, usually phosphines, with at least one π-acceptor ligand such as CO to moderate BD or strong trans-effect ligands such as hydride. The first H_2 complexes stabilized only by nitrogen-donor ancillary ligands were $Tp^*RhH_2(H_2)$, and, at about the same time, $[Os(NH_3)_5(H_2)]^{2+}$ and its ethylenediamine analogues, which have very long d_{HH} (ca 1.34 Å), more characteristic of dihydrides. Complexes containing only aquo, CO, or hydrocarbon ligands are known, although they have not been isolated as solids because of thermal instability.

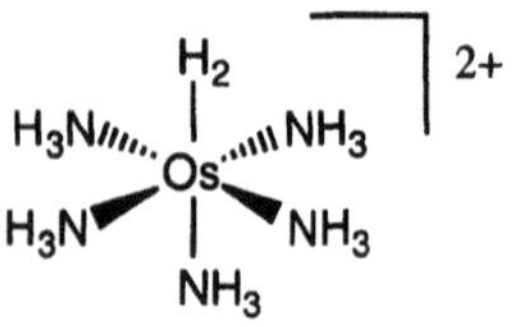

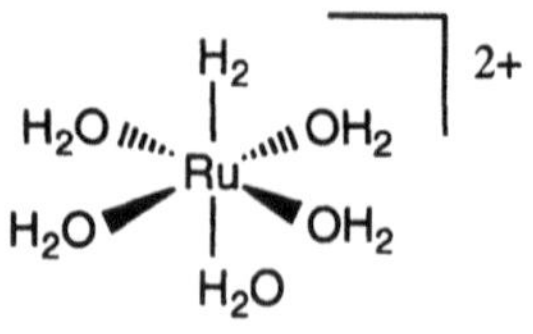

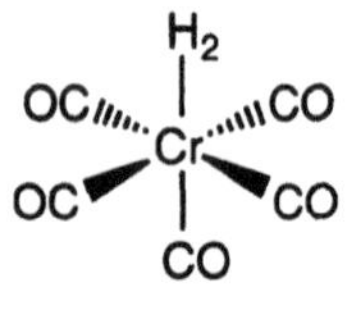

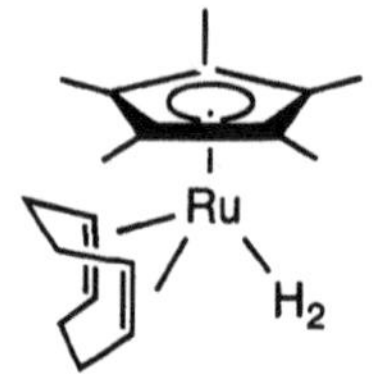

Several synthetic routes to H_2 complexes are available. The simplest method is reaction of H_2 gas with a coordinatively unsaturated complex such as $W(CO)_3(PR_3)_2$, the original method, as described in Chapter 2. This is analogous to the well-studied reversible addition of H_2 to Vaska's complex, $IrCl(CO)(PPh_3)_2$, except that for the latter OA to the dihydride occurs. A second, very common method of preparation is protonation of metal hydride complexes:

$$M-H \xrightarrow{H^+} [M-(H_2)]^+ \qquad (3.1)$$

In the late 1950s certain transition metal complexes were recognized to be basic and subject to protonation with acids. This method is an offshoot whereby protonation occurs at the basic hydride ligand and has been widely applicable because it does not require an unsaturated precursor that often either does not exist or is difficult to synthesize. Neutral polyhydride complexes L_nMH_x are often easy targets for protonation to cationic hydrido–H_2 complexes, $[L_nM(H_2)H_{x-1}]^+$, which are frequently more robust than complexes prepared from H_2 gas.

Table 3.1 lists the various generic types of H_2 complexes and primary literature references for the known variants of each class. These include complexes that may not be isolatable as solids at room temperature but are well-defined H_2 complexes from NMR or other evidence. The total number of known complexes has doubled since the latest review article in 1993,[120] and clearly the group 8 triad contains the overwhelming majority, with Ru and Os displaying the greatest variety of fragment types (especially Cp-type ligands[121]). The most common fragment is $[MH(H_2)P_4]^+$, where there are over 45 different variants, almost half of which are for Ru (P = phosphorus donor, primarily in a planar array). The series $[Os(H_2)(L)N_4]^{+/2+}$ (N_4 = $4NH_3$ or 2en) contains 27 members, which as in the previous series is ideal for correlating electronic and physical properties such as J_{HD}.[122] There are several isoelectronic series across the periodic table, e.g., $W(H_2)(CO)_3(PR_3)_2$, $[Re(H_2)(CO)_3(PR_3)_2]^+$, and $[Os(H_2)(MeCN)_3(PR_3)_2]^{+2}$. Note that the last does not contain π-acceptor CO ligands, which generally stabilize H_2 coordination against OA to hydride ligands, but instead the dipositive charge on M reduces BD that otherwise might promote OA (see Chapter 4). Highly electrophilic cationic metals are thus excellent targets for the design of σ complexes because increased σ donation to M stabilizes the bonding but can never cause the σ bond to rupture.

Many variations of "half-sandwich"-type complexes are known for Cp, Tp, and triazacyclononane (Cn) ligands, including the entire series for Ru^{II}. Because the Cn

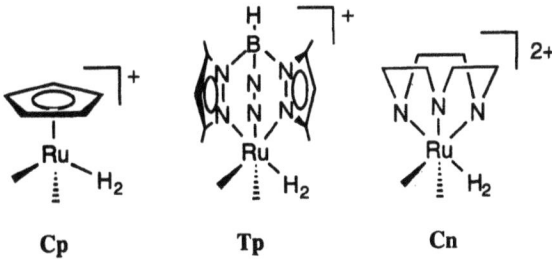

Cp   Tp   Cn

ligand is neutral, the corresponding Ru complex is dicationic but can still be prepared by protonation of the hydride monocation with strong acids.[83] As will be discussed in Chapter 9, the H_2 ligands in these and related cationic complexes are quite acidic, especially in the more electrophilic dicationic species such as the Cn complexes.

Very few bis(H_2) complexes are known:

$[Os(H_2)_2(MeCN)_2(P^iPr_3)_2]^{+2}$, $[OsH_3(H_2)_2(P^iPr_3)_2]^+$, $[IrH_2(H_2)_2(PCy_3)_2]^+$,

$RuH_2(H_2)_2(PR_3)_2$, (R = Cy, ^iPr) and $Tp^*RuH(H_2)_2$

Only the last two types have been isolated as solids, and the X-ray structure of $RuH_2(H_2)_2(PCy_3)_2$ shows unstretched cis H_2 ligands as in structure **2** below (X = H). The novel, X-ray-characterized 16e species $RuHX(H_2)(PCy_3)_2$ (X = Cl, I) add a second H_2 ligand in equilibrium fashion [Eq. (3.2), observable only in solution].[65]

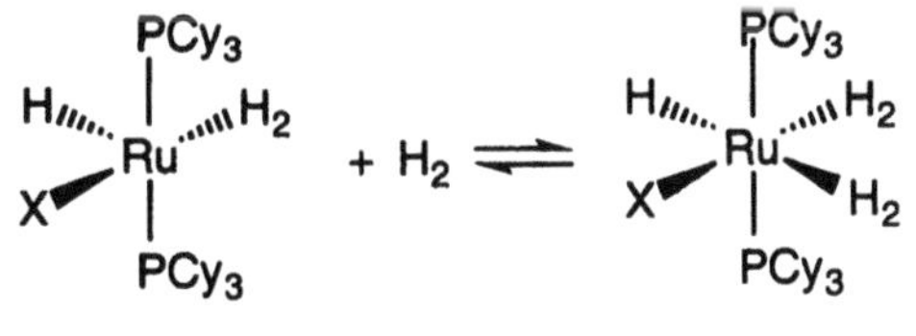

$$H_{\cdots}\overset{PCy_3}{\underset{PCy_3}{\overset{|}{\underset{|}{Ru}}}}\cdots H_2 \;+\; H_2 \;\rightleftharpoons\; H_{\cdots}\overset{PCy_3}{\underset{PCy_3}{\overset{|}{\underset{|}{Ru}}}}\overset{\cdots H_2}{\underset{H_2}{}} \qquad (3.2)$$

X   X

(1)   (2)

A complex with *two different* η^2-coordinated σ bonds is known, $RuH_2(\eta^2-H_2)$-$(\eta^2\text{-}SiHPh_3)(PCy_3)_2$, **3**, wherein the bulky phosphines are in an unusual cis configuration (P–Ru–P = 109.7°) apparently because favorable attractive interactions between the Si atom and the hydride ligands (see Chapter 4) overcome steric repulsion between the phosphines.[64]

Very few polynuclear complexes are known and these are primarily dinuclear hydride- and/or halide-bridged Ru, Os, and Ir complexes containing H_2 bound to only one of the metals (Table 3.1). Bridging H_2 ligands have not been definitively

Table 3.1. Stable H_2 Complexes by Generic Type[a]

Configuration	Complex	Reference
Group 5		
$M^{III}\ d^2$	$[Cp'_2M(H_2)(L)]^+$	Nb (L = P, CNR); Ta (L = CO): 2
Group 6		
$M^0\ d^6$	$M(H_2)(CO)_3(P)_2$	Cr: 3; Mo: 4; W: 4, 5
	$Mo(H_2)(L)(PP)_2$	6 (L = CO); 7 (L = CNR)
Group 7		
$M^I\ d^6$	$CpMn(H_2)(CO)_2$	8
	$[Mn(H_2)(CO)_n(P)_{5-n}]^+$	n = 1–3: 10
	$MnH(H_2)(PP)_2$	11
	$MCl(H_2)(PP)_2$	Tc: 12; Re: 13
	$ReCl(H_2)(P)_4$	14
	$[ReH(H_2)(CO)(NO)(P)_2]^+$	15
	$ReBr_2(H_2)(NO)(P)_2$	16
	$[Re(H_2)(CO)_n(P)_{5-n}]^+$	n = 1–4: 15, 17–21
	$[Re(H_2)(CNR)_n(P)_{5-n}]^+$	n = 3, 5: 20, 22
	$[Re(H_2)(NNN)(P)(PF_3)]^+$	23
	$[Cp*Re(H_2)(CO)(NO)]^+$	24
$M^{III}\ d^4$	$[ReH_2(H_2)(CO)(P)_3]^+$	25
	$Re^{III}(H_2)(N_3N_F)$	26
$M^V\ d^2$	$ReH_5(H_2)(P)_2$	27
	$[ReH_4(H_2)(PPP)]^+$	28
	$[ReH_{8-2n}(H_2)_n(P)_2]^+$	29
Group 8		
$M^{II}\ d^6$	$[MH(H_2)(P)_4]^+$	Fe, Ru, Os: 30, 31
	$[MX(H_2)(P)_4]^+$	Ru, Os; X = Cl, Br, I, SEt: 191
	$[MH(H_2)(PP)_2]^+$	Fe, Ru, Os: 32–37
	$[MH(H_2)(PPPP)]^+$	Fe, Ru, Os: 38–41
	$[MX(H_2)(PP)_2]^+$	Fe, Ru, Os; X = Cl, CN, CNBF$_3$: 42–44
	$[MCl(H_2)(L)(PNP)]^+$	Ru, Os; L = CO, P: 45
	$[M(H_2)(L)(PP)_2]^{2+}$	Fe, Ru, Os; L = CO, CNH: 43, 46, 47
	$MH_2(H_2)(P)_3$	Fe, Ru: 48–53
	$RuH_2(H_2)_2(P)_2$	51, 54
	$MH_2(H_2)(CO)(P)_2$	Ru, Os: 55–57
	$[OsH_2(H_2)(NO)(P)_2]^+$	58
	$MHX(H_2)(CO)(P)_2$	Ru, Os; X = Cl, I, C≡CR: 55–57, 59, 60
	$MCl_2(H_2)(P)_2(L)$	Ru: 61; Os: 57 (L = CO), 62 (L = pyrazole)
	$[RuH(H_2)(PPP)(L)]^+$	63 (L = CO, P)
	$RuH_2(H_2)(SiHR_3)(P)_2$	64
	$RuHX(H_2)(P)_2$	X = Cl, I, SR: 65, 66
	$RuH(\eta^2\text{-}YZ)(H_2)(P)_2$	67, 68 (YZ$^-$ = chelate)
	$Ru(\eta^2\text{-}YZ)_2(H_2)(P)_2$	68
	$Ru(H_2)(P)("S_4")$	69
	$Ru[\eta^3\text{-}Sn_3Ph_6(\mu\text{-}OMe)_2]^-$$(H_2)(CO)(P)$	70
	$[Ru(C≡CR)(H_2)(PP)_2]^+$	71
	$[Cp'M(H_2)(PP)]^+$	Fe, Ru: 72–74
	$[Cp'M(H_2)(CO)_n(P)_{2-n}]^+$	Fe: 75 (n = 1); Ru: 73, 76 (n = 0–2) Os: 77 (n = 2)
	$[Cp'Ru(H_2)(L)]^+$	78 (L = COD, tmeda)
	$Tp'RuH(H_2)_2$	79
	$Tp'RuH(H_2)(L)$	79, 80 (L = P, py, THT, NHEt$_2$)

Table 3.1. Continued

Configuration	Complex	Reference
	[Tp'M(H$_2$)(L)(L')]$^+$	Ru, Os: 81–83 (L, L' = P, CO, MeCN, H$_2$O)
	[CnRu(H$_2$)(L)(L')]$^{2+}$	83 (L, L' = P, CO)
	OsClX(H$_2$)(P)$_2$(L)	62, 84 (X = Hbim, η^2-S$_2$CR), 85 (X = SiEt$_3$, L = CO)
	Os(OSCR)$_2$(H$_2$)(P)$_2$	84
	M(OEP)(H$_2$)(L)	Ru, Os: 86, 87 (L = THF, Im)
	[OsH(H$_2$)(CO)(P)$_2$(L)]$^+$	88 (L = vacant or H$_2$O)
	[OsX(H$_2$)(P)$_2$(L)]$^+$	62, 89–91 (X = OAc, L = P; X = N,S-donor, L = CO)
	[Ru(H$_2$O)$_5$(H$_2$)]$^{2+}$	92
	[Os(NH$_3$)$_4$(H$_2$)(L)]$^{+/2+}$	93–95 (L = halide, O or N donor, carbene)
	[Os(en)$_2$(H$_2$)(L)]$^{+/2+}$	95, 96 [L = halide, OAc, O or N donor, M(CN)$_6^{4-}$]
	[Os(H$_2$)(MeCN)$_n$(P)$_{5-n}$]$^{2+}$	97 (n = 3), 98 (n = 1)
	[Os(H$_2$)$_2$(MeCN)$_2$(P)$_2$]$^{2+}$	97
	[M(H$_2$)(bipy)(CO)(P)$_2$]$^{2+}$	Ru, Os: 98
	[Os(H$_2$)(bipy)$_2$(CO)]$^{2+}$	98
Dinuclear		
	Ru$_2$H(μ-X)$_3$(H$_2$)(P)$_4$	51, 54, 99 (X = H/Cl)
	Ru$_2$Cl(μ-Cl)$_3$(H$_2$)(PP)$_2$	100
	Ru$_2$Cl$_2$(μ-H)$_2$(H$_2$)$_2$(PP)$_2$	101
	[Ru(H$_2$)(P)$_2$(μ-X)$_3$(μ-CO)ReH(P)$_2$]$^+$	102
	Ru(H$_2$)(CO)$_2$(P)[Sn$_3$Ph$_6$]	102
	OsCl(H$_2$)(P)$_2$(μ-Hbim)MCl(cod)	62 (M = Rh, Ir)
M^{IV} d^4	[RuH$_3$(H$_2$)(P)$_3$]$^+$	103
	[Cp*MH$_2$(H$_2$)(P)]$^+$	104 (M = Ru, Os)
	OsH$_2$(X)(Y)(H$_2$)(P)$_2$	105 (X,Y = H, halide)
	[OsH$_3$(H$_2$)$_2$(P)$_2$]$^+$	97
Group 9		
M^I d^8	Rh(H$_2$)(P–C–P)	106
M^{III} d^6	Tp'RhH$_2$(H$_2$)	107
	[Tp'MH(H$_2$)(P)]$^+$	108 (M = Rh, Ir)
	[IrH(LL)(H$_2$)(P)$_2$]$^+$	109, 110 (LL = bq or diphpyH)
	[IrH$_2$(H$_2$)$_2$(P)$_2$]$^+$	109
	[IrH$_2$(H$_2$)(P)$_2$]$^+$	111
	[IrH$_2$(H$_2$)(P)$_3$]$^+$	112
	[MH$_2$(H$_2$)(PPP)]$^+$	113 (M = Rh, Ir)
	IrH$_2$X(H$_2$)(P)$_2$	114 (X = Cl, Br, I)
	IrX$_2$H(H$_2$)(P)$_2$	115 (X = Cl, Br)
Dinuclear		
	Ir$_2$H$_3$(μ-H)(H$_2$)(μ-pz)$_2$(P)$_2$	116
Group 10		
M^{II} d^8	[PtX(H$_2$)(P)$_2$]$^+$	117, 118 (X = H, Me, Ph)
Lanthanide		
M^{II}	Cp$_2^*$Eu(H$_2$)	119

[a] Abbreviations: P, PP, PPP and PPPP = mono-, bi-, tri- and tetra-dentate phosphorus donors, including phosphites; NNN = tridentate amine; X = halide or anionic ligand; Cp' = Cp, Cp*, or other cyclopentadienyl derivative. See List of Abbreviations for others.

proven by diffraction methods, although there is NMR evidence in a porphyrin system (see Chapter 5). It is at times extremely difficult to determine conclusively

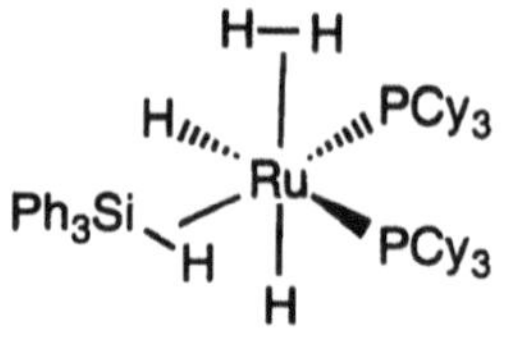

(3)

whether or not a complex contains hydride or η^2-H_2 (or how many of each). This is a problem especially in polyhydride complexes that contain both classical hydrides and η^2-H_2 that undergo dynamic exchange even at the lowest temperature accessible by solution NMR. The classic example is $RuH_2(H_2)(PPh_3)_3$, which had long been thought to contain η^2-H_2 (only evidence: solution NMR T_1) but had defied attempts to prove it definitively by diffraction methods.[48,123] Solid state NMR finally showed the d_{HH} to be 0.93 Å.[124] Another good example is $[Cp^*Ru^{IV}H_2(H_2)(PPh_3)]^+$, which could not be distinguished from a bis(H_2) species, $[Cp^*Ru^{II}(H_2)_2(PPh_3)]^+$.[104] Not surprisingly, as shown by Heinekey,[125] there have been cases of misassignment, even for complexes containing only two H atoms on M. Table 3.2 lists complexes that possibly contain coordinated H_2 and/or have d_{HH} in the "gray zone" (1.4–1.6 Å) between formulation as H_2 or dihydride complexes. In this regard, $Re^{III}(N_3N_F)(H_2)$ is a rare example of a relatively high-valent M^{III} system that

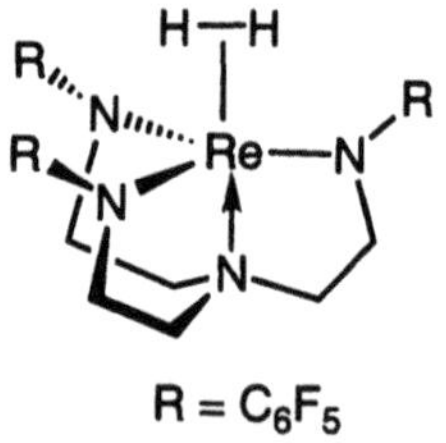

$R = C_6F_5$

can be considered to be a diamagnetic 18e H_2 complex.[26] Although the structure has not been determined, the value of J_{HD}, 17 Hz, for the HD complex is still within the range for an H_2 complex, although the low value indicates an elongated H–H bond (~1.15 Å; see Chapter 5). The "high" oxidation state of M (nearly always 0–II for H_2 complexes) and the amido ligand set are both unusual for any type of σ complex.

η^2-H_2 may also exist in solutions of organometallic complexes as equilibrium or transient species that cannot be observed spectroscopically. In the solid state, transition metal cations such as Ni^+ supported on silica and zeolitic materials show interaction with H_2,[145,146] and the rotational dynamics of H_2 absorbed on CoNa–A zeolite have been studied by inelastic neutron scattering (INS; see Chapter 6).[147] Other weak interactions of H_2 with surface species, bare metal ions, and main-group Lewis acids/bases are known and will be discussed at the end of this chapter and in Sections 14.3 and 14.4. It is proposed that short d_{HH}, as low as 1.5 Å ("hydrogen pairing"), are present in certain intermetallic rare-earth hydrides as shown in Table

Table 3.2. Complexes Possibly Containing H_2 Ligands and/or Having $d_{HH} < 1.6$ Å

Complex	Evidence	Reference
$\{Cp_2Ta[P(OMe)_3]H_2\}^+$	Equilibrium?	126
$Cp_2MH(H_2)^+$ (M = Mo, W)	Detected transient	127
$[MoH_4(H_2)(dppe)]^+$	Detected transient	128
$[WH_3(H_2)(PMePh_2)_4]^+$	Detected transient	129
$[WH(H_2)_2(PMePh_2)_4]^+$	Detected transient	129
$[WH_3Cl_2(PMe_2Ph)_4]^+$		130
$CpRe(CO)H_2$	Equilibrium? $J_{HD} = 6.5$	40
$ReH_7(\eta^2\text{-triphos})$	$\{T_1\}_{min} = 37$ ms	131
$ReH_5(PPh_3)_2(L)$	$\{T_1\}_{min} = 76$ ms	132
$ReH_3(\eta^3\text{-triphos})(L)$	$\{T_1\}_{min} = 60\text{--}66$ ms	131
$Cp^*FeH_3(PR_3)$	Equilibrium? $J_{HD} = 4.8$	133
$Cp^*RuH_3(PR_3)$		134
$[Ru(\eta^6\text{-}C_6H_6)(H_2)(CH_3CN)]^{+2}$		135
$Ru(H_2)(CO)_2(PBu^n_3)$	Intermediate?	136
$[RuX(H_2)(P)_2(CO)]^+$	Unstable	91
$Ru_2H(\mu\text{-H})_3(H_2)(P^iPr_3)_4$	$T_1 = 110$ ms	66
$[OsH(H_2)(tetraphos)_2]^+$	$T_1 = 160$ ms	137
$[OsH_3(H_2)(PR_3)_3]^+$	$T_1 = 30\text{--}82$ ms	48, 130, 138
$[OsH_3(H_2)(PPhMe_2)_3]^+$	$d_{HH} = 1.49$ (4) Å (neutron)	138–140
$[OsH(H_2)(MeCN)_2(PPr^i_3)_2]^+$	$\{T_1\}_{min} = 65.5$ ms	97
$Cp^*Rh(SiEt_3)H_3$	Transient?	141
$RhCl_2H(H_2)(PR_3)_2$	Undetected transient	142
$Tp'IrH_4$	Equilibrium? $J_{HD} = 17$?	143
$IrH_2Cl_2(PR_3)_2$	Equilibrium?	144
$Pt(Cy_2PC_2H_4PCy_2)(H_2)$	NMR intermediate	148
$CeNiInH_{1.0}$ (bulk hydride)	Pake doublet in ^{1}H NMR	149
$PrNiInH_{1.29}$ (bulk hydride)	Pake doublet in ^{1}H NMR	150

3.2. The observation of a Pake doublet at 140 K gives a d_{HH} of 1.48 ± 0.02 Å in $CeNiInH_{1.0}$, suggesting that the hydrogens may occupy nearest-neighbor tetrahedral sites separated by about 1.4 Å (2.1 Å had generally been believed to be the closest possible spacing in metal hydrides).[149] This phenomenon has been studied intensively both experimentally by methodologies such as solid state NMR[150] and theoretically.[151]

3.1.2. General Properties

The properties of H_2 complexes vary tremendously, depending on the degree of activation of the H_2 ligand toward the dihydride form, i.e., d_{HH}. True H_2 complexes with $d_{HH} < 0.9$ Å typically have labile H_2 ligands that readily exchange with D_2 and in some cases impart isotopic scrambling to produce HD (see Chapter 9). Atmospheric N_2 can even displace the H_2 ligand in these complexes. Most H_2 complexes are air-sensitive, reacting with oxygen to cause decomposition or, very rarely, O_2 binding. The exceptions tend to be cationic species of late transition metals such as $[IrH(H_2)(PPh_3)_2(bq)]^+$, $[RuCl(H_2)(PP)_2]^+$, and $[PtH(H_2)(P^iPr_3)_2]^+$. The last is stable to air even in solution (although it is thermally unstable above $-30°C$). Thus H_2 complexes are best prepared, handled, and stored under hydrogen-enriched atmospheres of rare gases such as Ar or He. Occasionally the solid complexes can be handled under N_2 or even briefly in air although it is often necessary to use an Ar-flushed glove bag filled with an Ar–H_2 (or D_2) mixture, e.g., when preparing Nujol-mull IR samples of H_2 or D_2 complexes. Stability to air increases toward the later transition elements, down the group, and for complexes that are more hydridic in character (longer d_{HH}). A trace amount of water in the atmosphere or solvent is usually not a problem if excess H_2 is present since H_2 competes favorably with H_2O binding.

Another common but key feature is lability of the H_2 ligand, which has two important connotations, namely *reversibility* and *ease of displacement* by other ligands. Reversibility in the strictest sense means that the H_2 can be removed *in vacuo*, by passage of an inert gas over the complex or by heating, either in solution or solid states, to regenerate a stable precursor that readds H_2 for at least several cycles without major degradation. This property was found for the original W complex and is obviously more common for the complexes prepared from H_2 gas, which are shown in Table 3.3 (though not all such complexes show facile reversibility). Often the solid will have a measurable H_2 dissociation pressure [~ 10 Torr for $W(H_2)(CO)_3(P^iPr_3)_2$], necessitating an H_2-enriched atmosphere over the complex at all times. Major color changes can occur on H_2 loss *in vacuo* and readdition on exposure to H_2 gas and are usually rapid, even in the solid. This is often an easy, valuable (and visually impressive) test of reversibility. Morris has tabulated the stability of a wide variety of H_2 complexes in both solution and solid states.[152]

Facile displacement of η^2-H_2 by more strongly bound ligands can occur for both the above cases and for systems that do not bind H_2 reversibly.[152] For group 6 and certain other complexes, this includes coordinating solvents such as THF and acetonitrile. The $M(H_2)(CO)_3(PR_3)_2$ series also decomposes in halogenated solvents such as dichloromethane, severely limiting the solvents available for use to hydrocarbons for the most part. However, some complexes are quite robust and stable to H_2 loss even on heating in coordinating solvents. In the Fe complex (4), the H_2 is so strongly bound that when it is used as a hydrogenation catalyst for alkynes to alkenes, a free coordination site for the incoming alkyne is provided by detachment of a phosphine arm instead of H_2 loss.[153] However catalysis by the Ru analogue

(4)

goes by usual H_2 loss,[154] which illustrates the difficulty in predicting stability. Normally a second-row M gives more stable complexes than a first-row M, but the situation is reversed for the group 8 metals apparently because INS experiments (see Section 6.2.3.2) and other evidence show that Fe is the better backbonder

Table 3.3. Complexes Prepared by Reversible Addition of H_2 to a Known Precursor Complex

Complex[a]	Precursor structure	H_2 lability	Reference
$M(H_2)(CO)_3(PR_3)_2$ (M = Cr, Mo, W)	Agostic	V high-med	3–5
trans-$Mo(H_2)(CO)(PP)_2$	Agostic	Med	6
$[Mn(H_2)(CO)_3(P)_2]^+$	Agostic,[b] solvento[c]	High	9
trans-$[Mn(H_2)(CO)(PP)_2]^+$	Agostic	High	9
trans-$[Mn(H_2)(CO)\{P(OR)_3\}_4]^+$	Agostic?	High	10
$Tc(H_2)Cl(dppe)_2$	Trig bipy	Med	12
$[Re(H_2)(CO)_3(PR_3)_2]^+$	Agostic	Med	15, 19, 20
$[Re(H_2)(CO)_4(PR_3)]^+$	Solvento	High	21
$[Re(H_2)(CO)_2(triphos)_2]^+$	Agostic	Med	18
$[CpRu(tmeda)(H_2)]^+$	2-Leg piano stool	Low	78
$[Ru(H_2)H(PP)_2]^+$		Med-high	32, 34, 36, 37
$[M(H_2)(CN)(PP)_2]^+$ (M = Fe, Ru)	Anion-coord	Med	47
$[M(H_2)(L)(PP)_2]^{2+}$ (M = group 8; L = CO, CNH)	Anion-coord	Med	43, 46, 47
$Ru(H_2)H_2(CO)(P^tBu_2Me)_2$	Sq pyr	High	56
$[Ru(H_2)Cl(PP)_2]^+$	Trig bipy	V high-med	35, 42, 43
$Ru(H_2)Cl_2(P-N)(PR_3)$	Sq pyr	High	61
$M(H_2)Cl(H)(CO)(P^iPr_3)_2$ (M = Ru, Os)		Med	55, 59
$(H_2)(dppb)Ru(\mu\text{-Cl})_3RuCl(dppb)$	Dimer	High	100
$[Os(H_2)Cl(PP)_2]^+$	Trig bipy	Low	43, 44
$OsH_3Cl(H_2)(P^iPr_3)_2$	Distorted Oct	Low	105
$OsH_2(X)(Y)(H_2)(PPr^i_3)_2$ (X, Y = Cl, Br, I)	Distorted Oct	Med	105
$Ir(H_2)H_2Cl(PR_3)_2$	Trig bipy	V high	114
trans-$Ir(H_2)HX_2(PR_3)_2$ (X = Cl, Br)	Sq pyr?	V high	115
$Ir(H_2)(H)(diphpyH)(PR_3)_2$	Agostic	Med	110
$[PtH(H_2)(PR_3)_2]^+$	Anion/solvento	V high	117, 118

[a] Abbreviations: P-N = o-diphenylphosphino-N,N dimethylaniline; diphpyH = 2,6-diarylpyridine.
[b] P = PCy_3.
[c] P = $P\{(OCH_2)_3CMe\}_2$

here.[152,155] The "cis-effect" between the hydride and H_2 ligands (see Section 4.11) may also contribute to the overall and relative stabilities.[155] Thus the general wisdom in M–H_2 chemistry is to carefully plan and monitor the environment around the complexes at all times. η^2-H_2 is normally a fragile entity in true H_2 complexes and can easily slip away even in solid complexes, sometimes without notice. By analogy, most of these principles also apply to other types of σ complexes.

The photochemical stability of H_2 complexes has not been well studied, but H_2 dissociation on exposure to visible light has been commonly observed in matrix-isolated species (see below). Reversibly bound isolable complexes can also be susceptible to decomposition. Yellow $W(H_2)(CO)_3(PR_3)_2$ darkens in color very slowly in artificial light and is severely damaged by low-power Ar laser light even at liquid-nitrogen temperature.

The electrochemistry of H_2 complexes has also not been widely studied and its application is limited to cyclic voltammetric determinations. Oxidation of H_2 complexes is much more common than reduction because the majority of these complexes are low-valent. Reversible redox systems are quite rare and include ReCl(H$_2$)(PMePh$_2$)$_4$[156] and [Os(H$_2$)(NH$_3$)$_5$]$^+$,[93] which show respective $E_{1/2}$ values of -0.07 and 0.58 V in organic solvents. In the latter case, oxidation is irreversible in acetone because the resulting Os^{III}–H$_2$ complex reduces acetone to isopropanol, an unusual case in which oxidation transforms a complex into a better reducing agent. Irreversible systems that show primarily anodic peaks are summarized by Jessop and Morris.[152] About the only complex to be reduced electrochemically is [FeH(H$_2$)(pp$_3$)]$^+$, which goes to FeH$_2$(pp$_3$) irreversibly.[157]

3.1.3. Complexes Synthesized by Addition of H$_2$ Gas to an Unsaturated Precursor

A common method of preparation is the reaction of H$_2$ gas at a pressure of about 1 atm with a coordinatively unsaturated precursor complex, ML$_n$ [Eq. (3.3)]:

$$L_nM + H_2 \rightleftharpoons L_nM(H_2) \tag{3.3}$$

The precursor complex can be a formally 16e species possessing an agostic C–H interaction[158] that is in effect displaced by the incoming H$_2$ ligand [Eq. (3.4)]. The

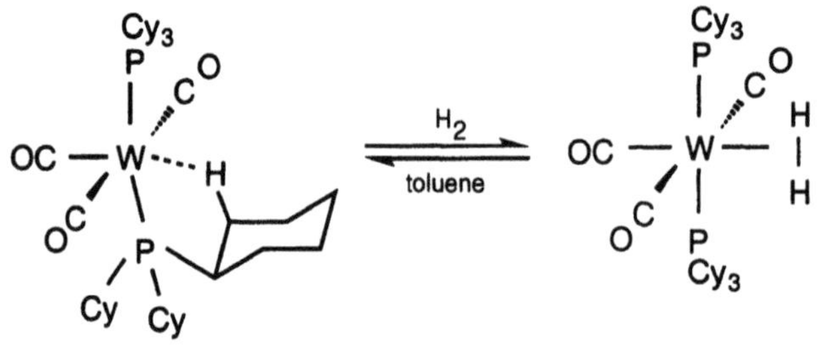

$$(3.4)$$

agostic interaction can readily displace the η^2-H$_2$ if excess H$_2$ is not present, facilitating the reversibility of the binding. This is the case for the original H$_2$ complexes, M(H$_2$)(CO)$_3$(PR$_3$)$_2$ (M = Cr, Mo, W) and certain others formed directly by *reversible* addition of H$_2$ gas to an isolated, formally unsaturated, precursor complex (Table 3.3). Virtually all of the precursors are "operationally unsaturated," i.e., formally 16e species stabilized by agostic interactions, π donation from halide ligands, or hydride ligands (see Section 12.2.3). In a few cases the precursor has an anion such as triflate or solvent (e.g., CH$_2$Cl$_2$) occupying the coordination site that can reversibly be displaced by H$_2$. The percentage of H$_2$ complexes synthesized by H$_2$ addition to precursors is actually surprisingly small (~10–15%). The reactions are generally carried out in noncoordinating or weakly coordinating organic solvents such as toluene or CH$_2$Cl$_2$, although solid-gas reactions can also be used.[78,114] Low-coordinating anions such as BAr$_f$ are often needed to stabilize a cationic M and prevent anion binding to M, especially for M = Mn, Re in Table 3.3. For example, the complex [Re(H$_2$)(CO)$_3$(PCy$_3$)$_2$]$^+$ with a BF$_4$ anion loses H$_2$ at low temperatures but the complex with less coordinating BAr$_f$ can be isolated as a solid at room temperature.[20]

The group 6 metal fragments are highly interesting both sterically and electronically in that they bind almost any molecule that can fit in between the bulky PR_3 except CO_2 and hydrocarbons (ethylene does bind similarly to H_2).[158] The Mo-diphosphine system was the first to show coordination of H–H, Si–H (see Chapter 11), and agostic C–H bonds (see Chapter 12).[6,159] In actual practice, $M(H_2)(CO)_3(PR_3)_2$ is prepared *in situ* under an H_2 atmosphere, e.g., as in Eq. (3.5).

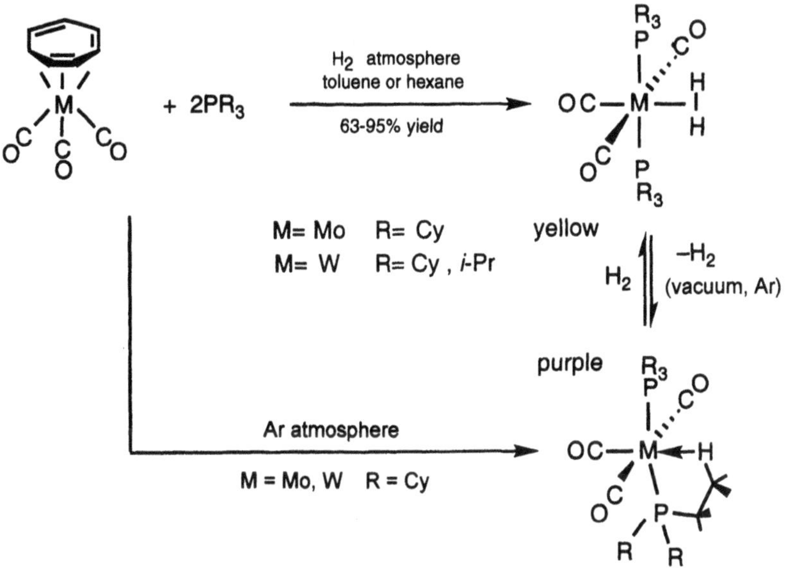

$$\begin{array}{c} \text{(reaction scheme 3.5)} \end{array} \tag{3.5}$$

M = Mo   R = Cy   yellow
M = W   R = Cy, *i*-Pr

purple

Ar atmosphere
M = Mo, W   R = Cy

If the reaction shown in Eq. (3.5) is carried out under Ar, the deep purple agostic complex results, which can easily be isolated and conveniently used to add the various isotopomers of hydrogen for NMR and IR studies. The H_2 addition can easily be monitored by the color change to yellow from the intense dark color that is often characteristic of unsaturated species. For R = ^{i}Pr, however, the agostic complex cannot be isolated directly in this manner. It is necessary to first synthesize the H_2 complex as in Eq. (3.5), and then remove the H_2 *in vacuo* in a high-boiling-point alkane such as nonane. The preparations and stabilities of each variant of a M–L fragment type can thus range far more than one might expect. The stabilities to H_2 dissociation, shown in Figure 3.1, serve to illustrate the characteristic high sensitivity of σ complexes to even minor changes in their electronic and stereochemical environments. Even the solubilities of the H_2 complexes vary greatly: the PCy_3 complexes are barely soluble in toluene, while the P^iPr_3 analogues are very soluble in hexane. The Mo complexes lose some H_2 when dissolved even under an atmosphere of H_2 (pressurization to several atmospheres is required to prevent dissociation).

The Cr complexes are among the most unstable H_2 complexes isolatable as solids at 25°C.[3] The deep-blue precursor, $Cr(CO)_3(PCy_3)_2$, was prepared initially by Hoff in THF solvent under Ar using $Cr(CO)_3$(naphthalene) instead of $M(CO)_3$(cycloheptatriene) as the starting material in Eq. (3.5).[160] In solution this complex had coordinated H_2 (or N_2) only at pressures greater than 10 atm, yet it

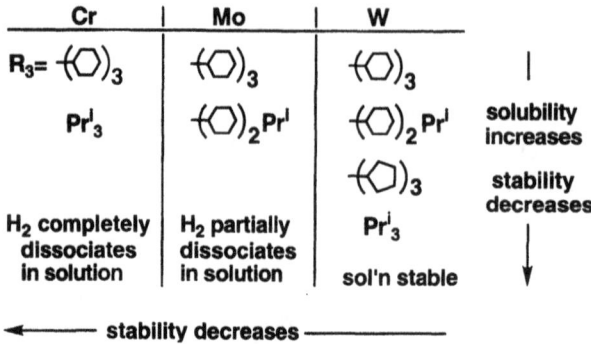

Figure 3.1. Stability of $M(H_2)(CO)_3(PR_3)_2$ complexes.

was later found[3] that solid $Cr(H_2)(CO)_3(PCy_3)_2$ precipitates from solution when a reaction according to Eq. (3.5) is carried out under only 1 atm of H_2! The H_2 complex is stable under H_2, but immediately on dissolving in toluene, loses all bound H_2 as H_2 gas, which vigorously effervesces from solution like CO_2 in carbonated water to give a deep-blue solution of $Cr(CO)_3(PCy_3)_2$. Such a large difference in stability between solution and solid states is rare in chemistry. It appears that coordinated H_2 can effectively be "trapped" in the solid state, possibly as a result of product solubility differences. The H_2 is less able to escape out of the crystal lattice than from the solution phase. This is reasonable in that the H_2 is not merely leaving the coordination site in these complexes; the whole molecule must rearrange itself to give back the agostic interaction with more acute P–Cr–P, Cr–P–C, and P–C–C bond angles. When placed into solution, $Cr(CO)_3(PCy_3)_2(H_2)$ would be more flexible in its ability to move a cyclohexyl ring into position to displace the H_2. Also, in toluene, transient solvent binding might induce rapid H_2 loss kinetically by mass action effects, although hydrocarbon binding could never actually be observed by NMR for any of these group 6 systems, even at low temperatures. Evidence for H_2 substitution by hydrocarbon solvents (toluene or even hexane) is seen for $IrXH_2(H_2)(P^iPr_3)_2$ (X = Cl, Br, I), which, as the Cr complex, readily liberates H_2 on dissolution in hydrocarbons.[161] Here agostic interactions do not appear to compete for the binding site, although the π-donating ability of the halide helps stabilize the formally 16e five-coordinate product of H_2 loss.

In addition to mass action effects, entropy effects are also often critical in determining relative stabilities of these weak complexes because enthalpies of ligand binding can be as low as 15 kcal/mol for M–H_2 or even lower for alkane complexes (see Chapter 7). This is particularly true when σ ligands are competing for binding sites against external ligands such as H_2O and N_2 and at the same time against intramolecular agostic interactions that are favored entropically by about 10 kcal/mol.

Other complexes prepared according to Eq. (3.1) are listed in Table 3.3 along with the structure of the precursor complex if known. Several 16e precursors have true five-coordinate geometries without agostic interactions. The H_2 is reversibly bound in all of these species and, as for the Cr complex, is extremely labile in the Ir complexes, even in the solid state. Unlike the group 6 complexes, [Ru(H$_2$)H(diphosphine)$_2$]$^+$ is *less* stable to H$_2$ loss than its first-row Fe congener. This has been attributed to greater BD for Fe than Ru.

3.1.4. Complexes Prepared from H$_2$ Gas by Ligand Displacement or Reduction

A related method of synthesis from H$_2$ gas involves displacement of a labile ligand

$$L_nML' + H_2 \rightleftarrows L_nM(H_2) + L' \qquad (3.6)$$

Neutral ligands L' that have been displaced include H$_2$O,[92,109] N$_2$,[41,162] NH$_3$,[163] CH$_2$Cl$_2$,[9,21,118] and PMe$_2$Ph.[31] One of the simplest conceivable H$_2$ complexes, [Ru(H$_2$O)$_5$(H$_2$)]$^{2+}$, is formed by displacement of an aquo ligand from the hexaquo complex by pressurized H$_2$ in aqueous solution.[92] Although it cannot be isolated, NMR indicates it has d_{HH} of 0.90 Å on the basis of the observed J_{HD} of 31.2 Hz (Chapter 5). Displacement of a charged ligand, X$^-$, by H$_2$ has occasionally been employed for synthesis:

$$L_nMX + H_2 \rightleftarrows [L_nM(H_2)]^+ + X^- \qquad (3.7)$$

Complexes prepared as in Eq. (3.7) are

[Re(H$_2$)Cp*(CO)NO]$^+$,[24] [M(H$_2$)H(depe)$_2$]$^+$, M = Fe, Ru, Os,[164]

[M(H$_2$)Cl(depe)$_2$]$^+$, M = Ru, Os,[42] [Ru(H$_2$)H(dcype)$_2$]$^+$,[35] and

[Os(H$_2$)H(CO)(P^{i}Pr$_3$)$_2$]$^+$ (for X = BH$_4^-$).[88]

Often a group 1 metal cation such as Na$^+$ or alternatively Tl$^+$ is present to precipitate with the anion. A Re complex has recently been found wherein H$_2$ directly displaces a normally strongly bound chloride ligand without such help[22]:

$$Re(CN^tBu)_3(PCy_3)_2Cl + H_2 \rightleftarrows [Re(CN^tBu)_3(PCy_3)_2(H_2)]Cl \qquad (3.8)$$

The syntheses of polyhydride complexes containing η^2-H$_2$, such as RuH$_2$(H$_2$)(PPh$_3$)$_3$, can be accomplished by hydride reduction according to[165]

$$L_nMX_m + mH^- + H_2 \rightarrow L_nM(H_2)H_m + mX^- \qquad (3.9)$$

Common sources of hydride in Eq. (3.9) are NaH, NaBH$_4$, LiAlH$_4$, and the anion X$^-$ is usually chloride or bromide. Complexes include ReH$_7$(PR$_3$)$_2$,[48] [Fe(H$_2$)H(pp$_3$)]$^+$,[39] M(H$_2$)H$_2$(PR$_3$)$_3$ (M = Fe, Ru),[49,123] Ru(H$_2$)H$_2$(cyttp),[53] and Tp*Rh(H$_2$)H$_2$.[107]

3.1.5. Protonation of a Hydride Complex

A very common and convenient method of preparation of H_2 complexes is the addition of H^+ to a hydride or polyhydride complex [Eq. (3.10)]. In most cases the

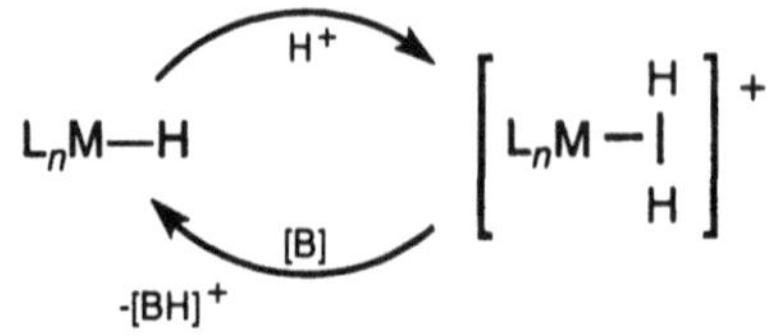

$$L_nM-H \underset{-[BH]^+}{\overset{H^+}{\underset{[B]}{\rightleftarrows}}} \left[L_nM-\overset{H}{\underset{H}{|}} \right]^+ \qquad (3.10)$$

resulting complex is cationic, and most of the cationic species in Table 3.1 are prepared in this way (some can be prepared from H_2 gas as in Table 3.3). The proton source can range from strong acids such as $HBF_4 \cdot Et_2O$ or triflic acid to very weak acids, even alcohols. The reactions are usually carried out below room temperature (ca $-60°C$), especially with strong acids, which often need to have low-interacting anions such as BF_4 or BAr_f. This method was first employed by Crabtree in 1985 by reaction of $IrH_2(PPh_3)_2(bq)$ with $PhCH(SO_2CF_3)_2$,[109] and a variety of H_2 complexes too numerous to list in detail in Table 3.1 have been prepared by protonation. For instance, the large class of half-sandwich complexes, $[Cp' M(H_2)(L)(L')]^+$ (M = Fe, Ru, Os; Cp' = cyclopentadienyl derivative), have all been prepared by protonation. Review articles list these as well as most other stable and unstable H_2 complexes synthesized prior to 1993.[120,121] A more recent review by Kuhlman discusses site selectivity of protonation of metal hydride–halide complexes.[166]

The attack of the proton is generally directly at the M–H bond, i.e., M is not protonated and then later transfers H^+ to the hydride.[74,109] An initial hydrogen bonding interaction may facilitate the protonation (see Section 7.3 for kinetic studies), which usually gives the H_2 complex as the product even when the dihydride may be the thermodynamically favored product [Eq. (3.11)].[167]

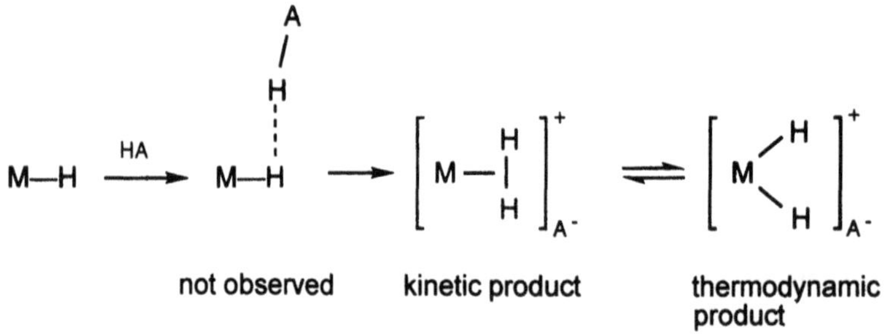

$$M-H \xrightarrow{HA} M-H \cdots H-A \longrightarrow \left[M-\overset{H}{\underset{H}{|}} \right]^+_{A^-} \rightleftharpoons \left[M\overset{H}{\underset{H}{<}} \right]^+_{A^-} \qquad (3.11)$$

not observed   kinetic product   thermodynamic product

Normally the protonation is carried out at low temperatures to give a $[M-H_2]^+$ complex, but on warming there is sometimes rearrangement to a dihydride or equilibrium mixture. Occasionally the product is unstable toward loss of H_2 and coordination of anion or solvent (S) if the electronics and thermodynamics of the system do not favor H_2 binding [Eq. (3.12)].

The stability of H_2 complexes prepared by protonation thus varies greatly: Some are stable only below room temperature and cannot be isolated as solids and

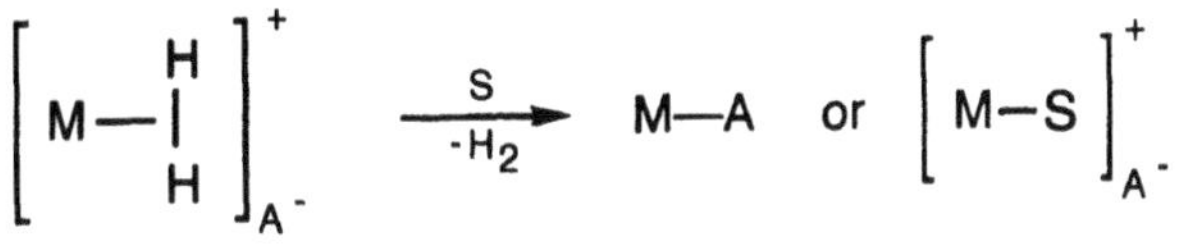

$$\left[\begin{array}{c} H \\ M-| \\ H \end{array}\right]^+_{A^-} \xrightarrow[-H_2]{S} M-A \text{ or } \left[M-S\right]^+_{A^-} \quad (3.12)$$

others are among the most robust H_2 complexes known. Generally the lability of a H_2-hydride system increases upon protonation or multiple protonation. Thus $M(dppe)_2$ (M = Ni, Pd, Pt) had been reported in 1966 to give a dication on double protonation, presumably via a monohydride and an unstable H_2 complex, which readily loses H_2[168]:

$$M(dppe)_2 \xrightarrow{HClO_4} [MH(dppe)_2][ClO_4] \xrightarrow{HClO_4} [M(H_2)(dppe)_2][HClO_4]_2$$

$$\xrightarrow{-H_2} [M(dppe)_2][HClO_4]_2 \quad (3.13)$$

Needless to say, complexes formed by protonation, especially where HA is a strong acid, are readily *deprotonated*, even by bases as weak as diethyl ether, and are highly sensitive to solvent media and trace water. These properties relate in large measure to the high acidity of certain H_2 complexes, which can have pK_a as low as -6, e.g., when generated from triflic acid (see Chapter 9).

3.1.6. Other Methods of Preparation

Some less common preparations have been reported. The reduction of complexes of Re^V or Os^{III} in the presence of a source of protons and electrons (H^+ and Mg or Na) gives the complexes $ReCl(H_2)(PMePh_2)_4$[14] and $[Os(H_2)(NH_3)_5]^{2+}$, respectively. The latter and its ethylenediamine (en) congeners are unique in containing pure σ-donor ligands, and a large series of such complexes has been prepared with a variety of L.[93-96] The dipositive charge is rare among H_2 complexes and undoubtedly is responsible for arresting OA. However, the d_{HH} is very

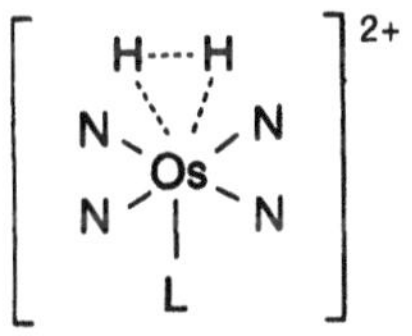

$$\left[\begin{array}{c} H\cdots H \\ N\diagdown \diagup N \\ Os \\ N\diagup \diagdown N \\ | \\ L \end{array}\right]^{2+}$$

long, ca 1.35 Å, in these species, indicating they are closer to being dihydrides. The reaction of $Ru(cod)(cot)$ with PCy_3 and H_2 gives $RuH_2(H_2)_2(PCy_3)_2$,[51] and protonation of $[RuH_5(P^iPr_3)_2]^-$ gives $RuH_2(H_2)_2(P^iPr_3)_2$.[54] These are among only a handful of well-characterized complexes that contain more than one η^2-H_2 ($TpRuH(H_2)_2$ derivatives[79] are the only others isolated as solids).

Decomposition of OsH(η^2-H$_2$BH$_2$)(CO)(P^{i}Pr$_3$)$_2$ in methanol produced OsH$_2$(H$_2$)(CO)(P^{i}P$_3$)$_2$,[167] which despite its wide use as a hydrogen transfer catalyst was not determined to have η^2-H$_2$ until nearly 10 years later,[57] yet another example that illustrates how difficult it can be to prove the presence of H$_2$ ligands! Another unusual synthesis involves hydrogenation of an ethylene complex either in solution or even in the solid state at 60°C [Eq. (3.14)][113]

$$[\text{IrH}_2(\text{triphos})(\text{C}_2\text{H}_4)]^+ + 2\text{H}_2 \rightarrow [\text{IrH}_2(\text{triphos})(\text{H}_2)]^+ + \text{C}_2\text{H}_6 \qquad (3.14)$$

3.2. DIHYDROGEN COMPLEXES UNSTABLE AT ROOM TEMPERATURE

Many H$_2$ complexes are unstable at room temperature, particularly those formed by protonation [Eqs. (3.8) and (3.9)]. In numerous examples, they can still be studied by low-temperature NMR methodologies and determined to have η^2-H$_2$ by measurement of J_{HD} and T_1. Virtually any metal system that eliminates H$_2$ gas via any route (e.g., protonation, photolysis, heating) must do so by a transient H$_2$ complex as demanded by the principle of microscopic reversibility. Obviously the transient will have widely varying degrees of instability roughly corresponding to the various points along the reaction coordinate toward OA along which H$_2$ complexes can be arrested. The sections below describe identification of H$_2$ complexes by non-NMR methods.

3.2.1. Organometallic Complexes Observed at Low Temperature in Rare Gas or Other Media

The first spectroscopic evidence for H$_2$ coordination was obtained in matrix-isolated Cr(CO)$_5$(H$_2$) by Sweany[172] (see Chapter 2) virtually at the same time as that for W(CO)$_3$(PR$_3$)$_2$(H$_2$). The investigations of low-temperature stable H$_2$ complexes (Table 3.4) in solid or liquid rare-gas media have continued to constitute a subdiscipline that goes hand in hand with studies of stable complexes as shown in reviews by Sweany (1992) and Poliakoff (1998).[174] In most cases the preparations involve photochemical displacement of CO either in a rare-gas matrix or in liquid Xe:

$$\text{L}_x\text{M(CO)}_n + \text{H}_2 \xrightarrow[\text{2-200 K}]{h\nu, -\text{CO}} \text{L}_x\text{M(CO)}_{n-1}(\text{H}_2) \qquad (3.15)$$

The most intensely studied species are the group 6 pentacarbonyls, M(CO)$_5$(H$_2$), which have been observed in rare-gas matrices, in liquid Xe solutions at −70°C (a very useful medium), in alkane solvents, and in the gas phase. Perhaps the most novel preparation is photolysis of the hexacarbonyls impregnated in polyethylene disks under H$_2$ or N$_2$ pressures to give M(CO)$_{6-n}$(L)$_n$, where $n = 1,2$ for L = H$_2$ and 1–4 for N$_2$.[175] Reactivity follows the order Mo > Cr > W, and H$_2$ can displace coordinated N$_2$ in the polyethylene systems. In all media, vibrational spectroscopy provides evidence for H$_2$ rather than dihydride binding, and the H–H, H–D, and

Table 3.4. Low-Temperature-Stable H_2 Coordination to Organometallic Fragments

Complex	Conditions	Reference
$V(CO)_5(H_2)$	Heptane, Xe soln	170
$Cp'M(CO)_3(H_2)$ M = V, Nb	Heptane, Xe soln	171
$M(CO)_5(H_2)$ M = Cr, Mo, W	Matrix; heptane, Xe soln; gas phase, PE[a]	172–176
$M(CO)_4(H_2)_2$	Matrix, Xe soln, PE	172, 173, 175
$M(CO)_{5-n}(olefin)_n(H_2)$ $n = 1,2$ M = Cr, Mo, W	Xe soln	177
$Cr(arene)(CO)_n(H_2)$ $n = 1,2$	Xe soln, gas phase	176, 178
$Mo(arene)(CO)_3(H_2)$	Matrix	179
$CpMH(CO)_2(H_2)$ M = Mo, W	Matrix	180
$MnX(CO)_4(H_2)$ X = Cl, Br	Matrix	181
$Fe(CO)(NO)_2(H_2)$	Xe soln	182
$Fe(C_4H_4)(CO)_2(H_2)$	Xe soln	176
$Ru(OEP)(H_2)$	THF soln	87
$Co(CO)_2(NO)(H_2)$	Xe soln	182
$CoX(H_2)(CO)_3$ X = H, Me	Matrix	183
$CpIr(CO)(H_2)$	Matrix	192
$Ni(CO)_3(H_2)$	Matrix	184
$CuCl(H_2)$	Ar matrix	185

[a]PE = polyethylene disk.

D–D stretching modes are often observed because of the clear spectroscopic window in rare-gas media (see Chapter 8). This is usually not a trivial measurement, however, as will be discussed later.

In nearly all cases, these complexes decompose rapidly and irreversibly at or near room temperature because of the weak H_2 binding on such CO-rich metals, where there is less BD. Their instability is exacerbated because the 16e product of H_2 dissociation is extremely reactive since it is not stabilized by internal agostic C–H interactions or solvent binding (hydrocarbon solvents are even more weakly bound than H_2). The rate of room-temperature dissociation of H_2 from $Cr(CO)_5(H_2)$ in hexane is actually slower than that for many stable species (see Chapter 7). Thus this complex and others like it might otherwise be stable under H_2. One such complex initially presumed to be unstable, $CpMn(H_2)(CO)_2$, has in fact been isolated as a solid from supercritical CO_2 ($scCO_2$) at room temperature in a flow reactor by rapid expansion of the $scCO_2$.[8] This complex, shown in Eq. (3.16), is more robust than

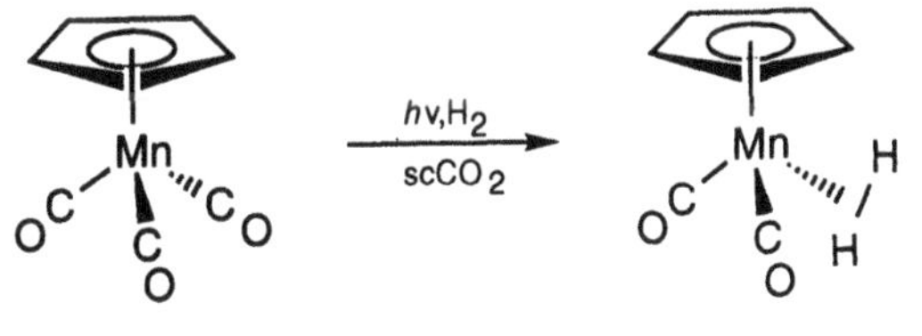

$$Cp(OC)_2Mn \xrightarrow{h\nu, H_2}_{scCO_2} Cp(OC)_2Mn(H_2) \qquad (3.16)$$

originally thought: Displacement of the η^2-H_2 by N_2 (500 psi) or ethylene requires 2 h in sc CO_2. $CpMn(H_2)(CO)_2$ is one of the simplest stable H_2 complexes and has by far the lowest molecular weight (178) and highest percentage of H_2 by weight

(1.1%). Analogues with Cp* and N_2, C_2H_4, and η^2-SiHEt$_3$ ligands have also been prepared, and interchange of these labile ligands can be promoted.[8]

3.2.2. Binding of H_2 to Bare Metal Atoms, Ions, and Surfaces

H_2 has been found to molecularly bind to metal surfaces such as Ni(510), metal atoms or cations, and small metal atom clusters (e.g., $Cu_2(H_2)_2$, $Cu_2(H_2)_3$, $Cu_3(H_2)$, $Fe_x(H_2)$ ($x = 3$ or 4) at low temperatures (Table 3.5). The evidence again is entirely spectroscopic, primarily vibrational and mass spectroscopy. H_2 is believed to be bound in η^2 fashion on the stepped edges of the Ni(510) surface, which are coordinatively unsaturated (Figure 3.2). Electron energy loss spectroscopy (EELS) at 100 K shows several bands comparable to those for organometallic H_2 complexes (see Chapter 8). No such chemisorption is observed on the flat Ni(100) surface, which lacks the residual unfilled d states at the step sites that bind the H_2. Undoubtedly H_2 coordination as in Figure 3.2 is the first step in the dissociation of H_2 on metal surfaces to form hydrides and is followed by rapid splitting of H–H analogous to OA in homogeneous solution activation.

The bare metal cationic species, $[M(H_2)_n]^+$ (M = first-row metals; see Section 4.13), are formed and studied by, e.g., electron-impact ionization of organometallic precursors such as $CpCo(CO)_2$, injection of the resulting Co^+ into a reaction cell containing H_2, and mass spectrometric analysis. Alternatively, the metal ions can be produced by sputtering them off a metal cathode in a flow tube where H_2 (or other small molecules) are added downstream in a guided ion-beam tandem mass spectrometer. These experiments are useful for determining M–H_2 binding energies on extremely electrophilic fragments and comparing them to those for CH_4 and other weak ligands (see Chapter 7).

Diatomic and triatomic Cu and Pd formed by vaporization reacts with up to three H_2 molecules to form complexes in Ar matrices at 7–15 K. Analogous reac-

Table 3.5. H_2 Bound to Bare Metal Species, Surfaces, or Clusters

Complex	Conditions	Reference
$[M(H_2)_n]^+$	Mass spectroscopy, surface ionization	a
Ni(510) surface	EELS	186
$Ni^{n+}(H_2)$(zeolite)	Esr, INS evidence	145
$Ni(SiO_2)(H_2)_n^+$	Esr evidence	146
$Pd(H_2)_x$ ($x = 1-3$)	Kr, Ar, or Xe matrix	187, 190
$Pd_2(H_2)$	Ar matrix	190
$Cu_2H_2(H_2)_x$ ($x = 1, 2$)	Ar matrix	188
$Cu_3(H_2)$	Ar matrix	188
$Fe_x(H_2)$ ($x = 3, 4$)	Ar matrix	188
$M_x(H_2)_y$; $M_2H_x(H_2)_y$ (M = Fe, Co, Ni)	Ar, Kr, and/or Xe matrices	193
Cr_2O_3 surface	Adsorption studies	189
CoNa-A zeolite(H_2)	INS	147

[a] See Section 4.13.

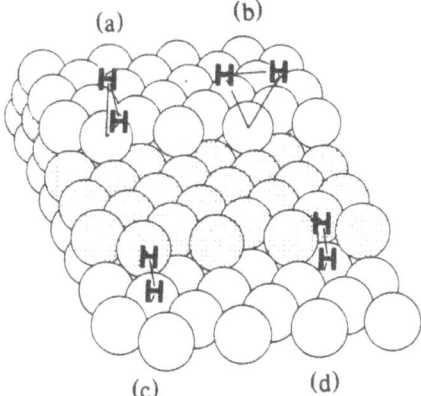

Figure 3.2. Model of the Ni(510) surface, showing some different adsorption sites (a)–(d). Case (a) is consistent with the EEL spectra. Reprinted with permission from Martensson et al.[186] Copyright 1986 by the American Physical Society.

tion of H_2 with Fe clusters forms only Fe_3 or Fe_4 hydrides (Fe_2 is unreactive). Both η^1 and η^2 bound H_2 is claimed to be observed in $M_x(H_2)_y$ and $M_2H_x(H_2)_y$, which has terminal and bridging hydrides.[193] Main-group species such as alkali halides, boron hydrides, and Lewis bases interact very weakly with H_2 at low temperatures (see Section 4.14).

REFERENCES

1. Hoff, C. D. *Coord. Chem. Rev.* **2000**, *206/207*, 451, and references therein.
2. Jalon, F. A.; Otero, A.; Manzano, B. R.; Villasenor, E.; Chaudret, B. *J. Am. Chem. Soc.* **1995**, *117*, 10123; Sabo-Etienne, S.; Chaudret, B.; el Makarim, H. A.; Barthelat, J.-C.; Daudey, J.-P.; Moise, C.; Leblanc, J.-C. *J. Am. Chem. Soc.* **1994**, *116*, 9335; Antinolo, A.; Carrillo-Hermosilla, F.; Fajardo, M.; Garcia-Yuste, S.; Otero, A.; Camanyes, S.; Maseras, F.; Moreno, M.; Lledos, A.; Lluch, J. M. *J. Am. Chem. Soc.* **1997**, *119*, 6107; Antinolo, A.; Carillo-Hermosilla, F.; Fajardo, M.; Fernandez-Baeza, J.; Garcia-Yuste, S.; Otero, A. *Coord. Chem. Rev.* **1999**, *193–195*, 43.
3. Kubas, G. J.; Nelson, J. E.; Bryan, J. C.; Eckert, J.; Wisniewski, L.; Zilm, K. *Inorg. Chem.* **1994**, *33*, 2954.
4. Kubas, G. J.; Unkefer, C. J; Swanson, B. I.; Fukushima, E. *J. Am. Chem. Soc.* **1986**, *108*, 7000.
5. Khalsa, G. R. K.; Kubas, G. J.; Unkefer, C. J.; Van Der Sluys, L. S.; Kubat-Martin, K. A. *J. Am. Chem. Soc.* **1990**, *112*, 3855.
6. Kubas, G. J.; Burns, C. J.; Eckert, J.; Johnson, S.; Larson, A. C.; Vergamini, P. J.; Unkefer, C. J.; Khalsa, G. R. K.; Jackson, S. A.; Eisenstein, O., *J. Am. Chem. Soc.* **1993**, *115*, 569. Luo, X.-L.; Kubas, G. J.; Burns, C. J.; Eckert, J. *Inorg. Chem.* **1994**, *33*, 5219.
7. Seino, H.; Arita, C.; Nonokawa, D.; Nakamura, G.; Harada, Y.; Mizobe, Y.; Hidai, M. *Organometallics* **1999**, *18*, 4165.
8. Banister, J. A.; Lee, P. D.; Poliakoff, M. *Organometallics* **1995**, *14*, 3876; Lee, P. D.; King, J. L.; Seebald, S.; Poliakoff, M. *Organometallics* **1998**, *17*, 524.
9. King, W. A.; Luo, X.-L.; Scott, B. L.; Kubas, G. J.; Zilm, K. W. *J. Am. Chem. Soc.* **1996**, *118*, 6782; King, W. A.; Scott, B. L.; Eckert, J. Kubas, G. J. *Inorg. Chem.* **1999**, *38*, 1069; Toupadakis, A.; Kubas, G. J.; King, W. A.; Scott, B. L.; Huhmann-Vincent, J. *Organometallics* **1998**, *17*, 5315; Fang, X.; Huhmann-Vincent, J.; Scott, B. L.; Kubas, G. J. *J. Organometal. Chem.*, **2000**, *609*, 95.

10. Albertin, G.; Antoniutti, S.; Bettiol, M.; Bordignon, E.; Busatto, F. *Organometallics* **1997**, *16*, 4959.
11. Perthuisot, C.; Fan, M.; Jones, W. D. *Organometallics* **1992**, *11*, 3622.
12. Burrell, A. K.; Bryan, J. C.; Kubas, G. J. *J. Am. Chem. Soc.* **1994**, *116*, 1575.
13. Kohli, M.; Lewis, D. J.; Luck, R. L.; Silverton, J. V.; Sylla, K. *Inorg. Chem.* **1994**, *33*, 879.
14. Cotton, F. A.; Luck, R. L. *J. Chem. Soc., Chem. Commun.* **1988**, 1277.
15. Gusev, D. G.; Nietlispach, D.; Eremenko, I. L.; Berke, H. *Inorg. Chem.* **1993**, *32*, 3628; Albertin, G.; Antoniutti, S.; Garcia-Fontan, S.; Carballo, R.; Padoan, F. *J. Chem. Soc., Dalton Trans.* **1998**, 2071.
16. Gusev, D.; Llamazares, A.; Artus, G.; Jacobsen, H.; Berke, H. *Organometallics* **1999**, *18*, 75.
17. Luo, X.-L.; Michos, D.; Crabtree, R. H. *Organometallics* **1992**, *11*, 237.
18. Bianchini, C.; Marchi, A.; Marvelli, L.; Peruzzini, M.; Romerosa, A.; Rossi, R.; Vacca, A. *Organometallics* **1995**, *14*, 3203.
19. Heinekey, D. M.; Schomber, B. M.; Radzewich, C. E. *J. Am. Chem. Soc.* **1994**, *116*, 4515.
20. Heinekey, D. M.; Radzewich, C. E.; Voges, M. H.; Schomber, B. M. *J. Am. Chem. Soc.* **1997**, *119*, 4172.
21. Huhmann-Vincent, J.; Scott, B. L.; Kubas, G. J. *J. Am. Chem. Soc.* **1998**, *120*, 6808.
22. Heinekey, D. M.; Voges, M. H.; Barnhart, D. M. *J. Am. Chem. Soc.* **1996**, *118*, 10792.
23. Chin, R. M.; Barrera, J.; Dubois, R. H.; Helberg, L. E.; Sabat, M.; Bartucz, T. Y.; Lough, A. J.; Morris, R. H.; Harman, W. D. *Inorg. Chem.* **1997**, *36*, 3553.
24. Chinn, M. S.; Heinekey, D. M.; Payne, N. G.; Sofield, C. D. *Organometallics* **1989**, *8*, 1824.
25. Luo, X.-L.; Crabtree, R. H. *J. Am. Chem. Soc.* **1990**, *112*, 6912.
26. Reid, S. M.; Neuner, B.; Schrock, R. R.; Davis, W. M. *Organometallics* **1998**, *17*, 4077.
27. Brammer, L.; Howard, J. A.; Johnson, O.; Koetzle, T. F.; Spencer, J. L.; Stringer, A. M. *J. Chem. Soc., Chem. Commun.* **1991**, 241.
28. Kim, Y.; Deng, H.; Meek, D. W.; Wojcicki, A. *J. Am. Chem. Soc.* **1990**, *112*, 2798; Luo, X.-L.; Crabtree, R. H. *J. Chem. Soc., Dalton Trans.* **1991**, 587.
29. Fontaine, X. L. R.; Fowles, E. H.; Shaw, B. L. *J. Chem. Soc., Chem. Commun.* **1988**, 482; Fontaine, X. L. R.; Layzell, T. P.; Shaw, B. L. *J. Chem. Soc., Dalton Trans.* **1994**, 917.
30. Albertin, G.; Antoniutti, S.; Bordignon, E. *J. Am. Chem. Soc.* **1989**, *111*, 2072; Arliguie, T.; Chaudret, B. *J. Chem. Soc., Chem. Commun.* **1989**, 155; Gusev, D. G.; Hubener, R.; Burger, P.; Orama, O.; Berke, H. *J. Am. Chem. Soc.* **1997**, *119*, 3716; Hills, A.; Hughes, D. L.; Jimenez-Tenorio, M.; Leigh, G. J. *J. Organomet. Chem.* **1990**, *391*, C41; Osakada, K.; Ohshiro, K.; Yamamoto, A. *Organometallics* **1991**, *10*, 404; Lough, A. J.; Morris, R. H.; Ricciuto, L.; Schleis, T. *Inorg. Chim. Acta* **1998**, *270*, 238; Albertin, G.; Antoniutti, S.; Baldan, D.; Bordignon, E. *Inorg. Chem.* **1995**, *34*, 6205.
31. Amendola, P.; Antoniutti, S.; Albertin, G.; Bordignon, E. *Inorg. Chem.* **1990**, *29*, 318.
32. Morris, R. H.; Sawyer, J. F.; Shiralian, M.; Zubkowski, J. D. *J. Am. Chem. Soc.* **1985**, *107*, 5581.
33. Bautista, M.; Earl, K. A.; Morris, R. H.; N. J.; Sella, A. *J. Am. Chem. Soc.* **1987**, *109*, 3780; Cappellani, E. P.; Drouin, S. D.; Jia, G.; Maltby, P. A.; Morris, R. H.; Schweitzer, C. T. *J. Am. Chem. Soc.* **1994**, *116*, 3375; Bautista, M. T.; Earl, K. A.; Morris, R. H. *Inorg. Chem.* **1988**, *27*, 1124; Baker, M. V.; Field, L. D. *J. Organomet. Chem.* **1988**, *354*, 351; Field, L. D.; Hambley, T. W.; Yau, B. C. K. *Inorg. Chem.* **1994**, *33*, 2009; Ogasawara, M.; Saburi, M. *J. Organometal. Chem.* **1994**, *482*, 7; Tsukahara, T.; Kawano, H.; Ishii, Y.; Takahashi, T.; Saburi, M.; Uchida, Y.; Akutagawa, S. *Chem. Lett.* **1988**, 2055; Kranenburg, M.; Kramer, P. C. J.; van Leeuwen, P. W. N. M.; Chaudret, B. *Chem. Commun.* **1997**, 373; Earl, K. A.; Jia, G.; Maltby, P. A.; Morris, R. H. *J. Am. Chem. Soc.* **1991**, *113*, 3027.
34. Saburi, M.; Aoyagi, K.; Takahashi, T.; Uchida, Y. *Chem. Lett.* **1990**, 601.
35. Mezzetti, A.; Del Zotto, A.; Rigo, P.; Farnetti, E. *J. Chem. Soc., Dalton Trans.* **1991**, 1525.
36. Jimenez-Tenorio, M.; Puerta, M. C.; Valerga, P. *J. Am. Chem. Soc.* **1993**, *115*, 9794.
37. Schlaf, M.; Lough, A. J.; Morris, R. H. *Organometallics* **1997**, *16*, 1253.
38. Bautista, M. T.; Earl, K. A.; Maltby, P. A.; Morris, R. H.; Schweitzer, C. T. *Can. J. Chem.* **1994**, *72*, 547; Bianchini, C.; Perez, P. J.; Peruzzini, M.; Zanobini, F.; Vacca, A. *Inorg. Chem.* **1991**, *30*, 279.
39. Bianchini, C.; Peruzzini, M.; Zanobini, F. *J. Organomet. Chem.* **1988**, *354*, C19.
40. Casey, C. P.; Tanke, R. S.; Hazin, P. N.; Kemnitz, C. R.; McMahon, R. J. *Inorg. Chem.* **1992**, *31*, 5474.
41. Bianchini, C.; Linn, K.; Masi, D.; Peruzzini, M.; Polo, A.; Vacca, A.; Zanobini, F. *Inorg. Chem.* **1993**, *32*, 2366.
42. Cappellani, E. P.; Maltby, P. A.; Morris, R. H.; Schweitzer, C. T.; Steele, M. R. *Inorg. Chem.* **1989**, *28*, 4437; Chin, B.; Lough, A. J.; Morris, R. H.; Schweitzer, C.; D'Agostino, C. *Inorg. Chem.* **1994**, *33*, 6278; Amrhein, P. I.; Drouin, S. D.; Forde, C. E.; Lough, A. J.; Morris, R. H. *J. Chem. Soc. Chem.*

Commun. **1996**, 1665; Rocchini, E.; Rigo, P.; Mezzetti, A.; Stephen, T.; Morris, R. H.; Lough, A. J.; Forde, C. E.; Fong, T. P.; Drouin, S. D. *J. Chem. Soc., Dalton Trans.* **2000**, 3591.
43. Rocchini, E.; Mezzetti, A.; Ruegger, H.; Burckhardt, U.; Gramlich, V.; Del Zotto, A.; Martinuzzi, P.; Rigo, P. *Inorg. Chem.* **1997**, *36*, 711.
44. Maltby, P. A.; Schlaf, M.; Steinbeck, M.; Lough, A. J.; Morris, R. H.; Klooster, W. T.; Koetzle, T. F.; Srivastava, R. C. *J. Am. Chem. Soc.* **1996**, *118*, 5396.
45. Jia, G.; Lee, H. M.; Williams, I. D.; Lau, C. P.; Chen, Y. *Organometallics* **1997**, *16*, 3941.
46. Forde, C. E.; Landau, S. E.; Morris, R. H. *J. Chem. Soc., Dalton Trans.* **1997**, 1663.
47. Fong, T. P.; Lough, A. J.; Morris, R. H.; Mezzetti, A.; Rocchini, E.; Rigo, P. *J. Chem. Soc., Dalton Trans.* **1998**, 2111; Fong, T. P.; Forde, C. E.; Lough, A. J.; Morris, R. H.; Rigo, P.; Rocchini, E.; Stephan, T. *J. Chem. Soc., Dalton Trans.* **1999**, 4475.
48. Hamilton, D. G.; Crabtree, R. H. *J. Am. Chem. Soc.* **1988**, *110*, 4126.
49. Van Der Sluys, L. S.; Eckert, J.; Eisenstein, O.; Hall, J. H.; Huffman, J. C.; Jackson, S. A.; Koetzle, T. F.; Kubas, G. J.; Vergamini, P. J.; Caulton K. G. *J. Am. Chem. Soc.* **1990**, *112*, 4831.
50. Albeniz, M. J.; Buil, M. L.; Esteruelas, M. A.; Lopez, A. M.; Oro, L. A.; Zeier, B. *Organometallics* **1994**, *13*, 3746.
51. Arliguie, T.; Chaudret, B.; Morris, R. H.; Sella, A. *Inorg. Chem.* **1988**, *27*, 598; Busch, S.; Leitner, W. *Chem. Comm.* **1999**, 2305.
52. Kohlmann, W.; Werner, H. *Z. Naturforsch. B* **1993**, *48b*, 1499; Bardajf, M.; Caminade, A.-M.; Majoral, J.-P.; Chaudret, B. *Organometallics* **1997**, *16*, 3489.
53. Jia, G.; Meek, D. W. *J. Am. Chem. Soc.* **1989**, *111*, 757.
54. Abdur-Rashid, K.; Gusev, D. G.; Lough, A. J.; Morris, R. H. *Organometallics* **2000**, 19, 1652.
55. Gusev, D. G.; Vymenits, A. B.; Bakhmutov, V. I. *Inorg. Chem.* **1992**, *31*, 1.
56. Poulton, J. T.; Sigala, M. P.; Eisenstein, O.; Caulton, K. G. *Inorg. Chem.* **1993**, *32*, 5490; Heyn, R. H.; Macgregor, S. A.; Nadasdi, T. T.; Ogasawara, M.; Esenstein, O.; Caulton, K. G. *Inorg. Chim. Acta* **1997**, *259*, 5.
57. Gusev, D. G.; Kuhlman, R. L.; Renkema, K. H.; Eisenstein, O.; Caulton, K. G. *Inorg. Chem.* **1996**, *35*, 6775.
58. Yandulov, D. V.; Streib W. E.; Caulton, K. G. *Inorg. Chim. Acta* **1998**, *280*, 125.
59. Esteruelas, M. A.; Sola, E.; Oro, L. A.; Meyer, U.; Werner, H. *Angew. Chem. Int. Ed. Engl.* **1988**, *27*, 1563.
60. Espuelas, J.; Esteruelas, M. A.; Lahoz, F. J.; Oro, L. A.; Valero, C. *Organometallics* **1993**, *12*, 663.
61. Mudalige, D. C.; Rettig, S. J.; James, B. R.; Cullen, W. R. *J. Chem. Soc., Chem. Commun.* **1993**, 830.
62. Esteruelas, M. A.; Lahoz, F. J.; Oro, L. A.; Onate, E.; Ruiz, N. *Inorg. Chem.* **1994**, *33*, 787.
63. Michos, D.; Luo, X.-L.; Crabtree, R. H. *Inorg. Chem.* **1992**, *31*, 4245.
64. Hussein, K.; Marsden, C. J.; Barthelat, J.-C.; Rodriguez, V.; Conjero, S.; Sabo-Etienne, S.; Donnadieu, B.; Chaudret, B. *Chem. Comm.* **1999**, 1315.
65. Chaudret, B.; Chung, G.; Eisenstein, O.; Jackson, S. A.; Lahoz, F. J.; Lopez, J. A. *J. Am. Chem. Soc.* **1991**, *113*, 2314; Christ, M. L.; Sabo-Etienne, S.; Chaudret, B. *Organometallics* **1994**, *13*, 3800.
66. Burrow, T.; Sabo-Etienne, S.; Chaudret, B. *Inorg. Chem.* **1995**, *34*, 2470; Wilhelm, T., E.; Belderrain, T. R.; Brown, S. N.; Grubbs, R. H. *Organometallics* **1997**, *16*, 3867.
67. Christ, M. L.; Sabo-Etienne, S.; Chung, G.; Chaudret, B. *Inorg. Chem.* **1994**, *33*, 5316; Guari, Y.; Sabo-Etienne, S.; Chaudret, B. *Organometallics* **1996**, *15*, 3471; Guari, Y.; Sabo-Etienne, S.; Chaudret, B. *J. Am. Chem. Soc.* **1998**, *120*, 4228.
68. Chung, G.; Arliguie, T.; Chaudret, B. *New J. Chem.* **1992**, *16*, 369.
69. Sellmann, D.; Gottschalk-Gaudig, T.; Heinemann, F. W. *Inorg. Chem.* **1998**, *37*, 3982.
70. Buil, M. L.; Esteruelas, M. A.; Lahoz, F. J.; Onate, E.; Oro, L. A. *J. Am. Chem. Soc.* **1995**, *117*, 3619.
71. Jimenez-Tenorio, M.; Puerta, M. C.; Valerga, P. *J. Chem. Soc., Chem. Commun.* **1993**, 1750.
72. Hamon, P.; Toupet, L.; Hamon, J.-R.; Lapinte, C. *Organometallics* **1992**, *11*, 1429; Jia, G.; Ng, W. S.; Yao, J.; Lau, C.-P.; Chen, Y. *Organometallics* **1996**, *15*, 5039; Conroy-Lewis, F. M.; Simpson, S. J. *J. Chem. Soc., Chem. Commun.* **1987**, 1675; Jia, G.; Morris, R. H. *J. Am. Chem. Soc.* **1991**, *113*, 875. Jia, G.; Lough, A. J.; Morris, R. H. *Organometallics* **1992**, *11*, 161. de los Rios, I.; Jimenez-Tenorio, M.; Padilla, J.; Puerta, M. C.; Valerga, P. *Organometallics* **1996**, *15*, 4565.
73. Chinn, M. S.; Heinekey, D. M.; *J. Am. Chem. Soc.* **1990**, *112*, 5166.
74. Chinn, M. S.; Heinekey, D. M. *J. Am. Chem. Soc.* **1987**, *109*, 5865.
75. Scharrer, E.; Chang, S.; Brookhart, M. *Organometallics* **1995**, *14*, 5686.

76. Chinn, M. S.; Heinekey, D. M.; Payne, N. G.; Sofield, C. D. *Organometallics* **1989**, *8*, 1824.
77. Bullock, R. M.; Song, J.-S.; Szalda, D. J. *Organometallics* **1996**, *15*, 2504.
78. Jia, G.; Ng, W. S.; Lau, C. P. *Organometallics* **1998**, *17*, 4538; Gemel, C.; Huffman, J. C.; Caulton, K. G.; Mauthner, K.; Kirchner, K. *J. Organomet. Chem.* **2000**, *593–594*, 342.
79. Moreno, B.; Sabo-Etienne, S.; Chaudret, B.; Rodriguez, A.; Jalon, F.; Trofimenko, S. *J. Am. Chem. Soc.* **1995**, *117*, 7441.
80. Halcrow, M. A.; Chaudret, B.; Trofimenko, S. *J. Chem. Soc., Chem. Commun.* **1993**, 465; Chen, Y.; Chan, W. C.; Lau, C. P.; Chu, H. S.; Lee, H. L.; Jia, G. *Organometallics* **1997**, *16*, 1241.
81. Chan, W.-C.; Lau, C.-P.; Chen, Y.; Fang, Y.-Q.; Ng, S.-M.; Jia, G. *Organometallics* **1997**, *16*, 34; Jimenez-Tenorio, M.; Jimenez-Tenorio, M. A.; Puerta, M. C.; Valerga, P. *Inorg. Chim. Acta.* **1997** *259*, 77.
82. Bohanna, C.; Esteruelas, M. A.; Gomez, A. V.; Lopez, A. M.; Martinez, M.-P. *Organometallics* **1997**, *16*, 4464.
83. Ng, S.-M.; Fang, Y. Q.; Lau, C.-P.; Wong, W. T.; Jia, G. *Organometallics* **1998**, *17*, 2052.
84. Esteruelas, M. A.; Oro, L. A.; Ruiz, N. *Inorg. Chem.* **1993**, *32*, 3793.
85. Esteruelas, M. A.; Oro, L. A.; Valero, C. *Organometallics* **1991**, *10*, 462.
86. Collman, J. P.; Hutchison, J. E.; Wagenknecht, P. S.; Lewis, N. S.; Lopez, M. A.; Guilard, R. *J. Am. Chem. Soc.* **1990**, *112*, 8206.
87. Collman, J. P.; Wagenknecht, P.; Hutchison, J. E.; Lewis, N. S.; Lopez, M. A.; Guilard; R.; L'Her, M.; Bothner-By, A. A.; Mishra, P. K. *J. Am. Chem. Soc.* **1992**, *114*, 5654.
88. Esteruelas, M. A.; Garcia, M. P.; Lopez, A. M.; Oro, L. A.; Ruiz, N.; Schlunken, C.; Valero, C.; Werner, H. *Inorg. Chem.* **1992**, *31*, 5580.
89. Siedle, A. R.; Newmark, R. A.; Korba, G. A.; Pignolet, L. H.; Boyle, P. D. *Inorg. Chem.* **1988**, *27*, 1593.
90. Albeniz, M. J.; Buil, M. L.; Esteruelas, M. A.; Lopez, A. M.; Oro, L. A.; Zeier, B. *Organometallics* **1994**, *13*, 3746.
91. Schlaf, M.; Lough, A. J.; Morris, R. H. *Organometallics* **1996**, *15*, 4423.
92. Aebischer, N. Frey, U.; Merbach, A. E. *Chem. Comm.* **1998**, 2303.
93. Harman, W. D.; Taube, H. *J. Am. Chem. Soc.* **1990**, *112*, 2261.
94. Nunes, F. S.; Taube, H. *Inorg. Chem.* **1994**, *33*, 3111, 3116; Li, Z.-W.; Taube, H. *J. Am. Chem. Soc.* **1994**, *116*, 11584.
95. Li, Z.-W.; Taube, H. *J. Am. Chem. Soc.* **1991**, *113*, 8946.
96. Lin, P.; Hasegawa, T.; Parkin, S.; Taube, H. *J. Am. Chem. Soc.* **1992**, *114*, 2712; Hasegawa, T.; Li, Z.; Parkin, S.; Hope, H.; McMullan, R. K.; Koetzle, T. F.; Taube, H. *J. Am. Chem. Soc.* **1994**, *116*, 4352; Li, Z.-W.; Yeh, A.; Taube, H. *Inorg. Chem.* **1994**, *33*, 2874.
97. Smith, K.-T.; Tilset, M.; Kuhlman, R.; Caulton, K. G. *J. Am. Chem. Soc.* **1995**, *117*, 9473.
98. Heinekey, D. M.; Luther, T. A. *Inorg. Chem.* **1996**, *35*, 4396; Luther, T. A.; Heinekey D. M. *Inorg. Chem.* **1998**, *37*, 127.
99. Hampton, C. R. S. M.; Butler, I. R.; Cullen, W. R.; James, B. R.; Charland, J.-P.; Simpson, J. *Inorg. Chem.* **1992**, *31*, 5509.
100. Joshi, A. M.; James, B. R. *J. Chem. Soc., Chem. Commun.* **1989**, 1785; Chau, D. E. K.-Y.; James, B. R. *Inorg. Chim. Acta* **1995**, *240*, 419.
101. Bianchini, C.; Barbaro, P.; Scapacci, G.; Zanobini, F. *Organometallics* **2000**, *19*, 2450.
102. He, Z.; Nefedov, S.; Lugan, N.; Neibecker, D.; Mathieu, R. *Organometallics* **1993**, *12*, 3837.
103. Halpern, J.; Cai, L.; Desrosiers, P. J.; Lin, Z.; *J. Chem. Soc., Dalton Trans.* **1991**, 717.
104. Zlota, A. A.; Tilset, M.; Caulton, K. G. *Inorg. Chem.* **1993**, *32*, 3816; Gross, C. L.; Young, D. M.; Schultz, A. J.; Girolami, G. S. *J. Chem. Soc., Dalton Trans.* **1997**, 3081.
105. Gusev, D. G.; Kuznetsov, V. F.; Eremenko, I. L.; Berke, H. *J. Am. Chem. Soc.* **1993**, *115*, 5831; Kuhlman, R. L.; Gusev, D. G.; Eremenko, I. L.; Berke, H.; Huffman, J. C.; Caulton, K. G. *J. Organometal. Chem.* **1997**, *536–537*, 139.
106. Vigalok, A.; Ben-David, Y.; Milstein, D. *Organometallics* **1996**, *15*, 1839.
107. Bucher, U. E.; Lengweiler, T.; Nanz, D.; von Philipsborn, W.; Venanzi, L. M. *Angew. Chem. Int. Ed. Engl.* **1990**, *29*, 548.
108. Heinekey, D. M.; Oldham, W. J., Jr. *J. Am. Chem. Soc.* **1994**, *116*, 3137; Oldham, W. J., Jr.; Hinkle, A. S.; Heinekey, D. M. *J. Am. Chem. Soc.* **1997**, *119*, 11028.
109. Crabtree, R. H.; Lavin, M. *J. Chem. Soc.*, Chem. Commun. **1985**, 794; Crabtree, R. H.; Lavin, M.; Bonneviot, L. *J. Am. Chem. Soc.* **1986**, *108*, 4032.

110. Albeniz, A. C.; Schulte, G.; Crabtree, R. H. *Organometallics* **1992**, *11*, 242.
111. Cooper, A. C.; Caulton, K. G. *Inorg. Chem.* **1998**, *37*, 5938; Cooper, A. C.; Eisensein, O.; Caulton, K. G. *New. J. Chem.* **1998**, *22*, 307.
112. Lundquist, E. G.; Huffman, J. C.; Folting, K.; Caulton, K. G. *Angew. Chem. Int. Ed. Engl.* **1988**, *27*, 1165.
113. Bianchini, C.; Moneti, S.; Peruzzini, M.; Vizza, F. *Inorg. Chem.* **1997**, *36*, 5818; Bakhmutov, V. I.; Bianchini, C.; Peruzzini, M.; Vizza, F.; Vorontsov, E. V. *Inorg. Chem.* **2000**, *38*, 1655.
114. Mediati, M.; Tachibana, G. N.; Jensen, C. M. *Inorg. Chem.* **1992**, *31*, 1827; Le-Husebo, T.; Jensen, C. M. *Inorg. Chem.* **1993**, *32*, 3797; Zidan, R. A.; Rocheleau, R. E. *J. Mater. Res.* **1999**, *14*, 286.
115. Gusev, D. G.; Bakhmutov, V. I.; Grushin, V. V.; Volpin, M. E. *Inorg. Chim. Acta* **1990**, *177*, 115; Albinati, A.; Bakhmutov, V. I.; Caulton, K. G.; Clot, E.; Eckert, J.; Eisenstein, O.; Gusev, D. G.; Grushin, V. V.; Hauger, B. E.; Klooster, W. T.; Koetzle, T. F.; McMullan, R. K.; O'Loughlin, T. J.; Pelissier, M.; Ricci, J. S.; Sigalas, M. P.; Vymenits, A. B. *J. Am. Chem. Soc.* **1993**, *115*, 7300; Bakhmutov, V. I.; Vymenits, A. B.; Grushin, V. V. *Inorg. Chem.* **1994**, *33*, 4413.
116. Sola, E.; Bakhmutov, V. I.; Torres, F.; Elduque, A.; Lopez, J. A.; Lahoz, F. J.; Werner, H.; Oro, L. A. *Organometallics* **1998**, *17*, 683.
117. Gusev, D. G.; Notheis, J. U.; Rambo, J. R.; Hauger, B. E.; Eisenstein, O.; Caulton, K. G. *J. Am. Chem. Soc.* **1994**, *116*, 7409; Stahl, S. S.; Labinger, J. A.; Bercaw, J. E. *Inorg. Chem.* **1998**, *37*, 2422.
118. Butts, M. D.; Kubas, G. J.; Scott, B. L. *J. Am. Chem. Soc.* **1996**, *118*, 11831.
119. Nolan, S. P.; Marks, T. J. *J. Am. Chem. Soc.* **1989**, *111*, 8538.
120. Heinekey, D. M.; Oldham, W. J., Jr. *Chem. Rev.* **1993**, 93, 913.
121. Jia, G.; Lau, C. P. *J. Organomet. Chem.* **1998**, *565*, 37.
122. Bacskay, G. B.; Bytheway, I.; Hush, N. S., *J. Am. Chem. Soc.* **1996**, *118*, 3753.
123. Ashworth, T. V.; Singleton, E. *J. Chem. Soc., Chem. Commun.* **1976**, *705*; Crabtree, R. H.; Hamilton, D. G. *J. Am. Chem. Soc.* **1986**, *108*, 3124.
124. Wisniewski, L.; Zilm, K. W.; Kubas, G. J.; Van der Sluys, L., unpublished results **1993**.
125. Heinekey, D. M.; Liegeois, A.; van Roon, M. *J. Am. Chem. Soc.* **1994**, *116*, 8388.
126. Leboeuf, J.-F.; Lavastre, O.; Leblanc, J.-C.; Moise, C. *J. Organomet. Chem.* **1991**, *418*, 359; Chaudret, B.; Limbach, H. H.; Moise, C. *C.R. Acad. Sci. Paris* **1992**, *315*, 533.
127. Henderson, R. A.; Oglieve, K. E. *J. Chem. Soc., Dalton Trans.* **1993**, 3431.
128. Henderson, R. A. *J. Chem. Soc., Chem. Commun.* **1987**, 1670.
129. Oglieve, K. E.; Henderson, R. A. *J. Chem. Soc., Chem. Commun.* **1992**, 441.
130. Rothfuss, H.; Gusev, D. G.; Caulton, K. G. *Inorg. Chem.* **1995**, *34*, 2894.
131. Bianchini, C.; Peruzzini, M.; Zanobini, F.; Magon, L.; Marvelli, L.; Rossi, R. *J. Organomet. Chem.* **1993**, *451*, 97.
132. Loza, M. L.; de Gala, S. R.; Crabtree, R. H. *Inorg. Chem.* **1994**, *33*, 5073.
133. Paciello, R. O.; Manriquez, J. M.; Bercaw, J. E. *Organometallics* **1990**, *9*, 260.
134. Arliguie, T.; Chaudret, B.; Devillers, J.; Poilblanc, R. *C.R. Acad. Sci. (Paris)*, Series II **1987**, *305*, 1523; Arliguie, T.; Border, C.; Chaudret, B.; Devillers, J.; Poilblanc, R. *Organometallics* **1989**, *8*, 1308.
135. Chan, W.-C.; Lau, C.-P.; Cheng, L.; Leung, Y.-S. *J. Organometal. Chem.* **1994**, *488*, 103.
136. Frediani, P.; Salvini, A.; Bianchi, M.; Piacenti, F. *J. Organomet. Chem.* **1993**, *454*, C19.
137. Bautista, M. T.; Earl, K. A.; Maltby, P. A.; Morris, R. H. *J. Am. Chem. Soc.* **1988**, *110*, 4056.
138. Johnson, T. J.; Huffman, J. C.; Caulton, K. G.; Jackson, S. A.; Eisenstein, O. *Organometallics* **1989**, *8*, 2073.
139. Rottink, M. K.; Angelici, R. J. *Inorg. Chem.* **1993**, *32*, 3282.
140. Johnson, T. J.; Albinati, A.; Koetzle, T. F.; Ricci, J.; Eisenstein, O.; Huffman, J. C.; Caulton, K. G. *Inorg. Chem.* **1994**, 33, 4966.
141. Ruiz, J.; Mann, B. E.; Spencer, C. M.; Taylor, B. F.; Maitlis, P. M. *J. Chem. Soc., Dalton Trans.* **1987**, 1963.
142. Gusev, D. G.; Bakhmutov, V. I.; Grushin, V. V.; Vol'pin, M. E. *Inorg. Chim. Acta* **1990**, *175*, 19.
143. Paneque, M.; Poveda, M. L.; Taboada, S. *J. Am. Chem. Soc.* **1994**, *116*, 4519.
144. Mura, P.; Segre, A.; Sostero, S. *Inorg. Chem.* **1989**, *28*, 2853; Bergamini, P.; Sostero, S.; Traverso, O.; Mura, P.; Segre, A. *J. Chem. Soc., Dalton Trans.* **1989**, 2367.
145. Michalik, J.; Narayana, M.; Kevan, L. *J. Phys. Chem.* **1984**, *88*, 5236.
146. Bonneviot, L.; Cai, F. X.; M.; Kermarec, M.; Legendre, O.; Lepetit, C.; Olivier, D. *J. Phys. Chem.* **1987**, *91*, 5912; Carriat, J.-Y.; Lepetit, C.; Kermarec, M.; Che, M. *J. Phys. Chem. B* **1998**, *102*, 3742.

147. Nicol, J. M.; Eckert, J.; Howard, J. *J. Phys. Chem.* **1988**, *92*, 7117.
148. Clark, H. C.; Hampden Smith, M. J. *J. Am. Chem. Soc.* **1986**, *108*, 3829.
149. Ghoshray, K.; Bandyopadhyay, B.; Sen, M.; Ghoshray, A.; Chatterjee, N. *Phys. Rev. B* **1993**, *47*, 8277.
150. Sen, M.; Ghoshray, A.; Ghoshray, K.; Sil, S.; Chatterjee, N. *Phys. Rev. B* **1996**, *53*, 14345.
151. Halet, J.-F.; Saillard, J.-Y.; Koudou, C.; Minot, C.; Nomikou, Z.; Hoffmann, R.; Demangeat, C. *Chem. Mater.* **1992**, *4*, 153, and references therein.
152. Jessop, P. G.; Morris, R. H. *Coord. Chem. Rev.* **1992**, *121*, 155.
153. Bianchini, C.; Meli, A.; Peruzzini, M.; Vizza, F.; Zanobini, F.; Frediana, P. *Organometallics* **1989**, *8*, 2080; Bianchini, C.; Meli, A.; Peruzzini, M.; Frediani, P.; Bohanna, C.; Esteruelas, M. A.; Oro, L. A. *Organometallics* **1992**, *11*, 138.
154. Bianchini, C.; Bohanna, C.; Esteruelas, M. A.; Frediani, P.; Meli, A.; Oro, L. A.; Peruzzini, M.; *Organometallics* **1992**, *11*, 3837.
155. Bianchini, C.; Masi, D.; Peruzzini, M.; Casarin, M.; Maccato, C.; Rizzi, G. A. *Inorg. Chem.* **1997**, *36*, 1061.
156. Cotton, F. A.; Luck, R. L. *Inorg. Chem.* **1989**, *28*, 2181.
157. Bianchini, C.; Laschi, F.; Peruzzini, M.; Ottaviani, F. M.; Vacca, A.; Zanello, P. *Inorg. Chem.* **1990**, *29*, 3394.
158. Wasserman. H. J.; Kubas, G. J.; Ryan, R. R.; *J. Am. Chem. Soc.* **1986**, *108*, 2294; Butts, M. D.; Kubas, G. J.; Luo, X.-L.; Bryan, J. C. *Inorg. Chem.*, **1997**, *36*, 3341.
159. Luo, X.-L.; Kubas, G. J.; Bryan, J. C.; Burns, C. J.; Unkefer, C. J. *J. Am. Chem. Soc.* **1994**, *116*, 10312.
160. Gonzalez, A. A.; Mukerjee, S. L.; Chou, S.-L.; Zhang, K.; Hoff, C. D. *J. Am. Chem. Soc.* **1988**, *110*, 4419.
161. Lee, D. W.; Jensen, C. M. *J. Am. Chem. Soc.* **1996**, *118*, 8749.
162. A. A. Gonzalez and C. D. Hoff, *Inorg. Chem.* **1989**, *28*, 4295.
163. Sweany R. L.; Moroz, A. *J. Am. Chem. Soc.* **1989**, *111*, 3577.
164. Bautista, M. T.; Cappellani, E. P.; Drouin, S. D.; Morris, R. H.; Schweitzer, C. T.; Sella, A.; Zubkowski, J. *J. Am. Chem. Soc.* **1991**, *113*, 4876.
165. Hlatky, G. G.; Crabtree, R. H. *Coord. Chem. Rev.* **1985**, *65*, 1.
166. Kuhlman, R. *Coord. Chem. Rev.* **1997**, *167*, 205.
167. Crabtree, R. H. *J. Organomet. Chem.* **1998**, *577*, 111.
168. Cariati, F.; Ugo, R.; Bonati, F. *Inorg. Chem.* **1966**, *5*, 1128.
169. Werner, H.; Esteruelas, M. A.; Meyer, U.; Wrackmeyer, B. *Chem. Ber.* **1987**, *120*, 11.
170. George, M. W.; Haward, M. T.; Hamley, P.; Hughes, C.; Johnson, F. P. A.; Popov, V. K.; Poliakoff, M. *J. Am. Chem. Soc.* **1993**, *115*, 2286.
171. Haward, M. T.; George, M. W.; Howdle, S. M.; Poliakoff, M. *J. Chem. Soc., Chem. Commun.* **1990**, 913; Haward, M. T.; George, M. W.; Hamley, P.; Poliakoff, M. *J. Chem. Soc., Chem. Commun.* **1991**, 1101; Johnson, F. P. A.; George, M. W.; Bagratashvili, V. N.; Vereshchagina, L. N.; Poliakoff, M. *Mendeleev Commun.* **1991**, 26; George, M. W.; Haward, M. T.; Hamley, P. A.; Hughes, C.; Johnson, F. P. A.; Popov, V. K.; Poliakoff, M. *J. Am. Chem. Soc.* **1993**, *115*, 2286.
172. Sweany, R. L. *J. Am. Chem. Soc.* **1985**, *107*, 2374.
173. Upmacis, R. K.; Poliakoff, M.; Turner, J. J. *J. Am. Chem. Soc.* **1986**, *108*, 3645.
174. Church, S. P.; Grevels, F.-W.; Hermann, H.; Shaffner, K. *J. Chem. Soc., Chem. Commun.* **1985**, 30; Upmacis, R. K.; Gadd, G. E.; Poliakoff, M.; Simpson, M. B.; Turner, J. J.; Whyman, R.; Simpson, A. F. *J. Chem. Soc., Chem. Commun.* **1985**, 27; Andrea, R. R.; Vuurman, M. A.; Stufkens, D. J.; Oskam, A. *Recl. Trav. Chim. Pays-Bas* **1986**, *105*, 372; Ishikawa, Y.; Weersink, R. A.; Hackett, P. A.; Rayner, D. M. *Chem. Phys. Lett.* **1987**, *142*, 271; Ishikawa, Y.; Hackett, P. A.; Rayner, D. M. *J. Phys. Chem.* **1989**, *93*, 652; Sweany, R. L. In *Transition Metal Hydrides*, Dedieu, A., ed., VCH: New York, 1992, pp. 65–101. Wells, J. R.; House, P. G.; Weitz, E. *J. Phys. Chem.* **1994**, *98*, 8343; Walsh, E. F.; Popov, V. K.; George, M. W.; Poliakoff, M. *J. Phys. Chem.* **1995**, *99*, 12016; Poliakoff, M.; George, M. W. *J. Phys. Organic Chem.* **1998**, *11*, 589.
175. Goff, S. E. J.; Nolan T. F.; George, M. W.; Poliakoff, M. *Organometallics* **1998**, *17*, 2730.
176. Howdle, S. M.; Healy, M. A.; Poliakoff, M. *J. Am. Chem. Soc.* **1990**, *112*, 4804.
177. Jackson, S. A.; Upmacis, R. K.; Poliakoff, M.; Turner, J. J.; Burdett, J. K.; Greves, F.-W. *J. Chem. Soc., Chem. Commun.* **1987**, 678; Jackson, S. A.; Hodges, P. M.; Poliakoff, M.; Turner, J. J.; Grevels, F.-W. *J. Am. Chem. Soc.* **1990**, *112*, 1221; Hodges, P. M.; Jackson, S. A.; Jacke, J.; Poliakoff, M.; Turner, J. J.; Grevels, F.-W. *J. Am. Chem. Soc.* **1990**, *112*, 1234.

178. Zheng, Y.; Wang, W.; Lin, J.; She, Y.; Fu, K.-J. *J. Phys. Chem.* **1992**, *96*, 9821.
179. Grinval'd, I. I.; Lokshin, B. V.; Rudnevskii, N. K.; Mar'in, V. P. *Dokl. Acad. Nauk SSSR* **1988**, *298*, 1142.
180. Sweany, R. L. *J. Am. Chem. Soc.* **1986**, *108*, 6986; Sweany, R. L. *Organometallics* **1986**, *5*, 387.
181. Sweany, R. L.; Watzke, D. *Organometallics* **1997**, *16*, 1037.
182. Gadd, G. E.; Upmacis, R. K.; Poliakoff, M.; Turner, J. J. *J. Am. Chem. Soc.* **1986**, *108*, 2547.
183. Sweany, R. L.; Russell, F. N. *Organometallics* **1988**, *7*, 719.
184. Sweany, R. L.; Moroz, A. *J. Am. Chem. Soc.* **1989**, *111*, 3577.
185. Plitt, H. S.; Bar, M. R.; Ahlrichs, R.; Schnockel, H. *Angew. Chem. Int. Ed. Engl.* **1991**, *30*, 832.
186. Martensson, A.-S.; Nyberg, C.; Andersson, S. *Phys. Rev. Lett.* **1986**, *57*, 2045; Martensson, A.-S.; Nyberg, C.; Andersson, S. *Surface Sci.* **1988**, *205*, 12.
187. Ozin, G. A.; Garcia-Prieto, J. *J. Am. Chem. Soc.* **1986**, *108*, 3099.
188. Hauge, R. H.; Margrave, J. L.; Kafafi, Z. H. *NATO ASI Ser., Ser. B* **1987**, *158* (*Phys. Chem. Small Clusters*), 787.
189. Burwell, R. L., Jr.; Stec, K. S. *J. Coll. Interface Sci.* **1977**, *58*, 54.
190. Andrews, L.; Manceron, L.; Alikhani, M. E.; Wang, X. *J. Am. Chem. Soc.* **2000**, *122*, 11011.
191. Albertini, G.; Antoniutti, S.; Bordignon, E.; Pegoraro, M. *J. Chem. Soc., Dalton Trans.* **2000**, 3575.
192. Bloyce, P. E.; Rest, A. J.; Whitwell, I.; Graham, W. A. G.; Holmes-Smith, R. *J. Chem. Soc., Chem. Commun.* **1988**, 846.
193. Park, M. A.; Hauge, R. H.; Margrave, J. L. in *2nd International Conference on Low Temperature Chemistry*, Durig, J. R.; Klabunde, K. J., eds, BkMk Press: Kansas City, Mo., 1996, pp. 109–112.

4

Bonding and Activation of Dihydrogen and σ Ligands: Theory versus Experiment

4.1. INTRODUCTION

H_2 and alkanes were among the last remaining small molecules not to have been found to form stable metal complexes. Complexes of CO, NO_x, SO_x, O_2, CS_2, N_2, N_2O, CO_2, and H_2S had all been synthesized at least a decade or so before H_2 complexes. Unlike H_2, all of these contain nonbonding or π electrons for the metal to clasp, so the nature of the chemical bonding was not novel. The beauty of H_2 coordination was the absolute proof it provided that the interaction of a *bonding* electron pair with a transition M is strong enough (with help from $M \rightarrow H_2\sigma^*$

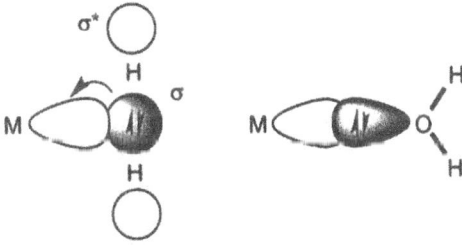

backdonation) to hold an external ligand to a metal. Internal agostic interactions[1] of C–H bonds with M (see Chapter 12) had been found shortly before the discovery of H_2 complexes but had not yet been well defined, considering their weak nature

and entropic advantage over external ligand binding. The hydrogen of the C–H bond is closer to M than the carbon, which added ambiguity to the nature of the interaction.

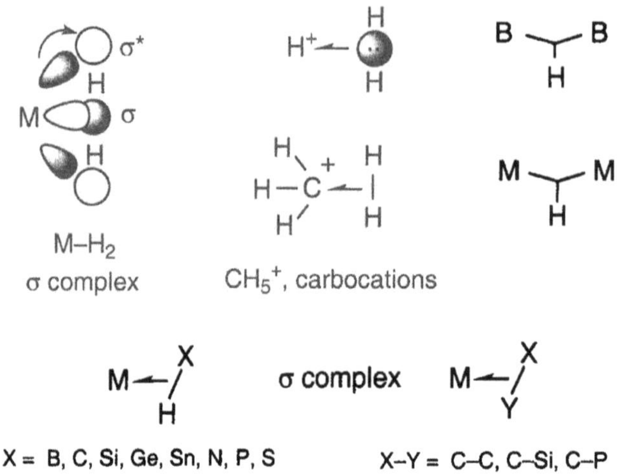

Figure 4.1. Electron-deficient bonding in chemistry.

The notion of coordinating molecules containing only "inert" σ bonds to M defied conventional bonding principles. Clearly a "nonclassical" form of bonding, e.g., *3c-2e bonding* had to join M to H_2 or other σ bonds, as for boron hydrides, carbocations, and H_3^+ (Figure 4.1). Indeed CH_5^+ is now considered[2] to be an extremely dynamic H_2 complex of CH_3^+, which bears an *isolobal* relationship[3] to a transition metal center. Even a bridging hydride should be viewed as rapidly

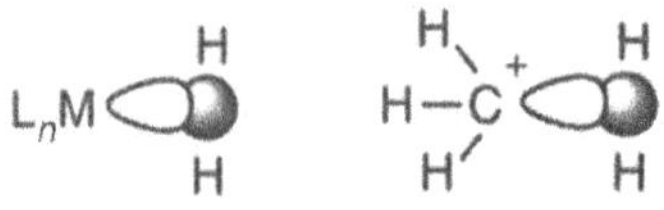

interconverting tautomers of a metal coordinating to an M–H bond (Figure 4.1). Despite all of these foreshadowings, the high stability of M–H_2 bonding was initially astonishing. Several review articles focus at least partially on the theoretical aspects of H–H and C–H bond coordination and activation,[4–8] including five in a special volume of *Chemical Reviews* devoted to computational transition metal chemistry.[9] In comparison to main-group electron-deficient bonding, transition metal σ-bond complexes are stabilized by M–L *backdonation* (BD) that is analogous to that in metal–olefin π coordination in the Dewar–Chatt–Duncanson model (Figure 4.2). This is shown by the absence of stable H_2 complexes of metals with d^0 electronic configurations or of strong main-group Lewis acids such as BX_3 where there are no d electrons to participate in BD. However *too much* BD destabilizes bound H_2 and other σ coordinated ligands toward OA, which is a key step in chemical reaction of these molecules. Thus a balance of σ donation and BD is needed to coordinate X–Y bonds to a metal.

Theoretical calculations provide some measure of this balance, and BD is much stronger than early estimates indicated, even for minimally activated H_2 complexes with short d_{HH} (0.85–0.9 Å). H_2 will thus stably bind to highly electrophilic cationic

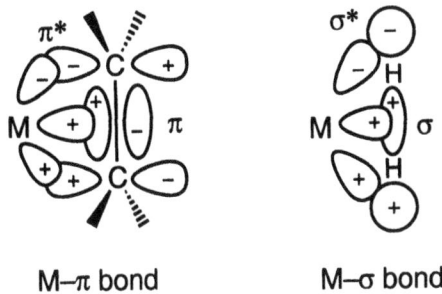

Figure 4.2. Comparison of M–H$_2$ coordination to Dewar–Chatt–Duncanson model for M–L π bonding.

complexes where the H$_2$ becomes a better σ donor in lieu of reduced BD, some of which must remain to stabilize the complex. H$_2$ is extremely versatile and can coordinate or oxidatively add to virtually any unsaturated metal fragment and *is perhaps the ideal ligand because it can act as both a Lewis acid (π acceptor) and a Lewis base (σ donor)*. The factors that can stabilize H$_2$ and other σ complexes over OA are: (1) electron-withdrawing ancillary ligands such as CO, particularly trans to the σ ligand; (2) positive charge, i.e., cationic rather than neutral complexes; (3) less electron-rich first- or second-row M; and (4) orbital hybridization, i.e., octahedral coordination and d^6 M. Electrophilic fragments favor σ coordination, although highly electrophilic cationic centers generally give weak coordination and/or promote heterolytic cleavage of the X–H bond.

The majority of H$_2$ complexes are of the d^6 octahedral type $[M(H_2)L_xX_y]^{z+}$, where the sum of neutral and anionic ligands, $x + y$, is five. Dihydride complexes of this type are also very stable when $x + y$ is four: e.g., FeH$_2$(CO)$_4$ prefers to be an octahedral dihydride rather than the five-coordinate Fe(H$_2$)(CO)$_4$ expected based on its electron-poorness. The seemingly simple H$_2$ ligand has proven to be much more sophisticated and complex than could have been imagined, and this chapter will address the bonding issues, emphasizing calculational analyses.

4.2. DEVELOPMENT OF THE THEORETICAL ASPECTS OF METAL–DIHYDROGEN BONDING AND CALCULATIONAL METHODOLOGIES

The existence of a complex between H$_2$ and a metal is one of the few notable examples in all of science of the nearly simultaneous and independent derivation of theory and fact. Neither the theoreticians nor the experimentalists involved were aware of the seminal research being carried out in their respective fields during 1979–1983 when H$_2$ activation was initially studied computationally and W(CO)$_3$(P^{i}Pr$_3$)$_2$(H$_2$) was first prepared and later proven to contain a H$_2$ ligand (see Chapter 2). This is marvelously exemplified by the classic theoretical paper by Saillard and Hoffmann in 1984 presenting extended Huckel calculations on the bonding of H$_2$ and CH$_4$ to metal fragments[10] that appeared only months after the seminal publication describing the W–H$_2$ complex. This and several other mid-1980s papers (see below) gave valuable early insights into M–H$_2$ bonding modeled after M–ethylene coordination. Much like the experimental discovery process, theoretical analysis of H$_2$ as a ligand developed over several years. The interplay between theory and experiment has continued hand in hand to this day as one of the most valuable synergistic relations in all of chemistry. There have been well over a hundred publications on computations of M–H$_2$ bonding and dozens more where experiment and theory are intertwined.

As discussed in an excellent review by Maseras *et. al.*,[9] the study of H$_2$ on metal complexes, including classical hydride ligands, has many advantages:

1. H is the smallest ligand and the only one to form a pure single bond to a metal, i.e., a crucial computational benchmark σ ligand.
2. The H atom is easy to locate computationally but difficult experimentally; even neutron diffraction (see Chapter 5) often cannot pinpoint the exact position of the hydrogens in H$_2$ ligands. Computational crystallography is a low-cost, high-quality technique for structural location and prediction of classical versus nonclassical structures.
3. Historically, the hydride ligand has offered most of the unforeseen, paradigm-shifting findings. Obviously the discovery of the H$_2$ ligand ranks at the top, but also surprising is the nearly boundless dynamics of H$_2$ and H ligands on metals discussed below and in Chapter 6.
4. M–H$_2$ is the ideal model for the binding/activation of other inert σ bonds such as M–alkane. Comparisons of the activation of H–H, Si–H, and C–H bonds will be drawn throughout this book.

The initial bonding concepts for M–H$_2$ coordination remain widely accepted, and extensive quantum-chemical calculations have provided increasingly accurate quantitative descriptions beginning about 1987 (earlier *ab initio* computations were limited to the lower-quality RHF level). The most recent calculations often show differences of only 0.01 Å between computed and experimental parameters such as d_{HH}. Many types of calculational methods have been employed to represent M–H$_2$ coordination, but density functional theory (DFT)[11] has emerged as the leading methodology, allowing the treatment of increasingly complex systems with excellent success. For example, a complex containing the 26-atom Tp ligand, TpRhH$_2$(H$_2$), has been fully analyzed calculationally by DFT, and this a structure that would have been unthinkable to model only a few years earlier.

Nonetheless theoretical analysis remains challenging because the M–H$_2$ binding energy is low and much effort is required for proper modeling. It is now apparent that the Hartree–Fock approximation underestimates the strength of the M–H$_2$ interaction. In all cases the electronic influence of the ancillary ligands such as phosphines is central, and use of PH$_3$, the "theoreticians phosphine," to model PR$_3$ can lead to systematic errors. On the other hand, useful information can sometimes be obtained. If the calculated (PH$_3$) and experimental (PR$_3$) structures are similar, the size of R or the precise electronics of the ligands must not be responsible for an observed structure/property. Often the substitution is reasonable electronically: an error of about 2 kcal/mol favoring the nonclassical isomer has been estimated for certain calculations.[6] DFT calculations show that larger variations in energy are possible using PH_3 rather than PMe_3, although this substitution does not significantly affect calculated structural parameters such as d_{MH}.[12] For studies of $IrXH_2(H_2)(PR_3)_2$ (X = Cl, Br, I; R = H, Me), Clot and Eisenstein have compared the two most widely used modern levels of theory in such inorganic systems, DFT (B3LYP) and *ab initio* MO perturbation theory (MP2).[13]

$$\begin{array}{c} R_3P \\ H\cdots\overset{|}{Ir}\cdots H \\ X\overset{|}{\diagup}\overset{|}{\diagdown}H \\ H \\ R_3P \end{array}$$

In general MP2 overestimates the energy of $M-H_2$ interaction while B3LYP underestimates it. Thus variations in method may have a severe impact on systems that border between weakly bound H_2 and elongated H_2 complexes ($d_{HH} > 1$ Å) approaching dihydrides. MP2 may give results that are more consistent with the latter while B3LYP tends to draw the system toward the former. A striking example is $IrCl_2H(H_2)(P^iPr_3)_2$ with a 1.1 Å neutron d_{HH}.[14] The MP2 distance is overestimated at 1.4 Å,[14] while for B3LYP it is 0.98 Å.[15] This may be a consequence of the highly delocalized bonding in stretched H_2 complexes with exceptionally flat potential energy surfaces (see Section 4.7.4). In regard to the most crucial structural parameter, d_{HH}, the MP2 value calculated for $IrIH_2(H_2)(PMe_3)_2$ is 0.851 Å while the experimental neutron distance is 0.856 Å for $IrIH_2(H_2)(P^iPr_3)_2$.[16] This agreement is much closer than can normally be expected or found, as neutron distances generally need correction for rapid rotation of the H_2 ligand that foreshortens observed d_{HH} (see Chapter 5).

4.3. THEORETICAL STUDIES OF OXIDATIVE ADDITION OF DIHYDROGEN

Qualitative ideas of catalytic H_2 activation on metals were devised in the late 1950s in connection with the formation of transition states or intermediates prior to OA of H_2 to hydride complexes (see Chapter 2). Surprisingly, there was no molecular orbital analysis of this intriguing theoretical problem until Dedieu carried one out in 1979.[17] Both extended Huckel and *ab initio* Hartree–Fock calculations were carried out on H_2 addition to square-planar d^8 $RhCl(PH_3)_3$, a model for the well-known Wilkinson catalyst in which the phosphine is PPh_3. In this 16e complex, the H_2 approaches the filled d_{z^2} metal orbital. Calculations indicate that at the beginning of the reaction, end-on (η^1) approach of H–H is preferred over side-on (η^2) approach (Eq. 4.1).

In Eq. (4.1) the antibonding σ^* orbital of H_2 and the antibonding combination $d_{z^2}-\sigma(H_2)$ mix to give rise to a $d_{z^2}-\sigma^* + \sigma$ nonbonding level. When the distance between the reactants shortens, the geometry of the $Rh-H_2$ moiety gradually changes from η^1 to η^2 via asymmetrically bound η^2-H_2. Crabtree also proposed such an attack of H_2 on Ir, much like his later model for C–H bond activation (see

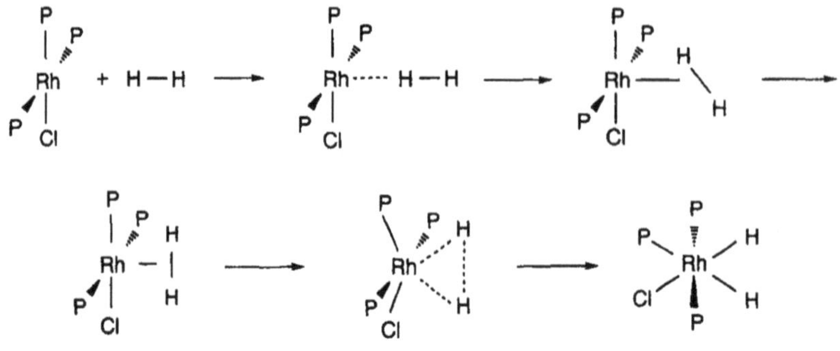

$$\text{(4.1)}$$

Section 12.2.1).[18] The H–H bond then symmetrically binds side-on and cleaves to give the *cis* hydride isomer that SCF calculations had shown to be the most stable

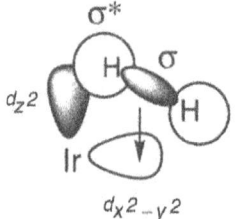

[Eq. (4.1)], which also was in agreement with experimental results. However, no true five-coordinate η^2-H_2 intermediate was found along the potential surface, i.e., H–H bond rupture to form a *cis* dihydride was not arrested along the reaction coordinate. $RhCl(PPh_3)_3$ does not form a stable H_2 complex, so this result propagated the belief that H_2 coordination could only be transitory.

As shown in the MO diagram, the σ orbital of H_2 is destabilized and the σ^* antibonding orbital is stabilized because d_{HH} has increased (Figure 4.3). It is important to note that the interaction of H_2 σ^* with the filled d_{yz} orbital adds to the

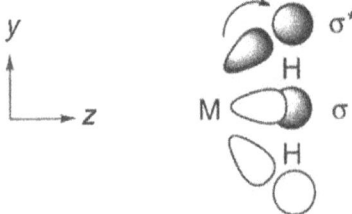

stabilization. As has already been emphasized, this *BD is critical to the binding and cleavage of* H_2. The lower the H_2 σ^* level becomes, the more d_{HH} increases. Furthermore, distorting the $RhCl(PH_3)_3$ fragment as shown below reduces steric repulsion with the incoming H_2 and destabilizes and hybridizes the d_{yz} orbital in a such as way that its overlap with σ^* is increased. The overlap increases from 0.17 to 0.27 when the angle Cl–Rh–P decreases from 180 to 130°. Therefore, owing to greater overlap and better energy match, BD is enhanced. Even minor skeletal

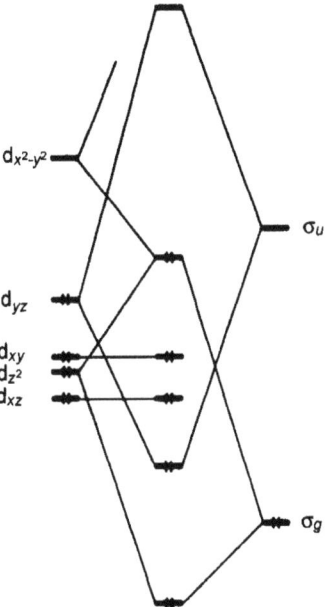

Figure 4.3. Qualitative molecular orbital diagram for interaction of H_2 and $RhCl(PH_3)_3$. Reprinted with permission from Dedieu.[17] Copyright 1979 American Chemical Society.

distortions were subsequently shown to affect the degree of BD, as reflected by measurements of the barrier to rotation of H_2 in bidentate phosphine complexes *trans*-$Mo(H_2)(CO)(Ph_2PC_2H_4PPh_2)_2$ (see Section 6.2.3.4).

Subsequent investigations support these early studies, and Hoffmann shows that side-on approach of H_2 to both $Cr(CO)_5$ and $[Rh(CO)_4]^+$ fragments gives a deeper energy minimum than end-on.[10] In the square planar Rh case, η^2 approach is again greatly stabilized only if two of the carbonyls bend back as for the $RhCl(PH_3)_3$ fragment above. The situation changes dramatically for approach of C–H bonds in CH_4 (see Chapter 12), where steric factors and orbital mismatch are a problem, and a pure η^2 approach becomes strongly repulsive. Modern DFT computations of OA

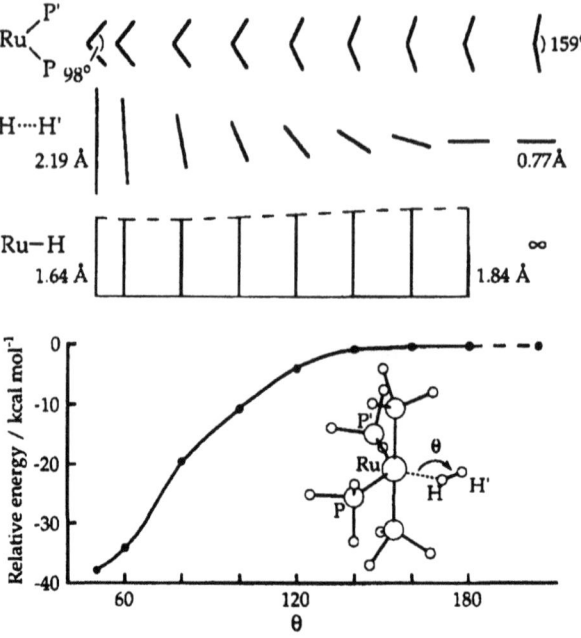

Figure 4.4. Reaction profile of geometrical changes for the $\eta^1-\eta^2$ swing of H_2 on $[Ru(PH_3)_4]$. Energies are relative to the sum of the isolated reactants set to zero, as indicated at the extreme right of the profile. Reprinted with permission from Macgregor *et al.* Copyright 1998 American Chemical Society.

of H_2 on $[M(PH_3)_4]$ for M = Fe, Ru, Rh^+ show that the optimum reaction coordinate involves an η^1 approach early in the reaction, followed by H_2 swinging around to an η^2 conformation (Figure 4.4).[19] A steep drop in energy begins when d_{RuH} approaches 1.77 Å and $\theta = 140°$, whereupon the P–Ru–P angle decreases rapidly. However, d_{HH} does not change much until $\theta = 100°$ (Ru–H = 1.65 Å, near its final value of 1.64 Å). Only then is there elongation of d_{HH}, corresponding to a late transition state often found for σ-bond activation processes. Most significantly, the $Ru(PH_3)_4 + H_2$ reaction is highly exothermic (37 kcal/mol) and proceeds without an activation barrier because of the donor/acceptor characteristics of the metal species.

By contrast the reaction profiles for OA of H_2 to *less electron-rich* cationic $[Rh(PH_3)_4]^+$ and $Ru(CO)_4$[20] fragments have small but distinct activation barriers (1–4 kcal/mol) and are less favorable thermodynamically (ca 15 kcal/mol for Rh). A plateau corresponding to $Ru(H_2)(CO)_4$ with an elongated d_{HH} of 1.00 Å was found, attesting to the crucial influence of the ancillary ligands and charge on H_2 activation, which will be discussed in depth below. BD to such electrophilic fragments is reduced, which stabilizes the σ complex. As will be shown, many other computations on the addition of σ ligands to unsaturated fragments demonstrate the existence of potential minima for σ complexes of, e.g., H_2 and CH_4 prior to full OA. Both Hall and Dedieu provided excellent overviews of OA and reductive elimination (RE) of H–H, C–H, and other σ bonds.[9]

4.4. PALLADIUM (DIHYDROGEN)—THE FIRST PREDICTION OF A STABLE DIHYDROGEN COMPLEX

The first quantum-mechanical calculations that showed that a M–H$_2$ interaction could be *stable* were described by Bagatur'yants.[21–23] In 1980 he hypothesized in 1980 a general bonding interaction of the two-electron type between low-lying localized orbitals of the σ core of ligands such as C–H and unoccupied diffuse orbitals of the transition metal.[21] A rough analysis indicated that the energy of the interaction could reach 10–20 kcal/mol, which agreed well with subsequent experimental and calculated values for M–H$_2$ and related σ-ligand binding. Semiempirical CNDO quantum-mechanical studies of the coordination of H$_2$ to model complexes of Pd^0 and Pt^0 such as M(PH$_3$)$_n$ (n = 1-3) led to proposing a molecular H$_2$ complex to explain reversible binding of H$_2$ by certain Pd complexes.[22] More reliable (but rudimentary by modern standards) nonempirical computations on H$_2$ interacting with a bare Pd atom showed a minimum for an η^2-H$_2$ structure on the potential curve at a Pd–H$_2$ distance of 2.05 Å, assuming an invariant d_{HH} of 0.74 Å (that in free H$_2$).[4,23] The calculated bond energy was 6.4 kcal/mol, and end-on H$_2$ bonding

$$Pd \xrightarrow{2.05 \text{Å}} \begin{array}{c} H \\ | \\ H \end{array} \quad 0.74 \text{Å}$$

was less favored. The critical finding was that BD from Pt $4d_{xy}$ to H$_2$ σ^* is a very significant bonding component. Calculations using a minimal basis set predicted that BD is apparently less than the σ donation from H$_2$, with the resulting charge on the H$_2$ being about 0.08 e. An expanded basis set showed that BD actually somewhat exceeds the σ donation, and the charge on H$_2$ is about -0.07 e. However, it is known that charge is only a very approximate reflection of the electron distribution, eliciting debate on the relative strengths of BD versus σ donation from H$_2$. However, as will be discussed below, recent advances indicate that the *BD component can be separated from the σ-donation component and can equal or exceed it in H$_2$ complexes containing* electron-donating ancillary ligands.

As discussed by Dedieu, who has reviewed theoretical aspects of Pd and Pt chemistry,[9] the Pd–H$_2$ system was refined by several investigators,[24–27] and almost identical results were obtained by Nakatsuji using a relativistic effect core potential.[24] Somewhat higher bond energies, about 14 kcal/mol for η^2-H$_2$ and 10 kcal/mol for η^1-H$_2$, were determined when electron correlation was taken into account by the CI method.[26] Despite discrepancies in the bond energies, the conclusion was that a Pd–H$_2$ complex should exist with d_{PdH} of 1.67–2.05 Å, H–Pd–H angles of 20–30°, and d_{HH} up to 0.81 Å, which is only slightly less than in L$_n$M–H$_2$ complexes. Balasubramanian found a second minimum corresponding to the OA product that was somewhat higher in energy (5.8 kcal/mol) and with a very small barrier for the reductive elimination back to the H$_2$ complex.[27] In contrast, reaction of H$_2$ + Pt leads directly to the very stable OA product, because of the different ground state of Pt.[27] Pt with its d^9s^1 ground state is ideally suited for making two Pt–H covalent bonds in PtH$_2$, but Pd with its d^{10}

ground state can easily form a Lewis acid/Lewis base complex between the empty 5s orbital and the σ_g orbital of H_2.

Direct experimental confirmation came in the form of low-temperature matrix-isolated $Pd(H_2)$ complexes studied spectroscopically by Ozin in 1986 and recently by Andrews (see Chapter 8).[28] Evidence for η^2-H_2 was obtained in a Xe matrix of Pd atoms, although Ozin suggested that η^1-H_2 may be present in Kr. This is a rare

```
        H
        |
Pd ←— |           Pd ←— H—H
        |
        H

solid Xe, 12 K    solid Kr, 12 K
```

experimental claim for η^1-H_2 (see also Section 3.2.2), which could be within 4 kcal/mol of being as stable as η^2-H_2 for $Pd(H_2)$. However, Andrews's studies in Ar matrices assign the IR bands found for $Pd(\eta^1$-$H_2)$ to be due to higher $Pd(H_2)_{2,3}$ complexes. His DFT calculations on $Pd(\eta^2$-$H_2)$ show $d_{HH} = 0.85$ and $H_2 \to Pd$ σ donation versus BD of 0.17 and 0.23 e, respectively.

Computations on $Ni(H_2)$ systems show η^2-H_2 geometry with $d_{NiH} = 1.82$ Å and $d_{HH} = 0.79$ Å.[29] This triplet state is stable by 9 kcal/mol with respect to dissociation to Ni plus H_2. Once again experimental confirmation soon followed in the form of vibrational spectroscopic evidence for H_2 binding to the stepped edges of a Ni(510) surface (see Chapter 8).[30] Although metal surfaces normally cleave H_2 to form hydrides, M atoms on this surface are effectively electronically unsaturated and bind H_2 molecularly.

Calculationally, when H_2 approaches small clusters such as M_2 or M_3, the nature of M (Pd or Pt) is important, and H_2 is bound and activated at one M for the Pt dimer.[31] The H–H bond is first broken and then a hydride migrates to the

```
       H                    H           H
       |                     \           \
0.85   | —Pt—Pt    →         Pt—Pt   →   Pt—Pt
       |                     /              \
       H                    H                H
```

other Pt with little barrier. However, for Pd_2, a molecular complex is not seen calculationally, and the metals "collaborate" to break the H–H bond without barrier on the singlet state to form a hydride-bridged dimer. IR studies of thermally

```
     H
     |
     H
     ↓                        H
                             / \
  Pd——Pd        →         Pd———Pd
                             \ /
                              H
```

evaporated Pd atoms in Ar matrices by Andrews confirm that one Pd atom cannot activate H_2 but that two can split H_2 with no activation energy.[28] A Pd–Pd(H_2) species is suggested to form and photochemically convert to $Pd(\mu$-$H)_2Pd$. There are also differences for CH_4 activation on Pd_n versus Pt_n. Such differences in σ-bond activation on second-versus third-row metals is a common theme in mononuclear organometallic complexes as well.

4.5. THEORETICAL STUDIES OF $M(CO)_3(PR_3)_2(H_2)$ AND BONDING MODEL FOR DIHYDROGEN COORDINATION

Not surprisingly, the first experimentally established H_2 complex to be studied theoretically was $W(CO)_3(PR_3)_2(H_2)$. A theoretician colleague at Los Alamos, Jeff Hay, had previously studied H_2 reaction with a $Pt(PH_3)_2$ fragment by *ab initio* methods.[32] Although he did not find a stable Pt–H_2 complex, his work paved the way for *ab initio* treatment of $W(CO)_3(PH_3)_2(H_2)$ at the Hartoch–Fock level.[33] A comparison of $W(CO)_3(PH_3)_2(H_2)$ with CO-free $W(PH_3)_5(H_2)$ led to important insights on H_2 binding and cleavage.[34] Hay showed that the bonding for an H_2 complex is described by the singlet state of M whereas the bonding in a comparable dihydride is described by the triplet state interacting with two hydrides. Extended Huckel calculations on similar systems also illuminated factors that stabilize H_2 versus dihydride coordination.[35]

Hay's calculations utilized the Dewar–Chatt–Duncanson type of bonding model with bond lengths for the $W(CO)_3(PH_3)_2$ fragment based on diffraction data for $W(CO)_3(P^iPr_3)_2(H_2)$. The η^2-H_2 geometry, where the H_2 is parallel to the P–W–P axis (**1**) rather than perpendicular (**1′**), is most favored, and a bond energy

(**1**)

(**1′**)

(**2**)

(**3**)

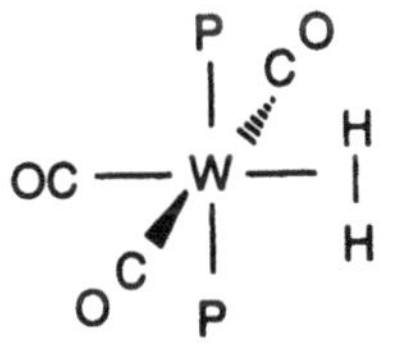

of 16.8 kcal/mol is calculated (Table 4.1). The dihydride forms (**2** and **3**) are less stable by 10–17 kcal/mol (Figure 4.5, upper) although the difference is much less (~1 kcal/mol) in the actual complex in the solution phase, which shows an equilibrium between **1** and a dihydride tautomer (see Section 7.4). The primary reason for this large discrepancy is that the geometries **2** and **3** used at the time to

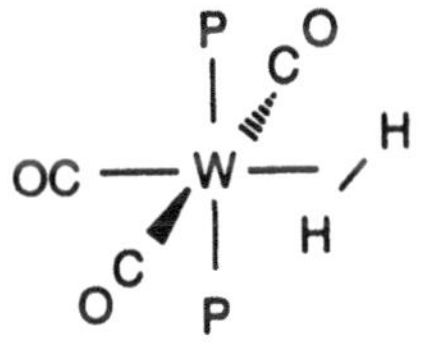

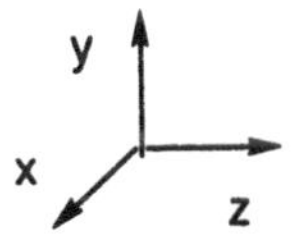

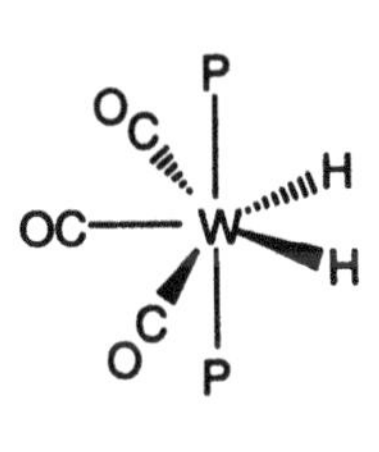

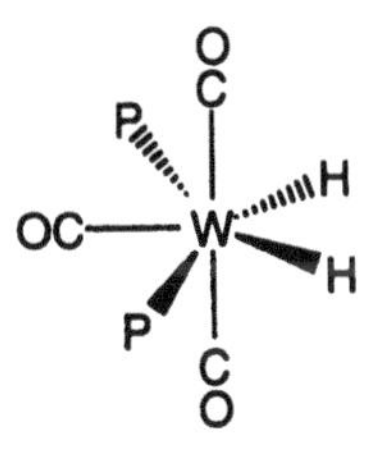

Table 4.1. Calculated Relative Energies of W(CO)$_3$(PH$_3$)$_2$(H$_2$)

Fragment + H$_2$	0.0
η^2-H$_2$ complex (1)	−16.8
η^2-H$_2$ complex (1′)	−16.5
η^1-H$_2$ complex	−5.7
Dihydride complex (2)	−6.8
Dihydride complex (3)	0.0

represent the dihydride form have cis hydrides, whereas it is now apparent that they are distal (see below).

The computed value of d_{WH} in 1 is 2.15 Å, which is much higher than the experimental value (1.89 Å), while d_{HH}, 0.80 Å, is close to the neutron diffraction value, 0.82 Å, although significantly lower than the more accurate solid state NMR value, 0.89 Å. Vibrational frequencies are another measure of the M–H$_2$ interaction (see Chapter 8): ν_{HH} in W(CO)$_3$(P^{i}Pr$_3$)$_2$(H$_2$) (2695 cm^{-1} exptl, 2474 cm^{-1}, calcd) is nearly half of that in free H$_2$ (4401 cm^{-1} exptl, 4561 cm^{-1}, calcd). The bonding energy for η^1-H$_2$ is only 5.7 kcal/mol, indicative of a stronger preference for η^2-H$_2$ in comparison to Pd(H$_2$). In regard to the two possible H$_2$ orientations in 1 and 1′ structure (1′) is less stable by only 0.3 kcal/mol, although subsequent experimental and theoretical determinations of the barrier to rotation of H$_2$ indicate that this energy difference is about 2 kcal/mol (see below and Chapter 6). The extended Huckel calculations of Jean[35] on W(CO)$_3$(H)$_2$(H$_2$) actually give a value very close to this (2.1 kcal/mol), favoring a structure analogous to 1 over 1′.

The preference for the orientation as in 1 is consistent with the fact that the d_π orbitals in the W(CO)$_3$(PR$_3$)$_2$ fragments have the ordering $d_{xz} < d_{yz}$ if the PR$_3$ ligands are chosen to lie along the y axis. This ordering arises from the greater stabilization of d_{xz} afforded by the CO π^* orbitals relative to the d_{yz} orbitals. In other words there is less competition for BD with the strongly π-accepting cis CO ligands when the H$_2$ is aligned with the better donor ligands (phosphines), which is the observed geometry of most H$_2$ complexes. This is similar to trans ethylene ligands coordinating 90° to each other to avoid competing for π-BD. The crystal structure of the W–C$_2$H$_4$ complex shows that ethylene also coordinates along the P–W–P axis.[36] The electronic preference for this must be strong to counterbalance the steric demand of forcing the ethylene to align toward the bulky tricyclohexylphosphines. As will be shown below, H$_2$ is a strong π acceptor like C$_2$H$_4$.

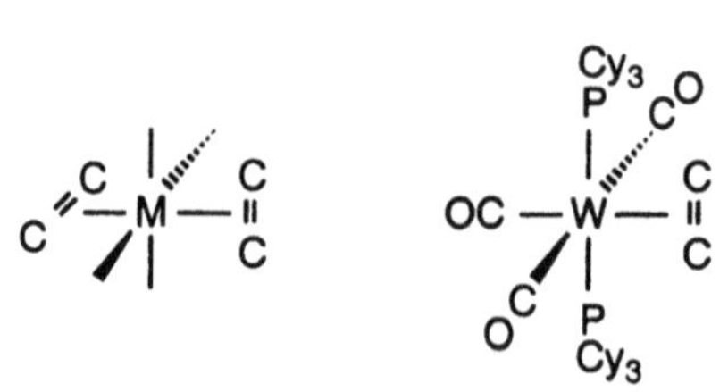

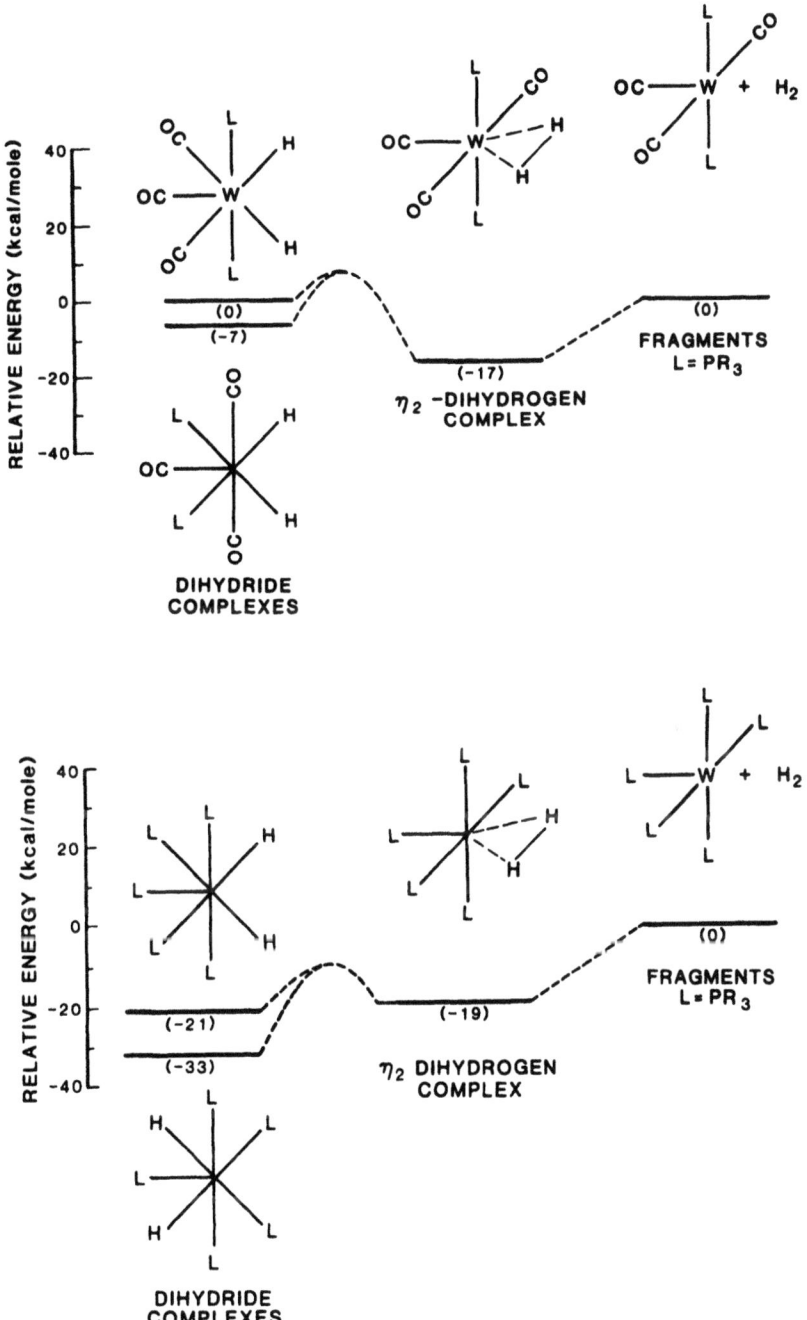

Figure 4.5. Calculated relative energies for W(CO)$_3$(PH$_3$)$_2$(H$_2$) and W(PH$_3$)$_5$(H$_2$) species. Reprinted with permission from Hay.[34] Copyright 1987 American Chemical Society.

Replacing CO by PR_3 both experimentally and theoretically leads to the seven-coordinate *dihydride* structure, $WH_2(PR_3)_5$, which is 3 kcal/mol more stable than the H_2 complex (Figure 4.5, lower) because of increased BD. If the d_{xz} orbital is to interact with the σ^* H_2 orbital, the presence of the CO ligands serves to decrease this interaction as the d levels are stabilized by BD. Replacing the COs by PR_3 destabilizes the orbital and promotes BD, leading to H–H bond rupture, especially when the trans CO is replaced.

Although the proximal dihydride is initially formed in a least-motion addition of H_2 to WL_5, further rearrangement can occur to give more stable forms (seven-coordinate complexes are fluxional). The distal form (**5**) lies 12 kcal/mol lower than

(4)  **(5)**

4 and 14 kcal/mol lower than η^2-H_2. Structure **5** is also less sterically crowded and indeed is the most stable form in solution for the known fluxional complex $CrH_2[P(OMe)_3]_5$.[37] OA of H_2 to $W[P(OMe)_3]_5$ has been reported,[38] and the X-ray structure of $MoH_2(PMe_3)_5$ is directly analogous to **5**.[39]

Modern DFT and *ab initio* studies have since been applied to H_2 binding and activation on the models $M(CO)_3(PH_3)_2$ (M = Cr, Mo, W) and $Mo(CO)_n(PH_3)_{5-n}$ (n = 1, 3, 5).[10,40–43] The calculated geometries and d_{MH} and d_{HH} are consistent with solid state NMR data for the known $M(H_2)(CO)_3(PR_3)_2$ and $Mo(H_2)(CO)(dppe)_2$ complexes, with deviations of approximately 0.03 Å (Table 4.2). By comparison, the value calculated for d_{HH}, 0.81 Å, in $W(CO)_3(PH_3)_2(H_2)$ by Hay using the original Hartree–Fock method was too short. A valuable computational result is that the lowest-energy structure for the dihydride tautomer in solution equilibrium with $W(CO)_3(PR_3)_2(H_2)$ has the trigonal bipyramidal geometry **6** shown in Eq. (4.2).[42,43]

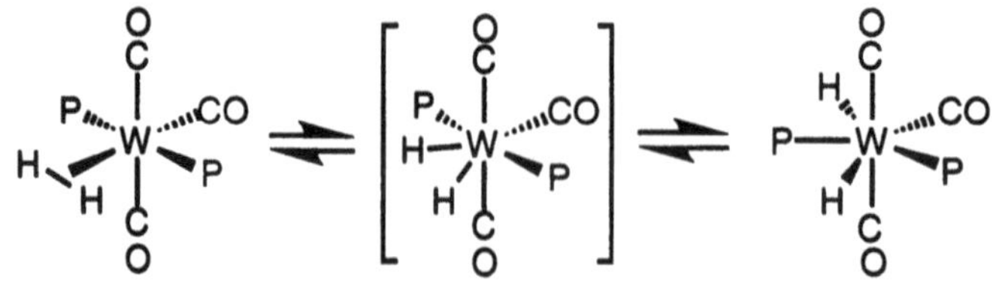

(4.2)

(1) **(3)** **(6)**

The first step presumably involves cleavage of H–H to give **3** followed by rearrangement to **6** by well-established mechanisms.[44] The enthalpy difference between **1** and **6** is 1.29 kcal/mol in favor of **1**, in excellent agreement with experiment (1.2–1.5 kcal/mol for R = ^{i}Pr or Cy; see Chapter 7). The actual structure of $WH_2(CO)_3(PR_3)_2$ is unknown and was originally proposed to be a capped

Table 4.2. Comparison between Calculated and Observed Bond Distances (Å) and Dissociation Energies (D_e, kcal/mol) for $M(CO)_nP_{5-n}(H_2)$ Complexes ($n = 0, 1, 3, 5$)

Fragment	d_{MH}			d_{HH}			D_e		
	calcd[a]	calcd[b]	exptl[c]	calcd[a]	calcd	exptl[d]	calcd[a]	calcd[b]	exptl
$Cr(CO)_5$	1.808			0.794			17.9		15 ± 1
$Mo(CO)_5$	2.006	1.899		0.787	0.824		15.7	19.6	
$W(CO)_5$	1.969			0.802			19.1		>16
$Cr(CO)_3P_2$	1.782	1.857	1.75X	0.808	0.822	0.85	16.9	21.3	~17[e]
$Mo(CO)_3P_2$	1.965	1.897		0.804	0.848	0.87	17.1	19.2	~17[e]
$W(CO)_3P_2$	1.918	1.872	1.89n	0.832	0.862	0.89	21.3	20.9	~19[e]
$Mo(CO)P_4$		1.896	1.92n		0.855	0.88			
CrP_5	1.730			0.829			18.7		
MoP_5	1.854			0.858			23.5		
WP_5	1.823			0.911			29.8		

[a] CCSD(T)/B3LYP level.[43]
[b] Relativistic NL-SCF + QR level.[41]
[c] X = X-ray, n = neutron.
[d] Solid state ^{1}H NMR, estimated deviation: 0.01 Å (see Chapter 5).
[e] Assuming a 10 kcal/mol correction for agostic interaction (see Chapter 7).

octahedron. The lower-energy geometry in **6** is consistent with the NMR finding of inequivalent H atoms and PR_3's and is consistent with the structures of $MoH_2(CO)(depe)_2$ (**9**, below) and similar seven-coordinate dihydrides. Thus the exact ligand arrangement is critical in the energetics of dihydrogen-dihydride systems and by analogy in the activation of all σ bonds.

Energy profiles associated with H–H bond-breaking in W complexes nicely illustrate the three possible scenarios encountered in the problem of η^2-H_2 versus dihydride. For each complex in Figure 4.6, d_{HH} is varied from 0.8 to 2.0 Å by steps of 0.2 Å with all other geometric parameters optimized at the B3LYP level at each fixed d_{HH}. In the electron-poor pentacarbonyl the stretching of H–H to 1.8 Å costs 14.6 kcal/mol, and there is no minimum associated with a cis dihydride structure. This is reversed for the electron-rich $W(PH_3)_5$ with greater BD; the dihydride is now lower in energy and no stable M–H_2 is found. The flatness of the profile indicates that the cis dihydride $WH_2(PH_3)_5$ is not the experimental structure but is most likely a trigonal bipyramidal (TBP) structure such as **6** as known for the Mo analogue. Finally, $W(CO)_3(PH_3)_2(H_2)$ gives an intermediate case with the two forms close in energy and a very low barrier for OA as found experimentally. The orientation of the H_2 is important, and when the H–H bond is in the plane of the carbonyls (rotated 90° from that in **1**, corresponding to $W(CO)_3(PH_3)_2(H_2)$-B in Figure 4.6), OA is much more difficult and similar to that in $W(H_2)(CO)_5$ where H_2 must compete with CO ligands for BD. OA is also more facile for third-row W than for the Cr or Mo tricarbonyls for which dihydride isomers are not detected. Hay's singlet-versus-triplet description[34] for $M(\eta^2$-$H_2)$ versus MH_2 led to a relationship that analyzes the trends in $MH_2(CO)_n(PH_3)_{5-n}$ with the help of a thermodynamic cycle (Scheme 1), where D_e(M–H) is the M–H bond energy, D_e(M–H_2) is the M–H_2 energy, D_e(H–H) is the H–H energy (104.2 kcal/mol) and $\Delta E_{S/T}$ is the

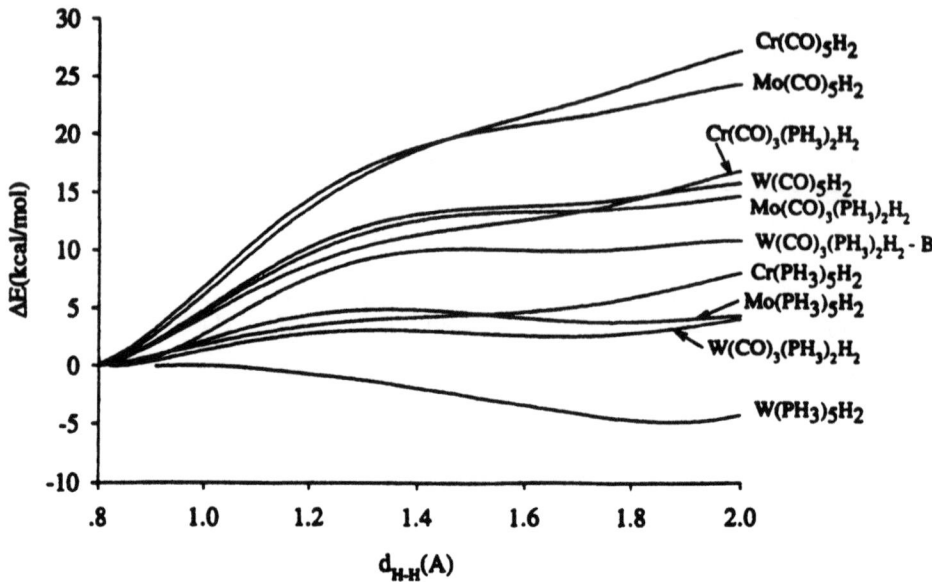

Figure 4.6. Energy profiles associated with the H–H bond-breaking in the $M(CO)_n(PH_3)_{5-n}(H_2)$ complexes (M = Cr, Mo, W; n = 0, 3, 5). Reprinted with permission from Tomas *et al.* Copyright 1998 American Chemical Society.

singlet/triplet energy gap of the fragment[43]:

$$\Delta E_{\text{dihydride/dihydrogen}} = D_e(M-H_2) + \Delta E_{S/T} + D_e(H-H) - 2D_e(M-H) \quad (4.3)$$

There is a linear relationship between

$$\Delta E_{\text{dihydride/dihydrogen}} \quad \text{and} \quad 2D_e(M-H) - D_e(M-H_2) - \Delta E_{S/T}$$

giving a straight-line plot with a slope of −1 passing near to the value for $D_e(H-H)$ (105.7 kcal/mol) at $\Delta E_{\text{dihydride/dihydrogen}} = 0$ (Figure 4.7). The different behavior of the

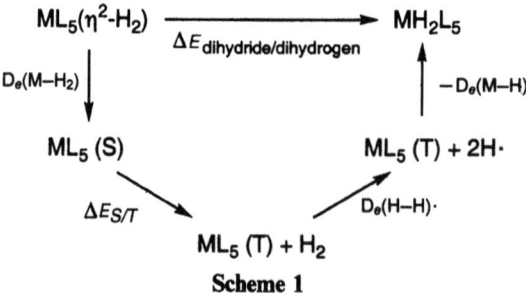

Scheme 1

various complexes upon H_2 addition is clearly apparent on this straight-line correlation. For the five complexes on the left side, the addition is arrested at the η^2-H_2 stage, while for the three complexes in the middle of the line there can be an

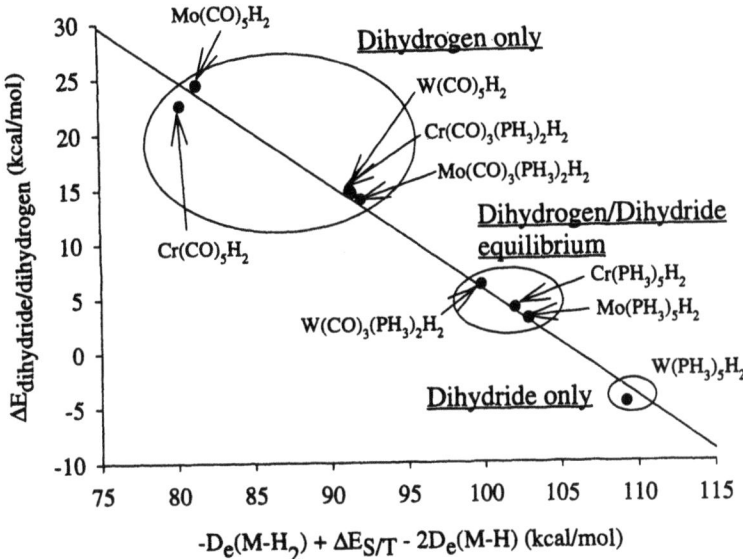

Figure 4.7. Plot of $\Delta E_{\text{dihydride/dihydrogen}}$ vs. $-D_e(\text{M-H}_2) + \Delta E_{S/T} - 2D_e(\text{M-H})$. Reprinted with permission from Tomas *et al*. Copyright 1998 American Chemical Society.

equilibrium between H_2 and dihydride forms. For $W(PH_3)_5(H_2)$, OA is complete and only dihydride is observed. Each of the three well-delineated regions associated with the three possible behaviors is represented by one or several systems. For a given set of ligands, the smallest $\Delta E_{S/T}$ and $D_e(\text{M-H})$ are found for Cr while for Mo both terms are larger. The similar behavior of Cr and Mo complexes stems from offsetting factors while going from Mo to W increases $D_e(\text{M-H})$ but $\Delta E_{S/T}$ is lower, which favors the dihydride.

Experimentally d_{HH} from the X-ray structure of $Cr(CO)_3(P^iPr_3)_2(H_2)$ is much shorter than the calculated d_{HH} (Table 4.2) because of the usual uncertainty in X-ray location. However, d_{HH} determined by solid state NMR (see Chapter 5) track surprisingly well with the DFT values (consistently 0.025 Å lower). This might not have been expected, considering the large ligand changes in going from $Mo(CO)_3(PR_3)_2(H_2)$ to $Mo(CO)(dppe)_2(H_2)$. These species turn out to have surprisingly similar d_{HH} despite the higher degree of BD in the latter. As will be shown below, the presence of the strong π-acceptor CO trans to H_2 greatly moderates the lengthening of d_{HH}.

4.6. RELATIVE STRENGTHS OF DIHYDROGEN AS A π ACCEPTOR AND AS A DONOR LIGAND COMPARED TO OTHER LIGANDS

If BD is so crucial, how do we know it exists and can we measure it quantitatively? The existence of $M \rightarrow H_2$ σ^* BD was definitively proven by inelastic neutron scattering measurements of a barrier to rotation of H_2, which would not

Table 4.3. Axial CO Stretching Frequencies (cm^{-1}) for Complexes Containing H$_2$ versus Other Ligands

L	W(CO)$_5$(L)	W(CO)$_3$(PCy$_3$)$_2$(L)	Mo(CO)(dppe)$_2$(L)
SO$_2$	2002	1873	1901
H$_2$	1971	1843	1815
C$_2$H$_4$	1973	1834	1813
N$_2$	1961	1835	1809
Ar/agostic[a]	1932	1797	1723
CH$_4$	1926		
Pyridine	1921	1757	1718
NR$_3$	1919[b]	1788[c]	1723[b]

[a] L = argon (matrix) for W(CO)$_5$(L); agostic C–H for others.
[b] NR$_3$ = NEt$_2$H.
[c] NR$_3$ = NH$_2$Bu.

exist if only σ donation from H$_2$ occurs (see Chapter 6). The question then becomes why are cationic H$_2$ complexes often as stable as their isoelectronic neutral analogues? This was demonstrated by Heinekey's studies of [Re(CO)$_3$(PR$_3$)$_2$(H$_2$)]$^+$ versus W(CO)$_3$(PR$_3$)$_2$(H$_2$), where the more electrophilic cation binds H$_2$ as strongly as the neutral species wherein BD is greater.[45] Another vivid example of this phenomenon is the series L$_n$Mo, [L$_n$Mn]$^+$, and [L$_n$Fe]$^{2+}$ [L$_n$ = (H$_2$)(CO)(PP)$_2$] in which the H$_2$ is most strongly bound in the dication (see Section 7.2). The absolute and relative amounts of electron donation and BD are difficult to gauge quantitatively either theoretically or experimentally. One of the first experimental clues to the relative strength of BD was a comparison of ν_{CO} in W(CO)$_5$(L).[46,47] The L = H$_2$ complex prepared in liquid Xe had the highest ν_{CO}, on a par with those for L = ethylene and N$_2$, and it was concluded that "the W(d_π) → H$_2$ (σ_u^*) interaction is stronger than expected theoretically" (alluding to the calculations by Hay[33] and Hoffmann[3]).[46] Gas-phase studies of the H$_2$ and N$_2$ complexes even led to the conclusion that this is the *main* interaction,[47] although later theoretical studies (see below) showed this to be an overstatement. The similarity of H$_2$, N$_2$, and C$_2$H$_4$ coordination holds true in the W(CO)$_3$(PR$_3$)$_2$(L)[48] and Mo(CO)(dppe)$_2$(L)[49–51] systems, for which the axial (trans to L) ν_{CO} are compared to those in W(CO)$_5$(L)[46,47,52,53] in Table 4.3. BD to L weakens BD from M to the trans CO, which raises ν_{CO}. These effects, while remarkably consistent, are all relative and nonquantitative. In all cases the complex with the strongest π acceptor, SO$_2$, gives the highest ν_{CO}, while pure σ donors such as alkylamines give the lowest. Complexes with the somewhat weaker π acceptors H$_2$, ethylene, and N$_2$ all have ν_{CO} within 10 cm^{-1} of each other. It is important to note that the complexes with C–H bond interactions, either the matrix-isolated CH$_4$ complex[53] or the agostic complexes, have ν_{CO} characteristic of pure σ donors. In W(CO)$_5$(L), ν_{CO} for L = CH$_4$ is even smaller than that for Ar, which is among the weakest pure σ donors. Overall the data suggest that *alkanes are poor backbonders* compared to H$_2$ and silanes, which oxidatively add to W(CO)$_3$(PR$_3$)$_2$. This matches Hoffmann's analysis showing that overlap between M d orbitals and σ* is small for C–H but large for H–H,[10] which helps explain why M–alkane binding is much weaker than M–H$_2$ binding.

Another experimental probe of BD versus σ donation is Mossbauer spectroscopy of a series of iron complexes.[54] Plots of isomer shift versus quadrupole splitting for [FeH(PP)(L)]$^+$ for L such as H_2, N_2, CO, MeCN, and Cl show that the H_2 complexes fell way off the straight line defined by the σ donors, more so than even CO. It was concluded that π bonding in H_2 complexes is very significant, more so than in N_2 or CO complexes. Theoretical analyses concur that H_2 is a much better π-acceptor ligand than early calculations showed. More quantitative measures of BD are provided by charge decomposition analysis (CDA) and extended transition state (ETS) analysis.[41,55–60]

In Frenking's CDA studies, the canonical (natural) MOs of the complex are expressed in terms of the MOs of appropriate fragments.[9,55–57] The MOs of the complexes, including group 6 species with multiple H_2, $M(CO)_{6-x}(H_2)_x$ ($x = 1-3$), are formed by a linear combination of the MOs of $M(CO)_5$ and H_2 in the geometry of, e.g., $M(CO)_5(H_2)$. The interaction is broken down into three components: charge donation (d), BD (b), and repulsive polarization (r). Table 4.4 shows each contribution at the MP2/II level for $M(CO)_5(H_2)$, trans-$M(CO)_4(H_2)_2$, and $W(CO)_5(L)$ for ligands with varying donor–acceptor properties. CDA indicates that CO is well balanced, both a good σ donor and a strong π acceptor, consistent with its ability to bind to most metal fragments (in this case electron-poor fragments). Cyanide is a powerful donor but a weak acceptor while N_2 is the opposite, a very poor donor and a moderate acceptor. Table 4.4 shows that H_2 is a slightly better acceptor than N_2 but, unlike the latter, is a strong donor. This is beautifully corroborated experimentally by small molecule interactions with the strong electrophile $[Mn(CO)_3(PCy_3)_2]^+$, which binds H_2 weakly but does not bind N_2 even at low temperature.[61] For $W(CO)_5(H_2)$, donation from H_2 (0.349 e) is greater than BD (0.129 e), as expected for this related electron-poor system. In the trans bis(H_2) complexes, BD is increased and the M–H_2 bonds are stronger than in $M(CO)_5(H_2)$.[56] Change in M is generally less significant, although Mo gives less BD. The negative sign of r indicates that charge is depleted from the overlap area of the occupied orbitals of the fragments. Depletion is large for CO and N_2, but much less for H_2.

In Ziegler's ETS method based on DFT, the M–H_2 bond energy can be decomposed into steric and orbital interaction terms plus a term for the energy

Table 4.4. Charge Decomposition Analysis of $M(CO)_5(L)$ and trans-$M(CO)_4(H_2)_2$ Complexes (Electron Units)

M	L	d	b	r
Cr	H_2	0.393	0.143	−0.147
Mo	H_2	0.315	0.105	−0.117
W	H_2	0.349	0.129	−0.105
W	CO	0.315	0.233	−0.278
W	N_2	0.027	0.107	−0.252
W	CN$^-$	0.488	0.024	−0.241
W	PH_3	0.278	0.091	−0.297
$Cr(CO)_4(H_2)_2$		0.277	0.209	−0.322
$Mo(CO)_4(H_2)_2$		0.362	0.149	−0.169
$W(CO)_4(H_2)_2$		0.483	0.191	−0.157

Table 4.5. Extended Transition State Analysis of Donation versus Backbonding Energies (kcal/mol) for Group 6 Dihydrogen Complexes

Complex	$-E_D$	$-E_{BD}$	Bond energy
$Mo(CO)_5(H_2)$	23.9	11.9	19.6
$Mo(CO)_3(PH_3)_2(H_2)$	17.6	18.9	19.2
$Mo(CO)(PH_3)_4(H_2)$	13.2	21.2	18.9
$Cr(CO)_3(PH_3)_2(H_2)$	16.9	17.9	21.3
$Mo(CO)_3(PH_3)_2(H_2)$	17.6	18.9	19.2
$W(CO)_3(PH_3)_2(H_2)$	18.2	19.5	20.9

required to relax the structures of the free fragments to the geometry of the combined molecule.[41,59] As for CDA, the orbital interaction can be further separated into donation (E_D), BD (E_{BD}), and other terms (all kcal/mol) [Eq. (4.4)]. Table 4.5 shows

$$M-H_2 \text{ bond energy} = E_D + E_{BD} + E_{other} \qquad (4.4)$$

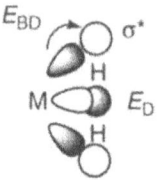

that the dependence of E_D versus E_{BD} on the ligand set is quite high, while variation with M is low. Overall the BD is much higher than earlier calculations suggested and can be *greater than σ donation*. Even for H_2 binding to the electron-poor $Mo(CO)_5$ fragment, E_{BD} is still one-third of the orbital energy. For the complex that models $Mo(CO)(H_2)(dppe)_2$, E_{BD} is nearly two-thirds of the orbital energy. Such substantial values for E_{BD} are supported by the observation of barriers to H_2 rotation as high as 11 kcal/mol in some systems (see Chapter 6), which implies at least that much BD energy.

The sum of E_D plus E_{BD} remains nearly constant, varying from -34.4 to -36.5 kcal/mol for the Mo complexes. E_{other} takes into account steric factors, and is generally 13–17 kcal/mol, giving net $M-H_2$ energies of 19–21 kcal/mol, close to experiment, e.g., 19.4 kcal/mol for $Mo(CO)_5(H_2)$. For very electrophilic centers, *loss in BD is almost completely offset by increased E_D from H_2 to the electron-poor center.* The $M-H_2$ energy for electron-poor $Mo(CO)_5(H_2)$ is surprisingly similar to that for the more electron-rich, isolable, phosphine complexes. *H_2 is the perfect ligand because it is effectively amphoteric like CO and SO_2 and is perhaps the most adaptable "weak" ligand*, reacting with virtually every unsaturated M fragment. From Table 4.3, SO_2 is a stronger acceptor, similar to CO, but H_2 is not far back, closely followed by ethylene and N_2. As pointed out by Hoffmann,[62] the reason CO is an excellent, ubiquitous ligand is the balance between its good donor/acceptor capabilities and its innate stability. H_2 offers the same advantages, on a smaller energy scale.

Table 4.6. Decomposition of the
$MH_2(PH_3)_3$–H_2 Interaction Energies (kcal/mol)

M	$-E_D$	$-E_{BD}$
Fe	13.8	13.4
Ru	12.3	14.2
Os (relativistic)	10.5	15.5
Os (nonrelativistic)	12.1	10.8

Similar energy decomposition calculations for M–CO reveal increased σ donation and decreased BD as electrophilicity of M increases, especially in cationic systems (nonclassical CO complexes).[62–64] The computed dissociation energy of the first CO from $[M(CO)_6]^n$ (M = group 4–9 metal; n = −2 to +3) is unexpectedly *higher* for the cationic complexes because σ donation from CO increases,[64] as it does for H_2.

For group 8 $MH_2(PH_3)_3$ fragments, E_{BD} can again be larger than E_D (Table 4.6).[59] $OsH_4(PMe_2Ph)_3$ is a hydride, and *relativistic effects* are crucial in calculations

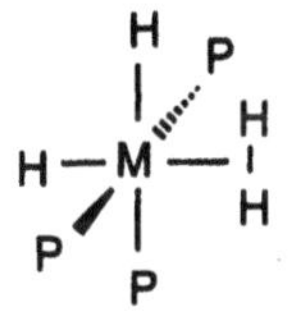

M= Fe, Ru, Os

of stabilities of $MH_2(H_2)$ versus MH_4 structures. At nonrelativistic levels, $MH_2(H_2)$ is the most stable, but this is reversed when relativistic effects are considered. Table 4.6 shows that E_{BD} is higher than E_D for relativistic calculations (less important for the lighter Fe and Ru congeners). Both d orbitals interacting with H_2 are raised in energy by relativity, which increases BD but reduces the energy for $H_2 \to$ Os donation. The relativistic increase in energy of the d orbitals results from contraction of the s and p core orbitals, which reduces the effective nuclear charge felt by the electrons in the Os 5d shell.[65,66] The same effect increases the bond strength between a heavy metal and acceptors such as CO and C_2H_4[67] and is caused by the relativistic mass increase of electrons moving with high instantaneous velocities near nuclei.[68]

4.7. DIHYDROGEN VERSUS DIHYDRIDE COORDINATION: ACTIVATION OF DIHYDROGEN TOWARD OXIDATIVE ADDITION

Computations on real M–L systems containing H_2 such as $W(CO)_3(PR_3)_2(H_2)$ rather than metal atoms such as $Pd(H_2)$ allowed variation of the ancillary L that led to crucial insight on the electronic factors favoring η^2-H_2 over dihydride. This was vital because steric congestion in $W(CO)_3(PR_3)_2(H_2)$ was initially believed to

Table 4.7. Examples of Effects of Variation of Metal,
Ligand, or Charge on Dihydrogen versus
Polyhydride Coordination

Dihydrogen complex	Polyhydride complex
$Mo(H_2)(CO)(dppe)_2$	$WH_2(CO)(dppe)_2$
	$MoH_2(CO)(depe)_2$
$CpMn(H_2)(CO)_2$	$CpReH_2(CO)_2$
$TcCl(H_2)(dppe)_2$	$TcH_3(dppe)_2$
$[Fe(H_2)(PH_3)_5]^{2+}$ (calcd)	$MoH_2(PR_3)_5$
$[MH(H_2)(PR_3)_4]^+$ (M = Fe, Ru)	$[OsH_3(PPh_3)_4]^+$
$MH_2(H_2)(PR_3)_3$ (M = Fe, Ru)	$OsH_4(R_3)_3$
$TpRuH(H_2)(PR_3)$	$CpRuH_3(PR_3)$
$[TpOs(H_2)(PPh_3)_2]^+$	$[CpOsH_2(PPh_3)_2]^+$
$[PtH(H_2)(PR_3)_2]^+$	$IrH_3(PH_3)_2$ (calcd)

stabilize H_2 as a ligand. Once again experiment and theory went hand in hand during the mid-1980s in defining H_2 activation by examining the effects of fine-tuning the electronics at M (Table 4.7). Computations showed that replacing acceptors such as CO by donors, i.e., increasing the basicity of M, promotes η^2-H_2 cleavage. This was dramatically demonstrated experimentally in $Mo(CO)(R_2PC_2H_4PR_2)_2(H_2)$ whereby merely changing R controlled whether a H_2 (**8**) or dihydride (**9**) complex was stable.[51] The more-electron-donating alkyl diphos-

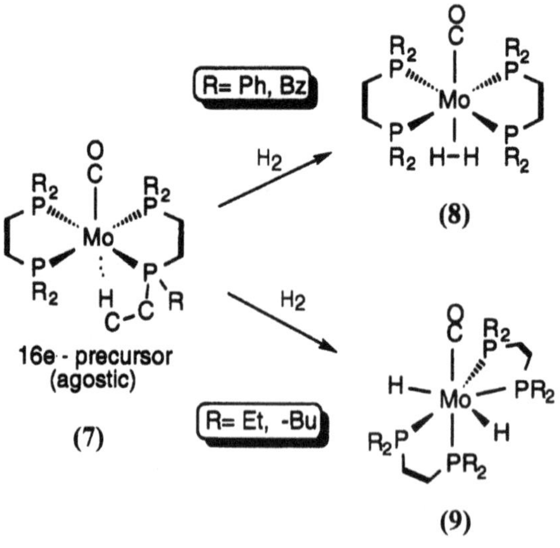

$$(4.5)$$

phines such as depe (R = Et) lead to increased BD, ultimately causing H–H rupture in **9**. Electronic rather than steric factors are crucial in stabilizing H_2 versus dihydride coordination since the phosphines with R = ^{i}Bu and phenyl (dppe) are similar in size. However, as will be shown in Section 4.7.1, which discusses the strong influence of the trans ligand, steric effects cannot be ignored. Changing M in

Mo(CO)(dppe)$_2$ to W also leads to dihydride formation[69] because W is a better backbonder than Mo (third-row metals have more diffuse d orbitals). Numerous examples of fine-tuning of H$_2$ versus hydride coordination are known in group 5–10 systems (Table 4.7) and provide excellent probes of electronics such as BD capability at specific fragments and stereoelectronic ligand effects. An example of the latter is the stabilization of η^2-H$_2$ and η^2-silane complexes over their dihydrido and hydrido(silyl) forms by Tp ligands relative to Cp analogues (see Sections 4.9 and 11.3.1).

4.7.1. The Crucial Influence of the Ligand trans to H$_2$

H$_2$ coordination chemistry affords a grand opportunity to study how metals activate and oxidatively add H$_2$ as a function of M, L, and charge. Conversely, H$_2$ binding can be used to probe site-specific electronics on a wide variety of fragments. W(CO)$_3$(PR$_3$)$_2$(H$_2$) and [FeH(H$_2$)(dppe)$_2$]$^+$ have trans to H$_2$ either the strong acceptor CO or the high-trans-effect hydride ligand. Their d_{HH} are shorter than 0.9 Å and their linearly related J_{HD} values (see Chapter 5) are greater than 30 Hz, indicative of true H$_2$ complexes. Widely varying H$_2$ complexes with different ligand sets can show much longer d_{HH} and lower J_{HD}, 11–26 Hz. Although stretching of d_{HH} can generally be rationalized by increased BD, some structure-bonding aspects of H$_2$ coordination remain unaccountable. As shown above, Mo(CO)(dppe)$_2$(H$_2$) is calculated to have high BD, yet it has one of the highest J_{HD} and shortest d_{HH} known. This lack of observable activation of the H–H toward OA is even more puzzling when compared to the cationic Mn congeners that actually have longer d_{HH} and no proclivity to cleave H$_2$ on increasing the donor strength of the cis phosphines (see Section 4.7.2).

The above observations can be explained in terms of the *trans effect*, i.e., the electronic influence of the ligand trans to the ligand of interest, which is crucial in coordination chemistry.[70] *The nature of the ligand trans to H$_2$ is the most important factor in determining where an H$_2$ complex lies along the reaction coordinate toward OA.*[14,57,71] For example, CO greatly reduces BD and has a powerful leveling effect that may even be underestimated in theoretical analyses. As seen from Table 4.8, d_{HH} *is normally shorter than 0.9 Å and J_{HD} is less than 30 Hz in complexes with CO trans to H$_2$, regardless of ligand set or overall charge.* Conversely, complexes with mild σ-donor ligands such as H$_2$O trans to H$_2$ or π donors such as Cl, have elongated H–H bonds (0.96–1.34 Å) and J_{HD} from 9–28 Hz because of increased BD. A good comparison is between the group 6 and 7 congeners, Mo(CO)(H$_2$)(dppe)$_2$ with J_{HD} = 34 Hz and d_{HH} = 0.88 Å and TcCl(H$_2$)(dppe)$_2$, where d_{HH} jumps to 1.08 Å from solution NMR evidence (1.44 Å, calculated[11]). For [Os(H$_2$)(L)(PP)$_2$]$^{2+}$ J_{HD} is 32 Hz for L = CO versus 14 Hz for L = H$_2$O.

If the trans ligand is a strong σ donor such as a hydride, there is a powerful trans labilizing effect that reduces donation from H$_2$, which once again weakens M–H$_2$ binding and contracts d_{HH}. Even relatively electron-rich neutral complexes with hydride trans to H$_2$ such as OsHCl(H$_2$)(CO)(PR$_3$)$_2$ and *trans*-IrCl$_2$H(H$_2$)(PR$_3$)$_2$ (**10**) often bind H$_2$ more weakly than comparable electrophilic cationic systems which rely on enhanced σ donation from H$_2$ for stability. For the isomer with Cl trans to H$_2$, IrCl$_2$H(H$_2$)(PR$_3$)$_2$ (**11**), J_{HD} decreases dramatically to

Table 4.8. H–D Coupling Constants and d_{HH} for H_2 Complexes of Selected Octahedral Transition Metal Fragments

Metal fragment	J_{HD}, Hz	d_{HH}, Å	Reference
trans CO Ligands			
$M(CO)_3(PR_3)_2$ (M = Cr, Mo, W[c])	33.5–35	0.84–0.89[a,b,e]	74
$Mo(CO)(PP)_2$	30–34	0.85–0.94[a,b,e]	50, 75
$[Mn(CO)(dppe)_2]^+$	32	0.89[a]	76
$[Mn(CO)(depe)_2]^+$	33	0.87–0.89[b]	70
$[Mn(CO)_3(PR_3)_2]^+$	33	0.87–0.89[b]	61
$[Re(CO)_4(PR_3)]^+$	34	0.85–0.87[b]	77
$[Re(CO)_3(PR_3)_2]^+$	32–33	0.87–0.90[b]	45
$[Re(CO)_2(PR_3)_3]^{+c}$	31	0.90–0.92[b]	78
$[Re(CO)(P\{(OR)_3\}_4]^{+c}$	33	0.87–0.89[b]	79
$[M(CO)(PP)_2]^{2+}$ (M = Fe, Ru, Os)	32–34	0.85–0.90[b]	80, 81
trans Cl or Other Donor Ligands			
$MCl(dppe)_2$ (M = Tc, Re)		1.08–1.21[d]	82
$ReCl(PR_3)_4$		1.17[e]	83
cis-$[Re(PR_3)_4(CO)]^+$	27.7	0.96–0.97[b]	84
$[RuCl(PP)_2]^+$	16–26	0.99–1.17[b]	85, 86
$[OsCl(PP)_2]^+$	10–14	1.19–1.27[b,d,e]	81, 86–88
$[Ru(H_2O)_5]^{2+}$	31.2	0.90–0.92[b]	89
$[Os(H_2O)(PP)_2]^{2+}$	14	1.19[b]	90
$[Os(L)(NH_3)_4]^{+,2+}$	9–20	1.09–1.34[b,e]	91
cis-$IrCl_2H(PR_3)_2$	12	1.11[e]	14
$RuIH(PR_3)_2$		1.03[e]	165
trans Hydride			
$[MH(PP)_2]^+$ (M = Fe, Ru)	28–33.5	0.86–0.97[b,d,e]	92
$[OsH(PP)_2]^+$	11–28	0.97–1.26[b,d,e]	92
$MHCl(CO)(PR_3)_2$ (M = Ru,Os)	31–34.5	0.84–0.92[b]	93
trans-$IrCl_2H(PR_3)_2$	34	0.85–0.87[b]	14
$IrIH_2(PR_3)_2$		0.86[e]	16

[a]Measured by solid state NMR.[74]
[b]Calculated from and bracketed by the empirical relationships, $d_{HH} = 1.42 - 0.0167 J_{HD}$[88] and $d_{HH} = 1.44 - 0.0168 J_{HD}$.[94]
[c]In equilibrium with dihydride tautomer in solution.
[d]Calculated from T_1 data.
[e]X-ray or neutron diffraction.

12 Hz and *ab initio* calculations (R = H) show a spectacular increase in d_{HH} from 0.81 to 1.4 Å (1.11 Å from neutron data for R = ^{i}Pr) on going from **10** to **11**.[14] Clearly the important concept is that *the influence of the trans ligand on H_2 activation is*

$J_{HD} = 34$ $J_{HD} = 12$
(10)   **(11)**

generally far greater than that of the cis ligand set, particularly in cations where there is less BD. This huge dependence on fragment stereochemistry can be expected to extend to other σ complexes, including alkane complexes where BD to C–H is much lower. It is crucial that the exact stereochemistry of a complex be known in order to understand σ-bond coordination. Although spectroscopy indicates that $TcH_3(dppe)_2$ is a trihydride (Table 4.7), its crystal structure and geometry are unknown.

Complexes related to **10**, such as

$$[IrH_2(H_2)(P^tBu_2Ph)_2]^+ \quad \text{and} \quad RuH_2(CO)(P^tBu_2Me)_2$$

are true *unsaturated 16e complexes that do not have a stablizing π-donor ligand* as in similar five-coordinate $RuHI(H_2)(PR_3)_2$.[72,73] DFT calculations on the Ir species show H_2 trans to H and a short d_{HH}, 0.80 Å (the experimental d_{HH} is computed from

ground state 0.0 Kcal/mol   transition state +6.9 kcal/mol

$J_{HD} = 34$ Hz to be 0.85 Å as in **10**). A transition state for hydride exchange is shown by calculations to have a Y-shaped geometry with an acute angle between the hydrides as computed[14] for $IrH_2Cl(PH_3)_2$. The binding energy is only 12 kcal/mol, and although this may be higher in the actual complex, it is remarkable that such a structure is energetically accessible in the absence of a π-donor ligand. However, in lieu of π donors, these isolable 16e species are stabilized without the need for an agostic C–H interaction by strong σ donation from the hydrides (see Section 12.2.3).[73] They can further relieve unsaturation by adding H_2 or other L, although binding is usually weak. A bis-H_2 complex can even be formed on H_2 addition to the Ir complex and shows rich dynamic behavior featuring independent hydrogen exchange and isotopic scrambling (see Chapter 6). It is interesting that silanes undergo OA on $RuH_2(CO)(P^tBu_2Me)_2$, in line with the higher susceptibility of Si–H to cleavage (see Chapter 11).

Calculations (MP2) on group 6 *trans*-$M(CO)_4(H_2)(L)$ well support the experimental results for a large variety of trans L.[57] For M = W, d_{HH} is 0.81 Å for L = CO, rising to 0.86 Å for L = H and stretching all the way to 1.69 Å for Cl, indicative of OA. In the last case the basicity of M is crucial since the Cr and Mo species remain H_2 adducts with $d_{HH} = 0.98$ and 0.89 Å. respectively. Computed M–H_2 bond dissociation energies decrease with increasing acceptor strength of L and correlate well with d_{HH}. d_{HH} in $[Os(NH_3)_4(L^z)(H_2)]^{z+2}$ where H_2 is trans to L^z depend not only on L^z but also on the method of calculation.[95–97] d_{HH} is long, ranging from 1.0 to 1.4 Å, and DFT methods consistently give geometries with shorter d_{HH} and longer d_{MH} distances than those at the MP2 level. When L^z is a strong π donor, the two methods predict d_{HH} near 1.3 Å and agree within 0.1 Å. However, when L^z is a π acceptor, DFT gives about 1.0 Å and differs by 0.3 Å from MP2. For elongated d_{HH}, the potential energy surfaces are very flat with respect to v_{HH} (see Section 4.7.4.), and the energy differences between the DFT and MP2 geometries are very small, about 3 kcal/mol.

The above rationalizations are based solely on electronic arguments, but steric effects should be taken into consideration. This appears to be the case in trans-Mo(L)(H$_2$)(dppe)$_2$ for L = CO [Eq. (4.5)] versus recently characterized analogues with L = CNR [Eq. (4.6)].[98] Isocyanides are better donors and weaker acceptors

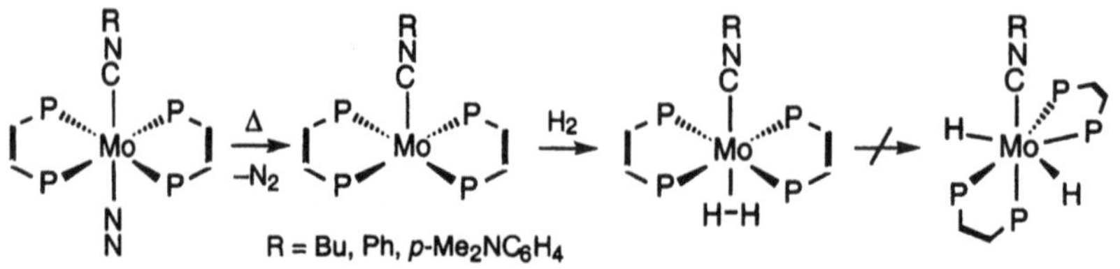

$$\text{(4.6)}$$

R = Bu, Ph, p-Me$_2$NC$_6$H$_4$

than CO and should have favored dihydride formation in Eq. (4.6), which is poised on the brink of OA. However, even for R = alkyl, the H$_2$ complex forms, and NMR T_1 values do not vary significantly from those for the trans CO analogue. As a sensitive gauge of the electronic richness of the M, ν_{NN} for

Mo[CN(p-Me$_2$NC$_6$H$_4$)](N$_2$)(dppe)$_2$

is 2018 cm^{-1}, significantly below that for

Mo(CO)(N$_2$)(dppe)$_2$

(2090 cm^{-1}) and in the range for dihydride formation or an elongated H$_2$ ligand (see Section 4.7.5). The crystal structure of

Mo[CN(p-Me$_2$NC$_6$H$_4$)](H$_2$)(dppe)$_2$

(Figure 4.8) shows that the molecule is very crowded [with a bent C–N–R (148°)], which could inhibit rearrangement to the dihydride.

4.7.2. Other Ligand/Metal/Charge Effects on H$_2$ Binding and Activation: Why FeH$_2$(CO)$_4$ Is a Dihydride

When d_{HH} and J_{HD} couplings for group 6–8 H$_2$ complexes with trans CO ligands are compared (Table 4.8), a surprising consistency is observed. Because H$_2$ bonding is governed by both the Lewis acid strength of M for σ bonding (donation from H$_2$) and the BD ability of M (H$_2$ as an acid), the observed d_{HH} results from a balance of these two components. The observation that isoelectronic but increasingly electron-poor neutral, cationic, and dicationic H$_2$ complexes (**12–14**) all possess comparable J_{HD} and d_{HH} raises an intriguing question. Why do highly electrophilic Mn$^+$ and Fe^{2+} systems with poor BD ability have d_{HH} similar to neutral Mo and W systems with much greater BD, the bonding component that had been thought to be the main control of d_{HH}? One explanation is that the *activation of η^2-H$_2$ in the more electropositive systems is occurring primarily via increased σ donation from H$_2$ as compared to the more electron-rich systems.*

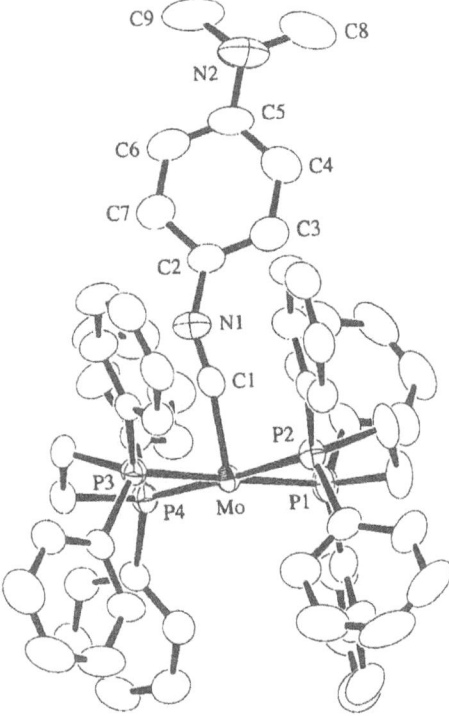

Figure 4.8. ORTEP diagram of Mo[CN(p-Me$_2$NC$_6$H$_4$)](H$_2$)(dppe)$_2$ showing steric crowding (hydrogen atoms omitted for clarity; H$_2$ ligand not located). Reprinted with permission from Seino et al.[98] Copyright 1999 American Chemical Society.

As discussed in Section 4.6, theoretical studies indicate that the σ donation and BD components vary greatly depending on the electron donating or withdrawing ability of the ligand sets, and the effect of M is much smaller. It is important to note that the sum of the bonding energies of the two components remains nearly constant in a particular system such as M(CO)$_n$(PH$_3$)$_{5-n}$(H$_2$). The calculations show that for

GROUP 6 (12)

GROUP 7 (13)

GROUP 8 (14) M = Fe, Ru, Os

highly electrophilic M *the loss in BD is almost completely compensated for by increased σ donation from H$_2$ to the electron-poor M.* Thus the extremely electron-poor Fe *dication* (**14**),[80] which is stable *in vacuo* in contrast to the Mo^0 and Mn$^+$

analogues (**12,13**), actually binds H_2 more strongly than the neutral complexes that are stabilized by BD. *Ab initio* results (P. J. Hay, unpublished) for **12** versus **13** indicate that the binding energy for the Mn cation is slightly higher than for the neutral Mo analogue (14.8 versus 13.6 kcal/mol).

The consistency in H–H activation on both cationic and neutral fragments is astonishing, especially the narrow J_{HD} range of 32–34 Hz for most of the complexes listed in Table 4.8. d_{HH} for the cationic systems generally lie in a narrow range, 0.86–0.90 Å near the calculated d_{HH} of 0.87 in H_3^+,[99] which is an ideal model for H_2 activation strictly through σ-orbital interactions. Nonetheless cationic H_2 complexes still contain some BD because ν_{CO} for **13** is significantly higher than those for the precursor complexes in Eq. (4.7) that contain multiple agostic interactions. The H–H bond is a much stronger acceptor compared to C–H in agostic interactions and alkane complexes because orbital mismatch for $M(\eta^2\text{-CH})$ greatly

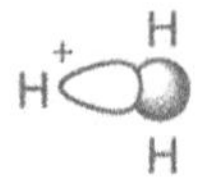

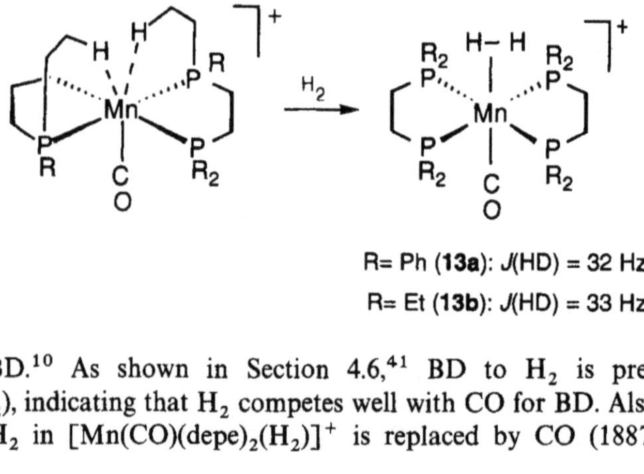

R= Ph (**13a**): $J(HD)$ = 32 Hz
R= Et (**13b**): $J(HD)$ = 33 Hz

(4.7)

diminishes BD.[10] As shown in Section 4.6,[41] BD to H_2 is present even in $Mo(CO)_5(H_2)$, indicating that H_2 competes well with CO for BD. Also ν_{CO} changes little when H_2 in $[Mn(CO)(depe)_2(H_2)]^+$ is replaced by CO (1887 versus 1888 cm^{-1}).

H_2 binds molecularly to $Mo(CO)(dppe)_2$ but undergoes OA on more electron-rich $Mo(CO)(depe)_2$ [Eq. 4.5)]. However, **13** binds H_2 molecularly for both the dppe and depe congeners [Eq. 4.7)] because the positive charge disfavors H–H bond rupture. What is surprising is the slightly lower J_{HD} and longer d_{HH} in **13a** than in **13b** *despite the greater donor ability of depe in **13b** and expected increase in BD ability of the M center that normally elongates or cleaves H–H*. This same counter-intuitive pattern is also seen in $[RuH(H_2)(PP)_2]^+$ despite electrochemical properties and pK_a of $\eta^2\text{-}H_2$ showing increased electron richness at Ru for depe.[100,101] In the series $[Re(H_2)(CO)_n\{P(OEt)_3\}_{5-n}]^+$, J_{HD} decreases from 33 to 30 Hz as n increases from 1 to 3,[79] an emphatic reversal of the expected trend because CO is a stronger acceptor than a phosphite. *The cis-ligands have little systematic effect in these cations*, especially when compared to neutral complexes at the brink of OA such as $Mo(CO)(H_2)(PP)_2$, where they can induce H–H cleavage. The lowest J_{HD} observed for any complex with CO trans to H_2 is 30 Hz for PP = $Bz_2PC_2H_4PBz_2$. Placement of a strong acceptor such as CO trans to an open coordination site in a cationic group 5-10 system may be predicted to favor formation of a σ complex for any X–H, regardless of the nature of M or the cis ligands.

As a dramatic example of the influence of metal charge, $MoH_2(PR_3)_5$ (**15**) is a dihydride while $[Fe(H_2)(PH_3)_5]^{2+}$ (**16**) is calculated to be an unactivated H_2 complex with a very short d_{HH} (0.744 Å).[102] One glaring exception to this principle

P
H,,,|,,,,,P
P——Mo
H/ `P
   P
(**15**)

P   2+
P,,,,|,,,,P
   Fe
H/   `P
  H   P
(**16**)

is $[CpWH_2(CO)_3]^+$, which by all accounts (cationic, containing mostly CO ligands) should be an H_2 complex.[103] However, as will be discussed below, complexes with such "piano-stool" geometries are unpredictable in the way they coordinate hydrogen, and $CpReH_2(CO)_2$ is also a hydride. Ironically, the first characterized hydride complex, $FeH_2(CO)_4$, would be expected to be an H_2 complex. However, recent microwave measurements in the gas phase confirm that the separation between hydrogens is about 2.2 Å and 2.36 Å in the Ru analogue, i.e., cis dihydrides.[104] Calculations do indicate, however, that an H_2 species is located in a shallow local minimum along the reaction pathway toward OA of H_2 to the $Fe(CO)_4$ fragment (Figure 4.9).[105] The DFT-derived d_{HH} for $Fe(H_2)(CO)_4$ is 0.924 Å, lengthening to 1.109 Å in the transition state. However, no spectroscopic evidence for $Fe(H_2)(CO)_4$ in equilibrium with the dihydride form could be found because of the low activation barrier (<4 kcal/mol) for breaking an H–H bond.[106] The reaction will pass through

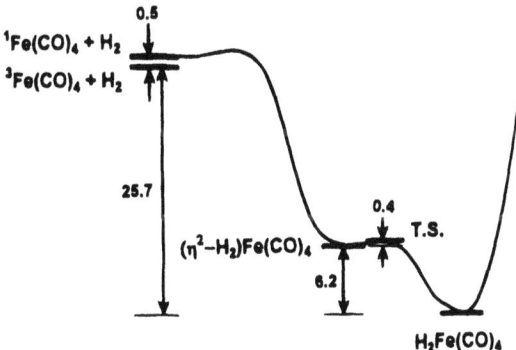

Figure 4.9. A diagram of the change in energy (kcal/mol) for the $H_2 + Fe(CO)_4$ system along the reaction coordinate based on BP86/II calculations. Reprinted with permission from Wang and Weitz.[105] Copyright 1997 American Chemical Society.

this local minimum directly to the $H_2Fe(CO)_4$ product with a favorable octahedral structure with cis hydrides. The PES for the H–Fe–H bending angle clearly shows anharmonicity with a very flat region from about 28 to 42° (local minimum near 28°) corresponding to d_{HH} of roughly 0.8–1.5 Å[104] typical of elongated H_2 complexes (see Section 4.7.4). Calculations on gas-phase reactions of $Fe(CO)_5$ with OH^- also identify $(\eta^2\text{-}H_2)Fe(CO)_4$ as a probable intermediate in the water–gas shift reaction catalyzed by iron carbonyl (see Section 9.1).

Microwave experiments also show that $OsH_2(CO)_4$ is a dihydride,[107] and the PES for $\angle$ HOsH is essentially harmonic without the flat region, while the Ru species shows intermediate behavior.[104] However, the reaction profile for OA of H_2 to $Os(CO)_4$ does show a midpoint plateau corresponding to formation of an elongated H_2 complex.[20] The nature of the electronic state of the complex may play a large role, as will be discussed below for H_2 addition to Fe atoms. $M(CO)_4(H_2)$ is destabilized by repulsion between a filled d orbital and the filled $\sigma(H_2)$ orbital and by poor BD from the low-energy d orbitals.[20]

4.7.3. The Reaction Coordinate for OA: The Point at Which The H–H Bond Is Broken

At the beginning of the 1990s polyhydride complexes were classified into two groups: classical hydrides with $d_{HH} < 1.6$ Å and nonclassical H_2 complexes with d_{HH} ranging from 0.8–1.0 Å. However, complexes with intermediate d_{HH} were subsequently observed by neutron diffraction and were labeled "elongated" or "stretched" H_2 complexes. A near *continuum of d_{HH}* now exists (0.8 to 1.5 Å and beyond; see Chapter 5) depending on the M–L fragment.

$$\underset{\text{"true" dihydrogen}}{M{\overset{H}{\underset{H}{\cdots\mid\cdots}}}\;0.8\text{--}1.0\text{ Å}} \longrightarrow \underset{\text{stretched dihydrogen}}{M{\overset{H}{\underset{H}{\cdots\mid\cdots}}}\;1.0\text{--}1.2} \longrightarrow \underset{\text{compressed dihydride}}{M{\overset{H}{\underset{H}{\diagup\!\!\mid\!\!\diagdown}}}\;1.2\text{--}1.6} \longrightarrow \underset{\text{dihydride}}{M{\overset{H}{\underset{H}{\diagup\diagdown}}}\;< 1.6}$$

(4.8)

The arbitrary bonding depictions and distance ranges in Eq. (4.8) essentially represent points along the reaction coordinate for OA of H_2. However, the bonding electrons are difficult to locate experimentally, and often d_{HH} cannot be accurately determined (see Chapter 5). True H_2 complexes (sometimes referred to as Kubas complexes) lie at one extreme and represent H_2 coordination to electrophilic M and/or complexes where H_2 is trans to CO or other strong trans-effect ligands. "Compressed dihydrides" lie at the other end and are essentially hydrides with weak H···H attractions. Most H_2 complexes have intermediate d_{HH} (ca 0.9–1.1 Å), and complexes with $d_{HH} > 1$ Å can have extremely delocalized M–H_2 bonding. Vibrational analysis of $W(CO)_3(PCy_3)_2(H_2)$ indicates a high degree of mixing between W–H and H–H modes (see Chapter 8), suggestive of a triangulo structure as depicted for a stretched H_2 complex even though d_{HH} is fairly short, 0.89 Å, and J_{HD} is quite high, 33 Hz. This complex undergoes equilibrium OA [Eq. (4.2)], and the main reason that d_{HH} is short in the H$_2$ tautomer seems to be the trans CO ligand. Thus judging the degree of H$_2$ activation based solely on d_{HH} or J_{HD} is often not meaningful and is sometimes paradoxical.

A difficult question to answer is: *At what point can the H–H bond considered to be broken?* Maseras et al.[15] provide theoretical guidance by applying Bader's atoms-in-molecules formalism[108] to determine where chemical bonds exist in H$_2$ complexes with varying d_{HH}. This model provides one of the most powerful tools available for the analysis of theoretical results and can be applied to any computational method that provides the electron density of the system. Basically, atomic nuclei correspond to electron density maxima, and a bond path (characterized by a bond critical point) can be defined between each bound pair of atoms. Four model complexes (Figure 4.10) corresponding approximately to each of the designations in Eq. (4.8) are optimized geometrically[15] using DFT[109] (which gives better results than MP2[110]). The four complexes are W(H$_2$)(CO)$_3$(PH$_3$)$_2$ (**1b**), IrHCl$_2$(H$_2$)(PH$_3$)$_2$ (**2b**, which models **11**), [Os(H$_2$)(NH$_2$(CH$_2$)$_2$NH$_2$)$_2$(HCO$_2$)]$^+$ (**3b**), and the classical hydride OsH$_4$(PH$_3$)$_3$ (**4b**). Table 4.9 compares calculated d_{HH} and d_{MH} with neutron values for the actual complexes. Electron density contour maps around the M(H···H) triangle (Figures 4.11–4.13) show that complexes with d_{HH} of 0.82 Å modeled by **1b** correspond to a true H$_2$ complex, i.e., a clear bond path connects the hydrogens (Figure 4.11). The W atom is not connected separately to each of the H atoms but rather directly to its bond critical point, giving a T-shaped bond structure. This is similar to the minimum energy structure of CH$_5^+$, which corresponds to CH$_3^+$ interacting with H$_2$ (see Section 4.1). However, complexes with d_{HH} near 1.3 Å and beyond such as in [OsH$_2$(en)$_2$(acetate)]$^+$ (neutron value, 1.34 Å), which are modeled by **3b** (calculated value, 1.43 Å), are best viewed as dihydrides with shortened (compressed) H···H distances. The electron density contour (Figure 4.13) is very similar in appearance to that for the hydride OsH$_4$(PH$_3$)$_3$. This can also be clearly seen in the plot of the Laplacian of charge density ($\nabla^2\rho$) in Figure 4.14, which shows two separated charge concentrations associated with each of the hydride ligands in **3b**. In contrast **2b** shows a region of charge concentration around the H$_2$ ligand with a distribution similar to that of the H$_2$ molecule. The actual complex, IrHCl$_2$(H$_2$)(P^{i}Pr$_3$)$_2$ (**11**) contains a stretched H$_2$ with d_{HH} = 1.11 Å and is difficult to model theoretically. The usually reliable MP2 methodology is unable to locate this species as a local minimum on the potential hypersurface (optimized d_{HH} = 1.4 Å; see Section 4.7.1). However

Table 4.9. B3LYP-Optimized Values of d_{HH} and d_{MH} (Å) for Model Complexes **1b–4b**, Compared with Neutron Diffraction Data for Actual Complexes

	1b	2b	3b	4b
H–H B3LYP	0.818	0.984	1.428	1.861
Neutron	0.82	1.11	1.34	1.84
M–H B3LYP	1.934	1.651	1.606	1.641
Neutron	1.95	1.54	1.60	1.65

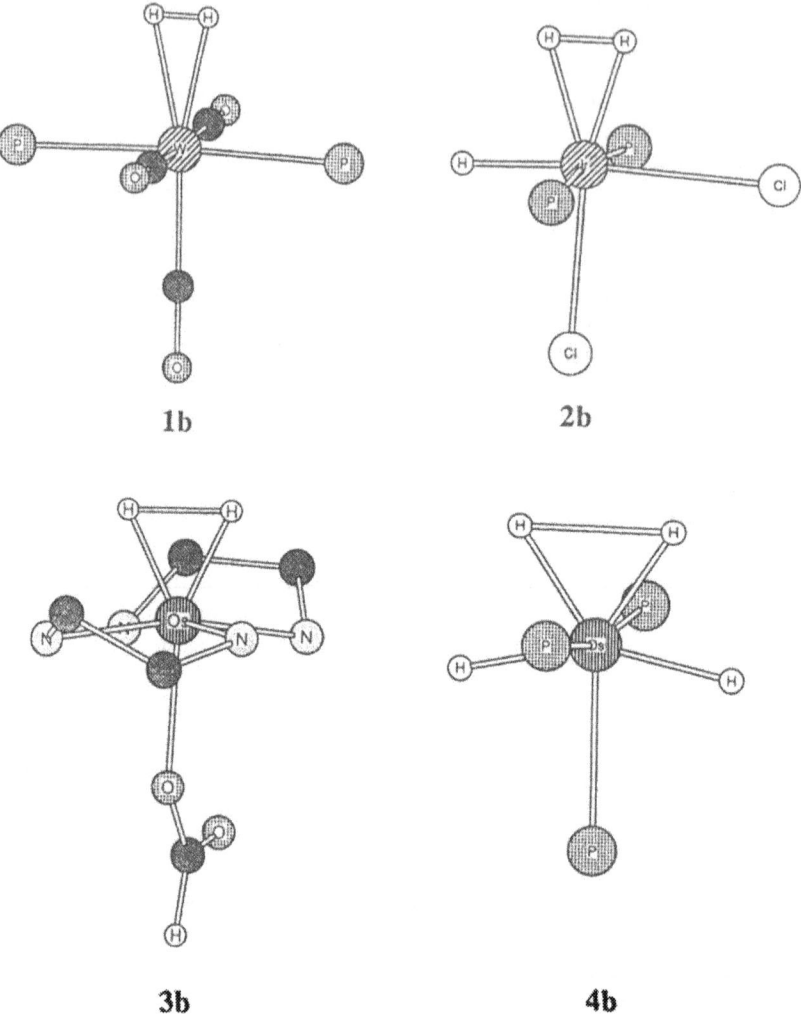

Figure 4.10. B3LYP-optimized geometries of model complexes **1b–4b**. Most of atoms not directly attached to M are omitted for clarity. Reprinted with permission from Maseras *et al.*[15] Copyright 1996 American Chemical Society.

B3LYP gives d_{HH} closer to the neutron value, albeit still about 0.1 Å shorter (Table 4.9). Complexes with d_{HH} of 1.0–1.1 Å can still be marginally considered to be H_2 complexes, but those with $d_{HH} > 1.35$ Å are in the realm of hydrides.

Another useful criterion employed by Hush[111] is the concept of *bond index*, p_{HH}, which for H_2 is a function of the 1s orbital structure:

$$p_{HH}^{1/2} = \alpha^2 - \beta^2 \qquad (4.9)$$

p_{HH}, also termed the bond order, is unity for a free molecule and essentially zero for

Figure 4.11. Electron isodensity contour map of the optimized system **1b**. The lines plotted correspond to the values of 0.0001, 0.001, 0.01, 0.02, 0.04, 0.06, 0.08, 0.10, 0.13, 0.16, 0.19, 0.22, 0.26, 0.30, 0.35, 0.40, 0.70, and 1.00 au (also for Figures 4.12–4.14). Reprinted with permission from Maseras *et al.*[15] Copyright 1996 American Chemical Society.

a dihydride. The H–H bond is weakened by a combination of σ donation to M (which reduces α) and BD from M (which increases β). This can be related via Eq. (4.10) to the reduced internuclear separation, $d'_{HH} = d_{HH}/d^0_{HH}$, where d^0_{HH} is the free-molecule equilibrium separation, 0.74 Å.

$$d'_{HH} = a + bp_{HH} \qquad (4.10)$$

Hush's calculations on complexes similar to **3b** suggest values of a and b very close to 2 and -1, respectively, and the expression simplifies to

$$d'_{HH} = 2 - p_{HH} \qquad (4.11)$$

By this criterion the H–H bond can be considered "broken" when d_{HH} is twice the free molecule separation, 1.48 Å, at which point p_{HH} becomes zero. This, however, is

Figure 4.12. Electron isodensity contour map of the optimized system **2b**. Reprinted with permission from Maseras *et al.*[15] Copyright 1996 American Chemical Society.

again only a rough guideline. Also according to Hush, the reduced value of $J'_{HD} = J_{HD}/J^0_{HD}$, where $J^0_{HD} = 43$ Hz in free H_2, is a direct measure of the bond index p_{HH}. For example, for $[Os(NH_3)_4(H_2)(H_2O)]^{2+}$ the observed J_{HD} of 8.1 Hz implies an H–H bond index of 0.19 (8.1 Hz/43 Hz) and a complementary Os–H bond index p_{MH} of 0.81. By contrast, the J_{HD} of 35 Hz reported for $Cr(H_2)(CO)_3(P^iPr_3)_2$ indicates an H–H bond index of 0.81, with a Cr–H bond index of 0.19, the reverse of the situation for the Os complex. This concept is useful only as a qualitative guide because J_{HD} and d_{HH} can be deceptive in judging the closeness of H_2 to cleavage.

The correlation of bond valence and bond length as applied above can be extended further according to the concept of Dunitz and Burgi of mapping a series of crystal structures as pathways of chemical reacions (Section 12.2.1). Hush derived the expression $p_{HH} + p_{MH} = 1$, and Grundemann *et al.*[112] extended it to give Eqs. (4.12)–(4.14).

$$p_{HH} = \exp[-(d_{HH} - d^0_{HH})/b_{HH}] \quad (4.12)$$

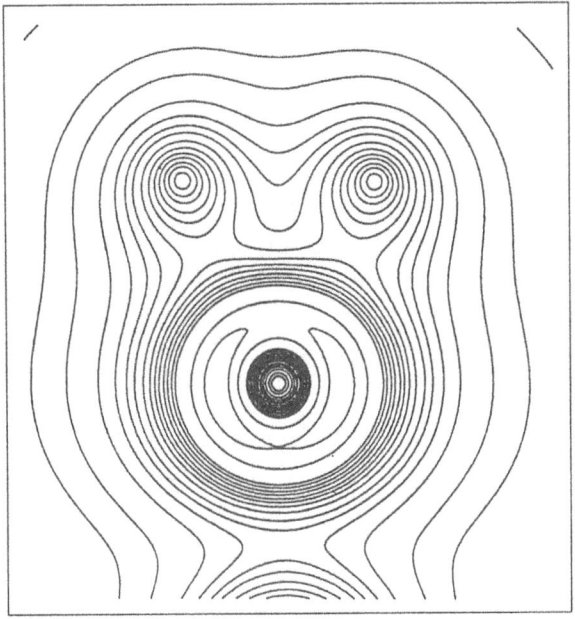

Figure 4.13. Electron isodensity contour map of the optimized system **3b**. Reprinted with permission from Maseras et al.[15] Copyright 1996 American Chemical Society.

$$p_{MH} = \exp[-(d_{MH} - d^0_{MH})/b_{MH}] \qquad (4.13)$$

$$d_{HH} = d^0_{HH} - b_{HH}\ln\{1 - \exp[-(d_{MH} - d^0_{MH})/b_{MH}]\} \qquad (4.14)$$

These equations are based on that for hydrogen bonds, and the decay parameters $b_{MH} = b_{HH}$ can accordingly be set to 0.404 Å. In Figure 4.15a, Eq. (4.14) is plotted as the solid line, which depicts the M–H$_2$ plane, i.e., the half-hydrogen distance as a function of the distance between M and the center of the H–H bond, with M at the origin. Figure 4.15b includes the corresponding bond order changes. The crystal structure data (shown as points in Figure 4.15a) shows a continuous transition from free H$_2$ to dihydrides, which makes it difficult to delineate the exact borders between dihydride, elongated H$_2$ complexes, and true H$_2$ complexes. This excellent correlation between calculated and experimental data represents snapshots of the pathway for approach of H$_2$ to M, much like that proposed by Crabtree for the approach of CH groups to M (see Section 12.2.1).

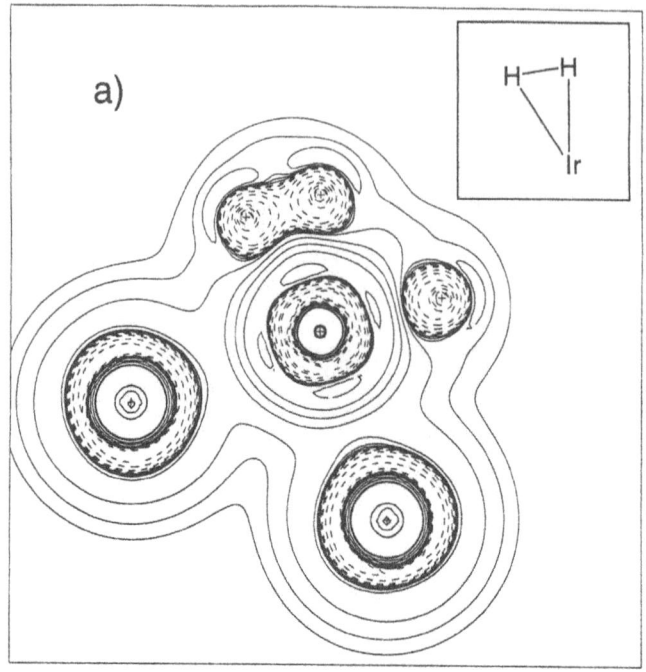
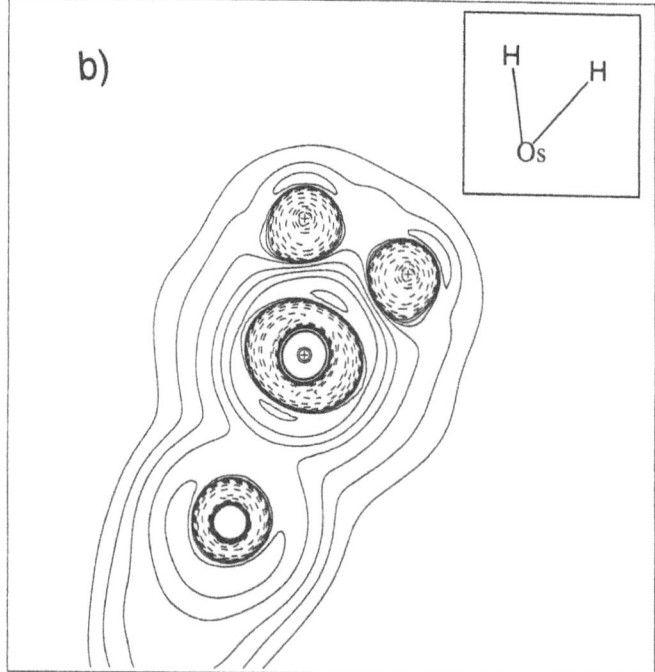

Figure 4.14. Plots of $\nabla^2\rho$ for complexes **2b** (a) and **3b** (b). Solid lines are for $\nabla^2\rho > 0$ (regions of charge depletion); dashed lines are for $\nabla^2\rho < 0$ (regions of charge concentration). Reprinted with permission from Maseras *et al.*[15] Copyright 1996 American Chemical Society.

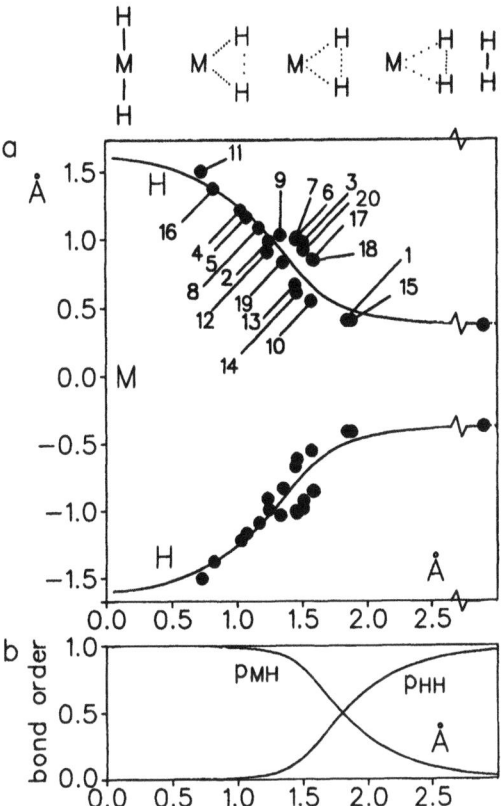

Figure 4.15. (a) Approach of H_2 to M (origin) according to the valence bond order concept. The solid line was calculated using Eq. (4.14) as described in the text. The numbered points (identified in Grundemann et al.[112]) represent neutron diffraction data for H_2 complexes (see Chapter 5) and hydrides. (b) Valence bond orders p_{MH} and p_{HH} as a function of the distance between M and the H_2 center. Reprinted with permission from Grundemann et al.[112] Copyright 1999 American Chemical Society.

What force is pulling the hydrogens closer to each other than observed in "normal" dihydrides ($d_{HH} > 1.6$ Å)? Even before H_2 complexes were discovered, $X\alpha$ calculations performed in 1978 by Ginsberg showed that significant H···H interactions occur even at d_{HH} up to 1.9 Å and were believed to stabilize high-coordination-number polyhydrides such as $[ReH_9]^{2-}$ and $[ReH_8(PR_3)]^-$.[113] Eisenstein also showed that weak long-range attractions exist between well separated cis hydride ligands.[114] Extended Huckel calculations on cis-$OsH_2(CO)_4$ show a very small positive Mulliken overlap population (MOP) of 0.004 at $d_{HH} = 2.33$ Å. This is far from a chemical bond (MOP is typically 0.5 in an H_2 complex) but is characteristic of hydrogen atoms bound directly to M. The bond order, p_{HH}, as calculated in Eq. (4.12) is still not zero (0.025) even for $d_{HH} = 2.2$ Å.[112] Isolobal "organic hydrides" such as CH_4 or methyl ligands show only small *negative* MOP between their hydrogens. The movement of two hydrogens on M toward each other produces an H_2 complex, as shown in a plot of MOP between the hydrogens (Figure 4.16). There

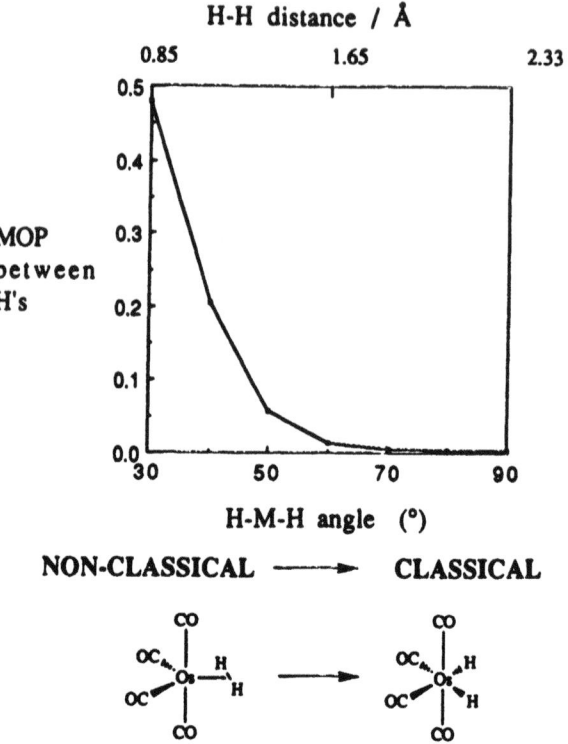

Figure 4.16. Plot of MOP between the two H's in *cis*-OsH$_2$(CO)$_4$ versus the distance and angle between them. Reprinted with permission from Jackson and Eisenstein.[114] Copyright 1990 American Chemical Society.

is a sharp drop-off in MOP as d_{HH} increases from 0.85 Å, and at $d_{HH} > 1.2$ Å less than a quarter "bond" might be postulated to exist.

The reaction coordinate for OA and the attendant d_{HH} might appear to depend simply on the degree of M → H$_2$ σ* BD, but this is deceptive. For example, why are d_{HH} not stretched past 0.9 Å in group 6 complexes close to OA, such as Mo(CO)(dppe)$_2$(H$_2$), for which the BD energy is calculated to be nearly twice that for σ donation? One answer is that such complexes with relatively short d_{HH} may actually lie further along the reaction coordinate than one might think. This is supported by normal coordinate analysis of W(CO)$_3$(PCy$_3$)$_2$(H$_2$), which shows that the force constant for v_{HH} is reduced by a factor of four from that in free H$_2$ and is similar to that for v_{WH} (see Chapter 8). The H–H and W–H$_2$ vibrational modes are highly coupled and are not pure normal modes that monotonically decrease and increase in frequency with the strength of the M–H$_2$ interaction. Thus H–H bond breaking and W–H bond formation are intertwined and well on their way here, although this is not reflected in d_{HH} and d_{MH} relative to other H$_2$ complexes. Rather than stretching like a rubber band, the H–H bond here *seems* to snap suddenly like a taut rope for the d^6 group 6 metal systems, whereas it appears to be much more flexible for later transition metals. However, even in the group 6 complexes the H–H bond can be stretched or shrunk with little energy cost and is probably delocalized. In solution, the activated W–H$_2$ system manifests a *tautomeric equilibrium* between a σ complex with a short H–H bond and the dihydride form [Eq. (4.2)]. The reasons why some systems show this equilibrium and others exhibit bond elongation are unclear and will be discussed below.

The above principles when modified proportionately extend to silane or other σ complexes. Thus a "stretched" Si–H bond (2.10 Å) is known that roughly corresponds to a d_{HH} of 1.05 Å (see Chapter 11). No matter how much analysis is undertaken, the point at which to draw broken or unbroken lines representing partial versus full bonds and even categorizations such as σ complexes, stretched σ complexes, and so forth will always be debatable, as it must be for continuum-like behavior. As will be shown below, elongated σ complexes present unique dynamical features that further emphasize the extraordinary complexity of σ complexes.

4.7.4. Elongated Dihydrogen Complexes: Extraordinarily Delocalized Dynamic Systems

Despite the massive number of theoretical studies on H$_2$ complexes, there are few that properly model elongated H$_2$ complexes.[6,15,95–97,115,116] These complexes are notorious for the problems encountered at sophisticated theoretical levels when trying to reproduce the geometry of the elongated H$_2$ ligand. Decomposition of the components of the interaction energy between [Os(NH$_3$)$_4$Cl]$^+$ and H$_2$ at the equilibrium geometry of [Os(NH$_3$)$_4$Cl(H···H)]$^+$ (d_{HH} = 1.40 Å) demonstrates that BD significantly increases in stretched H$_2$ systems and is close to the limiting case for formation of two M–H bonds.[97] Thus a strong bond between M and H$_2$ results, ranging from 23 to 45 kcal/mol in B3LYP calculations,[97,115,116] much stronger than that calculated in true H$_2$ complexes (15–20 kcal/mol).[40,43,55,57,59] In both cases the calculated binding energies correlate reasonably well with d_{HH}.

The rotational orientation of the H$_2$ with respect to cis ligands can be crucial and lead to huge differences in calculated d_{HH}. DFT computations dramatically show this for ReBr$_2$(H$_2$)(NO)(P^{i}Pr$_3$)$_2$ for which d_{HH} is 1.21–1.28 Å from NMR measurements (see Chapter 5, H$_2$ not located in X-ray structure).[117] This remarkable

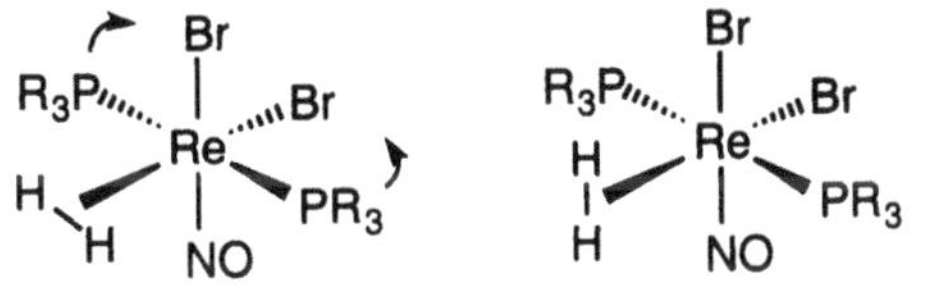

R =	d_{HH}	calcd d_{HH}
^{i}Pr:	1.285 Å	0.844 Å
Me:	1.426 Å	0.839 Å
H:	1.331 Å	0.838 Å

difference stems from the competition for BD with the powerful π acceptor nitrosyl ligand when the H$_2$ is rotated into the Br–Re–NO plane (note that the calculations are now sophisticated enough to include the actual isopropyl groups, not just PH$_3$

models). There is significant dependence on R for the elongated system, which partially arises from the steric bulk of the PR_3. The large P^iPr_3 inhibits the bending away of PR_3 illustrated above that favors BD to H_2 (see Section 4.3). The calculated $\angle$ PReP is 176° for R = Pr but can bend back to 162° for smaller R = Me, allowing even further elongation of the H–H bond to 1.426 Å. The PH_3 ligands can bend further (157°) but are not as strongly donating as the alkylphosphines, offsetting the effect.

The shape of the PES along the d_{HH} coordinate in elongated H_2 complexes is also of crucial importance in understanding the structure and bonding in elongated complexes. If these complexes are considered to be species where scission of H_2 is arrested at an intermediate stage between the $M(\eta^2$-$H_2)$ and MH_2 structures, the potential energy curve as a function of d_{HH} should have one minimum at the equilibrium distance. Perhaps the most accurate picture is based on the *rapid motion of two hydrogen atoms on a very flat PES with an exceptionally shallow minimum.* Hush and later Lluch and Lledos showed that the PES for the stretch of the H–H bond is exceptionally flat regardless of the method of calculation.[97,115,116] In relation to Hush's studies of $[Os(NH_3)_4(L)(H\cdots H)]^+$, Figure 4.17 shows a qualitative representation of the potential energy of H–H stretching in hypothetical structures for an incipient weak Os–H_2 complex (equilibrium d_{HH} = 0.74 Å) and a corresponding cis dihydride (1.8 Å).[97] In panels A and B, the minima of the diabatic curves are close in energy, with only weak electronic coupling leading to an avoided crossing and hence a low binding energy with a double minimum. In A, the H_2 complex is energetically preferred as in $W(CO)_3(PH_3)_2(H_2)$ while in B the dihydride tautomer is favored as in calculations for $W(PH_3)_5(H_2)$.[34] Panel C illustrates the situation for the known highly elongated Os complexes: very strong coupling at the crossing point leading to a single-minimum surface with an equilibrium d_{HH} very different from that of the diabatic curves. The binding energy is large as a result and the d_{HH} are calculated to be 1.3–1.4 Å as found experimentally. Further lowering of curve II with respect to I leads to increased coupling and ultimately a single-minimum dihydride description.

The possibility that elongated H_2 complexes could also be described in such a fashion was suggested by Morris and Koetzle in their insightful NMR and neutron studies of *trans*-$[OsCl(H_2)(dppe)_2]^+$ (see Chapter 5),[88] and subsequent computations suggest that the delocalization is even larger than had been imagined. The most definitive theoretical investigations of this phenomenon are by Lluch and Lledos, who combine DFT with quantum nuclear motion calculations of $[Cp^*Ru(H_2)(dppm)]^+$ and *trans*-$[OsCl(H_2)(dppe)_2]^+$ to obtain the nuclear vibrational energy levels.[115] These are rare cases of stretched H_2 complexes for which both a precise structure obtained by neutron studies (d_{HH} = 1.10 and 1.22 Å, respectively) and detailed temperature dependence of J_{HD} are available.[88,118] Furthermore the Ru complex (though not the Os complex) exists in solution equilibrium with the trans dihydride isomer in 2:1 ratio (see Section 7.4).

For $[Ru(H\cdots H)(C_5H_5)(H_2PCH_2PH_2)]^+$ (Figure 4.18), the H_2 structure is calculated to be more stable by 4.1 kcal/mol, in accord with the equilibrium that favors the H_2 tautomer. The energies of dissociation of H_2 from $Cp^*Ru(H_2)(dppm)]^+$ and $OsCl(H_2)(dppe)_2]^+$ are 22.7 and 34.9 kcal/mol, respectively, denoting that the M–H_2 bond is relatively strong and the H–H bond is weak. However, there

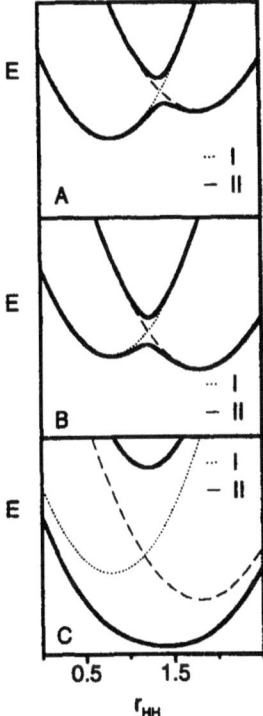

Figure 4.17. Qualitative representation of the potential energy of H–H stretching in hypothetical structures for an Os–H_2 complex (I, left-side curves) and a cis-dihydride (II, right-side curves). A and B correspond to small electronic coupling where both tautomers may exist, and C represents strong coupling leading to an elongated H_2 complex. Reprinted with permission from Craw et al.[97] Copyright 1994 American Chemical Society.

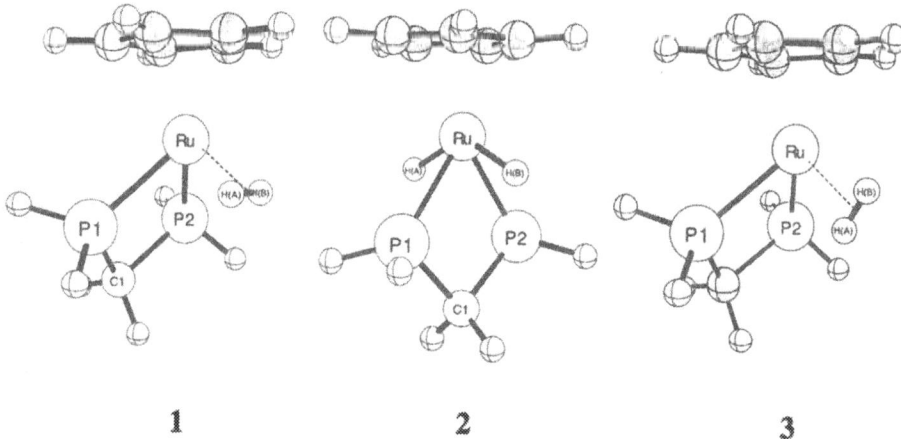

Figure 4.18. Optimized structures for the η^2-H_2 (left) and trans dihydride (middle) model complexes, and the transition-state structure for the rotation of η^2-H_2 of the H_2 complex (right). Reprinted with permission from Gelabert et al.[115] Copyright 1997 American Chemical Society.

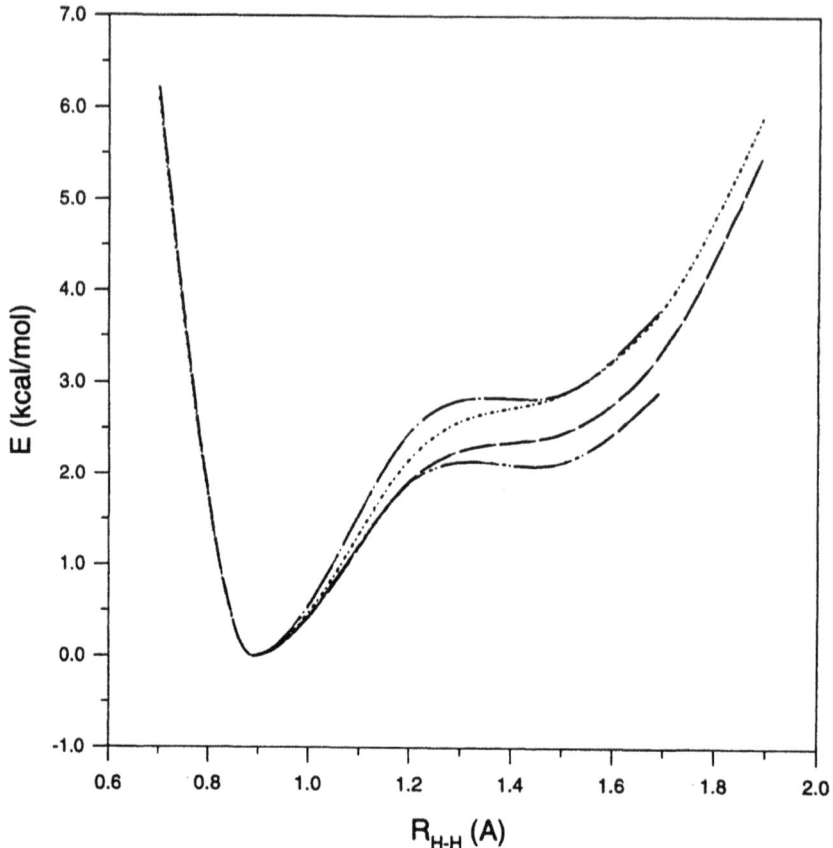

Figure 4.19. Energy profiles for lengthening of H–H while relaxing the rest of the structure in the model complex, [Ru(H···H)(C$_5$H$_5$)(H$_2$PCH$_2$PH$_2$)]$^+$, at different calculational levels. Reprinted with permission from Gelabert et al.[115] Copyright 1997 American Chemical Society.

is unexpectedly poor agreement between the experimental and computed values of d_{HH} and d_{MH}. The computations indicate that in electronic structure terms the most stable structure is an unstretched dihydrogen complex ($d_{HH} = 0.888$ Å) for the Ru complex. For the Os complex, the energy minimum gives a d_{HH} of 1.071 Å, a distance characteristic of an elongated complex but still significantly shorter than the neutron value for [OsCl(H$_2$)(dppe)]$^+$ of 1.22(3) Å. The monodimensional potential energy profile for lengthening the H–H bond for the model complex for [Cp*Ru(H$_2$)(dppm)]$^+$ is very flat and anharmonic (Figure 4.19) The "shoulder" at $d_{HH} > 1.4$ Å can be viewed as an incipient minimum corresponding to a nonexistent cis dihydride structure (the trans dihydride structure in Figure 4.18 is calculated).

Thus even sophisticated calculations cannot explain what is seen experimentally. One of the consequences of the quantum nature of nuclei is that the nuclei are not fixed so the overall structure of a molecule is vibrating even at 0 K. Therefore nuclear motion calculations are important for understanding elongated

M–H$_2$ systems. By solving the Schrödinger equation for the motion of nuclei on a reduced PES of two dimensions corresponding to the H–H and Ru–H$_2$ stretches, d_{HH} in the ground vibrational state is calculated to be 1.02 Å, much closer to the neutron value [1.10(2) Å] (Figure 4.20). It is important to note that the vibrational motion calculations show that the wave function here and for the Os complex is highly delocalized across a broad valley that lies oblique to the axes when compared to that for unstretched W(CO)$_3$(PR$_3$)$_2$(H$_2$), which is fairly parallel to the W–H$_2$ axis with curvature along the H–H direction as d_{MH} shortens (Figure 4.21). The crucial implication of these results is that the η^2-H$_2$ *ligand is greatly delocalized and cannot be envisaged as a fixed, rigid unit in elongated complexes.* The PES for the H–H stretch is so flat for the Os complex that the stretch of this bond can *traverse the entire distance range from 0.85 Å to 1.60 Å with attendant variations in d_{MH} at an energy cost of merely 1 kcal/mol*! The motion of the hydrogens is approximated in cartoonlike fashion in **17** and can be likened to that of two spring-linked soccer balls rapidly traversing opposite sides of a conical whirlpool.

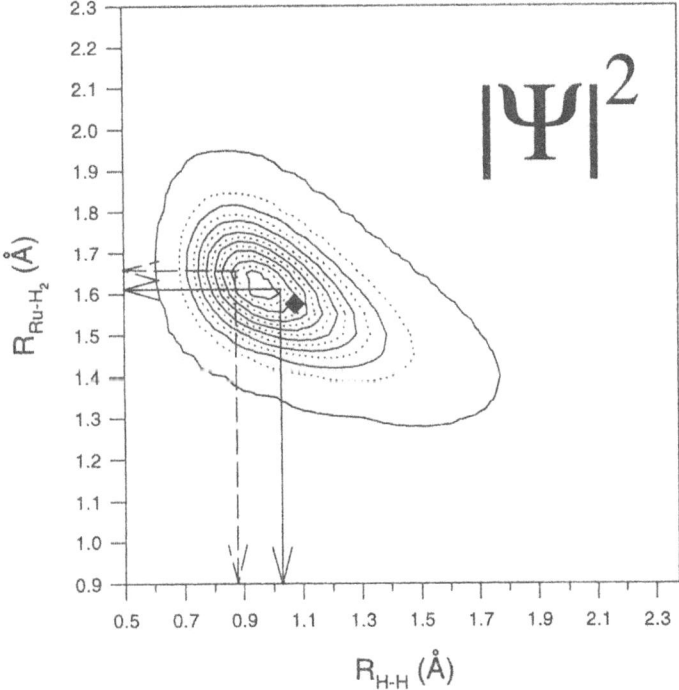

Figure 4.20. Probability density plot of the vibrational ground-state wave function of complex [Ru(H···H)(C$_5$H$_5$)(H$_2$PCH$_2$PH$_2$)]$^+$ as a contour plot. The dashed arrow indicates the position of the minimum in the potential energy surface (R$_{H-H}$ = 0.89 Å, R$_{Ru-H_2}$ = 1.66 Å), the solid arrow that of the expectation values for this vibrational state (R$_{H-H}$ = 1.02 Å, R$_{Ru-H_2}$ = 1.61 Å), and the square mark, that of the experimentally reported data from neutron diffraction (R$_{H-H}$ = 1.10 Å, R$_{Ru-H_2}$ = 1.58 Å). Reprinted with permission from Gelabert et al.[115] Copyright 1997 American Chemical Society.

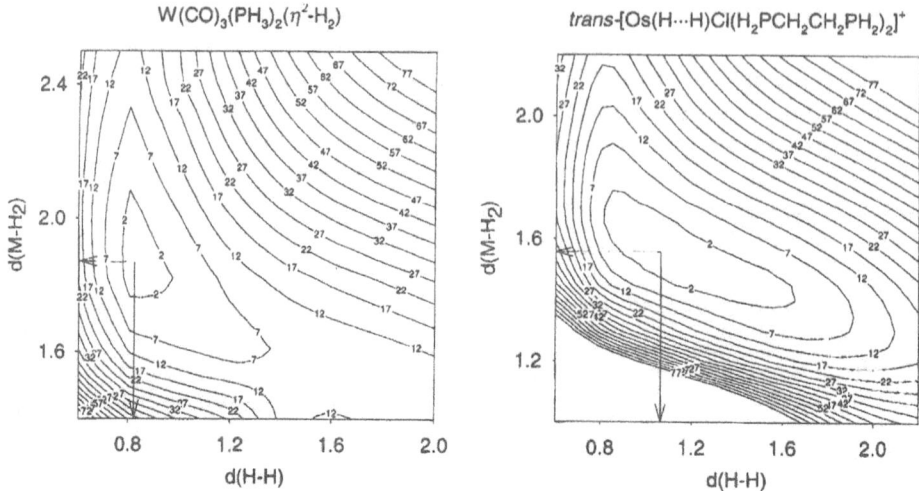

Figure 4.21. Left: Contour plot of the two-dimensional PES for $W(CO)_3(PH_3)_2(\eta^2\text{-}H_2)$. Energy contours appear every 5 kcal/mol. The arrows indicate the position of the minimum energy structure ($d_{HH} = 0.832$ Å, $d_{W-H_2} = 1.872$ Å). Right: Same for $[OsCl(H\cdots H)(H_2PCH_2CH_2PH_2)_2]^+$ ($d_{HH} = 1.071$ Å, $d_{W-H_2} = 1.567$ Å. Reprinted with permission from Torres et al.[115] Copyright 2000 American Chemical Society.

It is astonishing that a bond as strong as H–H can be weakened so as to be lengthened by 0.8 Å without a significant rise in energy. The ν_{HH} and ν_{MH_2} vibrational stretches in fact lose their meaning and the "normal" stretching modes have to be

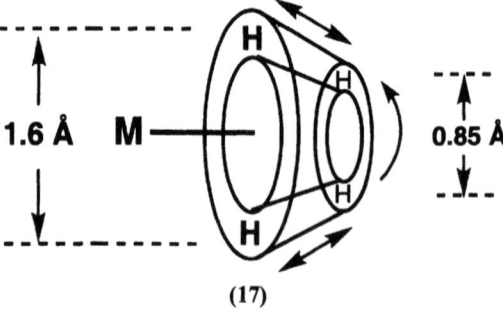

(17)

redefined (see Section 8.4) to one along the arrows shown in **17** (low-energy mode) and one orthogonal to it (high-energy mode). The soft vibrational mode parallels the reaction coordinate for OA. The hydrogens are also undergoing much slower librational/rotational motion with energy barriers of approximately 4 kcal/mol, so the delocalization is occurring along an annular "whirlpool" surface (neglecting M–H$_2$ deformational motion).

The variation of J_{HD} with small temperature changes and the remarkably high isotope dependence of d_{HH} for $[Cp^*Ru(H_2)(dppm)]^+$ and $[OsCl(H_2)(dppe)_2]^+$ offers experimental verification of the flat PES (see Section 5.4). Thus, the initial picture that stretched complexes represent arrested states (with stationary structures) along the OA pathway should be revised to two hydrogen atoms moving almost freely in a large region within the coordination sphere of the metal. This behavior would logically be expected to be present in unstretched H_2 complexes as well, although to a lesser degree. The existence of solution equilibria between distinct dihydrogen–dihydride tautomers even for complexes with relatively short d_{HH} such as $W(CO)_3(PR_3)_2(H_2)$ can be understood as a manifestation of this delocalization, although a double-minimum potential well is present here. The question remains why such equilibria occur for some systems, e.g., $[Cp^*Ru(H_2)(dppm)]^+$, and not others ($[OsCl(H_2)(dppe)_2]^+$).

One answer to the above is that forces other than the electronics at M must be considered, e.g., steric factors, structural rearrangement barriers for six versus seven coordination, or overall bond energetics. Sizable rearrangements of the coordination sphere take place for the $Mo(CO)(PP)_2$ system on OA of H_2 [Eq. (4.5)]. The two hydride ligands are *distal* to each other in the pentagonal plane of **9** and cis to the

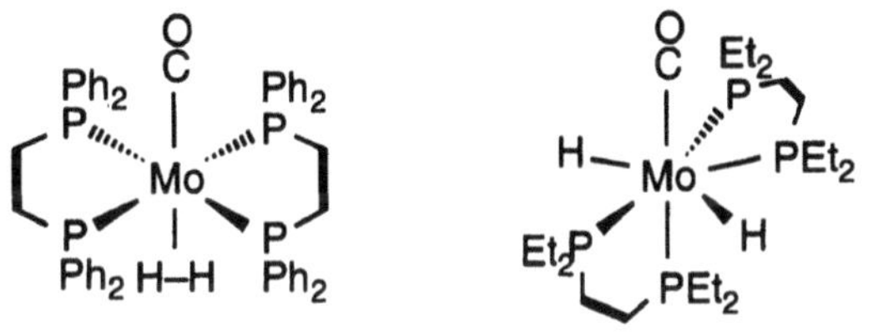

(8)   (9)

CO but in **8** the hydrogens are proximal and trans to CO. These rearrangement energies and other forces cannot easily be quantified or even identified as being present in many cases. The role of ligand rearrangement prior or subsequent to OA is not well known, and perhaps isomerization to give H_2 cis to CO precedes OA. This would parallel the situation for Si–H and Ge–H bond coordination and cleavage on this Mo fragment (see Chapter 11).

As shown in Scheme 2, SiH_4 and organosilanes always bind cis to CO, and the isomer with silane trans to CO is never observed presumably because the two good acceptor ligands (CO and silane) prefer to be cis to each other electronically [cf trans to cis isomerization of $Mo(CO)_2(PP)_2$]. For the H_2 case, the cis isomer would immediately undergo OA (H_2 now trans to a donor), explaining why it is not observed. However, Si–H and Ge–H coordinate cis to CO regardless of phosphine, perhaps because of the better acceptor strength of silanes and germanes. But why doesn't the X–H bond immediately cleave here similar to H–H? Steric factors may come into play, and hybrid quantum-mechanical and molecular mechanics calculations on $[RuH(H_2)(PP)_2]^+$ resolve some of the complexity of bis(diphosphine) systems.[119] The strong trans-effect hydride and the positive charge result in similar H_2 activation as in the Mo system (facile exchange between hydride and H_2 also occurs here; see Section 4.9). The trans H_2 complex is favored by 7–11 kcal/mol over both the cis isomer and a pentagonal bipyramid dihydride structure. When

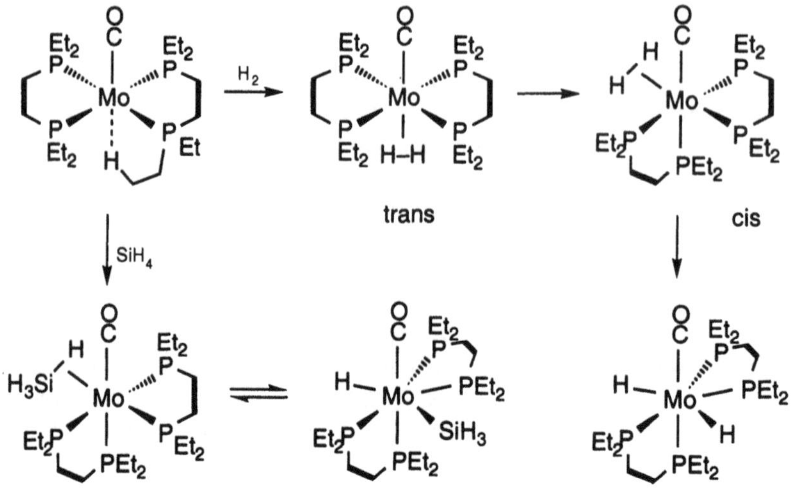

Scheme 2

electronic energies are added, the trans H_2 complex is still the absolute minimum by 17 kcal/mol for the dppe complex. Diphosphines with large bite angles such as diop make the system more sensitive to steric effects that favor the cis orientation also seen experimentally for cis-$[RuH(H_2)(thixantphos)_2]^+$ (bite angle of the free ligand is 103°).[120]

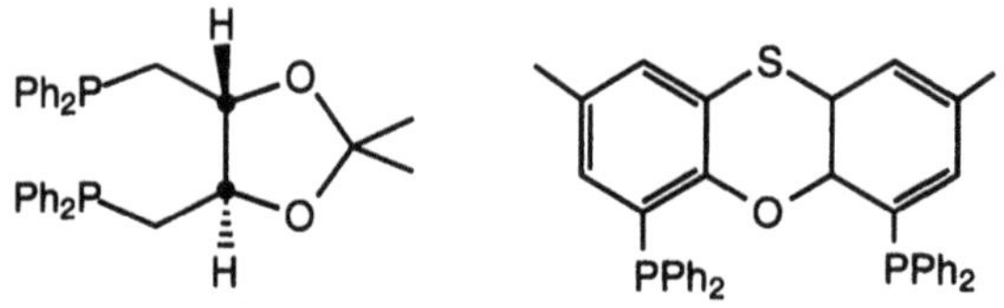

diop   thixantphos

Calculations show that distortion of the octahedral geometry induced by the rigid phosphine also leads to less optimal orbital overlap, which noticeably weakens the M–H_2 interaction compared to less sterically demanding diphosphines. Steric factors can indirectly influence the electronic properties of the complex, which further emphasizes the delicate nature of bonding of H_2 to a transition metal. For silane and other σ ligands that are much bulkier than H_2, steric influences are magnified and perhaps favor the cis (CO)(silane) orientation as well as disfavoring Si–H cleavage to a silyl ligand, which is really huge in size compared to a hydride.

4.7.5. Further Aspects of Electronic and Steric Control of Dihydrogen versus Dihydride Bonding

Extended Huckel calculations on interactions of η^2-H_2 with d^6 ML_5, d^8 ML_4, and d^{10} ML_3 fragments all show that replacement of σ-donor L by π-acceptor L weakens the M–H_2 interaction somewhat and inhibits OA.[35,121,122] This is corroborated experimentally by the instability of H_2 complexes with only π acceptors such as $Cr(CO)_5(H_2)$,[123] $Fe(CO)(NO)_2(H_2)$,[124] and $Ni(CO)_3(H_2)$,[125] primarily because of lower BD. However, as discussed above, for highly electrophilic centers the loss of BD is offset by increased $H_2 \rightarrow M$ σ donation, and the $M-H_2$ binding energies for $M(CO)_5(H_2)$ are not significantly lower than those for stable H_2 complexes such as $W(CO)_3(PR_3)_2(H_2)$ (17–20 kcal/mol; see Chapter 7). Species such as $M(CO)_5(H_2)$ are unstable because dissociation of H_2 leads to highly unstable "naked" $M(CO)_5$ or solvent-bound species that instantly decompose, but corresponding 16e fragments containing phosphines are stabilized by agostic interactions. Thus kinetic instability rather than thermodynamic instability can occur for H_2 coordination to electron-poor M, at least in solution. Complexes such as $M(CO)_5(H_2)$ may be isolable in the solid state by special techniques such as were found for $CpMn(CO)_2(H_2)$.[126]

Since the ligand *trans* to H_2 has the greatest electronic effect on $M-H_2$ bonding, even a complex with mostly phosphine ligands with only one strong *trans* π acceptor can still be stabilized as an H_2 complex, e.g., *trans*-$Mo(CO)(PP)_2(H_2)$. By variation of M/L, it is possible to *electronically arrest* the reaction coordinate for OA of H_2 and examine it in fine detail, almost as if snapshots were being taken along the pathway.

$L_5 = (CO)_5$ unstable

$L_5 = (CO)_3(PR_3)_2$ stable

$L_5 = (PR_3)_5$ dihydride

Virtually every gradation between these forms has been observed structurally and spectroscopically by varying M/L/charge, as discussed above. A useful empirical correlation devised by Morris enables one to predict whether or not ML_n will bind H_2 or undergo OA by use of comparisons to the properties of the related *dinitrogen* complex, $M(N_2)L_n$.[127] N_2 is a π acceptor like H_2, and parameters related to the electron richness of M such as v_{NN} and electrochemical potentials can be used as indicators of π basicity of the binding site in $M(N_2)L_n$ for group 6 to 9 metals. When v_{NN} is less than 2060 cm^{-1} a dihydride $M(H)_2$ will generally form, but when it is greater than 2060 cm^{-1} a true (unstretched) H_2 complex will be favored. As seen in Figure 4.22, the correlation is excellent for group 6 and 7 complexes with trans CO and generally good for all $M-L$ fragments. At the time the correlation was made, it was believed that very-electron-poor fragments with $v_{NN} > 2160$ cm^{-1} could not bind H_2 in stable fashion, but this is no longer the case. $Cr(CO)_5(H_2)$ is unstable for reasons discussed above, and very electrophilic cationic complexes such as **13** with v_{NN} above the demarcation line are stable.[76] The delineation between true H_2 complexes and dihydrides is generally valid although

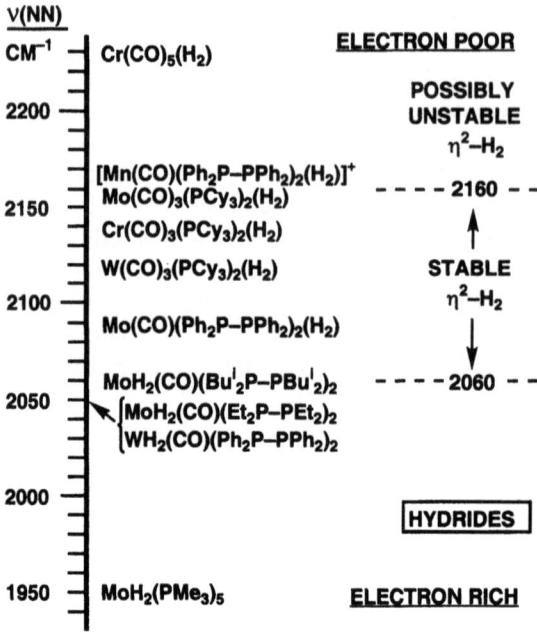

Figure 4.22. Correlation of stable dihydrogen versus dihydride binding with v_{N_2} of the corresponding N_2 complex.

Mo[CN(p-Me$_2$NC$_6$H$_4$)](H$_2$)(dppe)$_2$ has v_{NN} = 2018 cm^{-1} for the N$_2$ analogue (see Section 4.7.1). Steric factors may play a role, and d_{HH} may be elongated. Complexes with elongated H$_2$ ligands with small J_{HD} clearly can have v_{NN} < 2060 cm^{-1}, e.g., Re(H$_2$)(N$_3$N$_F$) with J_{HD} = 17 Hz (corresponding to d_{HH} ~ 1.15 Å), where v_{NN} is 2004 cm^{-1} for Re(N$_2$)(N$_3$N$_F$).[128]

This approach has been refined to include limits on the electrochemical potentials (Table 4.10). If $E_{1/2}$[M(N$_2$)L$_n^+$/M(N$_2$)L$_n$] < 0 V versus normal hydrogen electrode, H$_2$ should undergo OA on ML$_n$. If the H$_2$ is trans to a σ donor like a hydride and is on a 4d or 5d metal, splitting will occur at the more positive potential of 0.5 V. The strength of this method is that the potentials can be calculated from the structure of the complex by an additive ligand parameter method proposed by Morris.[129] In principle the stability of an H$_2$ complex with any combination of d^6 metal and five ancillary ligands can be predicted. However, if the corresponding dihydride is made particularly stable by the stereochemistry of the ligands then homolytic splitting might occur even when v_{NN} of the corresponding N$_2$ complex is greater than 2060 cm^{-1} and $E_{1/2}$ > 0.5 V. Complexes with piano-stool structures often are unpredictable (see below), e.g., [Cp*FeH$_2$(dippe)]$^+$ is a dihydride despite the high value (2112 cm^{-1}) of v_{NN} in [Cp*Fe(N$_2$)(dippe)]$^+$.[130] For both the v_{NN} and $E_{1/2}$ correlations in Table 4.10, the originally proposed upper boundaries for η^2-H$_2$ stability no longer generally apply. However, for many systems, e.g., neutral group 6 complexes and certain cationic species prepared by protonation, H$_2$ binding is too labile for isolation of a solid at room temperature. Regardless of stability, it is

Table 4.10. Properties of the Corresponding N_2 Complexes, $[M(N_2)X_nL_{5-n}]^{z+}$ That Indicate the Relative Stabilities of H_2 Complexes versus Dihydride Complexes or Coordinatively Unsaturated Species (H_2 Binding Too Weak)

	Too reducing	Correct energy of HOMO	Too oxidizing
N_2 trans to CO			
$E_{1/2}$ vs. NHE, V	<0	0–1	>0.5 (3d metal)
			>1.0 (4d, 5d)
v_{NN}, cm^{-1}	<2060	2060–2160	>2160
N_2 trans to σ donor			
$E_{1/2}$ vs. NHE, V	<0 (3d metal)	0–1.7 (3d)	>1.7 (3d)
	<0.5 (4d, 5d)	0.5–2.0 (4d, 5d)	>2.0 (4d, 5d)
v_{NN}, cm^{-1}	<2060	2060–2180	>2180
Corresponding complex	Dihydride $[MH_2X_nL_{5-n}]^{z+}$	Dihydrogen $[M(H_2)X_nL_{5-n}]^{z+}$	Coordinatively unsaturated $[MX_nL_{5-n}]^{z+}$

virtually certain that a "true" σ complex with a short d_{HH} (<1.0 Å) will be formed in the upper limits.

The stereochemistry of the ancillary L can influence whether an H_2 or dihydride complex is stable: OA of H_2 involves an increase by 1 in the M coordination number, and small L will favor the dihydride tautomer. However, large L can also accomplish this by preventing M from adopting the favored octahedral geometry, e.g., 1,1'-bis(diphenylphosphino)ferrocene (dppf) has a large bite angle of 106° that prevents octahedral coordination around Ru and promotes formation of a seven-coordinate trihydride complex, [Ru(H)$_3$(dppf)$_2$]$^+$,[131] instead of the usual trans-[RuH(H$_2$)(PP)$_2$]$^+$. As shown in Section 4.7.4., such phosphines with slightly smaller bite angles can also favor a cis H(H$_2$) geometry here. Piano-stool complexes also show a strong dependence on bite angles (see below). Apparently [OsH$_3$(PPh$_3$)$_4$]$^+$ does not contain η^2-H$_2$ because steric repulsions between the large PPh$_3$ disallow a regular octahedral structure.[132]

4.8. POLYHYDRIDE–DIHYDROGEN COMPLEXES

Up until now, we have concentrated mainly on complexes with only two hydrogens, but complexes with up to seven are known, and often it is these that show stretched H···H bonds. The factors governing the stabilities of classical versus nonclassical isomers in polyhydride complexes have been extensively studied theoretically by Hall and coworkers, who point out that electron correlation is important in computations but that has been recognized only since 1991.[6,133] For example, calculations by Hay in 1992 on ReH$_7$(PH$_3$)$_2$, showed the presence of one H$_2$ ligand with a short d_{HH} of 0.80 Å, but CI techniques, which include electron correlation, showed that only the classical heptahydride form.[134] The actual PPh$_3$ complex

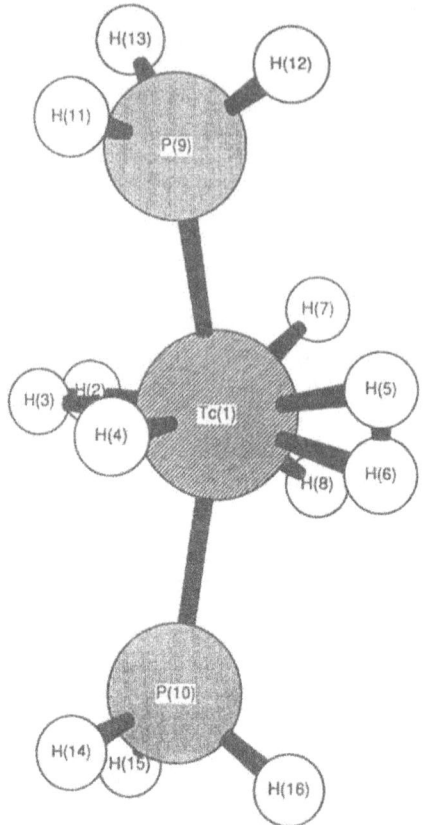

Figure 4.23. Structure corresponding to optimized geometry of the trans (bis-H_2) isomer of $TcH_7(PH_3)_2$. Reprinted with permission from Haynes et al.[134] Copyright 1992 American Chemical Society.

contains one long d_{HH} of 1.37 Å, much closer to the CI result. However, for the Tc analogue, seven-coordinate bis(H_2) structures (Figure 4.23) with short d_{HH} near 0.8 Å are calculated to be 2–12 kcal/mol lower in energy than the classical forms for both types of computations (mono-H_2 complex is unstable).

In addition, if electron correlation is included, calculations on

$$[Os(NH_3)_4(CH_3COO)(H_2)]^+$$

accurately predict a long d_{HH}, 1.39 Å, similar to that found by neutron diffraction for the ethylenediamine analogue, 1.34 Å.[97] Electron correlation plays such an important role in determining the geometry because the PES is very flat with respect to the d_{HH}. Comparisons of computed structures of model complexes IrH_5L_2 using different levels of electron correlation (CISD, CASSCF, MRCI, MP2, MP3, and MP4) versus the neutron structure of $IrH_5(P^iPr_3)_2$ show that MP2 gives the most economical and reliable determination of relative energies between the classical and nonclassical (nonexistent) isomers.[135] The usual replacement of PR_3 by PH_3 gives an error of 2–3 kcal/mol by overestimating the stability of the nonclassical isomer with respect to OA. The weaker σ-donor property of PH_3 compared to e.g., PMe_3 decreases the BD ability of M, which in turn decreases H–H activation. However, MP2 does not reproduce well structures with d_{HH} of 1–1.5 Å, as also found by Maseras, who utilized DFT methods as shown above.[15]

Periodic trends are established by MP2 calculations on various model complexes and energy differences between their classical and nonclassical isomers. For complexes with only phosphine coligands, $MH_x(PH_3)_y$ ($x = 2-7$, $y = 1-5$, $x + y = 7-8$), hydrides are preferred for M with more diffuse d orbitals, i.e., earlier M such as Mo. There is also a preference of third-row metals for the hydride isomer because of the increasingly diffuse nature of the d orbitals on descending the metal group. As shown in Figure 4.24, a diagonal line divides the metals into those that prefer hydride forms (left side of the line) from those that prefer forms containing one or more H_2 ligands (right side). For neutral complexes the line passes through Ru and Ir and for these metals both hydride and H_2 isomers may exist. For cationic complexes, the line shifts left because increasing the charge on a hydride contracts its d orbitals. The line position is also ligand-sensitive, since more basic PR_3 will favor the hydride form by increasing BD. For first-row metals, the division between hydride and H_2 forms is likely to shift left because the d orbitals are more contracted. For simple polyhydrides, $[MH_x]^n$ ($x = 3-9$), orbitally ranked symmetry analysis correctly predicts and rationalizes the geometries of over 100 complexes in comparison to their *ab initio* structures.[133] Such use of symmetry and group theory can rationalize unusual experimental geometries.

The accuracy of the predictions include the dual existence of H_2 complexes of Ru, e.g., $RuH_2(H_2)(PPh_3)_3$ along with hydrides, e.g., $RuH_2(PPh_3)_4$.[6] Also, most group 8 $[MH(H_2)(PR_3)_4]^+$ complexes contain H_2, although $[OsH_3(PPh_3)_4]^+$ is a trihydride,[136] while $[OsH(H_2)(dppe)_2]^+$ with *chelating* phosphines is a stretched H_2 complex with d_{HH} near 1.0 Å from solution NMR. Electronically these should be reversed since dppe is a better donor than PPh_3, again emphasizing that geometric considerations cannot be neglected. A good example of synergy between theory and experiment is $[OsH_5(PPh_3)_3]^+$, which has been studied extensively since its existence was first postulated in 1970.[137] Caulton was the first to elucidate spectroscopic

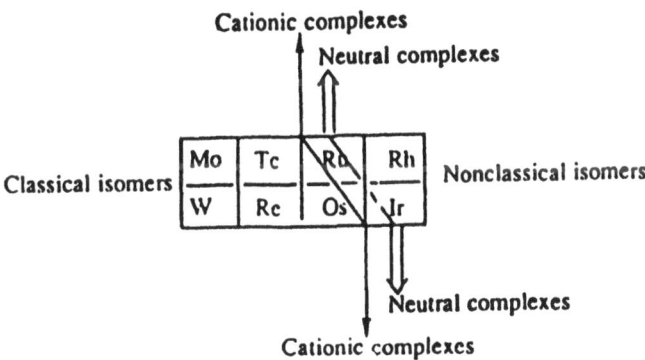

Figure 4.24. Periodic trends for neutral and cationic polyhydrides with only phosphine ligands.

data such as NMR spectra in 1984,[138] and this complex, formed by protonation of $OsH_4(PPh_3)_3$, was believed to be a pentahydride (the first H_2 complex had just been reported that year). In 1988 Hamilton and Crabtree[139] used this complex to test their NMR T_1 method (see Section 5.5) and concluded that it is actually $[OsH_3(H_2)(PPh_3)_3]^+$, as did Caulton in his followup paper.[140] However, Halpern's reassessment of the T_1 method indicated that both hydride and H_2 complexes were compatible with their data.[141] RHF calculations by Lin and Hall in 1992 then showed the pentahydride to be the most stable of three isomers considered, but by only 3.7 kcal/mol over $[OsH_3(H_2)(PPh_3)_3]^+$.[142] The next year *ab initio* MP2 computations by Maseras *et al.* on 20 possible isomers showed a dodecahedral pentahydride structure to be the most stable,[143] which was then found in the neutron structure (Figure 5.1), although somewhat distorted.[144]

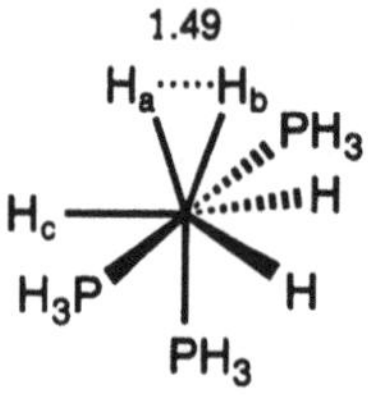

(18)

The story was not completely over, however, because the 1.49 Å distance between H_a and H_b in **18** is short for a hydride (other d_{HH} are longer than 1.75 Å), and H_a, H_b, and H_c are coplanar, both of which details were not reproduced by the elaborate calculations. Further calculations by Eisenstein rationalized the coplanarity as due to cis interactions (see Section 4.11), where the absence of an H–H bond does not affect the attraction between the hydrides. The reason that the H_a–H_b bond is essentially broken (and not an H_2 ligand) is the presence of sufficient (but unquantifiable) BD into σ^* H_a–H_b from a nonbonding orbital different in energy than other backdonating orbitals that apparently spread the other hydrides further apart.

One method for stabilizing an H_2 complex in a polyhydride system is protonation to a cationic species to lower the overall electron density at the metal. Although the above protonation of $OsH_4(PPh_3)_3$ failed to do this, the strategy is viable in many other systems and can even be effected by attachment of Lewis acids such as

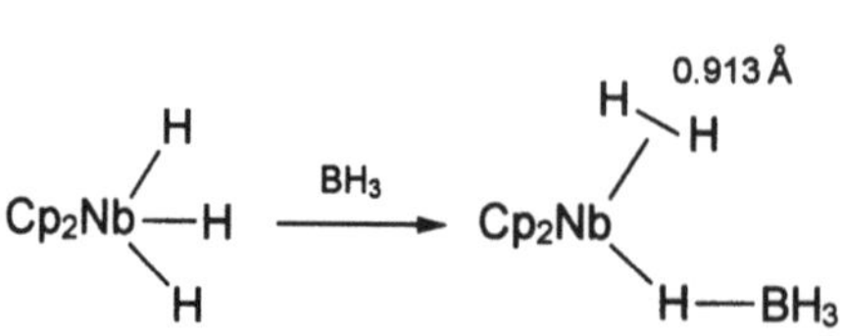

(4.15)

BF_3 or BH_3 to a hydride in, e.g., Cp_2NbH_3. Computations show that an H_2 complex results,[145] and again there is experimental support: Addition of HBR_2 lowers the temperature for H_2 evolution from Cp_2NbH_3 on heating,[146] consistent with increased η^2-H_2 character.

4.9. Cp AND Tp COMPLEXES, INCLUDING d^0 SYSTEMS

Many polyhydride complexes contain Cp ligands and have piano-stool geometries. Calculations by Lin and Hall on $CpMH_nL_{4-n}$ (M = Ru, Rh, Os, Ir; $n = 1-4$; L = PH_3) show that hydrides with four-legged piano-stool structures such as **19** and

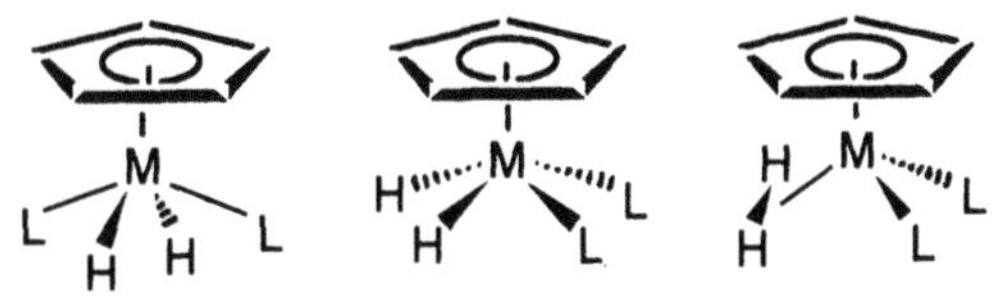

(19)   (20)   (21)

20 are generally preferred for both neutral and monocationic species.[6] The calculations indicate that Cp is a very strong σ and π donor, which increases BD favoring OA of hydrogen. Therefore the diagonal line dividing hydride and H_2 isomers (Figure 4.24) shifts toward late metals when three PR_3 are replaced by Cp.

Experimental results also show that hydrides predominate, although major exceptions are cationic complexes, e.g., $[CpM(H_2)L_2]^+$ (**21**) (M = Fe, Ru), as well as Ru analogues with Tp ligands discussed below.[147] Thus, although calculations on $[CpRuH_2(PH_3)_2]^+$ indicate that the dihydride form should be favored over the H_2 isomer by a hefty 12.5 kcal/mol, $[CpRuH_2(PPh_3)_2]^+$ is a H_2 complex. In some cases the reason the H_2 form is actually found appears to be steric, e.g., bidentate phosphines with small bite angles such as dppe or dppm may favor η^2-H_2. This is demonstrated by the work of Simpson [Eq. (4.16)] in which a dihydride forms for $n = 3$ (large bite angle) but an H_2 complex for $n = 1$.[148] For $n = 2$ a mixture of both

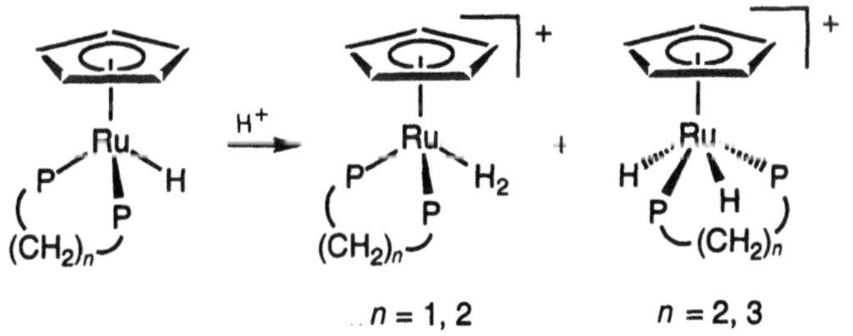

$$n = 1, 2 \qquad n = 2, 3 \tag{4.16}$$

tautomers forms, which Heinekey later found to be in equilibrium, i.e., thermodynamically controlled.[149] The kinetic product of protonation is likely to always be the H_2 complex (protonation at the hydride), but the sterics can control subsequent isomerization. Electronics are also of crucial importance in these finely balanced systems, e.g., the more electron-rich Cp* analogue is a mixture of tautomers for $n = 1$. $[CpRuH_2(PR_3)(CO)]^+$, where strong π acceptors such as CO or CNR replace one or both P donors is purely an H_2 complex. $CpMn(CO)_2(H_2)$ with two CO ligands is surprisingly *the only stable neutral $CpM(\eta^2$-$H_2)$-type complex*[126]; third-row $CpReH_2(CO)_2$ is a dihydride.[150] The thermally unstable H_2 and dihydride tautomers $CpNb(CO)_3(H_2)$ and $CpNb(H)_2(CO)_3$ are in equilibrium in liquid Xe.[151]

Apparently the Nb is not electron-rich enough to oxidatively add H_2 in these CO systems, but the Cp trans to H_2 is a strong enough donor to promote H–H cleavage. Energy profiles for OA of H_2 to one-legged piano stools of the type CpML (M = Rh, Ir; L = CO, PR_3) show a plateau at the midpoint of the reaction corresponding to formation of stretched H_2 complexes (d_{HH} = 1.04–1.12 Å) with formation energies of 14–24 kcal/mol, but have not been isolated.[20]

Polyhydrides containing higher coordination numbers $CpMH_nL_{6-n}$ (n = 5–6) do not show nonclassical structures experimentally, but calculations[6] indicate they could be stable for cationic complexes of late metals for n = 6, e.g., $[CpRh(H_2)_3]^{2+}$. As charge increases, d orbitals contract and are stabilized, which favors the nonclassical isomers: $[Cp*OsH_2(H_2)(PPh_3)]^+$ (**22**) contains elongated H_2 [d_{HH} = 1.014(11) Å neutron].[152] Complexes such as **22** with d^0 electronic configurations

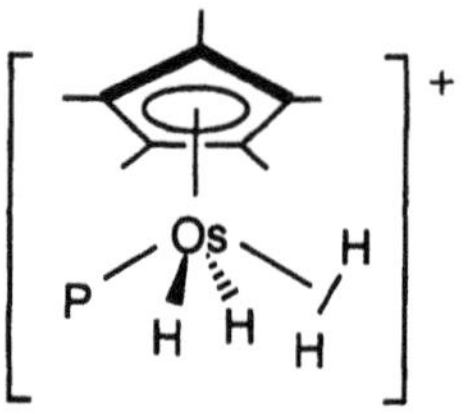

(22)

are of interest because there can be *no BD to H_2 to stabilize the complex*. Hall and Poli provide both experimental and computational evidence that d^0 $Cp*MH_5(PMe_3)$ is protonated to an unstable H_2 complex.[153] The more stable W

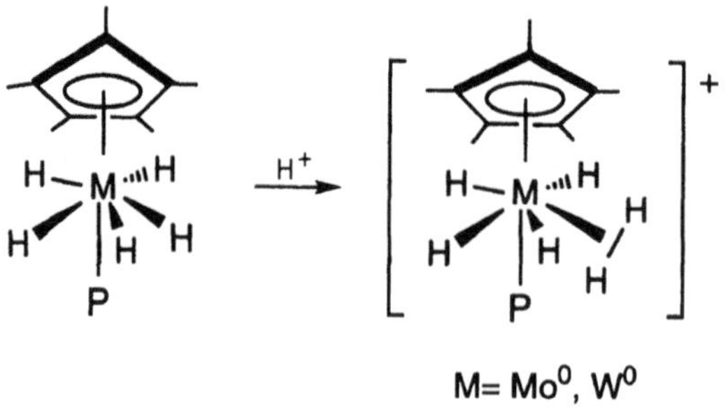

$$M = Mo^0, W^0 \qquad (4.17)$$

complex decomposes above −20°C and has a $T_{1(min)}$ of 51 ms at −90°C, indicative of H_2 coordination, but the H atoms exchange much too rapidly to give separate signals for the H_2 ligand or J_{HD} on deuterium labeling. Geometry optimizations on a $[CpMH_6(PH_3)]^+$ model show that the minimum energy structure of each system is the H_2 complex, and the potential bis(H_2) complex formed by double protonation is more than 14 kcal/mol less stable. d_{HH} for $[CpMH_4(H_2)(PH_3)]^+$ are 0.83–0.90 Å, close to that for H_3^+ (0.87 Å). Despite the lack of BD, high barriers to H_2 rotation (4.2 and 5.2 kcal/mol for M = Mo and W) are calculated, which are attributed to a strong *cis effect* (see Section 4.11).

Metallocene d^0 complexes of the early metals, which are very important catalysts for olefin polymerization, are not known to form H_2 complexes, but a Ti^{IV} metallocene H_2 complex is stable calculationally from Car–Parrinello *ab initio* molecular dynamics.[154] This methodology[155] can reveal novel reaction pathways that traditional "static" computations may overlook. An allyl H_2 complex computed to be stable in the gas phase at 300 K is located as a secondary minimum in β-elimination, the most common assumed unimolecular chain termination process in Ziegler–Natta ethylene polymerization. Static calculations show that the H_2

$$\text{agostic alkyl} \longrightarrow \text{olefin hydride (expected)} \longrightarrow \text{allyl dihydrogen} \tag{4.18}$$

complex is some 6 kcal/mol more stable than the olefin hydride and has a $Ti-H_2$ binding energy of 11 kcal/mol. This is the first indication experimentally or computationally of a group 4 H_2 complex and shows that a possible termination process in olefin polymerization could involve H_2 elimination. Experimental evidence for an H_2-producing side reaction in Zr and other metallocene catalysts now exists after the calculation brought it to light.[154]

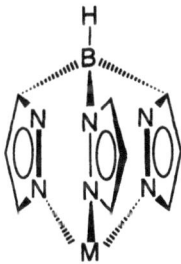

The Tp ligand (and variations thereof), has properties similar to Cp but has a higher tendency to stabilize H_2 versus dihydride coordination.[156,157] $CpRuH_3(PR_3)$ and Cp* congeners are trihydrides[158] but $TpRuH(H_2)(PCy_3)$ is an H_2 complex (see Table 4.7).[159] The electron-donating abilities and number of coordination sites occupied by these two ligands are analogous, but, unlike for Cp, the symmetry of the $M(\eta^3\text{-HBpz}_3)$ fragment ($\angle N-M-N$ near 90°) dictates the coordination mode of the other ligands of the complex. The TpM unit is strongly hybrid-biased to give and maintain *octahedral six-coordinate structures*, while the diffuse electron clouds of Cp are ineffective in promoting strong directional frontier orbitals. Therefore, processes involving coordination number increases such as OA are less likely for Tp complexes. Protonation of $TpOsHL_2$ gives an octahedral H_2 complex, $[TpOs(H_2)L_2]^+(L_2 = CO)(P^iPr_3)$ or $(PPh_3)_2)$ but the Cp analogue is a seven-coordinate dihydride.[156]

A DFT study of TpRhH$_2$(H$_2$) versus the experimentally unknown Cp analogue shows an octahedral dihydride–dihydrogen structure for Tp (d_{HH} = 0.836 Å) but a piano-stool tetrahydride for Cp, although CpRuH$_2$(H$_2$) is only slightly higher in energy (1.25 kcal/mol).[157] Known Tp*RhH$_2$(H$_2$) with 3,5-dimethyl groups on the pyrazoly rings has not been characterized by diffraction methods but contains an H$_2$

ligand with a roughly estimated d_{HH} of 0.94 Å based on inelastic neutron scattering studies of H$_2$ rotation (see Chapter 6).[160] Differences between Cp and Tp systems can also be ascribed to the somewhat stronger donating ability of Cp: d_{HH} in the calculated CpRhH$_2$(H$_2$) isomer is considerably longer (0.943 Å) than in the Tp analogue with less BD. This is reflected by a much lower rotational barrier calculated for the H$_2$ ligand (0.45 kcal/mol) compared to that in the Cp species (4.88 kcal/mol). The remarkable accuracy of DFT methods extends to rotational dynamics: The experimental barrier of 0.56(2) kcal/mol for Tp^{Me2}RhH$_2$(H$_2$) is in close agreement with that calculated.

4.10. POLYHYDRIDE COMPLEXES WITH CO VERSUS HALIDE LIGANDS

The principle that complexes containing strong acceptors such as CO favor the H$_2$ form over the dihydride can be extended to polyhydride systems as well.[6] For MH$_4$(CO)(PH$_3$)$_3$ both nonclassical H$_2$(H$_2$) and classical tetrahydride isomers are seen calculationally (MP2) for M = Mo, although the tetrahydride is preferred for M = W. For cations such as M = Tc$^+$, the nonclassical form is preferred, while for third-row Re$^+$ the tetrahydride is slightly favored. Once again the positive charge on the group 7 metals favors the H$_2$ isomer. In excellent agreement with theory, the replacement of a phosphine by CO in [ReH$_4$(PMe$_2$Ph)$_4$]$^+$, where only the classical isomer is seen experimentally,[161] gives a tautomeric mixture of both forms in solution[162]:

[ReH$_4$(CO)(PMe$_2$Ph)$_3$]$^+$ ↔ [ReH$_2$(H$_2$)(CO)(PMe$_2$Ph)$_3$]$^+$

Many varieties of H$_2$ complexes with halide ligands are known where the strong influence of the halide has been noted and extensively studied theoreti-

Table 4.11. Effect of Cl versus H Ancillary Ligands on H_2 Coordination in $OsCl_nH_{4-n}(P^iPr_3)$

n	d_{HH}, Å	$-\Delta G^0$ (H_2 binding)
0	1.65 (hydride)	Very high
1	0.95	High
2	1.25	Low

cally.[14,82,86,163–167] Although halides, particularly Cl, are strong σ-electron-withdrawing ligands, which confers cation-like character, their π-donating ability (Cl > Br > I)[168] generally elongates the X–H bond in σ ligands trans to it such as H_2 (Table 4.8) and agostic C–H (see Section 12.2.1). In trans-$[Os(X)(H_2)(NH_3)_4]^+$ for X = Cl, Br, and I, J_{HD} increases (10.2, 11.8, 12.5) and d_{HH} decreases as M → H_2 BD lessens because of reduced π donation from X.[91] For polyhydrides containing halides, the nature of the halide influences the energetics of reversible H_2 loss from $IrXH_2(H_2)(PR_3)_2$ even when the halide is cis to H_2 (see Chapter 7).[166,167] The situation can be quite complex however, as illustrated by H_2 addition to $OsCl_nH_{4-n}(P^iPr_3)_2$ (see Table 4.11).[164] H_2 coordination is weakened when Cl replaces hydride ligands, as can be seen by the trend in $-\Delta G^°$ of H_2 binding. This seems counterintuitive because more BD to H_2 is expected for $n = 2$, where the H_2 is stretched (d_{HH} = 1.25 Å), than for $n = 1$. However, the degree of Cl → M π bonding both stabilizes 16e complexes and destabilizes 18e complexes. While more Os–H bonding is lost when H_2 dissociates for $n = 2$ than for $n = 1$, proportionately more Os–Cl π bonding is created to compensate for the loss. As will be shown in Chapter 9, halides are subject to elimination as HX when cis to η^2-H_2, showing the substantial influence of cis interactions discussed below.

4.11. INTERACTION OF A COORDINATED σ BOND WITH A CIS LIGAND: THE CIS EFFECT IN HYDRIDE–DIHYDROGEN COMPLEXES

Heterolytic cleavage of H_2 can involve transfer of H^+ to a metal bound ligand cis to a transient H_2 ligand, while H^- binds to M.[169] Although the heterolytic

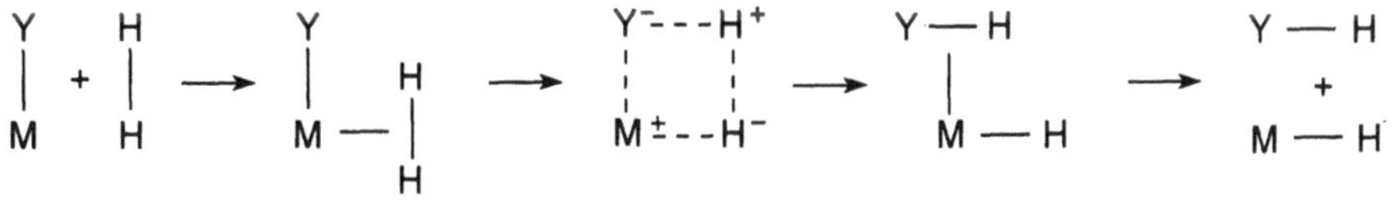

$$\begin{matrix} Y & H \\ | & + & | \\ M & H \end{matrix} \rightarrow \begin{matrix} Y & \\ | & H \\ M & — | \\ & H \end{matrix} \rightarrow \begin{matrix} Y^{-}\text{---}H^+ \\ | \\ M^{\pm}\text{--}H^- \end{matrix} \rightarrow \begin{matrix} Y — H \\ | \\ M — H \end{matrix} \rightarrow \begin{matrix} Y — H \\ + \\ M — H \end{matrix}$$

(4.19)

process in Eq. (4.19) is formally a concerted "ionic" splitting of H_2 as often illustrated by a four-center intermediate with partial charges, the mechanism does not have to involve such charge localization. In other words the two electrons originally present in the H–H bond do not necessarily both go into the newly formed M–H bond while a bare proton transfers onto Y$^-$ or, at the opposite extreme, onto an external base. As will be shown, the term σ-bond metathesis is actually a better description and comprises more transition states than the simple four-center intermediate shown above. An important concept is that the *mechanism involves at least transient coordination of* H_2 *or other σ ligand to M and cis to Y, and dissociation of transiently bound H–Y is the final step*. The key is interaction of the σ ligand with Y, leading to the four-center intermediate. This section will discuss the origin and theory behind this *cis interaction* as the first step in σ-bond metathesis and related atom-exchange processes.

Both experimentally and calculationally, complexes that contain a hydride cis to a H_2 ligand often show structural distortions and orientations of η^2-H_2 indicative of an interaction.[14,153,165,170–177] The barrier to H_2 rotation can be perturbed by the presence of a hydride cis to H_2 (see Chapter 6). From crystallography, the H_2 is eclipsed with the P–M–P axes in (A) and (B) in Scheme 3 but staggered with respect

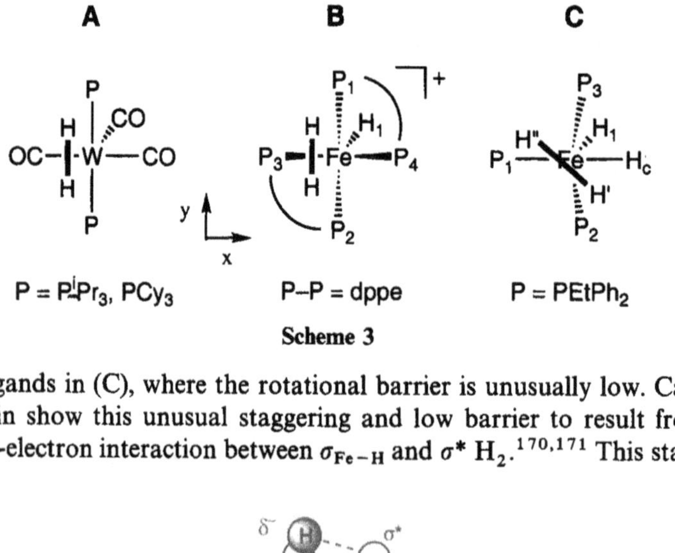

Scheme 3

to the cis ligands in (C), where the rotational barrier is unusually low. Calculations by Eisenstein show this unusual staggering and low barrier to result from the cis effect, a two-electron interaction between σ_{Fe-H} and σ^* H_2.[170,171] This stabilizes the

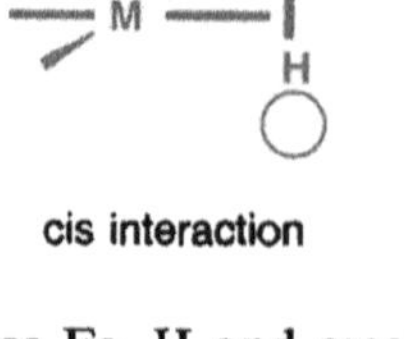

cis interaction

conformation where H–H eclipses Fe–H and creates a nascent bond between the closest nonbonded H centers. In the absence of a cis hydride, M → H_2 BD controls the H–H conformation, which optimally is eclipsed with the P–M–P axis (phosphines are good electron donors into the *d* orbitals that backbond). In (C), BD positions H_2 in the P_1–Fe–P_2 plane, but the cis effect favors eclipsing of H–H and Fe–H bonds (the orthogonal direction). The observed crystal structure reflects a balancing of the effects, and the H_2 lies in an intermediate position.[170]

The cis attraction is similar to other cis effects suggested to take place between an electron-rich L and a L containing acceptor orbitals and is also analogous to the anomeric effect in organic chemistry. The bonding is not covalent but rather an electrostatic attraction or a dipole/induced–dipole interaction between the partially positively charged H_2 that can be quite acidic (see Chapter 9) and the partially negatively charged hydride. The M–H bond is especially suited to this interaction because it is strongly localized onto the hydride s orbital and is high in energy owing to the elevated electron density on the hydride. Such an attraction was seen previously for the cis O–H and Ir–H bonds in $[IrH(OH)(PMe_3)_4]^+$, which show a 2.4 Å H···H separation in the neutron structure.[178]

δ^- δ^+
H-----H
P⋯⋯|⋯⋯O δ^-
P—Ir⁺—P
|
P
(23)

Structure **23** contains a form of *dihydrogen bonding* that is important in intramolecular heterolytic splitting of H_2 (see Section 9.4.4.1). Any interaction between the hydrogen of a σ ligand and a nearby hydride of the form η^2-X–H$^{\delta+}$···H$^{\delta-}$ can be considered a type of dihydrogen bond. The spherical nature of the hydride s orbital favors bond overlap with nearby vacant orbitals without severely stretching or breaking M–H, which is also advantageous in σ-bond metathesis reactions discussed below. *Ab initio* calculations by Eisenstein give a better understanding of this effect by examining the orientations and barrier to rotation of H_2 in (B) versus (C) in Scheme 3 (see also Chapter 6).[9,171] In (B), the orientation of H_2 should be controlled electronically principally by BD (no steric effects here). Calculations confirm this, giving a twofold barrier of 1.9 kcal/mol, close to the experimental value, of 2.4 kcal/mol. The optimal structure is the conformation with the H–H axis parallel to the P–Fe–P axis. For (C) the best conformation at the *SCF level* is found where the H_2 is closely aligned with the cis Fe–H bond and the top of the barrier height (1.6 kcal/mol) occurs when H_2 almost aligns with P_2–Fe–P_3. However at the *MCSCF level*, the preferred H–H orientation is rotated clockwise 64° away from alignment with P_1–Fe–H_c, which agrees well with the crystal structure (Scheme 3), and the top of the barrier (1.4 kcal/mol) is 90° beyond this. These two different types of calculations illustrate the two competing interactions present in (C): the Fe → H_2 BD, which prefers the H_2 to be perpendicular to Fe–H, and an electrostatic interaction (the cis effect), which favors coplanarity of H_2 and Fe–H. The latter effect, which manifests as a dipole/induced–dipole interaction, receives more emphasis in the SCF calculations. For MCSCF, the agreement with experiment is good because no ligand cis to H_2 is tilted toward it, so that all d orbitals are properly set for efficient BD.

When the hydride and H_2 are oriented cis in complex (B), *ab initio* and DFT calculations show significant preference for the H_2 in $[FeH(H_2)(PR_3)_4]^+$

(R = H,[12,173] Me[12]) to lie in the same plane as the hydride H_c and to lean toward it as if there is an attraction. The optimized geometry in **24** shows *asymmetry* in the Fe–H$_2$ bonding, which is not found in any other optimized structure. Remarkably, a subsequent X-ray structure[179] of [FeH(H$_2$)(PMe$_3$)$_4$]$^+$ shows the same asymmetry

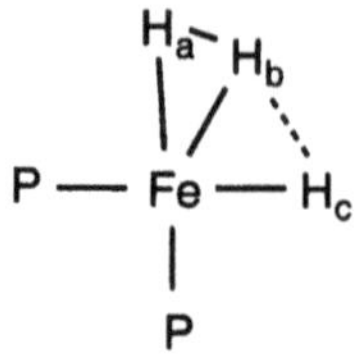

```
        Ha  Hb
         \ /
          ·
P ── Fe ── Hc
     |
     P
```

(24)

and similar bond distances and angles to those calculated (Table 4.12). The experimental d_{MH} are shorter than those calculated, especially for the hydride ligand, H_c, but values for these X-ray distances show the usual big errors. The distances calculated for the DFT study indicate little variation whether R = H or the real R = Me ligand, although it is much more important to use PMe$_3$ in assessing the relative energies of the various dihydrogen–hydride and trihydride isomers possible here.[12] The $H_b \cdots H_c$ distance is calculated to be 1.587 Å,[173] which is somewhat less than that found in hydrides. The Fe–H$_2$ bonding is stronger here than when H_a–H_b is oriented perpendicular to the plane (no cis interaction), as shown by the very short H_a–H_b distances for the latter case (0.743 Å). The energy difference of 7.0 kcal/mol favoring the in-plane geometry shown above over the perpendicular orientation of H$_2$ can be attributed only to cis attraction.[173]

DFT calculations on related tetradentate complexes,

[MH(H$_2$){P(CH$_2$CH$_2$PPh$_2$)$_3$}]$^+$

show a similar cis effect.[174] The H$_2$, which could not be located experimentally, is bound significantly more strongly to Fe (d_{HH} = 0.95 Å) than to Ru (d_{HH} = 0.89 Å)

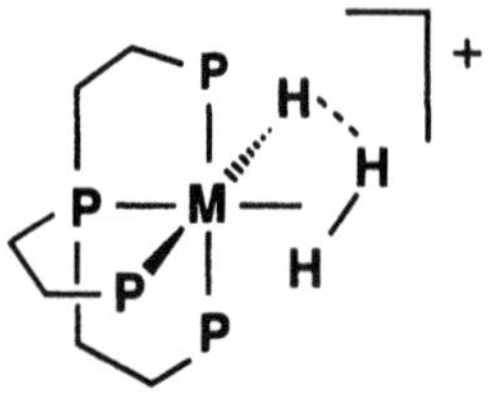

M= Fe, Ru

because of greater BD and also a stronger cis effect. The cis interaction is also believed to contribute to the higher stability of the Fe–H$_2$ complex as compared to the corresponding N$_2$ and ethylene complexes where a cis effect cannot operate. This extra stabilization may even influence reaction chemistry such as catalytic hydrogenation of alkynes to alkenes, where a binding site for the incoming alkyne is

Table 4.12. Calculated versus Experimental Distances and Angles for $[FeH(H_2)(PMe_3)_4]^+$

	Exptl	Ab initio[173]	DFT (PH_3)[12]	DFT (PMe_3)[12]
Fe–H_a	1.72(4)	1.743 Å	1.602	1.598
Fe–H_b	1.53(4)	1.619 Å	1.571	1.564
Fe–H_c	1.48(3)	1.657 Å	1.529	1.517
H_a–H_b	0.84(4)	0.786 Å	0.904	0.907
∠H_b–Fe–H_c	76(2)°	58°	68.7°	67.6°

provided by detachment of a phosphine arm from $[FeH(H_2)P(CH_2CH_2PPh_2)_3\}]^+$ rather than by loss of the tightly bound H_2.

The cis effect has also been observed experimentally and calculationally in several other Ru complexes, including bis(H_2) complexes. The extremely fluxional $RuH_2(H_2)_2(PCy_3)_2$, which has a low H_2 rotational barrier (1.1 kcal/mol; see Chapter 6)[175] has three isomeric structures (**25–27**) calculated to lie within an energy range of only 2 kcal/mol. Structure **28** with trans H_2 is 10–15 kcal/mol higher in energy.

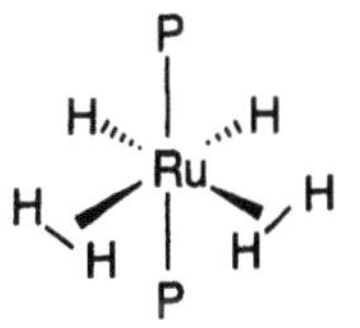

(25) 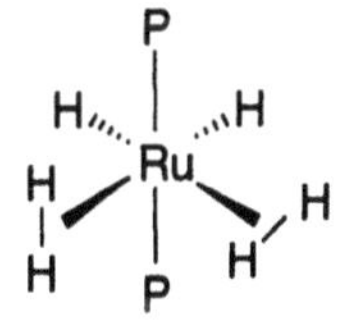   (26)

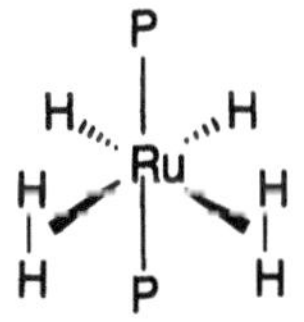

(27) 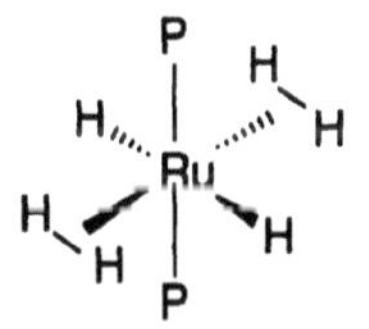   (28)

The lowest energy structure (**25**) unexpectedly has all the hydrogens in the same plane, which is evidence for cis interactions that could promote the facile intramolecular exchange in such systems (see Chapter 6). A subsequent X-ray structure (see Chapter 5) is consistent with **25**, showing d_{HH} near 0.85 Å just as computed (0.85 Å by DFT/B3LYP), another demonstration of the predictive power of computational chemistry.[177] A cis interaction is proposed between adjacent H_2 ligands rather than with the hydrides.

The *first complex with two different η^2-coordinated σ bonds,*

$$RuH_2(\eta^2\text{-}H_2)(\eta^2\text{-}SiHPh_3)(PCy_3)_2$$

is formed from $RuH_2(H_2)_2(PCy_3)_2$ by displacement of one H_2 by $SiHPh_3$ and shows essentially unstretched H–H and Si–H bonds. The bulky PCy_3 in **29** orient in an

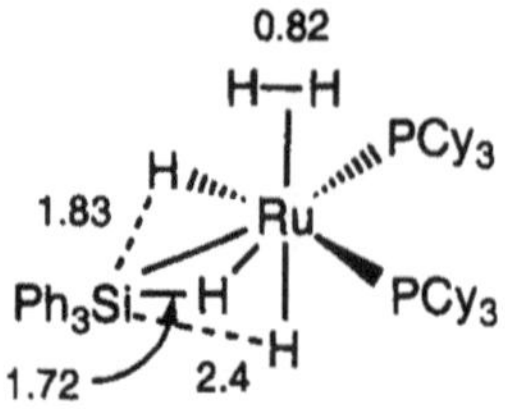

$$\begin{array}{c} 0.82 \\ H-H \\ 1.83 \quad H_{\prime\prime\prime\prime\prime}|\quad \prime\prime\prime\prime PCy_3 \\ Ph_3Si - H - Ru \\ 1.72 \quad 2.4 \quad H \quad PCy_3 \\ (\mathbf{29}) \end{array}$$

abnormal cis configuration apparently because favorable attractive cis interactions between the Si atom and the hydride ligands (see also Chapter 11) overcome steric repulsion between the phosphines.[180] The X-ray data and DFT/B3LYP calculations give distances between Si and the hydrides that are much shorter (2.116 and 2.071 Å, calcd) than the sum of the van der Waals radii of Si and H (3.3 Å). Also, Mulliken population analyses show positive overlap (values up to 0.09 versus 0.40 in free SiH_4). The lowest-energy isomer with trans phosphines is $RuH(SiH_3)(\eta^2-H_2)_2(PH_3)_2$ and is about 2 kcal/mol higher in energy.

The 16e H_2 complex $RuHI(H_2)(PCy_3)_2$ shows crystallographically that Ru–H is bent toward the H_2 ($\angle I$–Ru–H_1 = 99°).[172] The nonbonding $H_1 \cdots H_2$ distance is short [1.66(6) Å] because of cis interaction while H_2–H_3 is elongated (1.03 Å).

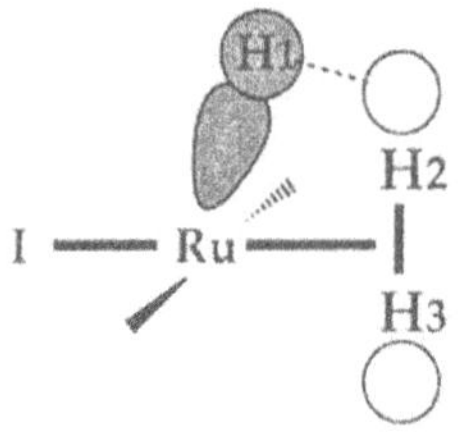

Extended Huckel calculations demonstrate that this structure is the minimum. The system $[Cp*OsH_2(PR_3)(H_2)]^+$ is remarkable in that a complete structural change occurs to give two cis interactions (H$\cdots$H–H$\cdots$H) on changing R from Ph to Cy. d_{HH} increases from 1.01 to 1.31 Å (PCy_3 is a better donor), the H_2 rotates 90°, and the H$\cdots H_2$ distances are 1.54(3) and 1.68(3) Å,[181] as in a putative H_4

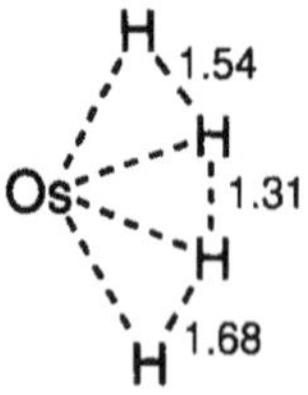

ligand (see Section 4.12). As noted above, H_2 binding in d^0 complexes such as $[CpWH_4(H_2)(PH_3)]^+$ that lack BD capability are stabilized by strong cis effects.[153]

4.12. INTRAMOLECULAR HYDROGEN EXCHANGE, POLYHYDROGEN COMPLEXES, AND σ-BOND METATHESIS

The above cis effect is significant because of its apparent role as the nascent interaction in intramolecular hydrogen exchange processes. The intermediate is essentially a trihydrogen complex, which was initially proposed by Brintzinger[182] as the key species in this direct hydrogen transfer process. The possible existence of η^3-H_3 ligands has been examined theoretically by Burdett[7,122,183] in his detailed studies of polyhydrogen species, H_n ($n = 3-13$) and also looked for in CpIr(PR$_3$)H$_3^+$ calculationally by Albright[176] and experimentally by Heinekey.[184] No evidence was found in either case for either an "open" (linear H_3^-) or "closed" (triangulo H_3^+) structure despite these tantalizing possibilities here (an H_2 species is located as a

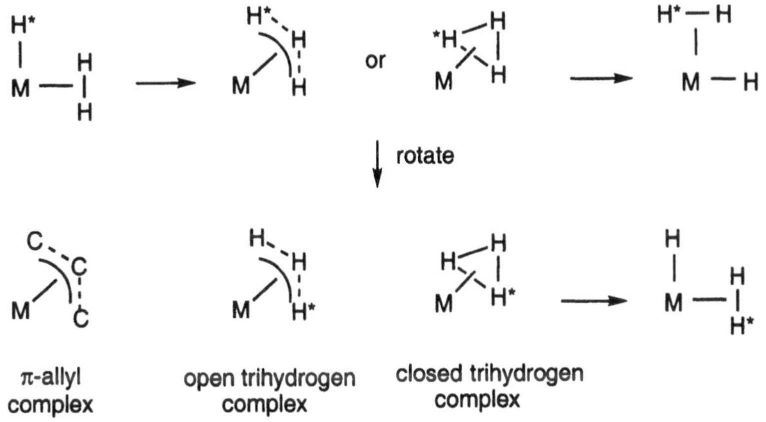

transition state in the H exchange process[176]). A trihydrogen complex has yet to be isolated, although there is experimental evidence for its intermediacy in a facile tautomerization reaction.[185] The ReH(H$_3$) species is estimated to be no more than 10 kcal/mol less stable than the ReH$_2$(H$_2$) complex. An open H$_4$ ligand may also be

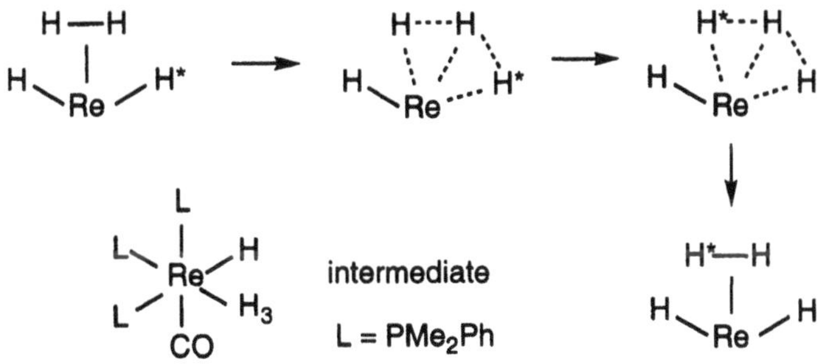

(4.20)

possible here and in other H$_4$ systems, at least as an entity with cis interactions as in [Cp*OsH$_2$(H$_2$)(PCy$_3$)]$^+$ above. On a Cr(CO)$_4$ fragment, an open H$_4$ unit with fixed 1.0-Å d_{HH} is calculated to be 6 kcal/mol more stable than an H$_3$(H) arrangement and 16 kcal/mol less stable than a bis(H$_2$) structure.[122]

Polyhydrogen species are known mass-spectrometer molecules, and H_3^+ has a closed (triangulo) structure with d_{HH} of 0.87 Å and is basically a H_2 complex of H^+ (isolobal with d^0 M; see Figure 4.1). However, a trihydrogen ligand is more likely to have an open linear structure best represented as H_3^- as supported by calculations. The energy for $[Ru(triangulo-H_3)(PH_3)_4]^+$ is calculated to lie well above that for the linear H_3 isomer and more than 50 kcal/mol above the trans-$[RuH(H_2)(PH_3)_4]^+$ energy.[119] The M–H_3 fragment with its kite-shaped "trefoil" topology can then effect transfer of hydrogen to the cis position to give η^2-H_2 where hydride was originally present. Facile exchange of hydrogen atoms H* with H, hence isotopic scrambling, can take place either by this route or by rotation of the H_3 followed by reformation of the original hydride–H_2 orientation. The H_3^- ligand is analogous to the π-allyl ligand in charge distribution and bonding properties. Each terminal H carries a charge of about -0.3 units, while the central H is about -0.02 units according to Brintzinger's early calculations.[182] Simplistically, H_3^- represents a bond/no-bond delocalized structure just as allyl represents single-bond/double-bond delocalization.

OA to a trihydride would also produce exchange, but this is a much higher energy path in *ab initio* calculations on $[FeH(H_2)(PH_3)_4]^+$.[186] The trihydride is 65.3 kcal/mol higher in energy than that of the reactants, which alone makes the existence of such an intermediate unlikely. By comparison the open direct transfer pathway

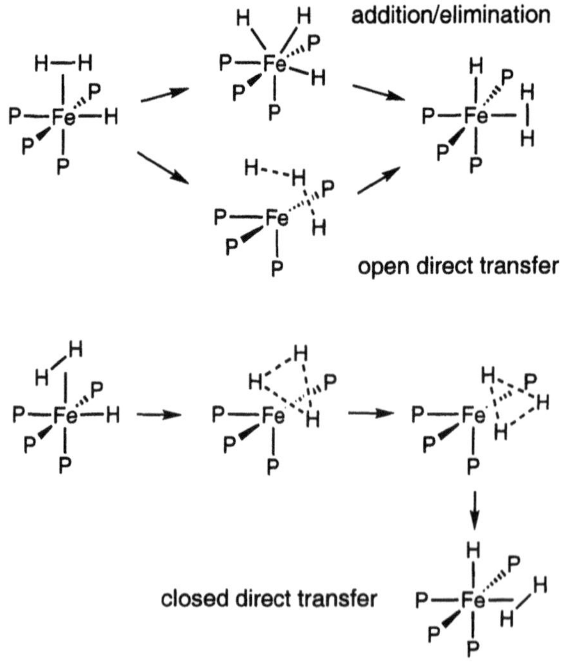

Scheme 4

for H exchange has an activation barrier of *only 3.2 kcal/mol* (the triangulo closed intermediate is much higher at 69 kcal/mol) (Scheme 4). DFT calculations on $[FeH(H_2)(PMe_3)_4]^+$ with the actual PMe_3 ligands instead of PH_3 models show an even lower barrier, ca 0.5 kcal/mol,[186] which is lower than the 2.3 kcal/mol barrier for the simple exchange reaction $H + H_2 \rightarrow H_2 + H$ from quantum Monte Carlo methods.[187] This low barrier corresponds well with the observation of extremely fast H/H_2 scrambling even at $-140°C$.[179]

Within thermodynamic limits, H* in Scheme 4 could be any other metal-bound atom (Z) such as a halide or a group such as alkyl, and H–H could be any other σ ligand (X–Y) such as alkane. This represents *σ-bond metathesis*,[188–192] a more general form of the above hydrogen exchange analogous to olefin metathesis. Mechanistically, two different kinds of σ-bond metathesis are possible. While the four-center intermediate in Eq. (4.21) looks to be similar to the trefoil intermediate in the hydride–H_2 exchange, the actual bonding and the overall pathway may differ

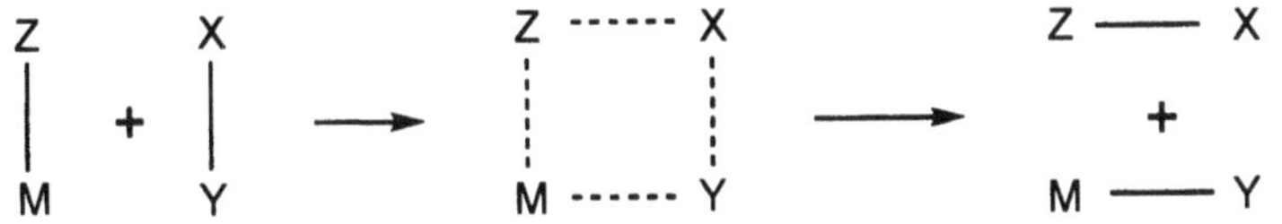

$$
\begin{array}{ccccccc}
Z & & X & & Z \cdots\cdots X & & Z \text{———} X \\
| & + & | & \longrightarrow & \vdots \quad\quad \vdots & \longrightarrow & \quad + \\
M & & Y & & M \cdots\cdots Y & & M \text{———} Y \\
\end{array}
\qquad (4.21)
$$

depending on the electronics of the system. A "traditional" mechanism of σ-bond metathesis is that shown above, which is similar to that postulated for heterolytic cleavage of H_2 (X and Y = H), where σ-complex intermediates were not involved (before H_2 complexes were discovered). This pathway works mainly with early transition metals, lanthanides, and actinides with little BD ability, e.g., complexes of the type Cp_2MR. The other potential mechanism is similar to the hydride–H_2 exchange [Eq. (4.20)], which does involve σ-complex intermediates, and this could occur on more electron-rich metals such as $[FeH(H_2)(PH_3)_4]^+$.

Theoretical studies of σ-bond metathesis processes include semiempirical,[182,193] *ab initio*,[194,195] and DFT[196–199] methods and provide acceptable explanations for the mechanisms, as discussed by Niu and Hall.[9] The results for the transition-state (TS) geometries are remarkably similar to those for the intramolecular hydrogen exchange [Eq. (4.20)]. The *ab initio* study by Goddard on a $[ScCl_2H] + H_2$ model gives a d_{HH} in the TS of 1.01 Å, which is very close to the d_{HH} of 1.04 Å calculated in $[FeH(H_2)(PH_3)_4]^+$.[194] The angles in the trefoil M–H_3 intermediates are also similar, which is surprising because of the different electronic situations in the Sc and Fe complexes. In order to understand this seeming contradiction, Maseras plotted the electron densities for the two systems and found differences in the topology of the bonds in the trihydrogen fragment and M (Figure 4.25).[186] In the Sc complex there is a four-membered ring defined by the three hydrogens and Sc because a ring point appears in the center and there are bond critical points[108] between all contiguous atoms. By contrast, such a ring does not exist for the Fe system, where the Fe is bound to the H_3 fragment by a *single* bond critical point through the central H. Thus the density plot for the Sc system corresponds to the traditional view of σ-bond metathesis [Eq. (4.21)], while for the Fe complex no four-membered ring exists but instead H_3 is attached to M. The key difference for this electronically is the electron count, which affects the occupation or vacancy of one of the molecular orbitals of the TS, specifically an antibonding orbital between M and the outer hydrogens. In the Sc case this orbital is unoccupied and M interacts with the outer hydrogens to give a four-membered ring. However, in d^6 $[FeH(H_2)(PH_3)_4]^+$ this

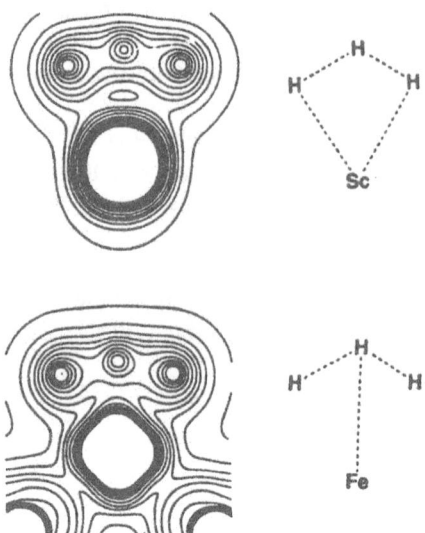

Figure 4.25. Electron isodensity contour maps for the transition states of the hydrogen transfer reaction in the [ScCl$_2$H] + H$_2$ (top) and [FeH(H$_2$)(PH$_3$)$_4$]$^+$ + H$_2$ (bottom) systems. The isodensity lines plotted correspond to the values of 0.0001, 0.01, 0.01, 0.02, 0.03, 0.08, 0.12, 0.16, 0.20, and 0.50 Å. Reprinted with permission from Maseras et al.[186] Copyright 1992 American Chemical Society.

orbital is occupied and the Fe is bound only to the central hydrogen. Calculations by Dedieu indicate that late M such as Rh and Pd also can provide σ-bond metathesis pathways for proton transfer from H$_2$ and CH$_4$, especially when involving Z with additional lone pairs such as OH or formate, e.g., **30**.[195] However, OA

(30)

pathways are energetically more favorable for Pt systems because Pt^{IV} is more accessible than Pd^{IV}, a common theme in C–H activation on these metals (see Section 12.4.3).

Ziegler has studied σ-bond metathesis of d^0 Cp$_2$Sc–R (R = H, Me, Et, Pr, vinyl, acetylide) with both H$_2$ and alkanes, using DFT methods where Cp rather than Cl or H is used to model the experimental[188] Cp* system.[197] A weak H$_2$ adduct forms in the early stages of the hydrogen exchange reaction with Cp$_2$ScD (using D to differentiate atoms):

$$\text{Cp}_2\text{Sc–D} + \text{H–H} \rightarrow \text{Cp}_2\text{Sc–H} + \text{H–D} \quad (4.22)$$

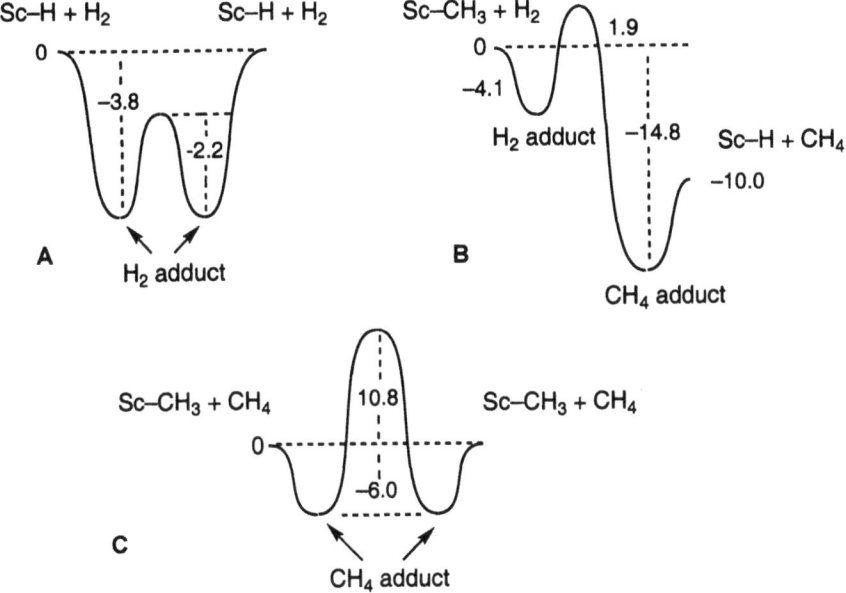

Figure 4.26. Energy profiles for σ-bond metathesis reactions in kcal/mol.

The structural parameters for the adduct and the four-center Sc–H$_3$ intermediate in the exchange process are shown below. d_{HH} is 0.81 Å in the transient adduct while the d_{ScH} are longer than the d_{ScH} calculated for Cp$_2$Sc–H (1.86 Å). The kite-shaped TS exhibits three Sc–H contacts nearly the same as the latter, and d_{HH} = 1.02 Å, similar to those in elongated H$_2$ complexes, i.e., a considerable amount of bond breaking/making. As shown in Figure 4.26A, the adduct formation energy is 3.8 kcal/mol, and even the four-center TS is slightly more stable than the reactants by 2.2 kcal/mol. The loss in H–H interaction energy on going from adduct to intermediate is compensated for by the formation of two new Sc–H linkages so that the TS is at nearly the same energy as the reactants. Thus the H exchange reaction has no barrier, consistent with the high experimental rates observed for H exchange (10^3 s^{-1} M^{-1} at −90°C).[188] A similar d^0 system [CpMH$_4$(H$_2$)(PH$_3$)]$^+$ (M = Mo, W) shows a low barrier of 4 kcal/mol and long d_{HH} (1.11–1.15 Å) for the H$_3$ ligand in the TS.[153]

For the metathesis of Cp$_2$Sc–CH$_3$ with H$_2$ to produce CH$_4$ and Cp$_2$Sc–H (referred to as hydrogenolysis), the profile in Figure 4.26B again passes from an initial H_2 adduct over a TS to a weak CH_4 adduct that readily converts to the final products [Eq (4.23)]. The barrier is small, 1.9 kcal/mol, and the overall exothermicity

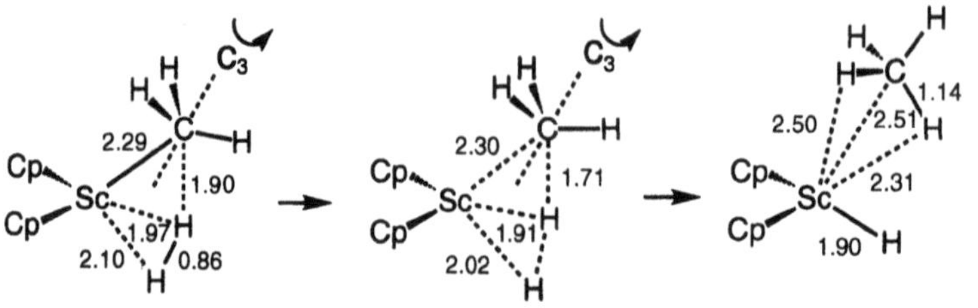

$$ (4.23) $$

is 10 kcal/mol. The TS structure is reached early on and is not changed much from the H_2 adduct. The C_3 axis of the methyl group is tilted away from the Sc–C bond vector toward the incoming H, which redirects the σ orbital on CH_3 toward H at the expense of losing some bonding interaction with the Sc-based orbitals. This weakens the Sc–C bond, unlike the situation for the Sc–H_a bond in the TS of the hydrogen exchange reaction. The spherical nature of the $1s_a$ orbital ensures strong Sc–H_a interaction while the H_a–H_b bond to the incoming H_b is formed, consistent with the lower barrier compared to hydrogenolysis of Cp_2Sc–CH_3. The small barrier calculated for hydrogenolysis (1.9 kcal/mol) is consistent with the fast experimental rates (4×10^{-1} s^{-1} M^{-1} at $-78°C$).[188]

The methane exchange reaction,

$$Cp_2M-{^*}CH_3 + CH_4 \rightarrow Cp_2M-CH_3 + {^*}CH_4 \qquad (4.24)$$

where M is lutetium, was a landmark discovery by Watson because it represented the *first known example of methane activation by an organometallic complex*.[189] Other alkyl exchange reactions have since been found for M = Sc,[188] Zr$^+$,[191] and the actinides Th and U.[190] As shown in Figure 4.26C, the process for M = Sc has a barrier of about 11 kcal/mol and the four-center TS has C_{2v} symmetry, where the activated C–H bond has been stretched to 1.33 Å. A methane adduct is formed in

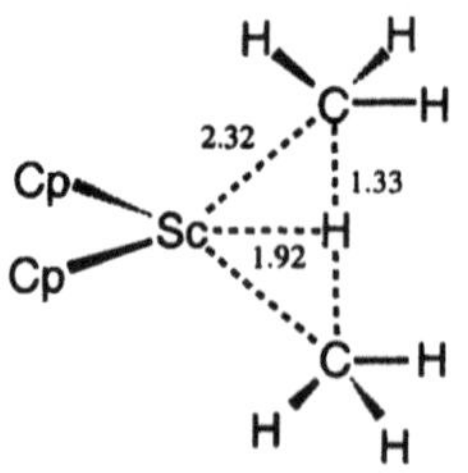

the early stages much as in the last step of the hydrogenolysis reaction [Eq. (4.23)]. The C–H activation here is experimentally much slower than the H–H activation processes in Figure 4.26. The activation energy observed by Watson[192] for the Lu system is 11.7 kcal/mol, which is remarkably close to that calculated for the Sc system (10.8 kcal/mole, Figure 4.26C).

The hydroformylation reaction in which hydrogen is transferred to an acyl ligand provides another potential example of a four-center intermediate. DFT

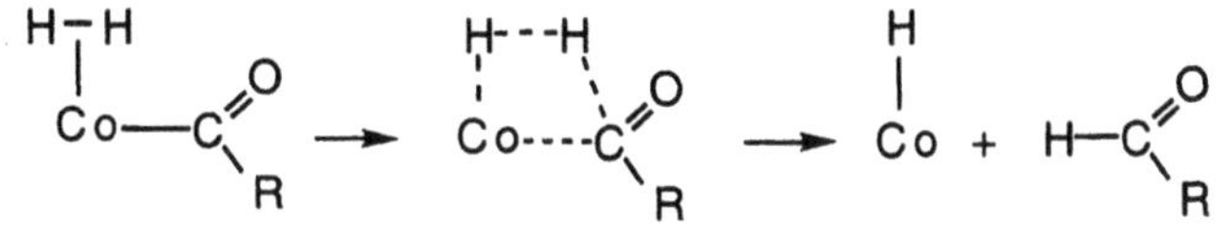

$$\begin{array}{c} H-H \\ | \\ Co-C \end{array} \begin{array}{c} O \\ \diagup \\ R \end{array} \rightarrow \begin{array}{c} H\cdots H \\ \vdots \quad \vdots \\ Co\cdots C \end{array} \begin{array}{c} O \\ \diagup \\ R \end{array} \rightarrow \begin{array}{c} H \\ | \\ Co \end{array} + H-C \begin{array}{c} O \\ \diagup \\ R \end{array} \quad (4.25)$$

calculations show that η^2-H_2 with $d_{HH} = 0.83$ Å binds initially and forms the four-center intermediate with $d_{HH} = 1.14$ Å and $d_{CH} = 1.37$ Å [Eq. (4.25)].[198] The σ^* orbital of H_2 can interact with the bonding Co–C orbital on $CH_3(O)C$–$Co(CO)_3$ much like in the cis interaction with a hydride. However, subsequent calculations in 1996 indicated an OA pathway that has a lower barrier (8.7 versus 16.8 kcal/mol).[198] Calculations on hydrogenation of CO_2 to formic acid catalyzed by cis-$RuH_2(PH_3)_4$ model complexes indicate a reaction mechanism involving a *six-membered* σ-bond metathesis.[199] Such a process directly relates to heterolytic splitting of H_2 by pendant ligands (see Section 9.4.4.1).

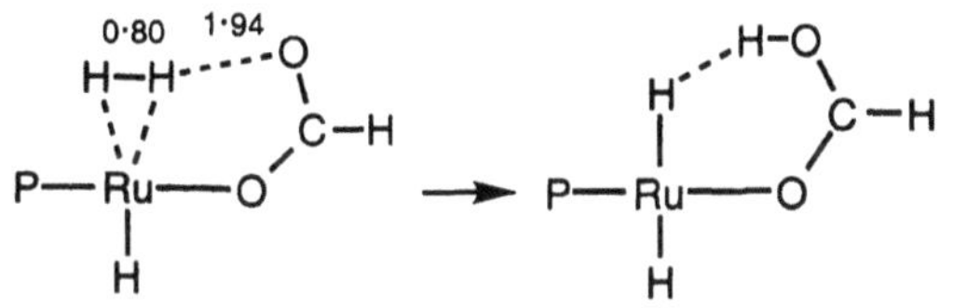

$$\begin{array}{c} 0{\cdot}80 \quad 1{\cdot}94 \\ H-H\cdots\cdots O \\ \diagdown \diagup \quad \diagdown \\ P-Ru-O \\ | \\ H \end{array} \begin{array}{c} \\ \\ C-H \\ \\ \end{array} \rightarrow \begin{array}{c} \quad\quad H-O \\ H \diagup \quad \diagdown \\ | \quad\quad C-H \\ P-Ru-O \diagup \\ | \\ H \end{array}$$

H_2 addition across M–L *multiple* bonds is not as common but has been studied by Rappe[200] by correlated *ab initio* calculations:

$$M{=}CR_2 + H_2 \rightarrow MH(CHR_2) \quad (4.26)$$

$$M{=}NX + H_2 \rightarrow MH(NHX) \quad (4.27)$$

$$M{-}O + H_2 \rightarrow MH(OH) \quad (4.28)$$

The model reaction is hydrogenation of a methylene complex to a H(Me) complex [Eq. (4.29)]. The reaction is quite exothermic by 21.5 kcal/mol at 300 K and the activation energy from a weakly bound H_2 adduct ($\Delta H = 3.8$ kcal/mol) is 10.5

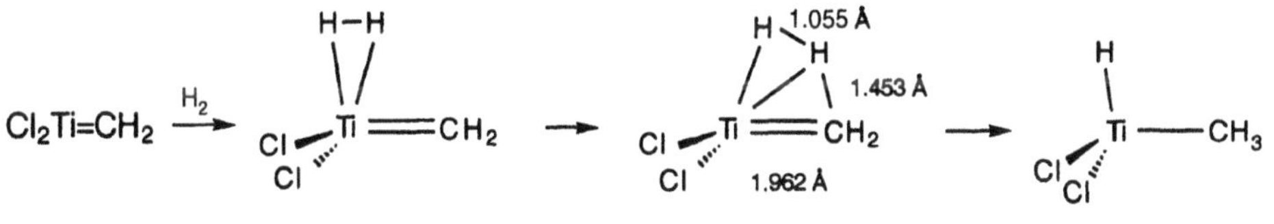

$$Cl_2Ti{=}CH_2 \xrightarrow{H_2} \begin{array}{c} H-H \\ \diagdown \diagup \\ Cl-Ti{=}CH_2 \\ | \\ Cl \end{array} \rightarrow \begin{array}{c} H\quad 1.055\,\text{Å} \\ \diagup \diagdown H \\ Cl-Ti{=}CH_2 \quad 1.453\,\text{Å} \\ | \quad\quad 1.962\,\text{Å} \\ Cl \end{array} \rightarrow \begin{array}{c} H \\ | \\ Ti-CH_3 \\ \diagup \\ Cl\; Cl \end{array}$$

(4.29)

kcal/mol. The barrier is small because of the ease of rehybridization of the Ti d_π orbital originally bonded to methylene, which can change its shape during the reaction and retain overlap with the methylene p orbital as it moves from carbon to H^a. The orbital motion along the reaction coordinate for Eq. (4.29) is shown in **31**.

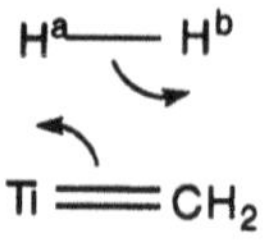

$$\text{Ti}=\text{CH}_2 \cdots \text{H}^a \text{---} \text{H}^b$$

(31)

The Ti–C π bond is transformed into a Ti–H^a σ bond analogously to the way M–H σ bonds rearrange, with remarkably similar low barriers. The TS geometry is similar to those above for four-center 2 + 2 reactions and also organic three-center reactions

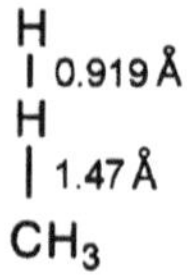

$$\begin{array}{c} \text{H} \\ | \; 0.919 \,\text{Å} \\ \text{H} \\ | \; 1.47 \,\text{Å} \\ \text{CH}_3 \end{array}$$

(32)

involving a radical and a σ bond (**32**). In **32** the $\angle$ H–H–C is 180° and 150° for **31**. Addition of the M orbital and the fourth electron perturbs the TS (**32**) very little.

4.13. DIHYDROGEN AND METHANE BINDING TO NAKED METAL IONS

A significant number of theoretical (and experimental) investigations of molecular H$_2$ binding have also been devoted to systems other than discrete transition metal complexes and early models such as Pd–H$_2$. A large class of "naked" metal cations [M(H$_2$)$_n$]$^+$, studied by ion-beam and mass spectrometric techniques give H$_2$ dissociation energies and are excellent systems for H$_2$ and alkane binding because of their high electrophilicity and reluctance to oxidatively add these molecules.[201] Neutral metals on surfaces nearly always transfer electrons to approaching H$_2$ molecules to split the H–H bond to give hydrides,

$$M + H_2 \rightarrow H\text{-}M\text{-}H \qquad (4.30)$$

analogous to excessive BD causing OA in complexes. However, when H$_2$ approaches a bare M$^+$, such donation is less energetically favorable because the second ionization potential of M$^+$ is rather high. Instead the cation polarizes the H$_2$ and the M$^+$–H$_2$ bonding takes on a dipole character. Calculations indicate that M$^+$ can in essence be "solvated" sequentially by up to ten H$_2$ molecules as in Eq. (4.31).[201]

Binding energies for all first-row clusters $[M(H_2)_n]^+$ ($n = 1-6$) and several small-molecule analogues have been determined by temperature-dependent equilibrium measurements[202–209] of mass-selected M^+ ions reacting with H_2 or by

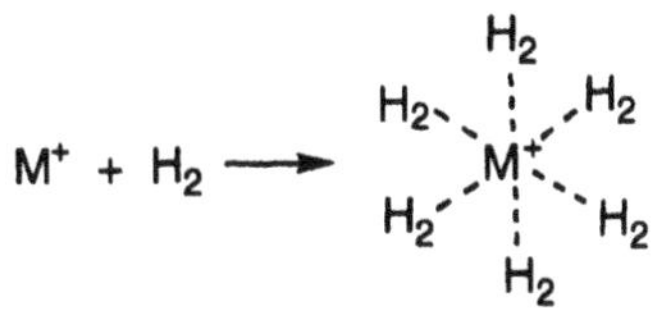

$$M^+ + H_2 \longrightarrow \begin{array}{c} H_2 \\ H_2 \diagdown \vdots \diagup H_2 \\ M^+ \\ H_2 \diagup \vdots \diagdown H_2 \\ H_2 \end{array} \quad (4.31)$$

collision-induced dissociation in a guided ion-beam mass spectrometer (Table 4.13).[210] These interactions have also been studied theoretically for Sc,[211] Ti,[207,212] V,[213,214] Cr,[215] Mn,[208,209] Fe,[216] Co,[214,217–219] Ni,[217] Cu,[209,214,217] and Zn.[208,209] Although noncovalent electrostatic interactions (charge-induced dipole and charge quadrupole) are present, they normally comprise a small fraction of the total bond strength because the purely electrostatic attraction in $[Na(H_2)_{1,2}]^+$ and $[K(H_2)_{1,2}]^+$ is only 1.3–2.5 kcal/mol.[206,209] The presence of covalent forces in the bonding is shown by the strong influence of the nature of M^+ on both bonding energies and structures. The four covalent forces include the main interaction: electron donation from the H_2 σ orbital to M^+ that stabilizes the ion charge. Most of this donation is to the M $4s$ orbital with a minor amount to a $3d$ orbital of proper symmetry. Secondly, some BD to the H σ^* orbital still occurs in the later M^+ with filled $3d$ orbitals, despite the highly electron-deficient M here. In ions with half-filled $3d$ σ orbitals, a hybridization between the $3d_{z^2}$ and the $4s$ orbital reduces on-axis Pauli repulsion. Lastly, minor contributions from hybridization with the $4p$ orbitals can occur, despite their significantly higher energy. The relative importance of these and the electrostatic factors depends strongly on the valence configuration of M^+.

Table 4.13. Comparison of Experimental Binding Energies (± 0.4–1.4 kcal/mol) for $[M(L)_{n-1}]^+ + L \rightarrow [M(L)_n]^+$ for L = H_2, CH_4, and Other Small Molecules

Ion	L	$n = 1$	$n = 2$	$n = 3$	$n = 4$
$[Ti(L)_n]^+$	H_2	10.0	9.7	9.3	8.5
	CH_4	16.7	17.4	6.7	
$[V(L)_n]^+$	H_2	10.2	10.7	8.8	9.0
$[Cr(L)_n]^+$	H_2	7.6	9.0	4.7	3.4
$[Mn(L)_n]^+$	H_2	1.90	1.65	1.4	1.2
$[Fe(L)_n]^+$	H_2	16.5	15.7	7.5	8.6
	CH_4	13.6	23.2	23.7	17.7
	N_2	12.9	19.8	10.8	13.6
	H_2O	30.6	39.2	18.2	19.6
	C_2H_4	34.5	36.1		
$[Co(L)_n]^+$	H_2	18.2	17.0	9.6	9.6
	CH_4	21.5	22.9	9.6	15.5
$[Ni(L)_n]^+$	H_2	17.3	17.6	11.3	7.1
$[Cu(L)_n]^+$	H_2	15.4	16.7	8.8	5.1
	CO	35.5	41.0		
$[Zn(L)_n]^+$	H_2	3.75	2.75	2.35	1.60

The observed binding energies for $[M(H_2)_n]^+$ as well as CH_4 analogues generally decrease with n, as shown in Table 4.13, which lists energies for $[M(L)_n]^+$ for $n = 1-4$, and L occupying octahedral sites. Computations show good agreement, e.g., in $[Ti(H_2)_n]^+$ bond energies at the DFT level are less than 1 kcal/mol lower than experimental values.[207] In general d_{HH} is near that in free H_2, 0.74–0.77 Å, for $n = 1-6$, although in some cases the distance can approach the 0.82 Å value seen crystallographically in organometallic complexes. For Sc^+, OA of H_2 occurs for $n = 1$, followed by molecular H_2 binding to give $[ScH_2(H_2)_n]^+$.[205] The bond strengths for $[M(H_2)_n]^+$ are greater for the later metals (Fe, Co, Ni) primarily because of greater BD and, secondarily, smaller ion size (much of the attraction is due to charge-induced dipole potential, which varies as $1/r^4$). The binding energies for Mn and Zn are by far the weakest because of repulsion between the singly occupied $4s$ orbital and the H_2 σ orbital.[208,209] All other first-row metals, in contrast, have a $3d^n$ valence electron configuration for the $[M(H_2)_n]^+$ species.

CID measurements for CH_4 binding to $[Co(CH_4)_n]^+$ exhibit parallel behavior to that for $[Co(H_2)_n]^+$ (Table 4.13).[220] Ab initio calculations show similar bond energies and predict that CH_4 binds in an η^2-H,H fashion. The trend in bond energies is rationalized by electronic changes at M (e.g., s–d hybridization) on coordination of the third and successive molecules. The different trends for the Fe^+ system for L binding are ascribed to changes in electronic structure of M with sequential coordination of ligands of varying field strengths.[221] The Ti^+ system is interesting in that when a third CH_4 is added, insertion of M into the C–H bond (OA) competes with σ binding.[222] The first two ligands provide a sufficiently strong ligand field that the low spin state is stabilized, which activates CH_4 better than the low-lying quartet state. Thus reactivity at metals can be tuned by altering the ligand shell surrounding them, just as in stable organometallic complexes. For the late metal Cu^+, the binding energy for the first CH_4 (21.4 kcal/mol) is calculated to be 50% higher than for H_2 (14.2 kcal/mol).[214] In contrast, $[M(SiH_4)_n]^+$ σ complexes are not obtained for any M, but rather Si–H cleavage and H-atom or H_2 dissociation occurs to give a variety of products, e.g., $[M(SiH_2)]^+ + H_2$. This is consistent with the tendency of silanes to undergo OA more easily than H_2 or alkanes.

It is most revealing that in all cases the *experimental and calculated binding energies of methane to positively charged metal ions are greater than those for H_2*! This is completely opposite to the situation for organometallic fragments, including cationic species studied thus far. The bare M^+ and the weakly coordinated $[M(L)_n]^+$ complexes are exceedingly electrophilic, which greatly promotes electron donation from the σ bond to M. As discussed above, this same effect is seen for cationic organometallic H_2 complexes that are poor backbonders, and M–H_2 complexes can be stabilized by strong σ donation with little BD equally as well as strong BD and less σ donation. BD is poor in alkane complexes however; hence there is less potential for "trade-off" in bonding components for CH_4 binding. Thus interaction of C–H bonds with highly electrophilic M could lead to stable alkane complexes purely by σ donation, and the bond energies for $[M(CH_4)_n]^+$ in Table 4.13 support this. The 15–18 kcal/mol energies for $n = 4$ (a reasonable approximation for a "real" fragment) may be sufficient to stabilize a methane complex toward observation near ambient temperature.

Table 4.14. Calculated Distances for $[FeH_2]^{+,0,-}$

System	d_{FeH}, Å	d_{HH}, Å	Binding energy (kcal/mol)
$[FeH_2]^+$	1.92	0.73	−33.8
$[FeH_2]^0$	2.01	0.77	−5.0
$[FeH_2]^-$	2.25	0.86	−42.4

The above complexation energies for alkanes on M^+ are larger than anticipated and suggest that the reason cations are good at activating C–H bonds toward cleavage is that the deep wells of the σ complexes provide the reactants with enough internal energy to overcome the barriers to insertion and elimination.[219] This is an advantage that neutral metal atoms do not possess, as further evidenced by the increase in binding energy with alkane size, which also correlates with increasing reactivity toward H_2 and CH_4 elimination from the alkane.[223] As will be shown in Chapter 12, highly electrophilic cationic centers can readily activate alkanes, even methane.

A comparison of the interaction of H_2 with Fe^0, Fe^+, and Fe^- atoms shows that positive charge on M favors η^2-H_2 binding while negative charge promotes OA to dihydride.[216] This corresponds well with organometallic systems where positive charge favors η^2-H_2 coordination. The H_2 binding energy for the positively charged molecule is much greater than for the neutral species. An energy barrier of 35 kcal/mol for H_2 OA on Fe^0 is calculated, but excitation to a quintuplet $3d^6 4s^1 4p^1$ state leads to OA without a barrier, as is experimentally known. This strong dependence on electronic state may relate to that for $FeH_2(CO)_4$, where H_2 is unexpectedly bound in dihydride form (see Section 4.7.2). Other calculations reiterate that metal cations bind H_2 with rather large binding energies while neutral metal atoms cleave H_2.[217–219] For neutral atoms the binding is caused by transfer of charge to the hydrogen, which limits the number of H atoms that can be bound. However, in the cations, the binding is due to polarization of the H_2 molecule, and a large number of H_2.

4.14. INTERACTION OF DIHYDROGEN WITH METAL SURFACES, METAL OXIDES AND HYDRIDES, AND NONTRANSITION METAL SYSTEMS

While the above ion species have been commonly observed spectroscopically, molecular binding of H_2 to metal surfaces and small metal clusters is rare, both experimentally and calculationally. Chemisorbed H_2 is observed on a stepped Ni(510) surface[224] and calculations for H_2 on a Ni_{13} cluster,[225] triatomic NiH_2,[225] and a Ni(100) surface[226] indicate that such molecularly bound states are possible, as are hydride states. For H_2 on Ni_{13}, d_{HH} is 0.89 Å and ν_{HH} is 2600 cm^{-1}, but no η^2-H_2 state is found on Cu(100) because of differences in $3d$ orbital occupation.[225] However, there is evidence for $Cu_2H_2(H_2)_x$ ($x = 1,2$) and $Cu_3(H_2)$ in an Ar matrix.[227]

Table 4.15. Parameters for H_2 Complexes Computed at the MP2 Level of Perturbation Theory[a]

Parameter	H_3^+	$Li(H_2)^+$	$Be(H_2)^{2+}$	$OBe(H_2)$
R, Å	0.753	1.983	1.553	1.531
d_{HH}, Å	0.870	0.745	0.809	0.772
v_{HH}, cm^{-1}	3481	4392	3642	4014
B.E., kcal/mol		6.3	54.9	15.4

[a] R is the distance between M and the H–H bond midpoint and B.E. is bound energy).

Calculations also predict H_2 binding to several other types of Lewis acidic sites, including nontransition metal cations and ionic solids such as BeO.[228–233] The simplest such species is H_3^+, formed by protonation of H_2 ($d_{HH} = 0.87$ Å). Similar

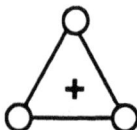

species are formed with M^+ with all outer electrons removed and include $Li(H_2)^+$ and $Be(H_2)^{2+}$ (Table 4.15).[228] $Be(H_2)^{2+}$ is much more stable than the Li complex because Be^{2+} can accommodate two electrons in degenerate $n = 2$ empty orbitals, and the energy of these lowest unoccupied molecular orbitals (LUMOs) lie closer to the energy of the occupied σ_g H_2 orbital. This extends to neutral complexes involving light metal atoms such as $OBe(H_2)$ and $SBe(H_2)$[230,231] or $F_2Mg(H_2)$[232] and its dimer,[233] where the "effective" positive charge on the M atom must be significant, e.g., metals with electronegative substituents such as O or F. Calculations[231] show that *monomeric* BeO is a substantially stronger Lewis acid than $AlCl_3$ (BeO is actually a polymeric solid like alumina). Theoretical and experimental evidence indicates that BH_5 exists as a very weak Lewis acid–base complex H_2-BH_3 with a very low dissociation energy of 1–5 kcal/mol, depending on methodology.[234–236] CDA results show that H_2 (and ethylene in C_2H_4-BH_3) are stronger donors than acceptors.[236] The barriers for hydrogen migration and rotation are very low, and the zero-point vibrational energy is similar to the binding energy so that H_2–BH_3 is barely a bound species. The dissociation energies for X_3B–H_2 (X = F, Cl) are even lower, 0.7–0.9 kcal/mol, indicative of van der Waals complexes.[236] Other weak interactions of H_2 with main-group species are known (Table 4.16), and help to define the Lewis acid–base strength of H_2 as a *pure σ donor or acceptor*. It is significant that complexes in which H_2 can act only as a pure Lewis base are unstable, attesting to the vital role of BD from metal d orbitals in stabilizing σ-ligand binding. Hypervalent main-group species such as CH_5^+ (33), CH_6^{2+} (34), CH_7^{3+}, $SiH_3(H_2)_2^+$, and analogous B and Al series starting with BH_6^+, AlH_4^+, and AlH_6^+ are rationalized theoretically as highly dynamic H_2 complexes of main group cations.

Transition metal oxides are vital heterogeneous catalysts and/or supports in many processes involving H_2, such as hydrotreatment of crude oils. Oxides studied

Table 4.16. Weak Lewis Acid–Base Interactions of H_2 with Main Group Compounds

Compound	Evidence	References
$SiH_3(H_2)_2^+$	IR	244
$Na^+/K^+ \cdot (H_2)_{1,2}$	Surface ionization	206
$Al^+ \cdot (H_2)$	Theory	245
$BX_3(H_2)$ (X = H, F, Cl)	Theoretical, solid Ar	234–236
$BH(H_2)$	Solid Ar	235
Lewis base–H_2	Solid Ar	247
Halide–H_2	Ar matrix	246
CH_5^+, BH_6^+, etc	Theory	2, 248
AlH_4^+, AlH_6^+, etc	Theory	249

theoretically include hematite (Fe_2O_3), modeled as a simple $Fe(\mu\text{-}O)_3Fe$ cluster with H_2 binding to an apical Fe.[237] The binding energy for $(Fe_2O_3)(H_2)$ is calculated by

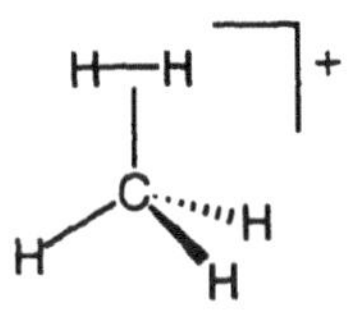

(33) 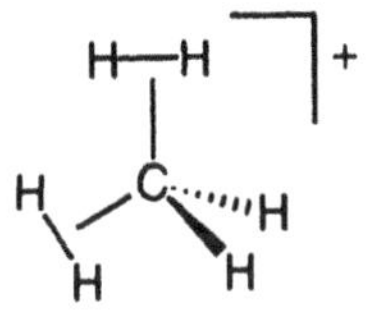   (34)

MINDO/SR methods to be relatively high, 37.6 kcal/mol, with $d_{HH} = 0.80$ Å, but placing a negative charge on the cluster decreases it to 10.1 kcal/mol and d_{HH} to 0.75 Å. This is unlike the situation for Fe atoms above (Table 4.14) because the negative charge on $[(Fe_2O_3)(H_2)]^-$ resides mainly on oxygen, reducing the Lewis acidity of Fe without increasing the BD that activates H_2 toward OA on Fe atoms. DFT studies of the reaction surface of $FeO^+ + H_2$ show $\eta^2\text{-}H_2$ on Fe with $d_{HH} = 0.77\text{–}0.81$ Å, depending on Fe spin state.[238]

Experimental counterparts for the above computations are rare because the surface of a metal oxide does not usually contain exposed unsaturated metal sites. Only very recently have coordinatively unsaturated sites (cus) been identified on an oxide surface: $RuO_2(110)$ can be seen to bind CO to Ru cus by scanning tunneling microscopy.[239] In 1969 Burwell proposed that dehydroxylated chromia (Cr_2O_3) contains cus, and that the Cr^{3+}(cus) and O^{2-}(cus) ion pairs chemisorb H_2 nondissociatively below $-130°C$, most probably at Cr via polarization adsorption at the ion pairs.[239] Pulses of D_2 at $-196°C$ completely and rapidly displace adsorbed

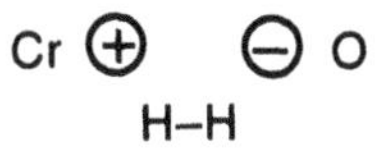

Cr ⊕   ⊖ O
    H–H

H_2 without formation of HD, although above $-163°C$ substantial HD is formed. This is consistent with molecular binding of H_2 to the metal at $-196°C$, with *heterolytic H_2 splitting* taking place on $Cr^{3+}\cdots O^{2-}$ at higher temperatures. A proposed mechanism for scrambling of $H_2 + D_2$ to HD involves a transient containing H^- associated with the Cr^{3+} and HD_2^+ with O^{2-}.

$$\begin{array}{c} D-D \\ \oplus \quad \ominus \\ H-H \end{array} \longrightarrow \oplus H^- \overset{D}{\underset{H}{\ominus}} D^+ \longrightarrow \begin{array}{c} H-D \\ \oplus \quad \ominus \\ H-D \end{array} \quad (4.32)$$

A reverse situation in Eq. (4.32) with HD_2^- associated with Cr^{3+} and H^+ with O^{2-} is also possible. Burwell points out that many other oxides adsorb and activate H_2 at low temperatures, including Co_3O_4, V_2O_3, MnO, and even main-group oxides such as MgO.[239] Various different types of adsorbed hydrogens have been observed on ZnO and alumina even at 25°C, some of which could be molecular. *Ab initio* studies of H_2 interaction and cleavage on a MgO surface use a cuboidal $(MgO)_4$ cluster as a model in which two types of interaction are found: η^1-H_2 on the oxygen site and η^2-H_2 at Mg.[240] Because the calculated d_{HH} (0.73 Å) in both cases is nearly the same as free H_2, the H_2 is most likely physisorbed. These weak complexes lead to a common TS featuring a bridging H_2 unit with $d_{HH} = 0.90$ Å, followed by

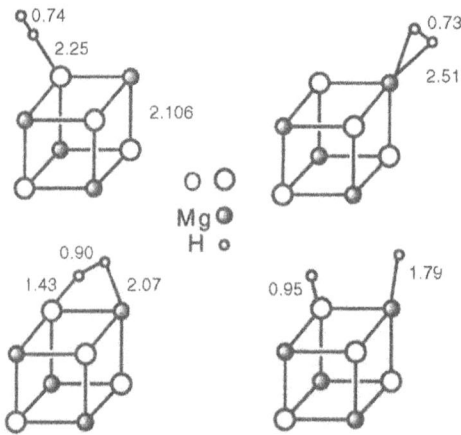

heterolytic cleavage of H_2. The estimated energies relative to the reactants are -2, $+2$, and -21 kcal/mol for the physisorbed complexes, the TS, and the product, respectively.

Calculations show that H_2 binds weakly to a large variety of binary hydrides (MH_n),[241,242] which have only rarely been observed, e.g., matrix-isolated $CrH_2 \cdot (H_2)$ (**35**).[243] The equilibrium structure for the $^5A_1$ ground state of **35** has a C_{2v} structure.[242] The binding energies for $MH_2 \cdot (H_2)$ decrease with increasing atomic number for M = Ti, V, and Cr, and BD is the dominant reason. Comparisons of calculated and experimental[243] vibrational frequencies support the existence of these species in matrices formed by cocondensation of M and H_2. Hydrogen exchange is calculated to occur on these systems via a trefoil-type M—H_3 transition state as in organometallic systems, which for alkali metal systems approximate ion pairs of M^+

and H_3^-.[241] The transition states for the exchange with group 3 transition metals have an energy of 8–10 kcal/mol relative to the reactants, which is lower than that

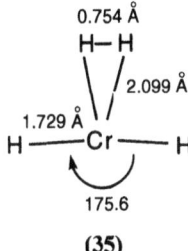

(35)

for the alkali metal systems (16–22 kcal/mol) and group 4 metal hydrides (32–46 kcal/mol).

REFERENCES

1. Brookhart, M.; Green, M. L. H. *J. Organomet. Chem.* **1983**, *250*, 395; Brookhart, M.; Green, M. L. H.; Wong, L.-L. *Prog. Inorg. Chem.* **1988**, *36*, 1.
2. Marx, D.; Parrinello, M. *Nature* **1995**, *375*, 216.
3. Elian, M.; Chen, M. M. L.; Mingos, D. M. P.; Hoffmann, R. *Inorg. Chem.* **1976**, *15*, 1148.
4. Ginsburg, A. G.; Bagaturyants, A. A. *Organometal. Chem. USSR* **1989**, *2*, 111.
5. Tsipis, C. A. *Coord. Chem. Rev.* **1991**, *108*, 163.
6. Lin, Z.; Hall, M. B. *Coord. Chem. Rev.* **1994**, *135/136*, 845.
7. Dedieu, A., ed. *Transition Metal Hydrides*, VCH: New York, **1992**.
8. van Leeuwen, P. W. N. M.; Morokuma, K.; van Lenthe, J. H., Eds., *Theoretical Aspects of Homogeneous Catalysis*, Kluwer Academic Publishers: Boston, 1995; Musaev, D. G.; Morokuma, K. *Adv. Chem. Phys.* **1996**, *95*, 61.
9. Maseras, F.; Lledos, A.; Clot, E.; Eisenstein, O., *Chem. Rev.* **2000**, *100*, 601; Niu, S.; Hall, M. B. *Chem. Rev.* **2000**, *100*, 353; Dedieu, A. *Chem. Rev.* **2000**, *100*, 543; Frenking, G.; Fröhlich, N. *Chem. Rev.* **2000**, *100*, 717; Torrent, M.; Solà, M.; Frenking, G. *Chem. Rev.* **2000**, *100*, 439.
10. Saillard, J.-Y.; Hoffmann, R. *J. Am. Chem. Soc.* **1984**, *106*, 2006.
11. Parr, R. G.; Yang, W. *Density Functional Theory of Atoms and Molecules*, Oxford University Press: Oxford, 1989.
12. Jacobsen, H.; Berke, H. *Chem. Eur. J.* **1997**, *3*, 881.
13. Clot, E.; Eisenstein, O. *J. Phys. Chem. A* **1998**, *102*, 3592.
14. Albinati, A.; Bakhmutov, V. I.; Caulton, K. G.; Clot, E.; Eckert, J.; Eisenstein, O.; Gusev, D. G.; Grushin, V. V.; Hauger, B. E.; Klooster, W. T.; Koetzle, T. F.; McMullan, R. K.; O'Loughlin, T. J.; Pelissier, M.; Ricci, J. S.; Sigalas, M. P.; Vymenits, A. B. *J. Am. Chem. Soc.* **1993**, *115*, 7300.
15. Maseras, F.; Lledos, A.; Costas, M.; Poblet, J. M. *Organometallics* **1996**, *15*, 2947.
16. Eckert, J.; Jensen, C. M.; Koetzle, T. F.; Le-Husebo, T.; Nicol, J.; Wu, P. *J. Am. Chem. Soc.* **1995**, *117*, 7271.
17. Dedieu, A.; Strich, A. *Inorg. Chem.* **1979**, *18*, 2940.
18. Crabtree, R. H.; Quirk, J. M. *J. Organomet. Chem.* **1980**, *199*, 99.
19. Macgregor S. A.; Eisenstein, O.; Whittlesey, M. K.; Perutz, R. N. *J. Chem. Soc., Dalton Trans.* **1998**, 291, and references therein.
20. Ziegler, T.; Tschinke, V.; Fan, L.; Becke, A. D. *J. Am. Chem. Soc.* **1989**, *111*, 9177.
21. Bagatur'yants, A. A.; Gritsenko, O. V.; Zhidomirov, G. M. *Russ. J. Phys. Chem.* **1980**, *54*, 2993.
22. Gritsenko, O. V.; Bagatur'yants, A. A.; Moiseev, I. I.; Kazanskii, V. B.; Kalechits, I. V. *Kinet. Katal.* **1980**, *21*, 632; **1981**, *22*, 354.
23. Bagatur'yants, A. A.; Anikin, N. A.; Zhidomirov, G. M.; Kazanskii, V. B. *Russ. J. Phys. Chem.* **1981**, *55*, 1157.

24. Nakatsuji, H.; Hada, M. *Croat. Chem. Acta* **1984**, *57*, 1371.
25. Blomberg, M. R. A.; Brandemark, U. B.; Petterson, L. G. M.; Siegbahn, P. E. M. *Int. J. Quantum Chem.* **1983**, *23*, 855; Brandemark, U. B.; Blomberg, M. R. A.; Petterson, L. G. M.; Siegbahn, P. E. M. *J. Phys. Chem.* **1984**, *88*, 4617.
26. Jarque, C.; Novaro, O.; Ruiz, M. E.; Garcia-Prieto, J. *J. Am. Chem. Soc.* **1986**, *108*, 3507.
27. Low, J. J.; Goddard, W. A., III *J. Am. Chem. Soc.* **1984**, *106*, 8321; Low, J. J.; Goddard, W. A., III *Organometallics* **1986**, *5*, 609; Nakatsuji, H.; Hada, M.; Yonezawa, T. *J. Am. Chem. Soc.* **1987**, *109*, 1902; Balasubramanian, K. *Chem. Phys.* **1988**, *88*, 6955.
28. Ozin, G. A.; Garcia-Prieto, J. *J. Am. Chem. Soc.* **1986**, *108*, 3099; Andrews, L.; Manceum, L.; Alikhani, M. E.; Wang, X. *J. Am. Chem. Soc.* **2000**, *122*, 11011.
29. Blomberg, M. R. A.; Siegbahn, P. E. M. *J. Chem. Phys.* **1983**, *78*, 986; Ruette, F.; Blyholder, G.; Head, J. *J. Chem. Phys.* **1984**, *80*, 2042.
30. Martensson, A.-S.; Nyberg, C.; Andersson, S. *Phys. Rev. Lett.* **1986**, *57*, 2045.
31. Cui, Q.; Musaev, D. G.; Morokuma, K. *J. Chem. Phys.* **1998**, *108*, 8418; *J. Phys. Chem. A* **1998**, *102*, 6373; Eremenko, I.; German, E. D.; Sheintuch, M. *J. Phys. Chem. A* **2000**, *104*, 8089.
32. Noell, J. O.; Hay, P. J. *J. Am. Chem. Soc.* **1982**, *104*, 4578.
33. Hay, P. J. *Chem. Phys. Lett.* 1984, *103*, 466.
34. Hay, P. J. *J. Am. Chem. Soc.* **1987**, *109*, 705.
35. Jean, Y.; Eisenstein, O.; Volatron, F.; Maouche, B.; Sefta, F. *J. Am. Chem. Soc.* **1986**, *108*, 6587.
36. Butts, M. D.; Kubas, G. J.; Luo, X.-L.; Bryan J. C. *Inorg. Chem.* **1997**, *36*, 3341.
37. Van-Catledge, F. A.; Ittel, S. D.; Jensen, J. P. *Organometallics* **1985**, *4*, 18.
38. Choi, H. W.; Muetterties, E. L. *J. Am. Chem. Soc.* **1982**, *104*, 153.
39. Lyons, D.; Wilkinson, G.; Thornton-Pett, M.; Hursthouse, M. B. *J. Chem. Soc., Dalton Trans*, **1984**, 695.
40. Eckert, J.; Kubas, G. J.; Hall, J. H.; Hay, P. J.; Boyle, C. M. *J. Am. Chem. Soc.* **1990**, *112*, 2324.
41. Li, J.; Ziegler, T. *Organometallics* **1996**, *15*, 3844.
42. Tomas, J.; Lledos, A.; Jean, Y. *Organometallics* **1998**, *17*, 190.
43. Tomas, J.; Lledos, A.; Jean, Y. *Organometallics* **1998**, *17*, 4932.
44. Drew, M. G. B. *Prog. Inorg. Chem.* **1977**, *23*, 67.
45. Heinekey, D. M.; Radzewich, C. E.; Voges, M. H.; Schomber, B. M. *J. Am. Chem. Soc.* **1997**, *119*, 4172.
46. Andrea, R. R.; Vuurman, M. A.; Stufkens, D. J.; Oskam, A. *Recl. Trav. Chim. Pays-Bas* **1986**, *105*, 372.
47. Ishikawa, Y.; Hackett, P. A.; Rayner, D. M. *J. Phys. Chem.* **1989**, *93*, 652.
48. Wasserman. H. J.; Kubas, G. J.; Ryan, R. R. *J. Am. Chem. Soc.* **1986**, *108*, 2294.
49. Tatsumi, T.; Tominaga, H.; Hidai, M.; Uchida, Y. *J. Organomet. Chem.* **1980**, *199*, 63.
50. Kubas, G. J.; Burns, C. J.; Eckert, J.; Johnson, S.; Larson, A. C.; Vergamini, P. J.; Unkefer, C. J.; Khalsa, G. R. K.; Jackson, S. A.; Eisenstein, O., *J. Am. Chem. Soc.* **1993**, *115*, 569.
51. Kubas, G. J.; Ryan, R. R.; Unkefer, C. J. *J. Am. Chem. Soc.* **1987**, *109*, 8113.
52. Schenk, W. A.; Baumann, F.-E. *Chem. Ber.* **1982**, *115*, 2615; Graham, M. A.; Poliakoff, M.; Turner, J. J. *J. Chem. Soc. A* **1979**, 2939; Brown, R. A.; Dobson, G. R. *Inorg. Chem. Acta* **1972**, *6*, 65; Stolz, I. W.; Dobson, G. R.; Sheline, R. K. *Inorg. Chem.* **1963**, *2*, 1264.
53. Perutz, R. N.; Turner, J. J. *Inorg. Chem.* **1979**, *18*, 2940. (d).
54. Morris, R. H.; Schlaf, M. *Inorg. Chem.* **1994**, *33*, 1725.
55. Dapprich, S.; Frenking, G. *Angew. Chem. Int. Ed. Engl.* **1995**, *34*, 354.
56. Dapprich, S.; Frenking, G. *Z. anorg. allg. Chem.* **1998**, *624*, 583.
57. Dapprich, S.; Frenking, G. *Organometallics* **1996**, *15*, 4547; Frenking, G.; Pidum, U. *J. Chem. Soc., Dalton Trans.* **1997**, 1653.
58. Bickelhaupt, F. M.; Baerends, E. J.; Ravenek, W. *Inorg. Chem.* **1990**, *29*, 350.
59. Li, J.; Dickson, R. M.; Ziegler, T. *J. Am. Chem. Soc.* **1995**, *117*, 11482, and references therein.
60. Maseras, F.; Li, X.-K.; Koga, N.; Morokuma, K. *J. Am. Chem. Soc.* **1993**, *115*, 10974.
61. Toupadakis, A.; Kubas, G. J.; King, W. A.; Scott, B. L.; Huhmann-Vincent, J. *Organometallics* **1998**, *17*, 5315.
62. Radius, U.; Bickelhaupt, F. M.; Ehlers, A. W.; Goldberg, N.; Hoffmann, R. *Inorg. Chem.* **1998**, *37*, 1080.
63. Ehlers, A. W.; Ruiz-Morales, Y.; Baerends, E. J.; Ziegler, T. *Inorg. Chem.* **1997**, *36*, 5031; Lupinetti, A. J.; Fau, S.; Frenking, G.; Strauss, S. H. *J. Phys. Chem.* **1997**, *101*, 9551.
64. Szilagyi, R. K.; Frenking, G. *Organometallics* **1997**, *16*, 4807; Diefenbach, A.; Bickelhaupt, F. M.; Frenking, G. *J. Am. Chem. Soc.* **2000**, *122*, 6449.

65. Pyykko, ; Declaux, J.-P. *Acc. Chem. Res.* **1979**, *12*, 276.
66. Schwarz, W. H. E.; van Wezenbeek, E. M.; Baerends, E. J.; Snijders, J. G. *J. Phys. B* **1989**, *22*, 1515.
67. Li, J.; Schreckenbach, G.; Ziegler, T. *J. Phys. Chem.* **1994**, *98*, 4838; *Inorg. Chem.* **1995**, *34*, 3245.
68. Ziegler, T.; Snijders, J. G.; Baerends, E. J. *J. Chem. Phys. Lett.* **1980**, *75*, 1; Ziegler, T.; Snijders, J. G.; Baerends, E. J.; Ros, P. *J. Chem. Phys.* **1981**, *74*, 1271.
69. Ishida, T.; Mizobe, Y,; Tanase, T.; Hidai, M. *J. Organomet. Chem.* **1991**, *409*, 355.
70. Coe, B. J.; Glenwright, S. J. *Coord. Chem. Rev.* **2000**, *203*, 5.
71. Schlaf, M.; Lough, A. J.; Maltby, P. A.; Morris, R. H. *Organometallics* **1996**, *15*, 2270, and references therein; King, W. A.; Scott, B. L.; Eckert, J.; Kubas, G. J. *Inorg. Chem.* **1999**, *38*, 1069.
72. Cooper, A. C.; Eisenstein, O.; Caulton, K. G. *New. J. Chem.* **1998**, *22*, 307.
73. Heyn, R. H.; Macgregor, S. A.; Nadasdi, T. T.; Ogasawara, M.; Esenstein, O.; Caulton, K. G. *Inorg. Chim. Acta* **1997**, *259*, 5.
74. Zilm, K. W.; Millar, J. M. *Adv. Mag. Opt. Reson.* **1990**, *15*, 163; Kubas, G. J.; Nelson, J. E.; Bryan, J. C.; Eckert, J.; Wisniewski, L.; Zilm, K. *Inorg. Chem.* **1994**, *33*, 2954; Kubas, G. J.; Unkefer, C. J; Swanson, B. I.; Fukushima, E. *J. Am. Chem. Soc.* **1986**, *108*, 7000.
75. Luo, X.-L.; Kubas, G. J.; Burns, C. J.; Eckert, J. *Inorg. Chem.* **1994**, *33*, 5219.
76. King, W. A.; Luo, X-L.; Scott, B. L.; Kubas, G. J.; Zilm, K. W. *J. Am. Chem. Soc.* **1996**, *118*, 6782.
77. Huhmann-Vincent, J.; Scott, B. L.; Kubas, G. J. *J. Am. Chem. Soc.* **1998**, *120*, 6808.
78. Luo, X.-L.; Michos, D.; Crabtree, R. H. *Organometallics* **1992**, *11*, 237.
79. Albertin, G.; Antoniutti, S.; Garcia-Fontan, S.; Carballo, R.; Padoan, F. *J. Chem. Soc., Dalton Trans.* **1998**, 2071.
80. Forde, C. E.; Landau, S. E.; Morris, R. H. *J. Chem. Soc., Dalton Trans.* **1997**, 1663.
81. Rocchini, E.; Mezzetti, A.; Rugger, H.; Burckhardt, U.; Gramlich, V.; Zotto, A. D.; Martinuzzi, P.; Rigo, P. *Inorg. Chem.* **1997**, *36*, 711.
82. Burrell, A. K.; Bryan, J. C.; Kubas, G. J. *J. Am. Chem. Soc.* **1994**, *116*, 1575; Kohli, M.; Lewis, D. J.; Luck, R. L.; Silverton, J. V.; Sylla, K. *Inorg. Chem.* **1994**, *33*, 879.
83. Cotton, F. A.; Luck, R. L. *Inorg. Chem.* **1991**, *30*, 767.
84. Gusev, D. G.; Nietlispach, D.; Eremenko, I. L.; Berke, H. *Inorg. Chem.* **1993**, *32*, 3628.
85. Mezzetti, A.; Del Zotto, A.; Rigo, P.; Farnetti, E. *J. Chem. Soc., Dalton Trans.* **1991**, 1525.
86. Chin, B.; Lough, A. J.; Morris, R. H.; Schweitzer, C.; D'Agostino, C. *Inorg. Chem.* **1994**, *33*, 6278.
87. Cappellani, E. P.; Maltby, P. A.; Morris, R. H.; Schweitzer, C. T.; Steele, M. R. *Inorg. Chem.* **1989**, *28*, 4437.
88. Maltby, P. A.; Schlaf, M.; Steinbeck, M.; Lough, A. J.; Morris, R. H.; Klooster, W. T.; Koetzle, T. F.; Srivastava, R. C. *J. Am. Chem. Soc.* **1996**, *118*, 5396.
89. Aebischer, N. Frey, U.; Merbach, A. E. *Chem. Comm.* **1998**, 2303.
90. Bartucz, T. Y.; Golombeck, A.; Lough, A. J.; Maltby, P. A.; Morris, R. H.; Ramachandran, R.; Schlaf, M. *Inorg. Chem.* **1998**, *37*, 1555.
91. Harman, W. D.; Taube, H. *J. Am. Chem. Soc.* **1990**, *112*, 2261; Li, Z.-W.; Taube, H. *J. Am. Chem. Soc.* **1991**, *113*, 8946.
92. Jessop, P. G.; Morris, R. H. *Coord. Chem. Rev.* **1992**, *121*, 155.
93. Esteruelas, M. A.; Sola, E.; Oro, L. A.; Meyer, U.; Werner, H. *Angew. Chem. Int. Ed. Engl.* **1988**, *27*, 1563; Gusev, D. G.; Vymenits, A. B.; Bakhmutov, V. I. *Inorg. Chem.* **1992**, *31*, 1.
94. Luther, T. A.; Heinekey, D. M. *Inorg. Chem.* **1998**, *37*, 127.
95. Bytheway, I.; Bacskay, G. B.; Hush, N. S. *J. Phys. Chem.* **1996**, *110*, 6023.
96. Bytheway, I.; Craw, I.; Bacskay, G. B.; Hush, N. S. *Adv. Chem. Ser.* **1997**, *253*, 21.
97. Craw, J. S.; Bacskay, G. B.; Hush, N. S. *J. Am. Chem. Soc.* **1994**, *116*, 5937.
98. Seino, H.; Arita, C.; Nonokawa, D.; Nakamura, G.; Harada, Y.; Mizobe, Y.; Hidai, M. *Organometallics* **1999**, *18*, 4165.
99. Pang T. *Chem. Phys. Lett.* **1994**, *228*, 555; Farizon, M.; Farizon-Mazuy, B.; de Castro Faria, N. V.; Chermette, H. *Chem. Phys. Lett.* **1991**, *177*, 451.
100. Bautista, M. T.; Cappellani, E. P.; Drouin, S. D.; Morris, R. H.; Schweitzer, C. T.; Sella, A.; Zubkowski, J. *J. Am. Chem. Soc.* **1991**, *113*, 4876.
101. Cappellani, E. P.; Drouin, S. D.; Jia, G.; Maltby, P. A.; Morris, R. H.; Schweitzer, C. T. *J. Am. Chem. Soc.* **1994**, *116*, 3375.
102. Maseras, F.; Duran, M.; Lledos, A.; Bertran, J. *J. Am. Chem. Soc.* **1991**, *113*, 2879.

103. Bullock, R. M.; Song, J.-S.; Szalda, D. J. *Organometallics* **1996**, *15*, 2504.
104. Drouin, B. J.; Kukolich, S. G. *J. Am. Chem. Soc.* **1998**, *120*, 6774; Lavaty, T. G.; Wikrent, P.; Drouin, B. J.; Kukolich, S. G. *J. Chem. Phys.* **1998**, *109*, 9473.
105. Wang, W.; Weitz, E. *J. Phys. Chem. A* **1997**, *101*, 2358.
106. Wang, W.; Narducci, A. A.; House, P. G.; Weitz, E. *J. Am. Chem. Soc.* **1996**, *118*, 8654.
107. Kukolich, S. G.; Sicafoose, S. M.; Breckenridge, S. M. *J. Am. Chem. Soc.* **1996**, *118*, 205.
108. Bader, R. F. W. *Atoms in Molecules: A Quantum Theory*, Clarendon: New York, 1990; Bader, R. F. W. *Acc. Chem. Res.* **1985**, *18*, 9.
109. Ziegler, T. *Chem. Rev.* **1991**, *91*, 651.
110. Moller, C.; Plesset, M. S. *Phys. Rev.* **1934**, *46*, 618.
111. Hush, N. S. *J. Am. Chem. Soc.* **1997**, *119*, 1717.
112. Grundemann, S.; Limbach, H.-H.; Buntkowsky, G.; Sabo-Etienne, S.; Chaudret, B.; *J. Phys. Chem. A* **1999**, *103*, 4752.
113. Ginsberg, A. P. *Adv. Chem. Ser.* **1978**, *167*, 201.
114. Jackson, S. A.; Eisenstein, O. *J. Am. Chem. Soc.* **1990**, *112*, 7203.
115. Gelabert, R,; Moreno, M.; Lluch, J. M.; Lledos, A. *J. Am. Chem. Soc.* **1997**, *119*, 9840; **1998**, *120*, 8168; Torres, L.; Gelabert, R.; Moreno, M.; Lluch, J. M. *J. Phys. Chem. A* **2000**, *104*, 7898.
116. Barea, G.; Esteruelas, M. A.; Lledos, A.; Lopez, A. M.; Onate, E.; Tolosa, J. I. *Organometallics* **1998**, *17*, 4065; Barea, G.; Esteruelas, M. A.; Lledos, A.; Lopez, A. M.; Tolosa, J. I. *Inorg. Chem.* **1998**, *37*, 5033; Craw, J. S.; Backsay, G. B.; Hush, N. S. *Inorg. Chem.* **1993**, *32*, 2230.
117. Gusev, D.; Llamazares, A.; Artus, G.; Jacobsen, H.; Berke, H. *Organometallics* **1999**, *18*, 75.
118. Klooster, W. T.; Koetzle, T. F.; Jia, G.; Fong, T. P.; Morris, R. H.; Albinati, A. *J. Am. Chem. Soc.* **1994**, *116*, 7677.
119. Maseras, F.; Koga, N.; Morokuma, K. *Organometallics* **1994**, 13, 4008.
120. Kranenburg, M.; Kramer, P. C. J.; van Leeuwen, P. W. N. M.; Chaudret, B. *Chem. Comm.* **1997**, 373.
121. Volatron, F.; Jean, Y.; Lledos, A. *New J. Chem.* **1987**, *11*, 651; Jean, Y.; Lledos, A.; Maouche, B.; Aiad, R. *J. Chim. Phys. Phys.-Chim. Biol.* **1987**, *84*, 805.
122. Burdett, J. K.; Phillips, J. R.; Pourian, M. R.; Poliakoff, M.; Turner, J. J.; Upmacis, R. *Inorg. Chem.* **1987**, *26*, 3054.
123. Upmacis, R. K.; Gadd, G. E.; Poliakoff, M.; Simpson, M. B.; Turner, J. J.; Whyman, R.; Simpson, A. F. *J. Chem. Soc., Chem. Commun.* **1985**, *27*; Church, S. P.; Grevels, F.-W.; Hermann, H.; Shaffner, K. *J. Chem. Soc., Chem. Commun.* **1985**, 30.
124. Gadd, G. E.; Upmacis, R. K.; Poliakoff, M.; Turner, J. J. *J. Am. Chem. Soc.* **1986**, *108*, 2547.
125. Sweany, R. L.; Moroz, A. *J. Am. Chem. Soc.* **1989**, *111*, 3577.
126. Banister, J. A.; Lee, P. D.; Poliakoff, M. *Organometallics* **1995**, *14*, 3876.
127. Morris, R. H.; Earl, K. A.; Luck, R. L.; Lazarowych, N. J.; Sella, A. *Inorg. Chem.* **1987**, *26*, 2674.
128. Reid, S. M.; Neuner, B.; Schrock, R. R.; Davis, W. M. *Organometallics* **1998**, *17*, 4077.
129. Lever, A. B. P. *Inorg. Chem.* **1990**, *29*, 1271; **1991**, *30*, 1980.
130. de la Jara Leal, A.; Jimenez Tenorio, M.; Puerta, M. C.; Valerga, P. *Organometallics* **1995**, *14*, 3839.
131. Saburi, M.; Aoyagi, K.; Kodama, T.; Takahashi, T.; Uchida, Y.; Kozawa, K.; Uchida, T. *Chem. Lett.*, **1990**, 1909.
132. Earl, K. A.; Jia, G.; Maltby, P. A.; Morris, R. H. *J. Am. Chem. Soc.* **1991**, *113*, 3027.
133. Bayse, C. A.; Hall, M. B. *J. Am. Chem. Soc.* **1999**, *121*, 1348.
134. Haynes, G. R.; Martin, R. L.; Hay, P. J. *J. Am. Chem. Soc.* **1992**, *114*, 28.
135. Lin, Z.; Hall, M. B. *J. Am. Chem. Soc.* **1992**, *114*, 2928.
136. Seidle, A. R.; Newmark, R. A.; Pignolet, L. H. *Inorg. Chem.* **1986**, *25*, 3412.
137. Douglas, P. G.; Shaw, B. L. *J. Chem. Soc. A* **1970**, 3334.
138. Bruno, J. W.; Huffman, J. C.; Caulton, K. G. *J. Am. Chem. Soc.* **1984**, *106*, 1663.
139. Hamilton, D. G.; Crabtree, R. H. *J. Am. Chem. Soc.* **1988**, *110*, 4126.
140. Johnson, T. J.; Huffman, J. C.; Caulton, K. G.; Jackson, S. A.; Eisenstein, O. *Organometallics* **1989**, *8*, 2073.
141. Desrosiers, P. J.; Cai, L.; Lin, Z.; Richards, R.; Halpern, J. *J. Am. Chem. Soc.* **1991**, *113*, 4173.
142. Lin, Z.; Hall, M. B. *J. Am. Chem. Soc.* **1992**, *114*, 6102.
143. Maseras, F.; Koga, N.; Morokuma, K. *J. Am. Chem. Soc.* **1993**, *115*, 8313.
144. Johnson, T. J.; Albinati, A.; Koetzle, T. F.; Ricci, J.; Eisenstein, O.; Huffman, J. C.; Caulton, K. G. *Inorg. Chem.* **1994**, *33*, 4966.

145. Camanyes, S.; Maseras, F.; Moreno, M.; Lledos, A.; Lluch, J. M.; Bertran J. *Angew. Chem. Int. Ed. Engl.* **1997**, *36*, 265; *Chem. Eur. J.* **1999**, *5*, 1166.
146. Hartwig, J. F.; De Gala, S. R. *J. Am. Chem. Soc.* **1994**, *116*, 3661.
147. Jia, G.; Lau, C. P. *J. Organomet. Chem.* **1998**, *565*, 37.
148. Conroy-Lewis, F. M.; Simpson, S. J. *J. Chem. Soc., Chem. Commun.* **1987**, 1675.
149. Chinn, M. S.; Heinekey, D. M. *J. Am. Chem. Soc.* **1990**, *112*, 5166.
150. Howdle, S. M.; Poliakoff, M. *J. Chem. Soc., Chem. Commun.* **1989**, 1099.
151. Haward, M. T.; George, M. W.; Hamley, P.; Poliakoff, M. *J. Chem. Soc., Chem. Commun.* **1991**, 1101.
152. Gross, C. L.; Young, D. M.; Schultz, A. J.; Girolami, G. S. *J. Chem. Soc., Dalton Trans.* **1997**, 3081.
153. Bayse, C. A.; Hall, M. B.; Pleune, B.; Poli, R. *Organometallics* **1998**, *17*, 4309.
154. Margl, P. M.; Woo, T. K.; Blochl, P. E.; Ziegler, T. *J. Am. Chem. Soc.* **1998**, *120*, 2174.
155. Car, R.; Parrinello, M. *Phys. Rev. Lett.* **1985**, *55*, 2471.
156. Bohanna, C.; Esteruelas, M. A.; Gomez, A. V.; Lopez, A. M.; Martinez, M.-P. *Organometallics* **1997**, *16*, 4464; Ng, W. S.; Jia, G.; Hung, M. Y.; Lau, C. P.; Wong, K. Y.; Wen, L.; *Organometallics* **1998**, *17*, 4556.
157. Gelabert, R.; Moreno, M.; Lluch, J. M.; Lledos, A. *Organometallics* **1997**, *16*, 3805.
158. Suzuki, H.; Lee, D. H.; Oshima, N.; Moro-oka, Y. *Organometallics* **1987**, *6*, 1569; Arliguie, T.; Border, C.; Chaudret, B.; Devillers, J.; Poilblanc, R. *Organometallics* **1989**, *8*, 1308; Heinekey, D. M.; Payne, N. G.; Sofield, C. D. *Organometallics* **1990**, *9*, 2643; Arliguie, T.; Chaudret, B.; Jalon, F. A.; Otero, A.; Lopez, J. A.; Lahoz, F. J. *Organometallics* **1991**, *10*, 1888.
159. Halcrow, M. A.; Chaudret, B.; Trofimenko, S. *J. Chem. Soc., Chem. Commun.* **1993**, 465.
160. Eckert, J.; Albinati, A.; Bucher, U. E.; Venanzi, L. M. *Inorg. Chem.* **1996**, *35*, 1292.
161. Lunder, D. M.; Green, M. A.; Streib, W. E.; Caulton, K. G. *Inorg. Chem.* **1989**, *28*, 4527.
162. Luo, X.-L.; Crabtree, R. H. *J. Am. Chem. Soc.* **1990**, *112*, 6912.
163. Eckert, J.; Jensen, C. M.; Jones, G.; Clot, E.; Eisenstein, O. *J. Am. Chem. Soc.* **1993**, *115*, 11056; Gusev, D. G.; Kuhlman, R. L.; Renkema, K. H.; Eisenstein, O.; Caulton, K. G. *Inorg. Chem.* **1996**, *35*, 6775; Grushin, V. V. *Acc. Chem. Res.* **1993**, *26*, 279; Christ, M. L.; Sabo-Etienne, S.; Chaudret, B. *Organometallics* **1994**, *13*, 3800; Gusev, D. G.; Kuznetsov, V. F.; Eremenko, I. L.; Berke, H. *J. Am. Chem. Soc.* **1993**, *115*, 5831; Kuhlman, R. *Coord. Chem. Rev.* **1997**, *167*, 205; Gusev, D. G.; Kuhlman, R.; Sini, G.; Eisenstein, O.; Caulton, K. G., in press.
164. Kuhlman, R. L.; Gusev, D. G.; Eremenko, I. L.; Berke, H.; Huffman, J. C.; Caulton, K. G. *J. Organometal. Chem.* **1997**, *536/537*, 139.
165. Chaudret, B.; Chung, G.; Eisenstein, O.; Jackson, S. A.; Lahoz, F. J.; Lopez, J. A. *J. Am. Chem. Soc.* **1991**, *113*, 2314.
166. Le-Husebo, T.; Jensen, C. M. *Inorg. Chem.* **1993**, *32*, 3797.
167. Hauger, B. E.; Gusev, D. G.; Caulton, K. G. *J. Am. Chem. Soc.* **1994**, *116*, 208.
168. Poulton, J. T.; Folting, K.; Streib, W. E.; Caulton, K. G. *Inorg. Chem.* **1992**, *31*, 3190; Caulton, K. G. *New J. Chem.* **1994**, *18*, 25.
169. Brothers, P. J. *Prog. Inorg. Chem.* **1981**, *28*, 1.
170. Van Der Sluys, L. S.; Eckert, J.; Eisenstein, O.; Hall, J. H.; Huffman, J. C.; Jackson, S. A.; Koetzle, T. F.; Kubas, G. J.; Vergamini, P. J.; Caulton K. G. *J. Am. Chem. Soc.* **1990**, *112*, 4831.
171. Jackson, S. A.; Eisenstein, O. *J. Am. Chem. Soc.* **1990**, *112*, 7203; Riehl, J.-F.; Pelissier, M.; Eisenstein, O. *Inorg. Chem.* **1992**, *31*, 3344.
172. Chaudret, B.; Chung, G.; Eisenstein, O.; Jackson, S. A.; Lahoz, F. J.; Lopez, J. A. *J. Am. Chem. Soc.* **1991**, *113*, 2314.
173. Maseras, F.; Duran, M.; Lledos, A.; Bertran, J. *J. Am. Chem. Soc.* **1991**, *113*, 2879.
174. Bianchini, C.; Masi, D.; Peruzzini, M.; Casarin, M.; Maccato, C.; Rizzi, G. A. *Inorg. Chem.* **1997**, *36*, 1061.
175. Rodriguez, V.; Sabo-Etienne, S.; Chaudret, B.; Thoburn, J.; Ulrich, S.; Limbach, H.-H. Eckert; J.; Barthelat, J.-C.; Hussein, K.; Marsden, C. J. *Inorg. Chem.* **1998**, *37*, 3475.
176. Soubra, C.; Chan, F.; Albright, T. A. *Inorg. Chim. Acta* **1998**, *272*, 95.
177. Borowski, A. F.; Donnadieu, B.; Daran, J.-C.; Sabo-Etienne, S.; Chaudret, B. *Chem. Comm.* **2000**, 543.
178. Stevens, R. C.; Bau, R.; Milstein, D.; Blum, O.; Koetzle, T. F. *J. Chem. Soc., Dalton Trans.* **1990**, 1429.
179. Gusev, D. G.; Hubener, R.; Burger, P.; Orama, O.; Berke, H. *J. Am. Chem. Soc.* **1997**, *119*, 3716.
180. Hussein, K.; Marsden, C. J.; Barthelat, J.-C.; Rodriguez, V.; Conjero, S.; Sabo-Etienne, S.; Donnadieu, B.; Chaudret, B. *Chem. Comm.* **1999**, 1315.

181. Girolami, G.; Schultz, A., private communication.
182. Brintzinger, H. H. *J. Organomet. Chem.* **1979**, *171*, 337.
183. Burdett, J. K.; Pourian, M. R. *Organometallics* **1987**, *6*, 1684; Burdett, J. K.; Pourian, M. R. *Inorg. Chem.* **1988**, *27*, 4445.
184. Heinekey, D. M.; Millar, J. M.; Koetzle, T. F.; Payne, N. G.; Zilm, K. W. *J. Am. Chem. Soc.* **1990**, *112*, 909.
185. Luo, X.-L.; Crabtree, R. H. *J. Am. Chem. Soc.* **1990**, *112*, 6912.
186. Maseras, F.; Duran, M.; Lledos, A.; Bertran, J. *J. Am. Chem. Soc.* **1992**, *114*, 2922.
187. Diedrich, D. L.; Anderson, J. B. *Science* **1992**, *258*, 786.
188. Thompson, M. E.; Baxter, S. M.; Bulls, A. R.; Burger, B. J.; Nolan, M. C.; Santarsiero, B. D.; Schaefer, W. P.; Bercaw, J. E. *J. Am. Chem. Soc.* **1987**, *109*, 203.
189. Watson, P. L.; Parshall, G. W. *Acc. Chem. Res.* **1985**, *18*, 51.
190. Jeske, G.; Lauke, H.; Mauermann, H.; Schumann, H.; Marks, T. J. *J. Am. Chem. Soc.* **1985**, *107*, 8111.
191. Christ, C. S., Jr.; Eyler, J. R.; Richardson, D. E. *J. Am. Chem. Soc.* **1988**, *110*, 4038; **1990**, *112*, 596.
192. Davis, J. A.; Watson, P. L.; Liebman, J. F.; Greenberg, A., eds. *Selective Hydrocarbon Activation*, VCH: New York, **1990**.
193. Rabaa, H.; Saillard, J.-Y.; Hoffmann, R. *J. Am. Chem. Soc.* **1986**, *108*, 4327.
194. Steigerwald, M. L.; Goddard, W. A., III. *J. Am. Chem. Soc.* **1984**, *106*, 308; Rappe, A. K. *Organometallics* **1990**, *9*, 466.
195. Dedieu, A.; Hutschka, F.; Milet, A. *ACS Symp. Ser.* **1999**, *721* (*Transition State Modeling for Catalysis*), 100–113; Milet, A.; Dedieu, A.; Kapteijn, G.; Van Koten, G. *Inorg. Chem.* **1997**, *36*, 3223.
196. Folga, E.; Ziegler, T. *Can. J. Chem.* **1992**, *70*, 333.
197. Ziegler, T.; Folga, E.; Berces, A. *J. Am. Chem. Soc.* **1993**, *115*, 636; see also Chapter 5 by Ziegler in reference 8.
198. Versluis, L.; Ziegler, T. *Organometallics* **1990**, *9*, 2985; Sola, M; Ziegler, T. *Organometallics* **1996**, *15*, 2611.
199. Dedieu, A.; Hutschka, F.; Milet, A. *ACS Symp. Ser.* **1999**, *721*(*Transition State Modeling for Catalysis*), 100–113; Musashi, Y.; Sakaki, S. *J. Am. Chem. Soc.* **2000**, *122*, 3867.
200. Rappe, A. K. *Organometallics* **1987**, *6*, 354.
201. Weisshaar, J. C. *Acc. Chem. Res.* **1993**, *26*, 213; Armentrout, P. B. *ibid.* **1995**, *28*, 430.
202. Kemper, P. R.; Bushnell, J. E.; von Helden; Bowers, M. T. *J. Phys. Chem.* **1993**, *97*, 52, 9821.
203. Kemper, P. R.; Bushnell, J. E.; Bowers, M. T. *J. Phys. Chem.* **1995**, *99*, 15602.
204. Bushnell, J. E.; Kemper, P. R.; Bowers, M. T. *J. Phys. Chem.* **1993**, *97*, 11628; Bushnell, J. E.; Kemper, P. R.; van Koppen, P.; Bowers, M. T. *J. Phys. Chem.* **1993**, *97*, 1810.
205. Bushnell, J. E.; Kemper, P. R.; Maitre, P.; Bowers, M. T. *J. Am. Chem. Soc.* **1994**, *116*, 9710.
206. Bushnell, J. E.; Kemper, P. R.; Bowers, M. T. *J. Phys. Chem.* **1994**, *98*, 2044.
207. Bushnell, J. E.; Maitre, P.; Kemper, P. R.; Bowers, M. T. *J. Chem. Phys.* **1997**, *106*, 10153.
208. Weis, P.; Kemper, P. R.; Bowers, M. T. *J. Phys. Chem. A* **1997**, *101*, 2809; Kemper, P. R.; Weis, P.; Bowers, M. T. *Chem. Phys. Lett.* **1998**, *293*, 503.
209. Kemper, P. R.; Weis, P.; Bowers, M. T.; Maitre, P. *J. Am. Chem. Soc.* **1998**, *120*, 13494.
210. Haynes, C. L.; Armentrout, P. B. *Chem. Phys. Lett.* **1996**, *249*, 64; Tjelta, B. L.; Armentrout, P. B. *J. Phys. Chem. A* **1997**, *101*, 2064; Sievers, M. R.; Jarvis, L. M.; Armentrout, P. B. *J. Am. Chem. Soc.* **1998**, *120*, 1891.
211. Alvarado-Swaisgood, A.; Harrison, J. F. *J. Phys. Chem.* **1985**, *89*, 5198; Rappe, A. K.; Upton, T. H. *J. Chem. Phys.* **1986**, *85*, 4400.
212. Mavridis, A.; Harrison, J. R. *J. Chem. Soc., Faraday Trans. 2* **1989**, *85*, 1391.
213. Niu, J.; Rao, B. K.; Khanna, S. N.; Jena, P. *Chem. Phys. Lett.* **1994**, *230*, 299; Maitre, P.; Bauschlicher, C. W., Jr. *J. Phys. Chem.* **1995**, *99*, 6836.
214. Maitre, P.; Bauschlicher, C. W., Jr. *J. Phys. Chem.* **1993**, *97*, 11912.
215. Rivera, M.; Harrison, J. F.; Alvarado-Swaisgood, A. *J. Phys. Chem.* **1990**, *94*, 6969.
216. Sanchez, M.; Ruette, F.; Hernandez, A. J. *J. Phys. Chem.* **1992**, *96*, 823.
217. Niu, J.; Rao, B. K.; Jena, P.; Manninen, M. *Phys. Rev. B* **1995**, *51*, 4475; Niu, J.; Rao, B. K.; Jena, P. *Phys. Rev. Lett.* **1992**, *68*, 2277.
218. Bauschlicher, C. W., Jr.; Maitre, P. *J. Phys. Chem.* **1995**, *99*, 5238.
219. Perry, J. K.; Ohanessian, G.; Goddard, W. A. III. *J. Phys. Chem.* **1993**, *97*, 5238.

220. Haynes, C. L.; Armentrout, P. B.; Perry, J. K.; Goddard, W. A. III. *J. Phys. Chem.* **1995**, *99*, 6340; Haynes, C. L.; Chen, Y.-M.; Armentrout, P. B. *J. Phys. Chem.* **1995**, *99*, 9110.
221. Schultz, R. H.; Haynes, C. L. *J. Phys. Chem.* **1993**, *97*, 596.
222. von Koppen, P. A. M.; Kemper, P. R.; Bushnell, J. E.; Bowers, M. T. *J. Am. Chem. Soc.* **1995**, *117*, 2098.
223. Tonkyn, R.; Ronan, M.; Weisshaar, J. C. *J. Phys. Chem.* **1988**, *92*, 92; Armentrout, P. B.; Beauchamp, J. L. *J. Am. Chem. Soc.* **1991**, *103*, 784; Jacobson, D. B.; Freiser, B. S. *J. Am. Chem. Soc.* **1983**, *105*, 5197.
224. Martensson, A.-S.; Nyberg, C.; Andersson, S. *Phys. Rev. Lett.* **1986**, *57*, 2045; Martensson, A.-S.; Nyberg, C.; Andersson, S. *Surf. Sci.* **1988**, *205*, 12.
225. Siegbahn, P.; Blomberg, M.; Panas, I.; Wahlgren, U. *Theor. Chim. Acta* **1989**, *75*, 143.
226. Li, J.; Schiott, B.; Hoffmann, R.; Proserpio, D. M. *J. Phys. Chem.* **1990**, *94*, 1554.
227. Hauge, R. H.; Margrave, J. L.; Kafafi, Z. H. *NATO ASI Ser., Ser. B* **1987**, *158* (*Phys. Chem. Small Clusters*), 787.
228. Nicolaides C. A.; Simandiras, E. D. *Comm. Inorg. Chem.* **1996**, *18*, 65, and references therein.
229. Blickensderfer, R. P.; Jordan, K. D.; Adams, N.; Breckenridge, W. H. *J. Phys. Chem.* **1982**, *86*, 1930; Curtiss, L. A.; Pople, J. A. *J. Phys. Chem.* **1988**, *92*, 894.
230. Nicolaides, C. A.; Valtazanos, P. *Chem. Phys. Lett.* **1990**, *174*, 489; Nicolaides, C. A.; Valtazanos, P. *Chem. Phys. Lett.* **1991**, *176*, 239; Valtazanos, P.; Nicolaides, C. A. *J. Chem. Phys.* **1993**, *98*, 549.
231. Frenking, G.; Dapprich, S.; Kohler, K. F.; Koch, W.; Collins, J. R. *Mol. Phys.* **1996**, *89*, 1245.
232. Nicolaides, C. A.; Simandiras, E. D. *Chem. Phys. Lett.* **1992**, *196*, 213.
233. Simandiras, E. D.; Nicolaides, C. A. *Chem. Phys. Lett.* **1994**, *223*, 233.
234. Watts, J. D.; Bartlett, R. J. *J. Am. Chem. Soc.* **1995**, *117*, 825.
235. Tague, T. J., Jr.; Andrews, L. *J. Am. Chem. Soc.* **1994**, *116*, 4970.
236. Fau, S.; Frenking, G. *Mol. Phys.* **1999**, *96*, 519.
237. Rodriguez, L. J.; Ruette, F.; Rosa-Brussin, M. *J. Mol. Catal.* **1990**, *62*, 199.
238. Fiedler, A.; Schroder, D.; Shaik, S.; Schwarz, H. *J. Am. Chem. Soc.* **1994**, *116*, 10734.
239. Over, H., Kim, Y. D.; Seitsonen, A. P.; Wendt, S.; Lundgren, E.; Schmid, M.; Varga, P.; Morgante, A.; Ertl, G. *Science* **2000**, *287*, 1474.
239. Burwell, R. L., Jr.; Haller, G. L.; Taylor, K. C.; Read, J. F. *Adv. Catal.* **1969**, *20*, 1; Burwell, R. L., Jr.; Stec, K. S. *J. Coll. Interface Sci.* **1977**, *58*, 54.
240. Sawabe, K.; Koga, N.; Morokuma, K.; Iwasawa, Y. *J. Chem. Phys.* **1992**, *97*, 6871.
241. Neuhaus, A. H.; Glendening, E. D.; Streitwieser, A. *Organometallics* **1996**, *15*, 3688.
242. Ma, B.; Collins, C. L.; Schaefer III, H. F., *J. Am. Chem. Soc.* **1996**, *118*, 870.
243. Xiao, Z. L.; Hauge, R. H.; Margrave, J. L. *J. Phys. Chem.* **1992**, *96*, 636.
244. Cao, Y.; Choi, J.-H.; Haas, B.-M.; Johnson, M. S.; Okumura, M. *J. Phys. Chem.* **1993**, *97*, 5215.
245. Niu, J.; Rao, B. K.; Jena, P. *Phys. Rev. B* **1995**, *51*, 4475.
246. Ogden, J. S.; Rest, A. J.; Sweany, R. L. *J. Phys. Chem.* **1995**, *99*, 8485; Sweany, R. L.; Ogden, J. S. *Inorg. Chem.* **1997**, *36*, 2523.
247. Moroz, A.; Sweany, R. L.; Whittenburg, S. L. *J. Phys. Chem.* **1990**, *94*, 1352.
248. Rasul, G.; Olah, G. A. *Inorg. Chem.* **1997**, *36*, 1278; Schreiner, P. R.; Kim, S.-J.; Schaefer, H. F.; Schleyer, P. v. R. *J. Chem. Phys.* **1993**, *99*, 3716.
249. Olah, G. A.; Rasul, G. *Inorg. Chem.* **1998**, *37*, 2047.

5

Structural and NMR Studies of Dihydrogen Complexes

5.1. INTRODUCTION

X-ray and neutron diffraction and NMR spectroscopy are the major techniques for determination of the structure of H_2 complexes, particularly H–H separation, by far the parameter of most interest. Because these methodologies are closely intertwined, they will be discussed in tandem. All the stable complexes studied to date feature symmetrically side-on (η^2) bound H_2 as in olefin binding in order to maximize backdonation (BD) from M. The existence of a bridging H_2 ligand has not been definitively proven crystallographically, although it has been proposed from NMR evidence in porphyrin systems as will be shown below. Because H_2 complexes represent points along the entire reaction coordinate to oxidative addition (OA), a remarkably full range of H–H distances (d_{HH}) has been observed, spanning 0.82 up to 1.6 Å, beyond which a complex is generally regarded to be a classical hydride. This represents over 100% elongation of the d_{HH} in free H_2 (0.74 Å) as compared to only 20% elongation on going from, e.g., a C≡C triple bond to a single bond. However, complexes with $d_{HH} \gg 1.0$ Å are more hydridelike in their properties and have highly delocalized bonding with very flat potential energy surfaces, as shown by both theoretical analyses (see Chapter 4) and NMR and diffraction data, which will be discussed below. A "true" dihydrogen complex thus can be considered to have $d_{HH} < 1.0$ Å, while complexes with d_{HH} much longer than this (certainly those with $d_{HH} > 1.3$ Å) can be thought of as dihydrides with weak H···H interactions ("compressed dihydrides"). In certain polyhydride complexes studied by neutron diffraction, two hydride ligands are observed to approach as closely as 1.6 Å in $OsH_6(P^iPr_3)_2$,[1] 1.49(4) Å in $[OsH_5(PPhMe_2)_3]^+$,[2] and 1.36(1) Å in $ReH_7[P(p\text{-}tolyl)_3]_2$.[3] Thus the dihydrogen versus dihydride "distinction" ("continuum" is a better word) is very subjective and blurred in this gray area. Although the d_{HH} are twice the normal bond length here, they are still shorter than the sum of the van der Waals radii. These stretched d_{HH} are not the results of solid state effects because they also persist in solution, e.g., solution NMR T_1 data show that the d_{HH} are shorter in $[OsH_5(PPhMe_2)_3]^+$ than in the tetrahydride precursor $OsH_4(PPhMe_2)_3$.[4] Arbitrarily, all complexes with $d_{HH} < 1.5$ Å will thus be included in discussions of H_2

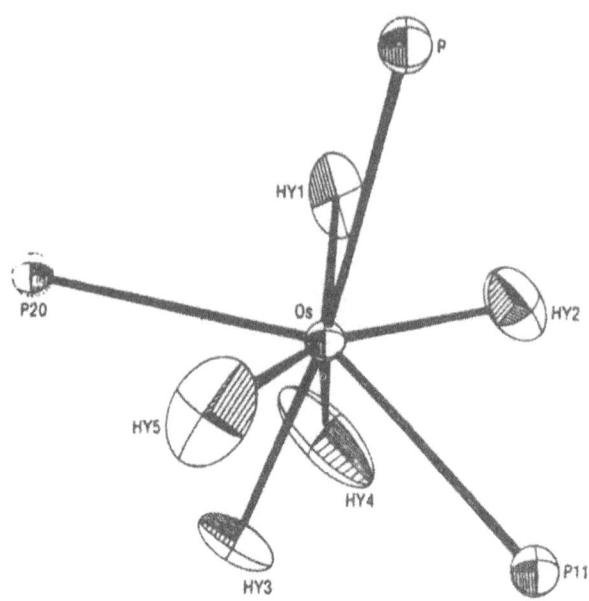

Figure 5.1. ORTEP drawing of [OsH$_5$(PPhMe$_2$)$_3$]$^+$ showing inner core atoms. Reprinted with permission from Johnson et al.[2] Copyright 1994 American Chemical Society.

complexes. However, as discussed in Chapter 4, theoretical calculations indicate that weak but measurable H $\cdots$ H interactions occur even at separations near 2 Å in classical polyhydrides.

The determination of d_{HH} and d_{MH} both accurately and precisely is nearly always a challenge. In certain cases, especially in polyhydride complexes, there is ambiguity as to whether H$_2$ ligands (elongated or otherwise) are really present, even in neutron diffraction structures. For example, as discussed in Section 4.8, [OsH$_5$(PPhMe$_2$)$_3$]$^+$ was originally formulated as an H$_2$ complex,[4,5] then calculationally as a pentahydride, until finally a neutron diffraction study at 11 K showed that it is indeed closer to a pentahydride with inequalities in d_{HH} (1.49 (HY3-HY4), 1.75 (HY1-HY2), and 1.98 Å (HY3-HY5) (Figure 5.1).[2] It took eight experimental and theoretical papers from six different research groups over a 25-year period to resolve the structure and bonding in a single complex! Thus it is not surprising that σ H$_2$ coordination was not identified until the 1980s. Locating hydrogen bound to heavy atoms by X-ray methods is a well-known problem, and even neutron diffraction is complicated by rapid rotation of η^2-H$_2$ (see Chapter 6), which foreshortens d_{HH}. Solid state NMR can be used to accurately determine d_{HH}, but with only moderate precision (± 0.01 Å). These values [e.g., 0.88 Å for Mo(CO)(H$_2$)(dppe)$_2$] are nearly always significantly longer (roughly 0.07 Å on average) than neutron values uncorrected for H$_2$ rotation. Solid state NMR directly measures the H-H internuclear separation (rotational and other dynamics are not factors) and can be a better gauge than neutron diffraction. However, there is only limited data here from the work of Zilm discussed below.

Solution ^{1}H NMR can be used to determine d_{HH} in solution by two different techniques involving measurement of either J_{HD} or relaxation time T_1. Comparison can then be made to solid state values, and usually there is close correlation between the different methodologies and solution versus solid state. However, determination of solution distances is not always straightforward and, as will be shown, can also be affected by rotational motion of the H_2 and other factors. The development of the T_1 methodology is a long and controversial saga. There is an overwhelming quantity of intricate NMR data for H_2 complexes because ^{1}H NMR is a universal characterization technique for several hundred known complexes. Thus comprehensive coverage will not be presented here (a thorough compilation of J_{HD} and T_1 data up till 1993 is given by Heinekey[6]), but important principles and newer results will be emphasized.

5.2. DIFFRACTION METHODS

X-ray diffraction rarely locates hydrogen atoms precisely, especially when they are bound to third-row metals. The problem is exacerbated for $M-H_2$ because much of the electron density resides between M and H_2 in the three-center bond, and there are no nonbonding electrons on H_2 to scatter X-rays. In one notable example, d_{HH} in $Cr(CO)_3(P^iPr_3)_2(H_2)$ was determined to be 0.67(5) Å,[7] which is 10% shorter than the distance in free H_2! As will be shown below, the distance measured by solid state NMR in the PCy_3 analogue is 0.85(1) Å, which is far truer and in line with that for the W analogue (0.89 Å, solid state NMR). The 0.85 Å d_{HH} is the shortest that has been accurately determined in a stable solid complex thus far and is also the shortest d_{HH} measured by solution NMR [in $Cr(CO)_3(P^iPr_3)_2(H_2)$], corresponding to $J_{HD} = 34$ Hz. Shorter X-ray and neutron diffraction values of 0.82 Å have been determined in, e.g., $W(CO)_3(P^iPr_3)_2(H_2)$ but corrections for H_2 rotation that typically increase these values by ca 0.07 Å often have to be applied, as will be seen below. The structures of H_2 complexes determined by diffraction techniques have been reviewed by both Bau and Koetzle[8] for neutron structures only and by Braga for X-ray and neutron.[9]

Hydrogen nuclei scatter neutrons better than almost any other nuclei, but there are limitations to this technique. Crystal size must be at least 1 mm^3, and available neutron sources are scarce. Also, because of the low energy barrier (0.5–3 kcal/mol) to H_2 rotation in most true H_2 complexes ($d_{HH} < 1.0$ Å) as discussed in Chapter 6, there is considerable librational motion even at temperatures as low as the 12 K typically used for neutron diffraction. This is graphically demonstrated in the structure of $Mo(CO)(dppe)_2(H_2) \cdot 4.5$ benzene by the large thermal ellipsoids elongated in the direction of H_2 rotation (Figure 5.2).[10] This motion gives an unusually large apparent bond shrinkage because the center of the thermal ellipsoid does not represent the position of the nucleus (the hydrogen motion traces out a "banana-like" shape improperly modeled by standard ellipsoids). The angular oscillations cause the maximum of an atomic peak in the electron density to be displaced toward the center of rotation (Scheme 1).[11]

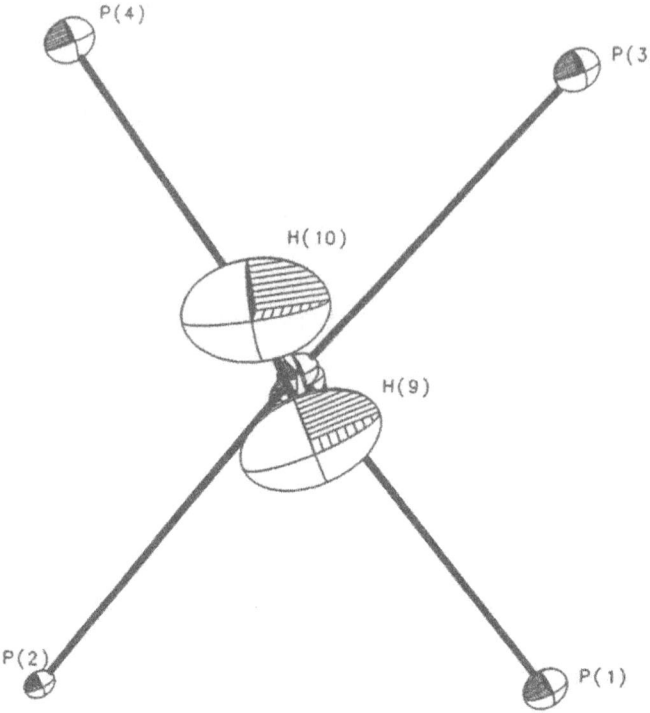

Figure 5.2. ORTEP drawing of Mo(CO)(dppe)$_2$(H$_2$)·4.5-benzene showing inner core atoms. Reprinted with permission from Kubas *et al.*[10]. Copyright 1993 American Chemical Society.

The refined distance H(9)–H(10) is 0.736(10) Å in Figure 5.2 (d''_{HH} in Scheme 1) is in fact shorter than the 0.74 Å value found in free H$_2$. This bond distance can be

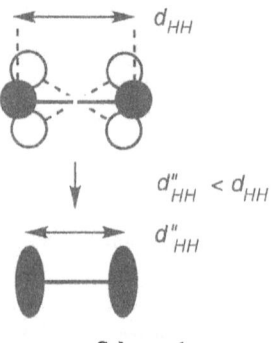

Scheme 1

corrected for thermal motion to give an estimate of the "true" d_{HH}, e.g., via the program THMA11.[12] However, the difficulty in subtracting the internal vibrational motion of the hydrogen atoms gives an overestimated bond length correction. The calculated corrected d_{HH} is 0.85–0.88 Å for $Mo(CO)(dppe)_2(H_2)$. If an overcorrection of 33–50% based on the inability to subtract the internal vibrational motion of the hydrogens is assumed, d_{HH} can only be estimated to lie somewhere in the range 0.80–0.85 Å,[10] an unsatisfactory answer to a critical question. The solid state NMR d_{HH} for the unsolvated complex is 0.88 Å, which is typical in octahedral d^6 complexes with CO or hydride trans to H_2. Thirteen complexes, mostly octahedral d^6 complexes, have been studied by neutron diffraction,[13] only three of which have d_{HH} corrected for libration.

A comparison of d_{HH} determined in the solid and solution states by different methodologies is given in Table 5.1 for complexes with $d_{HH} < 1.0$ Å and in Table 5.2 for "elongated" complexes with $d_{HH} > 1.0$. The range of d_{HH} is remarkable, corresponding to arrested points along the entire reaction coordinate for bond cleavage. A d_{HH} over 1.6 Å is generally observed to be present in dihydrides, but as discussed in Chapter 4 $d_{HH} > 1.0$ Å represent greater dihydride than dihydrogen character and the positions of the hydrogens are highly delocalized. The neutron structures, NMR studies (Section 5.4), and theoretical analyses of $[Os(en)_2(H_2)(acetate)]^+$ and trans-$[OsCl(H_2)(dppe)_2]^+$ were critical for uncovering the true nature of the bonding and dynamics in these species.

The nature of the ligand trans to H_2 can be the major determinant of d_{HH}. For CO or hydride trans to H_2, generally $d_{HH} < 0.9$ Å because the reduced BD to H_2 lessens activation toward OA. Nearly all the complexes in Table 5.1 have CO or H trans to H_2. Considering how numerous they are, only three complexes containing CO have been characterized by diffraction methods (often disorder between trans CO and H_2 ligands is a problem). d_{HH} in cationic complexes are generally shorter than or at least similar to those in neutral analogues, and changes in cis ligands often have little effect, especially in electrophilic cationic systems (see Table 4.8). It is clear from Table 5.1 that X-ray distances are normally shorter than the more trustworthy neutron and NMR values.

Normally d_{MH} are up to several tenths of an angstrom longer than for classical M–H, as might be expected. However, d_{MH} determined by X-ray once again have large uncertainties, and the number of more accurate neutron distances is not large enough for correlations. Furthermore, unlike d_{HH}, d_{MH} cannot readily be derived by NMR methods (see Section 5.5 for one example) and are not as useful for correlating degrees of H–H activation in complexes with different metals because of differences in the van der Waals radii. Within the same complex, d_{MH} can reflect the relative lability of the H_2 ligand. For example, the Ru–H_2 distance [1.81(2) Å] in the neutron structure of trans-$[RuH(H_2)(dppe)_2]^+$ is much longer than the Ru–$H_{hydride}$ distance [1.64(2) Å], consistent with weakly bound H_2. The H_2 readily dissociates in vacuo and d_{HH} corrected for H_2 libration is still relatively short, 0.94 Å. The difference in the M–H_2 [1.62(1) Å] and M–H distances [1.54(1) Å] in the analogous Fe complex is much less, corresponding to the lower lability of H_2. Complexes with elongated H–H bonds (>1.0 Å) and nonlabile H_2 generally have d_{MH} close to those in classical hydrides. For example, in cis-$IrCl_2H(H_2)(P^iPr_3)_2$ the Ir–H_2 distances are 1.550(17) and 1.537(19) Å while the Ir–$H_{hydride}$ distance is 1.584(13).[15] For $[Os(en)_2(H_2)(acetate)]^+$ the Os–H distances are quite short [1.59(1) and 1.60(1) Å], as much as 0.06 Å shorter than those in comparable third-row hydrides such as $OsH_4(PMe_2Ph)_3$ [1.659(8)].[8]

Table 5.1. Comparison of d_{HH} (Å) Determined by Diffraction and NMR Techniques for Complexes with $d_{HH} < 1.0$ Å

Complex	X-ray	Neutron[a]	Solid state NMR	Solution NMR From J_{HD}[b]	Solution NMR From $T_{1(min)}$[c]	Reference
$Cr(CO)_3(P^iPr_3)_2(H_2)$	0.67(5)		0.85[d]	0.84–0.85		7
$Mo(CO)(dppe)_2(H_2)$		0.736(10) 0.84[e]	0.88	0.85–0.87		10, 14
$W(CO)_3(P^iPr_3)_2(H_2)$	0.75(16)	0.82(1)	0.89	0.86–0.88	0.76/0.96	15–17
$[FeH(H_2)(dppe)_2]^+$	0.89(11)	0.816(16)	0.90	0.92–0.94		18
$[FeH(H_2)(dmpe)_2]^+$	0.86(12)				0.81	19
$[FeH(H_2)(PMe_3)]^{4+}$	0.84(4)					20
$FeH_2(H_2)(PEtPh_2)_3$		0.821(10)				21
$[RuH(H_2)(dppe)_2]^+$	0.83(8)	0.82(3) 0.94[e]		0.88–0.90	0.90(1)	22
$RuH_2(o-C_6H_5py)(P^iPr_3)_2]^+$	0.82(4),					85
$RuH_2(H_2)(SiHPh_3)-(PCy_3)_2$	0.82(2)					24
$[RuH(H_2)-(DuPHOS-Me)]^+$	0.8(2)			0.92–0.94	0.78/0.99	25
$Ru_2H(\mu-H)(\mu-Cl)_2(H_2)-(P-N)(PPh_3)_2$	0.80(6)					26
$[TpRu(dippe)_2(H_2)]^+$	0.71			0.92–0.94		27
$[OsH(H_2)(dppe)_2]^+$		0.97(2)		0.99–1.01	1.00/1.25	28
$[Os(H_2)(MeCN)(dppe)_2]^{+2}$	0.9(1)			1.06–1.08	0.94/1.19	28
$[OsCl(H_2)(H_2bim)-(P^iPr_3)_2]^+$	0.95(16), 1.02(13)[f]				0.95/1.19	29
$IrIH_2(H_2)(P^iPr_3)_2$			0.856			30
$IrBrH_2(H_2)(P^iPr_3)_2$			0.96			31

[a]Uncorrected for effects of librational motion.
[b]Using Eqs. (5.3) and (5.4).
[c]Assuming fast/slow rotation of the H_2 (for fast rotation if only one value is given).
[d]For the PCy_3 analogue.
[e]Neutron data corrected for H_2 librational motion.
[f]Values for two independent molecules in the asymmetric unit; H_2bim = 2, 2'-biimidazole.

5.3. SOLID STATE NMR

Solid state ^1H NMR is very effective in determining d_{HH}, as shown by Zilm and coworkers.[7,14,16,44,45] Only a small amount of solid in powder form is required to observe the η^2-H_2 signals using broad-line techniques (Figure 5.3). The basic principles are well established: Isolated pairs of nuclei in a rigid solid experience a mutual dipolar interaction directly proportional to the average of the inverse cube of the internuclear distance. For a powder sample a familiar Pake doublet line shape results (Figure 5.3C), which can be simulated to give internuclear distances within 1%. The patterns are quite sensitive to anisotropic motion, i.e., hindered rotation or torsion of the side-bound H_2 about the M–H_2 axis. This will not affect the H_2

Table 5.2. Comparison of d_{HH} (Å) Determined by Diffraction and NMR Techniques for Complexes with $d_{HH} > 1.0$ Å

Complex	X-ray	Neutron[a]	Solution NMR From J_{HD}[b]	From $T_{1(min)}$[c]	Reference
[Cp$_2$Ta(H$_2$)(CO)]$^+$	1.09(2)		0.96–0.98	1.06 ± 0.05	71
ReH$_5$(H$_2$)[P(p-tolyl)$_3$]$_2$		1.357(7)			3
[ReH$_4$(H$_2$)(Cyttp)]$^+$	1.08(5)				32
ReCl(H$_2$)(PMePh$_2$)$_4$	1.17(13)				33
ReBr$_2$(H$_2$)(NO)(P^{i}Pr$_3$)$_2$			1.21	1.27	34
[Ru(C≡CPh)(H$_2$)(dippe)$_2$]$^+$	1.1				35
RuHI(H$_2$)(PCy$_3$)$_2$	1.03(7)		1.19–1.21[d]	1.02(3)[d]	36
[Cp*Ru(dppm)(H$_2$)]$^+$		1.08(2) 1.10[e]	1.07–1.09	0.87/1.10	37
[Cp*OsH$_2$(H$_2$)(PPh$_3$)]$^+$	0.78(8)	1.01(1)	1.06–1.09	1.07[e]	38
[Cp*OsH$_2$(H$_2$)(AsPh$_3$)]$^+$		1.08(1)	1.15	1.15	39
[Cp*OsH$_2$(H$_2$)(PCy$_3$)]$^+$	1.14(7)	1.31(3)	1.09[f]	1.16[f]	39
trans-[OsCl(H$_2$)(dppe)$_2$]$^+$	1.11(6)	1.22(3)	1.19–1.21	1.08/1.35	40
trans-[OsBr(H$_2$)(dppe)$_2$]$^+$	1.13(8)		1.19–1.21	1.08/1.36	40
cis-OsCl$_2$(H$_2$)(Hpz)(P^{i}Pr$_3$)$_2$	1.3(1)			1.01/1.27	29
[Os(en)$_2$(H$_2$)(acetate)]$^+$		1.34(2)	1.27–1.29		41
[OsH$_3$(H$_2$)(PPhMe$_2$)$_3$]$^+$		1.49(4)			2
OsCl(H$_2$)(N–C)(P^{i}Pr$_3$)$_2$[g]	1.22(5)		1.31–1.33	1.08/1.36	42
cis-IrCl$_2$H(H$_2$)(P^{i}Pr$_3$)$_2$		1.11(3)	1.17–1.29[h]	1.07(1)/1.35(2)	43

[a] Uncorrected for effects of librational motion.
[b] Using Eqs. (5.3) and (5.4).
[c] Assuming fast/slow rotation of the H$_2$ (for fast rotation if only one value is given).
[d] For the P^{i}Pr$_3$ analogue.
[e] Neutron data corrected for H$_2$ librational motion.
[f] Not reliable because of exchange and interactions with cis hydrides.
[g] N–C is η^2 – NH=C(Ph)C$_6$H$_4$.
[h] J_{HD} is 12 ± 3 Hz.

dipolar splitting when this axis is parallel to the applied magnetic field. One pair of discontinuities in the Pake pattern is thus unaffected and is used to calculate d_{HH}. This discontinuity is temperature-independent (other temperature-dependent patterns can also exist, so variable-temperature measurements are needed).

The main problem is interference from the protons in ancillary ligands such as PCy$_3$, where 66 protons are present. These obscure the very wide Pake pattern for the H$_2$, but fortunately this can be averted by deuterating the ligands (Figure 5.3C) or using a hole-burning ligand suppression technique at 77 K to avoid costly deuteration. The latter method relies on the homogeneous line shape of the strongly coupled ligand protons, which can be saturated by application of a several-millisecond weak pulse. However, the inhomogeneous Pake doublet for η^2-H$_2$ is not saturated, a small hole is burned in the pattern, and the major part of the line shape is unaffected (Figure 5.3B). The first complex studied, W(CO)$_3$(PCy$_3$)$_2$(H$_2$), showed a d_{HH} of 0.890 ± 0.006 Å.[16] As will be discussed in Chapter 6, solid state NMR can also be used to study the dynamics of hydrogen exchange processes in e.g., IrClH$_2$(H$_2$)(P^{i}Pr$_3$)$_2$.[45]

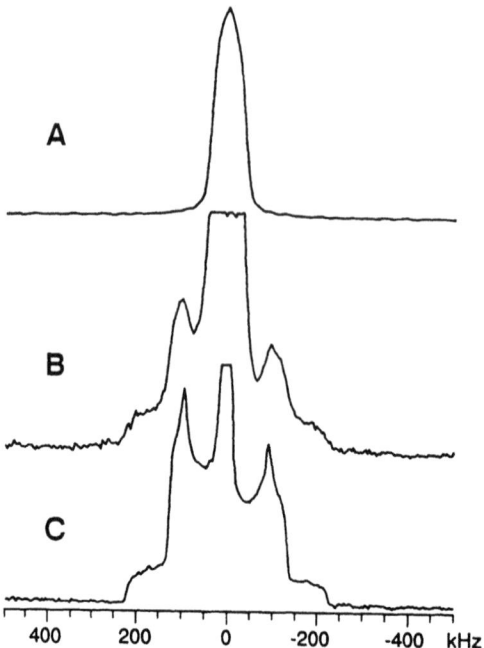

Figure 5.3. A: Solid state ^{1}H NMR spectrum for $W(CO)_3(PCy_3)_2(H_2)$ without suppression of the phosphine ligand signal. B: Same as A except using ligand suppression sequence. C: Same as B except using perdeuterated phosphine ligands. Reprinted with permission from Zilm et al.[16] Copyright 1986 American Chemical Society.

5.4. SOLUTION NMR: J_{HD} COUPLING AND ISOTOPE EFFECTS

As described in the discovery of H_2 complexes (see Chapter 2), the solution ^{1}H NMR spectra of η^2-H_2 normally gives broad uncoupled signals throughout a large range of chemical shifts (2.5 to −31 ppm) that overlap with those for classical hydrides. Among the several hundred H_2 complexes, $ReBr_2(H_2)(NO)(PCy_3)_2$ gives the furthest shift to low field (2.44 ppm) and has an elongated H–H (1.21–1.28 Å from NMR techniques described below).[34] If there are P-donor ligands in the complex, the J_{HP} between H_2 and ^{31}P nuclei is normally less than 6 Hz and unobserved. However, in a few cases larger couplings are found, usually for complexes with elongated d_{HH}.[46] In complexes containing both H_2 and hydride ligands, facile exchange between the hydrogens nearly always occurs, especially if these ligands are cis (see Chapter 6). Scrambling occurs via a M–H_3 transient, as shown in Eq. (5.1), where a D ligand differentiates the hydrogens. In most cases an

$$\begin{array}{c} D \\ | \\ M-\!\!\!\begin{array}{c} H \\ | \\ H \end{array} \end{array} \rightarrow \begin{array}{c} D\cdots H \\ M\cdots H \end{array} \rightarrow \begin{array}{c} H \\ | \\ M-\!\!\!\begin{array}{c} H \\ | \\ D \end{array} \end{array} \qquad (5.1)$$

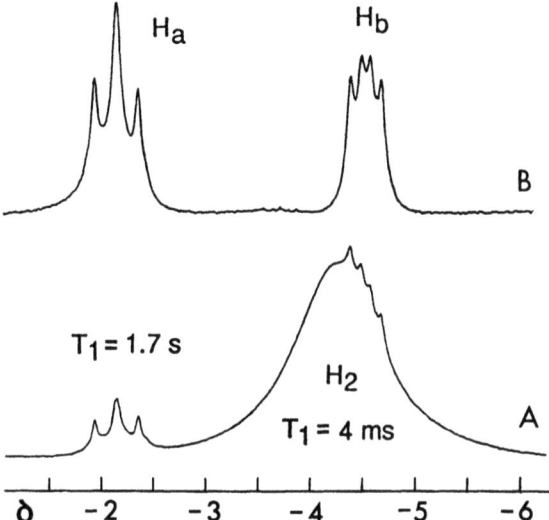

Figure 5.4. ^{1}H NMR spectrum for $W(CO)_3(P^iPr_3)_2(H_2)$ at -82 °C in toluene-d_8. A: normal spectrum showing broad resonance for the H_2 ligand, which obscures the upfield hydride (H_b) resonance. B: spectrum obtained with an inversion-recovery pulse sequence (180-τ-90) to null the interfering H_2 signal. Reprinted with permission from Kubas et al.[47] Copyright 1986 American Chemical Society.

averaged broad singlet results, which may not decoalesce at low temperatures; many complexes are highly fluxional even below -100°C (barriers as low as 1 kcal/mol).

Another dynamic process involves *tautomeric equilibria* between H_2 and dihydride forms in solution in about a dozen complexes, including $W(CO)_3(PR_3)_2(H_2)$ (**1**).[47] The dihydride tautomer (**2**) exists in about 20% concentration in Eq. (5.2) for

(5.2)

$J_{HD} = 34$ Hz

$\delta = -4.3$ (H_2)
$\delta = 33.5$ (P)
$T_1 = 4$ ms (H_2)

NMR
-92 °C
toluene-d_8

$\delta = -2.15$ (H_a), -4.5 (H_b)
$\delta = 30.8$ (P_a), 39.5 (P_b)
$T_1 = 1.67$ s (H_a and H_b)

$\nu_{HH} = 2695$ cm^{-1}
$\nu_{CO} = 1969, 1856$ cm^{-1}

IR
hexane

$\nu_{CO} = 1993, 1913,$
   $1867, 1828$ cm^{-1}

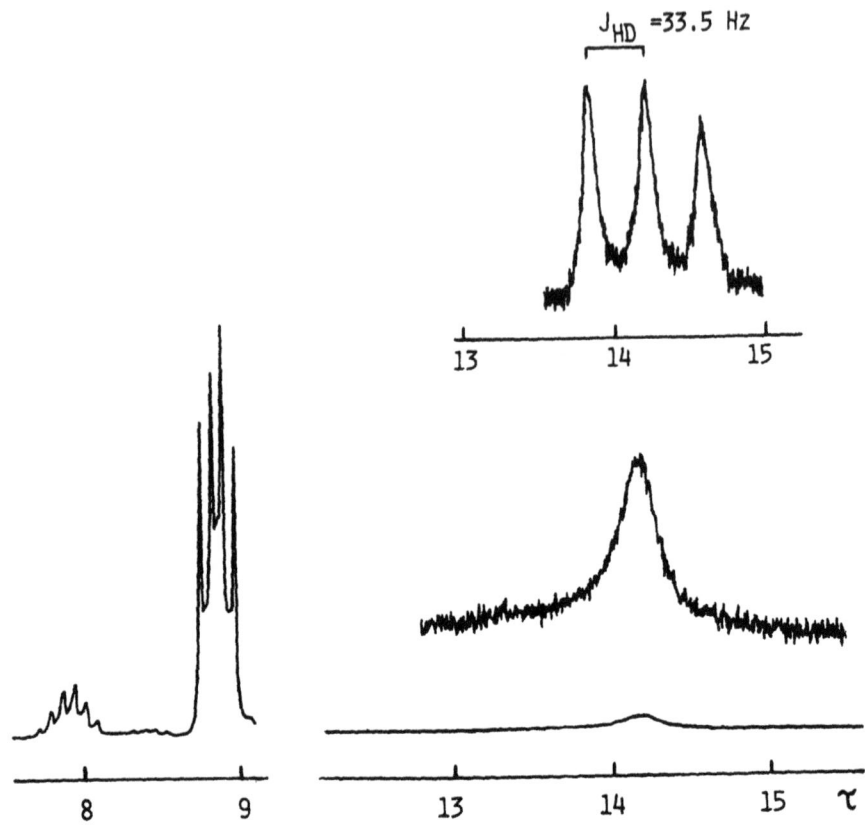

Figure 5.5. ^{1}H NMR spectrum in toluene-d_8 of $W(CO)_3(P^iPr_3)_2(H_2)$ and HD isotopomer (upper spectrum). Reprinted with permission from Kubas et al.[47] Copyright 1986 American Chemical Society.

$P = P^iPr_3$ and is not isolable but is detectable by IR and NMR, which show separate signals for the H_2 and hydride ligands at or below room temperature (Figure 5.4). Spin saturation transfer experiments using both ^{31}P and ^{1}H NMR confirm that these signals correspond to equilibrium species,[48] and the kinetic and thermodynamic parameters for this and other equilibria are discussed in Chapter 7. The relaxation time T_1 for the broad H_2 signal is much shorter than for the hydride signals, as will be described below. This difference can be used to suppress the H_2 signal by an inversion-recovery pulse sequence if it interferes with the hydride signals as in Figure 5.4. The inequivalency for both the hydride and phosphine signals in **2** is consistent with a calculated pentagonal bipyramid geometry (see Chapter 4), which is also observed crystallographically for $MoH_2(CO)(depe)_2$.

The single most important spectroscopic parameter is $^1J_{HD}$ for the HD isotopomer of an H_2 complex (J_{HD} without the superscript is normally used to refer to the one-bond coupling). The signal becomes a 1:1:1 triplet (D has spin 1) with much a narrower line width and is direct proof of the existence of an H_2 ligand since classical hydrides do not show significant $^1J_{HD}$ because no residual H–D bond is present. Figure 5.5 shows the first ^{1}H NMR spectrum of an HD complex, $W(CO)_3(P^iPr_3)_2(HD)$, which, owing to its high solubility, was recorded on a

90-MHz continuous-wave instrument ($\tau = 10 - \delta$).[47] J_{HD} for HD gas is 43 Hz, the maximum value ($d_{HD} = 0.74$ Å). A lower value represents a proportionately shorter d_{HD}. J_{HD} determined in solution correlates with d_{HH} in the solid state via the empirical relationships developed by both Morris[40] and Heinekey[49]:

$$d_{HH} = 1.42 - 0.0167 J_{HD} \text{ Å} \quad \text{[Morris]} \quad (5.3)$$
$$d_{HH} = 1.44 - 0.0168 J_{HD} \text{ Å} \quad \text{[Heinekey]} \quad (5.4)$$

The input data for formulating these expressions, which differ only slightly, include d_{HH} from both diffraction and solid state NMR measurements. Despite the inherent uncertainties of d_{HH} derived from X-ray and uncorrected neutron diffraction data, a plot of d_{HH} versus J_{HD} gives a straight line with little deviation (Figure 5.6). The intercept (1.44 Å) changes slightly but the slope is nearly identical in the more recent graphical analysis of Heinekey.[49] In Tables 5.1 and 5.2 the upper and lower limits of d_{HH} calculated from J_{HD} represent use of both Eqs. (5.3) and (5.4). It should be noted in this context that well-characterized classical hydrides can have small J_{HD} up to 3.3 Hz,[50] and use of Eq. (5.4) might suggest that 1.4 Å represents the maximum d_{HH} for η^2-H$_2$ ligands. For a large series of the elongated complexes $[Os(H \cdots H)(NH_3)_4(L)]^{z+}$ ($z = 1, 2$), low values of J_{HD} calculated by B3LYP methods are in good agreement with experimental values (8–20 Hz).[51] The computed J_{HD} inversely correlate well with calculated d_{HH}.

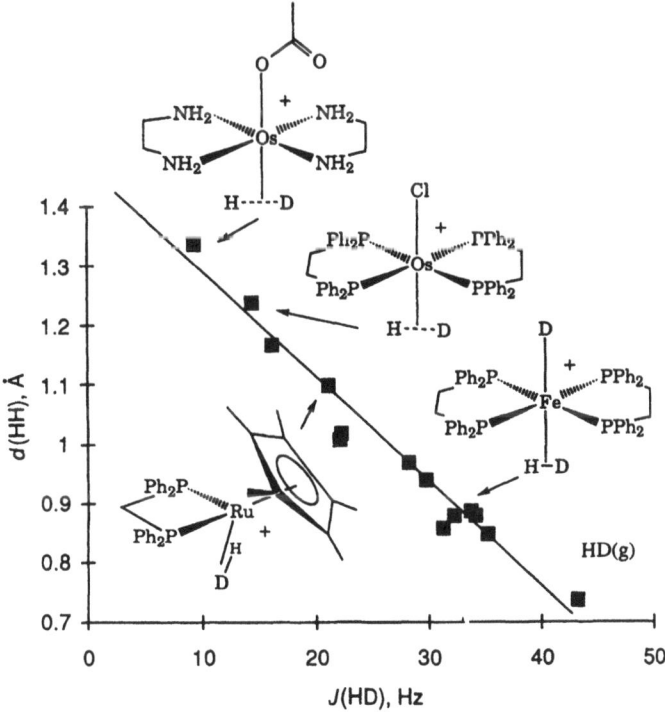

Figure 5.6. Plot of d_{HH} determined in the solid state versus J_{HD} measured for the corresponding HD complex in solution. Reprinted with author's permission from Morris, R. H. *Can. J. Chem.* **1996**, *74*, 1907.

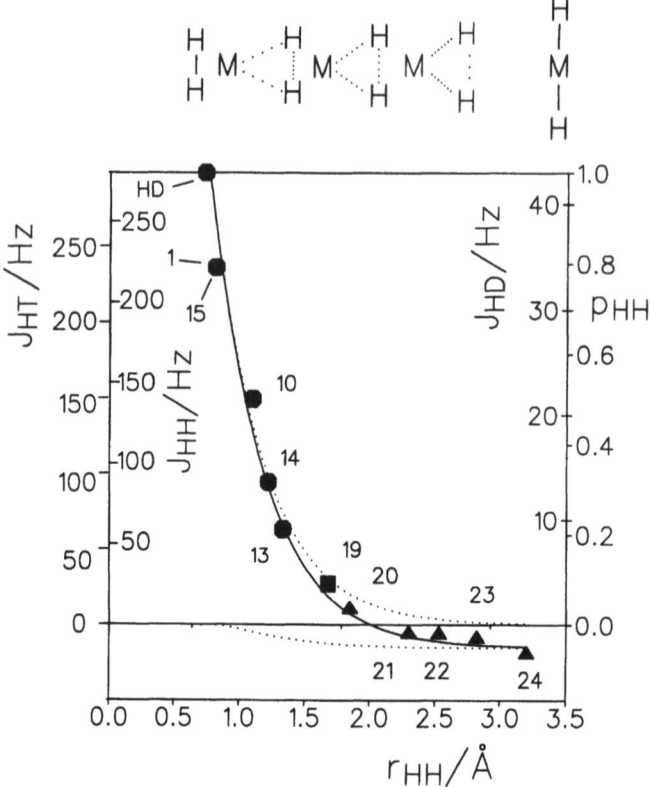

Figure 5.7. Correlation of the scalar coupling constants $J_{LL'}$, LL' = HH, HD, HT of transition metal dihydride–dihydrogen complexes with d_{HH}. Solid line: Eq. (5.6). Upper dotted line: first term of Eq. (5.6). Lower dotted line: second term of Eq. (5.6). For complexes with $d_{HH} > 2.0$ Å, negative signs for the scalar magnetic coupling constants were assumed. Reprinted with permission from Grundemann et al.[53] Copyright 1999 American Chemical Society.

According to Hush, J_{HD} is proportional to the H–H bond order, p_{HH}, described in Section 4.7.3.[52]

$$J_{HD} = J_{HD}^\circ p_{HH} \qquad J_{HD}^\circ = 43 \qquad (5.5)$$

This does not take into account magnetic couplings in trans dihydrides, where p_{HH} is near zero. Grundemann[53] adds a two-bond term proportional to the square of the M–H bond order, p_{MH}, to calculate $J_{LL'}$ for all isotopes L, L' = H, D, T (Eq. 5.6, where γ_L is the gyromagnetic ratio of L).

$$J_{LL'} = J_{HH}\gamma_L\gamma_{L'}/\gamma_H^2 = {}^1J_{LL'}^\circ p_{HH} + {}^2J_{LL'}^\circ(1 - p_{HH})^2 \qquad (5.6)$$

The validity of Eq. (5.6) is shown in Figure 5.7, where the solid line represents Eq. (5.6) and the points correspond to experimental data for the isotopic pairs HD (circles), HT (square), and HH (triangles). The d_{HH} are taken from neutron data (for $d_{HH} < 2$ Å, see below) or good-quality X-ray data (for $d_{HH} = 2$–3 Å). The agreement between experiment and theory is quite good and the one-bond term $^1J_{LL'}$ is proportional to the HH-bond order and dominates at $d_{HH} < 1.4$ Å (upper dotted line). At larger distances the two-bond order term (lower dotted line) leads to a change in the sign of the coupling constant. This provides a simple rationale for previous findings of inexplicably large coupling constants in some trans dihydrides with d_{HH} near 3.2 Å and small values of only a few hertz in octahedral cis dihydrides with d_{HH} near 2.2 Å. The determination of d_{HH} from Eq. (5.6) has the advantage that it is not affected by coherent or incoherent H_2 rotations, as in the case of the determination of d_{HH} from T_1 measurements (see below). The bond-valence/bond-order model (see Section 4.7.3), which is independent of M/L properties, well describes both the experimental hydrogen locations and the dependence of the coupling constants $J_{LL'}$ on d_{HH} and d_{MH}.

For very fluxional dihydrogen–polyhydride complexes, the use of deuterium labeling in NMR experiments is often critical in establishing the existence of η^2-H_2.[17,54] This is achieved by the observation of an averaged $J_{HD_{avg}}$ or a perturbation of chemical shift (see below). For complexes of the type $MH_x(H_2)_y$ containing rapidly exchanging hydrogens, an average J_{HD} is normally observed upon partial isotopic substitution by deuterium. J_{HD} for the η^2-H_2 can be extracted from this, but it is very dependent on assumptions made about other couplings and the structural model used.[17,54] The expression for an HD_{n-1} isotopomer is

$$^{avg}J_{HD} = n^{-1}\Sigma_i\chi_i\{\Sigma_j[J_{ij}/(n-1)]\} \tag{5.7}$$

where $i \neq j$ and χ_i is the likelihood that the proton occupies site i, and $\Sigma_i\chi_i = 1$. J_{ij} is the coupling between a proton in site i and a deuteron in site j. A major source of potential error is neglecting two-bond H–D couplings such as mutual couplings between the hydrides and/or hydride coupling to H_2 ligands. As an example, for an HD ligand in $OsHD_3(CO)(P^tBu_2Me)_2$, the J_{HD} calculated from the above expression neglecting $^2J_{ij}$ couplings involving hydrides is 25.2 Hz, which is simply six times the observed average J_{HD} of 4.2 Hz. However normal geminal H⋯H couplings in classical hydrides are near 6 Hz, which would give an H⋯D coupling of about 1 Hz (based on the gyromagnetic ratios of ^1H and ^2H isotopes). The inclusion of these couplings changes the extracted J_{HD} to 24.2 Hz. Thus, if such derived J_{HD} are used to calculate d_{HH}, it is important not to neglect two-bond H–D couplings. Also, there may be a small uncertainty in this procedure because the statistical weights will have some error (deuterium prefers to be in η^2-HD over M–D).

Tritium NMR of partially tritiated complexes has been carried out, and dynamically averaged J_{HT} occur at 51–65 Hz for $[TpIrH(H_2)(PR_3)]^+$.[55] The true J_{HT} should be about three times this value. Tritium (^3H) has a nuclear spin of $\frac{1}{2}$, the same as ^1H but unlike deuterium with spin equal to 1. Distinct resonances for the T_3, T_2H, and TH_2 isotopomers occur at -10 to -11 ppm, and the ^3H chemical shift for the T_3 isotopomer is the same as that for the H_3 isotopomer in the ^1H NMR (Figure 5.8). These relatively large shifts are a manifestation of *isotopic perturbation of resonance* (IPR),[56] which also occurs for deuterium versus protio isotopomers here and in other rapidly exchanging polyhydride complexes, including classical systems with inequivalent hydrides.[57–60] In the fast exchange of ^1H NMR spectra of isotopomers of nonclassical polyhydrides, each isotopomer shows a separate

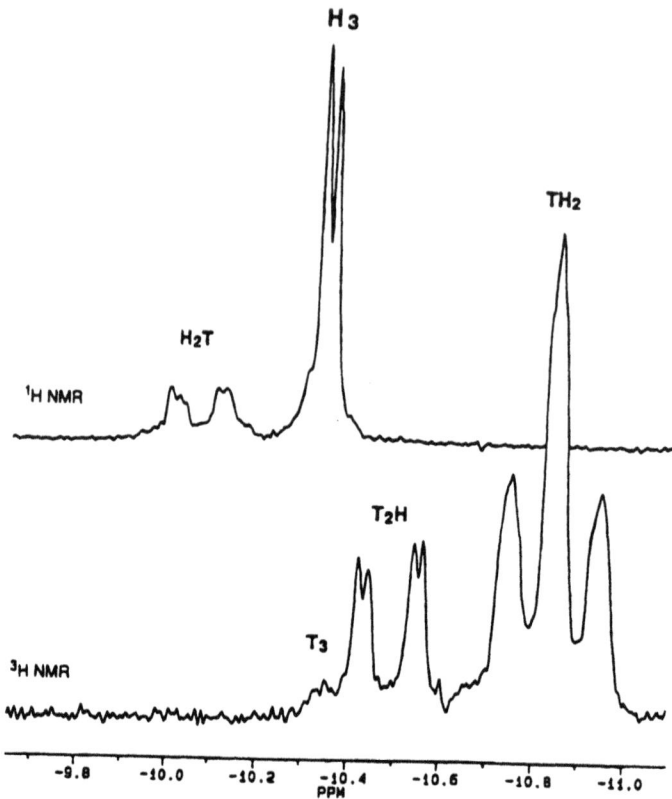

Figure 5.8. Top: ^{1}H NMR spectrum (CD$_2$Cl$_2$, 500 MHz) of a partially tritiated sample of [TpIrH(H$_2$)(PMe$_3$)]$^+$. Bottom: ^{3}H NMR spectrum (CD$_2$Cl$_2$, 533 MHz). Reprinted with permission from Oldham et al.[55] Copyright 1997 American Chemical Society.

hydride resonance provided that the M–H and M(H$_2$) sites have significantly different chemical shifts and that there is sizable deuterium fractionation between the sites. For example, the isotopomers of [TpIrH(H$_2$)(PMe$_3$)]$^+$ give an IPR shift of 0.228 ppm for $\delta(H_2D) - \delta(H_3)$ at 215 K, which decreases to 0.112 ppm at 281 K.[55] Over the same temperature range, the averaged J_{HD} varies from 6.87 to 7.63 Hz for the d_1 species, but is essentially invariant with temperature for the d_2 system (8.75 Hz). Thus there is a nonstatistical site preference for the deuterium isotope that varies with the degree of deuteration [Eqs. (5.8) and (5.9)].

The equilibrium constants determined in Eqs. (8) and (9) are actually Boltzmann factors (statistics not included), but they indicate that in this case the heavier isotope prefers to occupy the hydride site. In contrast, both Field and Crabtree had previously observed the opposite effect, e.g., $K = 1.3$ for ReD$_2$(HD) ↔ ReHD(D$_2$) in [ReH$_2$(H$_2$)(CO)(PMe$_2$Ph)$_3$]$^+$.[57,58] However, detailed analysis of chemical shift and coupling constant data as a function of temperature in the Ir complex has not been carried out in these systems. The isotope effect was simplistically interpreted to be a consequence of a greater vibrational zero-point energy difference between Re(η^2-HD) and Re(η^2-D$_2$) relative to Re–H and Re–D.[58] These systems are quite complex vibrationally as well as dynamically, however, and the isotopic preferences will be dictated by the changes in *all* of the force constants in both tautomers, anharmonicity in the M–H$_2$ vibrational modes, and relative H–H versus M–H bond strengths

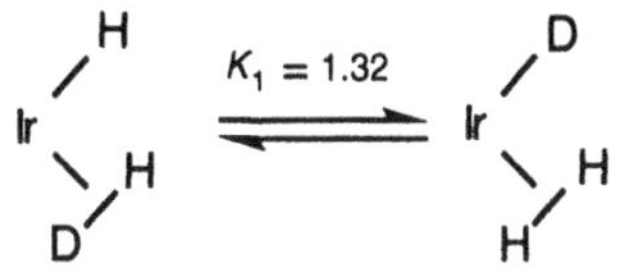

$$\text{Ir}\begin{pmatrix} H \\ H \\ D \end{pmatrix} \xrightleftharpoons{K_1 = 1.32} \text{Ir}\begin{pmatrix} D \\ H \\ H \end{pmatrix} \quad (5.8)$$

$$\text{Ir}\begin{pmatrix} H \\ D \\ D \end{pmatrix} \xrightleftharpoons{K_1 = 1.26} \text{Ir}\begin{pmatrix} D \\ H \\ D \end{pmatrix} \quad (5.9)$$

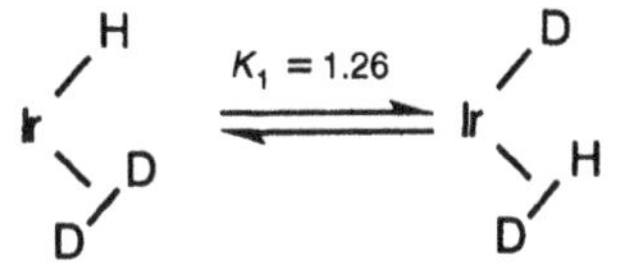

(Section 7.5.2). In regard to the latter, analysis of the NMR data for the Ir system provides the actual J_{HD}, approximately 25 Hz, for η^2-H$_2$, which along with a value of 22 ms for $T_{1(\min)}$ (see Section 5.5), is consistent with a long d_{HH} of about 1 Å. Thus the Ir–H bonds, particularly to the hydride, are undoubtedly stronger than the H–H bond, which would favor deuterium incorporation. The complex is close to a trihydride with inequivalent hydrides, and a tetrahydride with inequivalent hydrides also shows IPR.[60] By contrast the H–H bond in the Re complex, where H$_2$ is trans to CO, is much stronger (J_{HD} = 34 Hz), and hence is a true H$_2$ ligand. Systems such as these demand heroic efforts just to determine basic parameters such as d_{HH} and J_{HD} but provide a wealth of new information.

Unusual behavior in the temperature (and possibly field[37]) dependence of J_{HD} is found in complexes with elongated d_{HH}, [Cp*Ru(H···D)(dppm)]$^+$ and *trans*-[OsX(H···D)(dppe)]$^+$ (X = H, Cl), and gave initial indications of the delocalized bonding in these species.[37,40,61] In the OsCl species e.g., J_{HD} unexpectedly varied from 13.6 to 14.5 Hz depending on both temperature (253 to 308 K) and solvent.[40] Several different explanations evolved, including rapid temperature-dependent interconversion of H$_2$–dihydride tautomers, but these were discarded in favor of rapid motion of two hydrogen atoms in a flat potential energy surface with a shallow minimum at the neutron-diffraction-determined position of 1.2 Å. This

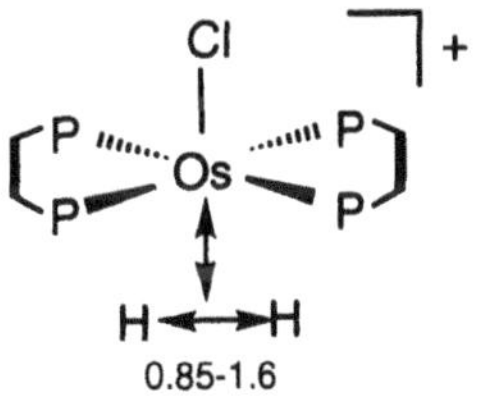

$$\begin{bmatrix} & \text{Cl} & \\ P\cdots & | & \cdots P \\ & \text{Os} & \\ P & & P \\ & \uparrow & \\ H & \leftrightarrow & H \\ & 0.85\text{--}1.6 & \end{bmatrix}^+$$

investigation led to important theoretical studies (see Section 4.7.4) that revealed the extraordinarily delocalized nature of the bonding in this case: d_{HH} can vary from 0.85 to 1.6 Å (with concomitant variation in d_{MH}) at a cost of only 1 kcal/mol! Subsequent NMR studies by Heinekey[37] of the HD, HT, and DT isotopomers of [Cp*Ru(H$_2$)(dppm)]$^+$ show remarkably high isotope and temperature dependence of d_{HH} (ranging from 1.037 Å for d_{DT} at 220 K to 1.092 Å for d_{HD} at 286 K) as determined by the various NMR J couplings. This is attributed to the extremely flat PES that defines the H–H and M–H interactions in this complex, which allows the ZPE differences among the various isotopomers to be directly reflected in d_{HH}. The striking change of d_{HH} with small changes in temperature is due to thermal population of vibrational excited states, which are only slightly higher in energy than the ground state, an unprecedented situation in a readily isolable molecule.

A small class of polyhydrides such as $[CpIrH_3L]^+$, $CpRuH_3L$, and Cp_2NbH_3 with hydrides that are separated by approximately 1.7 Å exhibit quantummechanical exchange coupling with extraordinarily large J_{HH} that can exceed 1000 Hz in some cases. As will be discussed in Chapter 6, quantum-mechanical tunneling of two protons through a vibrational potential surface accounts for quantum-mechanical exchange coupling; covalent bonding between the hydrogens is not needed.

Parahydrogen induced polarization (PHIP)[62] is useful in the detection of short-lived reaction intermediates in hydrogenation processes and is potentially useful in studying reactions of H_2 ligands. Apparently because of the rapid equilibrium with the dihydride tautomer in Eq. (5.2) the PHIP effect is not seen for para-H_2 addition to $W(CO)_3(PR_3)_2$ or $Mo(CO)(diphosphine)_2$ systems.[10] However, it provides crucial evidence in experiments demonstrating that η^2-H_2 can react directly with substrates bound to the same metal center without first undergoing OA (Section 9.5.1). The Mo–norbornadiene (NBD) complex in Eq. (5.10) forms an unstable H_2 complex on photolysis, which is proposed to directly transfer H_2 to NBD to form norbornene (NBN).[63] In situ NMR studies of the photocatalyzed

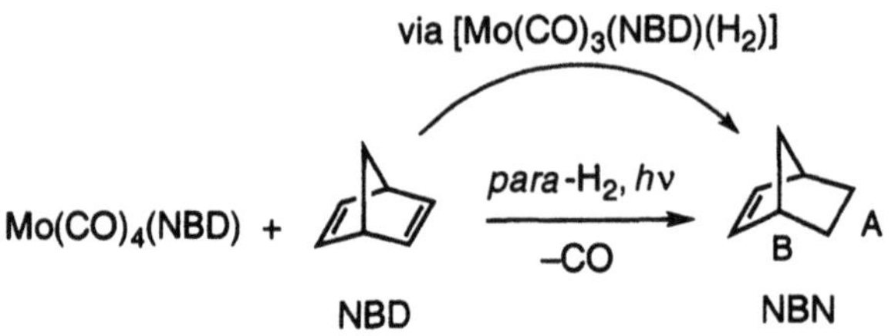

$$Mo(CO)_4(NBD) + \text{NBD} \xrightarrow[-CO]{para\text{-}H_2,\, h\nu} \text{NBN} \quad \text{via } [Mo(CO)_3(NBD)(H_2)] \tag{5.10}$$

reaction in Eq. (5.10) show that the nuclear spin polarization from para-H_2 is transferred to protons at the A and B positions of the NBN product. This can only occur if the H–H bond in the presumed $[Mo(CO)_3(NBD)(H_2)]$ intermediate is not first cleaved to dihydride ligands before the hydrogens are transferred to the coordinated NBD.

5.5. NMR RELAXATION TIME T_1, DEUTERIUM QUADRUPOLE COUPLING, AND EFFECTS OF DIHYDROGEN ROTATION

Another standard though more problematic method for determining the solution d_{HH} involves measuring the minimum value of the relaxation time, $T_{1(min)}$, for the H nuclei of η^2-H_2. This provides a reasonably accurate d_{HH} because dipolar relaxation of one H by its close neighbor, the partner H of the H_2 ligand, is dominant and T_1 is proportional to the sixth power of d_{HH}. In 1985 Crabtree attributed the broadness of NMR resonances for H_2 ligands to rapid dipole–dipole relaxation, which gives very small values for T_1 (normally < 50 ms compared to ≫100 ms for classical hydrides).[64] This became a useful criterion for η^2-H_2 versus dihydride complexes, although subsequent controversy necessitated extensive refinements over a period of many years. The theory for dipole–dipole relaxation shows that T_1 varies with temperature and goes through a minimum. Subsequently Crabtree and Hamilton developed a quantitative method for extracting d_{HH} from the experimentally determined $T_{1(min)}$.[5,65] Because T_1 depends on d_{HH}^6, it is extremely sensitive to the presence of hydrogens that are close together as in an H_2 complex, and values as low as 3 ms (at 300 MHz) are observed. Furthermore, the dipolar relaxation *usually* dominates the relaxation [e.g., >95% for $FeH_2(H_2)(PEtPh_2)_3$]. However, there can be several interfering factors, such as contribution to relaxation by nearby hydrogens on ligands, e.g. ortho H on arylphosphines.[66] More significant, as pointed out by Halpern,[46] is the fact that certain metals with a high gyromagnetic ratio, such as Co, Re, and Mn, can make substantial contributions, and in polyhydride complexes with cis hydrogens, all hydride–hydride interactions have to be considered. Corrections for this can be made in some cases, and measurement of $T_{1(min)}$ for both the H_2 and HD isotopomers can be used to cancel out relaxation caused by M. With the values of 15.2 and 116 ms determined for $[Mn(CO)(dppe)(H_2)]^+$ and its HD isotopomer, respectively, both d_{HH} [0.91(2) Å, cf 0.89(1) Å by solid state NMR[44]] and d_{MnH} [1.64(3) Å] can be calculated using this subtraction method.[67] The solution NMR determination of d_{HH} in polyhydrides of unknown structure is impractical, but qualitatively abnormally short $T_{1((min))}$ as in $[OsH_5(PPhMe_2)_3]^+$ (68 ms) shows evidence of a close H⋯H contact, which led to speculation that η^2-H_2 was present (see Section 4.8). Care must be exercised in applying this criterion because this value depends linearly on field strength (e.g. $T_{1(min)}$ = 50 ms at 200 MHz corresponds to 100 ms at 400 MHz). As will be shown below, $T_{1(min)}$ for ^2H NMR of polydeuterides *increases* on going from classical to nonclassical structures.

Complexes with a hydride ligand trans to η^2-H_2, as in $[MH(H_2)(PP)_2]^+$, give quite satisfactory results as shown in Tables 5.1 and 5.2. The fast internal rotation of η^2-H_2 that causes crystallographic problems can also affect the calculation of d_{HH} from $T_{1(min)}$ if the rotational rate is faster than molecular tumbling. Although the term "spinning" H_2 is often used, the H_2 does not really spin like a propeller but undergoes rapid libration combined with less frequent complete rotation. This reorientational motion can affect dipolar relaxation, which Morris addresses by viewing fast H_2 rotation much like methyl group rotation studied by Woessner, where d_{HH} calculated from $T_{1(min)}$ is corrected by a factor of 0.793.[68] The formulas for calculation are:

$$d_{HH}(\text{slow rotation}) = 5.81[T_{1(min)}/v]^{1/6}$$
$$d_{HH}(\text{fast rotation}) = 4.61[T_{1(min)}/v]^{1/6} \quad (5.11)$$
$$d_{HH}(\text{intermediate}) = C(\phi)[T_{1(min)}/v]^{1/6}$$

where v is the spectrometer frequency in megahertz and $C(\phi)$ is related to the average amplitude of the transitional libration as expressed by the angle ϕ.[66,68] This methodology can only set limits on d_{HH}, a longer one when the H_2 rotational frequency is lower than the spectrometer frequency and a shorter one when rotation is faster. Structural and spectroscopic data for a large series of cationic complexes, $[MH(H_2)(PP)_2]^+$ (3) (M = Fe, Ru) strongly reinforce this supposition.

$$\left[\begin{array}{c} R_2 \\ P_{\prime\prime\prime\prime} \\ \begin{array}{c} \end{array} \\ P \\ R_2 \end{array} \begin{array}{c} H \\ | \\ M \\ | \\ H_2 \end{array} \begin{array}{c} R_2 \\ \prime\prime P \\ \\ P \\ R_2 \end{array} \right]^+$$

(3)

Structure 3 shows large J_{HD} of 29.5–32.8 Hz, giving d_{HH} equal to 0.87–0.94 Å from Eqs. (5.3) and (5.4), which is consistent only with the d_{HH} calculated from the fast-rotation formula (0.86–0.90 Å versus 1.09–1.15 Å for slow rotation). The uncorrected neutron d_{HH} is 0.82(3) Å for $[RuH(H_2)(dppe)_2]^+$ (d_{HH} corrected for H_2 rotation is 0.94 Å). For many complexes, d_{HH} lies between d_{HH}(slow) and d_{HH}(fast).[68] $W(CO)_3(P^iPr_3)_2(H_2)$ clearly has a rapidly reorienting H_2 with rotational barrier near 2 kcal/mol in the solid state and d_{HH} of 0.89 Å. However as seen in Table 5.1, the fast reorientation correction gives a much too low d_{HH} of 0.76 Å, whereas no correction gives a value of 0.96 Å (the potential disparity between solution and solid state values must also be considered). To rationalize this as well as why some H_2 complexes require a correction and some do not, the character of H_2 reorientation must be examined, such as the angle of torsional libration, ϕ.[17,54,68] Inelastic neutron scattering studies generally show a double-minimum sinusoidal potential for rotation, but a smaller four-fold term is often added as a correction for nonsinusoidal behavior such as wobbling off the plane of rotation (see Chapter 6). However, in some cases, e.g., four identical ligands cis to H_2 as in $[MH(H_2)(PP)_2]$, a four-fold component to rotation truly exists (90° rotation) giving very low rotational barriers near 0.5 kcal/mol and a faster-spinning H_2.

Unlike most H_2 complexes, the metallocenes $[M(Cp')_2(H_2)(L)]^+$ (M = Nb, Ta; L = P-donor, CO, CNR) show *blocked rotation* of the H_2 ligand on the NMR timescale.[69–71] At low temperatures, decoalescence of signal is observed for the HD

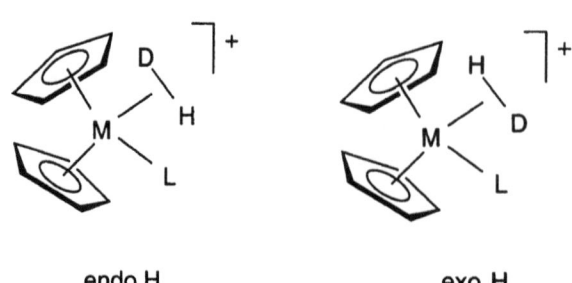

endo H   exo H

complex, although not for the HH species. For $[Nb(C_5H_4SiMe_3)_2(H_2)(PMe_2Ph)]^+$, normal values of $T_{1(min)}$ (19 ms at 188 K) and J_{HD} (15 Hz) are found, indicative of an elongated d_{HH} (1.17 Å). However, for the HD complex at 203 K, two different rotamers, endo H and exo H, are observed by ^1H NMR at 300 and 500 MHz. Very different J_{HP} (50.5 Hz for endo and less than 7 Hz for exo) are found, and when the temperature is raised, the rotamers interconvert by rotation of η^2-HD and coalescence occurs at 233 K. At 273 K, a single resonance appears with $J_{HD} = 16$ Hz and $J_{HP} = 26.4$ Hz. The free energy of activation of the H_2 internal rotation is 11.7 kcal/mol here and 8–12 kcal/mol both experimentally and calculationally for other Nb and Ta analogues.[69-71] The restricted rotation has also been seen in unstretched H_2 complexes such as $[Cp_2Ta(H_2)(CO)]^+$ ($J_{HD} = 27.5$ Hz) and has been linked to quantum-mechanical exchange coupling (see Section 6.3.4).[69,71] A nonmetallocene octahedral Ir system with elongated d_{HH} (1.22 Å) shows restricted H_2 rotation in the NMR.[42] Decoalescence occurs at 213 K and two signals are observed at 193 K for

the "frozen out" inequivalent hydrogens (Figure 5.9). The barrier to H_2 rotation is 12 kcal/mol as determined by NMR (11 kcal/mol from DFT calculations), which is much higher than in fast-rotating H_2 ligands with shorter d_{HH} (<3 kcal/mol from inelastic neutron scattering; see Chapter 6).

Because of the uncertainties in rotational effects discussed above, it is usually best to evaluate d_{HH} in solution by J_{HD} than by $T_{1(min)}$, although in some cases J_{HD} cannot easily be determined whereas $T_{1(min)}$ can. One advantage of T_1 is that it allows indirect detection of H_2 ligands that might be in fast exchange with other H nuclei or with free H_2. Measurement of anomalously short T_1 for the resonance of *dissolved* H_2 may thus be a sensitive indicator that an H_2 complex is present in low concentration. For the free H_2 in Eq. (5.12), T_1 is 1.2 s at 308 K (200 MHz) but is greatly reduced to 0.016 s at 193 K as the equilibrium shifts to the right and the H_2 resonance decreases from 4.4 ppm to 2.41 ppm.[72]

$$RuHCl(CO)(P^iPr_3)_2 + H_2 \rightleftharpoons RuHCl(H_2)(CO)(P^iPr_3)_2 \qquad (5.12)$$

The linewidth of the ^1H NMR signal for η^2-H_2 is generally larger than expected for reasons that are unclear.[73] Above the temperature corresponding to $T_{1(min)}$, T_1 should equal T_2, which determines the linewidth. Because T_1 should be the only significant contributor, a linewidth of $1/\pi T_1$ is expected, yet it is often substantially larger, e.g., 178 Hz at 253 K (300 MHz) for $[IrH(H_2)(PPh_3)_2(7,8\text{-benzoquinoline})]^+$ versus a calculated value of only 23 Hz from $T_1 = 14.1$ ms. Part of this (73 Hz) can be accounted for by exchange between the cis H_2 and hydride ligands, but the remaining 82 Hz has no obvious origin. The HD isotopomer has much narrower

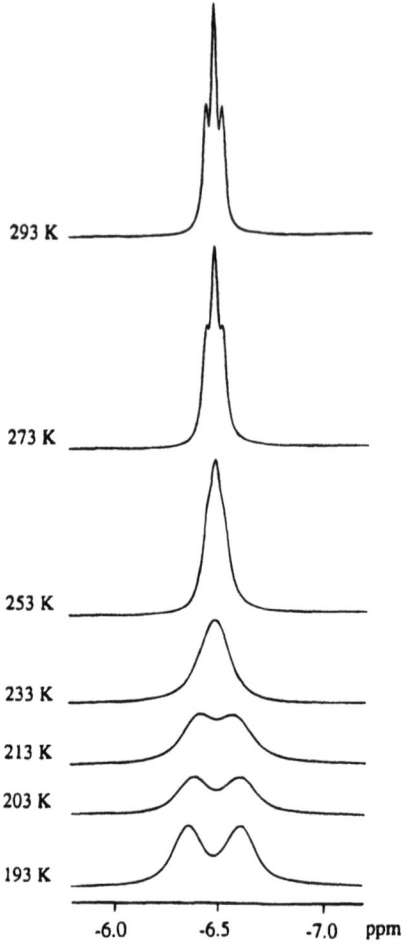

Figure 5.9. Variable temperature ^{1}H NMR spectra (300 MHz, toluene-d_8) of OsCl(H$_2$)(NH=C(Ph)-C$_6$H$_4$)(P^{i}Pr$_3$)$_2$. Reprinted with permission from Barea et al.[42] Copyright 1998 American Chemical Society.

bands, e.g., 24 Hz for each peak of the 1:1:1 triplet. The above extends to many other H$_2$ complexes, even those without cis hydrides or other exchangeable protons. Also the linewidth can increase sharply at low temperature for η^2-H$_2$ but not for η^2-HD, and the broadening is larger at 500 MHz than 300 MHz, possibly because H$_2$ rotation slows and quantum coupling effects make $J_{HH'}$ very large (see Chapter 6).

Another parameter that may be very relevant to the structural characterization of polyhydrido complexes in both solution and solid is the deuterium quadrupole coupling constant (C_Q).[47,74–78] Formation of a chemical bond creates a strong nonhomogeneous electric field along its direction (z axis) measurable by the electric field gradient, which at deuterium is expressed as C_Q and is affected by the X–D (X ≠ M) bonding interaction and other well-known factors. In particular, C_Q depends on the nuclear charge of the atom X, and rises as the X–D bond length decreases. Both theoretical and experimental methods have been developed to determine the C_Q of deuterium in a complex. MO calculations[75] on a $[Rb-D_2]^+$ model compound predict that C_Q increases from about 15 to 155 kHz when the Rb dihydride species is transformed into the η^2-H_2 derivative (C_Q is 227 kHz in free HD and D_2[76]). The electric field gradient is the sum of electronic and nuclear terms, and the latter is 404 kHz for free HD, where d_{HD} is 0.79 Å. When the bond is elongated to 0.85 Å, this contribution is reduced to 312 kHz so the C_Q of 155 kHz obtained calculationally for $[Rb-D_2]^+$ is reasonable.

C_Q can be determined from the quadrupole splitting (Δv_q) of deuterium from solid state ^2H NMR spectra,[47,77] but this is not universal because calculation of C_Q from Δv_q depends strongly on internal D_2 motion. For example, solid state ^2H NMR of $W(D_2)(CO)_3(P^iPr_3)_2$ shows a resonance with a quadrupole splitting of 62 kHz,[47] which gives C_Q = 165 kHz if fast D_2 rotational diffusion is assumed, but only about half this if not. Since librational motion occurs (averaged librational angle of 16° from solid state ^1H NMR), C_Q can only be estimated to lie somewhere between 83 and 165 kHz.

Bakhmutov and coworkers showed that variable-temperature ^2H-T_1 relaxation experiments in solution provide an alternative method for determining C_Q.[74] ^2H-$T_{1(min)}$ values of both D and D_2 ligands in $[pp_3M(D_2)D]^+$ (M = Ru, Os), $OsD(D_2)(CO)Cl(P^iPr_3)_2$, and $[Re(D_2)(PMe_3)_4(CO)]^+$ in CH_2Cl_2 can be measured using variable-temperature ^2H NMR. The ^2H-$T_{1(min)}$ values are 14–44 ms and do not readily correlate with the ^1H-$T_{1(min)}$, d_{HH}, or J_{HD} data for the H_2 and HD isotopomers. Various C_Q's for η^2-D_2 are determined from the $T_{1(min)}$ data using different models of internal D_2 motion, and the librational one provides the most realistic behavior. Accordingly, the calculated C_Q in the nonclassical complexes are 47–86 kHz or 56–101 kHz if α (the angle between the direction of the electric field gradient and the motion axis) is close to the magic angle.

For trans-$[M(D_2)(D)(dppe)_2]^+$ (M = Ru, Os) in solution, variable-temperature ^2H NMR gives much longer $T_{1(min)}$ values for the D_2 than the D ligands because of the rapid spinning of the D_2.[78] The ^2H-$T_{1(min)}$ for η^2-D_2 in [(triphos)-$RhD_2(D_2)]^+$, 1.34 s (average $T_{1(min)}$ = 0.0326 s) can be up to 80 times higher than for the deuterides in (triphos)RhD_3, 0.0165 s.[74] For the dppe system, C_Q for the terminal deuteride is calculated to be 79–81 kHz, and the motion-reduced coupling constant (C_Q^{eff}) for the D in η^2-D_2 is 19–31 kHz.[78] The actual C_Q for η^2-D_2 with short d_{DD} (<1 Å) should be at least $2C_Q^{eff}$ and greater than that of the terminal deuteride. This large difference demonstrates that the D_2 is rotating fast (61 MHz) and that the primary electric field gradient is not directed along the D–D vector but is tilted toward M. The tilt angles indicate that the major concentration of electron density is at the center of the MD_2 triangle for complexes with d_{HH} < 1 Å and that neither D–D nor M–D bonding is predominant, in agreement with bonding models. For $[Cp^*RuD_2(dppm)]^+$ with elongated d_{DD} (1.1 Å), C_Q^{eff} is 66 kHz, similar to the C_Q for its dideuteride tautomer, 71 kHz. Solid state NMR studies of $OsCl_2(D \cdots D)(CO)(P^iPr_3)_2$ and $[RuCl(D_2)(dppe)_2]^+$ give static C_Q of 120–167 and

107 ± 4 kHz respectively, the first such determination for D_2 complexes.[78] The barrier to twofold reorientation of the D_2 is 4.7 kcal/mol, the first such determination by a method other than inelastic neutron scattering (see Chapter 6). The first observation of quantum exchange coupling (see Section 6.3.4) between two deuterium nuclei is seen also.

5.6. H–H DISTANCE AND OTHER CRYSTALLOGRAPHIC AND STRUCTURAL ASPECTS OF DIHYDROGEN COORDINATION

The d_{HH} determined by solution NMR do not significantly vary from those in the solid state for the same complex. Theoretically calculated distances using modern approaches show remarkable agreement with experimental values (see Chapter 4). Models for complexes include not only PH_3 ligands but more realistic PMe_3 ligands,

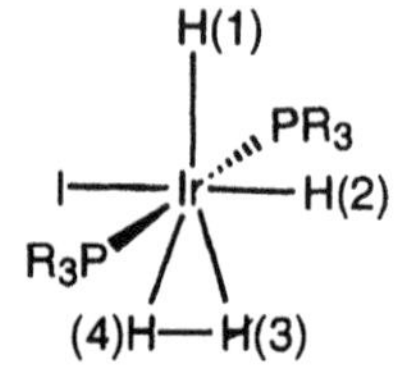

e.g., in calculations for $IrIH_2(H_2)(P^iPr_3)_2$.[79] Use of PMe_3 to model P^iPr_3 rather than PH_3 gives better results, and the calculated d_{HH} is only 0.006 Å longer than the neutron value of 0.856 Å.

In most cases, crystallographic d_{MH} for $M(\eta^2\text{-}H_2)$ are essentially equal, i.e., H_2 is bound symmetrically. Rare exceptions include $ReCl(H_2)(PMePh_2)_4$, with d_{ReH} = 1.49(9) and 1.98(9) Å,[33] although this is from X-ray data, and d_{HH} is long with a large standard deviation, 1.17(13) Å, i.e., closer to a dihydride with weak H ··· H interactions. A more genuine case of asymmetric binding of a true H_2 ligand was discussed in Chapter 4 for $[FeH(H_2)(PMe_3)_4]^+$, where Fe–H_a and Fe–H_b distances are 1.72(4) and 1.53(4) Å [d_{HH} = 0.84(4) Å].[80] Calculations give similar bond distances, and the asymmetry is due to *cis interaction* between the H_2 and H. Thus unless such a cis interaction is present, M–H_2 is symmetrical, as for M–ethylene.

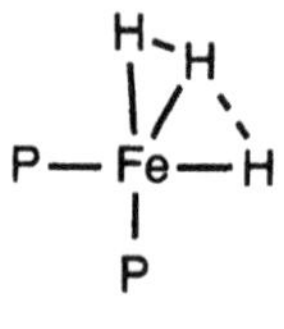

The existence of bridging H_2 ligands has not been definitively proven by diffraction methods, although there is NMR evidence for them in a porphyrin system. From NMR data (J_{HD} = 15 Hz, $T_{1(min)}$ of 66 ms at 200 MHz) $Ru_2(\mu\text{-}DPB)(H_2)(Im)_2$ is seen to have one H_2 between the two Ru, which are held in close proximity by the tethered DPB porphyrin rings (Im = imidazole).[81] Determination of the dipolar coupling in high-field NMR experiments suggests that H–H aligns perpendicular to the Ru–Ru vector in Eq. (5.13), but whether H_2 truly bridges or rapidly shuttles between Ru sites is unclear. H_2 binds reversibly, although the five-coordinate species slowly disproportionates to products that no longer bind H_2. Attempts to prepare Fe and Os analogues or the Ru complex with no axial ligands

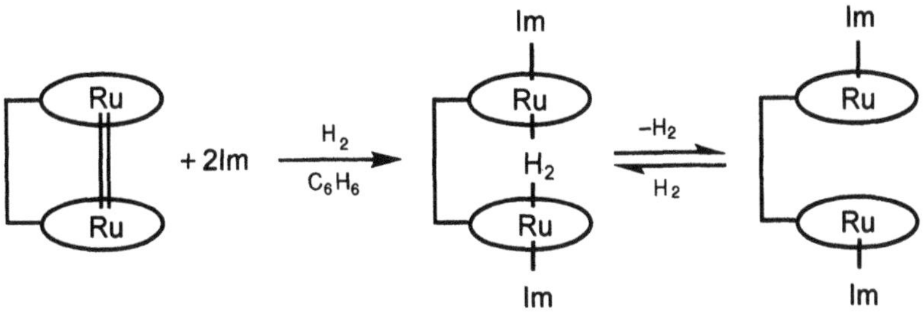

$$\text{(5.13)}$$

or with PPh_3 instead of sterically hindered Im failed, exemplifying the stringent stereoelectronic factors often required for stable σ-complex formation.

The crystal structure of a bis-(H_2) complex has now been determined. $RuH_2(H_2)_2(P^iPr_3)_2$ showed disorder but electron density in the difference Fourier map is consistent with the presence of either cis or trans H_2 ligands,[82] and $J_{HD} = 30 \pm 1$ Hz indicated $d_{HH} = 0.92 \pm 0.03$ Å. Also, v_{HH} is located at 2586 cm^{-1}. An X-ray structure of the PCy_3 analogue was also reported right after this, but disorder and other crystallographic problems were present (as reported in a *Chem. Comm. Corrigendum*).[23] The data give some indication that the H–H distance between the hydrogens in the H_2 ligands are close to 0.85 Å and that all hydrogens lie in the same plane as predicted theoretically (see Section 4.11). Cis interactions between the H_2 and/or hydride ligands may be present here, which would be consistent with the novel dynamics.[83] There is solution NMR evidence for a bis-trans-H_2 structure in an osmium pincer-ligand complex, $OsCl(H_2)_2(PCP)$, formed from reaction of $OsClH_2(PCP)$ under H_2 at $-70°C$ (PCP = [2,6-$(CH_2PBu_2^t)_2C_6H_3$]-).[84] However, on warming to $-30°C$ isomerization to a species with an H_2 ligand cis to two hydrides occurs. On further warming the latter loses H_2 and reversibly regenerates $OsClH_2(PCP)$.

Perhaps the most important structural determination since that of $W(CO)_3(PR_3)_2(H_2)$ was the neutron study of mer-$Fe(H_2)H_2(PEtPh_2)_3$, which in combination with inelastic neutron scattering data and theoretical calculations revealed that η^2-H_2 interacts with the hydride cis to it.[21] The Fe–hydride distances

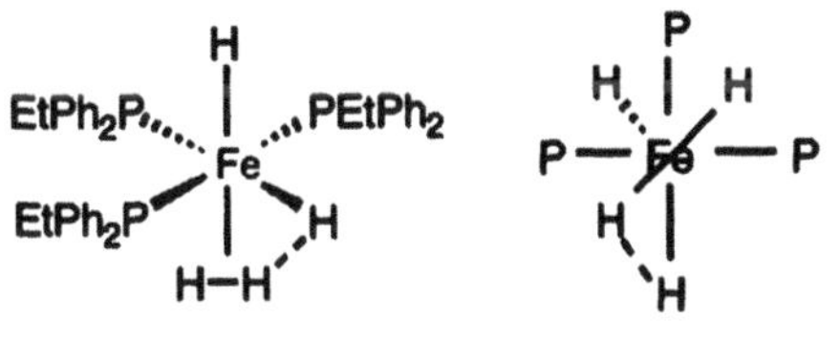

are 1.538(7) (trans to H_2) and 1.514(6) Å (trans to $PEtPh_2$), and the Fe–H (for H_2) distances are 1.607(8) and 1.576(9) Å. The d_{HH} in the η^2-H_2 is 0.821(10) Å, although it is undoubtedly closer to 0.9 Å because of bond foreshortening caused by H_2

rotation, as discussed above. Most significantly, the H–H bond orientation was unique: staggered with respect to the cis Fe–P and Fe–H axes [H–H eclipses equatorial ligands in all but one other case, Mo(CO)(dppe)$_2$(H$_2$), which has a very flat potential for H$_2$ rotation]. This is in direct opposition to both molecular mechanics and extended Huckel calculations, which predict a ground state structure where H–H eclipses the P–Fe–P direction based on more favorable backbonding. Inelastic neutron scattering studies reveal that the rotational barrier for η^2-H$_2$ is only about 1 kcal/mol, much lower than expected (see Chapter 6). All of these anomalies are explained by the attractive cis interaction between H$_2$ and hydride ligands, which overcomes the normal steric and electronic forces.

Hydrogen-bonding-type interactions can also exist between cis chloride and H$_2$ ligands, and these graphically illustrate the Brønsted acidity of η^2-H$_2$ and facilitate heterolytic cleavage of H$_2$ toward elimination of HCl (see Chapter 9). cis-IrHCl$_2$(H$_2$)(P^{i}Pr$_3$)$_2$ has both intramolecular and intermolecular hydrogen bonding to form an infinite chain structure, as shown by neutron diffraction.[43] The intermolecular separation between Cl and H is 2.64(2) Å compared to the sum of van der Waals radii of 2.7 Å. The intramolecular distance, Cl---H', is 2.65 Å compared to 2.76 Å between the noninteracting hydride and chloride ligands. Once again the H$_2$ ligand orients itself to maximize energetically favorable interactions with cis ligands since it easily could align parallel to P–M–P.

Clearly d_{HH} is the most important parameter for evaluating the degree of activation of the H–H bond along the reaction coordinate toward OA. The amount of BD the H$_2$ receives from M is the most important factor in elongating H–H and

the trans ligand is thus crucial, as discussed in Section 4.7.1. When strong π acceptors such as CO are trans to H$_2$, BD is greatly reduced and d_{HH} *is generally less than 0.9 Å*, even when the H$_2$ is at the brink of OA as in W(CO)$_3$(PR$_3$)$_2$(H$_2$). Here d_{HH} fails as an indicator of the apparent closeness of H$_2$ to OA as well as for Mo(CO)(dppe)$_2$(H$_2$), where merely changing the cis ligands to depe causes rupture of the short H–H bond (0.85–0.88 Å). Calculations show that BD is still extensive in these two systems (see Section 4.6) despite the enigmatic short d_{HH}. The presence of positive charge on M also reduces BD and keeps d_{HH} short, and many true H$_2$ complexes (d_{HH} < 1.0 Å) are cationic. Here, too, a trans CO ligand keeps d_{HH} below 0.9 Å, but there is surprisingly little difference between cationic and neutral congeners such as [M(CO)(dppe)$_2$(H$_2$)]$^{n+}$ (M = Mo, Mn$^+$, Fe^{2+}), where d_{HH} is near 0.88 Å and J_{HD} = 32–34 Hz. There is trade-off in the H$_2 \rightarrow$ M σ donation bonding component with the BD component, i.e., increased electrophilicity at cationic M increases σ donation at the expense of BD (the Fe *dication* actually has more tightly bound H$_2$). Thus, d_{HH} does not vary simply with changes in BD in this case, and the overall bonding situation must be considered. Increased σ donation from H$_2$ will also lengthen d_{HH} in lieu of decreased BD, although it alone cannot break H–H.

Conversely, the presence of a *donor* trans ligand and/or an overall donating ligand set can greatly increase d_{HH} to nearly that in classical hydrides. Substituting the trans CO with the π donor Cl stretches d_{HH} slightly beyond 1 Å for both neutral and cationic [MCl(dppe)$_2$(H$_2$)]$^{n+}$ (M = Tc, Ru$^+$). An even more dramatic example is [OsH$_2$(en)$_2$(acetate)]$^+$, which despite being cationic has a very long d_{HH} (neutron value is 1.34 Å, which is still marginally in the category of H$_2$ complexes).[17] Here the ligands are all pure σ donors, one of the rare H$_2$ complexes of this type. Os^{II} is also known to backbond strongly, so it seems clear here that BD freely stretches the H–H bond. This is in direct contrast to the group 6 complexes with trans CO, where the H–H bond suddenly snaps well before the H$_2$ becomes stretched along the reaction coordinate.

Complexes containing H$_2$ can show both intramolecular and intermolecular interactions of a hydrogen-bonding nature, as discussed by Braga and coworkers.[9] Often these involve contacts in the 2.2–2.6 Å range between hydrogens on phosphine ligands and the oxygens of CO ligands or the fluorines of anions such as BF$_4$. Except for IrHCl$_2$(H$_2$)(P^{i}Pr$_3$)$_2$ as noted above, the H$_2$ ligand itself only rarely is involved in intermolecular interactions because it is usually well-shielded by coligands. Intramolecular interactions can occur but these influence primarily the barrier to rotation of the H$_2$ (see Chapter 6).

REFERENCES

1. Howard, J. A. K.; Johnson, O.; Koetzle, T. F.; Spencer, J. L. *Inorg. Chem.* **1987**, *26*, 2930.
2. Johnson, T. J.; Albinati, A.; Koetzle, T. F.; Ricci, J.; Eisenstein, O.; Huffman, J. C.; Caulton, K. G. *Inorg. Chem.* **1994**, *33*, 4966.
3. Brammer, L.; Howard, J. A. K.; Johnson, O.; Koetzle, T. F.; Spencer, J. L.; Stringer, A. M. *J. Chem. Soc, Chem. Commun.* **1991**, 241.
4. Johnson, T. J.; Huffman, J. C.; Caulton, K. G.; Jackson, S. A.; Eisenstein, O. *Organometallics* **1989**, *8*, 2073.
5. Hamilton, D. G.; Crabtree, R. H. *J. Am. Chem. Soc.* **1988**, *110*, 4126.
6. Heinekey, D. M.; Oldham, W. J., Jr. *Chem. Rev.* **1993**, *93*, 913.
7. Kubas, G. J.; Nelson, J. E.; Bryan, J. C.; Eckert, J.; Wisniewski, L.; Zilm, K. *Inorg. Chem.* **1994**, *33*, 2954.
8. Bau, R.; Drabnis, M. H. *Inorg. Chim. Acta.* **1997**, *259*, 27; Koetzle, T. F. *Trans. Am. Cryst. Assoc.* **1997**, *31*, 57.
9. Braga, D.; De Leonardis, P.; Grepioni, F.; Tedesco, E. *Inorg. Chim. Acta* **1998**, *273*, 116.
10. Kubas, G. J.; Burns, C. J.; Eckert, J.; Johnson, S.; Larson, A. C.; Vergamini, P. J.; Unkefer, C. J.; Khalsa, G. R. K.; Jackson, S. A.; Eisenstein, O., *J. Am. Chem. Soc.* **1993**, *115*, 569.
11. Cruickshank, D. W. J. *Acta Crystallogr.* **1956**, *9*, 757; Busing, W. R.; Levy, H. A. *Acta Crystallogr.* **1964**, *17*, 142.
12. Maverick, E. F.; Trueblood K. N. THMA11-Program for Thermal Motion Analysis, UCLA, Los Angeles, CA 1988.
13. Koetzle, T. F. *Trans. Am. Crystallogr. Assoc.* **1997**, *31*, 57, and references therein.
14. Zilm, K. W.; Millar, J. M. *Adv. Mag. Opt. Reson.* **1990**, *15*, 163.
15. Kubas, G. J.; Ryan, R. R.; Swanson, B. I.; Vergamini, P. J.; Wasserman, H. J. *J. Am. Chem. Soc.* **1984**, *106*, 451.
16. Zilm, K. W.; Merrill, R. A.; Kummer, M. W.; Kubas, G. J. *J. Am. Chem. Soc.* **1986**, *108*, 7837.
17. Gusev, D. G.; Kuhlman, R. L.; Renkema, K. H.; Eisenstein, O.; Caulton, K. G. *Inorg. Chem.* **1996**, *35*, 6775.

18. Morris, R. H.; Sawyer, J. F.; Shiralian, M.; Zubkowski, J. D. *J. Am. Chem. Soc.* **1985**, *107*, 5581; Ricci, J. S.; Koetzle, T. F.; Bautista, M. T.; Hofstede, T. M.; Morris, R. H.; Sawyer, J. F. *J. Am. Chem. Soc.* **1989**, *111*, 8823.
19. Hills, A.; Hughes, D. L.; Jimenez-Tenorio, M.; Leigh, G. J.; Rowley, A.T. *J. Chem. Soc., Dalton Trans.* **1993**, 3041.
20. Gusev, D. G.; Hubener, R.; Burger, P.; Orama, O.; Berke, H. *J. Am. Chem. Soc.* **1997**, *119*, 3716.
21. Van Der Sluys, L. S.; Eckert, J.; Eisenstein, O.; Hall, J. H.; Huffman, J. C.; Jackson, S. A.; Koetzle, T. F.; Kubas, G. J.; Vergamini, P. J.; Caulton K. G. *J. Am. Chem. Soc.* **1990**, *112*, 4831.
22. Albinati, A.; Klooster, W. T.; Koetzle, T.F.; Fortin, J. B.; Ricci, J. S.; Eckert, J.; Fong, T. P.; Lough, A. J.; Morris, R. H.; Golombek, A. P. *Inorg. Chim. Acta* **1997**, *259*, 351.
23. Borowski, A. F.; Donnadieu, B.; Daran, J.-C.; Sabo-Etienne, S.; Chaudret, B. *Chem. Comm.* **2000**, 543. (See also *Corrigendum* published August 2, 2000).
24. Hussein, K.; Marsden, C. J.; Barthelat, J.-C.; Rodriguez, V.; Conjero, S.; Sabo-Etienne, S.; Donnadieu, B.; Chaudret, B. *Chem. Comm.* **1999**, 1315.
25. Schlaf, M.; Lough, A. J.; Morris, R. H. *Organometallics* **1997**, *16*, 1253.
26. Hampton, C. R. S. M.; Butler, I. R.; Cullen, W. R.; James, B. R.; Charland, J.-P.; Simpson, J. *Inorg. Chem.* **1992**, *31*, 5509.
27. Jimenez-Tenorio, M.; Jimenez-Tenorio, M. A.; Puerta, M. C.; Valerga, P. *Inorg. Chim. Acta.* **1997**, *259*, 77.
28. Schlaf, M.; Lough, A. J.; Maltby, P. A.; Morris, R.H. *Organometallics* **1996**, *15*, 2270.
29. Esteruelas, M. A.; Lahoz, F.J.; Oro, L. A.; Onate, E.; Ruiz, N. *Inorg. Chem.* **1994**, *33*, 787.
30. Eckert, J.; Jensen, C. M.; Koetzle, T. F.; Le-Husebo, T.; Nicol, J.; Wu, P. *J. Am. Chem. Soc.* **1995**, *117*, 7271.
31. Jensen, C. M.; Klooster, W. T.; Koetzle, T. F., unpublished data.
32. Kim, Y.; Deng, H.; Meek, D. W.; Wojcicki, A. *J. Am. Chem. Soc.* **1990**, *112*, 2798.
33. Cotton, F. A.; Luck, R. L. *Inorg. Chem.* **1991**, *30*, 767.
34. Gusev, D.; Llamazares, A.; Artus, G.; Jacobsen, H.; Berke, H. *Organometallics* **1999**, *18*, 75.
35. Jimenez-Tenorio, M.; Puerta, M. C.; Valerga, P. *J. Chem. Soc., Chem. Commun.* **1993**, 1750.
36. Chaudret, B.; Chung, G.; Eisenstein, O.; Jackson, S. A.; Lahoz, F. J.; Lopez, J. A. *J. Am. Chem. Soc.* **1991**, *113*, 2314; Burrow, T.; Sabo-Etienne, S.; Chaudret, B. *Inorg. Chem.* **1995**, *34*, 2470.
37. Klooster, W. T.; Koetzle, T. F.; Jia, G.; Fong, T. P.; Morris, R. H.; Albinati, A. *J. Am. Chem. Soc.* **1994**, *116*, 7677. Law, J.; Mellows, H.; Heinekey, D. M. *J Am. Chem. Soc.* **2001**, *123* (submitted).
38. Gross, C. L.; Young, D. M.; Schultz, A. J.; Girolami, G. S. *J. Chem. Soc. Dalton Trans.* **1997**, 3081.
39. Girolami, G. S.; Schultz, A. J.; private communication.
40. Maltby, P. A.; Schlaf, M.; Steinbeck, M.; Lough, A. J.; Morris, R. H.; Klooster, W. T.; Koetzle, T. F.; Srivastava, R. C. *J. Am. Chem. Soc.* **1996**, *118*, 5396.
41. Hasegawa, T.; Li, Z.; Parkin, S.; Hope, H.; McMullan, R. K.; Koetzle, T. F.; Taube, H. *J. Am. Chem. Soc.* **1994**, *116*, 4352.
42. Barea, G.; Esteruelas, M. A.; Lledos, A.; Lopez, A. M.; Onate, E.; Tolosa, J. I. *Organometallics* **1998**, *17*, 4065.
43. Albinati, A.; Bakhmutov, V. I.; Caulton, K. G.; Clot, E.; Eckert, J.; Eisenstein, O.; Gusev, D. G.; Grushin, V. V.; Hauger, B. E.; Klooster, W. T.; Koetzle, T. F.; McMullan, R. K.; O'Loughlin, T. J.; Pelissier, M.; Ricci, J. S.; Sigalas, M. P.; Vymenits, A. B. *J. Am. Chem. Soc.* **1993**, *115*, 7300; Bakhmutov, V. I.; Vorontsov, E. V.; Vymenits, A. B. *Inorg. Chem.* **1995**, *34*, 214.
44. King, W. A.; Luo, X-L.; Scott, B. L.; Kubas, G. J.; Zilm, K. W. *J. Am. Chem. Soc.* **1996**, *118*, 6782.
45. Wisniewski, L. L.; Mediati, M.; Jensen, C. M.; Zilm, K. W. *J. Am. Chem. Soc.* **1993**, *115*, 7533.
46. Cotton, F. A.; Luck, R. L. *J. Chem. Soc. Chem. Commun.* **1988**, 1277; Desrosiers, P. J.; Cai, L.; Lin, Z.; Richards, R.; Halpern, J. *J. Am. Chem. Soc.* **1991**, *113*, 4173.
47. Kubas, G. J.; Unkefer, C. J; Swanson, B. I.; Fukushima, E. *J. Am. Chem. Soc.* **1986**, *108*, 7000.
48. Khalsa, G. R. K.; Kubas, G. J.; Unkefer, C. J.; Van Der Sluys, L. S.; Kubat-Martin, K. A. *J. Am. Chem. Soc.* **1990**, *112*, 3855.
49. Luther, T. A.; Heinekey, D. M. *Inorg. Chem.* **1998**, *37*, 127.
50. Smith, K.-T.; Tilset, M.; Kuhlman, R.; Caulton, K.G. *J. Am. Chem. Soc.* **1995**, *117*, 9473, and references therein.
51. Bacskay, G. B.; Bytheway, I.; Hush, N. S. *J. Am. Chem. Soc.* **1996**, *118*, 3753.
52. Hush, N.S. *J. Am. Chem. Soc.* **1997**, *119*, 1717.

53. Grundemann, S.; Limbach, H.-H.; Buntkowsky, G.; Sabo-Etienne, S.; Chaudret, B. *J. Phys. Chem. A* **1999**, *103*, 4752.
54. Gusev, D. G.; Berke, H. *Chem. Ber.* **1996**, *129*, 1143.
55. Oldham, W. J., Jr.; Hinkle, A. S.; Heinekey, D. M. *J. Am. Chem. Soc.* **1997**, *119*, 11028.
56. Saunders, M.; Jaffe, M. H.; Vogel, P. *J. Am. Chem. Soc.* **1971**, *93*, 2558.
57. Bampos, N.; Field, L. D. *Inorg. Chem.* **1990**, *29*, 588.
58. Luo, X.-L.; Crabtree, R. H. *J. Am. Chem. Soc.* **1990**, *112*, 6912.
59. Bianchini, C.; Linn, K.; Masi, D.; Peruzzini, M.; Polo, A.; Vacca, A.; Zanobini, F. *Inorg. Chem.* **1993**, *32*, 2366.
60. Gutierrez,-Puebla, E.; Monge, A.; Paneque, M.; Poveda, M. L.; Taboada, S.; Trujillo, M.; Carmona, E. *J. Am. Chem. Soc.* **1999**, *121*, 346.
61. Earl, K. A.; Jia, G.; Maltby, P. A.; Morris, R. H. *J. Am. Chem. Soc.* **1991**, *113*, 3027.
62. Eisenberg, R. *Acc. Chem. Res.* **1991**, *24*, 110.
63. Thomas, A.; Haake, M.; Grevels, F. W.; Bargon, J. *Angew. Chem. Int. Ed. Engl.* **1994**, *33*, 755.
64. Crabtree, R. H.; Lavin, M. *J. Chem. Soc. Chem. Commun.* **1985**, 1661.
65. Crabtree, R. H. *Acc. Chem. Res.* **1990**, *23*, 95.
66. Bautista, M. T.; Earl, K. A.; Maltby, P. A.; Morris, R. H.; Schweitzer, C. T.; Sella, A. *J. Am. Chem. Soc.* **1988**, *110*, 7031; Morris, R. H. *Can. J. Chem.* **1996**, *74*, 1907.
67. King, W. A.; Scott, B. L.; Eckert, J.; Kubas, G. *J. Inorg. Chem.* **1999**, *38*, 1069.
68. Morris, R. H.; Wittebort, R. *J. Mag. Res. Chem.* **1997**, *35*, 243; Woessner, D. E. *J. Chem. Phys.* **1962**, *36*, 1; *37*, 647.
69. Antinolo, A.; Carrillo-Hermosilla, F.; Fajardo, M.; Garcia-Yuste, S.; Otero, A.; Camanyes, S.; Maseras, F.; Moreno, M.; Lledos, A.; Lluch, J. M. *J. Am. Chem. Soc.* **1997**, *119*, 6107.
70. Jalon, F. A.; Otero, A.; Manzano, B. R.; Villasenor, E.; Chaudret, B. *J. Am. Chem. Soc.* **1995**, *117*, 10123.
71. Sabo-Etienne, S.; Chaudret, B.; Abou el Makarim, H.; Barthelet, J.-C.; Ulrich, S.; Limbach, H.-H.; Moise, C. *J. Am. Chem. Soc.* **1995**, *117*, 11602; Sabo-Etienne, S.; Rodriguez, V.; Donnadieu, B.; Chaudret, B.; el Makarim, H. A.; Barthelat, J.-C.; Ulrich, S.; Limbach, H.-H.; Moïse, C. *New J. Chem.* **2001**, *25*, 55.
72. Gusev, D. G.; Vymenits, A. B.; Bakhmutov, V. I. *Inorg. Chem.* **1992**, *31*, 1.
73. Yao, W.; Faller, J. W.; Crabtree, R. H. *Inorg. Chim. Acta.* **1997**, *259*, 71.
74. Bakhmutov, V. I.; Bianchini, C.; Maseras, F.: Lledos, A.: Peruzzini, M.: Vorontsov, E. V. *Chem. Eur. J.* **1999**, *5*, 3318; Bakhmutov, V. I.; Bianchini, C.; Peruzzini, M.; Vizza, F.; Vorontsov, E. V. *Inorg. Chem.* **2000**, *38*, 1655.
75. Guo, K.; Jarrett, W. L.; Butler, L. G. *Inorg. Chem.* **1987**, *26*, 3001.
76. Butler, L. G.; Keiter, E. A. *J. Coord. Chem.* **1994**, *32*, 121.
77. Lambert, J. B.; Riddel, F. G., *The Multinuclear Approach to NMR Spectroscopy*, Reidel: Boston, 1982, pp. 151-167.
78. Facey, G. A.; Fong, T. P.; Gusev, D.; Macdonald, P. M.; Morris, R. H.; Schlaf, M.; Xu, W. *Can. J. Chem.* **1999**, *77*, 1899; Wehrmann, F.; Fong, T. P.; Morris, R. H.; Limbach, H H.; Buntkowsky, G. *Phys. Chem. Chem. Phys.* **1999**, *1*, 4033.
79. Li, S.; Hall, M. B.; Eckert, J.; Jensen, C. M.; Albinati, A. *J. Am. Chem. Soc.* **2000**, *122*, 2903.
80. Gusev, D. G.; Hubener, R.; Burger, P.; Orama, O.; Berke, H. *J. Am. Chem. Soc.* **1997**, *119*, 3716.
81. Collman, J. P.; Hutchison, J. E.; Wagenknecht, P. S.; Lewis, N. S.; Lopez, M. A.; Guilard, R. *J. Am. Chem. Soc.* **1990**, *112*, 8206; Collman, J. P.; Wagenknecht, P. S.; Hutchison, J. E.; Lewis, N. S.; Lopez, M. A.; Guilard; R.; L'Her, M.; Bothner-By, A. A.; Mishra, P. K. *J. Am. Chem. Soc.* **1992**, *114*, 5654.
82. Abdur-Rashid, K.; Gusev, D. G.; Lough, A. J.; Morris, R. H. *Organometallics* **2000**, *19*, 1652.
83. Sabo-Etienne, S.; Chaudret, B. *Coord. Chem. Rev.* **1998**, *178-180*, 381.
84. Gusev, D. G.; Dolgushin, F. M.; Antipin, M. Y. *Organometallics* **2001**, in press.
85. Toner, A. J.; Grundemann, S.; Clot, E.; Limbach, H.-H.; Donnadieu, B.; Sabo-Etienne, S.; Chaudret, B. *J. Am. Chem. Soc.* **2000**, *122*, 6777.

6

Intramolecular Dynamics of Dihydrogen–Hydride Ligand Systems: Hydrogen Rotation, Exchange, and Quantum-Mechanical Effects

6.1. INTRODUCTION

Transition metal complexes containing η^2-H_2 and hydride ligands are unquestionably the most dynamic ligand systems known. Classical polyhydride complexes have long been known to be stereochemically nonrigid in solution, as shown by investigations of group 6, 8, and 9 complexes of the type $MH_n(PR_3)_4$ ($n = 1, 2, 4$) by Meakin, Muetterties, and coworkers in the early 1970s.[1] All of these systems showed fluxional behavior, and low barriers for $n = 1$ or 2 were rationalized by a tetrahedral jump mechanism for rapid ligand exchange. Before H_2 coordination was recognized, the fluxionality of polyhydrides was viewed as isolated H atoms moving over the surface of the metal center. However, their association as H_2 ligands as an intermediate step is now a much more attractive possibility, e.g., for hydride site exchange in polyhydrides such as ML_4H_4 (M = Mo, W; L = P-donor), an intermediate with a geometry very much like *trans*-$M(H_2)_2L_4$ with elongated d_{HH} was considered possible even in 1973.[1] Many new examples of hydride fluxionality were

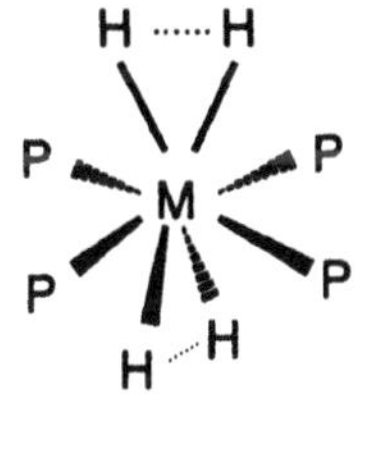

discovered subsequently, and the principal mechanistic aspects including systems containing η^2-H_2, have been reviewed.[2–4] As will be shown in this chapter, remarkably facile hydrogen site exchange between cis hydride and H_2 ligands can occur in the *solid state* at temperatures below 77 K with activation barriers as low as 1.5 kcal/mol!

For the H_2 ligand, additional solution dynamics can include equilibrium dissociation of H_2, equilibria between η^2-H_2–dihydride tautomers, ligand exchange via trihydrogen-like structures, and rotational motion of η^2-H_2 (Scheme 1). The first

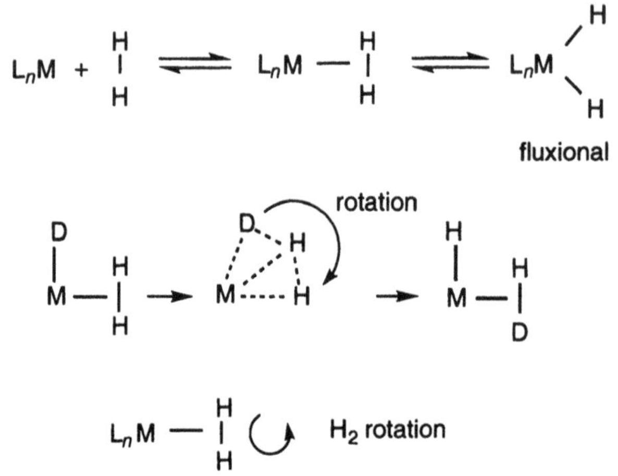

Scheme 1. Dynamic Behavior of M–H_2

set of equilibria essentially represents the reaction coordinate for H–H bond cleavage/formation, which in several systems takes place in solution at room temperature. In addition to or instead of this process, virtually all complexes with H_2 ligands cis to hydride undergo extremely facile ligand exchange with very low barriers of less than 5 kcal/mol. This cis effect was detailed in Chapter 4, and Jessop and Morris[3] provide an excellent review of exchange as well as of homolytic and heterolytic splitting of H_2 (see Chapter 9).

Finally, except for rare cases η^2-H_2 rotates rapidly (librational motion is more accurate) even in the solid state, further delocalizing the H atom positions over virtually the entire coordination sphere of a metal complex. In elongated H_2 complexes, the highly delocalized, rotating H_2 ligand can occupy a large volume as discussed in Section 4.7.4. One of the key diagnostics for coordination of H_2 is in fact the observation by inelastic neutron scattering (INS) of rotational transitions for η^2-H_2,[5–10] which cannot exist for classical hydrides. What is most important is that these extensive studies by Eckert and coworkers measure the barrier to rotation of H_2 ligands and consequently offer *direct experimental proof of* $M \to H_2$ *BD*. The rotational transitions and barrier are very sensitive to even minor changes in ligand environment about M and provide valuable insight into the reaction coordinate for splitting of H_2 on M and intramolecular interactions. A good example is the initial experimental evidence for interaction between η^2-H_2 and cis hydride ligand provided by INS, which will be discussed extensively later in this chapter.

Hydrogen reorientation among either identical or inequivalent sites is extremely complex because it can involve *quantum-mechanical* phenomena such as tunneling (in H$_2$ rotation) and exchange-coupling. The latter is an NMR effect in certain polyhydrides that undergo exchange of chemically inequivalent hydrogens, and will be discussed in Section 6.3.4.

6.2. DIHYDROGEN ROTATION IN M(η^2-H$_2$)

The H$_2$ ligand undergoes rapid two-dimensional hindered rotation about the M–H$_2$ axis, i.e., it spins (librates) in propeller-like fashion with little or no wobbling. The fact that there is a small barrier to rotation, ΔE, brought about by M → H$_2$ σ^*

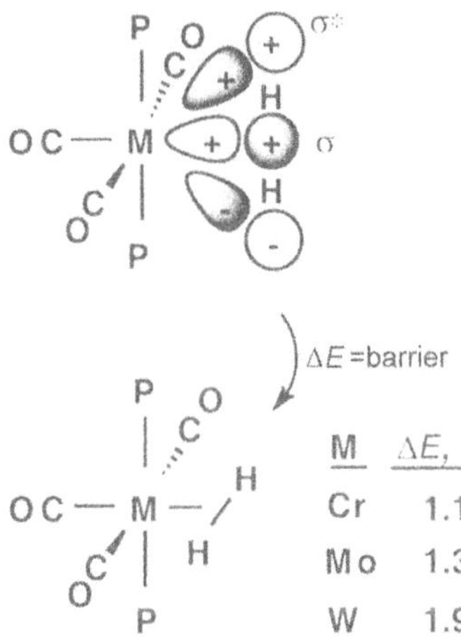

M	ΔE, kcal
Cr	1.17
Mo	1.32
W	1.90

backdonation (BD) is most significant. The σ donation from H$_2$ to M cannot give rise to a rotational barrier since it is completely isotropic about the M–H$_2$ bond. The barrier actually arises from the *disparity* in the BD energies from the *d* orbitals when H$_2$ is aligned parallel to P–M–P versus parallel to OC–M–CO, where BD is less (though not zero). ΔE varies with M and other factors and can be analyzed in terms of the BD and other forces that lead to it, both calculationally or by a series of experiments where M–L sets are varied. In most "true" H$_2$ complexes with $d_{HH} < 1.0$ Å, the barrier is only a few kcal/mol and observable only by neutron scattering methods. It can be as low as 0.5 kcal/mol for symmetrical ligand sets, e.g., all cis L are the same, but has never been measured to be zero because minor geometrical distortions are usually present. In the case of complexes with elongated H–H bonds[11] or where rotation is blocked as in [Cp$'_2$M(H$_2$)(L)]$^+$ (M = Nb, Ta) (see below), much higher barriers, up to 11 kcal/mol, are observed. For complexes with two H$_2$ ligands such as Cr(CO)$_4$(H$_2$)$_2$, a barrier of 4.8 kcal/mol is calculated for rotating the two H$_2$ ligands from their most stable orientation by 90°.[12] Interactions of η^2-H$_2$ with cis L can significantly lower the barriers, as will be shown below. INS is normally used for (and limited to) measurement of barriers lower than 3 kcal/mol, but solid state ^{2}H NMR can be used to determine higher barrriers. OsCl$_2$(D$_2$)(CO)(P^{i}Pr$_3$)$_2$ with a long d_{DD} of 1.1 Å gives a 4.7 kcal/mol barrier (see Section 5.5).

This hindered rotation of η^2-H_2 is governed by various forces, which can be divided into bonded (electronic) and nonbonded interactions ("steric" effects). The direct electronic interaction between M and H_2 results from overlap of the appropriate molecular orbitals. Nonbonded interactions such as van der Waals forces between the η^2-H_2 atoms and the other atoms on the molecule may vary as η^2-H_2 rotates. *Inter*molecular interactions should not contribute much to the barrier to rotation of η^2-H_2, as the metals are far apart. However, they may have a minor effect on the coordination geometry about M, which could in turn affect M–H_2 binding.

This section describes how information on M–H_2 binding can be extracted from series of measurements of the barrier to H_2 rotation. Background is given in Section 6.2.1 on the model for hindered rotation of a dumbbell molecule and the measurement of transitions of this hindered rotator by INS. Section 6.2.3 discusses the relationship of the rotational barrier to the various factors that give rise to it and the conclusions that can be drawn concerning M–H_2 binding.

6.2.1. Determination of the Barrier to Rotation of H_2

The geometry and height of the barrier can be derived by fitting the observed rotational transitions to a model for the barrier. The simplest possible model for the rotations of a dumbbell molecule is one of planar reorientation about an axis perpendicular to the midpoint of the H–H bond in a potential of twofold symmetry (Figure 6.1). More generally, terms with higher symmetry than twofold may be included in the Fourier expansion[13] of the rotational potential:

$$V(\phi) = \tfrac{1}{2} \sum_{2n} V_{2n}(1 - \cos 2n\phi) \tag{6.1}$$

where ϕ is the angle of rotation about this axis. While inspection of the structure of many H_2 complexes would suggest the presence of at least one additional set of potential minima (e.g., parallel to the OC–W–CO axis in Figure 6.1) theoretical calculations do not show the existence of a secondary set of potential minima.[6] The next term (V_4) in the above Fourier expansion of the potential is necessary to account for the temperature dependence of solid state NMR data.[14] However for the present purpose this term only slightly alters the shape of the potential well from purely twofold sinusoidal, which is indicated, e.g., in some of the theoretical treatments[15-17] of these systems.

The Schrödinger equation for this one-dimensional rotation[13] can be transformed into the Mathieu equation, and here (Figure 6.2, bottom) the H_2 performs large-amplitude librations within its potential minimum that are constrained to a plane perpendicular to the three-center M–H_2 bond. Strong support for this hypothesis is provided by solid state NMR, from which the out-of-plane librational amplitude can be estimated to be less than 6°.[14] In all other cases[18-20] where rotations of H_2 have been observed, both rotational degrees of freedom exist since the H_2 is not chemically bound. The energy level diagram for the twofold potential with one rotational degree of freedom is shown in Figure 6.1. The energy levels are given by BJ^2 for zero barrier height (B = rotational constant). The transition between the lowest two levels, i.e., $J = 0$ and $J = 1$, is akin to the ortho–para H_2

Intramolecular Dynamics of Dihydrogen–Hydride Ligand Systems

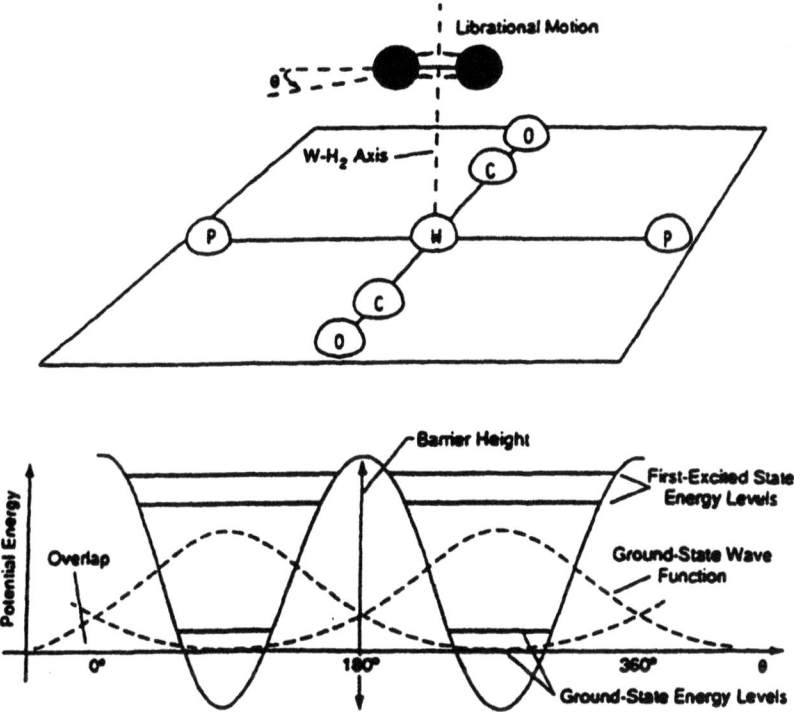

Figure 6.1. Model for the hindered rotation of the H_2 ligand in metal complexes. Schematic of H_2 rotation in $W(CO)_3(\eta^2\text{-}H_2)P_2$ about the axis from the W atom to the midpoint of the H–H bond in (top); double-minimum rotational potential with energy levels (not to scale) and wavefunctions (bottom). Reprinted with permission from Eckert and Kubas.[8] Copyright 1993 American Chemical Society.

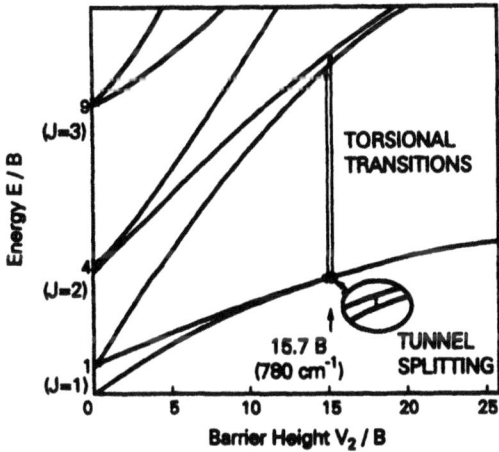

Figure 6.2. Energy-level diagram for rotation with one degree of freedom, ϕ, of a dumbbell molecule in a double-minimum potential $V_2(\phi)$. The transitions indicated are for $W(CO)_3(H_2)(PCy_3)_2$, where B is taken to be 49.5 cm^{-1}. Reprinted with permission from Eckert and Kubas.[8] Copyright 1993 American Chemical Society.

transition for (nearly) free hydrogen. Its rotation has two degrees of freedom and hence the energy levels are given by $BJ(J + 1)$. The ortho→para H_2 transition cannot be observed directly by optical methods, as it involves a change in the total nuclear spin of H_2, which is forbidden in optical spectroscopy.[18]

Application of a barrier to rotation rapidly decreases the separation between the lowest two rotational levels, which may then be viewed as a split librational ground state. Transitions within this ground state (Figure 6.1, bottom) as well as those to the excited librational state (often called torsions) may be observed by INS. The former occur by way of rotational tunneling[21] since the wave functions for the H_2 in the two wells 180° apart overlap. This rotational tunneling transition has an approximately exponential dependence on the barrier height and is therefore extremely sensitive to the latter. It is this property that is exploited to gain information on the origin of this barrier.

6.2.2. Inelastic Neutron Scattering

Both the rotational tunneling transition and the transitions to the first excited librational state can readily be observed by INS techniques.[5-10,21] Neutrons are extremely well suited as probes for molecular rotations when the motion involves mainly H atoms. These modes (torsions, librations) normally do not involve large changes in the polarizability or dipole moment of the molecule, and are therefore often difficult to observe by optical techniques. The INS studies allow observation of low-lying transitions within the ground librational state of the η^2-H_2 (tunnel splitting), which corresponds to the para ($I = 0$, $J = 0$) to ortho ($I = 1$, $J = 1$) transition for free H_2 (120 cm^{-1} in liquid hydrogen). This measurement can be performed without regard to other hydrogen-containing ligands that do not have observable excitations at low temperatures in the energy range of those of the H_2. Typical intensities of the tunneling peaks in these complexes are in fact about 0.25% or less than that of the intense central elastic peak (Figure 6.3).

INS measurements are typically carried out at about 5 K using approximately 1 g of polycrystalline H_2 complex sealed under inert atmosphere in aluminum or quartz sample holders. Boron-containing groups or anions such as BPh_4 can be a problem since boron is a strong absorber of neutrons. Low-frequency INS measurements are performed on the cold neutron time-of-flight spectrometers at the Institute Laue Langevin in Grenoble, France, and the Laboratoire Leon Brillouin. All the high-frequency data (>200 cm^{-1}) discussed below were obtained on the filter difference spectrometer at the Los Alamos Neutron Scattering Center. This measurement is only possible by use of a differential technique[22] involving subtraction of the spectra observed for a sample with a D_2-ligand (or another suitable "blank") from that of an identical sample with the H_2 ligand, which leaves only the vibrational modes for the M–H_2 fragment.

In most cases the ground-state rotational tunnel splitting, as well as the two transitions to the split excited librational state, are observed. Because the tunnel splittings (typically 1–10 cm^{-1}) can be measured with much better accuracy than the librational transitions, the value for the barrier height V_2 is usually extracted from the former. Higher terms in the expansion of the potential are introduced only if the librational transitions derived do not agree well with other observations. For

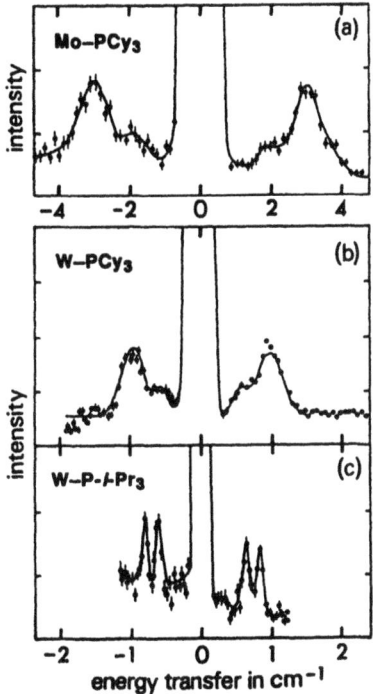

Figure 6.3. Rotational tunneling spectra for the series of complexes $M(H_2)(CO)_3(PR_3)_2$ where M = Mo and R = Cy (top) and M = W and R = Cy (middle) or ^{i}Pr (bottom). Note the change in energy scale between top and middle figures. Reprinted with permission from Eckert et al.[6] Copyright 1990 American Chemical Society.

example, analysis of the temperature dependence of solid state NMR data[14] has shown that for most of the systems studied a V_4 term improves the agreement with the NMR data significantly.

The use of INS for observing rotational tunneling transitions is limited by the energy resolution of the available spectrometers to studies involving molecules with a relatively large rotational constant B such as those in CH_4, $[NH_4]^+$, and methyl groups. Since H_2 has the largest value of B (59.6 cm^{-1}), it is the best candidate for the study of molecular rotation by neutron scattering. Because of the lengthening of the H–H bond upon activation, the value of B for the η^2-H_2 is reduced from that for free H_2 and is frequently not known. In cases where d_{HH} has been obtained by neutron diffraction, B is known, e.g., for $W(H_2)(CO)_3(PR_3)_2$ a B value of 49.5 cm^{-1} (B = 59.6 cm^{-1} for free H_2) can be derived from the d_{HH} (0.82 Å) determined by neutron diffraction. Prior to the discovery of H_2 complexes, however, the only systems known were those in which H_2 was either barely affected by its surroundings (physically bound) or dissociated into atomic hydrogen. The smallest splittings between the ortho- and para H_2 state that had previously been observed were 4.8–10.5 cm^{-1} for H_2 in K-intercalated graphite,[19] and 30.6 cm^{-1} for H_2 in Co ion-exchanged NaA zeolite.[20] In all likelihood in both of these cases H_2 is physisorbed as there was no indication of H–H bond activation. However, for $M(\eta^2$-$H_2)$, ground librational state splittings between 17 and 0.6 cm^{-1} are observed at temperatures as high as 200 K. The signals shift to lower energy and broaden but remain visible into the quasi-elastic scattering region, as will be discussed in Section 6.3.3 (see Figures 6.4 and 6.6). Observation of rotational tunneling, which is a *quantum-mechanical* phenomenon, at such a high temperature is extraordinary: in all previous studies of this type involving CH_3 or $[NH_4]^+$ the transition to classical behavior occurs well below 100 K. The barriers to rotation derived from the tunnel splitting by use of the simple model described in Section 6.2.1 range from 0.6 to 2.4 kcal/mol.

6.2.3. Origin of the Barrier to Rotation of η^2-H_2

6.2.3.1. Metal–Hydrogen Binding

As mentioned previously, the rotational barrier for H_2 is nearly completely intramolecular in origin—primarily the direct electronic interactions between H_2 and M and, to a lesser extent, the nonbonded interactions between the H atoms and the neighboring atoms on the same complex. The INS rotational tunneling spectra of several variants of $M(H_2)(CO)_3(PR_3)_2$ (M = group 6; R = Cy, ^{i}Pr) demonstrate the pronounced effect of the change in M (see Figure 6.3), as strongly supported by the theoretical calculations described below. The ratios of the tunnel splittings is approximately 1:3:5 for W:Mo:Cr, whereas variation of the phosphine shifts the mean tunneling transition by about 20%. As discussed earlier, the rotational barrier is mainly sensitive to BD and is affected by σ bonding only to the extent that it weakens and lengthens the H–H bond, thereby changing the value of B. However, the barrier to rotation is only a *relative* measure of the BD interaction because it results from the difference in energy between H_2 oriented in its minimum versus its maximum energy configurations, as is discussed in more detail below.

Electronic calculations support the above conclusion because calculated barrier heights, which do not include nonbonded interactions, show good agreement with experiment, as summarized in the review by Maseras *et. al.*[4] *Ab initio* and Fenske–Hall calculations performed on $M(H_2)(CO)_3(PR_3)_2$ (M = W, Mo and R = H, Me, ^{i}Pr, and Cy)[6,15] confirm the experimentally observed equilibrium orientation of η^2-H_2 along the P–M–P axis as well as give values for the barrier heights that are in remarkable agreement with those derived from INS data. For example, the barriers from the Fenske–Hall calculation[6] are 1.0 and 1.5 kcal/mol, respectively, for M = Mo, W and R = Cy compared with the revised experimental values of 1.3 and 1.9 kcal/mol discussed below. The energy barrier is a direct manifestation of the energy difference in the BD for H_2 aligned along the P–M–P axis versus the OC–M–CO axis, where BD is poorer because of competition from the strong π-acceptor CO ligands.

Virtually all the electronic calculations of barrier heights compare remarkably well with those derived from INS, which strongly supports the notion that the barrier arises from a variation in the overlap between $\sigma^*(H_2)$ and the relevant metal d orbitals upon rotation of η^2-H_2. The d orbital energies in turn are affected by at least three different factors: the nature of the M orbital structure, the effects of coligands on M (and thus on M–H_2 binding), and perturbations on the coordination geometry around M by counterions or solvent molecules via crystal-packing forces. The extent to which each of these contributions can be studied from barrier measurements will be described below.

6.2.3.2. Effect of the Metal Center

Systematic studies of the effect of M d orbitals on H_2 binding and the barrier to rotation have been carried out on group 6 and 8 complexes. Results for H_2 complexes with the same ligands but different M were compared in order to establish trends in $M-H_2$ bonding in general and, more specifically, the BD interaction to which the barrier is most sensitive. For $M(H_2)(CO)_3(PCy_3)_2$, the values for a simple twofold barrier increases down the group,[6,23,24] from 1.17 kcal/mol for the Cr analogue to 1.32 for Mo and 1.9 kcal/mol for W (revised values based on d_{HH} from solid state NMR for consistency[24]). The BD may therefore be said to increase in the order Cr < Mo < W. From this one might be tempted to conclude that H–H bond activation would increase in the same order (recall that the H–H bond of $W(CO)_3(PCy_3)_2(H_2)$ cleaves in solution to give equilibrium amounts of dihydride tautomer, while the Cr and Mo congeners do not). This might be indicated, e.g., by decreasing values of ν_{HH} in the order Cr < Mo < W. However, IR studies[25–27] of $M(CO)_5(H_2)$, show that ν_{HH} decreases in the order Mo > Cr > W (Table 6.1). Similarly, one might suppose that the strength of the M–H interaction would increase with an increasing amount of BD. Vibrational data (Table 6.1) for $\nu_{s(MH_2)}$ in the PCy_3 complexes, however, show that this mode has approximately the same value for the Cr and W complexes, and is significantly lower for the Mo analogue. Table 6.1 also shows that the enthalpies of H_2 binding for the series $M(CO)_3(PCy_3)_2(H_2)$ do not track well with ν_{MH_2} or the rotational barriers, where Mo is out of order.

These results in conjunction with solution stabilities (W < Mo < Cr) suggest a complicated picture of H_2 binding in terms of trends in physical properties versus electronic interaction down group 6. A major problem for correlating properties of the H_2 complexes is separating $H_2 \to M$ σ bonding effects from those of BD. The σ

Table 6.1. IR Frequencies, Rotational Barriers, and Enthalpies of Binding of Group 6 H_2 Complexes

Complex	ν_{HH}[a] (cm^{-1})	$\nu_{as(MH_2)}$[a]	$\nu_{s(MH_2)}$[a]	Barrier (kcal/mol)	$\Delta H_{binding}$[b] (kcal/mol)
$Cr(CO)_5(H_2)$	3030	1380	869,878	—	—
$Cr(CO)_3(PCy_3)_2(H_2)$	—	1540	950	1.17(10)	-7.3 ± 0.1
$Mo(CO)_5(H_2)$	3080	—	—	—	—
$Mo(CO)_3(PCy_3)_2(H_2)$	~2950[c]	~1420[c]	885	1.32(10)	-6.5 ± 0.2
$Mo(CO)(dppe)_2(H_2)$	2650	—	875	0.5(1), 0.6(1)[d]	—
$W(CO)_5(H_2)$	2711	—	919	—	—
$W(CO)_3(PCy_3)_2(H_2)$	2690	1568	951	1.9(1)	-9.4 ± 0.9

[a] In Nujol mulls for phosphine complexes and liquid Xe[26] or rare gas matrices[27] for the pentacarbonyls.
[b] Enthalpy of H_2 binding.[50]
[c] Estimated from observed D_2 isotopomer bands.
[d] For 2-toluene and 4.5-benzene solvates, respectively, assuming $d_{HH} = 0.85$ Å.

interaction is the major contributor to the total M–H$_2$ bond interaction, and as discussed in Chapter 7, H$_2 \to$ M σ-bonding strength does appear to correlate as Cr $\sim$ W > Mo in accord with the observed values of v_{MH_2} and $\Delta H_{binding}$ for M(CO)$_3$(PCy$_3$)$_2$(H$_2$). Calculations also indicate that Cr fragments are slightly better σ acceptors than W,[17] but the opposite is true for the backdonating ability of M, which correlates with the observed greater barrier height in the W complex. The stronger σ interaction in the Cr complex, on the other hand, would contribute to the total M–H$_2$ bond strength but would not affect the barrier to rotation because H$_2 \to$ M σ bonding is isotropic about the M–H$_2$. The effect of BD on v_{MH_2} is unclear. One would have anticipated v_{MH_2} to be significantly higher for W(CO)$_3$(PCy$_3$)$_2$(H$_2$) than for the weaker backbonder, Cr(CO)$_3$(PCy$_3$)$_2$(H$_2$), as found for Cr(CO)$_3$(PCy$_3$)$_2$(H$_2$) in comparison to Cr(CO)$_5$(H$_2$).

Some of these apparently conflicting observations can be explained by the same issues discussed concerning vibrational spectroscopy of the M–H$_2$ moiety (see Chapter 8). Mainly, the importance of v_{HH} as a good indicator of H–H bond activation and v_{MH_2} as a measure of M–H$_2$ binding energy is doubtful, because the H–H stretching coordinate involves a large degree of W–H stretching as well. Thus the frequencies are very mixed, e.g., the difference in v_{HH} between W(CO)$_5$(H$_2$) and W(CO)$_3$(PCy$_3$)$_2$(H$_2$) would have been expected to be much greater than 21 cm^{-1} because of far superior BD to H$_2$ in the more electron-rich phosphine complex. This is in marked contrast to end-on bound ligand stretches such as v_{NN} or v_{CO} that correlate well with the electron richness of M. N–N stretching involves little or no displacement of M and should therefore be a good measure of N–N bond activation.

Summarizing, it appears that in the series M(CO)$_3$(PR$_3$)$_2$(H$_2$) the M–H$_2$ σ interaction is weaker for Mo than for Cr or W, while BD is slightly better for Mo and much better for W than Cr based on rotational barriers. The very high lability of H$_2$ in the Cr analogue could be ascribed to this weaker BD, supporting the general notion that BD is more crucial than the M–H$_2$ σ interaction in influencing relative stabilities of complexes, d_{HH}, and possibly overall bond strengths (e.g., W–H$_2$ > Cr–H$_2$).

The barrier to H$_2$ rotation on group 8 metals has been determined only for the Fe and Ru complexes[5,28,29]:

[MH(H$_2$)(PP$_3$)]$^+$ (1) (M = Fe, Ru; PP$_3$ = P[CH$_2$CH$_2$PPh$_2$]$_3$),

[FeH(H$_2$)(dppe)$_2$]$^+$ (2) (dppe = Ph$_2$PCH$_2$CH$_2$PPh$_2$),

and

FeH$_2$(H$_2$)(PEtPh$_2$)$_3$ (3)

Surprisingly for **1** the barrier is higher in the Fe than the Ru complex, which suggests that *first-row Fe is a better backdonor to H$_2$ than second-row Ru*,[28] unlike for group 6 complexes. The v_{NN} of the N$_2$ analogues, i.e., 2110 and 2182 cm^{-1}, respectively, for the Fe and Ru complexes imply that more N–N bond activation is again present in the Fe compound on account of better BD, but in this case into the π^* orbital of N$_2$.

The barrier for **3** is only 1.1 kcal/mol, an important finding in regard to the cis interaction between the H$_2$ and hydride ligand here (see Sections 4.11 and 5.6). In **3** the H–H bond is staggered with respect to the cis Fe–P and Fe–H axes in opposition to both electronic and steric considerations. The influence of the d orbitals that favors eclipsing of the H–H bond with the P–Fe–P bond thus

competes with the attractive cis interaction. Consequently the electronic potential shows a broad minimum and the maximum is necessarily lower, as reflected by the rotational barrier, which is lower than it would be from either competing factor alone (in comparison to H_2 complexes without cis hydride attractions).

Rotational barriers are often lower than expected based on BD arguments for reasons other than the above. Theoretical considerations[30,31] show that the electronic barrier to H_2 rotation should be essentially zero if the ligands in the plane parallel to that of the rotation are highly symmetric, regardless of the strength of BD to H_2. Complexes of the type trans-$ML(H_2)(dppe)_2$ thus have low barriers (<1 kcal/mol for M = Mo; L = CO). However, the barrier for Tp*RhH$_2$(H$_2$) (4) is only 0.56(2) kcal/mol despite the low symmetry about the M–H$_2$ axis and the apparently weak H–H bond as judged by ν_{HH} = 2238 cm^{-1} and d_{HH} of 0.94 Å calculated from the value of the rotational constant, 37.1 cm^{-1}, from INS data.[32]

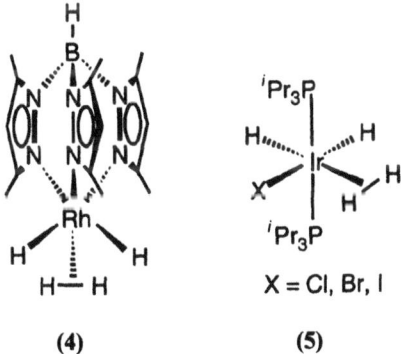

The most plausible explanation is that interaction of H_2 with the cis hydrides lowers the barrier (see Section 4.11). Another well-studied system discussed extensively in this chapter is IrXH$_2$(H$_2$)(P^{i}Pr$_3$)$_2$ (5), where the barrier for X = Cl is 0.51(2) kcal/mol, the lowest measured.[33,34] Ab initio calculations indicate that the energy difference between the two calculated H_2 orientations should be 2.2 kcal/mol, but better values are found using DFT methods (see Section 6.2.3.3). The experimental barrier has a very pronounced fourfold component (V_4), however, which may be related to the very rapid η^2-H_2–hydride exchange seen for this complex even in the solid state, the mechanism of which will be discussed in Section 6.3.2. Coupling between the rotational and exchange dynamics in 5 may play a role in the experimental observations because of the similar low energies of the processes (<2 kcal/mol). Quasi-elastic scattering attributable to the η^2-H_2–hydride site exchange is seen for the first time here (see Section 6.3.3).

Despite problems in certain situations, the barrier to H_2 rotation can still be a valuable *relative* measure of the degree of BD in complexes where the ligand set is the same and only the M is changed. When modifications are made to the ligand set, such comparisons are less useful unless only one ligand is changed with no geometric variations, as will be described in the following section.

6.2.3.3. Ligand Effects

The calculations described above deal with the direct electronic interaction between η^2-H_2 and M and therefore reflect the effect of the ancillary ligands L on the electronic state of M as well. For example, the more basic these L are the more electron density is shifted to M, which in turn raises the energy of the highest occupied d orbital in the complex and thereby increases the interaction with $\sigma^*(H_2)$. BD from M to η^2-H_2 may then become strong enough to cleave the H–H bond. A strong σ-donor ligand, on the other hand, may impede σ donation from H_2 to M and thus prevent H_2 binding, especially when it is trans to the H_2. Ligands that are π acceptors, e.g., CO, compete with BD and stabilize η^2-H_2 toward OA.

Careful analysis of these ligand effects in relation to measured barriers to H_2 rotation provides detailed information on the origin and strength of the critical $d_\pi(M)$–$\sigma^*(H_2)$ interaction. Measurements of the barriers for H_2 complexes with the same M but different L should relate the observed differences to the electronic effects of these L. An excellent case of such ligand effects studied theoretically (see Section 4.9) is $TpRhH_2(H_2)$ versus $CpRhH_2(H_2)$, where the barrier for the experimentally unknown complex with the more electron-donating Cp ligand is calculated to be 4.88 kcal/mol versus only 0.45 kcal/mol for the Tp derivative.[35] The computed value is remarkably close to the experimental value of 0.56 kcal/mol for the Tp' complex (**4**), which contains methyl groups on the pyrazole rings. DFT calculations on $TpRhH_2(H_2)$ give a rotational tunnel splitting of 9.2 cm^{-1} versus the 6.7(5) cm^{-1} found experimentally,[32] demonstrating their accuracy for obtaining dynamic and spectroscopic parameters for polyhydride complexes.

Morris has applied the concept of ligand additivity effects to derive generalizations of the stability of octahedral d^6 H_2 complexes (see Section 4.7.5).[36] Electrochemical parameters E_L for each type of L are added and a value for the electrochemical potential $E_{1/2}(d^5/d^6)$ is calculated by empirical formulas derived by Lever.[37] This quantity gives an overall measure of the electron-donating ability of the ligand set, and ranges of $E_{1/2}$ for stable H_2 binding may be defined. In order to assess the possible relevance of this overall measure of electron-donating ability of a M–L fragment to measurements of the barrier to rotation, data for three Fe, two Mo, and two Ru complexes with different sets of ligands are given in Table 6.2. The measured barriers are listed along with the sum of the E_L for the five ligands other than H_2, the computed $E_{1/2}$ for the N_2 complex, and the experimental value for v_{NN} if available. All of these are taken from papers by Morris.[3,36] Inspection of Table 6.2 reveals no obvious correlation between either the overall electron-donating ability of a M fragment as derived from the ligand additivity method or the values of v_{NN} and the measured barrier to H_2 rotation. There are several likely reasons for this, perhaps

Table. 6.2. Comparison of the Barrier to Rotation with Electrochemical Parameters and v_{NN}

	Barrier (kcal/mol)[a]	$\Sigma^S E_L{}^b$	$v_{NN}{}^c$ (cm^{-1})	$E_{1/2}$ (N$_2$ complex)
Fe				
[FeH(H$_2$)(dppe)$_2$]$^+$	1.8	1.04	2120	1.5
[FeH(H$_2$)(PP$_3$)]$^+$	2.1	0.8	2110	1.2
FeH$_2$(H$_2$)(PEtPh$_2$)$_3$	1.1	0.28	2058	0.6
Mo				
Mo(CO)(H$_2$)(dppe)$_2$	0.6–0.8	2.43	2120	0.1
Mo(CO)$_3$(H$_2$)PCy$_3$)$_2$	1.5–1.7	3.55	2159	0.9
Ru				
[CpRu(CO)(H$_2$)(PCy$_3$)]$^+$	(>3.5)	1.37	—	2.0
[RuH(H$_2$)(PP$_3$)]$^+$	1.4	0.8	2182	1.4
RuH$_2$(H$_2$)$_2$(PCy$_3$)$_2$	1.1	0.58	—	1.2

[a]Barrier heights for the Fe complexes were renormalized to the same value of rotational constant B (49.5 cm^{-1}) as determined from neutron diffraction measurements.
[b]Sum is over all the coligands of H$_2$.
[c]For the analogous N$_2$ complex.

the most important being that the apparent strong correlation between the electron-donating ability of a M fragment and v_{NN} is at best a measure of BD into the π^* orbital of N$_2$. Since that orbital is lower in energy and has a much greater spatial extent than σ^*(H$_2$), an analogous degree of BD in the H$_2$ analogue is not assured. Thus even though v_{NN} gives an indication of the *stability* of its M–H$_2$ counterpart, it evidently is not a good gauge of the amount of BD to σ^*(H$_2$).

Correlating the complexes in Table 6.2 is difficult because the fact that there are substantial geometric differences in the ligands sets. For the two Ru complexes, namely [RuH(H$_2$)(PP$_3$)]$^+$ (**1**) and RuH$_2$(H$_2$)$_2$(PCy$_3$)$_2$ (**6**), the latter has two H$_2$ and

```
          P
     H,,, | ,,,H
  H       Ru     H
    \     |     /
      H   |   H
          P
     | 1.4 Å |
         (6)
```

two hydride ligands. The rotational tunneling spectra show that the splitting for **6** is nearly twice that of **1**, and accordingly, the barrier to rotation is some 25% lower (Table 6.2).[38] Aside from the electronic effects of replacing two phosphorus donors with an H$_2$ and a hydride ligand, the reason for this no doubt relates to the presence of more extensive cis interactions, including that between the two η^2-H$_2$ that are proposed to be ~1.4 Å apart in the same plane in **6**. Another critical geometric effect relates to the fact that the barrier to rotation measures essentially a difference in BD for η^2-H_2 between its orientation at the maximum and minimum of the rotational potential function. If BD is strong, but does not differ much in these two orientations, as in 7, the rotational barrier may be quite low (0.6–0.8 kcal/mol in 7), provided the ligand set is symmetrical and not very distorted.

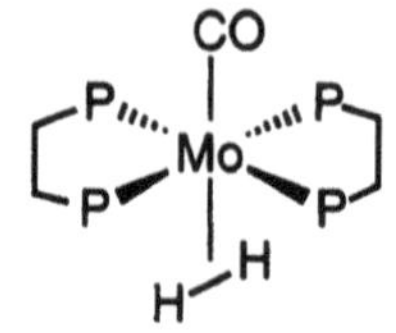

(7)

If the changes made to the ligands are not as drastic as going from the symmetrical FeH(dppe)$_2^+$ fragment to the FeH(PP$_3$)$^+$ fragment, then the barrier should be a good relative measure of the degree of BD, as it is for comparisons of the effect of changing the M keeping the ligands the same.[28] The IrXH$_2$(H$_2$)(P^{i}Pr$_3$)$_2$ system (5) where only X is varied does show small changes in barrier.[33,34] DFT calculations on IrXH$_2$(H$_2$)(PMe$_3$)$_2$ models reproduce the barriers from INS data quite well.[34]

X =	Cl	Br	I
Expt barrier, kcal/mol	0.51(3)	0.48(3)	1.0(4)
Calcd barrier, kcal/mol	0.37	0.42	0.66
d_{HH}, Å	0.78	0.82	0.86

The barriers for the PH$_3$ model are much higher (>2 kcal/mol), demonstrating a high sensitivity to the nature of the phosphine, which requires the use of a realistic phosphine such as PMe$_3$ in calculational estimates for the barrier. The increase in barrier for X = I may be evidence for increased BD. The d_{HH} as determined calculationally (Cl) or by neutron diffraction (Br, I) increases about 0.04 Å for each halide change, which might indicate increased BD down the series. However the possible coupling of rotational dynamics with highly dynamic site exchange between H$_2$ and hydride ligands (see below) may blur the correlation with INS data.

6.2.3.4. Intramolecular Interactions and Crystal Packing Effects

While the rotation barriers for highly symmetric complexes such as Mo(CO)(H$_2$)(dppe)$_2$ (7) can be extremely low (0.6 kcal/mol), they are not in fact zero. This is because X-ray data reveal that the structure of the MP$_4$ fragment is distorted with at least one of the P–M–P' axes bent back away from the H$_2$ (with which it is nearly parallel). The distortions rehybridize M orbitals such that the overlap with σ^*(H$_2$) does show a variation as the H$_2$ rotates in the plane parallel to that formed by the MP$_4$ fragment. In fact, the degree to which the P–M–P' angle deviates from 180° may be taken as an indicator of the strength of H_2 binding as discussed in Section 4.3.

This distortion about M is sensitive to steric effects or crystal-packing forces, e.g., the same complex can have slightly different molecular structures and therefore different barriers to H_2 rotation when crystallized with different counterions or lattice solvent molecules. The rotational tunnel splitting (17 cm^{-1}) is larger [and the barrier is thus lower (Table 6.1)] for Mo(CO)(H_2)(dppe)$_2$·2-toluene solvate than the 4.5-benzene solvate (13 cm^{-1}).[8,31] The X-ray data on these complexes supports the above hypothesis because the distortion about Mo is measurably larger for the benzene solvate. The P–Mo–P′ angles are 173.7(2)° and 174.5(1)° for the benzene and toluene solvates, respectively. The difference in the dihedral angles (the angle between the planes formed by the two different P–Mo–P′ axes) is more pronounced, namely 11.4° versus 5.7° in the benzene and toluene solvates, respectively. These data demonstrate how easily the bidentate ligands about M may be distorted and the remarkably high sensitivity of the rotational tunnel splitting to the small changes in geometry.

Because of slight variations in nonbonded interactions, crystallographic disorder can also result in small barrier differences that give two different tunnel splittings for the same complex. The doublet structure in the tunneling spectra of W(CO)$_3$(H_2)(P^{i}Pr$_3$)$_2$ (see Figure 6.3) probably occurs because one of the P^{i}Pr$_3$ has two possible (disordered) orientations. Complexes known from neutron diffraction not to have structural disorder do not show split tunneling peaks.[28,31] Another example of splitting is found in IrBrH$_2$(H_2)(P^{i}Pr$_3$)$_2$ (5), where tunneling transitions are seen at ± 19 and ± 29 cm^{-1} at 5 K.[34] The intensity of the latter shifts into the peak at 19 cm^{-1} on increasing temperature, and by 40 K only that peak remains. These splittings provide a built-in scale of the relative contribution of steric effects to the barrier to rotation. Based on this consideration and comparisons of experimental barrier heights with those from electronic calculations, one may conclude that the origin of the barrier is generally 60–70% electronic.

The nonbonded interactions between η^2-H_2 and its near-neighbor atoms, such as the H and C atoms on the phosphines, may be treated by the molecular mechanics calculation MM2.[39] The total energy is calculated from summing the pairwise Lennard-Jones interactions between the H atoms of the H_2 ligand and the other atoms on the molecule as a function of the rotation angle of the H_2. This procedure yields an orientational potential for the H_2 and thus a barrier height from "steric effects." This calculation does, however, overestimate the change in barrier height[6] on going from P^{i}Pr$_3$ to PCy$_3$ ligands in M(CO)$_3$(H_2)(PR$_3$)$_2$ by about a factor of four while giving reasonable values for the barrier heights (0.6 and 1.4 kcal/mol, respectively), for PCy$_3$ and P^{i}Pr$_3$ in W(CO)$_3$(H_2)(PR$_3$)$_2$, compared with the observed difference of 0.2 kcal/mol. This has been interpreted to imply that such "steric" effects are a less important contribution to the barrier than the MM2 calculation suggests.

Braga and coworkers have comprehensively examined the crystal structures of H_2 complexes for both intramolecular and intermolecular interactions and their effect on the rotational barrier.[40] Intermolecular contacts are extremely rare because the H_2 is well protected by ancillary ligands. The first and only well-documented example is IrHCl$_2$(H_2)(P^{i}Pr$_3$)$_2$, which has both intramolecular and intermolecular hydrogen bonding to form an infinite chain structure.[41] For **8** rotational tunneling transitions could not be seen out to 30 cm^{-1}, which is well above the largest known rotational tunnel splitting for H$_2$ complexes. Thus only a lower limit of 2.0 kcal/mol

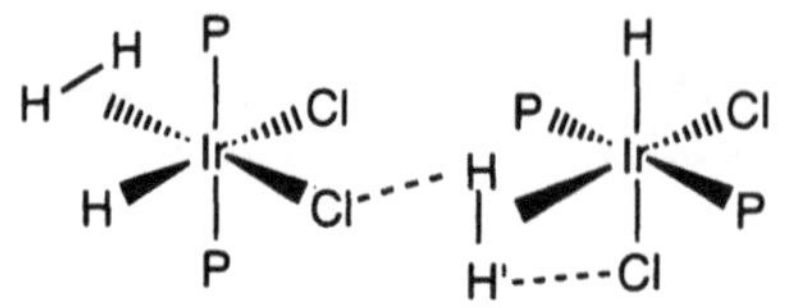

(8)

for the barrier could be estimated. The rotational barrier calculated by *ab initio* methods is 6.5 kcal/mol, which is consistent with the barrier being greatly raised by the hydrogen bonding.

6.2.4. Two-Dimensional Parameterized Model for H$_2$ Rotation

As described above, a one-dimensional model is used to derive a value of the barrier to H$_2$ rotation, where the H$_2$ is viewed as a planar rigid rotor whose equilibrium d_{HH} is assumed on the basis of other information. A new model refines this to explicitly include d_{HH} in the parameterization of the dynamics, and η^2-H$_2$ is considered to be a nonrigid planar rotor whose Hamiltonian is parameterized.[10] The parameters are optimized by means of a least-squares fit of the difference between the experimental and computed values for the INS transitions. This yields valuable structural information about the complex, in particular accurate equilibrium d_{HH} in excellent agreement with experimental data. In the new model, two hydrogen nuclei exchange their mutual positions in a potential created by a metal fragment. The coordinates used to locate the two hydrogen atoms with respect to the metal fragment (M) are shown in **9**.

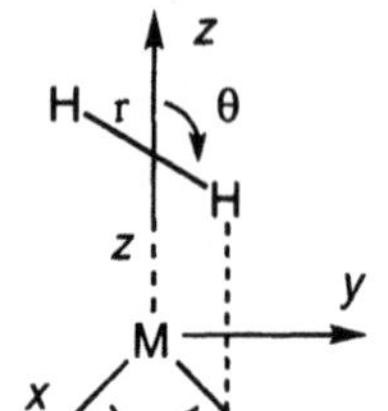

(9)

M can be considered to be a fixed reference frame wherein the H$_2$ is described by the usual spherical coordinates r, θ, and ϕ. The complete description of the exchange process involves the four dynamical coordinates r, θ, ϕ, and z, which can be simplified because presumably the H$_2$ remains perpendicular to the z axis during the rotation and is not displaced appreciably relative to M along z. In contrast to r and ϕ, the two coordinates z and θ do not play an important role in the dynamics of the exchange and are excluded from the dynamical model, which now depends only on r and ϕ, and thus is a two-dimensional (2D) model. The Schrödinger equation for a parameterized Hamiltonian is solved, and the reaction path Hamiltonian formalism coupled with sequential adiabatic reduction is used to fully display the particular physics of the problem. The vibrational states corresponding to the large-amplitude motion of the two H nuclei are obtained and the INS transitions can be computed.

Minimization of the least-squares fit of the differences between the calculated and observed transitions gives an independent determination of the four parameters introduced in the Hamiltonian. These parameters provide structural information about the M–H$_2$ interaction that is very difficult to obtain by *ab initio* methods. The 2D model is thus a better way of estimating d_{HH} in H$_2$ complexes with very high accuracy (0.001 Å) for the equilibrium d_{HH}, r_0. Some specific examples include IrClH$_2$(H$_2$)(P^{i}Pr$_3$)$_2$, where in the previous 1D model the d_{HH} has to be assumed. Here the choice is difficult, and a value of 0.82 Å was initially chosen because it was the distance initially observed by diffraction in true H$_2$ complexes (although the actual distances are somewhat longer). As discussed above, lack of agreement between the 1D model and experiment can be corrected by modification of the fixed d_{HH} and/or by addition of a V_4 term in the potential. In the 2D model, both contributions (influence of r and of the V_4 term) are automatically included, and the resulting r_0 is obtained independently by a fitting procedure. The calculated transitions are in excellent agreement with experiment, and the 2D model clearly improves the agreement over the 1D model. The d_{HH} obtained in the 2D model (0.782 Å) is almost identical to the one that had eventually been assumed in the 1D model (0.78 Å). This distance is very short for an H$_2$ complex and agrees with the observation that the binding energy of H$_2$ is quite low and the complex easily loses H$_2$. The frequency calculated for the ν_{HH} (3183 cm^{-1}) also indicates strong H–H interaction.

For three other systems studied, the 2D model also gives excellent results that compare well with experiment, e.g., d_{HH} in [FeH(H$_2$)(dppe)$_2$]$^+$ is 0.82 Å by neutron diffraction and is 0.822 Å with the 2D model. However, it is only useful if a sufficient number of INS transitions are observed. While instrumental limitations confine the observation of the tunneling transition to H$_2$ complexes with a rotational barrier of less than 3 kcal/mol, transitions to the torsional states can in principle be observed for any such compound. The 2D methodology could then be applied to any experiment in which the dynamic process involves permutation of two identical particles, e.g., the quantum exchange coupling phenomenon (see Section 6.3.4).

6.3. INTRAMOLECULAR HYDROGEN REARRANGEMENT AND EXCHANGE

In 1985 soon after the discovery of H$_2$ complexes the intramolecular site exchange of H atoms between H$_2$ and hydride ligands was found by the Crabtree and Morris research groups.[42] The ^{1}H NMR signals of the cis H$_2$ and hydride in [Ir(H$_2$)H(bq)(PPh$_3$)$_2$]$^+$ coalesce at 240 K because of exchange, and even the hydride trans to H$_2$ in [Fe(H$_2$)H(dppe)$_2$]$^+$ exchanges positions with the H atoms of η^2-H$_2$.

Many new examples include more sophisticated systems with four or more H-donor ligands that display very complex dynamic processes. The complexes encompass early to late metal species such as $Cr(CO)_4(H_2)_2$,[26] $[Cp*MoH_4(H_2)(L)]^+$,[43] cis-$[MH(H_2)(L)_4]^+$ (M = Fe, Ru)$_2$,[44] $[ReH_2(CO)(H_2)(L)_3]^+$,[45] cis, mer-$FeH_2(H_2)(L)_3$,[29] $[RuH(H_2)(CO)_2(L)_2]^+$,[46] and $[TpIrH(H_2)(L)]^+$,[47] (L = PR$_3$). *Ab initio* calculations show that various mechanisms are possible for the site exchange.[43,48] For example a RE/OA pathway through a bis(H$_2$) intermediate is proposed for the Re complex, but for cis-$[FeH(H_2)(L)_4]^+$ and the Cp*Mo species, the preferred pathway is via a M–(H$_3$) (trihydrogen) transition state. The essential features of fluxionality among hydride and H$_2$ ligands, including mechanistic aspects, have been well reviewed by Gusev and others.[2–4] The intramolecular dynamics will thus only be summarized here and will mainly focus on η^2-H$_2$ containing systems.

For the simple H$_2$/hydride situation, two general types of exchange mechanisms can be envisaged. The first is *dissociative* and involves homolysis of the H–H bond to produce a fluxional trihydride intermediate that facilitates intramolecular exchange of H atoms between either adjacent or distal H$_2$ and H ligands. Equations (6.2) and (6.3) are controlled by the same factors that affect the kinetics of homolytic

$$\begin{array}{c} H\!-\!H^* \\ | \\ L_nM\!-\!H \end{array} \rightleftharpoons \begin{array}{c} H \;\; H^* \\ \backslash \; / \\ L_nM\!-\!H \end{array} \rightleftharpoons \begin{array}{c} H \;\; H \\ \backslash \; / \\ L_nM\!-\!H^* \end{array} \rightleftharpoons \begin{array}{c} H\!-\!H \\ | \\ L_nM\!-\!H^* \end{array} \quad (6.2)$$

$$\begin{array}{c} H\!-\!H^* \\ | \\ L_nM \\ | \\ H \end{array} \rightleftharpoons \begin{array}{c} H \\ | \\ L_nM\!-\!H^* \\ | \\ H \end{array} \rightleftharpoons \begin{array}{c} H \\ | \\ L_nM\!-\!H \\ | \\ H^* \end{array} \rightleftharpoons \begin{array}{c} H\!-\!H \\ | \\ L_nM \\ | \\ H^* \end{array} \quad (6.3)$$

splitting of H$_2$. The second mechanism is *associative* and implies a trihydrogen intermediate or transition state such as that discussed extensively in Chapter 4.

$$M \begin{array}{c} H \cdots H \\ \; \; \; \; \; \vdots \\ \; \; \; H^* \end{array}$$

The complex $[Re(H_2)(H)_2(PMe_2Ph)_3(CO)]^+$ provides the first, and only good experimental evidence for an associative exchange mechanism that involves such a rotating H$_3$ intermediate.[45] The latter could occur in other complexes such as the previously discussed $FeH_2(H_2)(PEtPh_2)_3$, which contains a cis interaction between H$_2$ and a hydride that can be considered a nascent H$_3$ ligand. It is not possible to freeze out the J_{HD} for the M(H)$_2$(HD) isotopomer in such systems containing H$_2$ plus two or more hydrides, even at the lowest attainable temperature for solution NMR. However, even without NMR data, indirect evidence for unstable, fluxional H$_2$ and/or H$_3$ intermediates can be obtained merely from isotope exchange reactions using D_2 gas. Before $M-H_2$ was recognized, Brintzinger proposed in 1979 that a transient d^0 $Cp^*_2ZrH_2(D_2)$ mediated H–D exchange via the associative transition state species $[Cp^*_2Zr(H)(DDH)]^\ddagger$ as in Eq. (6.4).[49] Here the Zr

$$Cp^*_2Zr\begin{smallmatrix}H\\H\end{smallmatrix} \underset{-D_2}{\overset{D_2}{\rightleftharpoons}} \left[Cp^*_2Zr\begin{smallmatrix}H\\-D\\D\\H\end{smallmatrix}\right] \rightleftharpoons \left[Cp^*_2Zr\begin{smallmatrix}H\\D\\D\\H\end{smallmatrix}\right]^\ddagger \underset{HD}{\overset{-HD}{\rightleftharpoons}} Cp^*_2Zr\begin{smallmatrix}H\\D\end{smallmatrix}$$

(6.4)

center could not give a dissociative pathway because Zr^{VI} is unattainable. This was one of the first examples of σ-bond metathesis and postulation of a transient $M-H_3$ species (see Section 4.12). In the six-hydrogen system, $[CpMH_4(H_2)(PR_3)]^+$ (M = Mo, W), stretching the H_2 toward an adjacent hydride is a low-energy process that also leads to a transition state with H_3 character.[43] The calculated barriers for exchanges are thus only about 4 kcal/mol (R = H), in agreement with the inability to decoalesce the hydride NMR signals even at −133 K. Using ^{13}C solution NMR, Heinekey recently was able to measure the rate of cis $H-H_2$ exchange in $[RuH(H_2)(^{13}CO)_2(PCy_3)_2]^+$ at about $10^3 s^{-1}$ at 130 K, with $\Delta G^\ddagger_{120} = 5.5$ kcal/mol.[46] Remarkably, decoalescence of the averaged ^{13}CO signal does not occur until 130 K and there is rapid cleavage of the H–H bond even for this relatively unactivated H_2 complex ($d_{HH} = 0.9$ Å). This is consistent with a highly concerted exchange process, i.e., a Ru-trihydrogen-like transition state.

As Gusev and Berke remarked, "the finding of a universal rationale for all kinds of intramolecular ligand rearrangements is an impossible task." As a consequence, they highlight two types of principal motion that are often distinguishable in the dynamic behavior of metal hydrides: a migratory (M) type and a replacement (R) type (Scheme 2). In the M-type exchange one or more ligands migrate from their original inequivalent positions to give inversion of the entire structure. Subsequent replacement of the migrated ligands by each other in their former coordination sites does not occur. The R-type exchange involves a physical rearrangement of identical atoms or ligands that exchange their exact positional coordinates. Hydride H_a replaces H_b at the same time that H_b takes the former place of H_a. The simplest example of R exchange is rotation of η^2-H_2 as discussed above. The M and R mechanisms can be distinguished by NMR if they are not simultaneous events on the same timescale.

The R mechanism is important because polyhydrides can readily interchange cis hydrides by transient formation of H_2-like ligands where d_{HH} is shortened to 1.3–1.4 Å. For example, the 16e trihydrides $OsH_3X(P^iPr_3)_2$ (X = halide) have a structure with H_a and H_c exerting a strong mutual trans influence resulting in bending toward H_b [Eq. (6.5)]. Formation of an intermediate with an elongated η^2-H_2 allows H_c to interchange position with H_b, which has been studied both experimentally and theoretically.[51,52] The rate of site exchange increases slightly from X = I to X = Cl and the highest barrier, 8.8 kcal/mol ($\Delta G^\ddagger$ at 205 K), is for the former. These complexes display large *exchange couplings* between the hydrides (AB$_2$ patterns) with $J_{H_1-H_2}$ values of 920 (Cl), 550 (Br), and 280 Hz (I) at $-100°C$ (see

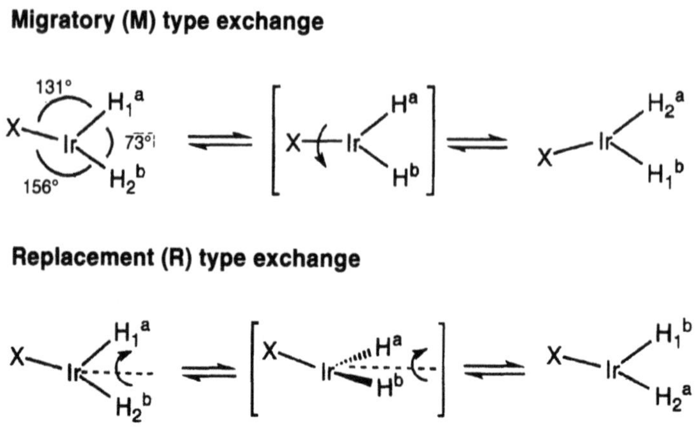

Migratory (M) type exchange

Replacement (R) type exchange

IrH$_2$X(P^{t}Bu$_2$Ph)$_2$ (X = Cl, Br, I)

Scheme 2

Section 6.3.4). The observations indicate that exchange couplings can operate between such hydrogens if they are involved in R-type exchange.[52]

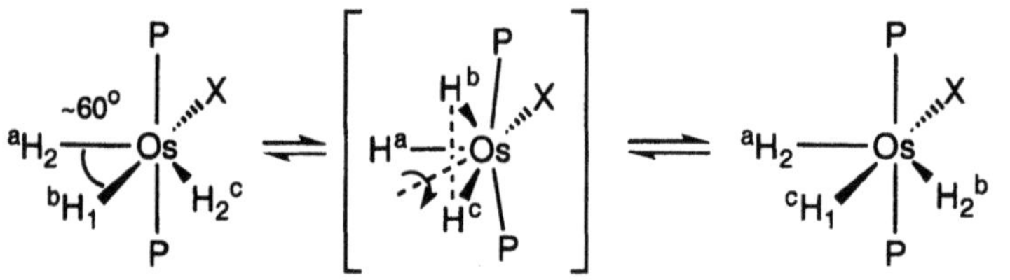

$$(6.5)$$

6.3.1. Extremely Facile Hydrogen Exchange in IrXH$_2$(H$_2$)(PR$_3$)$_2$ and Related Systems

Pseudooctahedral complexes MXH$_2$(H$_2$)L$_2$ with H$_2$ cis to a hydride are extremely fluxional and show M-type exchange. A well-studied case is IrClH$_2$(H$_2$)(P^{i}Pr$_3$)$_2$, where, in addition to the INS studies described above, solid state ^{1}H NMR studies by Zilm on a single crystal provided key initial information on the fluxional behavior.[53] A transition state with C_{2v} symmetry is attained in these systems by stretching the H–H bond followed by concerted migration of M-bound hydrogens. This transient structure inverts with H$_a$ and H$_b$ forming a new H$_2$ ligand, all of which happens in the equatorial plane of the molecule [Eq. (6.6)]. The NMR data indicate that the hydrogens remain as distinct pairs that do not cross the X–M–L plane, i.e., H$_c$ does not change with the H$_a$ site. Site exchange between H$_a$ and H$_b$ occurs via facile H$_2$ rotation. A combination of experimental and theoretical studies on $IrXH_2(H_2)(PR_3)_2$ has provided much insight into the mechanism of and energy barriers to exchange and attendant rotational dynamics of this

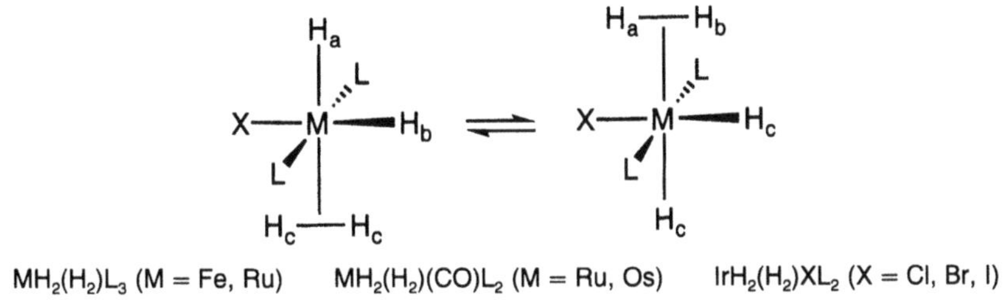

$$X-M(H_a)(L)(L')(H_b)(H_c-H_c) \rightleftharpoons X-M(H_a-H_b)(L)(L')(H_c)(H_c) \quad (6.6)$$

$MH_2(H_2)L_3$ (M = Fe, Ru) $MH_2(H_2)(CO)L_2$ (M = Ru, Os) $IrH_2(H_2)XL_2$ (X = Cl, Br, I)

system.[34] DFT calculations on model systems for X = Cl, Br, and I and R = H, Me determine the most favorable pathways and corresponding activation parameters for exchange as well as H_2 rotational barriers that are similar in energy. Several mechanisms for exchange are possible in $IrXH_2(H_2)(PR_3)_2$ (including a bis(H_2) intermediate), but calculations strongly support the OA/RE pathway through a tetrahydride intermediate as the most favored.

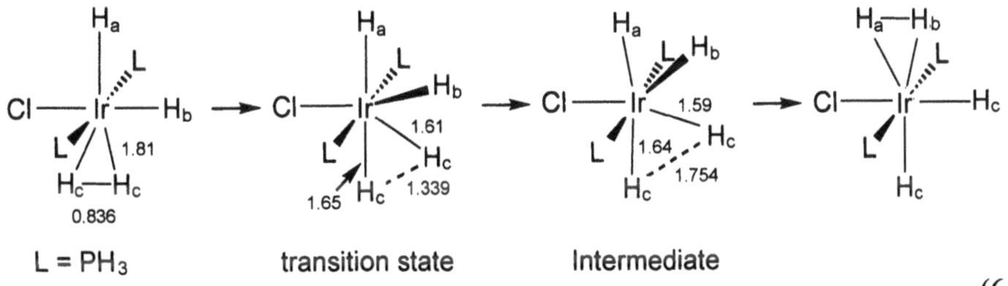

$$\text{Cl}-\text{Ir}(H_a)(L)(L')(H_b)(H_c-H_c)_{0.836}^{1.81} \rightarrow \text{Cl}-\text{Ir}(H_a)(L)(L')(H_b)_{1.61}(H_c)_{1.65}(H_c)_{1.339} \rightarrow \text{Cl}-\text{Ir}(H_a)(L)(L')(H_b)_{1.59}(H_c)_{1.64}(H_c)_{1.754} \rightarrow \text{Cl}-\text{Ir}(H_a-H_b)(L)(L')(H_c)(H_c)$$

L = PH_3 transition state   Intermediate

(6.7)

Use of the more realistic phosphine PMe_3 rather than PH_3 give exchange barriers of 1.9, 1.8, and 1.7 kcal/mol for X = Cl, Br, and I respectively, in excellent agreement with the experimental value obtained by quasi-elastic neutron scattering studies for X = Cl (1.5 kcal/mol) (see Section 6.3.3). This low barrier pathway is consistent with the original solid state ^{1}H NMR results that showed a barrier substantially less than 3 kcal/mol for the hydrogen exchange in $IrClH_2(H_2)(P^iPr_3)_2$ but a much higher barrier for hydrogens crossing the Cl–Ir–P plane.[53] This is a remarkably low barrier for a *solid state* process, a process involving considerable rearrangement yet facile enough to persist down to temperatures below 77 K. It is also significant that this is apparently a direct OA/RE process, one of several possible mechanisms that must be considered in the fluxional behavior of other polyhydride complexes. The tetrahydride intermediate here and intermediates like it may be difficult to detect because they are only slightly more stable (~ 1 kcal/mol) than the transition state, which bears a similarity to an elongated H_2 complex [Eq. (6.7)]. The H_2 complex "reactant" is itself only 4.3 kcal/mol lower in energy than the tetrahydride. The exchange barrier is only about 1 kcal/mol higher than that for H_2 rotation, and these highly dynamic process may be coupled to some extent.

A related complex, $[IrH_2(H_2)(P^tBu_2Ph)_2]^+$, is a rare coordinatively unsaturated complex (see Section 4.7.1) for which solution NMR indicates similar pairwise exchange between inequivalent hydrides.[54] Calculations show that two hydrides slide in the plane perpendicular to the P–Ir–P axis with no associated rotation of the H–Ir–H plane with respect to the rest of the complex (calcd vs. exptl barrier is 6.9 vs. 8.6 kcal/mol). For the bis-H_2 adduct, $[IrH_2(H_2)_2(P^tBu_2Ph)_2]^+$, two transition states corresponding to two different exchange processes have been identified, which separately lead to the H and H_2 exchange and H/D isotopic incorporation, where all six H atoms are involved. In related complexes $[IrH_2L'L_2]^+$ (L = PR_2Ph), the hydride exchange barriers show a strong dependence on the size of L' and L, especially when L' is another bulky phosphine.[55]

For $MH_2(H_2)XL_2$ congeners, d_{HH} in the intermediate for exchange should depend electronically on X whether or not the intermediate is actually similar to that for the Ir case. It has been an oft-repeated theme that CO in the trans position X favors formation of an η^2-H_2 with a short d_{HH}, as low as 0.84 in *ab initio* DFT calculations for $RuH_2(H_2)(CO)(PH_3)_2$.[56] The calculated energy difference for the transition state over the ground state (7.9 kcal/mol) is in excellent agreement with the experimental value ($\sim$8 kcal/mol[57]). For the opposite situation where X is the π donor Cl, the d_{HH} analogous to $H_c \cdots H_c$ is undoubtedly much longer, which is consistent with the fact that these hydrogens do not rotate on the much faster timescale of the migrational exchange in $IrClH_2(H_2)(P^iPr_3)_2$.

6.3.2. Other Systems that Exchange Hydrogens in H_2 and Hydride Ligands

The MH_4L_4 structural type is incredibly dynamic, displaying both M and R exchanges. The cationic complexes $[ReH_4(CO)L_3]^+$ (L = PMe_3, PMe_2Ph) exist in solution as two isomers **10** and **12** in equilibrium, each of which is highly fluxional.[45] The low-temperature ^1H NMR spectra display one exchange-averaged quartet in the hydride region for the tetrahydride **12** plus two decoalesced ReH_2 and $Re(H_2)$ resonances for the H_2 isomer **10**. ^{31}P NMR shows that the phosphine skeleton is rigid in **12** but very fluxional in **10**. Because these species are thermally unstable above −40°C, X-ray data are not available. A dodecahedral structure is probable for **12** from ^{13}C NMR evidence for the CO ligand.[45] However, the structure of **10** and the mechanism of exchange is controversial andtwo possibilities exist.[2] Crabtree proposed the pentagonal bipyramidal structure with H_2 in equatorial position shown in Eq. (6.8). A pseudooctahedral structure $[Re(H_3)(H)(PMe_2Ph)_3(CO)]^+$ with an H_3 ligand was proposed as the intermediate or transition state of the exchange reaction within the H_2-containing isomer. An H_3 intermediate was also proposed earlier by Bianchini in $[RuH(H_2)(PP_3)]^+$ containing a tripodal phosphine, although the evidence was less clear.[58] The $\Delta G^\ddagger$ of 9.9 kcal/mol for the rate of H-atom exchange in **14** is the lowest measured among nearly 30 complexes (including the Ru species) and is consistent with a facile associative process.[3] It has been noted that the H_3 intermediate would be no more than 10 kcal/mol less stable than the H_2–dihydride structure, making the isolation of an actual complex containing an H_3 ligand a realistic future goal.[45]

Intramolecular Dynamics of Dihydrogen–Hydride Ligand Systems

An alternate structure for $[ReH_2(H_2)(CO)(PR_3)_3]^+$ favored by Gusev and Berke[2] contains H_2 in the axial site trans to the strong acceptor CO, which would

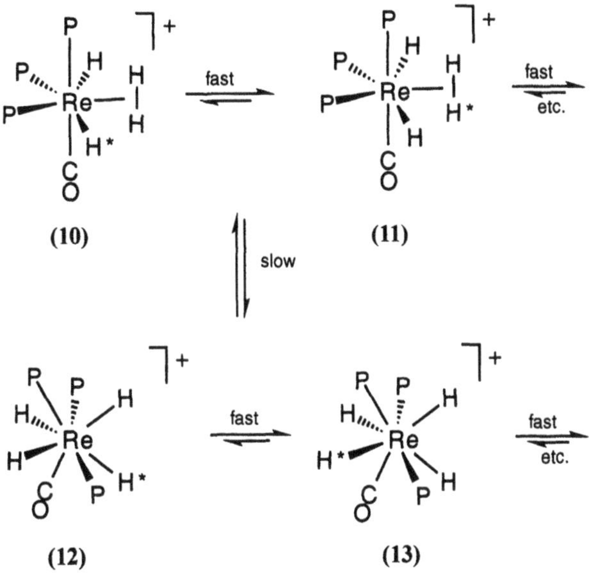

(10) ⇌ (11) fast; etc. fast

(12) ⇌ (13) fast; etc. fast

slow ⇌ between (10/11) and (12/13)

(6.8)

be consistent with the high value of J_{HD}, 34 Hz, because of the trans effect (cf. 27.7 Hz in $[Re(H_2)(CO)(PMe_3)_4]^+$ with H_2 cis to CO). In this case H-atom scrambling via a Re–H_3 transient is still possible.

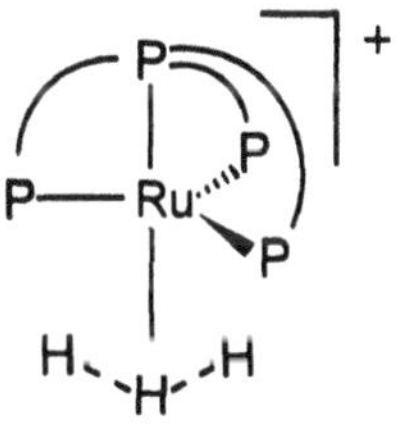

(14)

The complex $[Ir(H_2)_2H_2(PCy_3)_2]^+$ (15) is one of the few bis(H_2) complexes, and separate ^1H NMR resonances for the hydride and H_2 ligands (ratio 1:2) are observed at 188 K.[59] These peaks coalesce at 200 K, and Morris[3] calculates the $\Delta G^\ddagger$

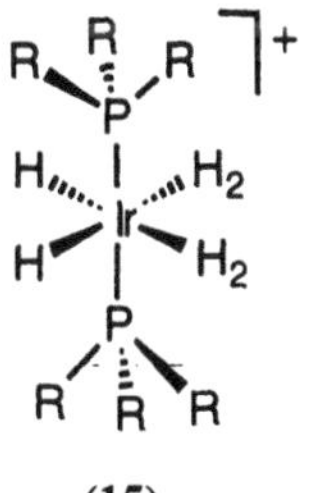

(15)

at this temperature to be 8.4 kcal/mol. Structure **15** presents another example of the very low barriers for exchange of H_2 and hydride ligands situated cis to each other around the equatorial plane of a complex. Chaudret's complex, $RuH_2(H_2)_2(PCy_3)_2$, another bis(H_2) complex, is also highly fluxional,[60] as is his $Tp^*RuH(H_2)_2$ complex (**16**) with the hydride and two η^2-H_2 residing on the same side of the complex.[61]

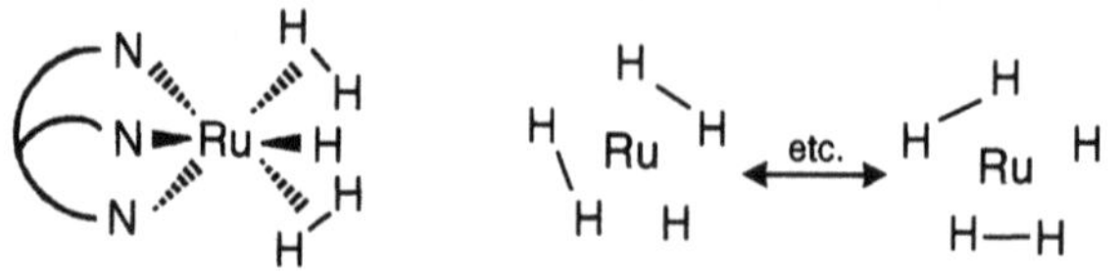

(16)

Although crystallographic evidence is unavailable, NMR data are compatible with averaging of the H positions in solution and cis interactions appear likely here. However, calculations indicate that the ground state structure is $H(H_2)_2$ rather than a "pentahydrogen ligand" analogous to Cp.

Exchange of H atoms between η^2-H_2 and a bridging hydride is seen whenever the two groups are cis to each other [Eq. (6.9)]. Complexes include

$$(L_2)(H_2)Ru(\mu\text{-H})(\mu\text{-Cl})_2RuH(PPh_3)_2 \quad (L_2 = FeCp(1,2\text{-}C_5H_3[CHMeNMe_2)(P^iPr_2)])^{62}$$

for which Morris uses rate data to calculate $\Delta G^\ddagger(293\ K) = 11.8$, $\Delta H^\ddagger = 15.3$ kcal/mol, and $\Delta S = 12$ cal/mol·K. Although this complex has a cis H_2/H interaction

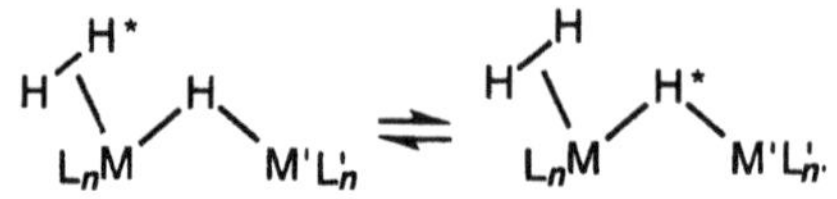

$$(6.9)$$

that might assist the exchange,[63] the $\Delta H^\ddagger$ value is still higher than those for mononuclear complexes, apparently because a hydride must shift from its stable bridging position. The complexes

$$[P(p\text{-tol})_3]_2(H_2)Ru(\mu\text{-H})(\mu\text{-Cl})_2RuH[P(p\text{-tol})_3]_2$$

and

$$(PCy_3)_2(H_2)Ru(\mu\text{-H})_3RuH(PCy_3)_2$$

are highly fluxional,[60,64] and exchange involves all hydrogens as rationalized by simultaneous OA of H_2 to give (H)Ru(μ-H) and reductive elimination of (μ-H)Ru(H*) to give Ru(HH*).

6.3.3. Quasi-Elastic Neutron Scattering Studies of H_2 Exchange with cis Hydrides

Quasi-elastic neutron scattering (QNS)[65] is potentially valuable for investigating the details of rapid H_2/hydride atom exchange. QNS is actually a form of INS in which experiments are carried out at higher temperatures in the regime where quantum-mechanical effects are in transition to classical dynamics (>100 K typically, but this varies). A complex with hydride(s) cis to H_2 should show increasingly strong interaction between the ligands as a function of temperature, including possibly exchange. The former affects the rate of rotation of the H_2 ligand as it becomes increasingly "aware" of the neighboring hydride's electrons. This is reflected in non-Arrhenius-like behavior of the quasi-elastic linewidth, i.e., broadening of the intense narrow elastic line. QNS data collected at temperatures up to 325 K on [FeH(H_2)(dppe)$_2$]BF$_4$, **2**; FeH$_2$(H_2)(PEtPh$_2$)$_2$, **3**; IrClH$_2$(H_2)(P^{i}Pr$_3$)$_2$, **5**[33,34]; and RuH$_2$(H_2)$_2$(PCy$_3$)$_2$, **6**,[38] show a temperature dependence of the spectral linewidth that can be fitted to an Arrhenius law that gives an activation energy for the rotation of η^2-H_2. The latter values for compounds **2**, **3**, and **5** were determined to be 30–50% of the experimentally determined barriers to rotation. This is very surprising as one would expect that at these temperatures the rotation would be essentially classical, i.e., thermally activated rotational hopping *over* the barrier. It is therefore apparent that *even at room temperature the rotational motion of η^2-H_2 is at least in part quantum-mechanical*. Similar effects are known for the translational diffusion of hydrogen in metals, where in many cases experimentally determined activation energies can be substantially lower than potential-well depths. Only the data for **2** could be fitted reasonably well to a model for stochastic rotation of a dumbbell molecule in a double-minimum potential, whereas the dynamics of H_2 in **3** and **5** appear to differ substantially from that of **2**. This difference may be attributable to interaction with the cis hydride in **3** and the very rapid exchange between hydride and η^2-H_2 that is known to occur in **5**.

The variable temperature QNS data for **5** represents a breakthrough in showing the *first experimental observation by INS of quasi-elastic scattering attributable to H_2/hydride site exchange and its associated activation energy.*[34] As seen in Figure 6.4, as the temperature is increased above 100 K the rotational tunneling transitions for **5** broaden, shift to slightly lower frequencies, and decrease in intensity while a very broad background appears beneath the peaks much as for **6**, as will be shown below. Additionally, the narrow elastic line broadens progressively, indicating that another dynamic process is now fast enough to be observable within the frequency window (i.e., energy resolution) provided by the spectrometer of about 2 cm^{-1} FWHM. Although the intensity of the quasi-elastic component is quite low, it can be extracted by fitting a Lorentzian convoluted with the measured Gaussian resolution function over this part of the spectrum, e.g., as shown in Figure 6.5 for the spectral line at 250 K in Figure 6.4. The extracted Lorentzian linewidths are fitted to an Arrhenius law to effectively provide an activation energy of 1.5(2) kcal/mol for the exchange. This remarkably low barrier closely matches the DFT calculated activation barrier for site exchange of 1.9 kcal/mol and is consistent with the mechanistic features as discussed above.

The bis(H_2) complex, RuH$_2$(H_2)$_2$(PCy$_3$)$_2$, **6**, is also extremely fluxional, and NMR studies in Freon solvent mixtures at temperatures as low as 143 K still give unresolvable spectra because of rapid exchange of hydrogens and a low H_2 rotational barrier.[38] In agreement with this, calculations show that this complex has three isomeric structures within an energy range of only 2 kcal/mol (see Chapter 4). The lowest-energy structure has all the H in the same plane, which is evidence for cis interaction of H_2 and hydride ligands, which would promote exchange. Equation

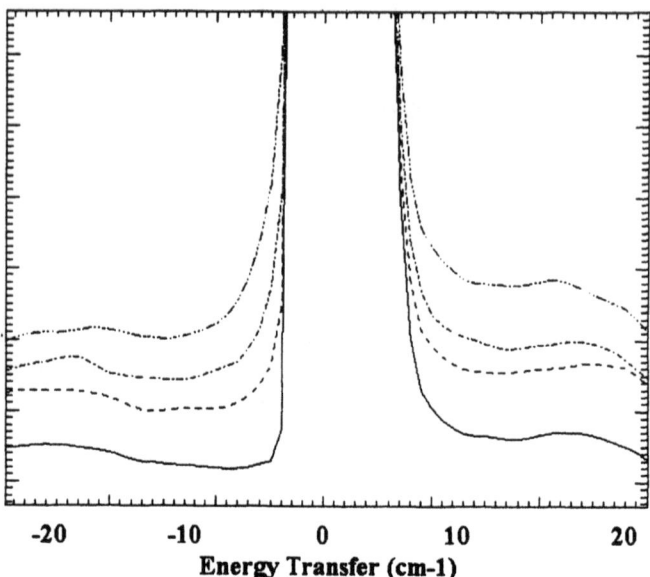

Figure 6.4. Temperature dependence of the quasi-elastic line (underneath the central inelastic peak) for $IrClH_2(H_2)(P^iPr_3)_2$ at 100, 175, 210, and 250 K (in upward direction). Broadening is evident above 175 K in the base of the line. Reprinted with permission from Li et al.[34] Copyright 2000 American Chemical Society.

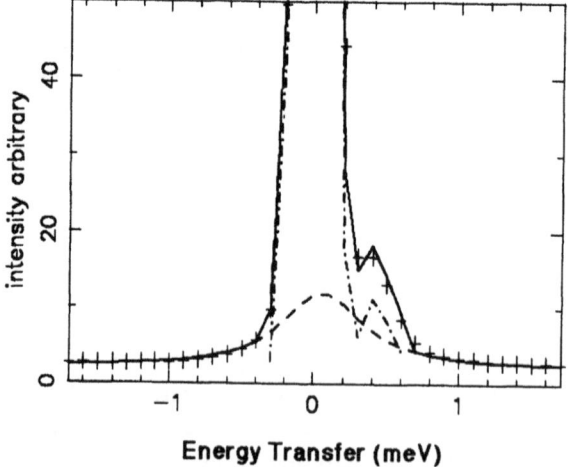

Figure 6.5. Example of a fit to the data in Figure 6.4 at 250 K. The line appears to be asymmetric because of a strong spectrometer-related false peak at about 2 cm^{-1}. Reprinted with permission from Li et al.[34] Copyright 2000 American Chemical Society.

(6.10) shows one of many possibilities for exchange that might, e.g., start with the known cis interaction between the η^2-H_2 and go through an OA/RE type mechanism

$$\begin{array}{c} H_b \\ | \\ H_a\text{---Ru---}H_c \\ L\,|\,\,\,\,\,H_d \\ H_f\text{---}H_e \end{array} \quad\longrightarrow\quad \begin{array}{c} H_b \\ H_a\diagdown | \diagup H_c \\ \text{Ru} \\ L\diagup \diagdown H_d \\ H_f\quad H_e \end{array} \quad\longrightarrow\quad \begin{array}{c} H_b \\ H_a\diagdown \diagup H_c \\ \text{Ru} \\ L\diagup \diagdown H_d \\ H_f\quad H_e \end{array} \quad (6.10)$$

(6)

as for **5**. The INS spectrum of **6** below 50 K consists of the usual pair of bands on either side of the elastic peak corresponding to a low barrier to rotation of 1.1 kcal/mol, which agrees with calculated barriers.[38] An interesting question is whether the two H_2 molecules rotate in the same or opposite sense, and theoretical studies indicate that the former occurs.

It is important to note that there is significant structure in the bands when compared with rotational tunneling lines in other H_2 complexes where molecular disorder is not known to be present. This structure may therefore be attributed to the effect of interactions between the two η^2-H_2. As the temperature is increased the rotational tunneling peaks shift to lower energy and broaden in the usual manner, where the width shows an Arrhenius-type temperature dependence with an activation energy of $\sim$0.4 kcal/mol (Figure 6.6). In addition, a broad quasi-elastic feature

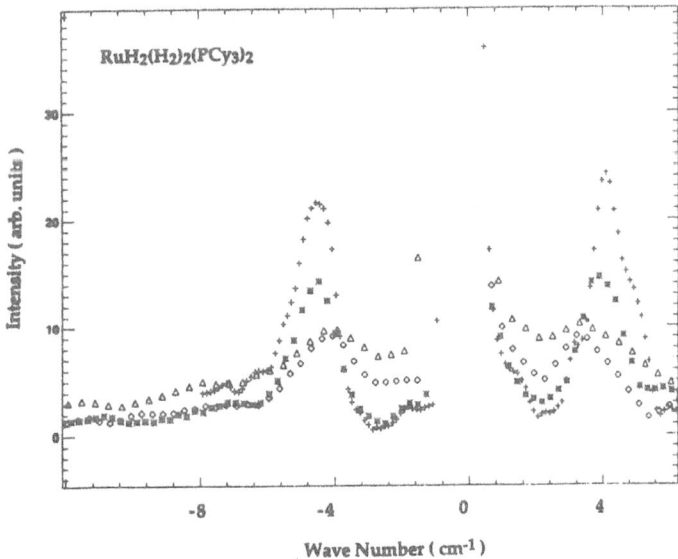

Figure 6.6. Temperature dependence of the INS spectrum of $RuH_2(H_2)_2(PCy_3)_2$: 5 K (+), 50 K(*), 75 K ($\triangle$), 100 K(O). Reprinted with permission from Rodriguez *et al.*[38] Copyright 1998 American Chemical Society.

appears below the inelastic peaks at temperature higher than 75 K, as for **5**. The rotational tunneling peaks coalesce into a broad quasi-elastic line below the elastic peak above 150 K. The width of this line continues to increase with temperature up to 250 K, the highest temperature reached in this experiment. The activation energy derived from the temperature dependence of the width of this quasi-elastic line is 0.25 kcal/mol, which suggests that the motion giving rise to this line may be described[65] as weakly hindered rotational diffusion of the hydrogen. In addition, a second, narrower quasi-elastic component is evident at 250 K with a FWHM of about 4 cm^{-1}.

The rationalization for the above temperature-dependent phenomena can be summarized as follows. The increase in width of the rotational tunneling transition lines as a function of temperature is the result of inhomogeneous broadening from coupling to the phonon bath and of the incoherent H_2 exchange processes observable by ^1H NMR and simulated for ^2H NMR of dideuteron systems.[66] The observed non-Arrhenius behavior (i.e., two different activation energies for different temperature regimes reported above) is consistent with the superposition of these processes. Note that both of these activation energies are well below the height of the barrier to rotation. Also, at the temperature at which the rotational tunneling lines coalesce into a quasi-elastic line (150 K) NMR data show rapid intramolecular H exchange even though the activation energy for the latter process is much higher than those observed by INS for the rotational motions. In the course of this interconversion of hydride and H_2 ligands, the latter presumably are subject to a variety of environments in addition to that of the octahedral sites. These considerations would account for the INS observations at 75 and 100 K, i.e., that the relatively sharp rotational tunneling bands arise from rotational transitions of H_2 located in well-defined potential wells, while the broad quasi-elastic component below the rotational tunneling peaks may be attributed to H_2 affected by the rapid intramolecular exchange of hydride and η^2-H_2.

6.3.4. Quantum-Mechanical Exchange Coupling and Hindered Rotational Phenomena

Several metallocene complexes such as $(C_5H_4SiMe_3)_2NbH_3$, $[CpIrH_3(PR_3)]^+$, and $Cp^*RuH_3(PR_3)$ were initially thought to have a trihydrogen structure (see Chapter 4) because they showed extraordinarily large (thousands of hertz), temperature-dependent J_{HH} in the ^1H NMR.[67-72] The spectra of these complexes displayed second-order AB_2 couplings that could result from an open $H_bH_aH_b$ ligand (**17**), but

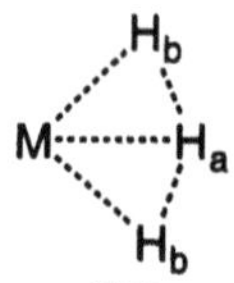

$$M \cdots \begin{matrix} H_b \\ H_a \\ H_b \end{matrix}$$

(**17**)

because they were greater than the coupling in H_2 gas (280 Hz) they could not be normal J_{HH}.[69] These were the first observations of such couplings in NMR for any molecular system, including organic and main-group compounds. The J_{HH} were also very unusual in that they increased with temperature and then vanished, giving spectra with A_3 ^{1}H spin systems characteristic of a highly fluxional trihydride. Heinekey and coworkers eventually showed that all these complexes are trihydrides with two relatively short d_{HH} [1.674(14) and 1.698(13) Å in the neutron structure for R = Me], as shown in **18**.[71]

(18)   (19)

Calculations suggest that this species exchanges a pair of H atoms through a tunneling path involving a transient H_2 complex,[73] and more recent calculations show that an H_2 complex **19** is a transition state in the H-exchange process with a barrier of 12.6 kcal/mol for R = H (with isolobal OH representing Cp).[74] A trihydrogen (η^3-H_3) complex as a transient was ruled out from this calculation. The shallow H–M–H vibrational potential wells here and in other species favor the occurrence of *quantum-mechanical exchange coupling* (QEC),[4,69-73,75-79] which rationalizes the large J_{HH} that do not originate from traditional magnetic interactions but result rather from exchange of equivalent H nuclei. Weitekamp and Zilm simultaneously proposed the QEC phenomenon as a manifestation of a tunnel effect between two equivalent configurations (Figure 6.7).[69] The permutation of two hydrides corresponds to a splitting of rovibrational levels detectable by NMR because of the extremely facile exchange between the configurations. Even though three hydrides are present, the tunnel effect involves only two H at a time and is thus

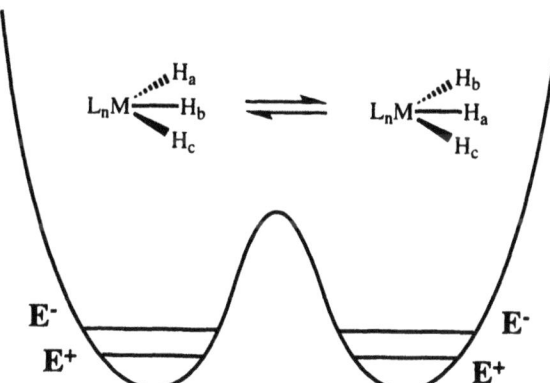

Figure 6.7. Tunnel effect between two equivalent configurations of a metal trihydride. Reprinted with permission from Maseras *et al.* Copyright 2000 American Chemical Society.

associated with a pairwise exchange leading to an AB NMR pattern. Such a phenomenon is extremely rare and has been seen previously for heavy particles (nonelectrons) only in studies of ^{3}He at cryogenic temperatures. The quantum nature of QEC is consistent with the dramatic decrease of the coupling constant upon incorporation of tritium, contrary to what is expected from the relative values of γ_T and γ_H.[71,79] Because the two particles are no longer identical, there are no longer any symmetry properties associated with the permutation operation, and therefore the observed AB pattern results only from the normal magnetic coupling. The QEC in polyhydride complexes that could have isomers with η^2-H_2 has prompted many studies on such systems and has been well reviewed by Sabo-Etienne and Chaudret, who have also been very active researchers in this field.[79]

Two seemingly disparate models for QEC have been proposed, both of which involve exchange of inequivalent H nuclei. One may involve thermal equilibrium between the ground state structure with inequivalent hydrides and an η^2-H_2 isomer that exchanges the two inequivalent nuclei as in **19** above.[73,78,80,81] Another model does not depend upon the existence of an H_2 ligand; rather the phenomenon is governed by a low barrier to H exchange through a pseudorotational process.[82] The T_1 values of all the trihydrides that display QEC are in fact consistent with trihydride rather than H_2-hydride structures. These models have been further refined and the present knowledge can be summarized as follows.[4,79]

1. A soft vibration exists in the polyhydride complex that combines in-plane and out-of-plane MH_2 bending to facilitate the approach of the two H that

$$L_nM\begin{matrix}H_a\\\\H_b\end{matrix} \longrightarrow \left[L_nM\begin{matrix}H_b\\H_a\end{matrix}\right]^{\ddagger} \longrightarrow L_nM\begin{matrix}H_b\\H_a\end{matrix} \qquad (6.11)$$

can come close enough to make an H_2 complex in the transition state. This is reminiscent of H_2 rotation in H_2 complexes but with higher barriers because the rotation is coupled with the approach of the two H.
2. A chemical path initiated by the vibration allows exchange via approach of the H within a distance that varies according to the complex but is shorter than the sum of the van der Waals radii. In some cases an H_2 ligand is clearly involved but in others (Ir and Os systems) it is not so clear, although short d_{HH} have been calculated.
3. A barrier between 8.7 and 16.7 kcal/mol to this chemical (classical) exchange must be present to be observable by NMR. If the barrier is too high the J_{HH} will be too small to be seen and if the barrier is too low, fluxional behavior (classical exchange, incoherent process[66]) will be observed rather than the coherent exchange process characteristic of QEC. The barriers can directly relate to the energetic cost for reaching an H_2 complex.
4. A rotational tunneling process through this barrier exists similar to that observed by INS for H_2 complexes and can be viewed as a transition between coordinated ortho- and para-hydrogen.

Another important characteristic of QEC is that the exchanging particles must have the same mass and spin, i.e., for mixtures of ^{1}H, ^{2}D, or ^{3}T there is negligible quantum-mechanical tunneling and the J_{HD} or J_{HT} have the usual scalar values. For example, [CpIr(D)(H)(D)(PPh$_3$)]$^+$ has no resolvable J_{HD} whereas [CpIrH$_3$(PPh$_3$)]$^+$ has a high J_{HH} of 397 Hz at 196 K owing to QEC.[71] Thus isotopic substitution is very diagnostic for QEC, as proposed by Heinekey.[72] In terms of electronic considerations, the exchange couplings apparently decrease as M becomes more electron-rich. Large QEC is favored by short ground state distances between two neighboring hydrides and low barriers for H–M–H bending motion, which promote quantum-tunneling.[81] Thus H–Os–H motions that would lead either to a M–H$_2$ structure or give a "turnstile" rotation of two H are not favored relative to the Ru analogue. The QEC effect can also be increased by reducing the electron density of M in metallocene trihydrides by attachment of Lewis acids or hydrogen-bond donors to one of the hydrides.[4,79,83] Even a slight shift of electron density away from M facilitates closer approach and exchange between the two "free hydrides."

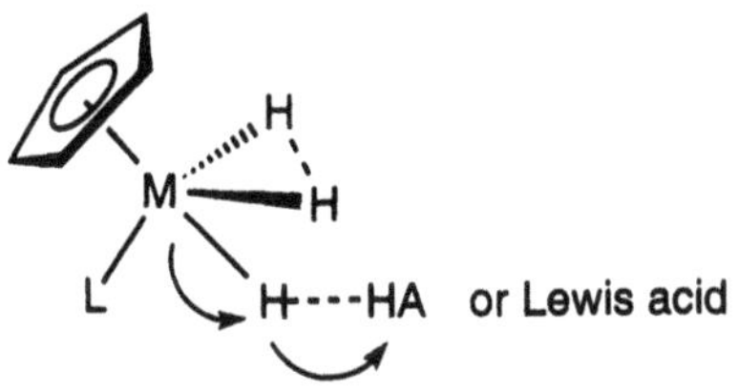

All the original complexes that showed QEC contained Cp-type ligands but new *electronically unsaturated* Os trihydrides, OsH$_3$X(P^{i}Pr$_3$)$_2$ (X = Cl, Br, I), and the first d^6 pseudooctahedral polyhydride, mer-RuH$_3$(NO)(P^{t}Bu$_2$Me)$_2$, are found by Caulton.[77] Experiment and theory show that an η^2-H$_2$ species mediates the exchange in the latter. Other non-Cp species include a particularly revealing complex, **20** [Eq. (6.12)] where a topological analysis (Bader's AIM method) of the

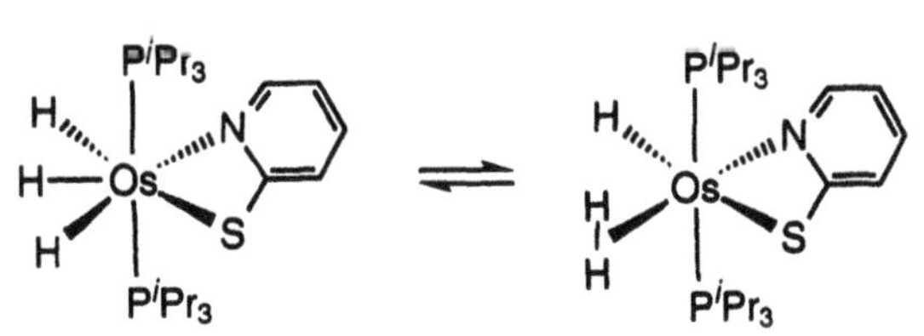

(6.12)

(20)   (21)

electron density indicates a transition state for H exchange that contains an η^2-H$_2$ (**21**, where PH$_3$ replaces P^{i}Pr$_3$).[78] The activation barrier for thermally activated site exchange going through transition state **21** is calculated to be 14.9 kcal/mol, and it is 16.2 kcal/mol for the isomer with H$_2$ trans to sulfur. This difference in barriers essentially represents the difference in energy needed for the exchanging H atoms to get close together as in Eq (6.11).

A unique case of QEC involving only a H$_2$ ligand is known. The complexes [Cp$_2'$M(H$_2$)(L)]$^+$ (M = Nb, Ta) show *blocked rotation of the H$_2$ ligand* in the NMR spectrum (see Section 5.5), and the free energy of activation of the H_2 internal rotation is determined to be 8–12 kcal/mol both experimentally and calculationally.[84–86] At 178 K, decoalescence of signal is observed for the HD complex but not for the HH species. This initially strongly suggested that rotational tunneling of η^2-H_2 is at the origin of QEC observed in d^2 transition metal polyhydrides, where only one orbital is available for BD.[85] The high rotational barrier can be attributed to the complete loss of BD on going from the global minimum to the transition state in which the H_2 is bound only by σ-donation from H_2 (calcd d_{HH} = 0.79 Å).[86]

With DFT calculations and a monodimensional rotational tunneling model, the absence of decoalescence for the HH isotopomer in $[Nb(Cp')_2(H_2)(CNR)]^+$ is shown to be due to the existence of a huge QEC of 10^6 Hz.[84] Conversely, for the HD species, the difference in zero-point energy corresponding to two inequivalent (H–D versus D–H) positions leads to a slight asymmetry that dramatically reduces QEC, allowing decoalescence to be seen. Thus the HD classical rotation and the quantum exchange processes will not be experimentally observed for this complex, whereas only the classical process for the HH isotopomer is quenched on the NMR timescale. This is an intermediate case between hydrogen rotation of H_2 complexes and the exchange of a pair of hydrides in polyhydride complexes. The former implies the exchange of H atoms via a low barrier (<3.5 kcal/mol) and a relatively short tunneling path, producing a splitting of the energy levels for the double-well potential observable by INS in the microwave region (frequencies of about 10^{10} Hz). For the latter, the QEC implies a high energy barrier and a very long tunneling path whereby both factors lower the splitting, which can be seen in the radiofrequency range (about 10^3 Hz). For the Nb complex, the energy barrier is high but the tunneling path is short, giving QEC of about 10^6 Hz, detectable in NMR spectra and preventing decoalescence of the H_2 signal. A similar effect is predicted for a case of a low barrier but a long tunneling path for H–H exchange.

In conclusion, QEC is probably present in all complexes with cis hydrides when the chemical path to exchange does not involve rearrangement of the heavy ligands, although it can only be observed for certain barriers to exchange. Theoretical studies including combined *ab initio*/dynamics studies[73,87] demonstrate that the two exchanging H need to get closer to the transition state than they are in the ground state. The occurrence of QEC in certain trihydrides probably can be related to the close proximity (∼1.6 Å) of their three H nuclei in the ground state geometry. Low energy barriers for permuting two H concomitant with small displacement of the heavy nuclei appear to be critical for observing large J_{AB}, but there is no specific requirement to reach a $MH(H_2)$ configuration. QEC has also been observed between deuteriums in $[RuCl(D_2)dppe)_2]^+$ by solid state deuterium NMR at very low temperature.[88] Despite its limited occurrence, QEC is an important new area for study because of the extremely high sensitivity of J to the electronics and coordination environment around a complex. One potential use is ultrasensitive detection of very weak interactions of metal centers with unconventional donors such as hydrocarbons.[77] For example, at low temperatures unsaturated $OsH_3Cl(P^iPr_3)_2$ forms weak solvent adducts, even with methylcyclohexane, that are detectable by large changes in exchange couplings.

REFERENCES

1. E. L. Muetterties, ed., *Transition Metal Hydrides*, Dekker: New York, 1971; Meakin, P.; Guggenberger, L. J.; Peet, W. G.; Muetterties, E. L.; Jesson, J. P. *J. Am. Chem. Soc.* **1973**, *95*, 1467, and references therein.
2. Gusev, D. G.; Berke, H. *Chem. Ber.* **1996**, *129*, 1143.
3. Jessop, P. G.; Morris, R. H. *Coord. Chem. Rev.* **1992**, *121*, 155.
4. Maseras, F.; Lledos, A.; Clot, E.; Eisenstein, O., *Chem. Rev.* **2000**, *100*, 601.
5. Eckert, J.; Kubas, G. J., Dianoux, A. J. *J. Chem Phys.* **1988**, *88*, 466; Eckert, J.; Blank, H.; Bautista, M. T.; Morris, R. H. *Inorg. Chem.* **1990**, 29747.
6. Eckert, J., Kubas, G. J.; Hall, J. H.; Hay, P. J.; Boyle, C. M. *J. Am. Chem. Soc.* **1990**, *112*, 2324.
7. Eckert, J. *Spectrochim. Acta* **1992**, *48A*, 363.
8. Eckert, J.; Kubas, G. J. *J. Chem. Phys.* **1993**, *97*, 2378.
9. Eckert, J. *Trans. Am. Crystallogr. Assoc.* **1997**, *31*, 45.
10. Clot, E.; Eckert, J. *J. Am. Chem. Soc.* **1999**, *121*, 8855.
11. Barea, G.; Esteruelas, M. A.; Lledos, A.; Lopez, A. M.; Onate, E.; Tolosa, J. I. *Organometallics* **1998**, *17*, 4065; Gelabert, R,; Moreno, M.; Lluch, J. M.; Lledos, A. *J. Am. Chem. Soc.* **1998**, *120*, 8168.
12. Pacchioni, G. *J. Am. Chem. Soc.* **1990**, *112*, 80.
13. Lewis, J., D.; Malloy, T., B.; Cao, T., H.; Laane, J. *J. Mol. Struct.* **1972**, *12*, 427.
14. Zilm, K. W., Merrill, R. A.; Kummer, M. W.; Kubas, G. J. *J. Am. Chem. Soc.* **1986**, *108*, 7837; Zilm, K. W.; Millar J. M. *Adv. Mag. Opt. Reson.* **1990**, *15*, 163.
15. Hay, P. J. *J. Am. Chem. Soc.* **1987**, *109*, 705.
16. Saillard, J.-Y.; Hoffmann, R. *J. Am. Chem. Soc.* **1984**, *106*, 2006.
17. Jean, Y.; Eisenstein, O.; Volatron, F.; Maouche, B.; Sefta, F. *J. Am. Chem. Soc.* **1986**, *108*, 6587.
18. Silvera, I. F. *Rev. Mod. Phys.* **1980**, *52*, 393.
19. Beaufils, J. P.; Crowley, T.; Rayment, R. K.; Thomas, R. K.; White, J. W. *Mol. Phys.* **1981**, *44*, 1257.
20. Nicol, J. M; Eckert, J.; Howard, J. *J. Phys. Chem.* **1988**, *92*, 7117.
21. Prager, M.; Heidemann, A. *Chem. Rev.* **1997**, *97*, 2933.
22. Eckert, J. *Physica*, **1986**, *136B*, 150.
23. Eckert, J; Kubas, G. J; White, R. P. *Inorg. Chem.* **1992**, *31*, 1550.
24. Kubas, G. J.; Nelson, J. E.; Bryan, J. C.; Eckert, J.; Wisniewski, L.; Zilm, K. *Inorg. Chem.* **1994**, *33*, 2954.
25. Gadd, G. E.; Upmacis, R. K.; Poliakoff, M.; Turner, J. J. *J. Am. Chem Soc.* **1986**, *108*, 2547.
26. Upmacis, R. K.; Poliakoff, M.; Turner J. J. *J. Am Chem. Soc.* **1986**, *108*, 3645.
27. Sweany, R. L. *J. Am. Chem. Soc.* **1990**, *107*, 2374.
28. Eckert, J.; Albinati, A; White, R. P.; Bianchini, C.; Peruzzini, M. *Inorg. Chem.* **1992**, *31*, 4241.
29. Van der Sluys, L. S.; Eckert, J.; Eisenstein, O.; Hall, J. H.; Huffman, J. C.; Jackson, S. A.; Koetzle, T. F.; Kubas, G. J.; Vergamini, P. J.; Caulton, K. G. *J. Am. Chem. Soc.* **1990**, *112*, 4831.
30. Maseras, M.; Duran, M.; Lledos, A.; Bertran, J. *J. Am. Chem. Soc.* **1991**, *111*, 2879.
31. Kubas, G., J.; Burns, C. J.; Eckert, J.; Johnson, S. W.; Larson, A. C.; Vergamini, P. J.; Unkefer, C. J.; Khalsa, G. R. K.; Jackson, S. A.; Eisenstein, O. *J. Am. Chem. Soc.* **1993**, *115*, 569.
32. Eckert, J.; Albinati, A.; Bucher, U. E.; Venanzi, L. M. *Inorg. Chem.* **1996**, *35*, 1292.
33. Eckert, J.; Jensen, C. M.; Jones, G.; Clot, E.; Eisenstein, O. *J. Am. Chem. Soc.* **1993**, *115*, 11056; Eckert, J.; Jensen, C. M.; Koetzle, T. F.; Le-Husebo, T.; Nicol, J.; Wu, P. *J. Am. Chem. Soc.* **1995**, *117*, 7271.
34. Li, S.; Hall, M. B.; Eckert, J.; Jensen, C. M.; Albinati, A., *J. Am. Chem. Soc.* **2000**, *122*, 2903.
35. Gelabert, R.; Moreno, M.; Lluch, J. M.; Lledos, A. *Organometallics* **1997**, *16*, 3805.
36. Morris, R., H. *Inorg. Chem.* **1992**, *31*, 1471.
37. Lever, A., B., P. *Inorg. Chem.* **1990**, *29*, 1271.; Lever, A., B., P. *Inorg. Chem.* **1991**, *30*, 1980.
38. Rodriguez, V.; Sabo-Etienne, S.; Chaudret, B.; Thoburn, J.; Ulrich, S.; Limbach, H.-H. Eckert; J.; Barthelat, J.-C.; Hussein, K.; Marsden, C. J. *Inorg. Chem.* **1998**, *37*, 3475; Borowski, A. F.; Donnadieu, B.; Daran, J.-C.; Sabo-Etienne, S.; Chaudret, B. *Chem. Comm.* **2000**, 543.
39. Burkert., U.; Allinger, N. L. *Molecular Mechanics*, ACS Monograph 177, American Chemical Society: Washington, D.C., 1982.
40. Braga, D.; De Leonardis, P.; Grepioni, F.; Tedesco, E. *Inorg. Chim. Acta* **1998**, *273*, 116.

41. Albinati, A.; Bakhmutov, V. I.; Caulton, K. G.; Clot, E.; Eckert, J.; Eisenstein, O.; Gusev, D. G.; Grushin, V. V.; Hauger, B. E.; Klooster, W. T.; Koetzle, T. F.; McMullan, R. K.; O'Loughlin, T. J.; Pelissier, M.; Ricci, J. S.; Sigalas, M. P.; Vymenits, A. B. *J. Am. Chem. Soc.* **1993**, *115*, 7300.
42. Crabtree, R. H.; Lavin, M. *J. Chem. Soc., Chem. Commun.* **1985**, 794; Morris, R. H.; Sawyer, J. F.; Shiralian, M.; Zubkowski, J. D. *J. Am. Chem. Soc.* **1985**, *107*, 5581.
43. Bayse, C. A.; Hall, M. B.; Pleune, B.; Poli, R. *Organometallics* **1998**, *17*, 4309.
44. Ricci, J. S.; Koetzle, T. F.; Bautista, M. T.; Hofstede, T. M.; Morris, R. H.; Sawyer, J. F. *J. Am. Chem. Soc.* **1989**, *111*, 8823; Gusev, D. G.; Hubener, R.; Burger, P.; Orama, O.; Berke, H. *J. Am. Chem. Soc.* **1997**, *119*, 3716.
45. Luo, X.-L.; Crabtree, R. H. *J. Am. Chem. Soc.* **1990**, *112*, 6912; Luo, X.-L.; Michos, D.; Crabtree, R. H. *Organometallics* **1992**, *11*, 237; Gusev, D. G.; Nietlispach, D.; Eremenko, I. L.; Berke, H. *Inorg. Chem.* **1993**, *32*, 3628.
46. Heinekey, D. M.; Mellows, H.; Pratum, T. *J. Am. Chem. Soc.* **2000**, *122*, 6498.
47. Oldham, W. J., Jr.; Hinkle, A. S.; Heinekey, D. M. *J. Am. Chem. Soc.* **1997**, *119*, 11028.
48. Maseras, F.; Duran, M.; Lledos, A.; Bertran, J. *J. Am. Chem. Soc.* **1992**, *114*, 2922; Lin, Z.; Hall, M. B. *J. Am. Chem. Soc.* **1994**, *116*, 4446.
49. H. H. Brintzinger, *J. Organomet. Chem.* **1979**, *171*, 337.
50. Gonzalez, A. A.; Zhang, K.; Mukerjee, S. L.; Hoff, C. D.; Khalsa, G. R. K.; Kubas, G. J. *ACS Symp. Ser.* **1990**, *428*, 133.
51. Gusev, D. G.; Kuhlman, R. L.; Sini, O.; Eisenstein, O.; Caulton, K. G. *J. Am. Chem. Soc.* **1994**, *116*, 2685.
52. Clot, E.; LeForestier, C.; Eisenstein, O.; Pelissier, M. *J. Am. Chem. Soc.* **1995**, *117*, 1797.
53. Wisniewski, L. L.; Mediati, M.; Jensen, C. M.; Zilm, K. W. *J. Am. Chem. Soc.* **1993**, *115*, 7533.
54. Cooper, A. C.; Eisensein, O.; Caulton, K. G. *New. J. Chem.* **1998**, *22*, 307.
55. Cooper, A. C.; Caulton, K. G. *Inorg. Chem.* **1998**, *37*, 5938.
56. Gusev, D. G.; Kuhlman, R. L.; Renkema, K. H.; Eisenstein, O.; Caulton, K. G. *Inorg. Chem.* **1996**, *35*, 6775.
57. Gusev, D. G.; Vymenits, A. B.; Bakhmutov, V. I. *Inorg. Chem.* **1992**, *31*, 1.
58. Bianchini, C.; Perez, P. J.; Peruzzini, M.; Zanobini, F.; Vacca, A. *Inorg. Chem.* **1991**, *30*, 279.
59. Lundquist, E. G.; Folting, K.; Streib, W. E.; Huffman, J. C.; Eisenstein, O.; Caulton, K. G. *J. Am. Chem. Soc.* **1990**, *112*, 855.
60. Arliguie, T.; Chaudret, B.; Morris, R. H.; Sella, A. *Inorg. Chem.* **1988**, *27*, 598.
61. Moreno, B.; Sabo-Etienne, S.; Chaudret, B.; Rodriguez, A.; Jalon, F.; Trofimenko, S. *J. Am. Chem. Soc.* **1995**, *117*, 7441.
62. Hampton, C.; Cullen, W. R.; James, B. R. Charland, J.-P. *J. Am. Chem. Soc.* **1988**, *110*, 6918.
63. Jackson, S. A.; Eisenstein, O. *Inorg. Chem.* **1990**, *29*, 3910.
64. Hampton, C.; Dekleva, T. W.; James, B. R.; Cullen, W. R. *Inorg. Chim. Acta* **1988**, *145*, 165.
65. Bee, M. *Quasielastic Neutron Scattering*: Adam Hilger: Bristol, 1988.
66. Limbach, H.-H.; Ulrich, S.; Gruendemann, S.; Buntkowsky, G.; Sabo-Etienne, S.; Chaudret, B.; Kubas, G. J.; Eckert, J. *J. Am. Chem. Soc.* **1998**, *120*, 7929; Buntkowsky, G.; Limbach, H.-H.; Wehrmann, F.; Sack, I.; Vieth, H.-M; Morris, R. H. *J. Phys. Chem. A.* **1997**, *101*, 4679.
67. Heinekey, D. M.; Payne, N. G.; Schulte, G. K. *J. Am. Chem. Soc.* **1988**, *110*, 2303.
68. Antinolo, A.; Chaudret, B.; Commenges, G.; Fajardo, M.; Jalon, F.; Morris, R. H.; Otero, A.; Schweitzer, C. T. *J. Chem. Soc., Chem. Commun.* **1988**, 1210.
69. Zilm, K. W.; Heinekey, D. M.; Millar, J. M.; Payne, N. G.; Demou, P., *J. Am. Chem. Soc.* **1989**, *111*, 3088; Jones, D. H.; Labinger, J. A.; Weitekamp, D. P. *J. Am. Chem. Soc.* **1989**, *111*, 3087.
70. Heinekey, D. M.; Payne, N. G.; Sofield, C. D. *Organometallics* **1990**, *9*, 2643; Heinekey, D. M. *J. Am. Chem. Soc.* **1991**, *113*, 6074; Zilm, K. W.; Heinekey, D. M.; Millar, J. M.; Payne, N. G.; Neshyba, S. P.; Duchamp, J. C.; Szczyrba, J. *J. Am. Chem. Soc.* **1990**, *112*, 920.
71. Heinekey, D. M.; Millar, J. M.; Koetzle, T. F.; Payne, N. G.; Zilm, K. W. *J. Am. Chem. Soc.* **1990**, *112*, 909.
72. Heinekey, D. M.; Harper, T. G. *Organometallics* **1991**, *10*, 2891.
73. Jarid, A.; Moreno, M.; Lledos, A.; Lluch, J .M.; Berttran, J. *J. Am. Chem. Soc.* **1995**, *117*, 1069.
74. Soubra, C.; Chan, F.; Albright, T. A. *Inorg. Chim. Acta* **1998**, *272*, 95.

75. Zilm, K. W.; Millar, J. M. *Adv. Mag. Reson.* **1990**, *15*, 163; Bowers, C. R.; Jones, D. H.; Kurur, N. D.; Labinger, J. A.; Pravica, M. G.; Weitekamp, D. P. *Adv. Mag. Reson.* **1990**, *14*, 269; Barthelat, J. C.; Chaudret, B.; Daudey, J. P.; De Loth, Ph.; Poilblanc, R. *J. Am. Chem. Soc.* **1991**, 113, 9896.
76. Heinekey, D. M.; Hinkle, A. S.; Close, J .D. *J. Am. Chem. Soc.* **1996**, *118*, 5353.
77. Kuhlman, R. L.; Clot, E.; LeForestier, C. L.; Streib, W. E.; Eisenstein, O.; Caulton, K. G. *J. Am. Chem. Soc.* **1997**, *119*, 10153; Yandulov, D. V.; Huang, D. Huffman, J. C.; Caulton, K. G. *Inorg. Chem.* **2000**, *39*, 1919.
78. Castillo, A.; Barea, G.; Esteruelas, M. A. Lahoz, F. J.; LLedós, A.; Maseras, F.; Modrego, J.; Oñate, E.; Oro, L. A.; Ruiz, N.; Sola, E. *Inorg. Chem.* **1999**, *38*, 1814.
79. Sabo-Etienne, S.; Chaudret, B. *Chem. Rev.* **1998**, *98*, 2077; Antinolo, A.; Carillo-Hermosilla, F.; Fajardo, M.; Fernandez-Baeza, J.; Garcia-Yuste, S.; Otero, A. *Coord. Chem. Rev.* **1999**, *193–195*, 43.
80. Limbach, H.-H.; Scherer, G.; Maurer, M.; Chaudret, B. *Angew. Chem. Int. Ed. Engl.* **1992**, *31*, 1369.
81. Camanyes, S.; Maseras, F.; Moreno, M.; Lledos, A.; Lluch, J. M.; Bertran, J. *J. Am. Chem. Soc.* **1996**, *18*, 4617.
82. Clot, E.; LeForestier, C. L.; Eisenstein, O.; Pelissier, M. *J. Am. Chem. Soc.* **1995**, *117*, 1797.
83. Camanyes, S.; Maseras, F.; Moreno, M.; Lledos, A.; Lluch, J. M.; Bertran, J. *Inorg. Chem.* **1998**, *37*, 2334.
84. Antinolo, A.; Carrillo-Hermosilla, F.; Fajardo, M.; Garcia-Yuste, S.; Otero, A.; Camanyes, S.; Maseras, F.; Moreno, M.; Lledos, A.; Lluch, J. M. *J. Am. Chem. Soc.* **1997**, *119*, 6107.
85. Jalon, F. A.; Otero, A.; Manzano, B. R.; Villasenor, E.; Chaudret, B. *J. Am. Chem. Soc.* **1995**, *117*, 10123.
86. Sabo-Etienne, S.; Chaudret, B.; Abou el Makarim, H.; Barthelet, J.-C.; Daudey, J.-C.; Ulrich, S.; Limbach, H.-H.; Moise, C. *J. Am. Chem. Soc.* **1995**, *117*, 11602; Sabo-Etienne, S.; Rodriguez, V.; Donnadieu, B.; Chaudret, B.; el Makarim, H. A.; Barthelat, J.-C.; Ulrich, S.; Limbach, H.-H.; Moïse, C. *New J. Chem.* **2001**, *25*, 55.
87. Jarid, A.; Moreno, M.; Lledos, A.; Lluch, J. M.; Bertran, J. *J. Am. Chem. Soc.* **1993**, *115*, 5861.
88. Wehrmann, F.; Fong, T. P.; Morris, R. H.; Limbach, H-H.; Buntkowsky, G. *Phys. Chem. Chem. Phys.* **1999**, *1*, 4033.

7

Thermodynamics, Kinetics, and Isotope Effects of the Binding and Cleavage of σ Ligands versus Classical Ligands

7.1. THERMODYNAMICS OF DIHYDROGEN AND σ BOND COORDINATION TO METAL CENTERS

Thermodynamic and kinetic parameters for coordination of H_2 are valuable for comparison with those for other σ-bonded ligands as well as classical ligands. They are especially useful for alkane complexes that have not been isolated but can be studied spectroscopically at low temperatures. The ligand strength of H_2 and presumably other σ X–H ligands varies markedly with the degree of activation of the X–H bond, which in turn depends primarily on the electronics at M. Complexes such as $W(CO)_5(H_2)$ readily dissociate H_2 well below room temperature while $[Os(NH_3)_5(H_2)]^{2+}$ can be considered to be hydridelike with very strongly bound H. The prototypical Kubas H_2 complex, $W(CO)_3(PCy_3)_2(H_2)$, is stable at 25°C but reversibly dissociates H_2 *in vacuo* and thus lies in between the extremes. This criterion of adduct formation between two species each capable of independent existence holds true for other genuine σ complexes as well. There can be an enormous dependence of H_2 binding strength on the nature of the coligands, L, whether they are π acceptors or σ donors, and whether they are cis or trans to η^2-H_2.

The thermodynamics of bonding are discussed below for σ complexes that are stable at 25°C, complexes that are thermally unstable but observable by IR, and also "bare metal" species that can only be observed by mass spectroscopy. The data are quite useful for understanding the electronic factors that stabilize the various types of σ-ligand coordination and may predict, e.g., which metal fragments are best for alkane binding.

7.1.1. Binding Energies of H_2 and σ Ligands to Stable Complexes

Solution calorimetric measurements on reactions of H_2 complexes and their precursor complexes were first carried out by Hoff and coworkers on the original

W–H$_2$ complex.[1,2]

$$W(CO)_3(PCy_3)_2 + py \rightarrow W(CO)_3(PCy_3)_2(py) \quad (7.1)$$

$$W(CO)_3(PCy_3)_2(H_2) + py \rightarrow W(CO)_3(PCy_3)_2(py) + H_2 \quad (7.2)$$

The enthalpy term for Eq. (7.1), $\Delta H°$, is -18.9 ± 0.4 kcal/mol in toluene, and that in Eq. (7.2) is -9.5 ± 0.5 kcal/mol under an H$_2$ atmosphere. The difference in enthalpies corresponds to the enthalpy of H$_2$ addition to W(CO)$_3$(PCy$_3$)$_2$, which is exothermic by 9.4 ± 0.9 kcal/mol. Note that these enthalpies are not the true binding energies because an agostic interaction is being displaced [Eq. (7.3)]. Thus the energy

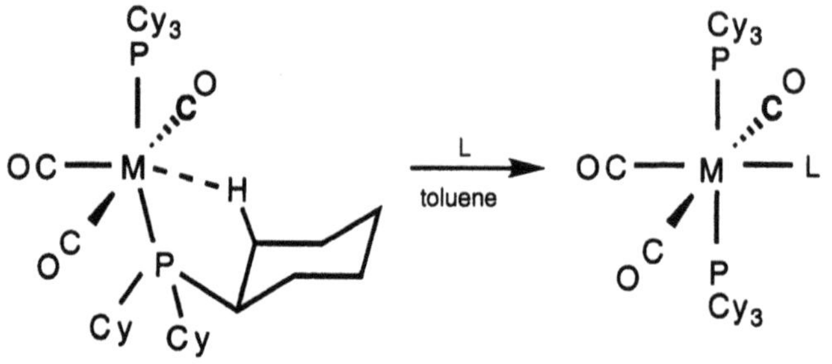

$$\text{(structure with agostic C-H)} \xrightarrow[\text{toluene}]{L} \text{(structure with L)} \quad (7.3)$$

of the agostic interaction should be added to the measured enthalpies to obtain the true binding energies. Unfortunately this can only be estimated to be on the order of 10–15 kcal/mol, based on the value determined by photoacoustic calorimetry for heptane binding to W(CO)$_5$, 13.4 ± 2.8 kcal/mol (see below).[3] Therefore, the binding energy of H$_2$ in W(CO)$_3$(PCy$_3$)$_2$(H$_2$) can best be approximated at 20 ± 7 kcal/mol. This agrees well with the values from theoretical calculations, 17–20 kcal/mol (see Chapter 4). These data can be used to estimate the average single W–H bond strength in W(CO)$_3$(PCy$_3$)$_2$(H$_2$), which is calculated to be one-half of the sum of the H–H bond strength (104 kcal/mol) plus 20 kcal/mol, or $124/2 = 62$ kcal/mol (± 3 kcal/mol). This is considerably lower than other estimates of W–H bond strengths in classical hydrides, e.g., 73 kcal/mol in Cp$_2$WH$_2$ and 81 kcal/mol in CpW(CO)$_3$H,[4] in line with the ease of H$_2$ dissociation in W(CO)$_3$(PCy$_3$)$_2$(H$_2$) versus the much higher stability of the hydrides, which can be sublimed *in vacuo*.

Another technique for measuring enthalpies of H$_2$ addition (or loss) is variable-temperature NMR in systems showing equilibrium binding of H$_2$ in solution.[5-11]

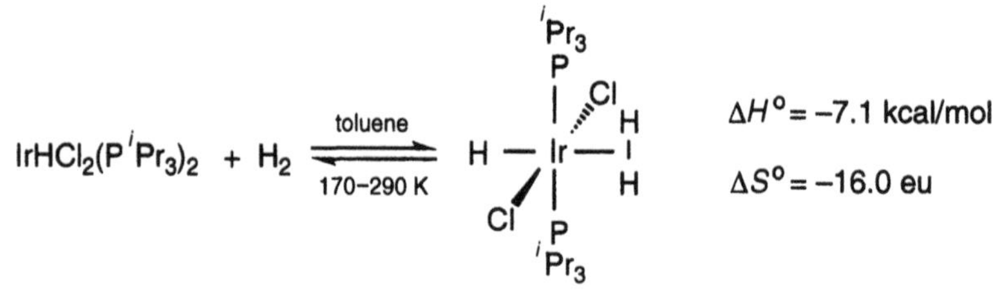

$$IrHCl_2(P^iPr_3)_2 + H_2 \xrightleftharpoons[170-290\,K]{\text{toluene}} \text{[Ir complex]} \quad \begin{array}{l} \Delta H° = -7.1 \text{ kcal/mol} \\ \Delta S° = -16.0 \text{ eu} \end{array} \quad (7.4)$$

Similar values are obtained for an analogue where one Cl is replaced by a hydride,[8] and measurements for $M(H_2)HCl(CO)(P^iPr_3)_2$ give $\Delta H° = -7.7 \pm 0.2$ kcal/mol and $\Delta S° = -23.2 \pm 1$ eu for $M = Ru$[6] and $\Delta H° = -14.1 \pm 0.5$ kcal/mol and $\Delta S° = -30 \pm 1$ eu for $M = Os$.[7] The values for the Ir and Ru complexes are somewhat smaller than those for $W(CO)_3(PCy_3)_2(H_2)$ and similar to those for the Cr and Mo analogues given below. The data are consistent with the high lability of the H_2 in all of these complexes and the generally higher stability of H_2 binding to a third-row element. The structures of the Ir, Ru, and Os precursor complexes are unknown, but the relatively low values of $\Delta H°$ indicate that, as for the W species, an energy of agostic interaction or structural reorganization must be added to give the true H_2 binding energies.

The π-donating halide coligands have a strong and complex influence on the energetics of H_2 binding, as discussed in Section 4.7.1. For example, η^2-H_2 is much more strongly bound when trans to halide than hydride because of increased BD. Calculations show that the bond dissociation energy (BDE) in the Ir complex in Eq. (7.4) ($d_{HH} = 0.81$ Å) is only 13.4 kcal/mol but 23.7 kcal/mol for the isomer with H_2 trans to Cl ($d_{HH} = 1.11$ Å).[12] The situation can be quite complex, however, as illustrated by the series $OsCl_nH_{6-n}(PR_3)_2$ (R = ^{i}Pr) (Table 7.1).[10] Experimentally, H_2 coordination becomes weaker with increasing n, as shown by $\Delta G°$. This seems counterintuitive because there is more BD to H_2 for $n = 2$ ($d_{HH} = 1.25$ Å). However, Cl $\rightarrow$ M π bonding both stabilizes 16e complexes and destabilizes 18e complexes. Thus, while more Os–H bonding is lost when H_2 dissociates for $n = 2$ than for $n = 1$, proportionately more Os–Cl π bonding is created to compensate for this loss. Exchange with free H_2 is much faster for $n = 1$ (seconds) than $n = 2$ (minutes). The main reason for the higher kinetic barrier in the latter is the significant changes in the heavy-atom positions between the unsaturated complex (P–Os–P = 112°) and the 18e complex (P–Os–P = 177°). In contrast, for $n = 1$, the PR_3 are trans for both $OsClH_3(PR_3)_2$ and $OsClH_3(PR_3)_2(H_2)$ and require almost no motion to bind H_2.

The effect of the nature of the halide (X) on the H_2 binding strength is small ($\Delta\Delta G° < 1$ kcal/mole) but complicated, following the stability trend Br > Cl > I for $OsX_2H_2(PR_3)_2(H_2)$ in solution at either $\pm 21°C$.[10] On the other hand, the ordering in stability found for $IrXH_2(PR_3)_2(H_2)$ is I > Br > Cl, which follows the halide σ-donor strength.[8,11] In general it appears that H_2 binding to complexes with heavier halides is favored enthalpically [$\Delta H°$ is most negative for $OsI_2H_2(PR_3)_2(H_2)$] but disfavored entropically [$\Delta S°$ is least positive for $OsI_2H_2(PR_3)_2(H_2)$ for steric crowding reasons)]. An abnormal order of stability can

Table 7.1. H_2 Binding Energy and H–H Distance in $OsCl_nH_{6-n}(PR_3)_2$

n	Shortest d_{HH}, Å	$-\Delta G°$ (H_2 binding)	$\Delta G^\ddagger$ (H_2 loss)
0	1.65 (hydride)	Very high	Very high
1	0.95 (H_2 complex)	High	Low
2	1.25 (H_2 complex)	Low	High

arise because of the low binding energies here where entropy can play an important role (see Section 7.2).

Few measurements of binding energies of σ ligands other than H₂ have been made on stable complexes. NMR measurements as above show that reaction of Ph₂SiH₂ with an agostic Mo complex is energetically similar to that for the H₂

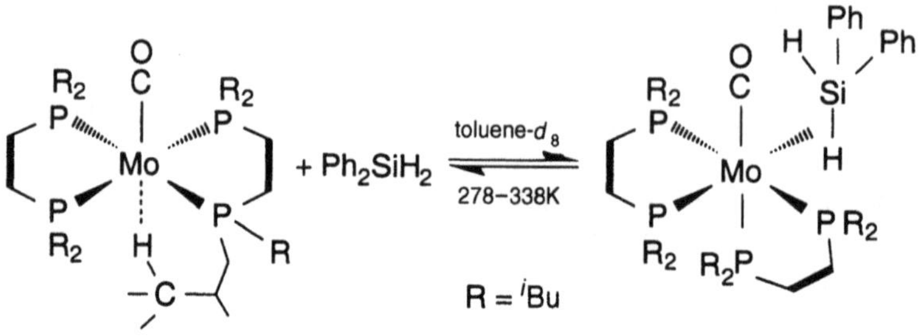

$$\Delta H^\circ = -12.8 \pm 1.2 \text{ kcal/mol}$$

$$\Delta S^\circ = -40 \pm 4 \text{ eu}$$

reactions [Eq. (7.5)].[13] The energy of the agostic interaction is again unknown but probably is about 10 kcal/mol.

7.1.2. Binding Energies of H₂ and σ Ligands to Thermally Unstable Metal Complexes

Several measurements of H₂ and alkane binding energies in thermally unstable complexes have been made, including studies in the *gas phase*. Gas-phase work has the advantage of being solvent-free and the production of very reactive intermediates is easier. Conversely it can be difficult to relate the results to solution systems because of the absence of solvation effects in the gas phase. Steric effects can also be much lower, especially for interactions with bare metal cations, M^+. The latter are excellent systems for molecular H₂ and alkane binding because of their high electrophilicity and reluctance to oxidatively add these molecules.[14] The study of metal ions is much easier than that of neutral atoms because mass-spectrometric methods permit extensive characterization of reaction intermediates and products. As discussed in Chapter 4, M–H₂ binding energies for first-row M–H₂ species, $[M(H_2)_n]^+$ ($n = 1$–6), vary greatly with M, but can be as high as 18.2 kcal/mol for $[Co(H_2)]^+$ using temperature-dependent equilibrium measurements of mass-selected M^+ ions reacting with H₂ (see Table 4.12).[15] Experiments using collision-induced dissociation (CID) in a guided ion-beam mass spectrometer show a similar value of 17.5 ± 2.3 kcal/mol.[16] A surprising result of this work is that in all cases the *binding energies of methane to positively charged metal ions, e.g., 15–18 kcal/mol for $[M(CH_4)_4]^+$ (M = Fe, Co), are greater than those for H₂*! This is completely opposite to the situation for organometallic fragments, including cationic species, studied thus far. The highly electrophilic $[M]^+$ (and to a lesser extent organometallic analogues) promote increased electron donation from the σ bond to M. As will be shown in Chapter 12, alkane activation occurs on highly electrophilic late-transition metal systems via transient σ-alkane complexes.

Reactions of small molecules with photochemically generated organometallics such as $M(CO)_5$ (M = Cr, W) can also be studied in the gas phase[17-19]:

$$Cr(CO)_6 \xrightarrow[-CO]{h\nu} Cr(CO)_5 \xrightarrow{H_2} Cr(CO)_5(H_2) \quad (7.6)$$

Use of time-resolved IR spectroscopy gives values for the binding energy of H_2 to 16e $Cr(CO)_5$ of 15.2 ± 1.3 kcal/mol[19] and at least 16 kcal/mol for M = W (Table 7.2).[17] It is interesting, that these binding energies are not significantly lower than those for stable H_2 complexes such as $W(CO)_3(PR_3)_2(H_2)$ (17-20 kcal/mol) yet the pentacarbonyls and similar complexes containing only π-acceptor ligands have not been isolated at room temperature. Presumably this is because any dissociation of H_2 leads to highly unstable $M(CO)_5$ whereas 16e fragments containing phosphines such as $M(CO)_3(PR_3)_2$ are stabilized by agostic interactions. It may be possible to trap $M(CO)_5(H_2)$ as a solid under the right conditions such as that for $CpMn(CO)_2(H_2)$, which has been isolated from liquid Xe.[20]

Alkanes show significantly weaker binding than H_2, and stronger bonds to $W(CO)_5$ are formed with longer-chain alkanes. Because the bond strengths correlate well with the ionization potential of the alkane C-H bond, it is reasonable that there is a simple dependence on the basicity of the alkane fragment (BD is of little consequence). There is similar behavior in binding of alkanes to $Cp^*Rh(CO)$ and polarizability of the alkanes can be correlated with binding strengths (see Section 12.4.2). Methane does not bind to either system but larger alkanes, especially cycloalkanes, coordinate increasingly more strongly, and $CpRe(CO)_2$(cyclopentane) is observable by NMR at low temperatures (see Section 12.3.1). Monofluoroalkanes such as CH_3F and C_2H_5F (12.2 ± 3 kcal/mol) are more strongly bound than alkanes to $W(CO)_5$, but no structural information is known here. On the other hand, *perfluoro*carbons such as perfluorodecalin bind more weakly than alkanes.[21]

Photoacoustic calorimetry (PAC) has been valuable for measuring enthalpy changes associated with rapid reactions of organometallic molecules in solution, particularly very reactive unsaturated species with weak ligands such as alkanes and H_2.[3,22,23] PAC utilizes a high-frequency microphone to measure acoustic waves generated during the absorption of a light pulse by the solution under study. The signal from the microphone can be used to calculate enthalpy changes associated with reactions, particularly those where CO photolytically dissociates to generate unsaturated species that rapidly coordinate weak donors such as hydrocarbon solvents. The reactions such as in Eq. (7.5) must be faster than the response time of the microphone, typically 1 or 2 μs, which limits this methodology. BDEs determined in heptane solution by Poliakoff[22-24] and Burkey[3] are summarized for various metals and ligands in Table 7.2 and compared to the BDE values determined in the gas phase.[17-19] The BDEs (ΔH_L) are actually calculated by measuring the overall enthalpy of displacement (ΔH_1) of CO by H_2 in e.g., Eq. (7.6) and subtracting it from the BDE of CO (ΔH_{CO}) in $Cr(CO)_6$:

$$\Delta H_L = \Delta H_{CO} - \Delta H_1$$

Table 7.2. Bond Dissociation Enthalpies (kcal/mol) of H_2, Alkanes, Silanes, and Other Ligands (L) from Various 16e Fragments Measured in Heptane Solution and Gas Phases

Fragment	Ligand, L	$\Delta H°$, solution	$\Delta H°$, gas
$Cr(CO)_5$	CH_4		8 ± 2
	Heptane	11.0 ± 1.0	
	Benzene		13.7 ± 0.8
	H_2	16.0 ± 1.0	15.2 ± 1.3
	$SiHEt_3$	21.1 ± 2.3	
	N_2	16.7 ± 1.0	
	C_2F_4		19.7 ± 1.4
	C_2H_4		24.7 ± 2.4
$Mo(CO)_5$	Heptane	15.3 ± 1.0	
	H_2	19.4 ± 1.0	
	$SiHEt_3$	22.0 ± 2.1	
	N_2	19.6 ± 1.0	
$W(CO)_5$	Xe	8.4 ± 0.2	8.4 ± 1.0
	CH_4, CF_4		<5
	CH_3F		11.2 ± 3
	Ethane		7.4 ± 2
	Propane		8.1 ± 2
	Butane		9.1 ± 3
	Hexane		10.8 ± 3
	Heptane		13.4 ± 2.1
	H_2		<16
	$SiHEt_3$	27.8 ± 2.3	
$CpV(CO)_3$	Heptane	9.9 ± 3.6	
	H_2	21.8 ± 4.8	
	N_2	28.7 ± 4.8	
$(Benzene)Cr(CO)_2$	Heptane	12	
	H_2	14.3	
	N_2	15.8	
	$SiHEt_3$	28	
$CpMn(CO)_2$	Heptane	10	
	Toluene	14.2 ± 0.8[a]	
	$SiHEt_3$	24.4	

[a]Bengali.[120]

The accuracy of the BDEs here depends on the accuracy of the ΔH_{CO} values, which may or may not be well determined or must even be estimated. For example, the BDEs listed in Table 7.2 for binding of heptane, H_2, and N_2 to $Cr(CO)_5$ determined by Poliakoff[23] are 2.6 kcal/mol lower than those reported in an earlier paper[22] because of a better estimate for ΔH_{CO} (36.0 kcal/mol). This value was also used for the (benzene)$Cr(CO)_2$ system for which ΔH_{CO} is unknown.[22]

There is little difference between H_2 and N_2 in their bonding energetics and also little dependence on M for both $M(CO)_5(L)$ and the $M(CO)_3(PR_3)_2(L)$ systems below. Unlike the situation for bare metal ions, alkanes are the weakest ligands on organometallic fragments (comparable to Xe binding,[25] Table 7.2), and the disparity with H_2 binding is greatest in the vanadium system. Silanes give the strongest interaction among the σ ligands, suggesting that complexes of the type $M(CO)_5$(silane) could be isolable. Further comparisons of σ-ligand bond strengths with those for classical ligands will be made below.

In summary, the binding energies of H_2 for "true" H_2 complexes are 15–25 kcal/mol and are comparable to the enthalpy of heterogeneous adsorption of H_2 onto the surface of many first-row metals. Thermodynamic measurements for complexes with elongated d_{HH} have not been performed, but binding energies would be expected to be significantly higher. Calculations for $[Os(H_2)(NH_3)_4L]^{n+}$ species indicate energies of 23–50 kcal/mol depending on L.[26] The ethylenediamine analogue with L = acetate has d_{HH} of 1.34 Å, and the calculated $\Delta H_{binding}$ is 44.5 kcal/mol, close to that for a classical hydride complex. The energies for alkane coordination in solution are much lower, about 9–13 kcal/mol, but can exceed those for H_2 on metal ions in the gas phase. Silane coordination is stronger than either alkane or H_2 binding [21–28 kcal/mol even on weakly coordinating $M(CO)_5$ fragments] because of superior BD and the greater electronegativity of Si. As will be shown in Chapter 13, borane binding (η^2-BH) is thermodynamically favored over silane binding in certain systems partly because the higher Lewis acidity of boron relative to silicon enhances M → XH back donation.

7.2. BINDING STRENGTH OF DIHYDROGEN AND σ LIGANDS VERSUS CLASSICAL LIGANDS; IMPORTANCE OF ENTROPY

Calorimetric measurements of the enthalpies of addition of various common ligands to $W(CO)_3(PCy_3)_2$ have been performed as for H_2 addition above [Eqs. (7.1)–(7.3)]. These values are given in Figure 7.1, and again the energy of the agostic interaction must be added to give the actual binding energies. H_2 is a stronger ligand than one might have imagined, much like N_2, with which it is electronically similar in terms of π-acceptor strength. However, as will be shown below, H_2 is a much better σ donor than N_2 and is *a more versatile ligand than any other weak ligand (and many strong ligands) in terms of the variety of L_nM fragments to which it binds*. As discussed in Chapter 4, H_2 can coordinate or oxidatively add to both highly electrophilic and electron-rich L_nM. Thus H_2 can be competitive with weak to moderately strong pure σ donors such as THF, water, and dichloromethane. Mass action effects are critical because H_2 can be displaced even by very weakly binding hydrocarbon solvents such as toluene or alkanes (see Chapter 9). However, elongated η^2-H_2 cannot easily be displaced even by moderate donors such as acetonitrile.

It is illuminating to compare the binding energy of H_2 to that for the aquo ligand, H_2O, the archetypal lone-pair donor in classical coordination chemistry. Addition of excess H_2O to a concentrated THF solution of $W(CO)_3(P^iPr_3)_2(H_2)$ gives instant vigorous effervescence of H_2, even under an H_2 atmosphere.[2] X-ray diffraction shows the product to be $W(CO)_3(P^iPr_3)_2(H_2O) \cdot$ THF, containing an H_2O ligand replacing the H_2 (Figure 7.2). The structure is also novel in that the H

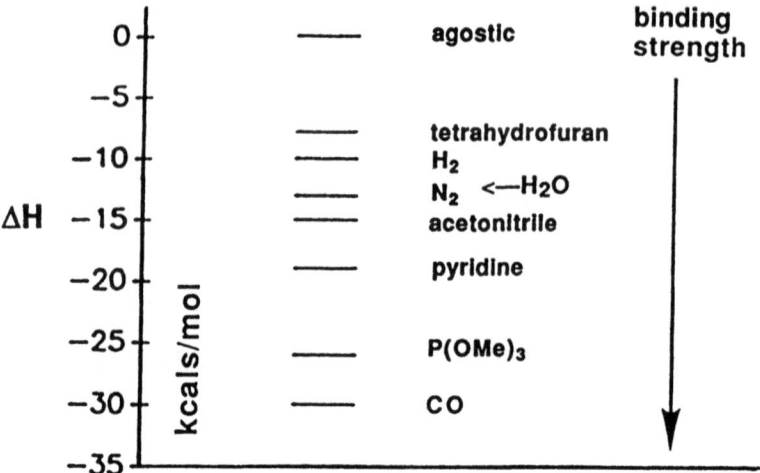

Figure 7.1. Comparison of enthalpies of addition of various ligands to $W(CO)_3(PCy_3)_2$ (binding strength increases going downward). Reprinted with permission from Gonzalez, A. A., et al. *ACS Symposium Series* **1990**, *428*, 133. Copyright 1990 American Chemical Society.

atoms on the aquo ligand hydrogen bond to the lattice THF oxygen atom and a CO oxygen on an adjacent molecule. Such hydrogen bonding in organometallic systems is becoming an increasingly recognized phenomenon, including unconventional systems with element–hydride bonds as proton acceptors (E–N⋯H–X) (see Section 9.4.4.1).[27] It is interesting to note that the aquo complex does not precipitate if

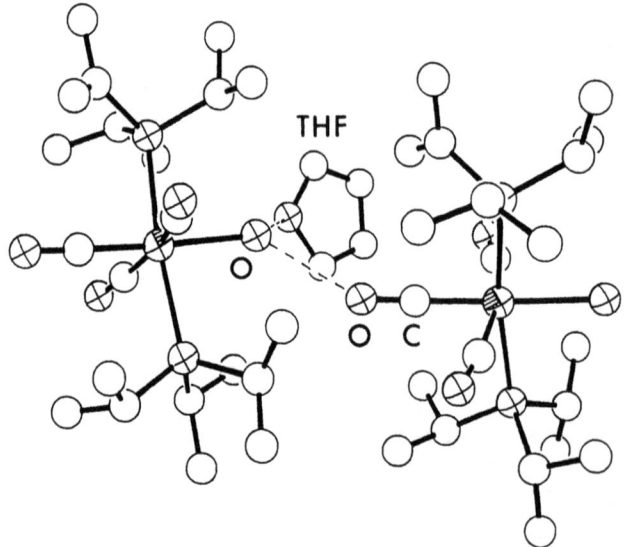

Figure 7.2. ORTEP of $W(CO)_3(P^iPr_3)_2(H_2O)\cdot THF$ showing hydrogen bonding interactions to THF and CO on neighboring molecule, which repeat to give chainlike linkages. Reprinted with permission from Kubas et al.[2] Copyright 1992 American Chemical Society.

addition of H_2O to $W(CO)_3(P^iPr_3)_2(H_2)$ is done in *hexane* under an H_2 atmosphere with a large excess of water present as an immiscible phase:

$$W(CO)_3(P^iPr_3)_2 + H_2O \xrightarrow[\text{hexane}]{H_2} \text{N.R.} \quad (7.7)$$

$$W(CO)_3(P^iPr_3)_2 + H_2O \xrightarrow[\text{hexane}]{Ar} W(CO)_3(P^iPr_3)_2(H_2O) \quad (7.8)$$

As soon as the H_2 atmosphere is replaced by Ar, the less soluble H_2O complex precipitates. Subsequent exposure to vacuum rapidly leads to dissociation of H_2O and precipitation of insoluble $W(CO)_3(P^iPr_3)_2$. This demonstrates the extremely delicate reversible nature of the H_2O and H_2 binding and indicates that *H_2 can compete thermodynamically and kinetically with H_2O as a ligand*. A major factor is mass action, i.e., concentration of unbound ligand in solution. In hexane the low solubility of H_2O limits its maximum concentration to the same order as that of dissolved H_2 (ca 0.005 M), as opposed to the situation in THF, where the high concentration of miscible H_2O overwhelms that of H_2. Other complexes demonstrating this effect are $[Ru\{HB(pz)_3\}(PPh_3)_2(H_2O)]^{+28}$ and $[Ru(H_2O)_6]^{2+}$, where H_2O can be displaced by H_2 under pressurized H_2 even in H_2O solution.[29] One of the first H_2 complexes, $[IrH(H_2)(PPh_3)_2(bq)]^+$, was prepared by displacement of H_2O under 1 atm of H_2 in organic solvents.[30]

The fact that H_2 and water can closely compete for the same binding site is relevant to biological activation of H_2 by metalloenzymes such as hydrogenase that have organometallic-like active sites with $OC-Fe-OH_2$ moieties (see Chapter 10). Reversible displacement of H_2O by H_2 followed by heterolytic cleavage of H_2 undoubtedly occurs here, and the thermodynamic data above show that H_2 binding should easily occur on large hydrophobic metalloenzyme sites, where the effective H_2O concentration is low. Catalytic heterolysis of H_2 and even of alkanes is known to occur in aqueous solutions of metal complexes. Of great interest are the Shilov-type systems for alkane functionalization, which utilize cationic Pt^{II} catalysts in *aqueous* acids (see Chapter 12).[31] It is most surprising that alkane binding is competitive with water, and as will be shown hydrogen bonding and entropic factors may play important roles in such weak ligand binding.

Equilibrium competitive binding between H_2O and H_2 is shown by variable-pressure IR spectra of a solution of $W(CO)_3(P^iPr_3)_2$ in 1% H_2O/THF.[2] H_2 pressures in excess of 1000 psi are required to quantitatively form $W(CO)_3(P^iPr_3)_2(H_2)$ at 25°C, as monitored by relative intensities of ν_{CO}. The equilibrium constants for displacement of H_2 by H_2O can be determined at pressures of several atm H_2 at 25 to -70°C. As the temperature is lowered, the peaks owing to the H_2 complex decrease and new peaks owing to the H_2O complex appear. The thermodynamic parameters for Eq. (7.9) are readily obtained from van't Hoff plots:

$$W(CO)_3(PR_3)_2(H_2) + H_2O + THF \rightleftarrows W(CO)_3(PR_3)_2(H_2O) \cdot THF + H_2 \quad (7.9)$$

R = Cy: $\Delta H = -2.8 \pm 0.1$ kcal/mol R = ^iPr: $\Delta H = -4.5 \pm 0.2$ kcal/mol
$\Delta S = -16.5 \pm 2.0$ cal/mol·deg $\Delta S = -18.8 \pm 2.0$ cal/mol·deg

Displacement of H_2 by water is exothermic by 3–4 kcal/mol, but hydrogen bonding between coordinated H_2O and solvent (THF) appears to play a role in the thermodynamics. The surprisingly high negative entropy change in Eq. (7.9) no doubt reflects free THF becoming bound (three particles converting to two). The unfavorable entropy of binding of H_2O is largely the reason why the equilibrium favors H_2 binding at room temperature and H_2O binding at low temperatures. ΔG_{298} can be calculated to be 1.1 kcal/mol for $R = {}^iPr$ and 2.1 kcal/mol for $R = Cy$, i.e., favoring the left side of Eq. (7.9). *Entropic factors can thus be critical in competition between weak ligands for binding sites*, as will be seen below for N_2 versus H_2 binding.

The enthalpies of binding of H_2O in Eq. (7.9) are relative to H_2 so it is of interest to determine the enthalpy of binding of H_2 to $W(CO)_3(P^iPr_3)_2$, which is measured directly to be -11.2 ± 0.5 kcal/mol in toluene at 20°C:

$$W(CO)_3(P^iPr_3)_2 + H_2 \rightarrow W(CO)_3(P^iPr_3)_2(H_2) \tag{7.10}$$

Indirect measurement of ΔH based on reactions with pyridine yields 10.4 ± 0.8 kcal/mol, giving an average value of 10.8 ± 1.0 kcal/mol that compares favorably with the ΔH of -9.4 ± 0.9 kcal/mol for $W(CO)_3(PCy_3)_2$.[1,2] These values reflect the enthalpy for net reaction with H_2 and do not incorporate enthalpies of tautomerization between η^2-H_2 (1) and dihydride (2) species. The major species present in

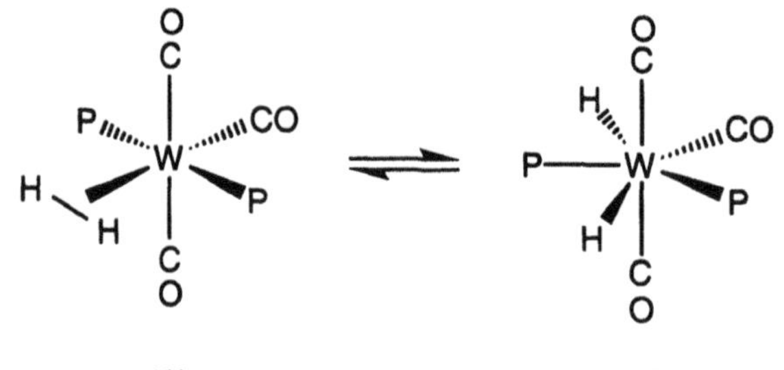

$$\tag{7.11}$$

(1)   (2)

toluene is the H_2 complex and the enthalpy of Eq. (7.11) in the forward direction is $+1.2 \pm 0.6$ kcal/mol for $R = {}^iPr$.[32] Thus the observed enthalpy of Eq. (7.10) includes very little contribution (~ 0.3 kcal/mol) from Eq. (7.11), where $K_{eq} = 0.25$ at 24°C.

The affinity of H_2 versus other ligands such as N_2 for L_nM varies and can be entropy-dependent (see below). In some cases N_2 is a better ligand than H_2 and sometimes the opposite is true, or N_2 does not bind at all. This is particularly true in electrophilic cationic fragments such as $[Mn(CO)_3(PCy_3)_2]^+$ (which contains an agostic interaction as shown in Eq. 7.12[33]) and the later transition elements, where Ir^{III} and Pt^{II} fragments such as $[PtH(PR_3)_2]^+$[34] weakly bind H_2 but not N_2. Surprisingly, no evidence for N_2 binding to $[Mn(CO)_3(PCy_3)_2]^+$ under 1 atm of $^{15}N_2$ is seen by ^{15}N solution NMR, even at $-58°C$, where entropy should favor binding of weak external ligands. In contrast, for $[Mn(CO)(dppe)_2(N_2)]^+$ the expected two resonances for end-on coordinated N_2 are observed at -73 to $25°C.^{35}$ Also, the Re congener, $[Re(CO)_3(PCy_3)_2]^+$, is 50% coordinated by N_2 at room temperature.[36] The explanation for these disparities goes

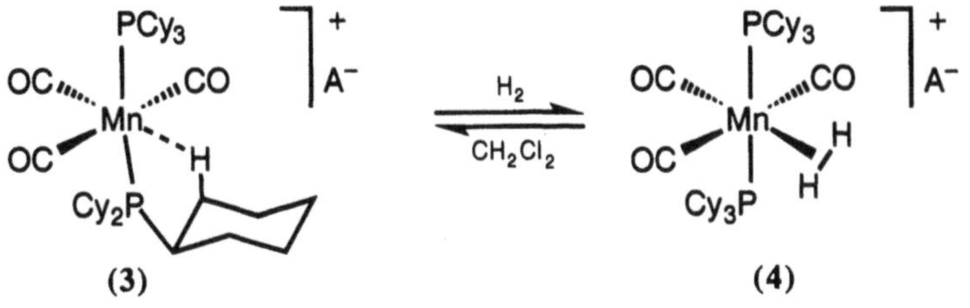

$$\begin{array}{c}\text{(3)} \end{array} \xrightleftharpoons[\text{CH}_2\text{Cl}_2]{\text{H}_2} \begin{array}{c}\text{(4)}\end{array} \quad (7.12)$$

beyond the simple rationale that third-row metals are better π donors than first-row metals and N_2 is a moderate π acceptor. It is quite informative to compare H_2, N_2, and other ligand coordination on a large array of cationic and neutral fragments. Table 7.3 lists 16e fragments, all of which bind H_2, in approximate order of decreasing electrophilicity and compares whether or not they bind N_2 and other common ligands discussed below. In some cases direct comparisons to the binding strength of H_2 can be made (denoted by "weaker" or "stronger"). In the neutral $M(CO)_3(PCy_3)_2$ system, coordination of H_2 and N_2 is very competitive and is nearly isoenergetic because of entropic effects, as will be discussed below. For the cationic

Table 7.3. Comparison of Small-Molecule Binding Relative[a] to H_2 on Various Metal Fragments in Approximate Order of Decreasing Electrophilicity

	←――― π-Acceptor strength				
Metal fragment	SO_2	Silanes	C_2H_4	N_2	Et_2O/CH_2Cl_2
$[trans\text{-}PtH(PR_3)_2]^+$	No			No	Stronger
$[Fe(CO)(dppe)_2]^{2+}$				No	
$[Re(CO)_4(PR_3)]^+$		Weak	Stronger	No	Stronger
$[Mn(CO)_3(PR_3)_2]^+$	Reversible	No ($-70°C$)	No ($-70°C$)	No ($-58°C$)	No
$[Re(CO)_3(PR_3)_2]^+$				Weaker	No
$[Re(CO)_2(triphos)]^+$				No (3 atm)	
$[Mn(CO)(dppe)_2]^+$	Reversible	No ($-70°C$)		Weaker	No
$[OsH(dppe)_2]^+$				Weaker	No
$[RuCl(dppe)_2]^+$				No	No
$Cr(CO)_5$	Strong	Stronger	Stronger[b]	Weaker	
$Cr(CO)_3(PR_3)_2$		No		Weaker[c]	No
$W(CO)_3(PR_3)_2$	Strong	OA[d]	Yes	Stronger	No
$Mo(CO)(dppe)_2$	Strong	Yes[e]	Yes	Stronger	No
$TcCl(dppe)_2$				Yes	No

[a]Weaker or stronger than H_2 in terms of ΔG of binding at room temperature. Terminology: strong = irreversible coordination; no = binding not observed at ambient temperature and pressure; yes = similar to H_2 binding but relative stabilities not determined; blank entry = not reported.
[b]$Cr(CO)_5(C_2H_4)$ has been isolated as a solid (Banister, J. A.; Lee, P. D.; Poliakoff, M. *Organometallics* 1995, *14*, 3876).
[c]At pressures above 1 atm.
[d]OA = oxidative addition.
[e]Complex rearranges to isomer with silane cis to CO.

[Mn(CO)$_3$(PCy$_3$)$_2$]$^+$ congener however, H$_2$ binding is clearly favored *enthalpically* over N$_2$ since at $-58°$C there is no N$_2$ coordination and complete H$_2$ coordination, while for Cr(CO)$_3$(PCy$_3$)$_2$ partial binding of both H$_2$ and N$_2$ is observed spectroscopically at about 2 atm. The entropy effect, even though greater at low temperatures, could not be responsible for this difference.

For [Mn(CO)(R$_2$PC$_2$H$_4$PR$_2$)$_2$]$^+$, direct competition studies show that H$_2$ binds more strongly than N$_2$. However, the thermodynamic difference is not as dramatic as for [Mn(CO)$_3$(PCy$_3$)$_2$]$^+$ since the complex with R = Et coordinates N$_2$ completely at 25°C (1 atm) while the less electron-rich R = Ph analogue still binds N$_2$ to the extent of 37%.[35] Since [Mn(CO)$_3$(PCy$_3$)$_2$]$^+$ is more electrophilic than the latter, the obvious trend is that *H$_2$ becomes an increasingly better ligand than N$_2$ as the electrophilicity of M increases and BD decreases.* Conversely, for more electron-rich systems such as Mo(CO)(dppe)$_2$, N$_2$ clearly binds more strongly than H$_2$.[37] The [FeH(dmpe)$_2$]$^+$ and FeH$_2$(PPh$_2$Et)$_3$ fragments also prefer N$_2$ and quantitatively exchange H$_2$ for it [Eq. (7.13)][38,39]:

$$2[FeH(H_2)(dmpe)_2]^+ + Mo(N_2)_2(dppe)_2 \rightarrow 2[FeH(N_2)(dmpe)_2]^+ + MoH_4(dppe)_2$$

(7.13)

This is surprising in that the N$_2$ is bound more tightly in the Mo complex than in the product but the reaction may be driven by the relative M–H bond energies [MoH$_4$(dppe)$_2$ is a tetrahydride with stronger M–H bonds].[38]

The disparities in relative coordinating abilities result from N$_2$ being a very poor σ donor[40-45] and a good, though slightly weaker, π acceptor like H$_2$,[41,43] as shown both theoretically and experimentally (mainly by Mossbauer studies). The data in Table 7.3 demonstrate that N$_2$ is a poorer σ donor than the very weak bases Et$_2$O and CH$_2$Cl$_2$, both of which form fairly robust complexes with the highly electrophilic [Re(CO)$_4$(PR$_3$)]$^+$ and [PtH(PR$_3$)$_2$]$^+$ fragments.[34,46] Theoretical calculations that include charge decomposition analysis of W(CO)$_5$L show that for L = H$_2$ the contribution from σ donation is 0.349e versus only 0.027e for N$_2$, while BD is 0.107e for N$_2$ versus 0.129e for H$_2$.[43] It is apparent that N$_2$ coordination must be stabilized by BD, even in *actinide* complexes[47] such as [{(NN$_3'$)U}$_2$(μ_2-η^2:η^2-N$_2$)] for which DFT calculations show that BD to side-on bound N$_2$ is the *only* significant U–N$_2$–U interaction.[44] Cationic organometallic electrophiles do not provide enough BD to stabilize N$_2$ binding, and N$_2$ cannot compensate for this loss by increasing σ donation to M as effectively as H$_2$ can. For the much more electron-rich Mo(CO)(dppe)$_2$ fragment on the other hand, BD to H$_2$ accounts for roughly two-thirds of the bond strength [versus two-thirds σ donation for Mo(CO)$_5$], showing how easily H$_2$ reverses its bonding capability (see Chapter 4). Reiterating, *nonclassically bound H$_2$ is a more versatile ligand than many classically coordinated ligands such as N$_2$ in its ability to adjust to a broader range of electronic situations.*

Even more surprising than N$_2$ versus H$_2$ binding, the normally strong ligand SO$_2$, which is a stronger π acceptor than CO, coordinates only very weakly to the Mn cations and not at all to [PtH(P^{i}Pr$_3$)$_2$]$^+$. Comparisons of η^2-H$_2$ to other types of ligands, including pure σ donors, further support the versatility of η^2-H$_2$, e.g., [Re(CO)$_4$(PR$_3$)]$^+$ gives isolable Et$_2$O and CH$_2$Cl$_2$ complexes that are more stable than the H$_2$ complex.[2] However, H$_2$ gains the bonding advantage over such weak σ donors on more electron-rich neutral systems such as M(CO)$_3$(PCy$_3$)$_2$ and Mo(CO)(R$_2$PC$_2$H$_4$PR$_2$)$_2$, which do not bind ethers or CH$_2$Cl$_2$ in stable fashion. In these cases, increased BD strengthens H$_2$ binding considerably, but pure σ bases cannot receive BD.

The main rivals to H$_2$ are ethylene and other olefins, which generally coordinate somewhat more strongly to the same fragments as H$_2$ and bind to M in the classic Dewar–Chatt–Duncanson π-bonding model to which M–H$_2$ bonding is comparable. Although ethylene does not coordinate to [Mn(CO)$_3$(PCy$_3$)$_2$]$^+$ even at −70°C, this may be a steric effect since olefins bind to a derivative with much less bulky tied-back phosphites (P(OCH$_2$)$_3$CMe)[48] and also to [Re(CO)$_4$(PR$_3$)]$^+$.[49]

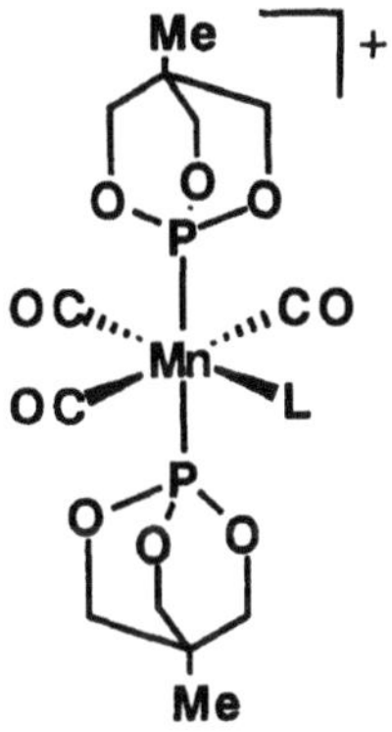

L = olefin, silane

Silanes also form σ complexes with a wide range of fragments, although these can be more prone to OA than H$_2$ because the Si–H bond is a slightly better acceptor than H–H (see Chapter 11). Silane coordination to first-row octahedral group 6 and 7 metals can be much weaker than to second- and third-row metals, again probably for steric reasons (smaller first-row metal). As for olefins, PhSiH$_3$ does not form a stable complex with [Mn(CO)$_3$(PCy$_3$)$_2$]$^+$ down to −75°C from ^{1}H and ^{31}P NMR evidence, but it does to the phosphite species at low temperatures. Similarly, silane coordination to [Mn(CO)(dppe)$_2$]$^+$ and even neutral Cr(CO)$_3$(PR$_3$)$_2$ is not observed,[35,50] but silanes bind to [Re(CO)$_4$(PR$_3$)]$^{+49}$ and even oxidatively add to W(CO)$_3$(PR$_3$)$_2$.[50] Silanes also readily undergo heterolytic cleavage at 25°C on these cationic species (see Chapter 11), so silane binding and activation is much more variable than that for H$_2$, which coordinates in a more predictable fashion. This is a result of the increased complexity of the R$_3$Si–H ligand, where substituents on Si influence activation, and reversal of trends can occur even on varying the electrophilicity of M (see Chapters 4 and 11).[51] Silanes and also N$_2$ and SO$_2$ coordinate as well or better than H$_2$ to CpMn(CO)$_2$, which is both less sterically crowded (piano-stool geometry) and less electrophilic than the other fragments in Table 7.3.

As shown in Chapter 4, the structure of the ligand trans to the ligand of interest is critical to the latter's bond strength and activation and in Table 7.3 the trans ligand is usually CO. However, even for systems with trans ligands that are primarily

Table 7.4. Equilibrium Parameters for Eq. (7.14)

L	K_{eq}, 298 K	ΔG, kcal/mol
H_2	8.08 ± 0.48	−1.24 ± 0.04
C_2H_4	0.070 ± 0.001	+1.57 ± 0.01
CO_2	0.007 ± 0.001	+2.97 ± 0.06

donors such as Cl or phosphine, e.g., [Re(CO)$_2$(triphos)]$^+$,[52] the above bonding trends still hold. Both H$_2$ and N$_2$ bind similarly to second-row TcCl(dppe)$_2$[53] while H$_2$ binding is again favored over N$_2$ on the more electrophilic [RuCl(dppe)$_2$]$^+$ cationic congener.[54] π-donating Cl generally has a low trans influence and is a weak-field ligand, and TcCl(dppe)$_2$(H$_2$) consequently has an elongated $d_{HH} > 1$ Å. Conversely, [PtH(P^{i}Pr$_3$)$_2$]$^+$ contains a hydride with strong trans-donor effect and does not coordinate N$_2$ yet binds H$_2$ at −40°C (H$_2$ dissociates on warming).[34] Even SO$_2$ does not bind to this feebly backdonating, highly electrophilic fragment at room temperature, although weak bases readily give isolable, air-stable [PtH(PR$_3$)$_2$(L)]$^+$ for L = Et$_2$O and CH$_2$Cl$_2$.

One exception to the general observation that N$_2$ is a slightly better ligand than H$_2$ on neutral complexes is a Rh complex with a pincer ligand that coordinates both H$_2$ and N$_2$ [Eq. (7.14)].[55] In cyclohexane solution, the H$_2$ binding is 1.24 kcal/mol more favorable than N$_2$ binding (Table 7.4), possibly because of the slightly better

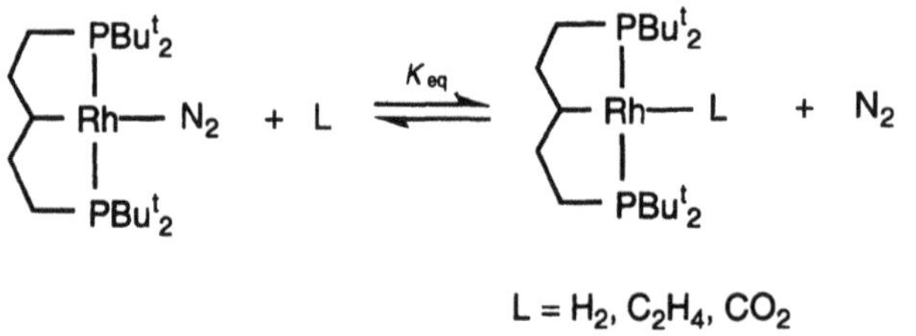

$$\{PBu^t_2\}Rh{-}N_2 + L \xrightleftharpoons{K_{eq}} \{PBu^t_2\}Rh{-}L + N_2 \quad (7.14)$$

L = H$_2$, C$_2$H$_4$, CO$_2$

BD ability of H$_2$ versus N$_2$ to the electron-rich Rh center here. Surprisingly, N$_2$ binds substantially more strongly than ethylene, which again may be a result of the steric demands of the bulky phosphines. Not surprisingly, CO$_2$ is the weakest ligand, as is usually observed in other systems. Very few metal fragments that bind H$_2$ also coordinate CO$_2$, and W(CO)$_3$(PR$_3$)$_2$ does not bind CO$_2$ even in liquid CO$_2$.[56] Dioxygen is also rarely found on the same fragment as H$_2$,[57] although the reason is that O$_2$ usually oxidizes the low-valent fragments that favor H$_2$ binding.

As seen above, entropy effects can be a key factor in H$_2$ versus H$_2$O coordination. This holds true for other situations involving weak ligand and agostic coordination where $\Delta H_{binding}$ and $T\Delta S$ are similar energetically.[58] From statistical thermodynamics, the *intramolecular* agostic interaction in Eq. (7.3) is favored entropically over external ligand binding by approximately 10 kcal/mol at 25°C. Thus stable binding of very weak ligands such as alkanes at room temperature is virtually impossible here since $\Delta H_{binding}$ is only about 10–15 kcal/mol, which leaves

Table 7.5. Thermodynamic Parameters for Binding of H_2 and N_2

$$M(CO)_3(PCy_3)_2 + L \rightarrow M(CO)_3(PCy_3)_2(L)$$

M	L	$\Delta H°$	$\Delta\Delta H°$	$\Delta S°$	$\Delta\Delta S°$	$\Delta\Delta G°_{298}$
Cr	N_2	-9.3 ± 0.2		-35.4 ± 2.3		
Cr	H_2	-7.3 ± 0.1	-2.0 ± 0.3	-25.6 ± 1.7	-9.8 ± 2.6	$+0.9$
Mo	N_2	-9.0 ± 0.6		-32.1 ± 3.2		
Mo	H_2	-6.5 ± 0.2	-2.5 ± 0.8	-23.8 ± 2.1	-8.3 ± 3.9	0.0
W	N_2	-13.5 ± 1.0				
W	H_2	-9.4 ± 0.9	-4.1 ± 0.4		-13.8 ± 3.5	-0.3

net ΔG of only 0 to -5 kcal/mol. Entropy effects can even include differences in the *absolute entropy* of small molecules. High pressures are needed to quantitatively bind H_2 and N_2 to $Cr(CO)_3(PCy_3)_2$ in solution,[59] and although N_2 is the stronger ligand enthalpically, the pressure required for N_2 binding to $W(CO)_3(PCy_3)_2$, 1500 psi, is paradoxically higher than that for H_2, 300 psi. Thermodynamic data for reaction of these molecules with group 6 $M(CO)_3(PCy_3)_2$ show that the preferred binding of H_2 is due to entropic rather than enthalpic factors (Table 7.5).[60] The enthalpies of binding are in the order $Cr \cong Mo < W$, in accord with the observed stronger binding of H_2 and N_2 to W. For each M, N_2 is preferred with regard to the enthalpy of binding but is disfavored with regard to the entropy of binding. The lower entropy of binding of H_2 compared to N_2 will influence the stability of the complexes as a function of T. For the displacement reaction [Eq. (7.15)], $K_{eq} = 1$ when $T = \Delta H/\Delta S$:

$$M(CO)_3(PCy_3)_2(N_2) + H_2 \rightleftarrows M(CO)_3(PCy_3)_2(H_2) + N_2 \quad (7.15)$$

Above this T the H_2 complex will be favored; below it the N_2 complex will be favored. These "crossover" temperatures are -70, 28, $46°C$ for Cr, Mo, W, respectively; the $\Delta\Delta G$ values at 25°C are given in Table 7.5.

What is the origin of the reduced entropy of H_2 binding? At first glance it might appear that since H_2 is side-on bound and rotates nearly freely about the M–H_2 axis, the lower entropy could be due to this additional degree of freedom not present for N_2 coordination. However, the real explanation is simpler: Compared to N_2, H_2 has less translational and rotational entropy to lose upon binding in the first place.[60] The entropy of gaseous ligand binding to a complex in solution is equal to the difference in total entropies of the species involved:

$$ML_n(soln) + H_2(g) \rightarrow ML_n(H_2)(soln) \quad (7.16)$$

Binding a gaseous ligand increases the total entropy of $ML_n(H_2)$ relative to ML_n but does so by a relatively minor amount compared to the entropy lost by the ligand.[61] On this basis the total entropy of exchange for Eq. (7.17) should depend primarily on the differences in absolute entropies for $N_2(g)$ and $H_2(g)$:

$$ML_n(N_2)(soln) + H_2(g) \rightarrow ML_n(H_2)(soln) + N_2(g) \quad (7.17)$$

The third-law entropies, $S°$, of the two gases can be calculated by using standard formulas of statistical thermodynamics.[62] At room temperature, the entropy is due exclusively to the translational and rotational components. Owing to its lower mass and moment of inertia, the absolute entropy of H_2 (31.2 cal/mol·deg) is 14.6 cal/mol·deg lower than that for N_2. If Eq. (7.17) is reexamined, it is clear that if the total entropies of the complexes in solution exactly canceled, the predicted entropy change would be 14.6 cal/mol·deg. This is reasonably close to the average values obtained for the three complexes: 11 ± 4 cal/mol·deg.

In summary, it is clear that the large entropy of ligand exchange is due to the fact that H_2 has the smallest absolute entropy ($S°$) of any diatomic gas. The net result is that at higher temperatures H_2 will be more competitive in binding relative to N_2 or other small molecules. This could have important consequences in catalysis, and this concept should apply to other weakly bound small molecules, including methane and most σ-bonded ligands. Binding of CH_4 ($S°$ is one of the lowest, 44.5 cal/mol·deg) should thus be entropically favored over coordination of virtually any other molecule except H_2, although enthalpically CH_4 is among the weakest polyatomic ligands known.

7.3. KINETICS OF FORMATION AND SUBSTITUTION OF DIHYDROGEN, ALKANE, AND RELATED σ LIGANDS

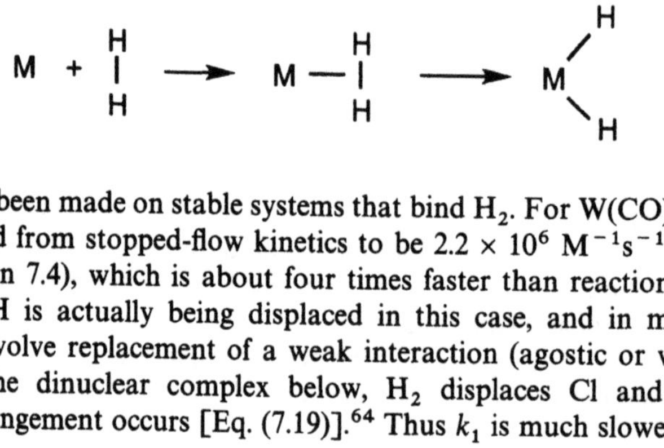

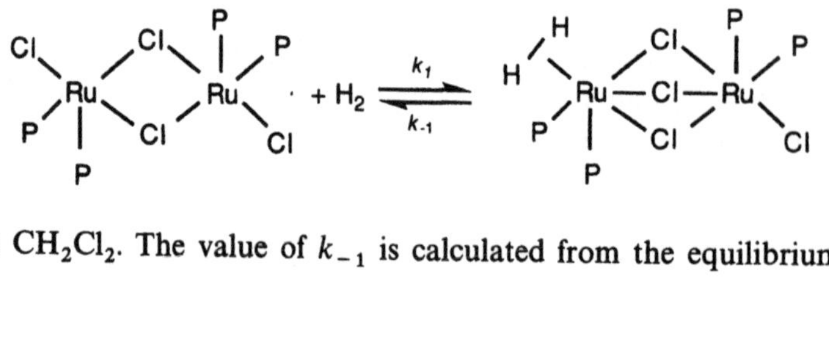

Oxidative addition of H_2 has long been recognized, but only relatively recently has it been realized that molecular coordination of H_2 lies along the reaction coordinate. Relatively few direct measurements of the rate of reaction of H_2 with metal complexes have been made on stable systems that bind H_2. For $W(CO)_3(PCy_3)_2$, the rate is calculated from stopped-flow kinetics to be 2.2×10^6 M^{-1}s^{-1} in toluene at 25°C (see Section 7.4), which is about four times faster than reaction with N_2.[1,63] An agostic C–H is actually being displaced in this case, and in most cases H_2 binding does involve replacement of a weak interaction (agostic or weakly bound solvent). For the dinuclear complex below, H_2 displaces Cl and considerable structural rearrangement occurs [Eq. (7.19)].[64] Thus k_1 is much slower, 32 M^{-1}s^{-1} at 25°C in CH_2Cl_2. The value of k_{-1} is calculated from the equilibrium constant to be 0.023 s^{-1}. The reaction is exothermic by about 14 kcal/mol and $\Delta S°$ is unexpectedly *positive* (ca 14 e.u.). Solvation of each five-coordinate Ru of the reactant by CH_2Cl_2 is rationalized to explain this (two solvent molecules released per dimer). This complex provides one of the few examples of *direct* transfer of hydrogen from η^2-H_2 to a substrate in a catalytic cyclic (see Chapter 9).

Kinetic studies of the protonation of metal hydrides with acids to form H_2 complexes of the type $[FeH(H_2)P_4]^+$ $[P_4 = $ 2dppe or $P(C_2H_4PPh_2)_3]$ show first-order dependence on concentration of both complex and acid.[65] This reaction in THF involves direct attack of acid at one hydride in FeH_2P_4, apparently by the mechanism in Eq. (7.20), where a hydrogen-bonding interaction (5) is likely to be

$$\begin{array}{c} Fe-H + HX \\ \downarrow \\ Fe-H\cdots HX \longrightarrow Fe\begin{array}{c}H\cdots\\ \ \ \ \ H\\ \ \ \ \ \ \ X\end{array} \longrightarrow Fe-\begin{array}{c}H\\ \vdots\\ H\cdots\\ \ \ \ \ X\end{array} \longrightarrow \left[Fe-\begin{array}{c}H\\ |\\ H\end{array}\right]^+ X^- \end{array} \quad (7.20)$$

(5) (6) (7) (8)

the first step. The second-order rate constants depend on both the nature of the acid and the complex and range from 1.7×10^{-4} (HBF_4) to 3.4×10^{-2} dm^3/mol·s (HBr) for the $P(C_2H_4PPh_2)_3$ complex and 9.7×10^{-3} to 1.48×10^{-1} dm^3/mol·s for the dppe complex (same acids). An inverse isotope effect is observed (see below), which indicates a late transition state similar to 7 in Eq. (7.20).

Many studies have been done on unstable or transient H_2 and other σ complexes using techniques such as flash photolysis. As in the above cases, H_2 displaces a weak ligand, which, e.g., can be an alkane solvent molecule[63]:

$$Cr(CO)_6 \xrightarrow{h\nu} Cr(CO)_5 \xrightarrow{C_6H_{12}} Cr(CO)_5(C_6H_{12}) \xrightarrow{H_2} Cr(CO)_5(H_2) \quad (7.21)$$

The initially formed $Cr(CO)_5$ reacts on the picosecond timescale to form the σ complex $Cr(CO)_5$(cyclohexane) in cyclohexane, where the strength of the Cr-alkane bond is about 10 kcal/mol (see Chapter 12). The presence of "token ligands" such as cyclohexane slows down the overall reactions of the naked unsaturated $Cr(CO)_5$ fragment, but reactions are still very fast. Formation of $Cr(CO)_5(H_2)$ occurs with $t_{1/2} = 36$ μs,[66] a rate similar to that for N_2 and CO addition to $Cr(CO)_5$(alkane). More recent and faster time-resolved infrared (TRIR) spectroscopic measurements show that $CpV(CO)_3(H_2)$ is formed from $CpV(CO)_4$ within 1 μs of the UV flash in heptane.[24] Second-order rate constants are given in Table 7.6 for this system and also Mn and Re analogues.[67,68] $CpV(CO)_3$ and $Cp*V(CO)_3$ are among the most reactive intermediates formed by loss of CO from a carbonyl complex. $CpV(CO)_3$ is about 100 times more reactive than its Nb and Ta analogues and nearly 1000 times more reactive than $CpMn(CO)_2$. Conversely, $CpRe(CO)_2$(heptane) is the longest-lived organometallic alkane complex in solution at room temperature (40 s^{-1} decay rate),[68] and this finding inspired the first direct observation of a similar alkane

Table 7.6. Second-Order Rate Constants k_2 for Reactions of Ligands with Metal-Heptane Complexes in Heptane at 298 K

Fragment	L	k_2, $M^{-1}s^{-1}$[a]
CpV(CO)$_3$ (heptane)	CO	1.3×10^8
	H$_2$	$2.1(5.7) \times 10^8$
	N$_2$	$1.5(3.4) \times 10^8$
	HSiEt$_3$	1.8×10^8
CpMn(CO)$_2$ (heptane)	CO	$8.1(16) \times 10^5$
	H$_2$	$8.8(17) \times 10^5$
	N$_2$	$4.7(11) \times 10^5$
	THF	4.4×10^6
	PPh$_3$	$9.0(8.8) \times 10^6$
CpRe(CO)$_2$ (heptane)	CO	2×10^3

[a] Values in parentheses are for Cp* derivatives.

complex by NMR (see Chapter 12). The Cp* derivatives are about twice as reactive as the less electron-rich Cp analogues, although steric factors may also affect these rates. Reaction rates for H$_2$ differ little from those for the much stronger CO ligand and are slightly faster than those for N$_2$. Surprisingly, the rates for phosphine reactions are fastest despite the greater bulk of PR$_3$.

Kinetic studies of H$_2$ dissociation and substitution rates show that the potential energy surfaces for these reactions vary dramatically even with minor changes in ancillary ligands or for isomers (Table 7.7).[5,69] Electronic effects, especially the influence of the trans ligand, appear to be more important than steric factors. For the Ir system, the cis isomer with H$_2$ trans to Cl has a strongly bound H$_2$ (d_{HH} = 1.11 Å) while the trans isomer with H$_2$ trans to H contains a weakly bound H$_2$ that dissociates nearly 10^5 times faster (see also Section 4.7.1). One of the few comprehensive quantitative studies of H$_2$ substitution reactions shows displacement of H$_2$ by L (MeCN, PhCN, Cl$^-$) from [MH(H$_2$)(P)$_4$]$^+$ (M = Fe, Ru, Os) is first-order in concentration of complex and zero order in L, i.e., a dissociative mechanism.[70]

Table 7.7. Kinetic Data for H$_2$ Dissociation in Solution at 273 K

Complex	k_1, s^{-1}	$\Delta G^{\ddagger}$, kcal/mol
W(H$_2$)(CO)$_3$(P^{i}Pr$_3$)$_2$	8	14.8
OsHCl(H$_2$)(CO)(P^{t}Bu$_2$Me)$_2$	2×10^4	10.6
OsHCl(H$_2$)(CO)(P^{i}Pr$_3$)$_2$	950	12.2
OsHI(H$_2$)(CO)(P^{i}Pr$_3$)$_2$	70	13.6
OsH$_2$(H$_2$)(CO)(P^{i}Pr$_3$)$_2$	0.002	19.6
cis-IrHCl$_2$(H$_2$)(P^{i}Pr$_3$)$_2$	<7.3[a]	>12.6
trans-IrHCl$_2$(H$_2$)(P^{i}Pr$_3$)$_2$	2.1×10^5[a]	8.9

[a] At 234 K.

Consistent with this is the independence of $\Delta H^{\ddagger}$ (~30 kcal/mol for reaction of $[\text{FeH}(\text{H}_2)(\text{pp})_2]^+$ with nitriles, near 300 K) and $\Delta S^{\ddagger}$ on the nature of L and the large positive values of $\Delta S^{\ddagger}$ (up to 19 cal/mol·deg). The rate constants at 298 K are near 4×10^4 s^{-1} for Fe complexes.

The rate of dissociation, hence lability, also increases dramatically with protonation for both H_2 and dihydride complexes [Eq. (7.22)].[71]

$$[\text{RuH}_3(\text{PPh}_3)_2]^- \xrightarrow{\text{H}^+} \text{RuH}_2(\text{H}_2)(\text{PPh}_3)_2 \xrightarrow{\text{H}^+} [\text{RuH}_3(\text{H}_2)(\text{PPh}_3)_2]^+ \quad (7.22)$$

Complex	k_1, s^{-1}	$\Delta H^{\ddagger}$, kcal/mol	$\Delta S^{\ddagger}$, e.u.
$[\text{RuH}_3(\text{PPh}_3)_2]^-$	6.8×10^{-6}	21.8 ± 1.4	-9 ± 4
$\text{RuH}_2(\text{H}_2)(\text{PPh}_3)_2$	2.1×10^0	17.9 ± 0.2	3 ± 1
$[\text{RuH}_3(\text{H}_2)(\text{PPh}_3)_2]^+$	3.6×10^3	8.8 ± 0.1	-12 ± 1

This is mainly an effect of the decreasing $\Delta H^{\ddagger}$, which in turn is due to decreased BD on protonation because of less available electron density. The same argument can be used for classical polyhydrides, where an η^2-H$_2$-like transition state would be stabilized on protonation. The rate of dissociation of H$_2$ from thermally unstable Cr(CO)$_5$(H$_2$) in hexane, 2.5 s^{-1},[66] is slower than might be expected by comparison with the data for the above stable species. However, the product of H$_2$ loss is extremely reactive and instantly decomposes because it is not stabilized by agostic interactions or a sixth ligand other than even more weakly bound hexane solvent. This complex and others like it might otherwise be stable under H$_2$ (see Chapter 3). The rate of dissociation of SiHEt$_3$ from Cr(CO)$_5$(SiHEt$_3$) is nearly identical to the H$_2$ complex, in keeping with the similarity of silane to H$_2$ coordination.[51] The dissociation is faster (143 s^{-1}) for the analogous SiHPh$_3$ complex and much faster for the SiHCl$_3$ complex (1.2×10^4 s^{-1}), demonstrating the importance of substituents at Si in the activation of Si-H bonds (see Chapter 11).

Reactions in the gas phase are also extremely fast, $1-2 \times 10^6$ Torr^{-1} s^{-1}, for W(CO)$_5$ plus H$_2$ or alkanes.[17-19] High reaction probabilities (0.03-0.14) imply that there is little or no barrier to the coordination of these molecules. Other photolytically generated organometallic fragments such as CpRh(CO)[72,73] show similar rate behavior to this and also to those determined in solution, i.e., reaction with H$_2$ is similar to CO and faster than N$_2$ (Table 7.8). In contrast, the reaction of H$_2$ with the RhCl(PPh$_3$)$_2$ fragment generated by flash photolysis gives a different type of kinetic behavior where the rate is much *slower* than with other ligands[74]:

$$\text{RhCl(CO)(PPh}_3)_2 \underset{k_{-1}}{\overset{h\nu}{\rightleftarrows}} \text{CO} + \text{RhCl(PPh}_3)_2 \xrightarrow{\text{H}_2}_{k_2} \text{RhH}_2\text{Cl(PPh}_3)_2 \quad (7.23)$$

The second-order rate $k_2 = 10^5$ M^{-1} s^{-1} seems fast but is 700 times slower than that for the backreaction with CO (k_{-1}), unlike the data in Table 7.8. In Eq. (7.23)

Table 7.8. Second-Order Rate Constants for Reactions of
H_2, N_2, CO, and Alkanes with Organometallic
Fragments in the Gas Phase

Fragment	Ligand	$k/10^9$
(η^6-C_6H_6)Cr(CO)$_2$	CO	6.3 ± 0.3 M^{-1}s^{-1}
	H_2	4.8 ± 0.2
	N_2	2.6 ± 0.2
CpMn(CO)$_2$	CO	0.59 ± 0.04
	H_2	0.52 ± 0.03
	N_2	0.37 ± 0.04
CpRh(CO)	CO	15 ± 3 cm^3/molecule s^{-1}
	H_2	16 ± 3
	CH_4	5.8 ± 3
	C_2H_6	18 ± 5
	c-C_6H_{12}	28 ± 9

complete OA of H_2 to form a dihydride is occurring, *and the lower rate is due to the barrier to breaking the H–H bond in the H_2 complex [Eq. (7.18)] rather than to binding of H_2.* This demonstrates the complexity of what on the surface appears to be simple H_2 addition to a M fragment. Section 7.4 addresses this further by comparing solution kinetics and thermodynamics of H_2 binding versus OA.

7.4. REACTION PROFILES AND KINETICS FOR COORDINATION AND OXIDATIVE ADDITION OF DIHYDROGEN AND OTHER σ LIGANDS

The thermodynamics for the binding and OA of σ-bonded ligands have been determined for H_2 and silanes, and are remarkably similar as will be shown below. In contrast the energy profile for alkane OA has a much shallower well for the intermediate σ complex. The existence of the remarkably facile and easily observable equilibria between H_2 and dihydride tautomers [Eq. (7.11)] in several systems allows study of the thermodynamics and kinetics of H–H bond cleavage in exquisite detail. Complexes showing this solution behavior and attendant thermodynamic data are given in Table 7.9. As expected for an equilibrium process involving intramolecular rearrangements, the values of $\Delta G°$, $\Delta H°$, and $\Delta S°$ are quite small. Either the H_2 or dihydride forms can be favored (positive or negative $\Delta G°$), and the entropy term can be positive or negative. For positive entropy changes, apparently more overall disorder in the seven-coordinate dihydride overcomes the loss of entropy (rotational freedom) of η^2-H_2 for M(H_2)L$_x$ → MH$_2$L$_x$. This favors the H_2 form at lower temperatures ($T\Delta S°$ becomes more positive and $\Delta G°$ becomes more negative). Steric interactions between the PR$_3$ and the Cp ring are important in determining the position of equilibria for the CpRu complexes. The substituents on phosphorus are closer to the Cp ring in the Ru(H_2) complex than in the transoid dihydride complex

Table 7.9. Thermodynamic Parameters for $M(H_2)L_x \rightarrow MH_2L_x$

Complex	Temperature, K	$\Delta G°$, kcal/mol	$\Delta H°$, kcal/mol	$\Delta S°$, e.u.	J_{HD}, Hz	Reference
$W(H_2)(CO)_3(P^iPr_3)_2$	298	0.80	1.2	1.2	33.5	32
$[Re(H_2)(CO)_2(PPhMe_2)_3]^+$	200	−0.1	−1.7	−8	31	75
$[ReH_2(H_2)(CO)(PPhMe_2)_3]^+$	213	−0.6	−1.1	−2.4	34	76
$[ReH_2(H_2)(CO)(PMe_3)_3]^+$	213	−0.6	−1.05	−2.3	33.6	77
$[ReH_2(H_2)(CO)\{P(OEt)_3\}_4]^+$	283	0.0	1.7	6.0	33	78
$[Cp*Re(H_2)(CO)(NO)]^+$	195	1.0			27	79
$[CpRu(H_2)(CO)(PCy_3)]^+$	298	2			28.5	80
$[CpRu(H_2)(dppe)]^+$	295	−0.4			24.9	81
$[CpRu(H_2)(dmdppe)]^+$	298	0.2	−0.92	−4.5	23.8	82
$[CpRu(H_2)(dmpe)]^+$	298	1.0			22	82
$[Cp*Ru(H_2)(dppm)]^+$	298	0.4			20.9	82
$Ru(H_2)(OCOCF_3)_2(PCy_3)_2$	243	−1				83
$Os(H_2)Cl(SiEt_3)(CO)(P^iPr_3)_2$	298	>1				84
$CpNb(H_2)(CO)_3$	228		0.8			85
$[CpOs(H_2)(CO)_2]^+$	193	<0				86

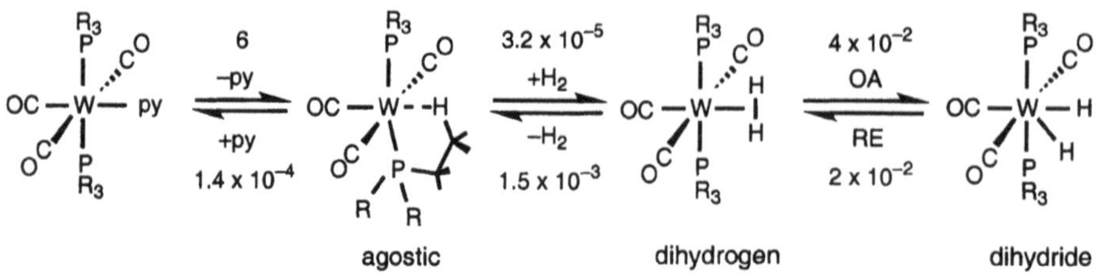

Figure 7.3. Stop-flow kinetic data (rates are $t_{1/2}$ in seconds) for substitution of H_2 in $W(CO)_3(PCy_3)_2$ (H_2) by pyridine (py). Reaction are shown in reverse order.

so that larger PR_3 favor the $Ru(H)_2$ form. ΔG decreases in the series $[CpRuH_2L_2]^+$ as L_2 increases in size from dmpe to dmdppe to dppe, the reverse of what would have been expected electronically.

The thermodynamic and kinetic reaction profile for molecular H_2 addition to $W(CO)_3(PR_3)_2$ and equilibrium H–H cleavage [Eqs. (7.11) and (7.18)] has been determined for R = Cy, ^{i}Pr.[1,63] The results of kinetic studies of displacement of H_2 by pyridine are given in Figure 7.3. In the first step of the reaction sequence shown in reverse, pyridine dissociates to generate a vacant site at M on the slow timescale of seconds. The agostic species $W(CO)_3(PCy_3)_2$ can then react with either pyridine or H_2 with $t_{1/2}$ of 140 and 32 μs, respectively ($k = 2.2 \times 10^6$ M^{-1}s^{-1} for H_2). If the H_2 complex is formed, it may dissociate H_2 and regenerate $W(CO)_3(PCy_3)_2$ within 1.5 ms ($k = 469$ s^{-1}) or undergo reversible OA ($K = \sim 0.25$ at 298 K) to form the dihydride tautomer with $t_{1/2} = 40$ ms. Under these conditions, the ratio of the rate of binding of H_2 to the rate of H_2 dissociation to the rate of OA is roughly 1200:25:1. The most surprising feature here is that *the rate of dissociation of H_2 is faster than the rate of OA by at least one order of magnitude.* Thus H_2 binds and dissociates many times prior to OA, which has vital importance in understanding σ-bond activation processes and attendant homogeneous catalytic reactions in general. *The barrier to breaking the σ bond in σ complexes is the dominant (and variable) factor in reaction rates rather than the binding of the σ ligand.* This will also be discussed in detail in Section 12.4 for C–H bond activation, where the key experiment for identifying σ complexes as intermediates in OA and RE of alkanes often involves intramolecular H/D exchange processes involving σ alkane complexes with a significant lifetime.

The complete reaction profile for H_2 addition to $W(CO)_3(PR_3)_2$ is shown in Figure 7.4 using a combination of data for complexes with R = Cy and ^{i}Pr obtained from different methodologies (calorimetric measurements, stopped-flow and NMR spin-saturation transfer kinetics). The enthalpy of activation, $\Delta H^{\ddagger}$, for loss of coordinated H_2 is 16.9 ± 2.2 kcal/mol, which implies a barrier of 6.9 ± 3.2 kcal/mol for the forward reaction between $W(CO)_3(PCy_3)_2$ and H_2, based on $\Delta H°$ measured for the latter reaction, 10.1 kcal/mol. The latter enthalpy is similar to the value of -12.8 kcal/mol for addition of an organosilane to $Mo(CO)(^iBu_2PC_2H_4P^iBu_2)_2$ [Eq. (7.5)], which also has an agostic interaction of 5–10 kcal/mol. These energies are much higher than that for N_2 addition to W, 4.3 ± 1.7 kcal/mol, and H_2 reac-

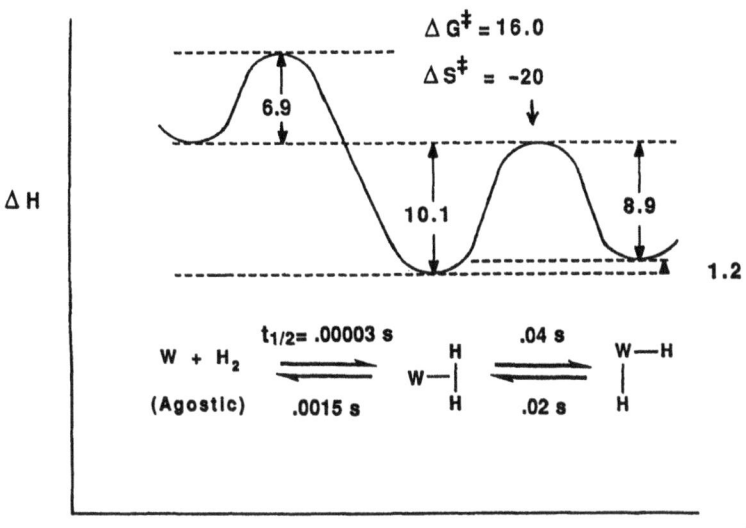

Figure 7.4. Thermodynamics and kinetics of $W(CO)_3(PR_3)_2 + H_2 \rightleftharpoons W(CO)_3(PR_3)_2(H_2) \rightleftharpoons WH_2(CO)_3(PR_3)_2$. The thermodynamic data for the first step is for $R = Cy$ and the second is for $R = {}^iPr$, and the kinetic data is for $R = Cy$. Reprinted with permission from Gonzalez, A. A., et al. *ACS Symposium Series* **1990**, *428*, 133. Copyright 1990 American Chemical Society.

tion is four times faster than for N_2, possibly due to a more favorable preexponential factor. A lower barrier of 0.5 ± 0.7 kcal/mol exists for H_2 addition to $OsHCl(CO)(P^iPr_3)_2$.[7] These barriers are in the range for diffusion-controlled reactions such as those in Table 7.7. Somewhat lower values of $\Delta H^\ddagger$ for H_2 loss are seen for $OsHCl(H_2)(CO)(P^iPr_3)_2$ (14.6 ± 0.2 kcal/mol) and $IrXH_2(H_2)(P^tBu_2Me)_2$ (X = halide; 9–11 kcal/mol).[7,8] The entropies of activation $\Delta S^\ddagger$ are quite low, 2–10 kcal/mol.

The second barrier in Figure 7.4 is for the actual splitting of the H–H bond in $W(CO)_3(P^iPr_3)_2(H_2)$ to give the dihydride and is based on thermodynamic parameters determined by NMR studies in toluene solution. $\Delta H^\ddagger$ for H–H cleavage is measured from Eyring plots of $\ln(k/T)$ versus $1/T$ to be 10.1 ± 1.8 kcal/mol and $\Delta G^\ddagger$ is 16.0 ± 2 kcal/mol at 298 K. By contrast $\Delta H^\ddagger$ for C–H splitting is only 4.5 kcal/mol (see Figure 12.11), representing a much shallower potential well for σ-alkane complexes, which undergo OA much faster than H_2. The ground state enthalpy difference in Figure 7.4 determined by Van't Hoff plots of $\ln K_{eq}$ versus $1/T$ favors formation of the H_2 complex by 1.2 ± 0.6 kcal/mol, with negligible $\Delta S°$ of 1 e.u. The equilibrium constants at 298 K for H–H rupture are 0.25 for $R = {}^iPr$, 0.29 for R = cyclopentyl, and 0.66 for $R = Cy$. Very similar data (Table 7.9) are seen for tautomeric conversions of

$$[ReH_2(H_2)(CO)(PMe_2Ph)_3]^+ \quad \text{and} \quad [Re(H_2)(CO)_2(PMe_2Ph)_3]^+$$

for which $\Delta G^\ddagger$ is 11.5 kcal/mol at 223 K in both cases. The profile for analogous

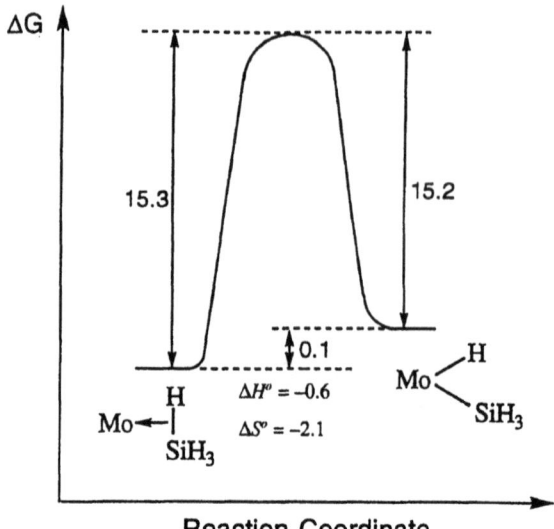

Figure 7.5. Energy profile for SiH$_4$ equilibrium cleavage on Mo(CO)(depe)$_2$ to give MoH(SiH$_3$)(CO)(depe)$_2$ at 333 K with energy values in kcal/mol.

SiH$_4$ equilibrium cleavage on Mo(CO)(depe)$_2$ to give

$$\text{Mo(SiH}_4\text{)(CO)(depe)}_2 \rightleftharpoons \text{MoH(SiH}_3\text{)(CO)(depe)}_2$$

in toluene (Figure 7.5) is remarkably similar to that for H$_2$ on W(CO)$_3$(PR$_3$)$_2$ in Figure 7.4.[87] In this system there is even less enthalpy difference (0.6 ± 0.2 kcal/mol) between the σ complex and the OA product, and the equilibrium constant is near unity at 298 K.

The rate of reaction of H$_2$ with the 17e metal-centered *radical*·CrCp*(CO)$_3$ to form HCrCp*(CO)$_3$ obeys the third-order rate law $d[P]/dt = k_{obs}[\cdot\text{Cr}]_2[\text{H}_2]$ in toluene solution.[88] In the range 20–60° $k_{obs} = 330 \pm 30$ M^{-2}s^{-1}, $\Delta H = 0 \pm 1$ kcal/mol, and $\Delta S = -47 \pm 3$ cal/mol·deg. The rate of OA is not inhibited by added pressure of CO. The rate of binding of D$_2$ is slower than that of H$_2$: $k_{H_2}/k_{D_2} = 1.18$. A mechanism consistent with the rate data involves formation of a 19e intermediate with bound H$_2$ that is attacked by a second radical in the rate determining step [Eq. (7.24)]. A complete reaction profile for hydrogenation of M–M bonded dimer

$$[\text{Cp*Cr(CO)}_3]_2 + \text{H}_2 \rightarrow 2\text{HCrCp*(CO)}_3$$

is shown in Figure 7.6. The transition state involves a ternary complex,

$$\text{L}_n\text{M}\cdots\text{H}_2\cdots\text{ML}_n$$

This is the first such experimentally determined reaction profile for hydrogenation of any M–M bonded complex. A similar reaction profile for a critical reaction in hydroformylation, Co$_2$(CO)$_8$ + H$_2$ → 2HCo(CO)$_4$, under high pressures of CO

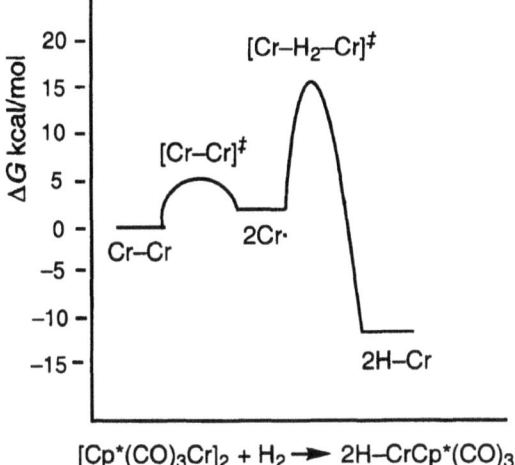

Figure 7.6. Reaction profile in toluene solution, 300 K. Overall thermochemistry for Eq. (7.24): $\Delta H° = 5$ kcal/mol; $\Delta S° = 20$ cal/mol·deg; $\Delta G° = -11$ kcal/mol.

can also be constructed, and here $\Delta G° = +2$ kcal/mol rather than -11 kcal/mol as for Eq. (7.24).

(7.24)

Theoretical and experimental studies indicate that in addition to electronic and steric effects, overall bond energetics can be crucial in OA,[89-91] e.g., the relative M–X energies can offset the X–H energy difference (H–H is 104 kcal/mol versus 72–100 kcal/mol for R_3Si–H, which varies with R[92]). Although little experimental data are known for M–Si bond strengths, theory provides some insight. *Ab initio* calculations by Morokuma of addition of SiH_4, H_2, and CH_4 to the CpRh(CO) fragment in the gas phase show that OA becomes more difficult as the respective X–H energy increases and the Rh–X energy decreases (73, 65, and 59 kcal/mol).[89] Potential energy profiles show no activation barrier for X = H, Si but a 6 kcal/mol barrier for X = C (Figure 7.7). Bergman has conducted extensive studies on OA of

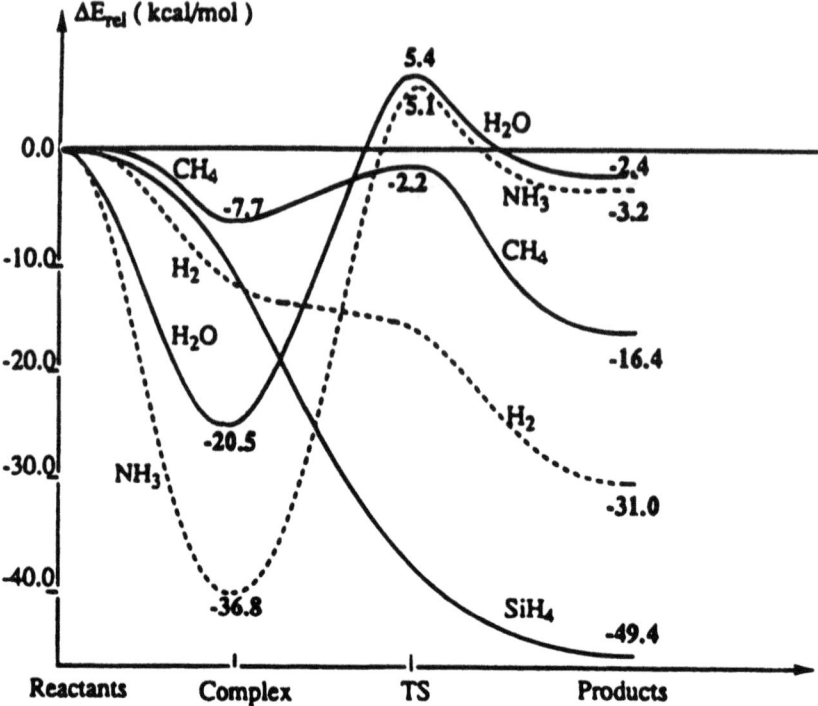

Figure 7.7. The potential energy profiles for the reaction of CpRh(CO) with various small molecules. Reprinted with permission from Musaev and Morokuma.[89] Copyright 1995 American Chemical Society.

alkane C–H bonds to photolytically generated CpRu(CO) (see Section 12.4.2).[73] The σ complex is too short-lived to be observed in solution, but a reaction barrier of 4.5 kcal/mol on going from the σ-cyclohexane complex to the OA product is measured in liquid Kr. A similar barrier is estimated for gas phase reaction, in good agreement with Morokuma's calculations.

A good comparison of H–H and C–H bond activation is shown in Eq. (7.25). The equilibrium lies to the left for basic phosphines (P^nBu_3 and PPh_2Me) but to the

right for less basic phosphines ($PAryl_3$). Crabtree suggests that this results from the preference of the basic metal in the P^nBu_3 complex for OA of the more electron-withdrawing aryl C–H group. For the $P(p\text{-tolyl})_3$ complex, the ΔH and ΔS are

1.6 ± 0.4 kcal/mol and 8 ± 2 cal/mol K. The positive entropy is attributed to the greater freedom of motion of the agostic group in the product compared to the Ir–aryl bond of the reactant. The small ΔH demonstrates that the agostic and H_2 ligands have similar binding strengths. For a related Ir complex (L = PPh$_3$), the bis(silane) form (9) is favored over silyl(H$_2$) isomers, indicating that either Si–H is a better ligand than H–H or that the Ir–H bond is much stronger than Ir–Si here (see Section 11.4.2).[91]

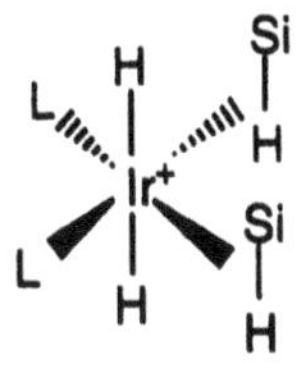

(9)

In conclusion, the η^2-H$_2$/dihydride equilibria have activation enthalpies ranging from 10 to 20 kcal/mol and negative activation entropies. The loss of the H–H bond could be offset by M–H bond-making so that this may not be an important enthalpic contribution to the barrier. The $\Delta H^\ddagger$ probably arises because the complex has to change its coordination number and geometry and concurrently the ancillary ligands have to shift position. Thus steric contributions to the barrier are important. A loss of rotational freedom of the H$_2$ is a good rationale for the negative $\Delta S^\ddagger$. In general, rate-constant determinations are difficult because of the instability of the complexes, the rapid relaxation of H$_2$ nuclei, and other dynamic exchange processes that complicate the analysis.

7.5. ISOTOPE EFFECTS IN σ-LIGAND COORDINATION AND OXIDATIVE ADDITION

Isotope effects can be extremely informative in organometallic chemistry, especially for M(H)(X) systems where X = H, C, Si, etc. An excellent review of this area is provided by Bullock.[93] Both kinetic and equilibrium (or thermodynamic) effects can provide crucial information about reaction mechanisms that is unavailable from other methods. Yet isotope effects often are poorly understood or even paradoxical at first glance. Unlike the situation in organic chemistry, the ability of metal sites (enzymes included) to reversibly coordinate substrates prior to rate-determining steps complicates isotope effect "rules" that were formulated (correctly) by organic chemists. For example, the nature of equilibrium isotope effects for H$_2$ versus D$_2$ addition to metal complexes has been understood only recently. The situation can become even more complex for σ ligands that can undergo homolytic or heterolytic cleavage, either of which can also be reversible. A "normal" isotope effect occurs when the rate of reaction of an unlabeled compound is faster than that for the corresponding labeled species, i.e., $k_H/k_D > 1$. It is "inverse" for $k_H/k_D < 1$, and this terminology also applies to equilibrium isotope effects (EIEs), K_H/K_D, which will be discussed first.

7.5.1. Equilibrium Isotope Effects for H_2 versus D_2 Binding

Deuterium EIEs have been observed for the reversible addition H_2 and D_2 to various complexes in solution to form either metal dihydride/dideuteride complexes[94-96] [Eqs. (7.26–7.27)] or H_2/D_2 complexes [Eqs. (7.28–7.29)].[6,8,97,98] The

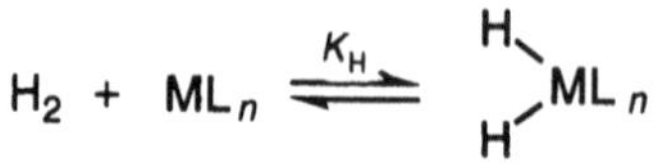

$$H_2 + ML_n \underset{}{\overset{K_H}{\rightleftharpoons}} \begin{array}{c} H \\ \diagdown \\ H \end{array}\!\!ML_n \qquad (7.26)$$

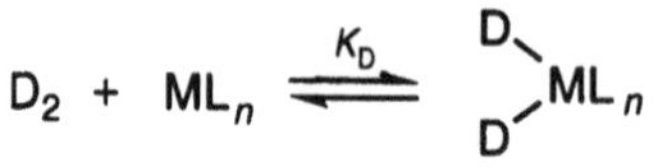

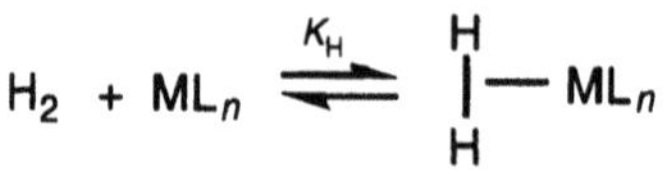

$$D_2 + ML_n \underset{}{\overset{K_D}{\rightleftharpoons}} \begin{array}{c} D \\ \diagdown \\ D \end{array}\!\!ML_n \qquad (7.27)$$

$$H_2 + ML_n \underset{}{\overset{K_H}{\rightleftharpoons}} \begin{array}{c} H \\ | \\ H \end{array}\!\!-ML_n \qquad (7.28)$$

$$D_2 + ML_n \underset{}{\overset{K_D}{\rightleftharpoons}} \begin{array}{c} D \\ | \\ D \end{array}\!\!-ML_n \qquad (7.29)$$

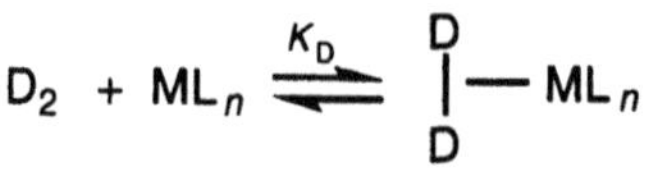

EIE values for H_2 versus D_2 addition are usually inverse over a broad temperature range (260–360 K), showing that counterintuitively D_2 binds more strongly than H_2. The values of K_H/K_D observed thus far are 0.36–0.77 for formation of H_2 complexes and 0.47–0.85 for complete OA. An inverse EIE (0.39) also occurs for protonation of metal hydrides to form η^2-H_2 in the reaction[99]:

$$Cp_2WH_2 + H^+ \rightarrow [Cp_2WH(H_2)]^+$$

For $Cr(CO)_3(PCy_3)_2$ the equilibria shown below are rapidly established under moderate pressures of H_2/D_2 (1–10 atm) in THF solution:

$$Cr(CO)_3(PCy_3)_2(soln) + H_2 \rightleftharpoons Cr(CO)_3(PCy_3)_2(H_2)(soln) \qquad (7.30)$$

$$Cr(CO)_3(PCy_3)_2(soln) + D_2 \rightleftharpoons Cr(CO)_3(PCy_3)_2(D_2)(soln) \qquad (7.31)$$

Equilibrium constants determined at 13–36°C by IR measurements of v_{CO} show that the thermochemical parameters for binding of H_2 are $\Delta H° = -6.8 \pm 0.5$ kcal/mol and $\Delta S° = -24.7 \pm 2.0$ e.u. (these newer values are more accurate than those in Table 7.5).[97] For binding of D_2, $\Delta H° = -8.6 \pm 0.5$ kcal/mol and $\Delta S° = -30.0 \pm 2.0$ e.u., i.e., D_2 binds better enthalpically ($\Delta\Delta H = 1.8$ kcal/mol), which easily overcomes the disfavored entropy of D_2 complexation ($\Delta\Delta S = 5.3$ cal/mol·deg). This domination of enthalpy over entropy is general to these primary isotope effects (unlike, e.g., H_2 versus N_2 binding), and thus EIEs are enthalpically driven. It should be kept in mind that EIEs are temperature-dependent because of the entropy differences.

EIE for the W analogue cannot be measured directly because of stronger W(H$_2$) bonding, but the equilibrium shown in Eq. (7.32) provides a means of determining accurate EIE values:

$$W(CO)_3(PCy_3)_2(N_2)(soln) + H_2(g) \rightleftharpoons W(CO)_3(PCy_3)_2(H_2)(soln) + N_2(g) \quad (7.32)$$

Spectroscopic measurements using calibrated H$_2$/N$_2$ and D$_2$/N$_2$ gas mixtures give $K_H/K_D = 0.70 \pm 0.15$ in THF solvent at 22°C.

Data for H$_2$ versus D$_2$ binding are reported for Ir and Os complexes that have both hydride and H$_2$ ligands, i.e., MH$_x$(H$_2$)L$_n$.[6-8] Lower K_H/K_D of 0.36–0.50 occur here, possibly because of a secondary isotope effect owing to the hydride ligands [values for MH$_x$(D$_2$)L$_n$ cannot be determined because of isotopic exchange, so values for MD$_x$(D$_2$)L$_n$ are used)]. D$_2$ loss from IrClD$_2$(D$_2$)(L)$_2$ is energetically some 1 kcal/mol higher than H$_2$ loss from IrClH$_2$(H$_2$)(L)$_2$ (L = P^{t}Bu$_2$Me).[8]

A large isotope effect (claimed to be ~ 50) is reported to favor D$_2$ binding versus H$_2$ on dehydroxylated chromia (Cr$_2$O$_3$) at -196°C.[100] The measurement involved passing a mixture of H$_2$/D$_2$ = 19 over activated chromia, sweeping with helium, and warming. Although the exact nature of the H$_2$ binding was not determined, it probably involves molecular binding to coordinatively unsaturated surface sites (see Section 4.14).

7.5.2. Origin of the Inverse EIE for H$_2$ and Other σ-Ligand Binding and EIE for OA

In view of the fact that the free D–D bond is 1.8 kcal/mol stronger than the H–H bond, one might naively expect H$_2$ to bind preferentially over D$_2$. If the W–(D$_2$) and W–(H$_2$) bonds were of equal strength in **10** and **10-d_2** in Eq. (7.33), the reaction would be predicted to be exothermic by -1.8 kcal/mol based on the fact

$$H_2 + \begin{array}{c} D \\ | \\ D \end{array} \!\!\to W(CO)_3L_2 \; \underset{}{\overset{K_H/K_D}{\rightleftharpoons}} \; D_2 + \begin{array}{c} H \\ | \\ H \end{array} \!\!\to W(CO)_3L_2 \quad (7.33)$$

$$(\textbf{10-}d_2) \qquad\qquad\qquad (\textbf{10})$$

that the D–D bond is that much stronger. The fact that ΔH in Eq. (7.33) is actually measured to be *endothermic* by this amount for the Cr complex implies that zero-point and excited state vibrational energies for the η^2-H$_2$ species determine the EIE.[97] Furthermore, one must consider more than just the lowering of ν_{HH} versus ν_{DD} on coordination. The favoring of the left side of Eq. (7.33) is the inverse of that predicted from simple changes in ν_{HH} alone, where deuterium should favor the stronger force constant, i.e., D$_2$ should prefer to remain unbound compared to H$_2$.

The EIE is calculated from molecular translational, rotational, and vibrational partition function ratios as described by Bigeleisen and Goeppert-Mayer[101]:

$$EIE = MMI \times EXC \times ZPE \quad (7.34)$$

The calculated EIE is the product of three factors: a rotational and translational factor containing the reduced (classical) rotational and translational partition function ratios of isotopic species (MMI); a factor accounting for contributions from excitations of vibrational energy levels (EXC); and a factor comprising zero-point energy contributions (ZPE). For H_2 complexes, all data concur that v_{HH} and v_{DD} (hence bond order) are lowered when H_2/D_2 binds to M (see Chapter 8). This should result in a "normal" equilibrium isotope effect *if changes in the HH(DD) force constant are the major contributor to the EIE*. However, as originally elucidated by Krogh-Jespersen and Goldman[96] and expanded by Bender and Kubas[97] to the case of H_2 complexes, contributions to the ZPE from new vibrational and rotational modes are of critical importance to EIEs when H_2 either coordinates or cleaves to dihydride. The change in ZPE for HH(DD) stretching mode contributes a large "normal" factor to the total EIE as expected. The calculated ZPE contribution would predict an EIE of about 3.2 for Eq. (7.34) if changes in HH(DD) stretching force constant were the sole contributors to the EIE. The five "new" vibrational normal modes in H_2 complexes (see Chapter 8) all contribute modest *inverse* EXC and ZPE factors to the EIE that when multiplied together overcome the strong "normal" ZPE component from v_{HH} (Table 7.10). These factors (ZPE = 0.216; EXC = 0.675) multiplied against the MMI term (5.77) predict an overall inverse EIE of 0.78 at 300 K for $W(CO)_3(PCy_3)_2(H_2)$ (this includes minor contributions from other modes that mix with those in Table 7.10).[97] This value agrees well with the experimental value of $K_H/K_D = 0.70 \pm 0.15$ in Eq. (7.32).

Nuclear motion quantum calculations (DVR methodology; see Section 8.4) give somewhat lower EIE values (0.53 at 300 K), but unexpectedly show *normal EIE for elongated H_2 complexes* such as

$[Cp^*Ru(H\cdots H)(dppm)]^+$ (1.22)   and $[Os(H\cdots H)Cl(dppe)_2]^+$ (1.69)

This is proposed to be due to the severe *anharmonicity* in the H_2-related vibrational modes in these complexes (see Section 8.4), which favors addition of H_2 over D_2. The ZPE factor is affected the most here, and corrections for anharmonicity must be

Table 7.10. Equilibrium Isotope Effect Contributions from Individual Modes for $H_2(D_2)$ Complexation to $W(CO)_3(PCy_3)_2(H_2)$ (6) at a Temperature of 300 K

Mode (sym)	$H_2(D_2)$, cm^{-1}	6(6-d_2), cm^{-1}	EXC[a]	ZPE[b]
v_{HH} (A$_1$)	4395(3118)	2690(1900)	1.000	3.215
v_{WH_2} (A$_1$)	—	953(703)	0.976	0.549
δ_{WH_2} (B$_1$)	—	640(442)	0.923	0.622
v_{WH_2} (B$_1$)	—	1575(1144)	0.996	0.356
δ_{WH_2} (B$_2$)	—	462(319)	0.879	0.710
τ_{WH_2} (A$_2$)	—	355(251)	0.856	0.780

[a] Π EXC = 0.675 (product of multiplication of all six EXC terms).
[b] Π ZPE = 0.216 (as above).

taken into account, especially for complexes with $d_{HH} > 1$ Å, although experimental confirmation is needed.[102]

OA of H_2 to WI_2 $(PMe_3)_4$ to give the dihydride $WH_2I_2(PMe_3)_4$ also shows an inverse EIE of 0.63(5) at 60°C because of the large number of isotope-sensitive vibrational modes in the product (two M—H stretching modes and four bending modes) compared to the H_2 reactant.[95] The calculated value is 0.73 using a similar approach to that above for formation of H_2 complexes and that by Krogh-Jespersen and Goldman for OA of H_2 to form $IrH_2Cl(CO)(PPh_3)_2$ (0.47; experimental, 0.55).[96] Because the MMI factors are similar for dihydride and H_2 complex formation, the inverse nature of the EIE may again be traced to the dominant zero-point energy term (ZPE), which for the dihydride is 0.10–0.17.

Because ΔZPE changes for the dihydride and H_2 cases are both referenced to free $H_2(D_2)$, this free-energy difference is due to changes in force constants, which favor D in the dihydride tautomer. This reasoning is consistent with an overall increase in the force constants (despite the weakening of the bound HH(DD) force constant) when the "loose" $M(H_2)$ fully adds. Indeed the preference of deuterium as the dideuteride tautomer (where it is more strongly bound to M) might seem to parallel the preference of D in the complex as opposed to the free ligand. Related to this is the tendency for D to concentrate in the hydride site in certain hydride (H_2) complexes [Eqs. (7.35) and (7.36)] versus in η^2-H_2 as discussed in Section 5.4.[103]

$$\text{Ir}\begin{array}{c}H\\ \diagup\\ \diagdown\\ D\end{array}\begin{array}{c}H\\ \diagdown\\ \diagup\\ D\end{array} \quad \xrightleftharpoons{K_1 = 1.32} \quad \text{Ir}\begin{array}{c}D\\ \diagup\\ \diagdown\\ H\end{array}\begin{array}{c}H\\ \diagdown\\ \diagup\\ H\end{array} \quad (7.35)$$

$$\text{Ir}\begin{array}{c}H\\ \diagup\\ \diagdown\\ D\end{array}\begin{array}{c}D\\ \diagdown\\ \diagup\\ D\end{array} \quad \xrightleftharpoons{K_1 = 1.26} \quad \text{Ir}\begin{array}{c}D\\ \diagup\\ \diagdown\\ D\end{array}\begin{array}{c}H\\ \diagdown\\ \diagup\\ D\end{array} \quad (7.36)$$

Ir = TpIr(PMe$_3$)

The equilibrium constants indicate that the heavier isotope prefers to occupy the hydride site here, but favors η^2-H_2 in $[ReH_2(H_2)(CO)(PR_3)_3]^+$. In the latter the isotope effect was interpreted to be a consequence of a greater vibrational zero-point energy difference between Re(η^2-HD) and Re(η^2-D_2) relative to Re–H and Re–D. The isotopic preferences will be dictated by the changes in *all* of the force constants in both tautomers and depend on relative H–H versus M–H bond strengths. The Ir system has a long d_{HH} of about 1 Å and the Ir–H bonds, particularly to the classical hydride, are undoubtedly stronger than the H–H bond, which would be expected to favor D incorporation. By contrast the H–H bond in the Re complex is much stronger ($J_{HD} = 34$ Hz); hence it is a "true" σ complex, which would bind D_2 more strongly than H_2. Also in support of this are K_H/K_D values that are 1.5 for tautomerization of $[ReH_2(H_2)(CO)(PMe_3)_3]^+$ to the tetrahydride (Table 7.9),[77] and the equilibrium between CpNb(CO)$_3$(H$_2$) and CpNb(CO)$_3$H$_2$ shows similar favoring of D in the nonclassical isomer. The $K_H/K_D = 0.20 \pm 0.05$ for equilibrium isomerization of [Cp$_2$WH(H$_2$)]$^+$ to [Cp$_2$WH$_3$]$^+$ (unisolated species formed by protonation of Cp$_2$WH$_2$ with HCl versus DCl) could reflect a more hydridic nature for this system, i.e., an H$_2$ ligand with a weak H–H bond.

The above considerations are also relevant to the complexation of alkenes,[104] alkanes,[94,96,105,106] and any σ ligand. The EIE is inverse (0.7) for reversible ethylene binding in Os$_2$(CO)$_8$(μ-C$_2$H$_4$). The isotope effect is normal for OA of alkanes (1.3–1.9 per C–H bond), both theoretically and experimentally. However, a large inverse EIE is measured for cyclohexane versus cyclohexane-d_{12} binding to Cp*Rh(CO)$_2$ and also to W(CO)$_5$ by Schultz (0.71, supported calculationally).[105] As in the situation for H$_2$ coordination, this is the opposite of what one would predict from simple changes in C–H stretching frequencies alone. Bergman rationalized the inverse effect by the σ-alkane complex gaining additional isotope-sensitive vibrational modes on coordination, thereby making the ZPE for the deuterated alkane complex lower. The situation thus undoubtedly closely parallels that established above for H$_2$ coordination. However, for cyclopentane coordination to CpRe(CO)$_2$ a normal effect of 1.33 is seen (see Section 12.3.1), an isotope effect that may parallel that calculated above for elongated H–H coordination (elongated C–H here?). The EIE for triethylsilane OA to Ir in Cp$_2$Ta(μ-CH$_2$)$_2$Ir(CO)$_2$ is inverse, with a large temperature dependence: 0.54 ± 0.04 at 273 K versus 0.76 ± 0.06 at 353 K.[94]

7.5.3. Kinetic Isotope Effects for H$_2$ and Alkane OA and Reductive Elimination

There are very limited data on kinetic isotope effects (KIE) for σ-ligand coordination/dissociation or σ-bond cleavage equilibria as shown in Eq. (7.37).

$$M + \begin{array}{c} H \\ | \\ H \end{array} \underset{k_{-1}}{\overset{k_1}{\rightleftarrows}} M - \begin{array}{c} H \\ | \\ H \end{array} \underset{k_{-2}}{\overset{k_2}{\rightleftarrows}} M \begin{array}{c} H \\ \diagup \\ \diagdown \\ H \end{array} \qquad (7.37)$$

For H$_2$ loss from the W(CO)$_3$(PCy$_3$)$_2$ fragment, $k_{-1} = 469$ s^{-1} for H$_2$ and 267 s^{-1} for D$_2$, giving $k^H_{(-1)}/k^D_{(-1)} = 1.7$.[63] Applying the EIE data above and the following expressions, one sets $k^H_{(1)}/k^D_{(1)} = 1.2$ for H$_2$ binding:

$$K_H/K_D = k^H_{(1)}/k^H_{(-1)} \times k^D_{(-1)}/k^D_{(1)} \qquad (7.38)$$

$$k^H_1/k^D_1 = K_H/K_D \times k^H_{-1}/k^D_{-1} = 0.7 \times 1.7 = 1.2 \qquad (7.39)$$

In comparison, the reaction below occurs 1.9 times as fast for H$_2$ as for D$_2$ (10^4 s^{-1})[66]:

$$\text{Cr(CO)}_5(\text{C}_6\text{H}_{12}) + \text{H}_2 \rightarrow \text{Cr(CO)}_5(\text{H}_2) + \text{C}_6\text{H}_{12} \qquad (7.40)$$

The subsequent rate of loss of H_2 (2.5 s^{-1}) is five times as fast as for D_2, consistent with stronger binding of D_2 over H_2. For OA of H_2 to RhCl(PPh$_3$)$_2$ as in Eq. (7.23), k_H/k_D is 1.5,[74] and for OA to Vaska's complex, it is smaller, 1.06.[107] For conversion of the dihydride to the H_2 complex in Eq. (7.37), $k_{-2} = 37$ s^{-1} for H_2 and 33 s^{-1} for D_2, giving $k_H/k_D = 1.08 \pm 0.04$ [the reverse reaction, OA of H_2, occurs about 50% slower ($k_2 = 18$ s^{-1})].[63] There is virtually no kinetic isotope effect for H–H bond formation (reductive elimination), which also probably applies to H–H cleavage (OA). The activation energies overlap within experimental error for the H_2 and D_2 systems in Eq. (7.37) for both the k_{-1} and k_{-2} steps.

In these cases it is the KIE for binding of H_2 in a preequilibrium step that determines the overall KIE for OA of H_2. In most cases H_2 coordination gives a normal KIE. Since the nature of the transition state is not known and can vary, it is not possible to draw conclusions about the slower rate of binding of D_2 versus H_2. Dissociation of H_2 can give either normal or inverse effects.[93] An overall inverse KIE for reductive elimination of H_2 from a dihydride is symptomatic of formation of an H_2 complex in an equilibrium step rather than in a single-step process.

As for RE of H_2, observation of similar inverse KIE (0.5–0.8) for alkane elimination from alkyl hydride complexes is evidence for unobserved σ complexes as

$$M\begin{matrix}R\\ \\H\end{matrix} \underset{k_{-1}}{\overset{k_1}{\rightleftharpoons}} \left[M-\begin{matrix}R\\|\\H\end{matrix}\right] \overset{k_2}{\longrightarrow} M + \begin{matrix}R\\|\\H\end{matrix} \qquad (7.41)$$

intermediates.[93,108–115] Complexes giving inverse overall k_H/k_D (values in parentheses) include Cp$_2$W(Me)H (0.75),[109] [Cp$_2$Re(Me)H]$^+$ (0.8),[112] Cp*Ir(PMe$_3$)(Cy)H (0.7),[110] Cp*Rh(PMe$_3$)(Et)H (0.5),[111] and [(tmeda)Pt(Me)(H)]Cl (0.29).[115] In single-step mechanisms, inverse isotope effects can occur when the transition state is productlike and the deuterium experiences a higher force constant in the product relative to that of the reactant, i.e., deuterium favors the stronger bond.[116] This effect can also result from an inverse *equilibrium* isotope effect (K_H/K_D) involving an intermediate with an appreciable lifetime that undergoes rapid and reversible migration and insertion prior to rate-limiting loss of alkane. Formation of a transient alkane complex from an alkyl hydride complex leads to significant C–H bond formation because the H or D atoms experience a higher force constant in a bond with C than with M. The inverse isotope effect then results from a buildup in the preequilibrium concentration of the deuterio isotopomer compared with the protio isotopomer, which in turn leads to a faster rate for reductive elimination for the deuterio species.[113] Another requirement for the inverse effect is that the nature of the C–H bond formed in the intermediate is not significantly altered in the rate-determining step in which alkane is lost, shown schematically in Figure 7.8.[117] Although these effects are measured as kinetic effects on the rates of alkane elimination, they are essentially EIE resulting from the preequilibrium step that gives an inverse effect. As for the H_2 case, the isotope effect on the loss of alkane from the alkane complex is expected to be small because the C–H bond is nearly completely formed in the σ complex.

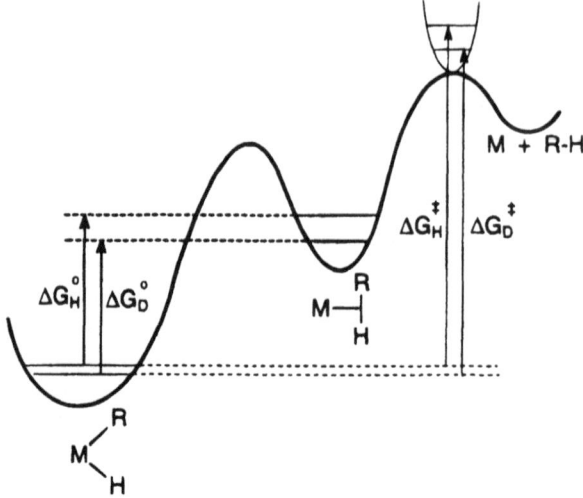

Figure 7.8. Reaction coordinate which depicts a larger zero-point energy difference between hydrogen and deuterium in the products relative to the reactants. The resulting equilibrium isotope effect is inverse, $K_H/K_D < 1$. Reprinted with permission from Wick et al.[117] Copyright 1999 American Chemical Society.

There are systems, particularly cis-L_2Pt(R)(H), where the isotope effect for Eq. (7.41) is normal (1.5–3.3).[118] Here, however, the alkyl hydride complexes are generally thermodynamically unstable, i.e., OA of C–H bonds is not thermodynamically favored. Parkin and Bercaw propose that σ complexes are intermediates in all of these elimination reactions, but the crucial factor is the energy of the transition state for dissociation of alkane relative to that of the transition state for σ-complex formation from the alkyl hydride.[113] In Eq. (7.41), then, an inverse effect will be seen when $k_{-1} \gg k_2$, whereas there will be a normal effect for $k_{-1} \ll k_2$.

OA of C–H bonds, the microscopic reverse of reductive elimination of C–H, must also proceed via a σ complex. There are abundant data to show that, as for H_2 additions, the kinetic isotope effects are normal for alkane OA. For example, reaction of Cp*Rh(PMe$_3$)(neopentyl)H with hexane versus hexane-d_{14} gives k_H/k_D = 1.2 ± 0.1 at 213 K,[111] and photolysis of Cp*Ir(PMe$_3$)H$_2$ gives k_H/k_D = 1.38 upon reaction with cyclohexane to form Cp*Ir(PMe$_3$)(cyclohexyl)H.[119]

Agostic C–H⋯M interactions bear considerable similarity to alkane complexes, and Hoff and Kubas show that the agostic species W(CO)$_3$(PCy$_3$)$_2$ is an intermediate in the ligand displacement reaction in Eq. (7.42).[1] A normal KIE effect of 1.2 is

(7.42)

observed for displacement of pyridine by $P(OMe)_3$ for the system with $P(C_6H_{11})_3$ ligands versus the same system with $P(C_6D_{11})_3$ ligands. This indicates that the agostic complex is an intermediate since no isotope effect would occur if loss of pyridine was completely dissociative.

Protonation of metal hydrides with HX acids to form H_2 ligands shows an inverse KIE of 0.21–0.64 for formation of $[FeH(H_2)P_4]^+$ (P_4 = 2dppe or $P(C_2H_4PPh_2)_3$). This reaction involves direct attack of HX at one hydride in FeH_2P_4 (see Section 7.3), and the inverse effect suggests that protonation occurs through a late transition state with a structure similar to that of the product H_2 complex.[65] Calculated values are similar in magnitude to the observed values, thus supporting the mechanism in Eq. (7.20).

REFERENCES

1. Gonzalez, A. A.; Zhang, K.; Nolan, S. P.; de la Vega, R. L.; Mukerjee, S. L.; Hoff, C. D.; Kubas, G. J. *Organometallics* **1988**, *7*, 2429; Gonzalez, A. A.; Zhang, K.; Mukerjee, S. L.; Hoff, C. D.; Khalsa, G. R. K.; Kubas, G. J. *ACS Symp. Ser.* **1990**, *428*, 133.
2. Kubas, G. J.; Burns, C. J.; Khalsa, G. R. K.; Van Der Sluys, L. S.; Kiss, G.; Hoff, C. D. *Organometallics* **1992**, *11*, 3390.
3. Morse, J. M.; Parker, G. H.; Burkey, T. J. *Organometallics* **1989**, *8*, 2471.; Burkey, T. J. *J. Am. Chem. Soc.* **1990**, *112*, 8329.
4. Calado, J. C. G.; Dias, A. R.; Salem, M. S.; Simoes, J. *J. Chem. Soc., Dalton Trans.* **1981**, *1174*; Nolan, S. P.; Lopez de la Vega, R.; Hoff, C. D. *J. Organometal. Chem.* **1986**, *315*, 187.
5. Bakhmutov, V. I.; Vorontsov, E. V.; Vymenits, A. B. *Inorg. Chem.* **1995**, *34*, 214.
6. Gusev, D. G.; Vymenits, A. B.; Bakhmutov, V. I. *Inorg. Chem.* **1992**, *31*, 1.
7. Bakhmutov, V. I.; Bertran, J.; Esteruelas, M.A.; Lledos, A.; Maseras, F.; Modrego, J.; Oro, L. A.; Sola, E. *Chem. Eur. J.* **1996**, *2*, 815.
8. Hauger, B. E.; Gusev, D. G.; Caulton, K. G. *J. Am. Chem. Soc.* **1994**, *116*, 208.
9. Gusev, D. G.; Kuznetsov, V. F.; Eremenko, I. L.; Berke, H. *J. Am. Chem. Soc.* **1993**, *115*, 5831.
10. Kuhlman, R. L.; Gusev, D. G.; Eremenko, I. L.; Berke, H.; Huffman, J. C.; Caulton, K. G. *J. Organometal. Chem.* **1997**, *536/537*, 139.
11. Le-Husebo, T.; Jensen, C. M. *Inorg. Chem.* **1993**, *32*, 3797.
12. Albinati, A.; Bakhmutov, V.I.; Caulton, K. G.; Clot, E.; Eckert, J.; Eisenstein, O.; Gusev, D. G.; Grushin, V. V.; Hauger, B. E.; Klooster, W. T.; Koetzle, T. F.; McMullan, R. K.; O'Loughlin, T. J.; Pelissier, M.; Ricci, J. S.; Sigalas, M. P.; Vymenits, A. B. *J. Am. Chem. Soc.* **1993**, *115*, 7300.
13. Luo, X.-L.; Kubas, G. J., unpublished results.
14. Weisshaar, J. C. *Acc. Chem. Res.* **1993**, *26*, 213; Armentrout, P. B. *Acc. Chem. Res.* **1995**, *28*, 430.
15. Bushnell, J. E.; Maitre, P.; Kemper, P. R.; Bowers, M. T. *J. Chem. Phys.* **1997**, *106*, 10153.
16. Haynes, C. L.; Armentrout, P. B. *Chem. Phys. Lett.* **1996**, *249*, 64.
17. Ishikawa, Y.; Hackett, P. A.; Rayner, D. M. *J. Phys. Chem.* **1989**, *93*, 652.
18. Brown, C. E.; Ishikawa, Y.; Hackett, P. A.; Rayner, D. M. *J. Am. Chem. Soc.* **1990**, *112*, 2530.
19. Wells, J. R.; House, P. G.; Weitz, E. *J. Phys. Chem.* **1994**, *98*, 8343.
20. Banister, J. A.; Lee, P. D.; Poliakoff, M. *Organometallics* **1995**, *14*, 3876.
21. Nayak, S. K.; Burkey, T. J. *Organometallics* **1991**, *10*, 3745; Kiplinger, J. L.; Richmond, T. G.; Osterberg, C. E. *Chem. Rev.* **1994**, *94*, 373.
22. Walsh, E. F.; Popov, V. K.; George, M. W.; Poliakoff, M. *J. Phys. Chem.* **1995**, *99*, 12016, and references therein.
23. Walsh, E. F.; George, M. W.; Goff, S.; Nikiforov, S. M. ; Popov, V. K.; Sun, X.-Z.; Poliakoff, M. *J. Phys. Chem.* **1996**, *100*, 19425.
24. George, M. W.; Haward, M. T.; Hamley, P. A.; Hughes, C.; Johnson, F. P. A.; Popov, V. K.; Poliakoff, M. *J. Am. Chem. Soc.* **1993**, *115*, 2286.

25. Wells, J. R.; Weitz, E. *J. Am. Chem. Soc.* **1992**, *114*, 2783; Weiller, B. H. *J. Am. Chem. Soc.* **1992**, *114*, 10910.
26. Bacskay, G. B.; Bytheway, I.; Hush, N. S., *J. Am. Chem. Soc.* **1996**, *118*, 3753.
27. Crabtree, R. H.; Siegbahn, P. E. M.; Eisenstein, O.; Rheingold, A. L.; Koetzle, T. F. *Acc. Chem. Res.* **1996**, *29*, 348.
28. Chan, W.-C.; Lau, C.-P.; Chen, Y.; Fang, Y.-Q.; Ng, S.-M.; Jia, G. *Organometallics* **1997**, *16*, 34.
29. Aebischer, N. Frey, U.; Merbach, A. E. *Chem. Comm.* **1998**, 2303.
30. Crabtree, R. H.; Lavin, M. *J. Chem. Soc., Chem. Commun.* **1985**, 794; Crabtree, R. H.; Lavin, M.; Bonneviot, L. J. *J. Am. Chem. Soc.* **1986**, *108*, 4032.
31. Shilov, A. E. In *Activation and Functionalization of Alkanes*, Hill, C. L., ed., Wiley: New York, 1989; Stahl, S. S.; Labinger, J. A.; Bercaw, J. E. *J. Am. Chem. Soc.* **1996**, *118*, 5961.
32. Khalsa, G. R. K.; Kubas, G. J.; Unkefer, C. J.; Van Der Sluys, L. S.; Kubat-Martin, K. A. *J. Am. Chem. Soc.* **1990**, *112*, 3855.
33. Toupadakis, A.; Kubas, G. J.; King, W. A.; Scott, B. L.; Huhmann-Vincent, J. *Organometallics* **1998**, *17*, 5315.
34. Butts, M. D.; Kubas, G. J.; Scott, B. L. *J. Am. Chem. Soc.* **1996**, *118*, 11831.
35. King, W. A.; Luo, X.-L.; Scott, B. L.; Kubas, G. J.; Zilm, K. W. *J. Am. Chem. Soc.* **1996**, *118*, 6782; King, W. A.; Scott, B. L.; Eckert, J. Kubas, G. J. *Inorg. Chem.* **1999**, *38*, 1069.
36. Heinekey, D. M.; Schomber, B. M.; Radzewich, C. E. *J. Am. Chem. Soc.* **1994**, *116*, 4515.
37. Kubas, G. J.; Burns, C. J.; Eckert, J.; Johnson, S.; Larson, A. C.; Vergamini, P. J.; Unkefer, C. J.; Khalsa, G. R. K.; Jackson, S. A.; Eisenstein, O. *J. Am. Chem. Soc.* **1993**, *115*, 569.
38. Hills, A.; Hughes, D. L.; Jimenez-Tenorio, M.; Leigh, G. J.; Rowley, A. T. *J. Chem. Soc., Dalton Trans.* **1993**, 3041.
39. Aresta, M.; Sacco, A. *Gazz. Chim. Ital.* **1972**, *102*, 755.
40. Chatt, J.; Dilworth, J. R.; Richards, R. L. *Chem. Rev.* **1978**, *78*, 589.
41. Bancroft, G. M.; Garrod, R. E.; Maddock, A. G.; Mays, M. J.; Prater, B. E. *J. Am. Chem. Soc.* **1972**, *94*, 647.
42. Morris, R. H.; Schlaf, M. *Inorg. Chem.* **1994**, *33*, 1725.
43. Dapprich, S.; Frenking, G. *Angew. Chem. Int. Ed. Engl.* **1995**, *34*, 354.
44. Kaltsoyannis, N.; Scott, P. *J. Chem. Soc., Chem. Commun.* **1998**, 1665.
45. Rosi, M.; Sgamellotti, A.; Tarantelli, F.; Floriani, C.; Cederbaum, L. S. *J. Chem. Soc., Dalton Trans.* **1989**, 33.
46. Huhmann-Vincent, J.; Scott, B. L.; Kubas, G. J., *J. Am. Chem. Soc.* **1998**, *120*, 6808.
47. Roussel, P.; Scott, P. *J. Am. Chem. Soc.* **1998**, *120*, 1070; Odom, A. L.; Arnold, P. L.; Cummins, C. C. *J. Am. Chem. Soc.* **1998**, *120*, 5836.
48. Fang, X.; Huhmann-Vincent, J.; Scott, B. F.; Kubas, G. J., *J. Organomet. Chem.* **2000**, *609*, 95.
49. Huhmann-Vincent, J.; Scott, B. L.; Kubas, G. J. *Inorg. Chim. Acta* **1999**, *294*, 240.
50. Butts, M. D.; Kubas, G. J.; Luo, X.-L.; Bryan J. C. *Inorg. Chem.* **1997**, *36*, 3341.
51. Zhang, S.; Dobson, G. R.; Brown, T. L. *J. Am. Chem. Soc.* **1991**, *113*, 6908.
52. Bianchini, C.; Marchi, A.; Marvelli, L.; Peruzzini, M.; Romerosa, A.; Rossi, R.; Vacca, A. *Organometallics* **1995**, *14*, 3203.
53. Burrell, A. K.; Bryan, J. C.; Kubas, G. J. *J. Am. Chem. Soc.* **1994**, *116*, 1575.
54. Chin, B.; Lough, A. J.; Morris, R. H.; Schweitzer, C.; D'Agostino, C. *Inorg. Chem.* **1994**, *33*, 6278.
55. Vigalok, A.; Ben-David, Y.; Milstein, D. *Organometallics* **1996**, *15*, 1839.
56. Mason, M. G.; Ibers, J. A. *J. Am. Chem. Soc.* **1982**, *104*, 5153.
57. Jimenez-Tenorio, M.; Puerta, M. C.; Valerga, P. *Inorg. Chem.* **1994**, *33*, 3515.
58. Minas da Piedade, M. E.; Martinho Simoes, J. A. *J. Organomet. Chem.* **1996**, *518*, 167.
59. Gonzalez, A. A.; Mukerjee, S. L.; Chou, S.-L.; Zhang, K.; Hoff, C. D. *J. Am. Chem. Soc.* **1988**, *110*, 4419.
60. Gonzalez, A. A.; Hoff, C. D. *Inorg. Chem.* **1989**, *28*, 4295.
61. Page, M. I. *Angew. Chem., Int. Ed. Engl.* **1977**, *16*, 449.
62. Stull, D. R.; Westrum, E. F., Jr.; Sinke, G. C. *The Chemical Thermodynamics of Organic Compounds*, Wiley: New York, 1969.
63. Zhang, K.; Gonzalez, A. A.; Hoff, C. D. *J. Am. Chem. Soc.* **1989**, *111*, 3627, and references therein.
64. Chau, D. E. K.-Y.; James, B. R. *Inorg. Chim. Acta* **1995**, *240*, 419.

65. Basallote, M. G.; Duran, J.; Fernandez-Trujillo, J.; Manez, M. A. *J. Chem. Soc., Dalton Trans.* **1998**, 2205; Basallote, M. G.; Duran, J.; Fernandez-Trujillo, J.; Manez, M. A.; Rodriguez de la Torre, J. *J. Chem. Soc., Dalton Trans.* **1998**, 745; Basallote, M. G.; Duran, J.; Fernandez-Trujillo, J.; Manez, M. A. *J. Organometal. Chem.* **2000**, *609*, 29.
66. Church, S. P.; Grevels, F.-W.; Hermann, H.; Shaffner, K. *J. Chem. Soc., Chem. Commun.* **1985**, 30.
67. Johnson, F. P. A.; George, M. W.; Bagratashvili, V. N.; Vereshchagina, L. N.; Poliakoff, M. *Mendeleev Commun.* **1991**, 26; Klassen, J. K.; Selke, M.; Sorensen, A. A.; Yang, G. K. *J. Am. Chem. Soc.* **1990**, *112*, 1267.
68. Sun, X.-Z.; Grills, D. C.; Nikiforov, S. M.; Poliakoff, M.; George, M. W. *J. Am. Chem. Soc.* **1997**, *119*, 7521.
69. Gusev, D. G.; Kuhlman, R. L.; Renkema, K. H.; Eisenstein, O.; Caulton, K. G. *Inorg. Chem.* **1996**, *35*, 6775.
70. Helleren, C. A.; Henderson, R. A.; Leigh, G. J. *J. Chem. Soc., Dalton Trans.* **1999**, 1213.
71. Halpern, J.; Cai, L.; Desrosiers, P. J.; Lin, Z. *J. Chem. Soc., Dalton Trans.* **1991**, 717.
72. Zheng, Y.; Wang, W.; Lin, J.; She, Y.; Fu, K.-J. *J. Phys. Chem.* **1992**, *96*, 9821.
73. Wasserman, E. P.; Moore, C. B.; Bergman, R. G. *Science* **1992**, *255*, 315.
74. Wink, D. A.; Ford, P. C. *J. Am. Chem. Soc.* **1987**, *109*, 436.
75. Luo, X.-L.; Crabtree, R. H. *J. Am. Chem. Soc.* **1990**, *112*, 6912.
76. Luo, X.-L.; Michos, D.; Crabtree, R. H. *Organometallics* **1992**, *11*, 237.
77. Gusev, D. G.; Nietlispach, D.; Eremenko, I. L.; Berke, H. *Inorg. Chem.* **1993**, *32*, 3628.
78. Albertin, G.; Antoniutti, S.; Garcia-Fontan, S.; Carballo, R.; Padoan, F. *J. Chem. Soc., Dalton Trans.* **1998**, 2071.
79. Chinn, M. S.; Heinekey, D. M.; Payne, N. G.; Sofield, C. D. *Organometallics* **1989**, *8*, 1824.
80. Chinn, M. S.; Heinekey, D. M. *J. Am. Chem. Soc.* **1987**, *109*, 5865.
81. Conroy-Lewis, F. M.; Simpson, S. J. *J. Chem. Soc., Chem. Commun.* **1987**, 1675; Jia, G.; Morris, R. H. *J. Am. Chem. Soc.* **1991**, *113*, 875.
82. Chinn, M. S.; Heinekey, D. M. *J. Am. Chem. Soc.* **1990**, *112*, 5166.
83. Arliguie, T.; Chaudret, B. *J. Chem. Soc., Chem. Commun.* **1989**, 155.
84. Esteruelas, M. A.; Oro, L. A.; Valero, C. *Organometallics* **1991**, *10*, 462.
85. Haward, M. T.; George, M. W.; Hamley, P.; Poliakoff, M. *J. Chem. Soc., Chem. Commun.* **1991**, 1101.
86. Bullock, R. M.; Song, J.-S.; Szalda, D. J. *Organometallics* **1996**, *15*, 2504.
87. Luo, X.-L.; Kubas, G. J.; Burns, C. J.; Bryan, J. C.; Unkefer, C. J. *J. Am. Chem. Soc.* **1995**, *117*, 1159; Luo, X.-L.; Kubas, G. J.; Unkefer, C. J., unpublished data.
88. Capps, K. B.; Bauer, A.; Kiss, G.; Hoff, C. D. *J. Organomet. Chem.* **1999**, *586*, 23.
89. Musaev, D.; Morokuma, K. *J. Am. Chem. Soc.* **1995**, *117*, 799.
90. Maseras, F.; Lledos, A. *Organometallics* **1996**, *15*, 1218; Esteruelas, M. A.; Oro, L. A.; Valero, C. *Organometallics* **1991**, *10*, 462; Poulton, J. T.; Sigala, M. P.; Eisenstein, O.; Caulton, K. G. *Inorg. Chem.* **1993**, *32*, 5490.
91. Luo, X.-L.; Crabtree, R. H. *J. Am. Chem. Soc.* **1989**, *111*, 2527.
92. Corey, J. Y.; Braddock-Wilking, J. *Chem. Rev.* **1999**, *99*, 175.
93. Bullock, R. M. In *Transition Metal Hydrides*, Dedieu, A., ed., VCH: New York, 1992, p. 263.
94. Hostetler, M. J.; Bergman, R. G. *J. Am. Chem. Soc.* **1992**, *114*, 7629.
95. Rabinovich, D.; Parkin, G. *J. Am. Chem. Soc.* **1993**, *115*, 353; Hascall, T.; Rabinovich, D.; Murphy, V. J.; Beachy, M. D.; Friesner, R. A.; Parkin, G. *J. Am. Chem. Soc.* **1999**, *1221*, 11402.
96. Abu-Hasanayn, F; Krogh-Jespersen, K.; Goldman, A. *J. Am. Chem. Soc.* **1993**, *115*, 8019.
97. Bender, B. R.; Kubas, G. J.; Jones, L. H.; Swanson, B. I.; Eckert, J.; Capps, K. B.; Hoff, C. D. *J. Am. Chem. Soc.* **1997**, *119*, 9179.
98. Gusev, D. G.; Bakhmutov, V. I.; Grushin, V. V.; Vol'pin, M. E. *Inorg. Chim. Acta* **1990**, *177*, 115.
99. Henderson, R. A.; Oglieve, K. E. *J. Chem. Soc., Dalton Trans.* **1993**, 3431.
100. Burwell, R. L., Jr.; Stec, K. S. *J. Coll. Interface Sci.* **1977**, *58*, 54.
101. Bigeleisen, J.; Goeppert-Mayer, M. *J. Chem. Phys.* **1947**, *15*, 261.
102. Torres, L.; Gelabert, R.; Moreno, M.; Lluch, J. M. *J. Phys. Chem. A* **2000**, *104*, 7898.
103. Heinekey, D. M.; Oldham, W. J., Jr. *J. Am. Chem. Soc.* **1994**, *116*, 3137; Oldham, W. J., Jr.; Hinkle, A. S.; Heinekey, D. M. *J. Am. Chem. Soc.* **1997**, *119*, 11028.
104. Bender, B. R. *J. Am. Chem. Soc.* **1995**, *117*, 11239.

105. Bengali, A. A.; Arndtsen, B. A.; Burger, P. M.; Schultz, R. H.; Weiller, Kyle, K. R.; Moore, C. B.; Bergman, R. G. *Pure Appl. Chem.* **1995**, *67*, 281; Schultz, R. H.; Bengali, A. A.; Tauber, M. J.; Weiller, B. H.; Wasserman, E. P.; Kyle, K. R.; Moore, C. B.; Bergman, R. G. *J. Am. Chem. Soc.* **1994**, *116*, 7369; Paur-Afshari, R.; Lin, J.; Schultz, R. H. *Organometallics* **2000**, *19*, 1682.
106. Bengali, A. A.; Schultz, R. H.; Moore, C. B.; Tauber, M. J.; Weiller, B. H.; Wasserman, E. P.; Kyle, K. R.; Bergman, R. G. *J. Am. Chem. Soc.* **1994**, *116*, 9585.
107. Zhou, P.; Vitale, A. A.; San Filippo, J.; Saunders, W. H. *J. Am. Chem. Soc.* **1985**, *107*, 8049.
108. Bullock, R. M.; Headford, C. E. L.; Kegley, S. E.; Norton, J. R. *J. Am. Chem. Soc.* **1985**, *107*, 8049.
109. Bullock, R. M.; Headford, C. E. L.; Hennessey, K. M.; Kegley, S. E.; Norton, J. R. *J. Am. Chem. Soc.* **1989**, *111*, 3897.
110. Buchanan, J. M.; Stryker, J. M.; Bergman, R. G. *J. Am. Chem. Soc.* **1986**, *108*, 1537.
111. Periana, R. A.; Bergman, R. G. *J. Am. Chem. Soc.* **1986**, *108*, 7332.
112. Gould, G. L.; Heinekey, D. M. *J. Am. Chem. Soc.* **1989**, *111*, 5502.
113. Parkin, G.; Bercaw, J. E. *Organometallics* **1989**, *8*, 1172.
114. Wang, C.; Ziller, J. W.; Flood, T. C. *J. Am. Chem. Soc.* **1995**, *117*, 1647.
115. Stahl, S. S.; Labinger, J. A.; Bercaw, J. E. *J. Am. Chem. Soc.* **1996**, *118*, 5961.
116. Halpern, J. *Pure Appl. Chem.* **1986**, *58*, 576.
117. Wick, D. D.; Reynolds, K. A.; Jones, W. D. *J. Am. Chem. Soc.* **1999**, *121*, 3974.
118. Abis, L.; Sen, A.; Halpern, J. *J. Am. Chem. Soc.* **1985**, *107*, 8049; Michelin, R. A.; Faglia, S.; Uguagliati, P. *Inorg. Chem.* **1983**, *22*, 1831; Hackett, M.; Ibers, J. A.; Whitesides, G. M. *J. Am. Chem. Soc.* **1988**, *110*, 1436.
119. Janowicz, A. H.; Bergman, R. G. *J. Am. Chem. Soc.* **1983**, *105*, 3929.
120. Bengali, A. *Organometallics* **2000**, *19*, 4000.

8

Vibrational Studies of Coordinated Dihydrogen

8.1. VIBRATIONAL MODES FOR η^2-H_2

The vibrational modes for M(η^2-H_2) are completely different from those for dihydrides, MH_2, and polyhydrides, which typically have only two fundamental modes: M–H stretches in the range 1700–2300 cm^{-1} and M–H deformations at 700–950 cm^{-1}.[1] When diatomic H_2 combines with a M–L fragment to form a η^2-H_2 ligand, in addition to ν_{HH} five "new" vibrational modes are created, which are related to the "lost" translational and rotational degrees of freedom for H_2 (Scheme 1).[1–3] The ν_{HH} is still present, but is shifted to a much lower frequency and, as will be shown, becomes highly coupled with a MH_2 mode. Thus six fundamental vibrational modes are expected to be formally isotope-sensitive: three stretches, ν_{HH}, $\nu_{as(MH_2)}$, $\nu_{s(MH_2)}$; two deformations, $\delta_{(MH_2)\text{in-plane}}$ and $\delta_{(MH_2)\text{out-of-plane}}$; and a torsion ($H_2$ rotation), τ_{H_2}. The bands shift hundreds of wavenumbers on isotopic substitution with D_2 or HD, which greatly facilitates their assignment (Figure 8.1). It is critical to note that the frequencies of the bands for the η^2-HD complexes are in between those for the η^2-HH and η^2-DD species and *not a superimposition of MH_2 and MD_2 bands* as is seen for classical hydrides. This is another valuable diagnostic for distinguishing H_2 versus dihydride coordination, although these vibrational modes are often difficult to observe: The entire set of bands has been identified only in the first H_2 complex, $W(CO)_3(PR_3)_2(H_2)$ (R = Cy, ^{i}Pr). All but $\nu_{s(MH_2)}$, observed in both the IR and Raman, are weak, and most of the bands tend to be obscured by other ligand modes. In the Nujol mull IR spectrum, four bands, ν_{HH} at 2690 cm^{-1}, $\nu_{as(MH_2)}$ at 1575 cm^{-1}, $\nu_{s(MH_2)}$ at 953 cm^{-1}, and $\delta_{(MH_2)\text{in-plane}}$ at 462 cm^{-1}, can be observed to shift to lower frequency for the D_2 analogue (Figure 8.2). The band at 442 cm^{-1} in the D_2 complex is assigned to $\delta_{(WD_2)\text{out-of-plane}}$. The modes for H_2 rotation about the M–H_2 axis, τ_{H_2} and also $\delta_{(MH_2)\text{out-of-plane}}$ near 640 cm^{-1} are observable only by inelastic neutron scattering (INS) methods (Figure 8.3; see also Chapter 6).

The frequency of most interest ν_{HH} is not formally forbidden in the IR of H_2 complexes, but is polarized along the direction of the M–H_2 bond in highly symmetric complexes. Therefore, intensity arises only from coupling of ν_{HH} with other modes of the same symmetry such as $\nu_{s(MH_2)}$ or ν_{CO} if CO is present, and ν_{HH}

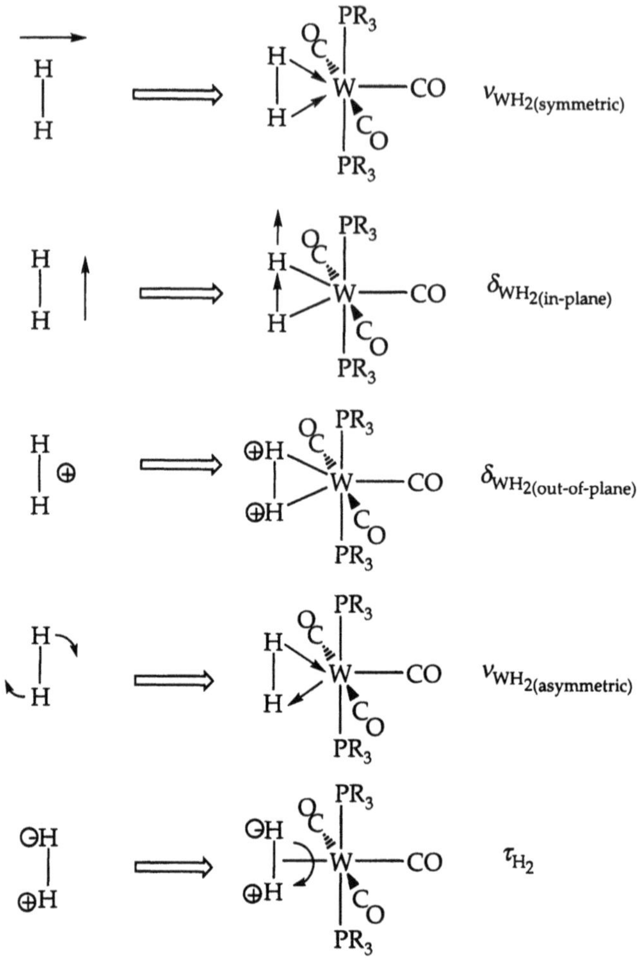

Scheme 1

is generally weak. Furthermore, its frequency varies tremendously and is often near the v_{CH} region, the worst possible position because most ancillary ligands such as phosphines have strong v_{CH}. Use of perdeuterated phosphine ligands to eliminate such interference enables location of v_{HH} in $W(CO)_3[P(C_6D_{11})_3]_2(H_2)$ as a broad, weak band at 2690 cm^{-1} (Figure 8.2). As shown Table 8.1, several other compounds, including surface and cluster species, exhibit v_{HH} in a range, 2080–3200 cm^{-1}, that is considerably lower than that for free H_2 gas (4300 cm^{-1}).

Low-temperature-stable species with only CO ligands, such as group 6 $M(CO)_5(H_2)$ generated photochemically in liquid rare gas solvents below $-70°C$, are favorable for study because of little solvent or coligand interference.[4] The v_{HH} bands are generally quite broad (<40 cm^{-1} FWHM) even at low temperatures (Figure 8.4). Several explanations are possible, including rapid hindered rotation of the H_2 ligand (or some other motion/behavior that dephases v_{HH}), "hot" band contributions, and a flat rotational potential.[22] The HD and DD isotopomers give

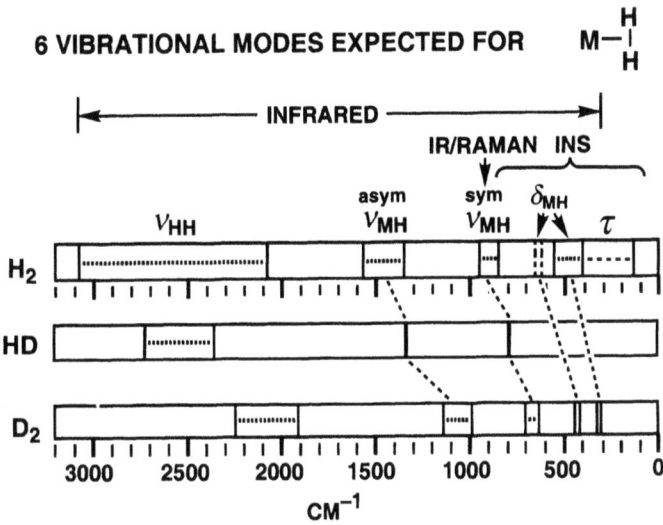

Figure 8.1. Ranges and shifts observed for vibrational modes in HH, HD, and DD complexes (INS = inelastic neutron scattering). The D_2 and HD complexes are primarily $M(CO)_3(PR_3)_2(H_2)$ (M = group 6).

increasingly narrower ν_{HD} and ν_{DD}. As can be seen in Figure 8.4, the nature of M greatly influences the band position, and third-row metals generally gives lower frequencies because the H–H bond here is weaker owing to higher BD. As will be shown, however, ν_{HH} is highly coupled with $M-H_2$ vibrational modes, and the degree of bond activation is difficult to correlate with vibrational frequencies.

Modes other than ν_{HH} have rarely been observed in room-temperature-stable complexes, partly because of interference from coligands or difficulty in assignment, especially if hydride ligands are also present. This is the case for $Tp^*RuH(H_2)_2$, which shows four unassignable bands at 458–834 cm^{-1}.[17] Such frequencies are also seen by Raman spectroscopy for $[CpRu(dppm)(H_2)]BF_4$, which has an elongated H–H bond (1.10 Å) and the lowest reported value for ν_{HH}, 2082 cm^{-1}.[18] As will be shown in Section 8.4, however, the high degree of mode-mixing in such elongated H_2 complexes, including especially $[Os(NH_3)_5(H_2)]^{2+}$, renders ν_{HH} and the other mode representations in Scheme 1 for true (unstretched) η^2-H_2 complexes meaningless and new vibrational modes must be defined. The photolytically generated pentacarbonyl and nitrosyl species and the compounds containing H_2 bound to a Ni(510) surface or to Pd atoms deposited on rare gas matrices at 7–12 K are among the few to display MH_2 modes aside from the group 6 complexes (Table 8.1). H_2 is believed to be bound in η^2 fashion on the stepped edges of the Ni surface (Figure 3.1) and electron energy loss spectroscopy (EELS) at 100 K shows several bands comparable to those for true H_2 complexes such as $W(CO)_3(PCy_3)_2(H_2)$.

As expected there is generally a strong dependence of ν_{HH} and ν_{MH_2} modes on both metal and ligand sets. One might anticipate a correlation of ν_{HH} with d_{HH} and the BD ability (electron-richness) of the M, as found for ν_{NN} and ν_{CO} in similar

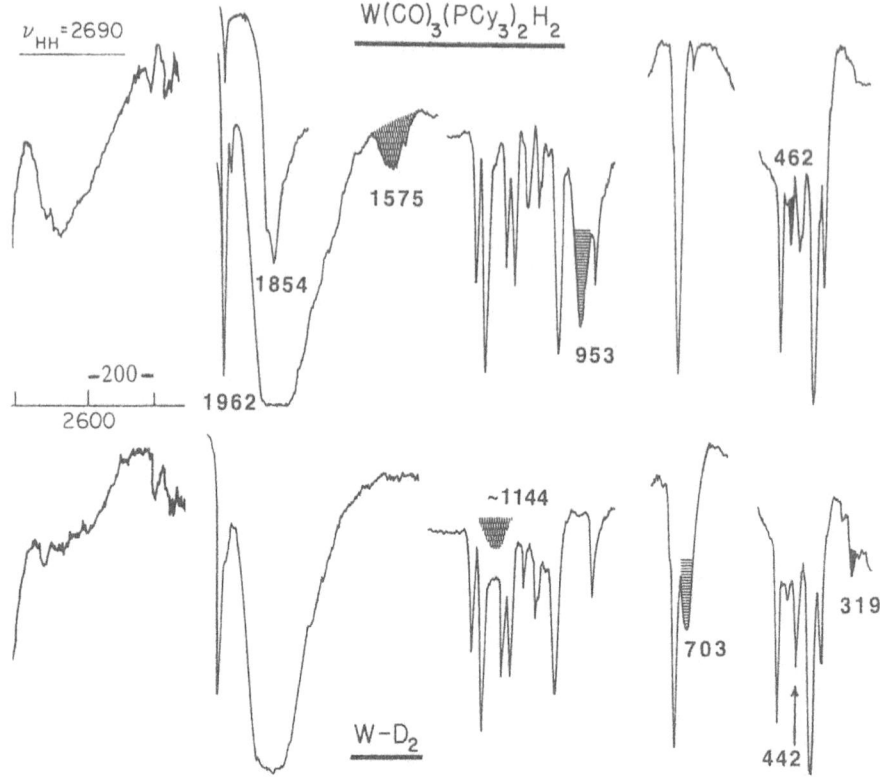

Figure 8.2. IR of W(CO)$_3$(PCy$_3$)$_2$(H$_2$) and D$_2$ analogue in Nujol mulls. Bands at 1962 and 1854 cm^{-1} are ν_{CO}. Spectra in ν_{HH} region (2690 cm^{-1}) are for P(C$_6$D$_{11}$)$_3$ complexes to reduce interference from ν_{CH} of phosphines. Reprinted with permission from Bender et al.[3]. Copyright 1997 American Chemical Society.

π-acceptor N$_2$ and CO ligands. However, this is not the case because of complexity of the bonding and mixing of ν_{HH} and ν_{MH_2} modes, as will be discussed in Section 8.4.

The lower-frequency deformations and torsions have been the least observed modes. INS is a powerful technique for locating such large-amplitude vibrations involving hydrogen. Because deuterium does not scatter as well, modes not due to η^2-H$_2$ can effectively be removed by subtracting the spectrum of the D$_2$ complex or a suitable "blank," e.g., a similar complex with a non-hydrogen-containing ligand in place of H$_2$. Excellent spectra of W(CO)$_3$(PR$_3$)$_2$(H$_2$) (Figure 8.3) and related complexes are obtained in the range 200–1000 cm^{-1} using neutron spectrometers at, e.g., Los Alamos National Laboratory (see Chapter 6). The torsional mode, τ_{WH_2}, appears in the INS difference spectrum as a split mode at 385 and 325 cm^{-1} resulting from transitions to two split excited librational states ($J = 1, 2$).[23] The two WH$_2$ deformational modes at 640 and 462 cm^{-1} are also seen in the INS as broad features. The WH$_2$ deformation around 640 cm^{-1} is obscured in the IR but the D$_2$ isotopomer gives a band at 442 cm^{-1} for the corresponding WD$_2$ deformation (Figure 8.2).

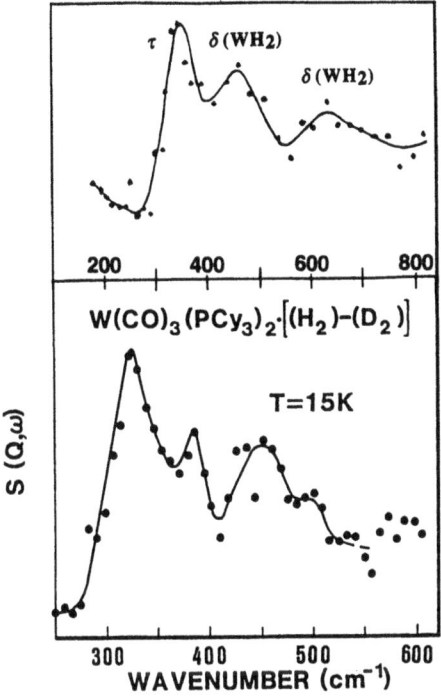

Figure 8.3. INS spectra of W(CO)$_3$(PCy$_3$)$_2$(H$_2$) at 15 K obtained as difference spectra by subtracting the spectrum of the D$_2$ complex from that of the H$_2$ complex. The lower spectrum with higher resolution shows splitting in the torsional mode (385/325 cm^{-1}). Reprinted with permission from Bender *et al.*[3] Copyright 1997 American Chemical Society.

In certain cases H–H and M–H vibrational frequencies have been calculated in order to assess the weakening of the H–H bond on coordination.[24] The reduction in frequency of the ν_{HH} over that in free H$_2$ generally agrees with experimental values, although these data are rather limited.

8.2. NORMAL COORDINATE ANALYSIS OF W(H$_2$)(CO)$_3$(PCy$_3$)$_2$

The normal coordinate analysis of W(H$_2$)(CO)$_3$(PCy$_3$)$_2$ (**1**) provides increased understanding of the M–H$_2$ interaction and the relation of H–H and MH$_2$ frequencies to bond strengths.[3] In addition to isotopic shifts of the six modes discussed above, several other metal–ligand modes also show small shifts to higher or lower wavenumber upon deuterium substitution, which result from vibrational coupling between modes that belong to the same symmetry block and are close in energy.[25] This is especially obvious in the Raman spectra of **1** and **1-d_2** below 650 cm^{-1}, e.g., a band at 400 cm^{-1} due mainly to $\nu_{as(WC)}$ shifts 13 cm^{-1} to *higher* frequency for the D$_2$ complex, presumably because of mixing with $\delta_{(WH_2)\text{out-of-plane}}$. Clearly the Cy groups should not show HH, HD, DD isotope shifts and are not

Table 8.1. IR Frequencies (cm^{-1}) for Nominal v_{HH} and v_{MH_2} Modes in H$_2$ Complexes Compared to d_{HH} (Å)

Complex	v_{HH}	$v_{as(MH_2)}$	$v_{s(MH_2)}$	δ_{MH_2}[a]	d_{HH}	Reference
CpV(CO)$_3$(H$_2$)	2642					10
CpNb(CO)$_3$(H$_2$)	2600					10
Cr(CO)$_5$(H$_2$)	3030	1380	869, 878			4
Cr(CO)$_3$(PCy$_3$)$_2$(H$_2$)		1540	950	563[b]	0.85	5
Mo(CO)$_5$(H$_2$)	3080					4
Mo(CO)$_3$(PCy$_3$)$_2$(H$_2$)	~2950[c]	~1420[c]	885	471	0.87	6
Mo(CO)(dppe)$_2$(H$_2$)	2650		875		0.88	7
W(CO)$_5$(H$_2$)	2711		919			4
W(CO)$_3$(P^{i}Pr$_3$)$_2$(H$_2$)	2695	1567	953	465	0.89	6
W(CO)$_3$(PCy$_3$)$_2$(H$_2$)	2690	1575	953	462	0.89	6
W(CO)$_3$(PCyp$_3$)$_2$(H$_2$)		1565	938			8
Fe(CO)(NO)$_2$(H$_2$)	2973	1374	~870			9
Co(CO)$_2$(NO)(H$_2$)	{3100, 2976}[d]	1345	868			9
FeH$_2$(H$_2$)(PEtPh$_2$)$_3$	2380		850	500, 405[e]	0.82	11
RuH$_2$(H$_2$)(PMe$_3$)$_3$	2360					12
RuH$_2$(H$_2$)$_2$(P^{i}Pr$_3$)$_2$	2568	1673	822[b]		0.92	33
Tp*RuH(H$_2$)$_2$	2361				0.90	17
Tp*RuH(H$_2$)(THT)	2250				0.89	17
[Os(NH$_3$)$_5$(H$_2$)]$^{2+}$	2231[b]				[1.34][f]	13
[CpRu(dppm)(H$_2$)]$^+$	2082[b]	1358[b]	679[b]	486, 397[b]	[1.10][g]	18
Tp*RhH$_2$(H$_2$)	2238				0.94[h]	14
Pd(H$_2$) (matrix)	2971	1507	950		0.85[h]	15, 34
Ni(510)-(H$_2$)[i]	3205	1185	670			16
Cu$_3$(H$_2$) (matrix)	3351, 3232					19

[a] In-plane deformation.
[b] Assignments unclear; in the case of the elongated Ru and Os complexes, these are highly mixed modes that could involve M–H modes (if present).
[c] Estimated from observed D$_2$ isotopomer bands.
[d] Split possibly by Fermi resonance.
[e] Assignment unclear (data from INS).
[f] For [Os(ethylenediamine)$_2$(H$_2$)(acetate)]$^+$.[21]
[g] For the Cp* analogue.[20]
[h] Calculated from inelastic neutron scattering data or DFT.
[i] Data from EELS spectroscopy.

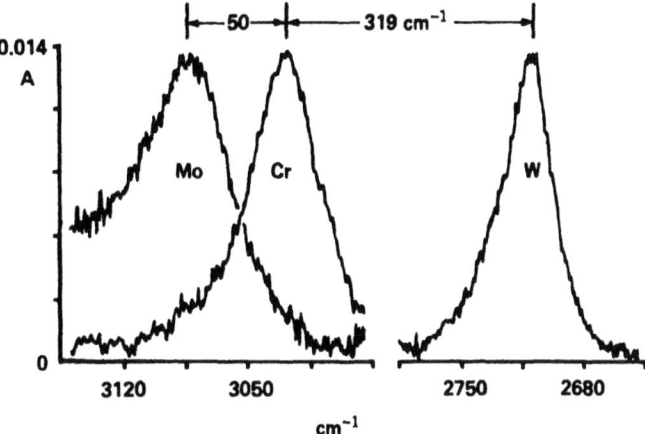

Figure 8.4. IR spectra (liquid Xe, −70°) showing the substantial wavenumber shift between bands assigned to v_{HH} in $M(CO)_5(H_2)$ for M = Cr, Mo, W. The three traces were recorded in separate experiments and the absorbance scale refers to the Mo system. Reprinted with permission from Upmacis et al.[4] Copyright 1986 American Chemical Society.

significantly coupled to vibrations other than Cy and P–Cy modes. Therefore only the "$W(H_2)(CO)_3P_2$" core fragment is treated in force field calculations.

The normal coordinate analysis includes 13 observed IR and Raman frequencies (W–P modes were not located). Obviously a large number of force constants are

(1)

fixed in the calculations, e.g., force constants of 2–3 are assumed for W–P, and for CO-related modes values are assumed that seem reasonable based on other systems, such as $W(CO)_6$.[26] The solution is given in Table 8.2, which summarizes only the data for H_2-related modes. The agreement between observed and calculated frequencies is not strong, being off by several wavenumbers in several cases. Altering the constrained constants somewhat could improve the situation, but this is not appropriate for such an ill-determined problem. Even a perfect fit would not be unique, but overall the analysis suffices to assign the observed vibrational modes for **1** and its isotopomers.

In order to understand M–H_2 bonding interactions, it is necessary to investigate the reasonableness and meaning of the force constants in Table 8.3 for the hydrogen-related modes. The first and most important mode is v_{HH}, and the value of 1.3 mdyn/Å for F_{HH} is much smaller than the value for free H_2 (5.7 mdyn/Å).[27] This is not too surprising, but multiplying the latter by the ratio of the square of

Table 8.2. Observed[a] and Calculated Vibrational Frequencies (cm^{-1}) and Mode Assignments for **1**, **1-d_2**, and **1-d_1**

Method/intensity	**1** obsd(calcd)	**1-d_2** obsd(calcd)	**1-d_1** obsd(calcd)	Assignment
IR, w	2690(2692.1)	~1900(1909.9)	2360(2357.8)	ν_{HH}
IR, R, m	953(949.0)	703(703.1)	791(799.8)	$\nu_{s(WH_2)}$
IR, w	1575(1574.7)	~1144(1136.1)	~1360(1357.9)	$\nu_{as(WH_2)}$
IR, w	462[b](456.2)	319(326.0)	360(368.7)	$\delta_{(WH_2)\text{in-plane}}$[c]
INS,[d] IR, w	640[e](640.0)	442[f](442.0)		$\delta_{(WH_2)\text{out-of-plane}}$
INS, m	385(325)			τ_{WH_2}

[a] Resolution: 2 cm^{-1}.
[b] Also observed in INS.
[c] This mode shows a greater observed isotope shift than calculated. It apparently arises from the $\delta_{(WH_2)\text{in-plane}}$ rocking coordinate coupled strongly with other coordinates.
[d] INS.
[e] Observed in INS only.
[f] Observed in IR only.

frequencies for bound and free H_2 indicates that F_{HH} should have only been lowered to 2.1 mdyn/Å:

$$5.7 \times (2690/4395)^2 = 5.7(0.37) = 2.1 \qquad (8.1)$$

The lower than calculated value of F_{HH} would suggest a longer d_{HH} than either the value of 0.82 Å observed by neutron diffraction in $W(CO)_3(P^iPr_3)_2(H_2)$ or 0.89Å determined by solid state NMR for both the R = ^{i}Pr and Cy species. An empirical correlation between bond length and force constant, known as Badger's rule, $k_e = b/(r_e - a)^3$, in fact predicts d_{HH} to be about 0.94 Å.[28] However, we see that ν_{HH} has considerable ν_{MH} character so it cannot be treated as an isolated HH mode. In complexes with more activated H_2 ligands with longer d_{HH}, this mixing of ν_{HH} and ν_{WH_2} is much more extensive and increases to the point at which these modes themselves no longer have meaning and must be redefined, as will be shown in Section 8.4. Thus the values of ν_{HH} in Table 8.1 are not reliable predictors of d_{HH} for H_2 complexes, especially for elongated complexes where the values listed for ν_{HH} and ν_{WH} effectively should be reversed. Similar situations arise for other π-acceptor

Table 8.3. Force Constants (mdynes/Å) for Hydrogen-Related Modes in $W(CO)_3P_2(H_2)$ and a Triatomic Model Complex

	$W(CO)_3P_2(H_2)$	$W(H_2)$
F_{HH}	1.32	1.46
$F_{s(WH)}$	1.46	—
$F_{as(WH)}$	1.42	—
$F_{WH,WH'}$	0.02	−0.05
$F_{HH,WH}$	0.67	0.62

ligands with a large degree of M → L backbonding such as metal–*ethylene* complexes: v_{CC} is not a reliable measure of d_{CC} because of normal mode coupling to same-symmetry C_2H_4 wagging and deformation modes.[29]

The WH stretching constant surprisingly is as large as that for the HH stretch, and the WH, WH' interaction is negligible. The HH, WH interaction force constant, $F_{HH,WH}$, is very large (0.67 mdyn/Å), indicating that stretching the HH bond leads to strengthening of the WH bonds, and vice versa. In hexacarbonyls such as $Cr(CO)_6$, $F_{CO,MC}$ is also large (0.61 mdyn/Å), but is notably smaller (0.37 mdyn/Å) in $[Fe(CO)_6]^{2+}$ because of reduced (but still substantial) BD from the highly electrophilic metals.[30] Stretching force constants for an *isolated* WH_2 group using the frequencies observed for the HH and WH stretches can also be calculated. The results are shown in Table 8.3 for comparison with the more complete treatment. Surprisingly the agreement is quite good; hence one can do rather well with the simplified treatment, though it does not tell us about the rest of the molecule.

8.3. BONDING CHARACTER OF DIHYDROGEN COMPLEXES SUGGESTED BY VIBRATIONAL ANALYSIS

As expected, there is generally a strong dependence of the modes on both M and L. However, vibrational analysis of a $M-\eta^2-H_2$ system is complicated by the 3c-2e bonding. The bonding is of the Dewar–Chatt–Duncanson type present in metal–olefin complexes, where there is a strong $M \to H_2$ σ^* component, E_{BD}, to the bonding in addition to electron donation, E_D, to the empty metal d orbital from

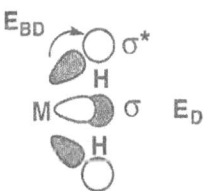

the H_2 electron pair (see Section 4.6). Calculations have shown that the E_{BD} component can be energetically as strong as or stronger than E_D. BD is much weaker in complexes with mostly π acceptors, although calculations show that E_{BD} in $Mo(CO)_5(H_2)$ still has about one-half the energy of E_D. One might anticipate a correlation of v_{HH} with the BD ability (electron-richness) of M, as found for v_{NN} and v_{CO} in similar π-acceptor N_2 and CO ligands. The decrease in v_{HH} on going from $Mo(CO)_5(H_2)$ to $Mo(CO)_3(PCy_3)_2(H_2)$ to $Mo(CO)(dppe)_2(H_2)$ at first glance does seem to reflect increased H–H bond-weakening by the increasingly electron-rich M. However, v_{HH} decreases in the order Mo > Cr > W for $M(CO)_5(H_2)$ (Figure 8.4), whereas Cr should be the worst backbonder and give the highest v_{HH}. The value of v_{HH} in $W(CO)_5(H_2)$ (2711 cm^{-1}) is far out of line with the much higher values in the Cr and Mo congeners (3030 and 3080 cm^{-1}) and also differs little (20 cm^{-1}) from that in vastly more electron-rich $W(CO)_3(PCy_3)_2(H_2)$ (**1**). The pentacarbonyl is unstable at room temperature and almost certainly has a much shorter d_{HH} and presumably a stronger H–H bond.

It is quite clear from the data in Table 8.1 that ν_{HH} does not correlate well with d_{HH} because of extensive mode-mixing. The normal coordinate analysis of $W(CO)_3(PCy_3)_2(H_2)$ treats the W–H$_2$ interaction as a triangulo system, i.e., where there are direct BD electronic interactions between W and H atoms (below, left) rather than as the strictly three-center bonding representation (below right). This is

confirmed by the fact that the WH stretching force constant is as large as that for the HH stretch and that the HH, HW interaction is very large, indicating that stretching the HH bond leads to strengthening of the HW bonds, and vice versa. This extensive mixing along with the reduction of the ν_{HH} force constant to one-fourth the value in free H$_2$ indicates that weakening of the H–H bond and formation of W–H bonds is already well along the reaction coordinate to OA in **1**. Furthermore, as the H–H bond becomes more activated on a metal fragment, the observed "ν_{HH}" mode will have increasing M–H character relative to H–H character. Upon H–H cleavage, this mode will then be assimilated into the M–H stretching mode, which can occur in the 1700–2300 cm^{-1} range. In the elongated H$_2$ complexes with d_{HH} of 1.1 Å or greater, the 2231 and 2082 cm^{-1} frequencies in Table 8.1 for the cationic Os and Ru complexes should be thought of more as ν_{OsH} than ν_{HH}, as will be shown in Section 8.4. The frequency of the nominal ν_{HH} mode decreases and in a sense crosses over with that for the increasing ν_{HH} on a metal fragment as the H–H bond is activated and then broken by changing the ligand environment. This is unprecedented in a vibrational spectroscopic analysis of a chemical system. It would be most interesting to study the gradual activation of an M–η^2-HD system since ν_{HD} and the two ν_{M-HD} modes (the frequencies for all of which are intermediate in value to those for the H$_2$ and D$_2$ isotopomers) must eventually transform into two widely separated M–H and M–D stretches. The question is how and at what point might this happen, given the continuum nature of the system. Section 8.4 will address this problem further for the realm of elongated H$_2$ complexes that lie further along the reaction coordinate.

The force constant analysis of **1** indicates that the H$_2$ ligand here is further activated toward OA than may have been previously thought. It has been a paradox that the d_{HH} in **1** or any of the group 6 complexes in Table 8.1 (0.85–0.89 Å, solid state NMR) are not as "stretched" as some of those found in later transition element complexes (1.0–1.5 Å), yet the H–H bond in **1** undergoes equilibrium cleavage in solution. Thus the observed d_{HH} may not always reflect the degree of "readiness to break," i.e., a very late transition state may exist. At the other end of the spectrum, for $W(CO)_5(H_2)$ and other complexes with extremely *weakly* bound H$_2$, the T-shaped entity pictured above with one internal coordinate, the H–H stretch, may be a more appropriate model for vibrational analysis. However, it is difficult to determine which analysis should be applied because some minimal BD and incipient

M–H bond formation will always be present even on highly electrophilic metals. As a further complication, the T-shaped bonding that represents the purely three-center interaction in σ complexes can be reasonably strong by itself in the absence of much BD. The interaction of H_2 with an electrophilic M has a much stronger E_D than an electron-rich M, and this can completely offset the lower E_{BD}. For example, a cationic H_2 complex such as $[Mn(CO)(dppe)_2(H_2)]^+$ can have properties, e.g., d_{HH}, that are remarkably similar to those of its isoelectronic neutral analogue, $Mo(CO)(dppe)_2(H_2)$.[31] In this case, the H–H bond here and in electron-poor $M(CO)_5(H_2)$ can be weakened to a considerable extent (i.e., a frequency much lower than 4300 cm^{-1} for free H_2) primarily by the E_D bonding component alone.

The lack of reliable correlation of vibrational versus other properties brought about by the bonding complexities extends to M–H_2 modes. These also show M/L dependence, though to a lesser degree than ν_{HH} (Table 8.1). More propitiously, $\nu_{s(MH_2)}$ is about 70 cm^{-1} lower in $Mo(CO)_3(PCy_3)_2(H_2)$ than in the W analogue, which combined with the higher ν_{HH} for the Mo complex suggests a weaker Mo–H_2 binding energy than for W–H_2, which is in line with the thermal stabilities. The ν_{MH_2} values for the unstable W, Fe, Co carbonyl complexes are also lower than those for the stable W complexes. However, one would have anticipated ν_{MH_2} to be appreciably higher for $W(CO)_3(PCy_3)_2(H_2)$ (1) than for its Cr analogue because of the higher M–H_2 binding energy measured for 1 and the far greater stability of 1 to H_2 dissociation in solution. This was not the case, however, and a minor ligand change in 1 from PCy_3 to tricyclopentylphosphine affected $\nu_{s(MH_2)}$ more than changing M. Also $\nu_{s(MH_2)}$ values are nearly identical for $Mo(CO)_3(PCy_3)_2(H_2)$ and $Mo(CO)(dppe)_2(H_2)$ though ν_{HH} is 300 cm^{-1} lower for the latter. As is the case for ν_{HH}, ν_{MH_2} is an inseparable measure of both HH and MH interactions, as will be further shown below.

8.4. HIGHLY MIXED H–H AND M–H_2 MODES IN ELONGATED DIHYDROGEN COMPLEXES: NEW NORMAL-MODE DEFINITIONS

Elongated H_2 complexes exhibit even greater mode-mixing to the extent that *new normal modes must be defined*. Raman studies of $[CpRu(dppm)(H_2)]^+$ by Chopra et al.[18] show unusually low values for all of the assigned modes in comparison to most of the other complexes. This might be expected for a weakened ν_{HH} but the ν_{MH_2} stretches would have been expected to rise with increasing M–H bond strength. As shown in the calculations by Lluch and Lledos (see Section 4.7.4), there is significant anharmonicity in the immediate neighborhood of the potential energy minimum, at least with respect to the H–H and Ru–H_2 stretches. This is central to an understanding of the true delocalized nature of the bonding in these types of complexes as revealed by their nuclear motion quantum calculations. These types of calculations, including discrete variable representation (DVR) analysis, reproduce accurate vibrational energy levels and wavefunctions without resorting to the harmonic approximation, which is not appropriate for an elongated H_2 complex.[32] The computations on a $[CpRu(H_2PCH_2PH_2)_2(H_2)]^+$ model complex determine the nuclear energy levels as a means of obtaining the vibrational energy levels.

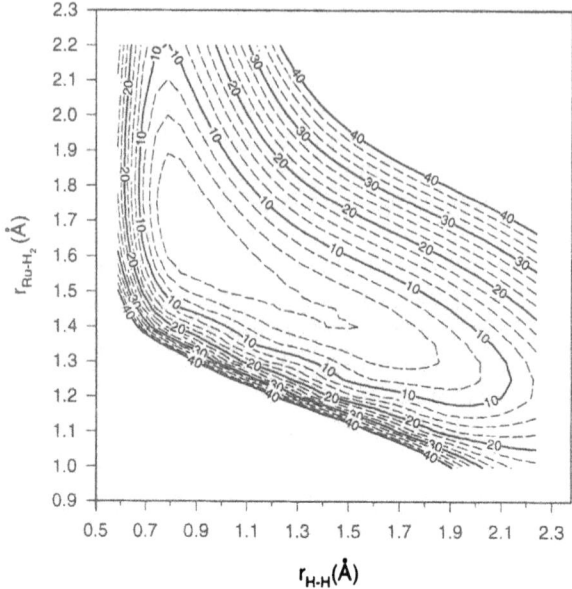

Figure 8.5. Contour plot of the potential energy surface used in the bidimensional calculations. Energy units are kcal/mol relative to the minimum in potential energy. Reprinted from Gelabert *et al.*[32] Copyright 1999, with permission from Elsevier Science.

A contour plot (Figure 8.5) of the two-dimensional PES (d_{HH} versus Ru–H$_2$ distance) similar to that shown in Figure 4.19 is very revealing. The valley for the minimum potential energy is oblique with respect to both v_{HH} and v_{RuH_2}, so that a normal mode analysis would lead to two normal modes that would not be a pure H–H stretch plus a pure Ru–H$_2$ stretch. What then are the "true" normal modes in this system? Because only a two-dimensional representation of the PES is calculated, only two normal modes can be described here, which cannot include the asymmetric modes (stretches or deformations). The "new" modes are represented qualitatively

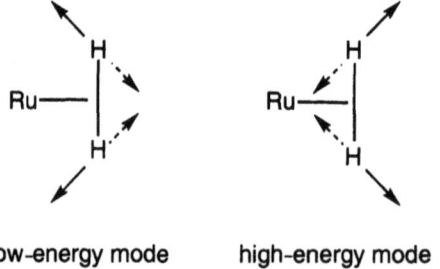

low-energy mode   high-energy mode

below. The first mode essentially parallels the oblique minimum-energy path that links the two exits of the potential energy valley in Figure 8.5. Along this normal mode the changes in energy are very damped; hence it is labeled as the low-energy mode, which from the DVR, analysis is calculated to occur at 555 cm^{-1}. This mode is essentially represented by the arrows along the sides of the whirlpool-like illustration (21) in Section 4.7.4 and directly follows the reaction coordinate for OA of H_2. Here the stretching of the H–H bond leads to shortening of d_{RuH} and thus strengthening of the Ru–H bonds. In the orthogonal high-energy mode, both the H–H bond and the Ru–H bonds stretch simultaneously, which costs much more energy: 2229 cm^{-1} from the DVR calculations. The experimental value of 2082 cm^{-1} in [CpRu(dppm)(H_2)]$^+$ would then correspond to the high-energy mode while the 679 cm^{-1} band relates to the 555 cm^{-1} value calculated for the low-energy mode. The high-frequency band can no longer be interpreted as a much weakened H–H stretch because it has a very significant component of Ru–H stretching and, in fact, must now essentially be regarded as representing primarily the Ru–H mode (mixed with some H–H stretch). Similarly, the low-frequency band should be described as a mode in which the H atoms separate from each other as they approach M, i.e., mainly an H–H stretch mixed with some Ru–H stretching. In other words, for elongated H_2 complexes *the assignments for the bands as v_{HH} and v_{RuH_2} have become in a sense reversed,* although they are still highly mixed.

Even simple unstretched H_2 complexes show anharmonicity in the M–H_2 potential function. In Pd(H_2), where d_{HH} is calculated to be 0.854, the $v_{s(MH_2)}/v_{s(MD_2)}$ ratio (950/714 = 1.331) is lower than the expected ratio, 1.414. The vibrational analysis of other types of σ-bond coordination would be expected to follow the same pattern of high degree of mode mixing and redefinition of normal modes for complexes with highly activated σ bonds at the brink of OA. Vibrational data for silane and related complexes are sparse, although the nominal stretching mode for the coordinated Si–H bond is observable by both IR and Raman (see Chapter 11). This mode will be highly mixed with M–H types of modes and will be similar to vibrational modes for μ-H ligands. The position of the $v_{M\cdots H-Si}$ frequency can be as high as 1890 cm^{-1} but is generally in the 1650–1800 cm^{-1} range.

REFERENCES

1. Sweany, R. L. In *Transition Metal Hydrides*, Dedieu, A., ed., VCH: New York, NY, 1992, pp. 65–101.
2. Kubas, G. J.; Ryan, R. R.; Swanson, B. I.; Vergamini, P. J.; Wasserman, H. J. *J. Am. Chem. Soc.* **1984**, *106*, 451.
3. Bender, B. R.; Kubas, G. J.; Jones, L. H.; Swanson, B. I.; Eckert, J.; Capps, K. B.; Hoff, C. D. *J. Am. Chem. Soc.* **1997**, *119*, 9179.
4. Upmacis, R. K.; Poliakoff, M.; Turner, J. J. *J. Am. Chem. Soc.* **1986**, *108*, 3645.
5. Kubas, G. J.; Nelson, J. E.; Bryan, J. C.; Eckert, J.; Wisniewski, L.; Zilm, K. *Inorg. Chem.* **1994**, *33*, 2954.
6. Kubas, G. J.; Unkefer, C. J; Swanson, B. I.; Fukushima, E. *J. Am. Chem. Soc.* **1986**, *108*, 7000.
7. Kubas, G. J.; Burns, C. J.; Eckert, J.; Johnson, S.; Larson, A. C.; Vergamini, P. J.; Unkefer, C. J.; Khalsa, G. R. K.; Jackson, S. A.; Eisenstein, O., *J. Am. Chem. Soc.* **1993**, *115*, 569.
8. Khalsa, G. R. K.; Kubas, G. J.; Unkefer, C. J.; Van Der Sluys, L. S.; Kubat-Martin, K. A. *J. Am. Chem. Soc.* **1990**, *112*, 3855.
9. Gadd, G. E.; Upmacis, R. K.; Poliakoff, M.; Turner, J. J. *J. Am. Chem. Soc.* **1986**, *108*, 2547.
10. George, M. W.; Haward, M. T.; Hamley, P. A.; Hughes, C.; Johnson, F. P. A.; Popov, V. K.; Poliakoff, M. *J. Am. Chem. Soc.* **1993**, *115*, 2286.

11. Van Der Sluys, L. S.; Eckert, J.; Eisenstein, O.; Hall, J. H.; Huffman, J. C.; Jackson, S. A.; Koetzle, T. F.; Kubas, G. J.; Vergamini, P. J.; Caulton K. G. *J. Am. Chem. Soc.* **1990**, *112*, 4831.
12. Kohlmann, W.; Werner, H. *Z. Naturforsch. B* **1993**, *48b*, 1499.
13. Harman, W. D.; Taube, H. *J. Am. Chem. Soc.* **1990**, *112*, 2261.
14. Eckert, J.; Albinati, A.; Bucher, U. E.; Venanzi, L. M. *Inorg. Chem.* **1996**, *35*, 1292.
15. Ozin, G. A.; Garcia-Prieto, J. *J. Am. Chem. Soc.* **1986**, *108*, 3099.
16. Martensson, A.-S.; Nyberg, C.; Andersson, S. *Phys. Rev. Lett.* **1986**, *57*, 2045.
17. Moreno, B.; Sabo-Etienne, S.; Chaudret, B.; Rodriguez, A.; Jalon, F.; Trofimenko, S. *J. Am. Chem. Soc.* **1995**, *117*, 7441.
18. Chopra, M.; Wong, K. F.; Jia, G.; Yu, N.-T. *J. Mol. Struct.* **1996**, *379*, 93.
19. Hauge, R. H.; Margrave, J. L.; Kafafi, Z. H. *NATO ASI Ser., Ser. B* **1987**, *158* (*Phys. Chem. Small Clusters*), 787.
20. Klooster, W. T.; Koetzle, T. F.; Jia, G.; Fong, T. P.; Morris, R. H.; Albinati, A. *J. Am. Chem. Soc.* **1994**, *116*, 7677.
21. Hasegawa, T.; Li, Z.; Parkin, S.; Hope, H.; McMullan, R. K.; Koetzle, T. F.; Taube, H. *J. Am. Chem. Soc.* **1994**, *116*, 4352.
22. Turner, J. J.; Poliakoff, M.; Howdle, S. M.; Jackson, S. A.; McLaughlin, J. G. *Faraday Trans.* **1988**, *86*, 271.
23. Eckert, J. *Spectrochim. Acta A* **1992**, *48A*, 363.
24. Hay, P. J. *J. Am. Chem. Soc.* **1987**, *109*, 705; Dapprich, S.; Frenking, G. *Angew. Chem. Int. Ed. Engl.* **1995**, *34*, 354.
25. Wilson, E. B. Jr.; Decius, J. C.; Cross, P. C. *Molecular Vibrations*, McGraw-Hill: New York, 1955, pp. 197–200.
26. Jones, L. H.; McDowell, R. S.; Goldblatt, M. *Inorg. Chem.* **1969**, *8*, 2349.
27. Levine, I. N. *Molecular Spectroscopy*, Wiley: New York, 1975, p. 160.
28. Badger, R. M. *J. Chem. Phys.* **1934**, *2*, 128.
29. Anson, C. E.; Sheppard, N.; Powell, D. B.; Bender, B. R.; Norton, J. R. *J. Chem. Soc., Faraday Trans.* **1994**, *90*, 1449, and references therein.
30. Bernhardt, E.; Bley, B.; Wartchow, R.; Willner, H.; Bill, E.; Kuhn, P.; Sham, I. H. T.; Bodenbinder, M.; Brochler, R.; Aubke, F. *J. Am. Chem. Soc.* **1999**, *121*, 7188.
31. King, W. A.; Luo, X.-L.; Scott, B. L.; Kubas, G. J.; Zilm, K. W. *J. Am. Chem. Soc.* **1996**, *118*, 6782.
32. Gelabert, R,; Moreno, M.; Lluch, J. M.; Lledos, A. *Chem. Phys.* **1999**, *241*, 155; Torres, L.; Gelabert, R.; Moreno, M.; Lluch, J. M. *J. Phys. Chem. A* **2000**, *104*, 7898.
33. Abdur-Rashid, K.; Gusev, D. G.; Lough, A. J.; Morris, R. H. *Organometallics* **2000**, *19*, 1652.
34. Andrews, L.; Manceron, L.; Alikhani, M. E.; Wang, X. *J. Am. Chem. Soc.* **2000**, *122*, 11011.

9

Reactions and Acidity of Dihydrogen Complexes

9.1. INTRODUCTION

The reactivity of H_2 complexes is dominated by oxidative addition (homolytic cleavage of H_2), deprotonation (heterolytic cleavage), and elimination of H_2. Complexes with H_2 ligands are exceedingly dynamic systems, particularly when hydride ligands are also present, and exhibit facile $M-H_2$ rotation, H-atom exchange, and fluxionality, which were all discussed in Chapter 6. It is important to note that H_2 ligands exhibit widely varying degrees of acidity, in some cases as strong as triflic acid, and thus are readily able to protonate weak bases. Jessop and Morris thoroughly review much of the reaction chemistry and dynamic properties.[1] A more recent review by Jia and Lau emphasizes the acidity and heterolytic cleavage reactions of H_2 ligands.[2] An overview of the general reactivity will be presented in this chapter, along with more recent specific reactions.

Dissociation or substitution of the often labile H_2 ligand is the most common reaction, and the widely varying solution and solid state stabilities of H_2 complexes have been tabulated.[1] Bound H_2, particularly in electrophilic cationic complexes and/or when trans to CO or a hydride, can readily be displaced by coordinating solvents and, remarkably, even hydrocarbon solvents such as toluene and alkanes.[3] The energetics of the reversible loss of H_2 from $IrIH_2(H_2)(P^iPr_3)_2$ are highly solvent-dependent, and even alkanes can compete with H_2 binding (Scheme 1), providing evidence that toluene and alkane complexes (see Chapter 12) can be of major thermodynamic importance. In cationic systems, η^2-H_2 is also subject to displacement by the anion, e.g., triflate (Scheme 2). Low-interacting anions are thus often crucial to the stability of weak H_2 and other σ complexes. On the other hand, many H_2 complexes further along the reaction coordinate to OA are quite robust and the H_2 cannot easily be removed, e.g., even in refluxing acetonitrile. Simple H_2 dissociation as found for $[RuH(H_2)(dppe)_2]^+$ is not necessarily the first step in substitution reactions, and concerted formation of an agostic complex (see Chapter 7) or an even more complicated mechanism can occur, e.g., opening up of a chelate ring in $[FeH(H_2)(dppe)_2]^+$.[4]

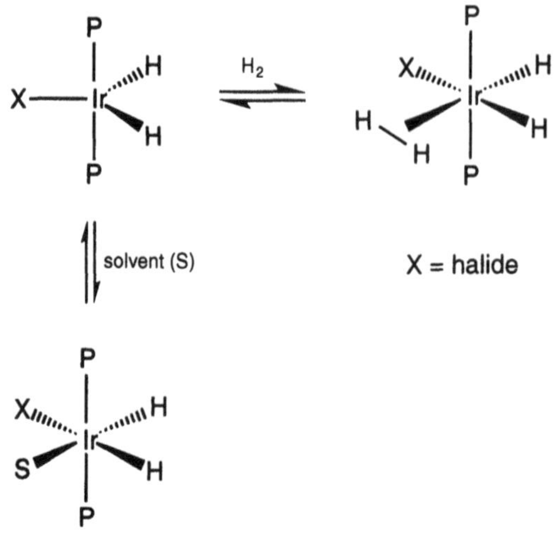

Scheme 1

Reductive elimination of H_2 from stable cis dihydride complexes can be photochemically promoted and undoubtedly $M-H_2$ intermediates are involved.[5] Theoretical studies[6] of gas-phase reactions of $Fe(CO)_5$ with OH^- indicate that H_2

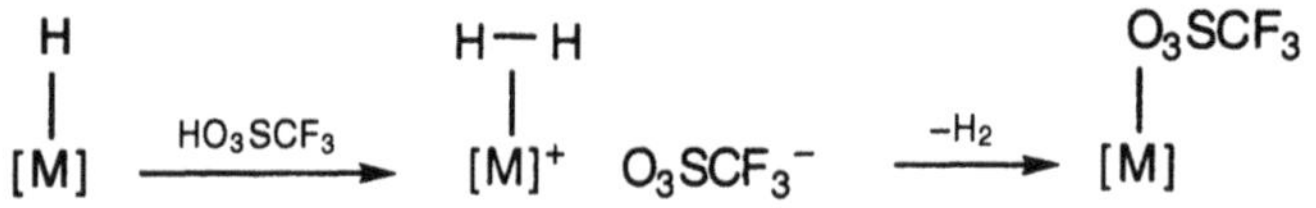

Scheme 2

complexes are probable intermediates in the heterogeneous catalyzed water/gas shift reaction, a key process in the chemical industry. Homogeneous catalysis by carbonyls such as $Fe(CO)_5$ takes place under mild conditions, and a key step is facile elimination of H_2 gas from $FeH_2(CO)_4$, a borderline H_2 complex (see Section 4.7.2).

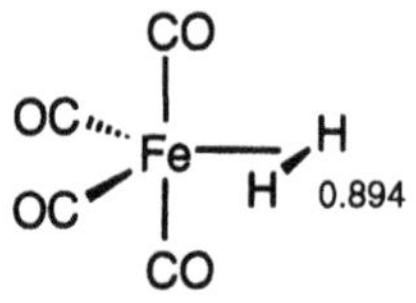

net reaction:

$CO + H_2O \rightarrow CO_2 + H_2$

Computations show that an H_2 complex lies along the reaction coordinate, and $W(H_2)(CO)_5$ may also be an intermediate in $W(CO)_6$-catalyzed reactions.[7] Examples of the significance of σ complexes in industrial and biological processes will be discussed below as well as in Chapter 10. Theoretical studies of a variety of hydrogenations and hydroborations have been reviewed by Frenking, and σ-complex intermediates are often identified along the reaction coordinate.[6]

The solid state reactivity of H_2 complexes has not been extensively studied, but significant differences occur for $RuH_2(H_2)_2(PCy_3)_2$, which is stable to 70°C, whereas

H_2 dissociates readily in solution.[8] Reaction with CO gives a mixture of species but not $RuH_2(CO)_2(PCy_3)_2$, which forms in solution reaction. Ethylene does not react, apparently because of steric hindrance, but does react in solution (see below).

H_2 coordination is important in isotopic exchange reactions and catalytic hydrogenation reactions, including enzymatic processes (discussed in Chapter 10 along with related activation of H_2 on sulfur ligands). *Coordination of H_2 not only tremendously raises its acidity but an H_2 complex can transfer a proton to a base more rapidly than a related dihydride complex.* Evidence for direct reactivity of η^2-H_2 in homogeneous catalytic hydrogenation is mounting, and the first direct evidence for the existence of dihydride complexes under catalytic hydrogenation conditions has been proven by Bargon and coworkers by para-H_2 enhanced ^1H NMR spectroscopy.[9] As always, the principles learned from H_2 coordination and activation apply to alkane and other σ ligands, and σ complexes are vital in the activation of C–H bonds (see Chapter 12).

9.2. HOMOLYTIC SPLITTING OF COORDINATED DIHYDROGEN

The H–H bond of an H_2 ligand can cleave in a homolytic fashion via two pathways: intramolecular [Eq. (9.1)] and intermolecular [Eq.(9.2)]. There are

$$L_nM-\begin{matrix}H\\|\\H\end{matrix} \rightleftharpoons L_nM\begin{matrix}H\\ \diagup \\ \diagdown \\ H\end{matrix} \qquad (9.1)$$

$$L_nM-\begin{matrix}H\\|\\H\end{matrix} + L_nM \rightleftharpoons 2L_nM-H \qquad (9.2)$$

very few cases of Eq. (9.2), mainly those involving metalloradicals such as in Eq. (7.24) and metalloporphyrin[10] systems. However, there are several mechanisms of homogeneous catalysis that involve reaction of H_2 with two metal centers to give two monohydrides. Formally Eq. (9.1) is a two-electron OA reaction where the oxidation state of the metal increases by two and is by far the most common reaction of H_2. Equation (9.1) is a key step in the mechanism of many hydrogenation catalysts such as Wilkinson's $RhCl(PPh_3)_3$, where presumably an unobserved Rh–H_2 intermediate rapidly gives the dihydride intermediate that has always been assumed to hydrogenate unsaturated substrates.

However, a crucial question arose when H_2 complexes were first discovered: "Does coordinated H_2 directly transfer H atoms to unsaturated coligands such as olefins without prior OA of the H–H bond, as in Eq. (9.3)?" Ironically this is still

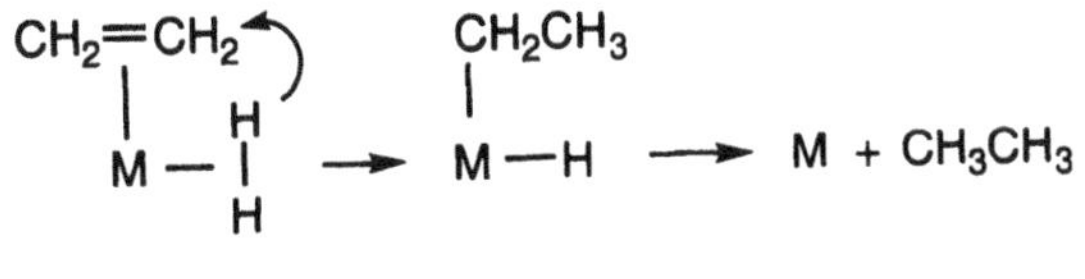

$$\begin{matrix} CH_2=CH_2 \\ | \quad H \\ M-| \\ \quad H \end{matrix} \longrightarrow \begin{matrix} CH_2CH_3 \\ | \\ M-H \end{matrix} \longrightarrow M + CH_3CH_3 \qquad (9.3)$$

difficult to prove conclusively, but evidence has been found, including catalytic systems (see Section 9.5.1). There is no question that η^2-H_2 transfers hydrogen in σ-bond metathesis and heterolytic cleavage processes, so that direct transfer in catalytic hydrogenation is quite reasonable. Equation (9.1) directly parallels equilibria that exist between other σ complexes of, e.g., alkanes and silanes, and their respective OA products. The electronic influences and theoretical basis for H–H activation and cleavage are extensively discussed in Chapter 4, and Figure 9.1 in Section 9.4.1 summarizes both the homolytic and heterolytic pathways and shows sample complexes.

It is remarkable that Eq. (9.1) is observed as a *tautomeric equilibrium* in solution in about a dozen types of H_2 complexes, including the first H_2 complex, $W(CO)_3(PR_3)_2(H_2)$ (1).[11] Structure 2 exists in about 20% concentration in Eq. (9.4)

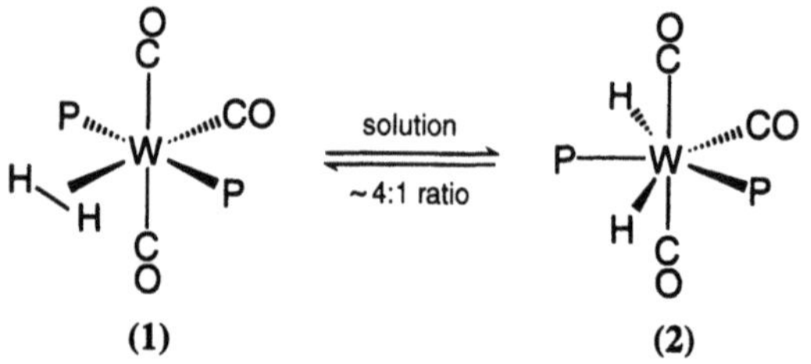

$$\text{(1)} \quad \rightleftharpoons \quad \text{(2)} \quad \quad (9.4)$$

for $P = P^iPr_3$ and is not isolable but is seen by NMR, which shows separate signals for the H_2 and hydride ligands (see Chapter 5). ^{31}P NMR and ^1H NMR are consistent with the distal hydride geometry for **2** and give thermodynamic parameters for the equilibria, summarized in Table 7.9 for all systems showing tautomeric equilibria. $\Delta G°$ is typically less than 1 kcal/mol positive or negative, i.e., either the H_2 or the dihydride tautomer is predominant. The crystal structure of the dihydride isomer (**4**) in a related cationic Re system (Eq. 9.5)] also shows the

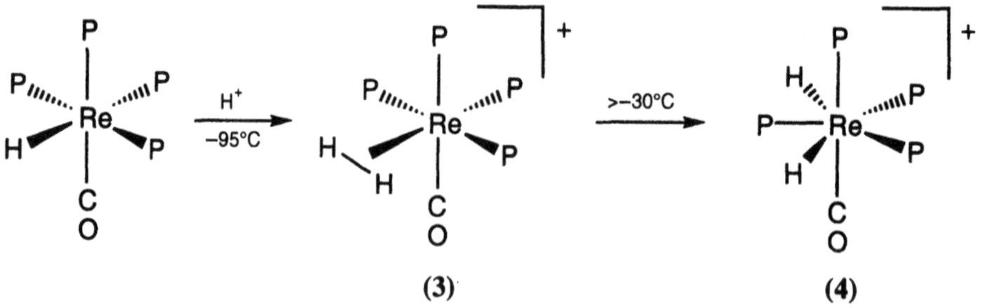

$$(9.5)$$

characteristic pentagonal bipyramidal geometry with distal hydrides observed in **2**. However, the transformation of **3** (kinetic product of protonation) to the much more thermodynamically stable **4** is an *irreversible* rather than an equilibrium process.[12]

$[ReH_4(CO)(PMe_2Ph)_3]^+$ is the only example of a pentagonal bipyramidal dihydrogen–dihydride complex (**5**) in equilibrium with a dodecahedral tetrahydride complex (**6**).[13] In this case **5** is at a lower concentration than **6**, but [**5**] increases at the expense of [6] as the temperature increases (negative ΔS as found for most such equilibria; see Chapter 7).

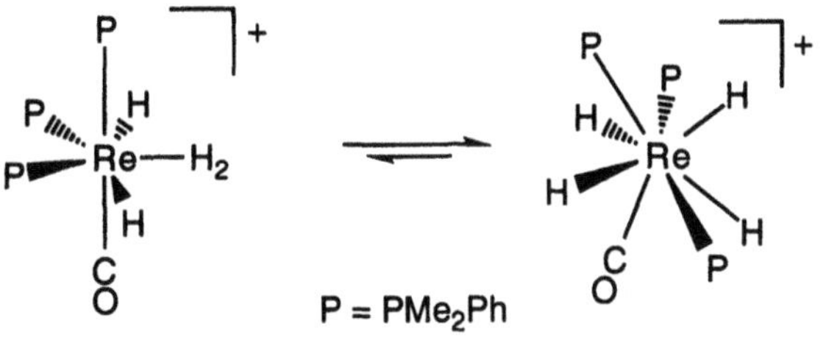

$$\text{(5)} \quad \rightleftharpoons \quad \text{(6)} \qquad (9.6)$$

P = PMe$_2$Ph

Solution IR of CO-containing complexes such as 1 also indicate the simultaneous existence of H$_2$ and dihydride forms by showing distinct ν_{CO} absorptions for each tautomer [Eq. (9.4)],[11] even for unstable photogenerated CpNb(CO)$_3$(H$_2$) and CpNb(H)$_2$(CO)$_3$ in liquid Xe, where NMR studies would be difficult.[14] The H$_2$ complex is thought to have a square-based–piano-stool geometry but the geometry of the dihydride is unknown. More recently a similar cationic system [Cp*Os(CO)$_2$(H$_2$)]$^+$ (7) prepared by protonation of the corresponding monohydride showed equilibrium exclusively with the trans dihydride isomer (8) (87% at $-80°C$).[15]

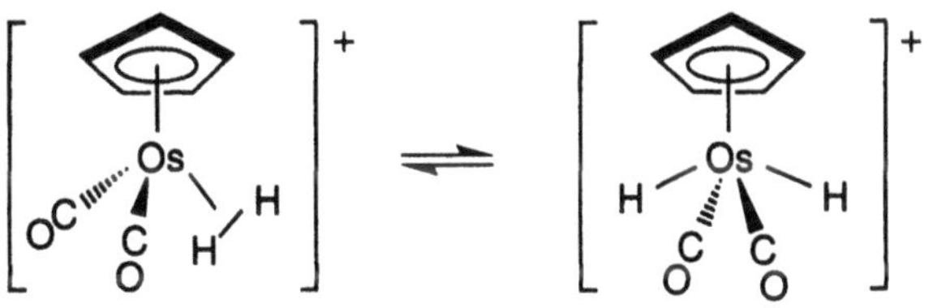

(7)   (8)

The factors that stabilize H$_2$ versus dihydride tautomers (see Chapter 4) relate to other OA reactions, e.g., cleavage of C–H and other σ bonds. If the MH$_2$ tautomer has strong M–H bonds, Eq. (9.1) will be pushed to the right, and M–H bond strength will generally increase down a column. A perfect example is group 5 CpM(CO)$_3$(H$_2$), which is an H$_2$ complex for V and a dihydride for Ta, while CpNb(H$_2$)(CO)$_3$ and CpNbH$_2$(CO)$_3$ coexist.[13] This trend is based on relativistic contraction of orbitals, which improves overlap of the vacant metal d orbital with the electrons in H$_2$. Because 5d metals also give the most stable H$_2$ complexes with respect to H$_2$ loss, Eq. (9.1) most often occurs in 5d M–H$_2$ complexes (W, Re, Os). The nature of the ancillary ligands can also shift Eq. (9.1) right or left, and the stereochemistry of the ancillary ligands can stabilize the dihydride form and shift Eq. (9.1) to the right. This might then give tautomeric equilibria not expected on the basis of electronic criteria such as electrochemical potentials

$$E_{1/2}\{M(N_2)L_n^+/M(N_2)L_n\}$$

for the N$_2$ analogues (see Section 4.7.5).

In summary, both neutral and cationic H_2 complexes with a relatively broad range of J_{HD} (21–34 Hz), d_{HH}, and thermal stability can have dihydride tautomers remarkably close in energy. Why do some complexes show this equilibria and others not, particularly those with stretched H–H bonds and thus highly activated H_2 ligands? Factors other than simple electronics, such as stereochemical requirements of the ancillary ligands, must play a important role, e.g., bulky ligands favor the dihydride form in complexes $[CpRuH_2L_2]^+$. Several complexes of Ru and Os with low J_{HD} of 10–21 Hz listed in Morris's review may have rapid equilibria that are difficult to identify and distinguish from elongated H···H bonds.[1] It is interesting to note that there are no examples of tautomeric equilibria for *first-row* transition metal complexes because seven-coordinate complexes are disfavored on a small 3*d* M, and M–H bonds in electrophilic 3*d* M systems are weak.

Elimination of H_2 from dihydrides is the microscopic reverse of H_2 splitting [Eq. (9.1)], and proceeds via a M–H_2 intermediate. The ease of H_2 loss is highly variable, and although $MoH_2(CO)(R_2PC_2H_4PR_2)_2$ (R = Et, ^iBu) (**9**) contains hydrides well separated by phosphines, it slowly dissociates H_2 *in vacuo* via **10** to form agostic **11**.[16] Well-known unsaturated systems that form a cis dihydride rather

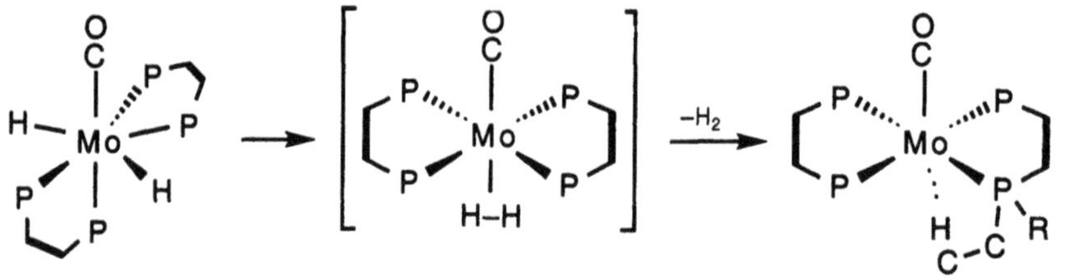

than an H_2 complex such as Vaska's complex can also show facile reversible addition and loss of H_2. On the other hand, a UV photon is usually required to accomplish this in very stable hydrides such as $CpIrH_2L$, $FeH_2(dmpe)_2$, and $MoH_4(dppe)_2$ that lie far to the right in Eq. (9.1) where $\Delta H > 25$ kcal/mol.[17] The photon excites an electron from a molecular orbital that is H–H antibonding (dihydride) to a MO that is H–H bonding and M–H antibonding (η^2-H_2 transient). In some cases unstable highly reactive unsaturated species such as CpIrL are generated, which can insert into the C–H bonds of unreactive alkanes including methane.[18]

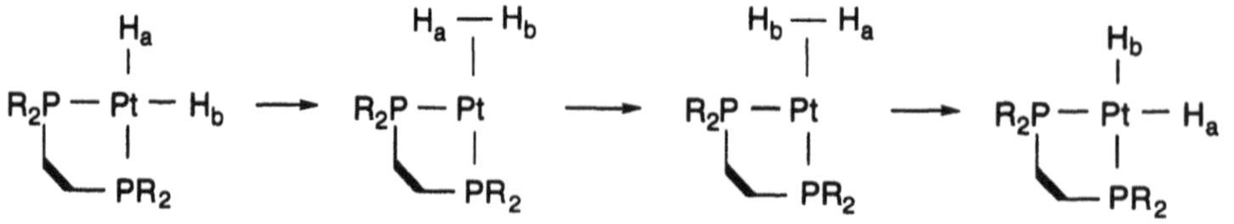

In many cases an unobserved H_2 complex is an intermediate in dihydrogen–dihydride interconversions where only the reactant or the product is observed as in Eq. (9.8) (R = Cy).[19] The scrambling of H atoms in $[M(H_2)H(L_2)_2]^+$, L_2 = dppe, depe, is postulated to occur via splitting of η^2-H_2 to an unobserved trihydride.[20] The activation parameters for M = Fe, Os ($\Delta H^\ddagger$ = 9–12 kcal/mol and $\Delta S^\ddagger$ = −13 to −2 e.u.) are similar to those for tautomeric H_2/dihydride equilibria. The trend that depe causes faster exchange than dppe for a given M is rationalized by the higher basicity of depe, which would favor formation of the trihydride intermediate.

Several well-known examples of complexes, particularly those of cobalt, split H_2 gas in an intermolecular homolytic fashion to give monohydrides[21]:

$$Co_2(CO)_8 + 0.5H_2 \rightarrow CoH(CO)_4 \tag{9.9}$$

$$Co(CN)_5^{3-} + 0.5H_2 \rightarrow CoH(CN)_5^{3-} \tag{9.10}$$

Equation (9.9) is crucial in catalytic hydroformylation of olefins, and Orchin and Rupilius were perhaps the first authors to properly illustrate the side-on bonded H_2 ligand in analogy to olefin binding and propose M → H_2 backbonding as an important component to H_2 activation.[22] η^2-H_2 intermediates have not been detected in such reactions, but calculations show that a *σ-bond metathesis* pathway is possible (see Section 4.12). An η^1 bridging H_2 is proposed in the transition state for addition of H_2 to the d^7 tetramesitylporphyrin complex Rh(tmp) to give d^6 RhH(tmp).[23]

There are a variety of reactions where monohydrides intermolecularly eliminate H_2 as the reverse of Eq. (9.2) in which a M–H_2 intermediate is not seen or is a transition state structure. The best characterized example is elimination of H_2 from d^5 octaethylporphyrin (oep) complexes[24]:

$$2RuH(oep)(thf) \rightarrow intermediate \rightarrow 2Ru(oep)(thf)_2 + H_2 \tag{9.11}$$

The intermediate is proposed to have a bridging H_2. A precedent is a similar diporphyrin complex for which NMR evidence suggests a μ-H_2 ligand (see Chapter 3). As discussed in the following section, these systems split H_2 heterolytically in the presence of strong bases.

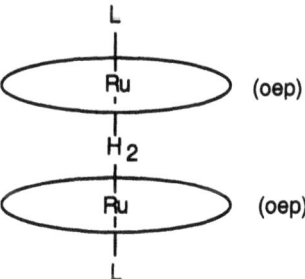

9.3. ISOTOPIC EXCHANGE AND OTHER INTRAMOLECULAR HYDROGEN EXCHANGE REACTIONS

Hydrogen-containing systems readily lend themselves to isotopic substitution or labeling by deuterium and tritium. This is most useful in IR and NMR spectroscopic studies, particularly for determining J_{HD}, which is often critical for the proof of H_2 coordination. Isotopic exchange of H_2 with hydrogen atoms in other molecules such as H_2O may find applications in hydrogen isotope separation chemistry and synthesis of isotopically labeled molecules.

Usually HD or D_2 ligands can be formed by direct addition to unsaturated precursors such as agostic complexes. In most cases a convenient precursor does not exist, and labeling can be done by facile exchange of the H_2 ligand with HD or D_2 gas, possibly combined with intramolecular isotopic scrambling [Eqs. (9.12)–(9.16)], or by adding a source of D^+ to a hydride complex [Eq. (9.17)][1]:

$$M(H_2)(H)L_n \xrightleftharpoons{D_2} M(D_2)(H)L_n \qquad (9.12)$$

$$M(D_2)(H)L_n \xrightleftharpoons{H_2} M(HD)(D)L_n \qquad (9.13)$$

$$M(HD)(D)L_n \xrightleftharpoons[HD]{D_2} M(DD)(D)L_n \qquad (9.14)$$

$$M(H_2)(H)L_n \xrightleftharpoons{HD} M(HD)(H)L_n \qquad (9.15)$$

$$M(HD)(H)L_n \xrightleftharpoons{H_2} M(H_2)(D)L_n \qquad (9.16)$$

$$MHL_n \xrightleftharpoons{D^+} [M(HD)L_n]^+ \qquad (9.17)$$

Intramolecular H/D exchange gives essentially a statistical mixture of isotopomers, but not always exactly statistical because deuterium usually prefers to be in the HD or DD site (see Chapter 6). Isotopomers can be detected by solution NMR or by IR in low-temperature matrices. Separate resonances for H_2 and hydride site isotopes are observed in the spectra of complexes when no intramolecular exchange occurs, but in cases where the reaction in Eq. (9.13) is a rapid one, only averaged chemical shifts and J_{HD} are observed (see Chapter 5). In the fast-exchange ^{1}H NMR spectra of isotopomers of nonclassical polyhydrides, a phenomenon called isotopic perturbation of resonance (IPR) occurs. For example, in a partially deuterated $MH(H_2)$ complex each isotopomer (H_3, DH_2, and HD_2) shows a separate hydride resonance for the species provided the M–H and $M(H_2)$ sites have significantly different chemical shifts and there is sizable deuterium fractionation between the sites (see Section 5.4).

Qualitative studies show that *intra*molecular exchange between isotopomers of group 8 *trans*-$[M(H_2)HL_4]^+$ is much faster than the *inter*molecular exchange of D_2 with $M(H_2)$ despite the trans disposition of the exchanging ligands. Lifetimes of $[MHD_2(depe)_2]^+$ undergoing exchange at 300 K according to Eq. (9.13) are approximately 0.1 s for Ru and 0.005 s for Fe and Os. The $t_{1/2}$ for replacement of the bound H atoms with D by reaction with D_2 gas according to Eqs. (9.12)–(9.16) is less than 5 min for Ru, 2 h for Fe, and 180 h for Os.[20] Conversely, some complexes with trans hydrogen ligands do not show any intramolecular H/D exchange, e.g., trans-$[Fe(HD)(H)(meso\text{-}tetraphos)]^+$, where tetraphos locks the octahedron into a rigid configuration.[25]

As shown above, H$_2$ complexes containing hydride ligands, M(H$_2$)H$_x$L$_n$, are usually effective catalysts for H$_2$/HD/D$_2$ scrambling, but several *coordinatively saturated H$_2$ complexes with no hydrides also catalyze exchange*. Reaction of unstable Cr(D$_2$)$_2$(CO)$_4$ in liquid Xe with H$_2$ in the presence of the more stable Cr(D$_2$)(CO)$_5$ gives D$_2$/HD/H$_2$ exchange as seen by formation of Cr(HD)(CO)$_5$.[26] Conversely, Cr(HD)$_2$(CO)$_4$ under HD gives η^2-H$_2$ and D$_2$:

$$Cr(HD)_2(CO)_4 \rightleftarrows Cr(H_2)(D_2)(CO)_4 \qquad (9.18)$$

However, Cr(H$_2$)(CO)$_5$ reacts with D$_2$ under pressure to give Cr(D$_2$)(CO)$_5$ and no Cr(HD)(CO)$_5$, apparently signifying that two coordinated molecules XY (X, Y = H, D) are required for H/D exchange. Calculations support several possible mechanisms, although transition states with closed H$_n$ polyhedra (triangle, square, or tetrahedron) or the Cr^{IV} tetrahydride Cr(H)$_4$(CO)$_4$ are too high in energy for this reaction.[27,28] More likely structures for the intermediate(s) of Eq. (9.18) are isotopomers of Cr(H$_2$)(H)$_2$(CO)$_4$ and/or Cr(HHHH)(CO)$_4$ with an open H$_4$ ligand (12) or Cr(H$_3$)H(CO)$_4$ with an open H$_3$ ligand (13).

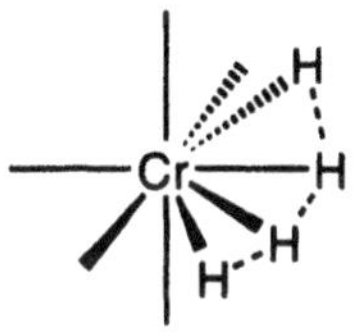

(12) 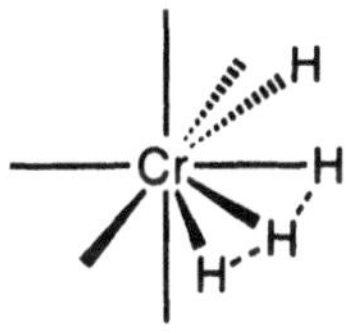   (13)

While the above exchange has several reasonable pathways, scrambling of D$_2$ with W(CO)$_3$(PR$_3$)$_2$(H$_2$) and a few other 18e complexes as in Eq. (9.19) is more enigmatic[11,16,29-31]:

$$D_2 + W(H_2)(CO)_3L_2 \rightleftarrows HD + W(HD)(CO)_3L_2 \rightleftarrows H_2 + W(D_2)(CO)_3L_2 \qquad (9.19)$$

Equimolar amounts of D$_2$ gas (1 atm) and the H$_2$ complexes give complete isotope equilibration *even in the solid state* within days for group 6 species or 12 h for [Re(CO)$_3$(PR$_3$)$_2$(H$_2$)]$^+$ in solution. Prior loss of CO or phosphine to allow D$_2$ into the coordination sphere followed by isotopic exchange as in Eq. (9.14) seems unlikely because ligand loss would be a high-energy process, especially in the solid. Scrambling involving ligand or solvent hydrogens has also been disproven, although intramolecular proton transfer from an H$_2$ ligand to an amido ligand is a plausible mechanism for a triamidoamine Re system.[31] Other possible mechanisms could involve seven- or eight-coordinate 20e intermediates such as a (H$_2$)(D$_2$) complex or a dihydride-dideuterium complex, WH$_2$(D$_2$)(CO)$_3$(PR$_3$)$_2$. The latter might lead to a transition state with coordinated H$_2$D as in 13, even in the solid. Although Cr(H$_2$)(CO)$_5$ does not exchange with D$_2$ to give HD, apparently because the dihydride is too high in energy,[28] the dihydride/dihydrogen tautomers for W(CO)$_3$(PR$_3$)$_2$(H$_2$) lie close in energy. Also, the latter is believed to undergo associative substitution reactions via a 20e intermediate.[32] However, no evidence exists for either the dihydride form in the solid state or seven- or eight-coordinate complexes of the type discussed here.

Trace quantities of adventitious water may lead to exchange since isotopic scrambling of $W(CO)_3(P^iPr_3)_2(D_2)$ with H_2O occurs in solution within days[33] or less for other complexes.[34-37] A reasonable mechanism for exchange for complexes with one open coordination site is deprotonation of η^2-H_2 by the weak base water followed by reprotonation with H_2DO^+. As will be discussed in Section 9.4, η^2-H_2

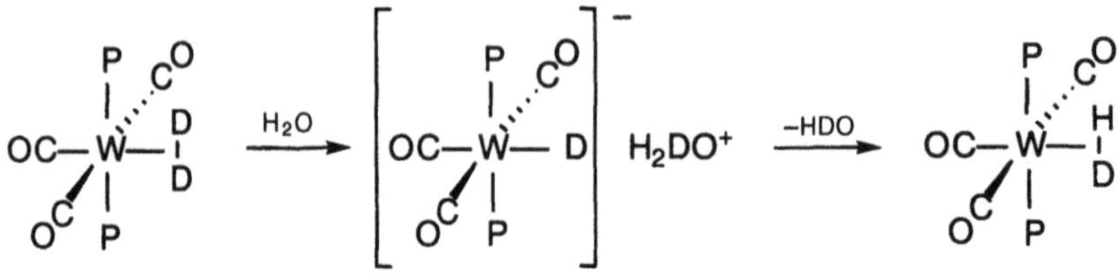

$$\text{OC--W--D} \xrightarrow{H_2O} [\text{OC--W--D}]^- H_2DO^+ \xrightarrow{-HDO} \text{OC--W--H} \quad (9.20)$$

can be quite acidic. For cationic complexes such as $[Os(H_2)(CH_3CN)(dppe)_2][BF_4]_2$ formed by protonation, reversible deprotonation by the anion to form equilibrium amounts of free HBF_4/DBF_4 is proposed to facilitate exchange [Eq. (9.21)].[30]

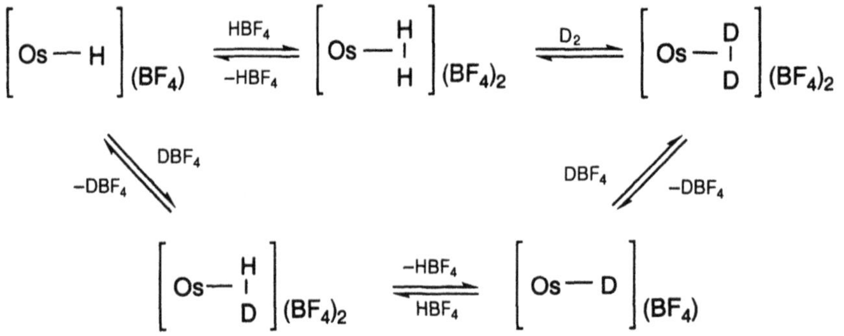

$$[Os-H](BF_4) \underset{-HBF_4}{\overset{HBF_4}{\rightleftharpoons}} \left[Os\begin{matrix}H\\|\\H\end{matrix}\right](BF_4)_2 \underset{}{\overset{D_2}{\rightleftharpoons}} \left[Os\begin{matrix}D\\|\\D\end{matrix}\right](BF_4)_2$$

$$\underset{-DBF_4}{\overset{DBF_4}{\searrow}} \qquad\qquad\qquad \underset{-DBF_4}{\overset{DBF_4}{\nearrow}} \qquad (9.21)$$

$$\left[Os\begin{matrix}H\\|\\D\end{matrix}\right](BF_4)_2 \underset{HBF_4}{\overset{-HBF_4}{\rightleftharpoons}} [Os-D](BF_4)$$

The isotopic exchange in CD_2Cl_2 is slow (days) as for the W system, and the deutero solvent does not become involved (see below). However in Eq. (9.20), much stronger bases than H_2O, such as alkoxides,[38] are required to deprotonate the W complex. Also the rate of H_2/D_2 exchange is much faster than H_2O/D_2 exchange, which is unlikely to occur as above in the solid state and is not seen for solid $W(CO)_3(P^iPr_3)_2(D_2)$ plus H_2O. This pathway could operate in solution for systems with more acidic η^2-H_2, but another explanation is needed for scrambling in group 6 complexes.

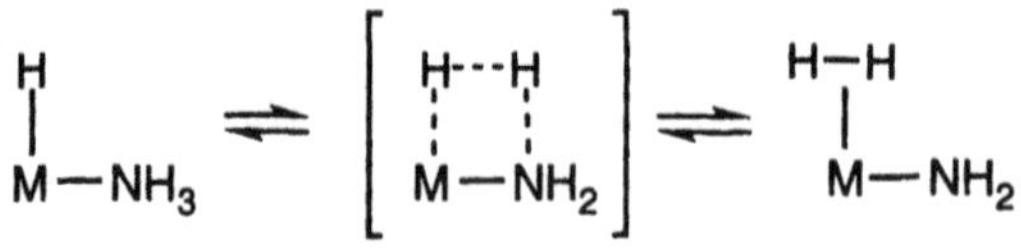

$$\begin{matrix}H\\|\\M-NH_3\end{matrix} \rightleftharpoons \left[\begin{matrix}H\cdots H\\ \vdots\\ M-NH_2\end{matrix}\right] \rightleftharpoons \begin{matrix}H-H\\|\\M-NH_2\end{matrix} \quad (9.22)$$

Hydrogen exchange between NH and CH bonds in neighboring ligands can occur via novel equilibria such as that in Eq. (9.22), which resembles σ-bond metathesis. An Ir complex in THF slowly exchanges the deuterium in both Ir–D and N–D bonds with protons from a cyclooctene ligand, which becomes deuterated.[39]

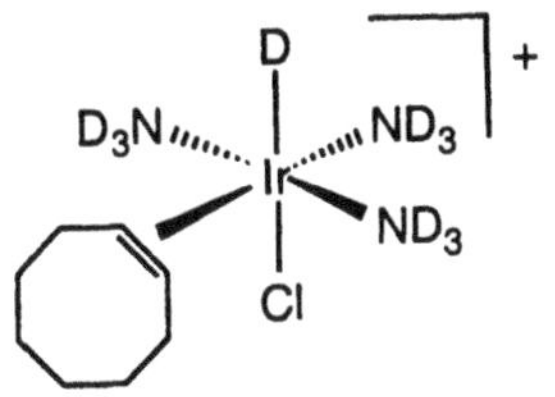

Another remarkable H/D exchange reaction that occurs even in the solid state involves intramolecular incorporation of deuterium into the PMe$_3$ ligands of [Fe(H$_2$)H(PMe$_3$)$_4$]$^+$, which may involve agostic interactions in which associative (σ-bond metathesis) or dissociative C–H bond activation occurs.[40]

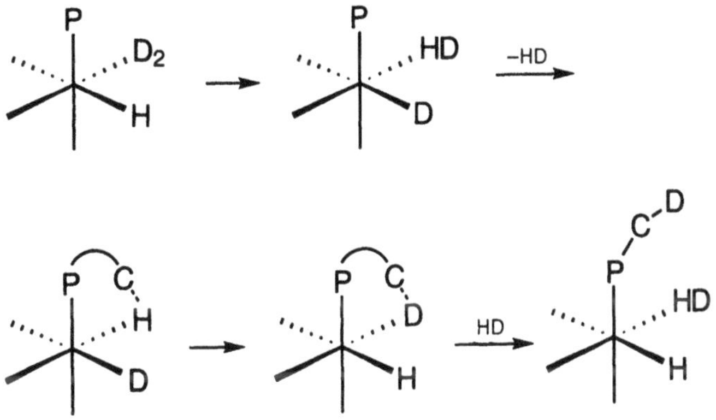

In solution, isotopic incorporation of deuterium from deuterated solvents into metal-bound hydrogen is common, e.g., reaction of acetone-d_6 and [RuCl(dppe)$_2$(H$_2$)]$^+$ or [OsH(H$_2$)(PP$_3$)]$^+$ gives the HD isotopomer in 20 min and the fully deuterated complexes in a few hours.[35,41] No reaction occurs between the Os complex and acetone even at reflux temperature; thus the isotopic exchange with acetone-d_6 does not proceed by H$_2$ dissociation at any stage.[35] The pK_a of [OsH(H$_2$)(PP$_3$)]$^+$ is estimated to be 12–15 (see Section 9.4.2), which represents a higher acidity than acetone (pK_a = 20). The isotopic exchange may therefore occur via a deprotonation/reprotonation pathway coupled with a keto-enol tautomerization.

Complexes with both hydride and H$_2$ ligands such as [Ir(H$_2$)H(bq)(PPh$_3$)$_2$]$^+$ and Ir$_2$H$_3$(μ-H)(H$_2$)(μ-Pz)$_2$(P^{i}Pr$_3$)$_2$ or unsaturated hydrides such as IrClH$_2$(P^{i}Pr$_3$)$_2$ are advantageous for isotopic exchange because of facile exchange of the bound H$_2$ with D$_2$ and substrates with exchangeable protons combined with low barriers to intramolecular exchange with the hydride ligand. The latter two complexes undergo H/D scrambling with toluene-d_8, which must bind to Ir by adding as a sixth ligand or displacing H$_2$.[42] For example, the cationic Ir species is an excellent catalyst for

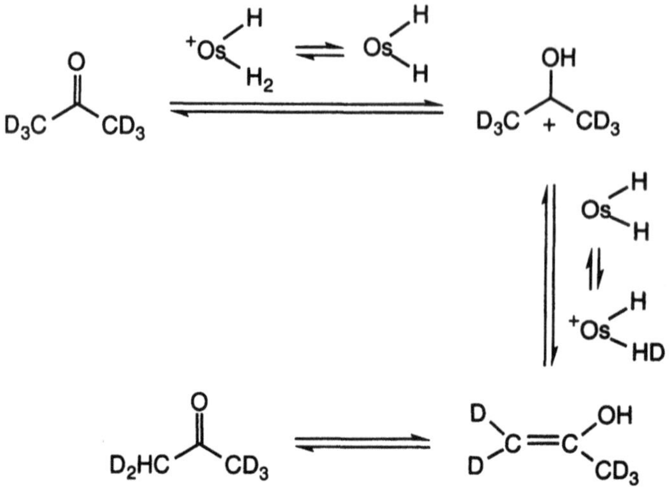

$$D_3C-\underset{\underset{CD_3}{\|}}{C}=O \xrightarrow{^+Os\overset{H}{\underset{H_2}{\diagdown}} \rightleftharpoons Os\overset{H}{\underset{H}{\diagdown}}\overset{H}{\diagdown}} D_3C-\underset{\underset{CD_3}{|}}{\overset{OH}{\overset{|}{C}}}$$

$$\Updownarrow Os\overset{H}{\underset{H}{\diagdown}} \qquad \Updownarrow ^+Os\overset{H}{\underset{HD}{\diagdown}} \qquad (9.23)$$

$$D_2HC-\underset{\underset{CD_3}{\|}}{\overset{O}{C}} \rightleftharpoons \underset{D}{\overset{D}{\diagdown}}C=C\underset{CD_3}{\overset{OH}{\diagup}}$$

deuterium incorporation into alcohols[34]:

$$ROH + D_2 \underset{R=Me, Et, {}^tBu}{\overset{[Ir(H_2)H(bq)(PPh_3)_2]^+}{\rightleftharpoons}} ROD + HD \qquad (9.24)$$

In addition to the deprotonation mechanism in Eq. (9.20), a mechanism involving exchange with the cis hydride is likely here.

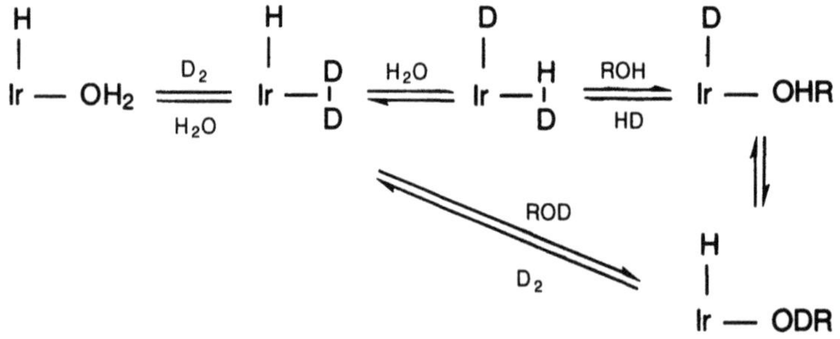

$$\underset{\text{Ir}-OH_2}{\overset{H}{\overset{|}{}}} \underset{H_2O}{\overset{D_2}{\rightleftharpoons}} \underset{\text{Ir}-D}{\overset{H}{\overset{|}{\underset{D}{|}}}} \overset{H_2O}{\rightleftharpoons} \underset{\text{Ir}-D}{\overset{D}{\overset{|}{\underset{D}{\overset{H}{|}}}}} \overset{ROH}{\underset{HD}{\rightleftharpoons}} \underset{\text{Ir}-OHR}{\overset{D}{\overset{|}{}}}$$

$$\underset{D_2}{\overset{ROD}{\rightleftharpoons}} \qquad \Updownarrow \qquad (9.25)$$

$$\underset{\text{Ir}-ODR}{\overset{H}{\overset{|}{}}}$$

9.4. HETEROLYTIC CLEAVAGE AND ACIDITY OF COORDINATED DIHYDROGEN

9.4.1. Introduction

One of the oldest, most significant and widespread reactions of coordinated H_2, heterolytic cleavage, is also the simplest and involves breaking the H–H bond into H^+ and H^- fragments (Figure 9.1). There are two ways this can take place on H_2 complexes generated either by addition of H_2 gas to unsaturated precursors or by protonation of a M–H bond [Eq. (9.26)]. The η^2-H_2 ligand becomes acidic, i.e., polarized toward $H^{\delta-}$–$H^{\delta+}$, where the highly mobile H^+ is ready to transfer.

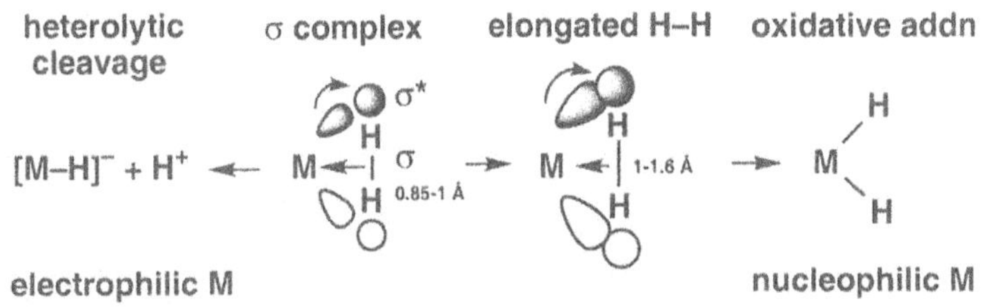

Reactions and Acidity of Dihydrogen Complexes

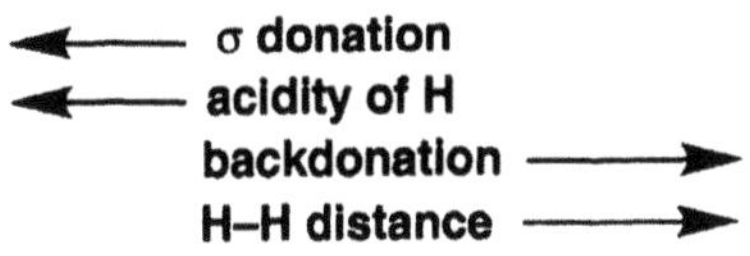

Figure 9.1. Summary of homolytic and heterolytic pathways for H_2 activation on transition metal centers, showing examples and reactivity and structural trends correlated with $H_2 \rightarrow M$ σ donation versus $M \rightarrow H_2$ backdonation.

Intermolecular heterolytic cleavage involves proton transfer to an external base B to give a metal hydride (H^- fragment) and the conjugate acid of the base, HB^+. This

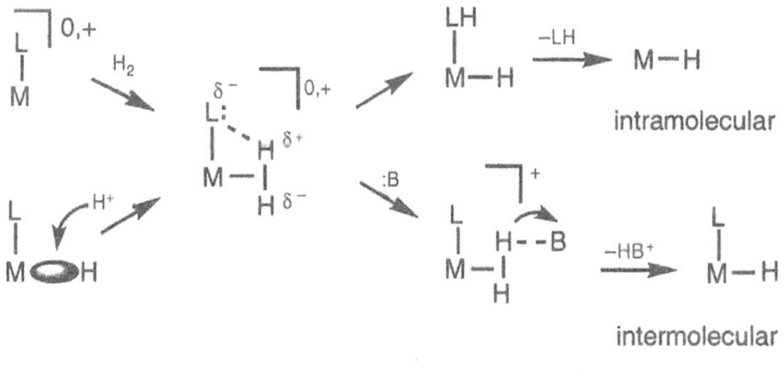

$$\text{(9.26)}$$

is essentially the reverse of the protonation reaction used to synthesize H_2 complexes, but also can occur for H_2 complexes that are generated from H_2, which is the most useful reaction because elemental H_2 can be turned into a strong acid on binding to electrophilic cationic complexes. *Intramolecular* heterolytic cleavage involves protonation of a cis ligand [Eq. (9.26)] or the counteranion of a cationic complex by an H_2 ligand:

$$[ML_x][A] + H_2 \rightleftarrows [M(H_2)L_x][A] \rightleftarrows M(H)L_x + HA \quad (9.27)$$

These reactions are also extremely important in H$_2$ activation and catalysis as will be shown.

Free H$_2$ is an extremely weak acid with a pK_a >49 in THF,[43] and heterolytic splitting of η^2-H$_2$ in relatively electron-rich neutral complexes is usually achieved only by strong bases. For example, copper alkoxides deprotonate W(CO)$_3$(PR$_3$)$_2$(H$_2$) and FeH$_2$(H$_2$)(PR$_3$)$_2$ to give heterobimetallic species with bridging hydrides.[38]

$$\text{FeH}_2(\text{H}_2)(\text{PR}_3)_2 \xrightarrow{\text{Cu(O}^t\text{Bu)(P')}} \text{bridging hydride complex} + {^t}\text{BuOH} \quad (9.28)$$

However, when H$_2$ is bound to a highly electrophilic cationic metal center, *the acidity of H$_2$ gas can be increased spectacularly, up to 40 orders of magnitude.* The pK_a of H$_2$ can become as low as −6 and thus *the acidity of η^2-H$_2$ becomes as strong as that in sulfuric or triflic acid.*

Crabtree first demonstrated heterolytic cleavage of η^2-H$_2$ by isotopic labeling studies to show that the H$_2$ in [IrH(H$_2$)(bq)(PPh$_3$)$_2$]$^+$ is deprotonated by LiR in preference to the hydride ligand.[44] A milder base, NEt$_3$, was shown by Chinn and Heinekey[45] to specifically deprotonate the η^2-H$_2$ tautomer in the equilibrium mixture (84:14 ratio of η^2-H$_2$ to dihydride form in Eq. (9.29):

$$[\text{CpRuH}_2(\text{dmpe})]^+ \rightleftarrows [\text{CpRu(H}_2)(\text{dmpe})]^+ + \text{NEt}_3 \rightleftarrows \text{CpRuH(dmpe)} + [\text{NEt}_3\text{H}]^+ \quad (9.29)$$

This indicates a pK_a of 17.6 in CH$_3$CN, and, more importantly, other NMR evidence (spin saturation transfer) shows that the H$_2$ tautomer is deprotonated more rapidly than the dihydride form, which shows a *greater kinetic acidity of the H$_2$ ligand* (the dihydride is actually a slightly stronger acid (pK_a of 16.8). A significant sidelight on this observation is that from the principle of microscopic reversibility, the reverse reaction, protonation of CpRuH(dmpe), must give only the H$_2$ complex as the initial kinetic product. Many H$_2$ complexes are synthesized by this type of reaction in which the M–H bond is apparently directly protonated, and any equilibration of dihydride and H$_2$ complexes must result from isomerization. It has long been argued that protonating a hydride ligand should be faster than protonating the metal because this involves less geometric and electronic rearrangement.[46] Qualititative evidence for this is provided by fast exchange between CpW(CO)$_2$(PMe$_3$)H, **14** (which exists in solution as cis and trans isomers) and triflic acid and by selective broadening of the hydride resonance during exchange with anilinium.[47] Isotopic exchange of **14** with a deuterated acid is a convenient measure of the rate of proton exchange. An unusually large equilibrium isotope effect (EIE) of 0.19 is observed, and a value of 0.18 is calculated by including the low-energy W–H bending modes (see Section 7.5.2). Rate data for isotopic exchange between **14** and N-d_1-4-*tert*-butyl-N,N-dimethylanilinium combined with the rate of deprotonation of **15** by 4-*tert*-butyl-N,N-dimethylaniline and a knowledge of all the pK_a's involved proves that hydride ligand protonation is faster than M protonation. The kinetic site of protonation for $CpW(CO)_2(PMe_3)H$ is H, even though the thermodynamic site of proton transfer is W (direct protonation at metal centers of M–H is also known to

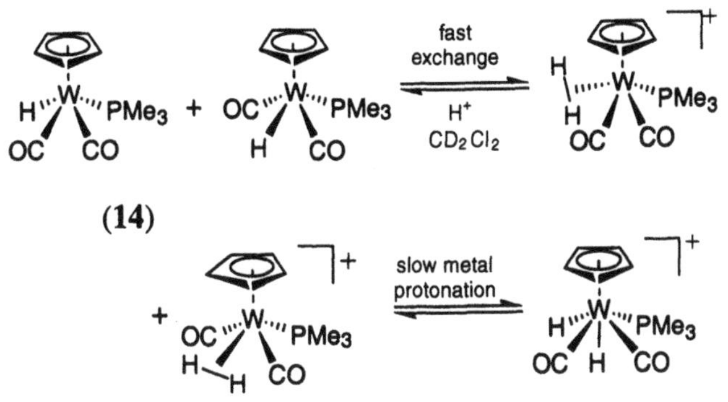

(14)

(15)

occur[48]). These reactions and the ability of acidic H_2 ligands to protonate substrates such as olefins and N_2 are relevant to processes such as ionic hydrogenation (see Section 9.5.3) and also the *structure and function of metalloenzymes* (see Chapter 10).

9.4.2. Thermodynamic and Kinetic Acidity of H_2 Ligands

The pK_a values of H_2 complexes are rarely measured in aqueous solution but are usually discussed on an adjusted pseudoaqueous scale for comparative purposes because pK_a values determined by solution equilibrium measurements are solvent-dependent. As discussed in reviews by Morris[1] and Jia[2] and subsequent work,[43,48,49] pK_a values are usually determined by NMR measurement of the concentrations of species in an equilibrium:

$$M(H_2)L_x + B \underset{}{\overset{K_{eq}}{\rightleftarrows}} [M(H)L_x]^- + BH^+$$

$$pK_a = pK_{eq} + pK_{BH^+}$$

(9.30)

with an external base such as a phosphine or amine. The pK_a of the conjugate acid of the base is known or readily measurable and covers a broad range, e.g., 10 to 20 for alcohols and 0 to 11 for protonated phosphines. A good polar, nonreactive solvent for pK_a measurements is THF and Morris has recently developed a wide-ranging pK_a^{THF} scale for dihydrogen and hydride complexes that can be related to other acidity scales (e.g., pK_a^{DMSO}) for other compounds as well.[43] The pK_a of an H_2 complex, which is a thermodynamic property, is determined from K_{eq} by the relationship in Eq. (9.30). The pK_a's of interconverting H_2 and dihydride tautomers,

$$M(H_2) \overset{K}{\rightleftarrows} M(H)_2$$

(9.31)

are related by

$$pK_a[M(H_2)] = pK_a[M(H)_2] - \log K \qquad (9.32)$$

Thus the predominant tautomer is always the weaker acid, and interestingly, the apparent pK_a of rapidly interconverting tautomers is larger than the pK_a of either tautomer alone.[48]

For a $MH(H_2)$ complex such as $[IrH(H_2)(bq)(PPh_3)_2]^+$, the acidities of the hydride and H_2 ligands are generally the same because deprotonation from either site results in the same thermodynamic product, MH_2. Selective deprotonation of the H_2 ligand over the hydride [Eq. 9.35 below] correlates with greater kinetic rather than thermodynamic acidity of bound H_2. In regard to correlating H–H bond weakening with increased acidity, it must be recalled that dihydrides are not necessarily more acidic than their dihydrogen tautomers. An increase in acidity from H–H bond weakening can be offset by a strengthening of the M–H bond, which lowers acidity. Furthermore, electron-deficient cationic H_2 complexes with strong short H–H bonds (<0.9 Å) and weakly bound H_2 such as $[Cp*Re(H_2)(CO)(NO)]^+$ and $[Re(H_2)(CO)_4(PR_3)]^+$ are among the most acidic complexes (Table 9.1). There is very little *overall* correlation of acidity with the degree of activation of the H–H bond as judged by J_{HD} (hence d_{HH}), although for a particular series of complexes such as $[CpRu(H_2)(diphosphine)]^+$ the pK_a usually decreases as J_{HD} increases (d_{HH} shortens).

Table 9.1. Reported pK_a Values (Pseudoaqueous Scale) and Corresponding J_{HD} of Selected H_2 Complexes, Emphasizing Highly Acidic Species[a]

Complex	pK_a	J_{HD}, Hz	Reference
$[Cp*Re(H_2)(CO)(NO)]^+$	-2	27	50
$[Re(H_2)(CO)_4(PPh_3)]^+$	-2 to 1	33.9	51
$[FeH(H_2)(depe)_2]^+$	~ 16	28	36
$[FeH(H_2)(dppe)_2]^+$	12.1	30	36
$[FeH(H_2)(dtfpe)_2]^+$	7.8	32	36
$RuH_2(H_2)(PPh_3)_3$	36		43
$[CpRu(H_2)(dmpe)]^+$	10.1	22.1	45
$[CpRuH_2(dppe)]^+$	7.5	Hydride	53
$[CpRu(H_2)(dppe)]^+$	7.0	24.9	53
$[CpRu(H_2)(dtfpe)]^+$	4.3	25.3	53
$[CpRu(H_2)(dfepe)]^+$	-5	29.1	68
$[TpRu(H_2)(dppe)]^+$	7.9	32.5	54
$[^{Me}CnRu(H_2)(dppe)]^{2+}$	3.8	29.4	54
$[^{Me}CnRu(H_2)(CO)(PPh_3)]^{2+}$	-2.6	31.0	54
$[OsH(H_2)(dppe)_2]^+$	13.6	25.5	36
$[OsCl(H_2)(dppe)_2]^+$	7.4	13.9	55
$[Os(CH_3CN)(H_2)(en)_2]^{2+}$	17^b	17.7	49
$[Os(CH_3CN)(H_2)(dppe)_2]^{2+}$	-2	21.4	49
$[Os(CO)(H_2)(dppp)_2]^{2+}$	-5.7	32.0	56
$[Os(CO)(H_2)(bpy)(PPh_3)_2]^{2+}$	-2 to -4.9	25.5	57
$[Os(CO)(H_2)(pyS)(PPh_3)_2]^+$	ca -2	21	58

[a] pyS = N,S-2-pyridinethiol anion; ^{Me}Cn = 1,4,7-trimethyl-1,4,7-triazacyclononane.
[b] Estimated from Eq. (9.34); error margin is ± 3.

Neutral complexes are typically not very acidic (pK_a's of 15–20) but positive charge and electron-withdrawing coligands such as CO, particularly when trans to H_2, give much more acidic complexes, e.g., the $[Os(L)(H_2)(dppe)_2]^{n+}$ and $[(Cp/Cn)Ru(H_2)(L)(L')]^{n+}$ series. One of the most acidic complexes, $[Os(CO)(H_2)(dppp)_2]^{2+}$ (estimated pK_a −5.7) contains both a strong π acceptor trans to H_2 and a *dipositive* charge. This unstable complex is formed by reaction of $[OsH(CO)(dppp)_2]^+$ with triflic acid (pK_a − 4.9), which is barely strong enough to protonate the already electron-poor hydride.[56] In these complexes, the M–H_2 bonding interaction is primarily σ donation from H_2 to the very electropositive metal, with little BD. Virtually all greatly acidic complexes ($pK_a < 0$) are either dicationic or else monocationic with strong acceptors trans to H_2. However, not every dicationic species is acidic, e.g., $[Os(CH_3CN)(H_2)(en)_2]^{2+}$ in which the cis amine ligands are stronger electron donors than dppe. The pK_a is estimated to be 17 ± 3 from $E_{1/2}$ data (see below), a difference of 19 units from $[Os(CH_3CN)(H_2)(dppe)_2]^{2+}$ ($pK_a = -2$) on variation of the cis ligands![49] Replacing just *one* cis ligand {PR_3 in $[Re(H_2)(CO)_3(PPh_3)_2]^+$ by CO to give $[Re(H_2)(CO)_4(PPh_3)]^+$} tremendously increases acidity (the tricarbonyl can be deprotonated only by relatively strong bases such as hindered amines). The influence of the cis ligands on pK_a is much greater than that on J_{HD}, which increases only 2–3.7 Hz in these examples, and other properties of H_2 ligands. Most of the highly acidic complexes are found for Re and especially Os, although less stable Ru analogues exist in some cases.[56,57,68]

The electrochemical potential $E_{1/2}$ for d^5/d^6 couples is a good diagnostic for electron density at a d^6 M, where a low $E_{1/2}$ corresponds to an electron-rich M. Based on the work of Tilset and Parker[59] for hydrides, Morris[1,36,52] has related the pK_a of a cationic H_2 complex $[M(H_2)L_5]^+$ to the electrochemical potential of the conjugate base MHL_5 by the thermochemical cycle:

$$[M(H_2)L_5]^+ \rightarrow MHL_5 + H^+ \qquad \Delta G = 2.301RT(pK_a) = 1.37pK_a$$
$$MHL_5 \rightarrow [MHL_5]^+ + e^- \qquad \Delta G = 23.1E_{1/2}(MH^+/MH)$$
$$H^+ + e^- \rightarrow H^- \qquad \Delta G = \text{constant}$$
$$\overline{[M(H_2)L_5]^+ \rightarrow [MHL_5]^+ + H^- \qquad \Delta G_{BDE} = 1.37pK_a + 23.1E_{1/2} + \text{constant}}$$

(9.33)

These terms combine to give the free energy of H–atom abstraction in Eq. (9.33) (BDE = bond dissociation energy), which can be used to calculate pseudoaqueous pK_a values by rearrangement, conversion to enthalpies instead of free energies,[59] and use of an empirical constant:

$$pK_a = 0.730\Delta H_{BDE} - 16.9E_{1/2} - 48 \qquad (9.34)$$

where $E_{1/2}$ is the reduction potential (in volts) for the MH/MH$^-$ (d^5/d^6) pair with reference to the $FeCp_2^+/FeCp_2$ potential. The constant of −48 derives from the ΔH_{BDE} (72 kcal/mol) and pK_a (6.3) of a reference complex, $[CpRuH_2(PPh_3)][BF_4]$. $E_{1/2}$ can be measured or estimated from Lever's additive electrochemical method,[60] but ΔH_{BDE} is difficult to measure or estimate and decreases as d_{HH} increases. Typical values for labile H_2 complexes are about 80 kcal/mol, but could range from 60–85 kcal/mol, which translates into a potential range of 18 units for the predicted pK_a in the absence of any guiding information.

The observed trends in pK_a with L variation in Table 9.1 are consistent with the trend that lower values of $E_{1/2}$, i.e., greater electron density at M, correspond to lower acidity. Thus the auxiliary ligand set plays a major role in determining acidities, as already noted above. For the $[CpRuH_2(PP)]^+$ system for both classical or nonclassical complexes, acidity increases with decreasing basicity of the diphosphines: dmpe < dppp < 2PPh$_3$ < dppe < dtfpe. Substitution of PPh$_3$ by CO decreases pK_a of related classical hydrides by 5–8 units [HCo(CO)$_4$ and HCo(CO)$_3$(PPh$_3$) have pK_a(MeCN) of 8.3 and 15.4, respectively[61]]. A similar effect is seen for H_2 complexes: pK_a drops 8.2 on going from $[TpRu(H_2)(PPh_3)_2]^+$ (7.6) to $[TpRu(H_2)(CO)(PPh_3)]^+$ (−0.6).[54] P-donors such as dppe give much more acidic complexes than corresponding N-donors such as en or bipy. In regard to variations with M, acidity generally decreases down the group for congeneric monohydrides because of increasing M–H bond strength. For H_2 complexes, only one complete series of complexes with reported pK_a values is known, $[MH(H_2)(dtfpe)_2]BF_4$, and the trend in acidity is Fe > Os > Ru.

The relative acidities of H_2 complexes can also be evaluated by their relative deprotonation energies (DP), which can be calculated by DFT methods. Computed values of DP for model diphosphine complexes $[M(H_2)(L)(H_2PC_2H_4PH_2)]^{n+}$ (M = Ru, Os) correlate well with their measured pK_a (Figure 9.2) and support the above general principles.[62] Complexes with neutral L have smaller DP values and therefore higher acidities than those with anionic L. Acidity increases in the

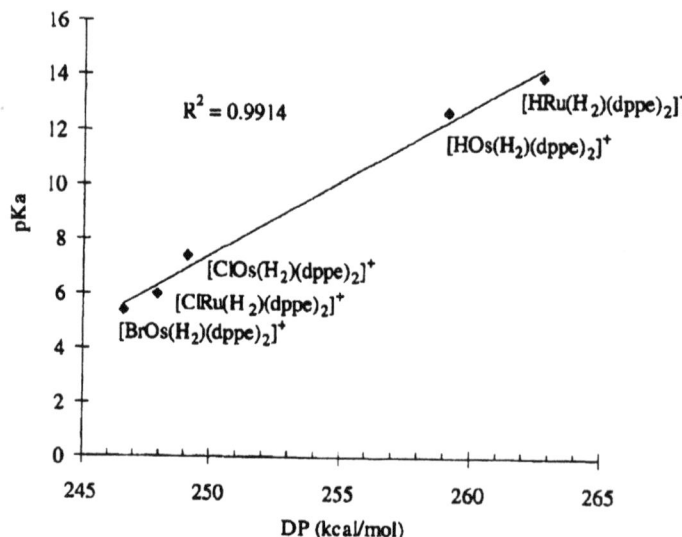

Figure 9.2. Plot of measured pK_a values against calculated DP energies. Values for dicationic complexes are not included because they are much more acidic and pK_a could be measured only approximately. Reprinted with permission from Xu et al.[62] Copyright 1999 American Chemical Society.

following orders, which relate to electron-donating versus electron withdawing properties of L:

Anionic ligands: $H^- < CH_3^- < F^- < CF_3^- < Cl^- < Br^-$

Neutral ligands: $NH_3 < NCH < PH_3 < CO$

The trend with M is generally normal and Os complexes have higher DP (less acidic) than Ru analogues. However when L is anionic and a strong donor, stabilization of the conjugate hydride makes the Ru species less acidic.

The hydride and H_2 ligands in $[IrH(H_2)(bq)(PPh_3)_2]^+$ and other $H(H_2)$ complexes have equal Brønsted acidity as discussed above, yet reactions with partially deuterated isotopomers show that η^2-H_2 is selectively deprotonated [Eq. (9.35)].[20,63] This again demonstrates the greater kinetic acidity of the H_2 ligand

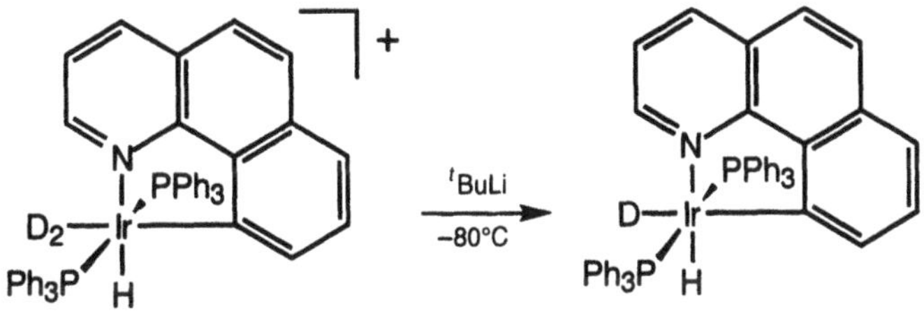

(9.35)

compared to hydride ligands as in the observation that the H_2 tautomer of $[CpRuH_2(diphosphine)]^+$ is more rapidly deprotonated than the more acidic dihydride form.[45,63] The main reason H_2 complexes have greater kinetic acidity than classical hydrides of similar structure is that deprotonation of an H_2 complex involves *no change in coordination number*. Also the η^2-H_2 can become polarized toward $H^{\delta-}$–$H^{\delta+}$, and H^+ is exceedingly mobile, especially for cationic complexes [Eq. (9.26)].

Surprisingly, certain complexes with pK_a much higher than that of H_3O^+ (1.74) such as $[TpRu(H_2)(MeCN)(PPh_3)]^+$ (8.9) are found to be deprotonated by water.[54] A base as weak as H_2O would not have been expected to deprotonate such complexes; displacement of H_2 to form $[TpRu(H_2O)(MeCN)(PPh_3)]^+$ would seem more likely. Solvation effects could play a role here since H_2O strongly solvates H^+. Also the difference in the pK_a of the H_2 complexes in aqueous solution may not actually be as large as the experimental pseudoaqueous values imply.

The higher kinetic acidity of H_2 complexes requires that the reverse reaction, protonation of a metal hydride, occur at H rather than at M, for which there is ample evidence. Actually "protonation at the hydride" is misleading because it is really the M–H *bond* that is protonated to form M–η^2-H_2, as pointed out in a review by Kuhlman that addresses site selectivity of protonation of hydride–halide complexes, MH(X).[64] Formal protonation of a hydride ligand would give M–η^1-H_2, which is not known to be stable. Proton transfer to *halide ligands* is quite rare because an acid with a lower pK_a than the *coordinated* HX produced is necessary. One example is protonation of *trans*-PtHX(P^{t}Bu$_3$)$_2$ with triflic acid, which gives *trans*-[PtH(η^2-H_2)(P^{t}Bu$_3$)$_2$][OTf] for X = H and an unstable species claimed to be trans-[PtH(η^1-ClH)(P^{t}Bu$_3$)$_2$][OTf] for X = Cl.[65] Thus in some cases protonolysis of halide is favored over hydride, particularly for nonoctahedral complexes. Hydrohalic acid ligands, which are not σ ligands because the halide is attached to M via

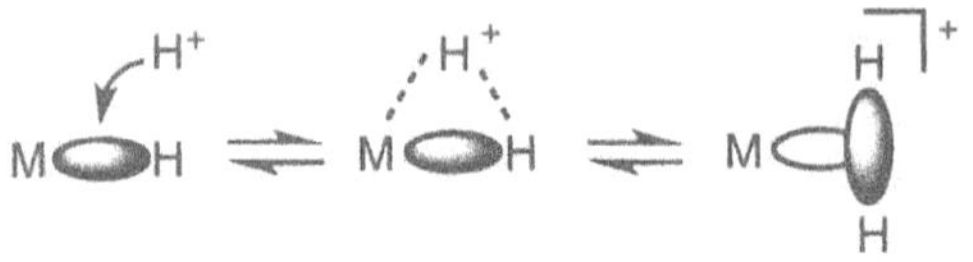

a lone pair, are elusive chemical entities that represent one of the last frontiers in small-molecule coordination.

9.4.3. Intermolecular Heterolytic Cleavage of Coordinated H$_2$

One of the best examples of intermolecular heterolytic cleavage of η^2-H$_2$ is the protonation of ethers by extremely electrophilic cationic H$_2$ complexes containing electron-withdrawing ligands such as CO[50,51]:

$$2\text{M-H} + 2\text{H}^+ \longrightarrow 2[\text{M-H}_2]^+ \xrightarrow{\text{Et}_2\text{O}} [\text{M}_2(\mu\text{-H})]^+ + \text{Et}_2\text{OH}^+ + \text{H}_2 \quad (9.36)$$

$$\text{M} = \text{Cp*Ru(CO)}_2, \text{Cp*Re(CO)(NO)}$$

$$2[\text{M}'\text{-CH}_2\text{Cl}_2]^+ + 2\text{H}_2 \longrightarrow 2[\text{M}'\text{-H}_2]^+ \xrightarrow{\text{Pr}_2\text{O}} [\text{M}'_2(\mu\text{-H})]^+ + \text{Pr}_2\text{OH}^+ + \text{H}_2 \quad (9.37)$$

$$\text{M}' = \textit{cis}\text{-Re(CO)}_4(\text{PR}_3)$$

In all cases, a μ-H dimer is the product even though the mononuclear hydride M-H is known in Eq. (9.36) and is used to generate the thermally unstable H$_2$ complex by protonation with HBF$_4$. M-H is not observed by NMR in Eq. (9.37), indicating strong thermodynamic preference for the μ-H dimer. Interestingly, the hydrogenase enzymes that heterolyically activate H$_2$ have dinuclear active sites that are capable of forming bridging hydrides by reversible protonation of M-M bonds (see Chapter 10). The pK_a of bound H$_2$ in Eqs. (9.36) and (9.37) can be estimated to be near -2 (the pK_a of Et$_2$OH$^+$ is -2.4 in sulfuric acid[66]), although the irreversible formation of the μ-H product provides a driving force for deprotonation that could raise the effective pK_a of the H$_2$ complex a few units. A notable difference between Eqs. (9.36) and (9.37) is that [Re(H$_2$)(CO)$_4$(PR$_3$)]$^+$ is *synthesized directly from reaction of H$_2$* with an isolable precursor,[51] while the Cp complexes are formed by protonation of a hydride with a strong acid.[50] As will be discussed below, only a few other examples of highly acidic η^2-H$_2$ directly generated from H$_2$ are known.[67-69] However, complexes formed by protonation such as [Cp*Re(H$_2$)(CO)(NO)]$^+$ can show exchange of η^2-H$_2$ with gaseous H$_2$.

In addition to the above simple stoichiometric reactions, intermolecular heterolysis is observed or postulated in several types of exchange that can be catalytic, e.g., exchange of protons between a hydride and a deutero acid (DA) can involve H_2 complexes:

$$M(H) + DA \rightleftarrows [M(HD)]^+ + A^- \rightleftarrows M(D) + HA \qquad (9.38)$$

$$M(H) + DA \rightleftarrows M(H)(DA) \rightleftarrows M(HD)(A) \rightleftarrows M(D)(HA) \rightleftarrows M(D) + HA \qquad (9.39)$$

One example of Eq. (9.38) is Eq. (9.40), where the intermediate H_2 complex $[FeH(H_2)(dmpe)_2]^+$ is known but not observed[70]:

$$FeH_2(dmpe)_2 + DC{\equiv}CPh \rightleftarrows FeH(D)(dmpe)_2 + HC{\equiv}CPh \qquad (9.40)$$

Similar Ru complexes with dppm ligands show equilibria between species observable by NMR [Eq. (9.41)].[71] This is also seen in $ROH{\cdots}HW(CO)_2(NO)(PR_3)_2$ and $ROH{\cdots}HRe(CO)_2(triphos)$ where $H{\cdots}H$ contacts are estimated from T_1 measurements to be 1.77 and 1.83 Å, respectively, within the 1.7–1.9 Å range for *intramolecular* interactions (shown below).[72] As in Eq. (9.41) the $ReH{\cdots}HOR$ adduct is in equilibrium with the $Re-H_2$ complex, which is thermodynamically more stable. For protonation of $RuH_2(dmpe)_2$ and $RuH_2(PMe_3)_4$ by alcohols and thiols, the H-bond adduct is not observed, and reaction proceeds to the H_2 adduct and beyond.[73]

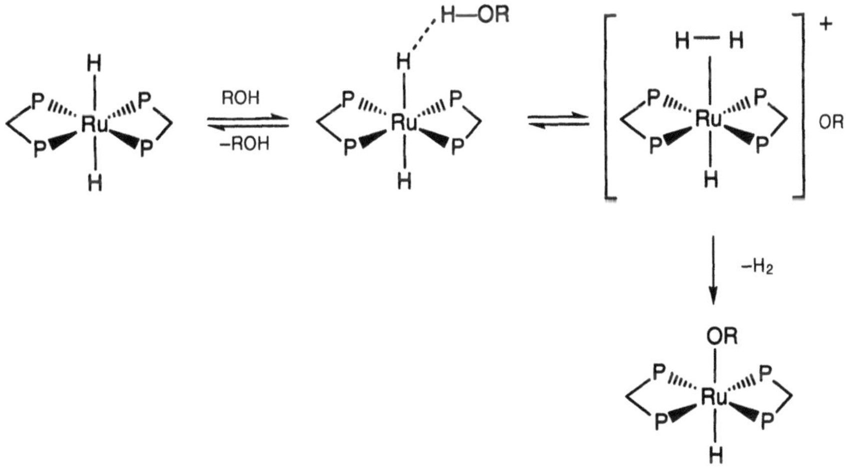

(9.41)

A proposed example of the mechanism in Eq. (9.39) is the deuteration of the hydride ligand of $[IrH(H_2O)(bq)(PR_3)_2]^+$ by reaction with D_2O.[74] Exchange of protons between an H_2 complex and ROH or H_2O can also occur via protonation of ROH/H_2O as in Eq. (9.20) in Section 9.3. M–HD complexes can be synthesized by the addition of D_2O/ROD to $M-H_2$[75] or the addition of H_2O/ROH to a $M-D_2$ complex.[33]

9.4.4. Intramolecular Heterolytic Cleavage

9.4.4.1. Proton Transfer to Ancillary Ligands and H_2 Bonding

As discussed in Chapter 2, *intramolecular* heterolytic cleavage of H_2 is one of the oldest reactions of H_2 and was among the first homogeneous catalytic conversions. η^2-H_2 and η^2-X–H can protonate a counteranion or a basic ligand, either at the M–L bond or at a ligand lone pair. Intramolecular cleavage of X–H is most likely to be essential in many industrial processes such as hydrodesulfurization and in the function of metalloenzymes such as hydrogenase and nitrogenase (see Chapter 10). In general, mechanisms of protonation of L include direct proton transfer via σ-bond metathesis [Eq. (9.42)] and external base catalysis [Eq. (9.43)], where B is a

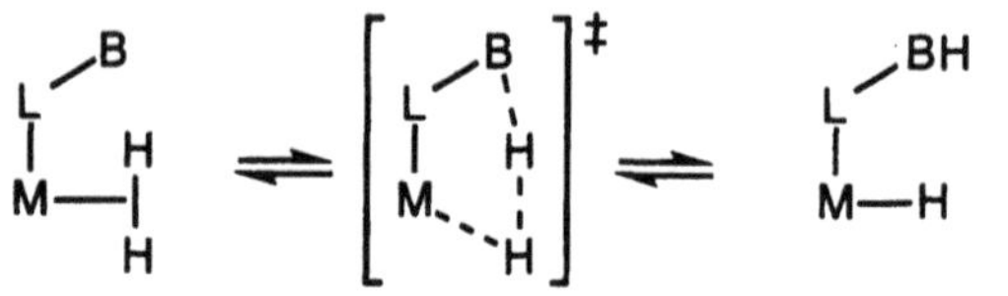

$$(9.42)$$

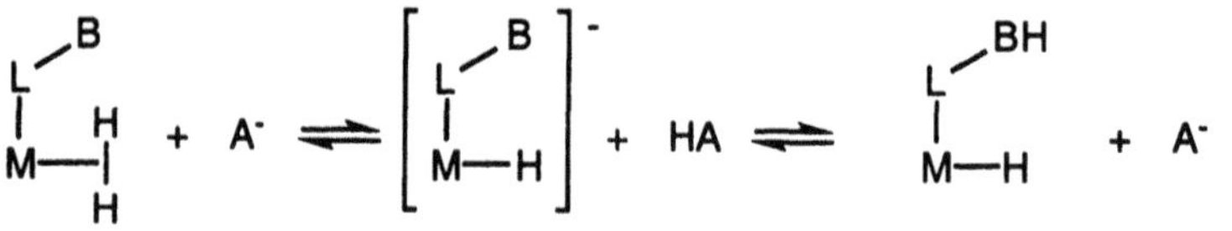

$$(9.43)$$

basic ligand site. Equation (9.43) facilitates protonation at any lone pair regardless of its distance from η^2-H_2 and is identifiable by an increase in the rate in basic solution.[76] An excellent example of Eq. (9.42) is the heterolytic cleavage of H_2 on an Ir complex with a pendant aniline [Eq. 9.44)].[77] The conversion is completely

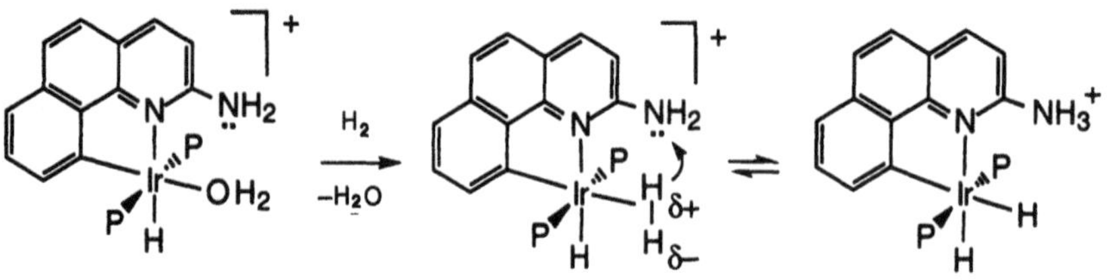

$$(9.44)$$

(16)   (17)   (18)

reversible by removing the H_2 gas from solution and is remarkably sensitive to phosphine basicity. For P = PPh$_3$, **17** is not observed but for the slightly more electron-donating PMePh$_2$, only **17** forms and no trace of the **17** ↔ **18** equilibrium is seen! Thus the pK_a of η^2-H_2 is very dependent on electronics at M: the acidity is higher for the slightly less electron-rich PPh$_3$ complex. Although **17** and **18** cannot be isolated, theoretical studies of a model system confirmed this effect and led to the experimental study of the phosphine variation because oft-used PH$_3$ is too basic to model PPh$_3$ here (calculated **17** is 12 kcal/mol more stable than **18**). Use of more-electron-withdrawing model phosphines PF$_x$H$_{3-x}$ is necessary computationally and shows that d_{HH} in the H$_2$ complex is short, ~0.85 Å, and varies little with x.

Equation (9.44) is facilitated by hydrogen-bonding interactions, e.g., Eq. (9.45), where the OH and IrH hydrogens scramble via rotation of the H$_2$ ligand in **20**. The

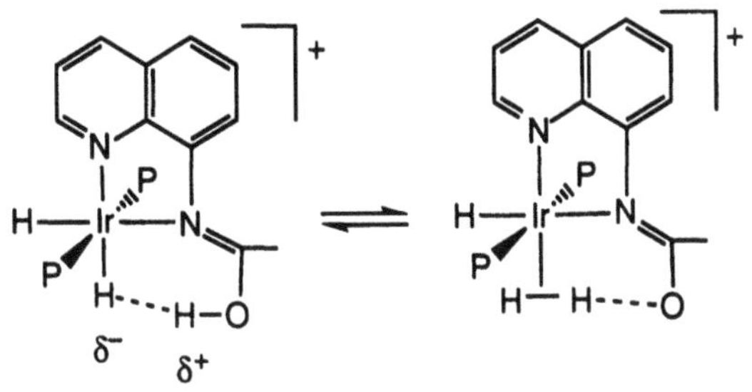

$$\tag{9.45}$$

(19)   (20)

H···H contacts (1.75–1.9Å) in **19** are closer than that between the cis O–H and Ir–H bonds in [IrH(OH)(PMe$_3$)$_4$]$^+$ [78] and are related to the cis effect discussed in Section 4.11. Equations (9.44) and (9.45) and related systems can be considered to be forms of *six-membered σ-bond metathesis* that is also calculated to occur for Ru-catalyzed hydrogenation of CO$_2$, which shows transfer of a proton from transient η^2-H$_2$ to oxygen (see Section 4.12). The H···H interactions here are referred to as "protonic-hydridic bonding" by Morris[79–82] and "dihydrogen bonding" by Crabtree,[83] who reviewed such *unconventional hydrogen bonds*, which include M–H···H–M′, M–H···H–X, and X–H···σ interactions in general (X = C, N, P, O, etc). These complexes represent intermediates in the heterolytic splitting of H$_2$ and illustrate both the basicity of the M–H bond and the acidity of η^2-H$_2$. The interactions can be comparable in strength to classical X–H···(lone-pair) hydrogen bonds (3–7 kcal/mol). In Eq. (9.44) d_{HH} in the theoretical model for **16** with PH$_3$ is calculated to be only 1.4 Å, in the range for stretched H$_2$ complexes, and increases to 1.6 Å for the PF$_3$ species.[77] Species such as those in Eq. (9.45) are capable of promoting catalytic processes, e.g., H/D exchange between D$_2$ and EtOH on a Ni complex containing ligands with pendant OH groups that exchange with η^2-D$_2$ by protonation/deprotonation [as with external H$_2$O in Eq. (9.20)] and then exchange with the ROH proton).[84] In the solid, anion interactions can give bifurcated hydrogen bonding as in **21**,[81] and crystalline polyhydride materials can self-assemble via proton–hydride bonds in, e.g., polyhydride anion pairing with cations containing H-bond donors.[82] Computationally the H···H bond in unconventional hydrogen

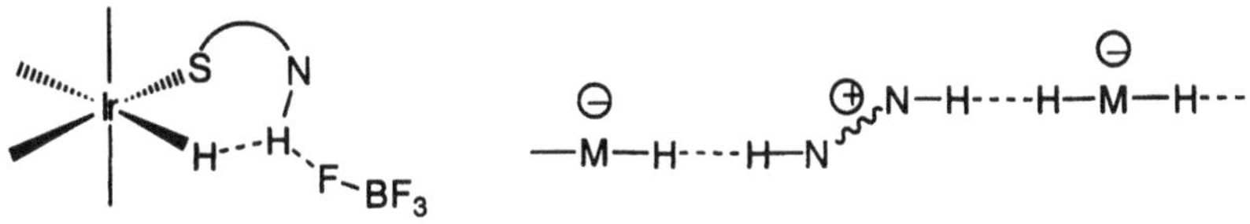

(21)

bonds possesses a large electrostatic component with a small but significant contribution from both charge transfer and polarization, as opposed to conventional hydrogen bonding, where polarization is minor.[85] The discovery of the dihydrogen bond and new findings in this area[86] has led to a significant rebirth of interest in hydrogen bonding in transition metal chemistry. Many aspects of chemistry are impacted here, including crystal engineering, organometallic architecture,[80,87] and proton transfer reactions.

Equation (9.46) represents the first direct observation of equilibrium between an acidic H_2 complex and a corresponding hydride complex with a protonated ancillary ligand.[58] Ancillary ligands with *no* lone pairs, such as hydride, alkyl, and silyl groups,

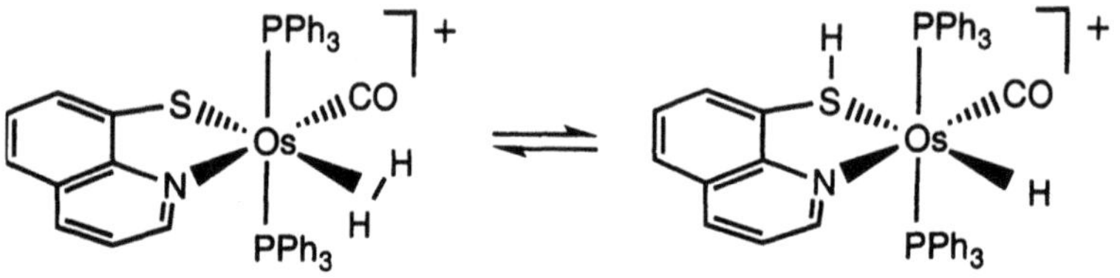

$$ \text{(structures)} \tag{9.46} $$

can be protonated at the M–L bond by mechanisms invoking external or internal base catalysis, OA, and σ-bond metathesis. Some of the theoretical and experimental features of these mechanisms have been discussed previously.[1] One interesting case that involves activation of an agostic C–H bond is the dynamic equilibrium shown in Eq. (9.47) as a four-center transition state mechanism, although no direct evidence for an H_2 intermediate is seen and other mechanisms are possible.[88]

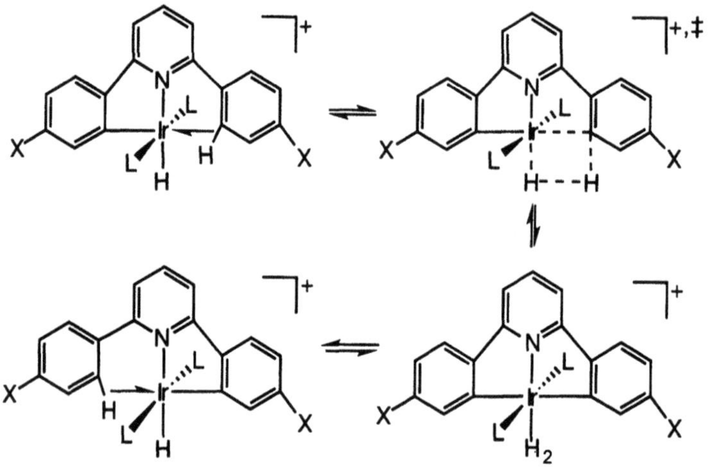

$$ \text{(structures)} \tag{9.47} $$

Intramolecular heterolysis of $L_nMX(H_2)$ to give HX (X = halide, usually Cl) is common and is useful for preparative and catalytic chemistry, e.g., a metal halide (including bridging X) can be converted to a metal hydride in the presence of base or under phase-transfer or high-pressure conditions:[41,64,89,90]

$$ L_nMCl + H_2 \rightarrow L_nMCl(H_2) \rightarrow L_nMH \text{ [or } L_nMH(H_2)] + HCl \tag{9.48} $$

A base or forcing conditions are not even required for a dehydrofluorination[90]:

$$[RuF(dppp)_2]^+ + H_2 \xrightarrow{CDCl_3} [RuH(H_2)(dppp)_2]^+ + HF \qquad (9.49)$$

Hydrogen-bonding interactions can exist between cis chloride and H_2 ligands that demonstrate the acidity of H_2 and facilitate HCl elimination. $IrHCl_2(H_2)(P^iPr_3)_2$ has both intramolecular and intermolecular hydrogen bonding to form an infinite chain structure, as shown by neutron diffraction (see Chapter 5). Such systems can lead to catalytic reactivity, as will be discussed below.

9.4.4.2. Proton Transfer to Anions

Even stronger acids than HCl can be eliminated by proton transfer from η^2-H_2 to the counteranion of highly electrophilic $[L_nM]^+$. One of the strongest acids known, *triflic acid*, CF_3SO_3H, can be eliminated from an H_2 complex formed from H_2 gas, as seen for reaction of a dicationic complex $[Ru(OTf)(CNH)(L)_2][OTf]$ (23, L = diphosphine) containing triflate anions and a protonated cyanide ligand, CNH.[68] The crystal structure of 22, which is a weak Brønsted base, shows a

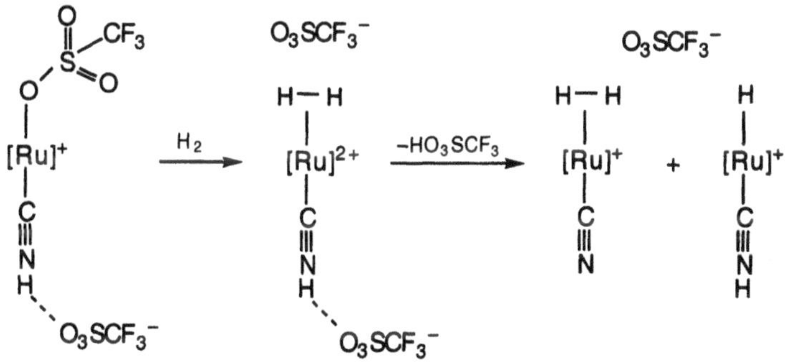

(9.50)

coordinated anion that is displaceable by H_2 to give dicationic 23. The product expected from heterolysis of H_2 is 25, but 24 is the thermodynamically stable product for L = dppe, while a mixture results for L = dppp. The isotopic exchange of $[Os(H_2)(CH_3CN)(dppe)_2][BF_4]_2$ with D_2 to form the HD isotopomer in Eq. (9.21) may involve similar deprotonation of η^2-D_2 by the anion to give DBF_4. The Fe and Os analogues of 23 and 24 are more stable with respect to H_2 displacement similar to related group 8 complexes. The thermodynamic protonation site on $MH(CN)(PP)_2$ can be hydride to give an η^2-H_2 tautomer (PP = depe) or CN to give a CNH tautomer (PP = dppe), but can in no case be M to give a dihydride. There is a delicate balance between the tautomers that can be altered merely by solvent changes and hydrogen-bonding properties of the anion. Migration of protons between H_2 and CNH ligands might be relevant in hydrogenases with Fe–CN linkages (see Section 10.1).

In Scheme 3, **29** is in nearly 1:1 equilibrium equilibrium with **30**, formed by methyl abstraction by $B(C_6F_5)_3$ to give the $MeB(C_6F_5)_3^-$ counterion. This indicates that the electrophilicity of the $[Re(CO)_4(PR_3)]^+$ fragment is similar to that of

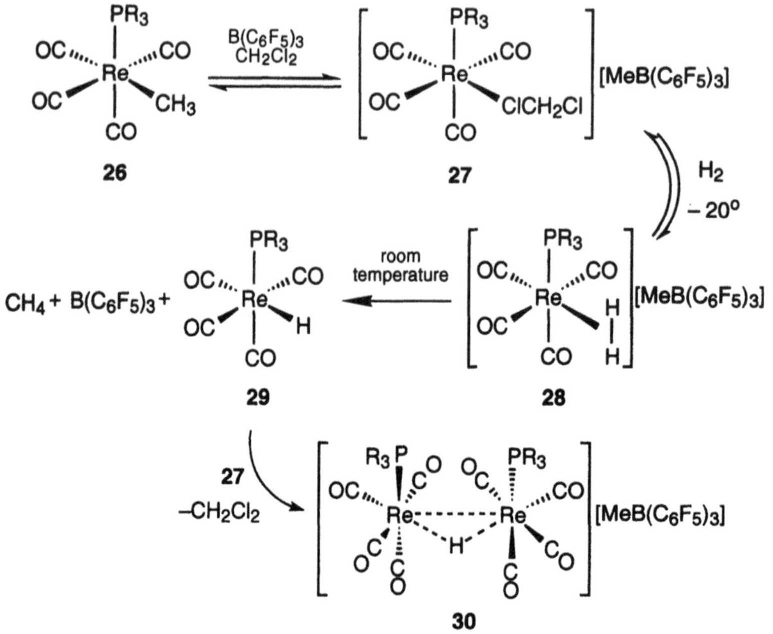

Scheme 3

$B(C_6F_5)_3$. Complex **30** reacts under H_2 atmosphere below room temperature to form equilibrium amounts (~5%) of **28**.[69] On warming, methane, $B(C_6F_5)_3$, and *cis*-$ReH(CO)_4(PR_3)$, **29**, form apparently by protonation of the anion $MeB(C_6F_5)_3$ by the acidic H_2 in **28**. As in Eq. (9.37) **29** is not observed by NMR, but presumably quickly reacts with unreacted **27** (or **28**) to form the hydride bridged dimer **27**, which is a "thermodynamic sink" in these systems. Such protonation of a borane anion has precedence as shown in Eq. (9.51), where the H_2 complex is also unstable and is generated from H_2 gas.[67]:

$$[IrH_2(H_2)(triphos)]^+[BPh_4]^- \rightarrow IrH_3(triphos) + BPh_3 + benzene \quad (9.51)$$

A mononuclear hydride results here (analogous to unobserved **29**). Another possible scenario in Scheme 3 is *intermolecular* heterolysis of H_2, e.g., protonation of the Me group in equilibrium quantities of **27** by the acidic H_2 in **28** to give CH_4, **27**, and **29**.

9.5. CATALYTIC HYDROGENATION AND RELATED REACTIONS

9.5.1. Direct Transfer of Hydrogen from H_2 Ligands

The most well-known reaction of hydrogen is catalytic hydrogenation, and H_2 complexes are always intermediates in these processes, even if only transients in

OA. Two key questions are whether η^2-H$_2$ *directly* transfers H atoms to substrates such as olefins [Eq. 9.3)], i.e., without cleavage to hydrides, and whether H$_2$ complexes have unusual or advantageous properties as catalysts. Examples of Eq. (9.3) or even M(olefin)(H$_2$) complexes are extremely rare. Unstable group 6 complexes M(CO)$_4$(norbornadiene)(H$_2$) detected by IR are proposed as intermediates for photocatalytic hydrogenation of norbornadiene involving direct reaction of H$_2$ ligands (see Section 5.4).[91] Evidence for such transfer of the hydrogens in H$_2$ ligand to cobound substrates is limited, however, partly because it is difficult to prove that cleavage of the H–H bond does not occur first. James[92] provides excellent kinetic, NMR, and tensiometric evidence for a dinuclear Ru–H$_2$ complex transferring both hydrogens to styrene in DMA solution [Eq. (9.52)], DMA = *N,N*-dimethylacetamide; PP = Ph$_2$P(CH$_2$)$_4$PPh$_2$). Although NMR data for the H$_2$

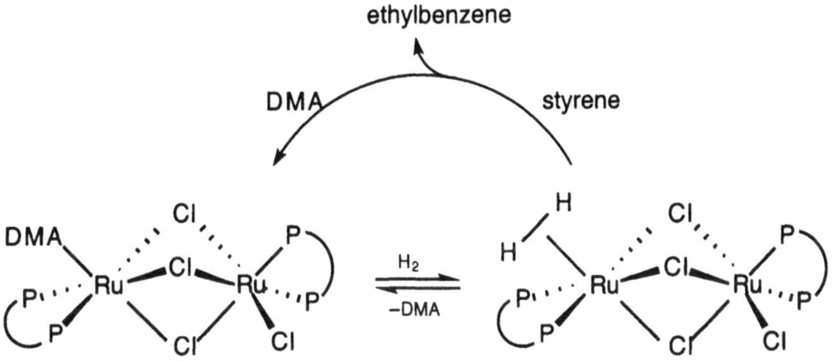

$$(9.52)$$

complex in the catalytic system is tenuous, it is a known complex[93] formed by a novel rearrangement reaction [Eq. (9.53)] that undoubtedly exists in equilibrium

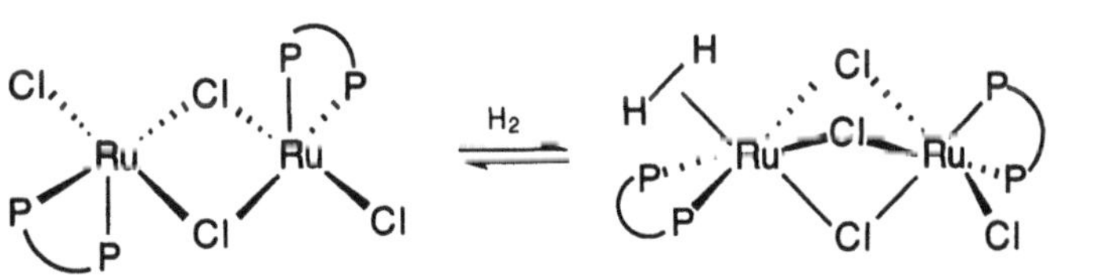

$$(9.53)$$

with the DMA solvent in Eq. (9.52). The first-order dependence on total Ru concentration, the first- to zero-order dependence on [styrene], and the fractional dependence on [H$_2$] are best interpreted in terms of the mechanism in Eq. (9.52). It is important to note that no hydride ligands are present in addition to or in equilibrium with η^2-H$_2$. Prior to this, catalytic hydrogenation of alkynes to alkenes using [FeH(H$_2$)(pp$_3$)]$^+$, which contains a hydride, was proposed by Bianchini to involve intramolecular transfer from η^2-H$_2$ to a σ-vinyl ligand [Eq. (9.54)].[94] The H$_2$ is not labile here and an open site is created by dissociation of a phosphine arm. The alkene elimination step is intramolecular heterolysis of η^2-H$_2$. Although the hydride apparently serves only to form the σ vinyl, there is ambiguity in the mechanism because, as will be discussed below, complexes with a hydride and labile H$_2$ can give different pathways. This system is notable in that Fe-based industrial homogeneous catalysts are very rare, although common in enzymes.

Espuelas *et al.* proposed that reaction of $OsH_2Cl_2(P^iPr_3)_2$ with $HC \equiv CR$ to give the carbyne complex $OsHCl_2(\equiv CCH_2R)(P^iPr_3)_2$ occurs via hydrogen transfer from H_2 to the β-carbon of the vinylidene ligand in the intermediate $OsCl_2(H_2)(C=CHR)(P^iPr_3)_2$.[94]

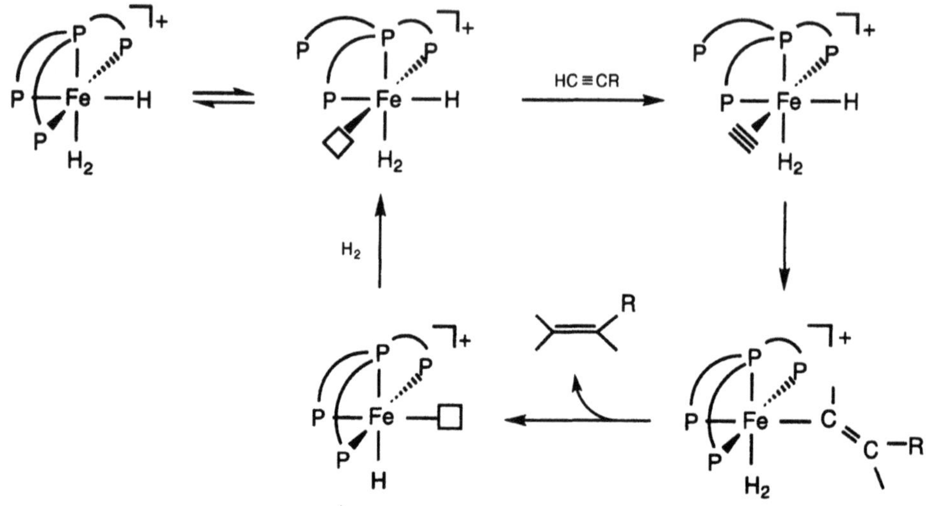

(9.54)

Protonation of an olefin–hydride complex to generate $M(olefin)(H_2)$ and observation of transfer of protons from η^2-H_2 to the olefin could prove direct reaction of H_2 ligands, as in Eq. (9.55), where NBD is seen to be hydrogenated to nortricyclene.[2,95] A dinuclear hydride with a vinylcyclopentadiene ligand is also

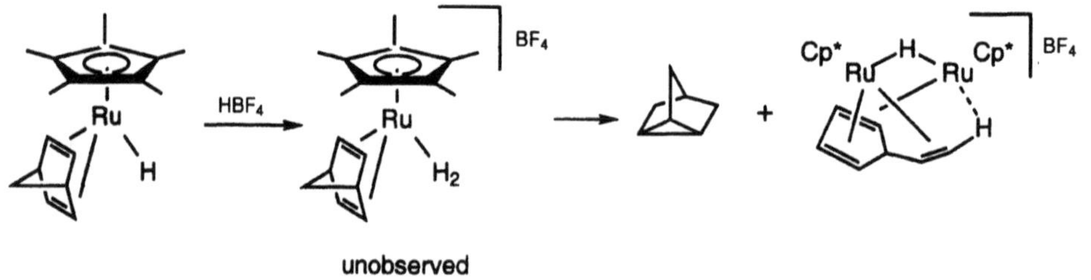

(9.55)

produced, possibly arising from reaction of unprotonated starting material. Although the Ru–H_2 complex is not detected, a similar COD complex is seen by NMR. Also displacement of the aquo ligand in $[Cp^*Ru(NBD)(H_2O)]^+$ by H_2 (50 atm) leads to nortricyclene. Equation (9.55) could be a form of ionic hydrogenation (see Section 9.5.3), and detailed mechanistic studies will be required to unravel whether direct transfer from η^2-H_2 to olefin occurs here.

A T-shaped 14e Rh^I complex containing a rigid P–C–P pincer ligand (see Chapter 7) reversibly binds H_2 (**31**) and CO_2 (**33**) giving net insertion of CO_2 into the H_2 complex. The hydrido-formate **32** quantitatively forms either upon treating **31** with CO_2 or **33** with H_2 at room temperature. In the latter case predissociation

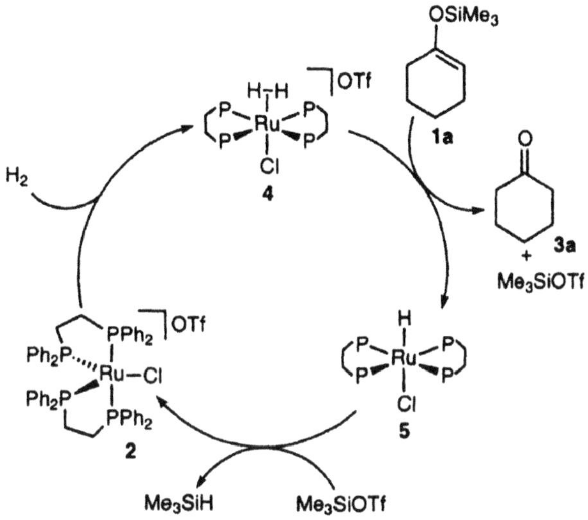

Figure 9.3. Mechanism for catalytic hydrogenation involving heterolytic cleavage of η^2-H_2.

of CO_2 from **33** presumably takes place, followed by insertion of the displaced CO_2 into **31**, although concerted H_2 addition to **33** cannot be excluded. Significantly, this is the first demonstration of reaction of an H_2 complex with CO_2 and one of the clearest examples of any type of direct reaction of η^2-H_2 since a dihydride tautomer is not observed here. Although the structure of **31** is not known, NMR and other evidence points to a true H_2 complex (T_1 = 30–60 ms, reversible binding). Reaction of a well-characterized CO_2 complex with H_2 to give insertion is also unprecedented.

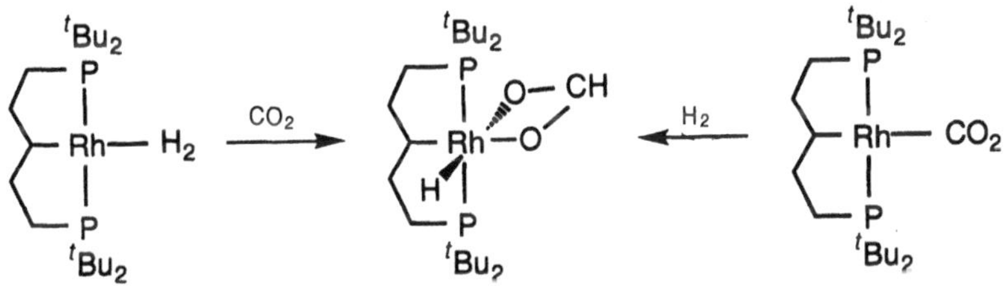

A novel catalytic hydrogenation process is proposed that involves heterolysis of H_2 on an acidic cationic Ru–H_2 complex.[96] Treatment of the silyl enol ether (**1a** in Figure 9.3) with H_2 in the presence of a catalytic amount of **2** in benzene at 50°C for 3 h gives the cyclohexanone **3a** and Me_3SiH in nearly quantitative yield. Based on studies using D_2, the initial step is thought to involve proton transfer from η^2-H_2 in **4** in Figure 9.3 to the oxygen atom of **1a** to give **3a**, the neutral hydride **5**, and Me_3SiOTf from the triflate anion. The latter can accept a hydride to form known **2**, which then converts back to **4** under H_2. A delicate balance between the acidity of η^2-H_2 in **4** and the nucleophilicity of the hydride **5** might be critical here.

9.5.2. H_2 Complexes as Precursors for Catalytic and Other Reactions

Many catalytic and stoichiometric reactions access H_2 complexes as *precursors*, as summarized in reviews by Morris[1] and Oro.[97] As Morris points out, it is remarkable that H_2 complexes serve as catalysts because efficient hydrogenation catalysts have at least two very labile accessible sites for simultaneous hydrogen and substrate binding and are usually electron-rich, which reduces the activation energy for hydride transfer. Stable H_2 complexes normally have converse properties, e.g., only one open coordination site (the H_2 binding site). In the above examples of direct H_2 ligand reactivity, open sites were created by unusual rearrangements or dissociations. However, the presence of a hydride ligand in addition to labile η^2-H_2 allows a straightforward catalytic pathway and virtually all catalytic systems possess $MH(H_2)$ formulations.

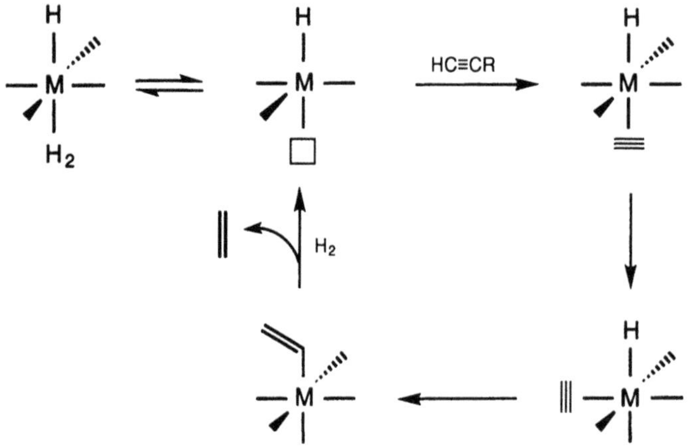

$$(9.56)$$

The loss of H_2 (usually trans to H) opens a site for binding of the unsaturated substrate, which in most cases is an alkyne [as in Eq. 9.56)], an alkene, or a ketone. The hydride migrates to substrate, followed by hydrogenolysis to final product, which could involve an η^2-H_2 intermediate and/or four-center transition state. The H_2 complex thus functions as a reservoir of unsaturation and stabilizes the catalyst system against decomposition. Because excess H_2 is usually present in hydrogenations, the η^2-H_2 behaves in a way that is similar to a weakly coordinated solvent molecule and can compete with solvent binding as in Eq. (9.52). Catalysts include both neutral and cationic systems with labile H_2 such as $OsHCl(H_2)(CO)P_2$,[97-99] $RuH_2(H_2)P_3$,[100-104] $RuH_2(H_2)_2P_2$,[8,103] $[MH(H_2)P_4]^+$ (M = Fe, Ru, Os),[94,105-107] $[IrH_2(H_2)P_3]^+$,[108,109] and $Tp*RuH_2(H_2)$[110] Some patented processes include hydrogenation of nitriles and nitro compounds, e.g., hydrogenation of adiponitrile to hexamethylenediamine in butylamine solvent containing $RuHCl(H_2)(PCy_3)_2$ catalyst under H_2 (860 kPa) at 60°C.[103] Loss of H_2 can give reactive 16e species that undergo stoichiometric reactions such as decarbonylation of alcohols [Eq. (9.57)] or addition of alkynes to $[MX(H_2)P_4]^+$ (M = Ru, Os) to form initially $[MX(=C=C(H)R)P_4]^+$ that can be deprotonated to acetylide complexes[104]:

$$RuH_2(H_2)P_3 \xrightarrow{-H_2} RuH_2P_3 \xrightarrow{ROH} RuH_2(CO)P_3 \qquad (9.57)$$

The $[MH(H_2)(pp_3)]^+$ system, which was studied extensively by Bianchini and coworkers,[94,105] highlights the importance of differences in M–H$_2$ bond strengths (Os > Fe > Ru) on the selective hydrogenation of alkynes to alkenes. The Fe congener is one of the most stable H$_2$ complexes and reactions with alkynes under 1 atm of H$_2$ produce only alkenes (with one exception) via the mechanism in Eq. (9.54), where the H$_2$ remains attached while the alkyne coordinates. In contrast, the complex $[RuH(H_2)(pp_3)]^+$ catalyzes dimerization of 1-alkynes to nearly 100% Z-1,4-disubstituted enynes.[107]

The complex $IrH_2Cl(P^iPr_3)_2$ behaves in a different manner to Eq. (9.56), and *heterolytically activates* the H$_2$ ligand as discussed above to form HCl and $IrH_3(P^iPr_3)_2$, where the known H$_2$ complex[111] is identified as an intermediate[112]:

$$IrH_2Cl(P^iPr_3)_2 \rightleftharpoons IrH_2Cl(H_2)(P^iPr_3)_2 \xrightleftharpoons{-HCl} IrH_3(P^iPr_3)_2 \qquad (9.58)$$

The unsaturated trihydride then catalyzes hydrogenation of benzylideneacetone to 4-phenylbutan-2-one in isopropanol at 60°C. $IrH_2Cl(H_2)(P^iPr_3)_2$ is closely related to the dichloro analogue $IrHCl_2(H_2)(P^iPr_3)_2$, which shows hydrogen bonding betweeen H$_2$ and Cl ligands, which could facilitate HCl loss.

Several of the precious-metal complexes are alkene hydrogenation catalysts including $OsHCl(H_2)(CO)(P^iPr_3)_2$, which converts styrene to ethylbenzene under 1 atm of H$_2$ at 60°C in isopropanol.[98] $[Ir(H_2)H_2(PMe_2Ph)_3]^+$ catalyzes room-temperature hydrogenation of ethylene,[109] and $RuH_2(H_2)(PPh_3)_3$ is a versatile catalyst for a variety of processes,[100-104] including hydrogenation of 9-methylanthracene and unactivated ketones. In all cases, facile H$_2$ loss from the complex is crucial in the mechanism, particularly for dehydrogenation of ethanol to acetaldehyde in ethanol/NaOH solution at 150°C catalyzed by $RuH_2(H_2)(PPh_3)_3$.[102] The catalyst precursor is actually $RuH_2(N_2)(PPh_3)_3$, which is rapidly converted

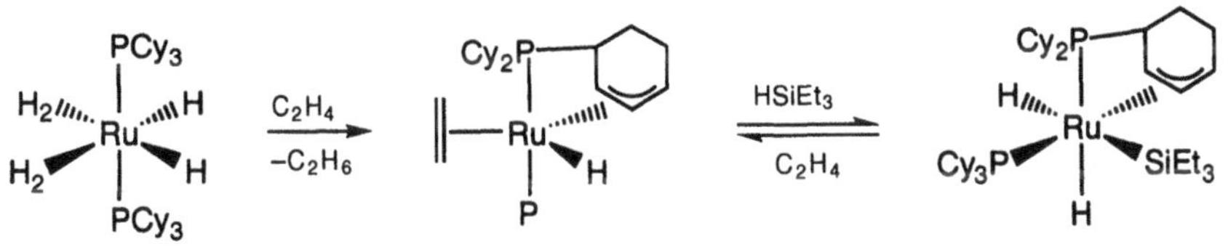

$$(34) \xrightarrow[-C_2H_6]{C_2H_4} (35) \xrightleftharpoons[C_2H_4]{HSiEt_3} (36) \qquad (9.59)$$

to the H$_2$ complex under these conditions. The versatile bis(H$_2$) complex $RuH_2(H_2)_2(PCy_3)_2$ (34), the chemistry of which has been reviewed,[8] catalyzes hydrogenation of arenes and the activated alkene, dimethylfumarate.[113] Reaction of

34 with ethylene leads to dehydrogenation of a Cy group on the phosphine to give 35 containing a cyclohexenyl group [Eq. (9.59)].[114] This unusual species is a highly efficient catalyst for selective dehydrogenative silylation of ethylene into the vinylsilane $CH_2=CHSiEt_3$. Thermal decomposition of solid 34 also leads to dehydrogenation of Cy rings to products such as $RuH_3(\eta^3-C_6H_8)PCy_2)(PCy_3)$ followed by formation of benzene.[8] Structure 34 catalyzes room-temperature addition of ethylene (20 atm) to acetophenone and benzophenone to give mono– and di-addition products, e.g., 2,2′-diethylbenzophenone[115]

$OsHCl(H_2)(CO)(P^iPr_3)_2$ is a very active, selective catalyst for silylation of acetylenes[99]:

$$HSiEt_3 + PhC\equiv CH \xrightarrow{60°C} PhHC=CH(SiEt_3) \quad \text{(trans and cis isomers)} \quad (9.60)$$

The η^2-H_2 or η^2-Si–H complexes could be relevant in Eq. (9.60), although the isolable complex $Os(SiEt_3)Cl(H_2)(CO)(P^iPr_3)_2$ proposed as an intermediate contains η^2-H_2, which is favored over the $OsH(\eta^2$-$HSiEt_3)$ structure in theoretical studies also (see Chapter 4). A related cationic complex $[OsH(H_2)(CO)(P^iPr_3)_2]^+$ acts as a template for a noncatalytic C–C coupling reaction between methyl propiolate and 1,1-diphenyl-2-propyn-1-ol.[116]

9.5.3. Ionic Hydrogenation and Zeolite-Catalyzed Hydrogenation

In 1989 Bullock reported the first examples of ionic hydrogenation wherein a mixture of a organometallic hydride such as $CpMoH(CO)_3$ and a strong acid such as HO_3SCF_3 reduces sterically hindered olefins to alkanes via protonation to carbocations followed by hydride transfer from the metal hydride.[117] Several other

$$\text{>C=C<} + CpM(CO)_3H \xrightarrow[-50°C, 5\ min]{HOTf} \text{>C–C<} \text{ (with H, H)} + CpM(CO)_3(OTf) \quad (9.61)$$

examples have since been reported, including hydrogenation of alkynes and ketones.[118–122] It is conceivable that an acidic H_2 (or dihydride) complex might be involved in the proton transfer step of some of these reactions [Eq. (9.62)]. Although $CpMH(CO)_3$ (M = Mo, W) gives dihydrides on protonation, $Cp*OsH(CO)_2$ gives an equilibrium mixture of dihydride and H_2 complexes.[120] The proposed intermediate ROH species, $[CpW(CO)_3(Me_2CHOH)][A]$ and $[ReH(CO)(NO)(PR_3)_2$-$(Me_2CHOH)][A]$ have not been isolated or observed by NMR (A = triflate).[121,122] The latter Re system is known to form a very acidic H_2 complex with pK_a near -1, and ROH is eliminated from the unstable ROH complex above 240 K to give the monohydride complex.[121] Although definitive proof that the organic substrates are directly protonated by M-bound hydrogen rather than the acid is lacking, Harman and Taube reported rapid reduction of acetone by electrochemically generated $[Os(H_2)(NH_3)_5]^{3+}$.[120] Since the Os^{II} complex $[Os(H_2)(NH_3)_5]^{2+}$ is inactive in

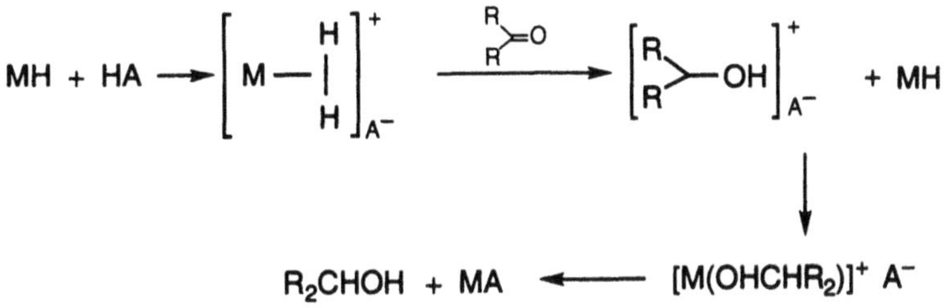

$$MH + HA \longrightarrow \left[M-\overset{H}{\underset{H}{|}} \right]^{+}_{A^{-}} \xrightarrow{\underset{R}{\overset{R}{>}}=O} \left[\underset{R}{\overset{R}{>}}-OH \right]^{+}_{A^{-}} + MH$$

$$R_2CHOH + MA \longleftarrow [M(OHCHR_2)]^{+} A^{-}$$

(9.62)

such a reaction and the Os^{III} complex is acidic, an ionic hydrogenation mechanism is proposed:

$$[Os(H_2)(NH_3)_5]^{3+} + Me_2CO \rightarrow [OsH(NH_3)_5]^{2+} + [Me_2C(OH)]^{+}$$
$$\rightarrow [Os(NH_3)_5(OHCHMe_2)]^{3+}$$

(9.63)

Whether or not catalytic cycles based on such ionic hydrogenation mechanisms can be devised remains to be seen.

There is a possibility that *transition-metal-free hydrogenation catalysts*, e.g., zeolites such as ZSM-5, can be devised.[123] Although experimental studies are few, calculations have been used to probe the zeolite-catalyzed hydrogenation of prototypical unsaturated molecules. Both Brønsted acid and alkali metal sites in model zeolites have been examined. For the hydrogenation of ethene, the barrier is predicted to be lowered by about 50% at the Brønsted acid sites and by about 40% at the alkali metal sites. A concerted mechanism with intermediates containing H_2 bridging zeolite oxygen and ethene carbon in end-on fashion is proposed. The barriers for the hydrogenation of formimine and formaldehyde are predicted to be lowered even more, with overall barriers of only 7 and 14 kcal/mol, respectively, at the Brønsted acid sites of the zeolites.

REFERENCES

1. Jessop, P. G.; Morris, R. H. *Coord. Chem. Rev.* **1992**, *121*, 155.
2. Jia, G.; Lau, C.-P. *Coord. Chem. Rev.* **1999**, *190–192*, 83.
3. Lee, D. W.; Jensen, C. M. *J. Am. Chem. Soc.* **1996**, *118*, 8749.
4. Basallote, M. G.; Duran, J.; Fernandez-Trujillo, J.; Gonzalez, G.; Manez, M. A.; Martinez, M. *Inorg. Chem.* **1998**, *37*, 1623; Basallote, M. G.; Duran, J.; Fernandez-Trujillo, J.; Maneg, M. A. *Inorg. Chem.* **1999**, *38*, 5067.
5. Perutz, R. N. *Pure Appl. Chem.* **1998**, *70*, 2211.
6. Torrent, M.; Sola, M.; Frenking, G. *Organometallics* **1999**, *18*, 2801; Torrent, M.; Solà, M.; Frenking, G. *Chem. Rev.* **2000**, *100*, 439.
7. King, R. B. *J. Organomet. Chem.* **1999**, *586*, 2.
8. Chaudret, B.; Dagnac, P.; Labroue, S.; Sabo-Etienne, S. *New J. Chem.* **1996**, *20*, 1137; Sabo-Etienne, S.; Chaudret, B. *Coord. Chem. Rev.* **1998**, *178–180*, 381.
9. Harthun, A.; Kadyrov, R.; Selke, R.; Bargon, J. *Angew. Chem. Int. Ed. Engl.* **1997**, *36*, 1103.
10. Wayland, B. B.; Ba, S.; Sherry, A. E. *Inorg Chem.* **1992**, *31*, 148.
11. Kubas, G. J.; Unkefer, C. J; Swanson, B. I.; Fukushima, E. *J. Am. Chem. Soc.* **1986**, *108*, 7000.

12. Gusev, D. G.; Nietlispach, D.; Eremenko, I. L.; Berke, H. *Inorg. Chem.* **1993**, *32*, 3628.
13. Luo, X.-L.; Crabtree, R. H. *J. Am. Chem. Soc.* **1990**, *112*, 6912.
14. Haward, M. T.; George, M. W.; Hamley, P.; Poliakoff, M. *J. Chem. Soc., Chem. Commun.* **1991**, 1101.
15. Bullock, R. M.; Song, J.-S.; Szalda, D. J. *Organometallics* **1996**, *15*, 2504.
16. Kubas, G. J.; Burns, C. J.; Eckert, J.; Johnson, S.; Larson, A. C.; Vergamini, P. J.; Unkefer, C. J.; Khalsa, G. R. K.; Jackson, S. A.; Eisenstein, O., *J. Am. Chem. Soc.* **1993**, *115*, 569.
17. Pierantozzi, R.; Geoffroy, G. L. *Inorg. Chem.* **1978**, *19*, 1821; Ziegler, T.; Tschinke, V.; Fan, L.; Becke, A. D. *J. Am. Chem. Soc.* **1989**, *111*, 9177; Geoffroy G. L.; Wrighton, M. S. *Organometallic Photochemistry*, Academic Press: New York, 1979; Veillard, A. *Chem. Phys. Lett.* **1990**, *170*, 441.
18. Bloyce, P. E.; Rest, A. J.; Whitwell, I.; Graham, W. A. G.; Holmes-Smith, R. *J. Chem. Soc., Chem. Commun.* **1988**, 846; Baker, M. V.; Field, L. D. *J. Am. Chem. Soc.* **1987**, *109*, 2825; Sponsler, M. B.; Weiller, B. H.; Stoutland, P. O.; Bergman, R. G. *J. Am. Chem. Soc.* **1989**, *111*, 6841.
19. Clark, H. C.; Hampden Smith, M. J. *J. Am. Chem. Soc.* **1986**, *108*, 3829.
20. Bautista, M. T.; Cappellani, E. P.; Drouin, S. D.; Morris, R. H.; Schweitzer, C. T.; Sella, A.; Zubkowski, J. *J. Am. Chem. Soc.* **1991**, *113*, 4876.
21. James, B. R. *Homogeneous Hydrogenation*, Wiley: New York, 1973.
22. Orchin, M.; Rupilius, W. *Catal. Rev.* **1972**, *6*, 85.
23. Wayland, B. B.; Ba, S.; Sherry, A. E. *Inorg. Chem.* **1992**, *31*, 148.
24. Collman, J. P.; Hutchison, J. E.; Wagenknecht, P. S.; Lewis, N. S.; Lopez, M. A.; Guilard, R. *J. Am. Chem. Soc.* **1990**, *112*, 8206.
25. Bautista, M. T.; Earl, K. A.; Maltby, P. A.; Morris, R.H. *J. Am. Chem. Soc.* **1988**, *110*, 4056.
26. Upmacis, R. K.; Poliakoff, M.; Turner, J. J. *J. Am. Chem. Soc.* **1986**, *108*, 3645.
27. Burdett, J. K.; Eisenstein. O.; Jackson, S. A. In A. Dedieu (ed.), *Transition Metal Hydrides*, VCH: New York, 1992, p. 149; Kober, E. M.; Hay, P. J. *The Challenge of d and f Electrons: Theory and Computation*, ACS Symp. Ser., 1989, p. 394.
28. Pacchioni, G. *J. Am. Chem. Soc.* **1990**, *112*, 80.
29. Heinekey, D. M.; Schomber, B. M.; Radzewich, C. E. *J. Am. Chem. Soc.* **1994**, *116*, 4515.
30. Schlaf, M.; Lough, A. J.; Maltby, P. A.; Morris, R. H. *Organometallics* **1996**, *15*, 2270.
31. Reid, S. M.; Neuner, B.; Schrock, R. R.; Davis, W. M. *Organometallics* **1998**, *17*, 4077.
32. Zhang, K.; Gonzalez, A. A.; Mukerjee, S. L.; Chou, S. J.; Hoff, C. D.; Kubat-Martin, K. A.; Barnhart, D.; Kubas, G. J. *J. Am. Chem. Soc.* **1991**, *113*, 9170.
33. Kubas, G. J.; Burns, C. J.; Khalsa, G. R. K.; Van Der Sluys, L. S.; Kiss, G.; Hoff, C. D. *Organometallics* **1992**, *11*, 3390.
34. Albeniz, A. C.; Heinekey, D. M.; Crabtree, R. H. *Inorg. Chem.* **1991**, *30*, 3632.
35. Bianchini, C.; Linn, K.; Masi, D.; Peruzzini, M.; Polo, A.; Vacca, A.; Zanobini, F. *Inorg. Chem.* **1993**, *32*, 2366.
36. Cappellani, E. P.; Drouin, S. D.; Jia, G.; Maltby, P. A.; Morris, R. H.; Schweitzer, C. T. *J. Am. Chem. Soc.* **1994**, *116*, 3375.
37. Sellmann, D.; Kappler, J.; Moll, M. *J. Am. Chem. Soc.* **1993**, *115*, 1830; Lau, C.-P.; Cheng, L. *J. Mol. Catal.* **1993**, *84*, 39; Hembre, R. T.; McQueen, S. *J. Am. Chem. Soc.* **1994**, *116*, 2141; Chan, W.-C.; Lau, C.-P.; Chen, Y.; Fang, Y.-Q.; Ng, S.-M.; Jia, G. *Organometallics* **1997**, *16*, 34.
38. Van Der Sluys, L. S.; Miller, M. M.; Kubas, G. J.; Caulton, K. G. *J. Am. Chem. Soc.* **1991**, *113*, 2513.
39. Koelliker, R.; Milstein, D. *J. Am. Chem. Soc.* **1991**, *113*, 8524.
40. Gusev, D. G.; Hubener, R.; Burger, P.; Orama, O.; Berke, H. *J. Am. Chem. Soc.* **1997**, *119*, 3716.
41. Chin, B.; Lough, A. J.; Morris, R. H.; Schweitzer, C.; D'Agostino, C. *Inorg. Chem.* **1994**, *33*, 6278.
42. Mediati, M.; Tachibana, G. N.; Jensen, C. M. *Inorg. Chem.* **1992**, *31*, 1827; Sola, E.; Bakhmutov, V. I.; Torres, F.; Elduque, A.; Lopez, J. A.; Lahoz, F. J.; Werner, H.; Oro, L. A.; *Organometallics* **1998**, *17*, 683.
43. Abdur-Rashid, K.; Fong, T. P.; Greaves, B.; Gusev, D. G.; Hinman, J. G.; Landau, S. E.; Lough, A. J.; Morris, R. H. *J. Am. Chem. Soc.* **2000**, *122*, 9155 and references therein.
44. Crabtree, R. H.; Lavin, M. *J. Chem. Soc., Chem. Commun.* **1985**, 794.
45. Chinn, M. S.; Heinekey, D. M. *J. Am. Chem. Soc.* **1987**, *109*, 5865.
46. Jordan, R. F.; Norton, J. R.; *J. Am. Chem. Soc.* **1982**, *104*, 1255–1263; Jordan, R. F.; Norton, J. R.; *ACS Symp. Ser.* **1982**, *198*, 403; Moore, E. J.; Sullivan, J. M.; Norton, J. R. *J. Am. Chem. Soc.* **1986**,

108, 2257–2263; Edidin, R. T.; Sullivan, J. M.; Norton, J. R. *J. Am. Chem. Soc.* **1987**, *109*, 3945–3953; *Inorganic Reactions and Methods*; Zuckerman, J. J., ed., VCH: New York, 1987, p. 207.

47. Papish, E. T. Rix, F.; Spetseris, N.; Norton, J. R.; Williams, R. D. *J. Am. Chem. Soc.* **2000**, *122*, 12235.
48. Kristjansdottir, S. S.; Norton, J. R. In *Transition Metal Hydrides*, Dedieu, A., ed., VCH: New York, NY, 1992, pp. 309–359 (proven case of protonation at metal; Scharrer, E.; Chang, S.; Brockhart, M. *Organometallics* **1995**, *14*, 5686).
49. Morris, R. H. *Can. J. Chem.* **1996**, *74*, 1907.
50. Chinn, M. S.; Heinekey, D. M.; Payne, N. G.; Sofield, C. D. *Organometallics* **1989**, *8*, 1824.
51. Huhmann-Vincent, J.; Scott, B. L.; Kubas, G. J., *J. Am. Chem. Soc.* **1998**, *120*, 6808.
52. Morris, R. H. *Inorg. Chem.* **1992**, *31*, 1471.
53. Jia, G.; Morris, R. H. *J. Am. Chem. Soc.* **1991**, *113*, 875.
54. Ng, S.-M.; Fang, Y. Q.; Lau, C.-P.; Wong, W. T.; Jia, G. *Organometallics* **1998**, *17*, 2052.
55. Maltby, P. A.; Schlaf, M.; Steinbeck, M.; Lough, A. J.; Morris, R. H.; Klooster, W. T.; Koetzle, T. F.; Srivastava, R. C. *J. Am. Chem. Soc.* **1996**, *118*, 5396.
56. Rocchini, E.; Mezzetti, A.; Ruegger, H.; Burckhardt, U.; Gramlich, V.; Del Zotto, A.; Martinuzzi, P.; Rigo, P. *Inorg. Chem.* **1997**, *36*, 711.
57. Luther, T. A.; Heinekey D. M. *Inorg. Chem.* **1998**, *37*, 127.
58. Schlaf, M.; Lough, A. J.; Morris, R. H. *Organometallics* **1996**, *15*, 4423.
59. Tilset, M.; Parker, V. D. *J. Am. Chem. Soc.* **1989**, *111*, 6711; *J. Am. Chem. Soc.* **1990**, *112*, 2843.
60. Lever, A. B. P. *Inorg. Chem.* **1990**, *29*, 1271.
61. Moore, E. J.; Sullivan, J. M.; Norton, J. R. *J. Am. Chem. Soc.* **1986**, *108*, 2257.
62. Xu, Z.; Bytheway, I.; Jia, G.; Lin, Z. *Organometallics* **1999**, *18*, 1761.
63. Jia, G.; Lough, A. J.; Morris, R. H. *Organometallics* **1992**, *11*, 161.
64. Kuhlman, R. *Coord. Chem. Rev.* **1997**, *167*, 205.
65. Hauger, B. E.; Gusev, D. G.; Caulton, K. G. *J. Am. Chem. Soc.* **1994**, *116*, 208.
66. Perdoncin, G.; Scorrano, G. *J. Am. Chem. Soc.* **1977**, *99*, 6983.
67. Bianchini, C.; Moneti, S.; Peruzzini, M.; Vizza, F. *Inorg. Chem.* **1997**, *36*, 5818.
68. Fong, T. P.; Lough, A. J.; Morris, R. H.; Mezzetti, A.; Rocchini, E.; Rigo, P. *J. Chem. Soc., Dalton Trans.* **1998**, 2111; Fong, T. P.; Forde, C. E.; Lough, A. J.; Morris, R. H.; Rigo, P.; Rocchini, E.; Stephan, T. *J. Chem. Soc., Dalton Trans.* **1999**, 4475; Ontko, A. C.; Houlis, J. F.; Schnabel, R. C.; Roddick, D. M.; Fong, T. P.; Lough, A. J.; Morris, R. H. *Organometallics* **1998**, *17*, 5467.
69. Huhmann-Vincent, J.; Scott, B. L.; Kubas, G. J., *Inorg. Chim. Acta* **1999**, *294*, 240.
70. Field, L. D.; George, A. V.; Hambley, T. W.; Malouf, E. Y.; Young, D. J. *J. Chem. Soc., Chem. Commun.* **1990**, 931.
71. Ayllon, J. A.; Gervaux, C.; Sabo-Etienne, S.; Chaudret, B. *Organometallics* **1997**, *16*, 2000.
72. Shubina, E. S.; Belkova, N. V.; Krylov, A. N.; Vorontsov, E. V.; Epstein, L. M.; Gusev, D. G.; Niedermann, M; Berke, H. *J. Am. Chem. Soc.* **1996**, *118*, 1105; Shubina, E. S.; Belkova, N. V.; Bakhmutova, E. V.; Vorontsov, E. V.; Bakhmutov, V. I.; Ionidis, A. V.; Bianchini, C.; Marvelli, L.; Peruzzini, M.; Epstein, L. M. *Inorg. Chim. Acta* **1998**, *280*, 302.
73. Field, L. D.; Hambley, T. W.; Yau, B. C. K. *Inorg. Chem.* **1994**, *33*, 2009; Osakada, K.; Ohshiro, K.; Yamamoto, A. *Organometallics* **1991**, *10*, 404.
74. Crabtree, R. H.; Lavin, M.; Bonneviot, L. J. *J. Am. Chem. Soc.* **1986**, *108*, 4032.
75. Baker, M. V.; Field, L. D.; Young, D. J. *J. Chem. Soc., Chem. Commun.* **1988**, 546; Chinn, M. S.; Heinekey, D. M. *J. Am. Chem. Soc.* **1990**, *112*, 5166.
76. Ryan, O. B.; Tilset, M. *J. Am. Chem. Soc.* **1991**, *113*, 9554; Ryan, O. B.; Tilset, M.; Parker, V. D. *Organometallics* **1991**, *10*, 298.
77. Lee, D.-H.; Patel, B. P.; Clot, E.; Eisenstein, O.; Crabtree, R. H. *Chem. Comm.* **1999**, 297.
78. Stevens, R. C.; Bau, R.; Milstein, D.; Blum, O.; Koetzle, T. F. *J. Chem. Soc., Dalton Trans.* **1990**, 1429.
79. Lough, A. J.; Park, S.; Ramachandran, R.; Morris, R. H. *J. Am. Chem. Soc.* **1994**, *116*, 8356.
80. Xu, W.; Lough, A. J.; Morris, R. H. *Inorg. Chem.* **1996**, *35*, 1549.
81. Park, S.; Lough, A. J.; Morris, R. H. *Inorg. Chem.* **1996**, *35*, 3001.
82. Abdur-Rashid, K.; Gusev, D. G.; Landau, S. E.; Lough, A. J.; Morris, R. H. *Organometallics* **2000**, *19*, 1652, and references therein.
83. Crabtree, R. H.; Siegbahn, P. E. M.; Eisenstein, O.; Rheingold, A. L.; Koetzle, T. F. *Acc. Chem. Res.* **1996**, *29*, 348; Crabtree, R. H.; Eisenstein, O.; Sini, G.; Peris, E. *J. Organometal. Chem.* **1998**, *567*, 7;

Bosque, R.; Maseras, F.; Eisenstein, O.; Patel, B. P.; Yao, W.; Crabtree, R. H.; *Inorg. Chem.* **1997**, *36*, 5505; Crabtree, R. H. *Science* **1998**, *282*, 2000.
84. Zimmer, M.; Schulte, G.; Luo, X.-L.; Crabtree, R. H. *Angew. Chem. Int. Ed. Engl.* **1991**, *30*, 193.
85. Orlova, G.; Scheiner, S. *J. Phys. Chem. A* **1998**, *102*, 260.
86. Buil, M. L.; Esteruelas, M. A.; Onate, E.; Ruiz, N. *Organometallics* **1998**, *17*, 3346; Chu, H. S.; Lau, C. P.; Wong, K. Y.; Wong, W. T. *Organometallics* **1998**, *17*, 2768; Ayllon, J. A.; Sayers, S.; Sabo-Etienne, S.; Donnadieu, B.; Chaudret, B.; Clot, E. *Organometallics* **1999**, *18*, 3981.
87. Braga, D.; Grepioni, F. *Coord. Chem. Rev.* **1999**, *183*, 19; Braga, D.; Grepioni, F.; Desiraju, G. R. *Chem. Rev.* **1998**, *98*, 1375; Braga, D.; Grepioni, F.; Tedesco, E.; Calhorda, M. J.; Lopes, P. E. M *New J. Chem.* **1999**, *23*, 219.
88. Albeniz, A. C.; Schulte, G.; Crabtree, R. H. *Organometallics* **1992**, *11*, 242.
89. Grushin, V. V. *Acc. Chem. Res.* **1993**, *26*, 279; Bianchini, C.; Barbaro, P.; Scapacci, G.; Zanobini, F. *Organometallics* **2000**, *19*, 2450.
90. Barthazy, P.; Stoop, R. M.; Worle, M.; Togni, A.; Mezzetti, A. *Organometallics* **2000**, *19*, 2844.
91. Jackson, S. A.; Hodges, P. M.; Poliakoff, M.; Turner, J. J.; Grevels, F.-W. *J. Am. Chem. Soc.* **1990**, *112*, 1221; Thomas, A.; Haake, M.; Grevels, F. W.; Bargon, J. *Angew. Chem. Int. Ed. Engl.* **1994**, *33*, 755.
92. Joshi, A. M.; MacFarlane, K. S.; James, B. R. *J. Organomet. Chem.* **1995**, *488*, 161.
93. Joshi, A. M.; James, B. R. *J. Chem. Soc., Chem. Commun.* **1989**, 1785; Chau, D. E. K.-Y.; James, B. R. *Inorg. Chim. Acta* **1995**, *240*, 419.
94. Bianchini, C.; Meli, A.; Peruzzini, M.; Frediani, P.; Bohanna, C.; Esteruelas, M. A.; Oro, L. A. *Organometallics* **1992**, *11*, 138; Espuelas, J.; Esteruelas, M. A.; Lahoz, F. J.; Oro, L. A.; Ruiz, N. *J. Am. Chem. Soc.* **1993**, *115*, 4683.
95. Jia, G.; Ng, W. S.; Lau, C. P. *Organometallics* **1998**, *17*, 4538.
96. Nishibayashi, Y.; Takei, I.; Hidai, M. *Angew. Chem. Int. Ed. Engl.* **1999**, *38*, 3047.
97. Esteruelas, M. A.; Oro, L. A. *Chem. Rev.* **1998**, 98, 577; Esteruelas, M. A.; Sola, E.; Oro, L. A.; Meyer, U.; Werner, H. *Angew. Chem. Int. Ed. Engl.* **1988**, *27*, 1563; Esteruelas, M. A.; Oro, L. A.; Valero, C. *Organometallics* **1992**, *11*, 3362.
98. Andriollo, A.; Esteruelas, M. A.; Meyer, U.; Oro, L. A.; Sanchez-Delgado, R. A.; Sola, E.; Valero, C.; Werner, H. *J. Am. Chem. Soc.* **1989**, *111*, 7431.
99. Esteruelas, M. A.; Oro, L. A.; Valero, C. *Organometallics* **1991**, *10*, 462.
100. Linn, D. E.; Halpern, J. *J. Am. Chem. Soc.* **1987**, *109*, 2969.
101. Lin, Y.; Zhou, Y. *J. Organomet. Chem.* **1990**, *381*, 135.
102. Morton, D.; Cole-Hamilton, D. J.; Utuk, I. D.; Paneque-Sosa, M.; Lopez-Poveda, M. *J. Chem. Soc., Dalton Trans.* **1989**, 489.
103. Beatty, R. P.; Paciello, R. A., Patents WO 9623802, 9623803, and 9623804, 1995.
104. Van Der Sluys, L. S. ; Kubas, G. J.; Caulton, K. G. *Organometallics* **1991**, *10*, 1033; Albertin, G.; Antoniutti, S.; Bordignon, E.; Pegoraro, M. *J. Chem. Soc., Dalton Trans.* **2000**, 3575.
105. Bianchini, C.; Meli, A.; Peruzzini, M.; Vizza, F.; Zanobini, F.; Frediana, P. *Organometallics* **1989**, *8*, 2080; Bianchini, C.; Farnetti, E.; Frediani, P.; Graziani, M.; Peruzzini, M.; Polo, A. *J. Chem. Soc., Chem. Commun.* **1991**, 1336; Bianchini, C.; Bohanna, C.; Esteruelas, M. A.; Frediani, P.; Meli, A.; Oro, L. A.; Peruzzini, M. *Organometallics* **1992**, *11*, 3837.
106. Albertin, G.; Amendola, P.; Antoniutti, S.; Ianelli, S.; Pelizzi, G.; Bordignon, E. *Organometallics* **1991**, *10*, 2876; Tsukahara, T.; Kawano, H.; Ishii, Y.; Takahashi, T.; Saburi, M.; Uchida, Y.; Akutagawa, S. *Chem. Lett.* **1988**, 2055; Saburi, M.; Takeuchi, H.; Ogasawara, M.; Tsukahara, T.; Ishii, Y.; Ikariya, T.; Takahashi, T.; Uchida, Y. *J. Organomet. Chem.* **1992**, *428*, 155; Mezzetti, A.; Del Zotto, A.; Rigo, P.; Farnetti, E. *J. Chem. Soc., Dalton Trans.* **1991**, 1525; Lough, A. J.; Morris, R. H.; Ricciuto, L.; Schleis, T. *Inorg. Chim. Acta* **1998**, *270*, 238.
107. Albertin, G.; Antoniutti, S.; Bordignon, E. *Proc. 21 Congr. Naz. di Chim. Inorg. Bressanone* **1991**.
108. Lundquist, E. G.; Huffman, J. C.; Folting, K.; Caulton, K. G. *Angew. Chem. Int. Ed. Engl.* **1988**, *27*, 1165; Marinelli, G.; Rachidi, I. E.-I.; Streib, W. E.; Eisenstein, O.; Caulton, K. G. *J. Am. Chem. Soc.* **1989**, *111*, 2346.
109. Lundquist, E. G.; Folting, K.; Streib, W. E.; Huffman, J. C.; Eisenstein, O.; Caulton, K. G. *J. Am. Chem. Soc.* **1990**, *112*, 855.
110. Vicente, C.; Shul'pin, G. B.; Moreno, B.; Sabo-Etienne, S.; Chaudret, B. *J. Mol. Catal., A* **1995**, *98*, L5.
111. Mediati, M.; Tachibana, G. N.; Jensen, C. M. *Inorg. Chem.* **1990**, *29*, 3.

112. Esteruelas, M. A.; Herrero, J.; Lopez, A. M.; Oro, L. A.; Schulz, M.; Werner, H. *Inorg. Chem.* **1992**, *31*, 4013.
113. Borowski, A. F.; Sabo-Etienne, S.; Christ, M. L.; Donnadieu, B.; Chaudret, B. *Organometallics* **1996**, *15*, 1427.
114. Christ, M. L.; Sabo-Etienne, S.; Chaudret, B. *Organometallics* **1995**, *14*, 1082.
115. Guari, Y.; Sabo-Etienne, S.; Chaudret, B. *J. Am. Chem. Soc.* **1998**, *120*, 4228.
116. Bohanna, C.; Callejas, B.; Edwards, A. J.; Esteruelas, M. A.; Lahoz, F. J.; Oro, L. A.; Ruiz, N.; Valero, C. *Organometallics* **1998**, *17*, 373.
117. Bullock, R. M.; Rappoli, B. J. *J. Chem. Soc., Chem. Commun.* **1989**, 1447.
118. Bullock, R. M.; Song, J.-S. *J. Am. Chem. Soc.* **1994**, *116*, 8602; Bullock, R. M.; Song, J.-S.; Szalda, D. J. *Organometallics* **1996**, *15*, 2504; Ito, T.; Koga, M.; Kurishima, S.; Natori, M.; Sekizuka, N.; Yoshioka, K. *J. Chem. Soc., Chem. Commun.* **1990**, 988.
119. Luan, L.; Song, J.-S.; Bullock, R. M. *J. Org. Chem.* **1995**, *60*, 7170.
120. Harman, W. D.; Taube, H. *J. Am. Chem. Soc.* **1990**, *112*, 2261.
121. Bakhmutov, V. I.; Vorontsov, E. V.; Antonov, D. M. *Inorg. Chim. Acta* **1998**, *278*, 122.
122. Song, J.-S.; Szalda, D.; Bullock, R. M.; Lawrie, C. J. C.; Rodkin, M. A.; Norton, J. R. *Angew. Chem. Int. Ed. Engl.* **1992**, *31*, 1233.
123. Senger, S.; Radom, L. *J. Am. Chem. Soc.* **2000**, *122*, 2613, and references therein.

10

Activation of Hydrogen and Related Small Molecules by Metalloenzymes and Sulfur Ligand Systems

10.1. HYDROGENASES: BIORGANOMETALLIC FORMATION AND SPLITTING OF DIHYDROGEN

10.1.1. Introduction

Recent developments in metalloenzyme and organometallic chemistry point to a growing link between these seemingly incongruous fields. The latter is characterized by "abiological" and often toxic ligands such as CO coordinated to reactive air-sensitive transition metal complexes that would appear to be abhorred by Nature. This notion has been radically altered by the recent discovery of CO and cyanide ligands bound to dinuclear iron sites in hydrogenase (H-ase) enzymes that have existed in certain microorganisms for over a billion years. Organometallic linkages (M–C bonds) were first recognized in biology in cobalamins in the early 1960s, giving birth to bioorganometallic chemistry.[1] Biological activation and production of "inert" σ-bonded compounds such as H_2 and CH_4 have been known for many decades, but the mechanism has always been a mystery. As originally noted by Crabtree,[2] several properties of η^2-H_2, such as its acidity and ability to compete with N_2 ligands, clearly must be considered in relation to the structure and function of enzymes such as hydrogenases (H-ases) and nitrogenases (N-ases). For example, these enzymes catalyze H/D exchange between H_2O and D_2, which an acidic H_2 ligand can easily promote (see Chapter 9). In order for this to occur, H_2 must bind competitively with water as well as atmospheric N_2, and this has been shown in previous chapters. The electronics at M must be just right: η^2-H_2 is a better ligand than N_2 on electrophilic M but if M is too electrophilic, water will bind more strongly than H_2. An organometallic biological active site with a mix of donor and acceptor ligands such as CO is advantageous here. Nature has found extremely efficient ways to use first-row metals such as Fe and Ni rather than the precious metals widely used as industrial catalysts. Although H-ases and N-ases often contain Ni (and in most cases N-ases contain Mo), *Fe* appears to be the site of small-molecule binding and activation. and enzymes that contain only Fe are known. The sites are usually polynuclear, and Fe–Fe bonding interactions appear to play a critical role in electron transfer.

The study of H-ases is a very active, vast research field (>3000 publications) that has been the subject of many reviews.[2-11] Only the aspects directly involving the biological activation and production of H_2 will be discussed here. Most notably, the active sites of H-ases *feature the first biological systems with CO and cyanide ligands*. These are coordinated to dinuclear M–M bonded centers, such as shown in **1**

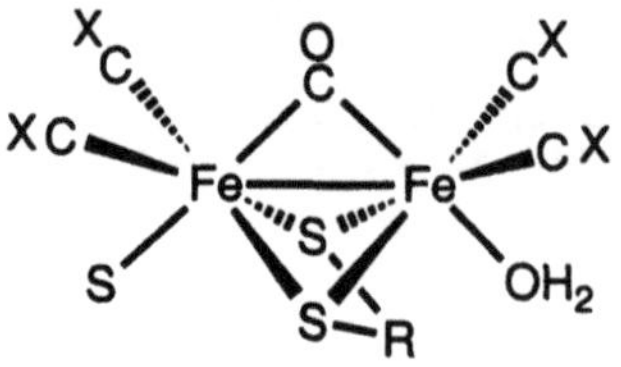

X = O or N

(1)

for an Fe-only H-ase. IR provided the first strong evidence that both ligands are present since protein crystallography often cannot distinguish CO from CN and other diatomic molecules. These ligands, especially CO, make the metals electrophilic, which favors η^2-H_2 binding and heterolytic cleavage in anaerobic enzymes. How CO and CN are formed and assembled in biological systems is unknown, but CO favors the reversible H_2 binding essential to the astoundingly high reaction rates of H-ases. Nature has been opportunistic in designing a nearly perfect organometallic site for H_2 activation, beating us to the punch 2–4 billion years ago when microorganisms with these metalloenzymes first evolved! Their organometallic-compound-like air sensitivity stems from the fact that the earth was then still in a reducing environment.[9] A large set of diatomic molecules, including H_2, N_2, NO, CO, CN, and O_2 all play key roles in biology as metabolites, substrates, signaling agents, toxins, and now as components of active-site structures.

H-ases are redox enzymes that catalyze *reversible* interconversion of H_2 and protons to either utilize H_2 as an energy source or dispose excess electrons as H_2:

$$H_2 \rightleftharpoons 2H^+ + 2e^- \tag{10.1}$$

Equation (10.1) is a rare true equilibrium process much like the hydrogen electrode; e.g., there is dependence on the H_2 pressure whether H_2 is produced or consumed. From isotope exchange evidence (such as the catalytic reaction shown in Eq. (10.2) wherein the HD/H_2 ratio is pH-dependent), it is inferred that the H_2 molecule is split heterolytically.

$$H_2 + D_2O \rightleftharpoons HD + HDO \tag{10.2}$$

The enzymes are crucial to hydrogen metabolism, which is essential to microorganisms of immense biotechnological interest, e.g., methanogenic, acetogenic, N-fixing, photosynthetic, and sulfate-reducing bacteria. The last, for instance, are strictly anaerobic and can grow on H_2 as an electron donor with sulfate and thiosulfate as terminal electron acceptors that are converted to sulfide:

$$H_2 + SO_4^{2-} \xrightarrow{\text{sulfate reducers}} H_2S \qquad (10.3)$$

$$H_2 + CO_2 \xrightarrow{\text{methanogens}} CH_4 \qquad (10.4)$$

$$H_2 + N_2 \xrightarrow{\text{nitrogen fixers}} NH_4^+ \qquad (10.5)$$

Sulfate-reducing bacteria corrode metallic iron just to acquire Fe for the active site! The reverse reaction, hydrogen production, occurs when the bacteria cells are growing, e.g., by fermentation of pyruvate, and this H_2 may be passed on to other organisms such as methanogens. The efficiency of H-ases in producing hydrogen is phenomenal: up to *10^4 turnovers/s per site* at 30°C, a mercurial rate rarely matched in industrial catalysis. As noted by Cammack[5] (and discounting practicality), this rate of H_2 production from 1 mol of enzyme would fill the airship *Graf Zeppelin* in 10 min and the liquid-hydrogen fuel tank of the space shuttle in 2 h!

As stated by Thauer,[8] hydrogen is an important intermediate in the degradation of organic matter by microorganisms in anoxic habitats such as freshwater/marine sediments and even in the gastrointestinal tract: hydrogen in the breath can be used to monitor intestinal malfunction.[4] Over 200 million tons of H_2 are estimated to be globally formed and consumed yearly. Despite this high rate, the steady state concentration of H_2 gas in most anoxic habitats is quite low (1–10 Pa), suggesting that H_2 formation rather than consumption is rate-limiting in the overall process. H_2 is also formed in aerobic environments as a side product of nitrogen fixation catalyzed by N-ase. The studies are important in that the active site is fairly well identified and could be useful in designing biomimetic catalysts for photovoltaic production of H_2 from water or new fuel cell technology.

Three basic types of H-ase active sites have been identified. The most prevalent contain Ni in combination with Fe, but a select few contain only Fe, and are classified as iron-only ([Fe]) H-ases. Selenium replaces sulfur in the Ni-bound cysteine in some [NiFe] H-ases, and one H-ase active site contains no metals at all! The activation of H_2 is as multifaceted in biology as it is in industrial catalysis. The general structural components involved in the H-ase mechanism are shown in Figure 10.1. Although the active site is deeply buried (e.g., about 30 Å from the protein surface), channels exist for both proton and H_2 diffusion. Amino acid residues carry protons away, and crystallographic and molecular dynamics analyses of Xe binding identify hydrophobic channels for H_2 gas ingress to or egress from the active site.[12] Concomitantly, the electrons hop across Fe–S cluster stepping stones spaced 10–15 Å apart to or from a docking site for e.g., a *c*-type cytochrome. The [NiFe] systems, which generally consume H_2 are less active but more resistant to oxidation than the anaerobic [Fe] enzymes, which most frequently produce H_2. The many [NiFe] H-ases have attracted the most attention because Ni is rarely observed in such a biological function, although a major question is the site of H_2 activation: Ni,

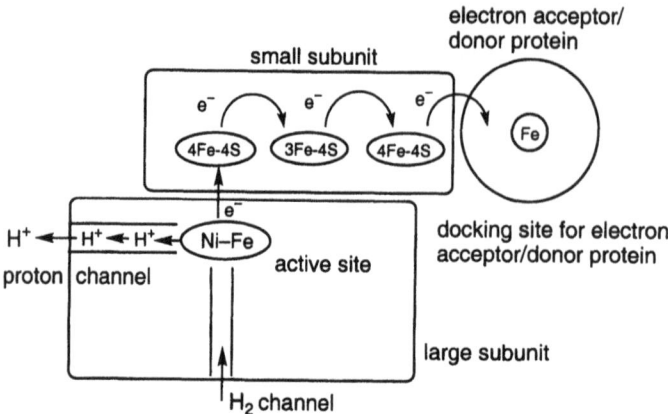

Figure 10.1. Structural components in the hydrogenase mechanism, shown for $H_2 \rightarrow 2H^+ + 2e^-$ in NiFe active sites. The reversible electron pathway is facilitated by three Fe–S clusters that act as stepping stones-to the surface of the small subunit, where there is a potential docking site for a *c*-type cytochrome. The proton channel is proposed to consist of a chain of proton-carrying amino acid residues leading to the surface of the large subunit. Channels for H_2 ingress/egress have been identified.

Fe, or possibly both. X-ray crystallography of both [NiFe] and [Fe] systems indicate that Fe is the primary site of H_2 binding and cleavage, but bridging hydride ligands are likely to be involved at some stage in the activation process. The presence of a bimetallic site in H-ases is intriguing in itself because H_2 is easily activated on a large array of mononuclear complexes without need for a second M. M–M bonds (Ni–Fe and Fe–Fe) are present in the H-ases and would be expected to serve a useful function in Nature. Other nonheme metalloenzymes such as N-ases also contain multiple M–M bonding interactions. Polynuclear systems reduce the severe electronic requirements and structural rearrangements that might be imposed at a single metal site.[9]

10.1.2. [NiFe] H-ases

The [NiFe] H-ases have different numbers of metal clusters depending on their origin, but always possess at least one NiFe-containing cluster considered as the probable H_2 binding site and one or more Fe–S clusters for electron transfer functions. Fe–S "cubane" clusters are ubiquitous throughout biology,[13] and are particularly important in H-ases and N-ases. As shown by EPR combined with isotopic substitution (^{61}Ni shows hyperfine couplings with ^{1}H) and X-ray absorption spectroscopy, H_2 activation takes place in the vicinity of Ni. This does not imply, however, that H_2 is activated exclusively at Ni, and as noted above, bridging hydrides could explain hydride coupling to ^{61}Ni. In the hydrogen activation process, these enzymes can be stabilized in up to six different redox states depending on conditions, e.g., whether H_2 is present. For example, *Desulfovibrio gigas* purified under aerobic conditions exists in oxidized forms (Ni–A and Ni–B) that are

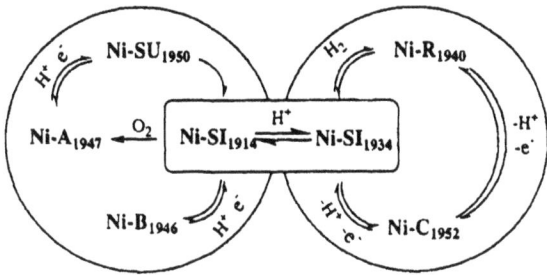

EPR signals: **Ni-A**, **Ni-B**, and **Ni-C**
EPR-silent: **Ni-SU**, **Ni-SI**, and **Ni-R**

Figure 10.2. Various states of the [Ni/Fe] H-ase active site and plausible catalytic cyclic. The experimental values of v_{CO} for the CO ligand are shown as subscripts.

catalytically inactive and have two distinct EPR signals, i.e., two paramagnetic Ni states (Figure 10.2). The "unready" Ni-A state reacts only after hours of contact with H_2 or strong reductants to give the "ready" Ni-B state. Their ratios can be changed by controlling oxidation of the reduced enzyme, and a typical aerobically purified sample contains about 40% Ni-A, 10% Ni-B, and a 50% fraction that does not give an EPR signal. Upon further hydrogen activation or reductive titration, the Ni-A and Ni-B signals vanish to first give the Ni-SI state (silent intermediate), then the catalytically active form Ni-C that binds H_2 in a light-sensitive fashion and shows yet a third EPR signal due to $S = \frac{1}{2}$ Ni^I. This form is present in all [NiFe] enzymes and it has been proposed that it contains either H_2 or hydride ligands. The photoreaction at temperatures less than 77 K (to an EPR-active Ni-C' form) shows a strong kinetic isotope effect, $k_H/k_D = 6$. Also, CO is a competitive inhibitor of H_2 binding, forming a bound Ni–CO state, which also leads to Ni-C' on photolysis. These experiments support the role of Ni-C in catalysis. Full reduction of samples again gives an EPR silent state termed Ni-R (reduced).

The oxidation and spin states of Ni and whether they are variable (e.g., existence of Ni^{III}/Ni^I redox couples) are controversial, although recent X-ray spectroscopy is consistent with Ni-A in *D. gigas* being covalent Ni^{III} and all reduced forms of the enzyme (Ni-R, Ni-SU exhibiting high spin Ni^{II}.[14] All forms of H-ases (except perhaps the oxidized form of an Fe hydrogenase below), apparently contain Fe^{II}, which is the d^6 electronic state most favored for H_2 binding, although Fe^{III} or even Fe^I is possible in dinuclear M–M bonded systems. In addition to the active site, *D. gigas* has two associated 4Fe–4S clusters with midpoint redox potentials (-290 mV and -340 mV) similar to those of the respective Ni-SI/Ni-C and Ni-C/Ni-R couples. Altogether, twelve microstates, forming three cycles, are possible.

The X-ray structure of the enzyme from *D. gigas* provided the first crucial details about the coordination environment around the metals even though it was for the inactive air-oxidized "unready" form.[15,16] The overall protein unit is a heterodimer with subunits of masses 60,000 and 30,000, with the former containing one Ni and one Fe, and the latter the Fe–S clusters (one 3Fe–4S and two 4Fe–4S). Until the structure determinations in 1995 (2.85 Å resolution) and 1996 (2.54 Å), it was believed that the active site of the enzyme contained Ni coordinated by either pure S ligands or mixed (S,N,O) coordination. However, an Fe atom is identified as being close to Ni using X-ray data collected to 3.00 Å resolution on both sides of the Fe absorption edge and confirmed by ENDOR using ^{57}Fe.[17] The 2.54 Å structure revealed a second unusual characteristic of the active site: the presence of three exogenous diatomic ligands bound to Fe that presumably give three very-high-frequency IR bands at 1800–2100 cm^{-1} similar to those reported earlier[18,19] for the [NiFe] H-ase of *Chromatium vinosum*.[16,20] The IR positions are dependent on the redox state of the H-ase, and no signals in this region are found in other enzymes. Subsequently, Happe *et al.* identified the ligands as two CN$^-$ and one CO after elegant investigation of band shifts and intensities in ^{13}C- and ^{15}N-enriched samples of *C. vinosum* combined with the release of appropriate quantities of CO and cyanide on denaturation of the enzyme.[21]

All evidence points to the Fe(CO)(CN)$_2$ group as shown in **2**, where the S ligands are endogenous cysteinyl sulfurs (or selenocysteinyl in rare cases) and X is

(2)

oxygen (H$_2$O, OH$^-$, or O^{2-}). v_{CN} bands at 2093 and 2083 cm^{-1} and v_{CO} at 1945 cm^{-1} clearly shifted down 31–47 cm^{-1} on isotopic labeling. CO and CN ligands were later confirmed by X-ray diffraction in the even more organometallic-like [Fe] H-ases discussed below, suggesting their presence in all metal-containing H-ases. It should again be emphasized that *CO and CN had never been found as intrinsic constituents of a prosthetic group in biology*. These ligands stabilize both low redox (-100 to -500 mV) and low spin states of Fe and control the electronics at Fe for H$_2$ binding and activation. The protonation states of the cysteinyl sulfurs (S$^-$ versus SH) are unknown, and hence also the metal oxidation states, but diamagnetic d^6 Fe^{II} is likely. The oxidation states for Ni^I, Ni^{II}, or Ni^{III} in the various forms of the enzyme (Figure 10.2) are more controversial however. The 2.9 Å separation between the Ni and Fe atoms indicates that a M–M bonding interaction is present, also an important feature of the [Fe] H-ases. In addition to the metals at the active site, the enzyme contains one 3Fe–4S cluster and two 4Fe–4S clusters that are likely to be involved in the transfer of the electrons emanating from the redox reaction at the active site.

The crystal structure (2.15 Å resolution) of a *reduced* [NiFeSe] H-ase *D. baculatum* provides insight into the actual catalytically active Ni-R or Ni-C states.[22] The overall architecture of the active site is very similar to that in *D. gigas* but with Se replacing one S. Most significantly, however, the putative oxo ligand X present in the unready oxidized form is absent, and the Fe–Ni distance is 0.4 Å shorter than in the above oxidized *D. gigas* enzyme. Structure (**3**) suggests that the closely spaced M may now be bridged by a hydride, which cannot be seen by X-ray but is supported calculationally, as will be discussed below. Higher-resolution (up to 1.4 Å) structural determinations by Higuchi[23] of the [NiFe] H-ase from *D. vulgaris* Miyazaki in both oxidized and reduced forms also confirm the coordination of Fe

(3)

by CO/CN ligands. It is proposed that one of the latter is replaced by an SO ligand in this enzyme based on X-ray electron density, but this has not been confirmed either spectroscopically or by other methods. The overall structures are similar to Ni–Fe separations of 2.55 Å in the inactive oxidized form (versus 2.9 Å in *D. gigas*) and 2.59 Å upon reduction with H_2. Once again the bridging X is removed on reduction with H_2 (in the presence of its electron carrier, methylviologen), but here H_2S is evolved, indicating that X is sulfido, S^{2-}.

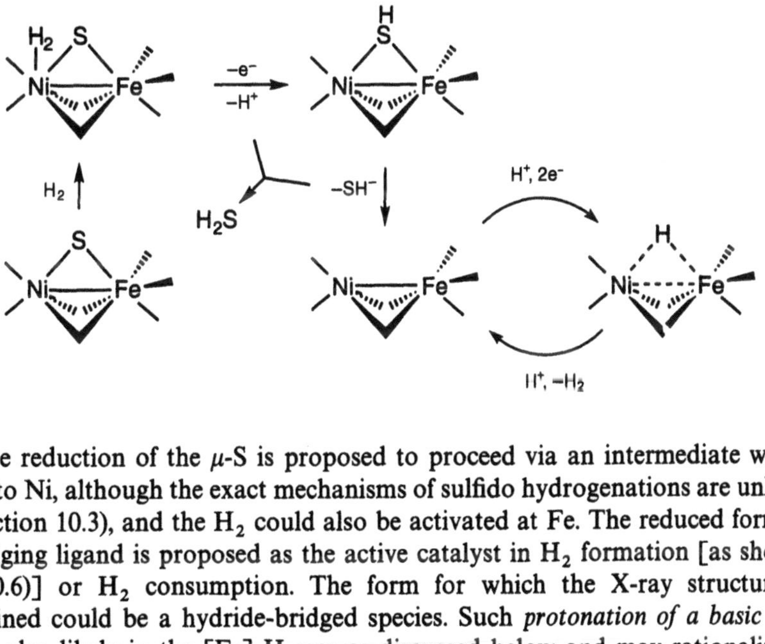

(10.6)

The reduction of the μ-S is proposed to proceed via an intermediate with H_2 bound to Ni, although the exact mechanisms of sulfido hydrogenations are unknown (see Section 10.3), and the H_2 could also be activated at Fe. The reduced form with no bridging ligand is proposed as the active catalyst in H_2 formation [as shown in Eq. (10.6)] or H_2 consumption. The form for which the X-ray structure was determined could be a hydride-bridged species. Such *protonation of a basic M–M bond* is also likely in the [Fe] H-ases as discussed below and may rationalize why two metal atoms are necessary when one would seem to suffice. The electronics, sterics, and certain other factors in H_2 activation at Fe should be unaffected by the presence of a second M. Organometallic complexes such as $[CpFe(CN)_2(CO)]^-$ and $[Fe_2(SR)_2(CO)_2(CN)]^{2-}$ synthesized as models for the Fe site show excellent agreement of ν_{CN} and ν_{CO} with those in *C. vinosum*, *D. gigas*, and *D. vulgaris* (Fe only).[9,24,25] Also, Cp*(dppf)RuH is an active "redox switch" catalyst for performing one-electron reductions with H_2.[26]

This very unusual (for biology) ligand set around Fe bears remarkable resemblance to many organometallic octahedral fragments that activate hydrogen. $Co(CN)_5^{3-}$ was one of the first organometallic complexes found to cleave H_2 homolytically [Eq. (9.10)], and $[Fe(H_2)(CN)(R_2PC_2H_4PR_2)_2]^+$ is known and can exist as an FeH(CNH) tautomer depending on R [see also Eq. (9.50)].[27] In H-ases the CN could be involved in proton transfer or important hydrogen-bonding interactions with protein components. Since CN is not a strong acceptor and is an excellent donor, it must be concluded that *CO is the crucial ligand in controlling the electronics of the system, specifically to increase the electrophilicity of the binding site to enhance both reversible molecular binding and heterolytic cleavage of H_2*. Remarkably, the highly electrophilic dicationic fragment $[Fe(CO)(R_2PC_2H_4PR_2)_2]^{2+}$ (R = Et, Ph) can still bind H_2 trans to CO in stable fashion via the enhanced σ donation from H_2, offsetting the greatly reduced backdonation (see Sections 4.6 and 4.7.2). This could be the case in the [Fe] H-ases in which both irons are surrounded by CO, including in one case a possible bridging CO. This would disfavor OA of H_2 to give nonlabile metal hydrides and increase the acidity of η^2-H_2 toward proton transfer to basic protein residues, such as cysteine, near the active site. The IR value of 1945 cm^{-1} believed to be due to CO in *C. vinosum* is characteristic of a fairly electrophilic metal center. An important experimental finding is that IR spectral changes occur when the H_2 atmosphere over the fully activated enzyme is replaced by CO gas. The ν_{CO} for the CO ligand that apparently binds to the site of H_2 activation is extremely high, 2060 cm^{-1},[19] and indicates a very electrophilic site. This site is even more electron-poor than those in $Fe(CO)_5$ ($\nu_{CO} = 2013$ cm^{-1}), and the acidity of η^2-H_2 bound here could far exceed normal physiological pH values.

DFT calculations starting from the X-ray structures and varying all plausible oxidation and spin states provide guidance for the mechanism of H_2 activation.[28-31] The first computations used a model $(CO)(CN)Fe(\mu-SH)_2Ni(SH)(SH_2)$ in which the overall charge is neutral in all species and S represents cysteinyl sulfur.[28] The bridging X was assumed to be removed (e.g., protonation of an OH$^-$ to labile H_2O) on reductive activation, and the M oxidation states of the "ready" form (Ni-B) were best represented by a spin-delocalized notation NiFe^{II,III}, i.e., no assiged individual oxidation states. The initial step of the mechanism involves heterolysis of Fe-bound H_2 in conjunction with CN temporarily moving to a bridging position (Ni–N≡C–Fe). This might rationalize the unusual presence of CN, although other ligands such as hydride or CO could also bridge the metals in the H_2 activation process. Energies are in general more critical tests of a model than are structures, and it is important that they match the experimental energetics of the H_2 reaction.[28] The activation of H_2 should have a barrier of about 10 kcal/mol, be slightly exothermic, and most likely include an H_2 complex along the reaction coordinate. Calculationally the only site to which H_2 binds significantly (3.1 kcal/mol) is the electrophilic Fe ($d_{HH} = 0.78$ Å), consistent with organometallic systems; the estimated barrier height for H–H cleavage is 8.7 kcal/mol.

Although these energetics appear proper, the calculations do not take into account the interactions of the active site with the protein environment that closely surrounds the CO/CN ligands with little freedom for movement.[30] Although this may not be the case for other H-ases (some [Fe] H-ases have only one point of attachment to the protein), terminal-bridging ligand reorientations might require large conformational changes with prohibitive energetics. Another possible mechanism (Scheme 1) features intramolecular heterolysis of H_2 on cysteine-530. Although not shown, H_2 initially is proposed to be split at Ni, which is pictured as

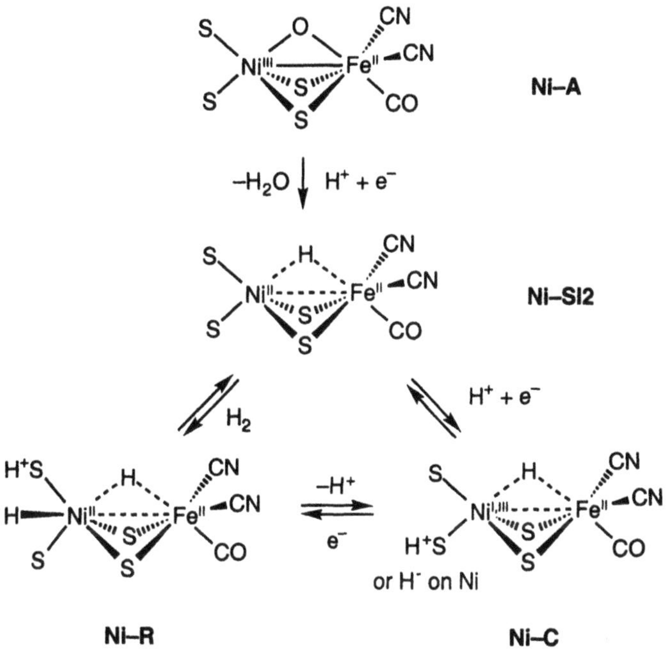

Scheme 1

containing both terminal and μ-hydrides in the Ni-R form. This is inspired by ENDOR studies that indicate that two types of exchangeable H nuclei are present in the vicinity of the Ni ligands in the Ni-C active form of a [NiFe] enzyme, consistent with μ-H.[32] In contrast, concurrent DFT calculations by Hall postulate Fe as the site of initial H_2 binding/heterolysis and incorporate monoanions as some of the key intermediates.[29] These computations do not take into account the protein backbone or hydrogen bonding of the CN to the protein known to be present,[16] potentially important considerations for future study. Optimized geometries (B3LYP) reveal that H_2 prefers to bind to Fe rather than Ni, and d_{HH} is 0.77 Å for a structure with H_2 trans to CO. The Fe^{II} center is perfectly configured for capture of H_2 as it diffuses to the active site, consistent with the existence of iron-only hydrogenases (see below). The H_2 coordination leads to an increase in d_{NiFe} with respect to that in Ni-SI. The proposed mechanism for H_2 activation is shown in Figure 10.3 and features hydride-bridged frameworks for the key intermediates that undoubtedly must be present on such dinuclear sites, as suggested by Fontecilla-Camps.[15] The calculated d_{NiFe} vary greatly in these species as shown, and this flexibility would be expected to facilitate both the electron and proton transfer processes (the M−M bond is a possible site for protonation). Although the proposed mechanisms may not be completely correct, the structure/bonding principles mirror those of H_2 activation on organometallic complexes.

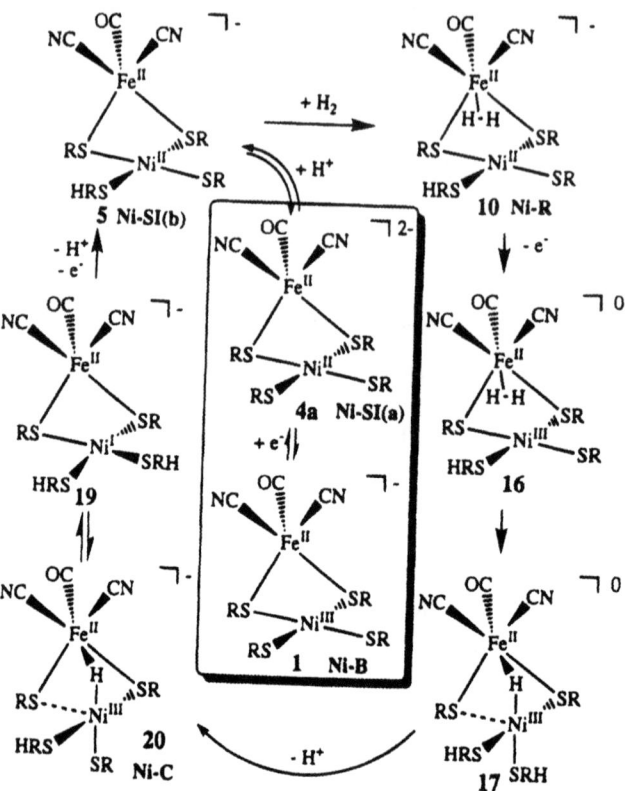

Figure 10.3. A possible mechanism for H-ase function as suggested by the calculations of Niu et al.[29]

10.1.3. A Metal-Free Hydrogenase

Remarkably, one H-ase found in methanogenic archaea, *Methanobacterium thermoautotrophicum*, does not contain transition metals at all![8] It catalyzes the reduction of a pterin compound by H_2 and also produces a proton, as a step in methane formation from CO_2 and H_2. One proposed mechanism is analogous to that of Olah[33] for the reversible formation of carbocations and H_2 from alkanes in superacid media, e.g., isobutane conversion [Eq. (10.7)]. However, the enzyme is

$$(CH_3)_3CH + H^+ \rightleftharpoons \left[(CH_3)_3C - \overset{H}{\underset{H}{|}} \right]^+ \rightleftharpoons (CH_3)_3C^+ + H_2 \quad (10.7)$$

active at pH as high as 7 partially because the positive charge is delocalized among conjugated N–C–N atoms rather than just at C, as shown by *ab initio* studies.[34,35] An electrophilic site modeled by the formamidinium ion in Eq. (10.8) is critical to

H_2 activation. The experimental free energy in the enzymatic process is very low (1.3 kcal/mol) as expected for a reversible system. The computed reaction enthalpies

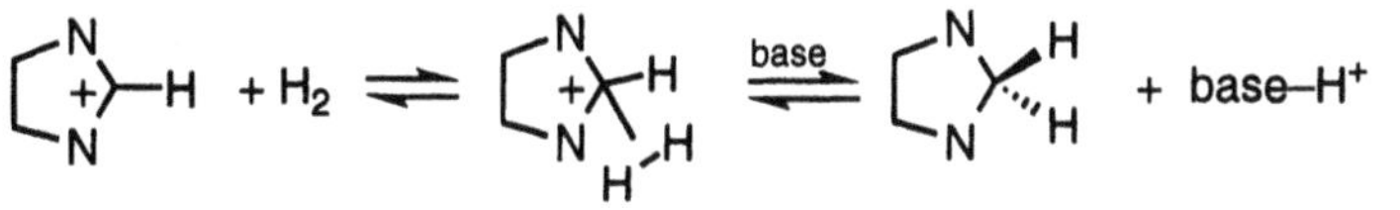

$$\begin{array}{c}\text{N}\\ \text{+}\!\!\!\!\diagup\!\!\!\text{H} \;+\; H_2 \end{array} \;\rightleftharpoons\; \begin{array}{c}\text{N}\\ \text{+}\!\!\!\!\diagup\!\!\!\text{H}\\ \text{N}\!\!\diagdown\!\!\text{H}\\ \text{H}\end{array} \;\underset{\text{base}}{\rightleftharpoons}\; \begin{array}{c}\text{N}\!\!\!\!\diagup\!\!\!\text{H}\\ \text{N}\!\!\diagdown\!\!\text{H}\end{array} \;+\; \text{base–H}^+ \qquad (10.8)$$

(in kcal/mol) are strongly dependent on the nature of the proton-accepting base, e.g., water (50.2) or its dimer (16.7), trimer (3.3), or tetramer (13.1). A primary amine such as the ω-amino group of a lysine residue is proposed as the actual primary proton acceptor (exothermic by 4.0 kcal/mol).[35]

10.1.4. [Fe] Hydrogenases: Highly Organometallic Active Sites

The above carbocation, R_3C^+, and electron-poor 16e metal centers such as $[Re(CO)_4(PR_3)]^+$ are strongly electrophilic, isolobal fragments. The CO ligands in H-ases would appear to be designed by Nature to increase the electrophilicity of the active site, thereby enhancing intramolecular heterolysis of H_2 as in carbonyl-rich $[Re(CO)_4(PR_3)(\eta^2\text{-}H_2)][MeB(C_6F_5)_3]$ illustrated in Scheme 3 in Chapter 9. Such electrophilic metal sites favor binding of H_2 over N_2, a potential competing ligand in metalloenzymes. The crystal structures of the Fe hydrogenases *Clostridium pasteurianum* (1.8 Å resolution)[36] and *Desulfovibrio desulfuricans* (1.6 Å)[37] point to a remarkable similarity between H_2 activation on organometallic centers and biological systems. Five CO and/or CN ligands are identified as being bound to a dinuclear Fe center in *C. pasteurianum*, including one in a bridging position along with two bridging SR ligands. An electron-transfer Fe_4S_4 "cubane" cluster is attached to Fe(1) via a cysteine thiol bridge as shown below in **4**, which represents one possible structure of the active site with one CN and CO on each Fe (alternatively two CO on Fe(2) would give a more electrophilic cationic Fe center analogous to **2**)

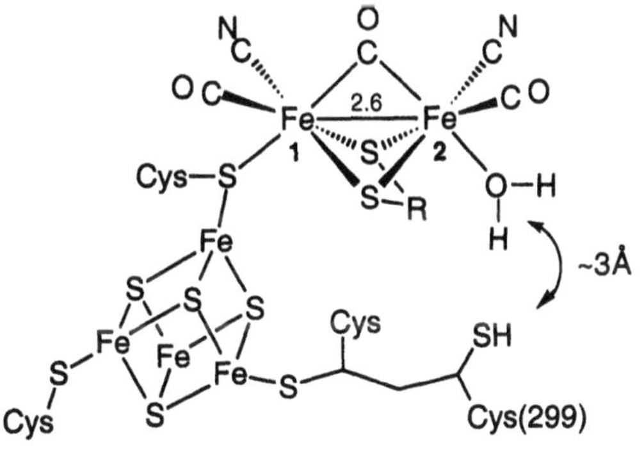

(4)

An Fe–Fe bond (2.6 Å) is present in both *C. pasteurianum* and *D. desulfuricans* and is typical of dithio-bridged organometallic Fe–Fe systems. The dinuclear Fe core is unusual in being low-spin with most of the ligands *exogenous*, the only attachment to the protein being through the cysteinyl sulfur bridging to the Fe_4S_4 cluster, i.e., a nearly independent organometallic complex. Mossbauer spectroscopy indicates that the Fe oxidation state is 2+ in the reduced form but $Fe^{II}Fe^{III}$ in the oxidized form.[38] The bridging diatomic ligand is probably CO and not CN, which does not bridge through carbon only (μ-CNH may be possible). This is crucial because it places CO trans to the aquo ligand located crystallographically on Fe(2), as in $W(CO)_3(PR_3)_2(H_2O)$, wherein H_2 can displace H_2O, and H_2 binding is favored by 1–2 kcal/mol over H_2O in terms of ΔG (see Chapter 7). In *C. pasteurianum* the probable site for H_2 binding/elimination is thus trans to μ-CO, which would stabilize σ-H_2 coordination, favor reversible binding and elimination of H_2, and promote heterolytic cleavage. Although CO ligands are crucial, additional CO is a known inhibitor of H_2 activation by the enzyme and irreversibly binds to the site occupied by H_2O [Eq. (10.9)] as shown crystallographically.[39] This mimics

$$\text{[structure]} \underset{h\nu}{\overset{CO}{\rightleftharpoons}} \text{[structure]} \overset{H_2}{\longrightarrow} N.R. \quad (10.9)$$

organometallic systems, where CO is also a much stronger ligand than H_2. Furthermore, X-ray diffraction studies of a single crystal of the CO adduct after photolysis shows dissociation of the CO and replacement by H_2O. The Fe–C distance to the μ-CO is significantly elongated when CO is trans to it, reflecting the strong competition for backbonding between the two trans CO ligands. The terminal CO trans to the μ-CO is more labile than the other CO ligands, which are trans to sulfur donors.

The more electrophilic cationic fragment $[Re(CO)_4(PR_3)]^+$ also coordinates H_2O trans to CO,[40] although the aquo ligand is less labile than in the neutral $W(CO)_3(PR_3)_2(H_2O)$ and appears to be more strongly bound than H_2. A binding energy of 23 kcal/mol is calculated in a model Fe^{II}–$Fe^{II}(H_2O)$ species for *D. vulgaris*, but reduction to Fe^{I}–Fe^{II} can release H_2O to make the site available for H_2 binding,[41] which may be key to activation of the oxidized inactive form. Addition of H_2 to a Fe^{I}–Fe^{II} species with an empty coordination site is computed to be exothermic by 6.1 kcal/mol ($d_{HH} = 0.824$ Å in the resulting η^2-H_2), and this EPR-active species is postulated to convert to an EPR silent Fe^{II}–$Fe^{II}(H_2)$ form via electron (or proton) transfer. Upon binding of H_2 to CO-rich, cationic complexes, the acidity of H_2 greatly increases, particularly when H_2 is trans to CO. Thus the $[Re–H_2]^+$ complex is much more easily deprotonated than the neutral W–H_2 species, which requires a strong base (see Section 9.4.1). In the hydrogenases, it is likely that the Fe active site is intermediate in electrophilicity. The active site of *D. desulfuricans* is similar to that of *C. pasteurianum*, but, in lieu of μ-CO, an oxygen donor such as H_2O or OH apparently bridges the irons (it could also be terminal) and Fe(2) is proposed to be coordinatively unsaturated.[37] A 1,3-propanedithiolate ligand bridges the Fe, similar to the thiols in **4** (where R could also be

$-CH_2CH_2CH_2-$). Assuming accurate crystallography, one explanation of the structural differences is that the two structures represent different oxidation states and that the open coordination site in *D. desulfuricans* is the potential site for H_2 binding (it may actually be occupied by H_2 since crystallization was carried out under H_2). Also, shifts of CO between terminal and bridging positions and similar ligand rearrangements are extremely facile in organometallic systems; thus in the enzyme mechanism H_2 and hydride ligands could be positioned trans to a variety at ligands in either bridging or terminal sites. Calculations (below) show that such transformations are nearly barrierless on models for the active site. Because of the many easily accessible ligand arrangements and strong trans-ligand influences, the active site is tremendously flexible for either consuming or releasing H_2, adjusting the acidity of η^2-H_2 for heterolysis, and attaining the low redox potentials typical of these active sites.

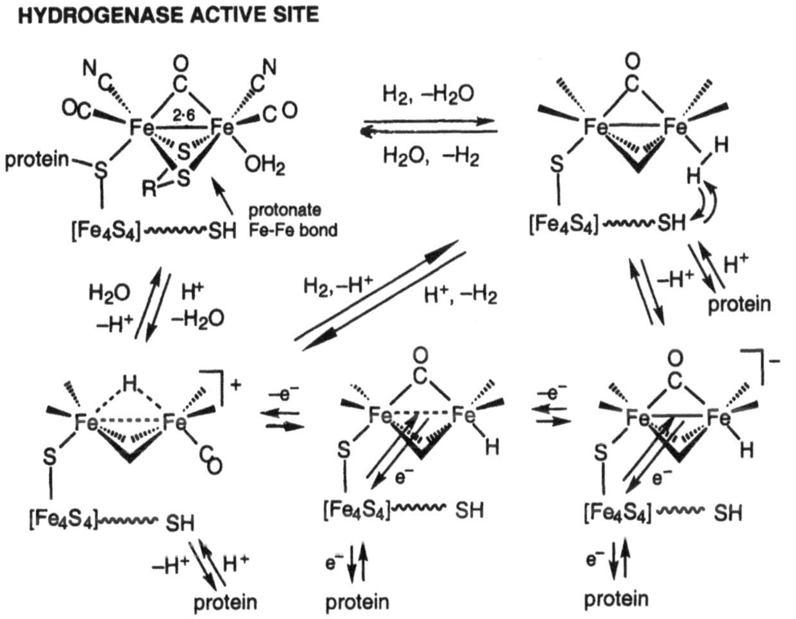

HYDROGENASE ACTIVE SITE

Scheme 2

Scheme 2 presents one (of many!) reasonable mechanism for reversible H_2 consumption/production on *C. pasteurianum*,[40] the basic principles of which could be applied to the function of other enzymes as well. Although the dinuclear center could be cationic (e.g., a CO replacing a CN) for ease of heterolysis of H_2, Scheme 2 assumes that, as generally believed, one CN and one CO is present on each Fe^{II} *and a low spin d^6 Fe^{II} octahedral configuration is well known to favor H_2 binding.* In the mechanism for H_2 consumption, an intermediate Fe–H_2 species could form by displacement of the H_2O ligand and transfer a proton to the cysteine-299 sulfur site conveniently located in close proximity to the H_2 binding site in 4. Both Crabtree and Morris have demonstrated that such intramolecular heterolytic cleavage readily occurs in inorganometallic complexes, as exemplified by the equilibrium proton transfer in Eq. (10.10), which is highly sensitive to the electronics at M and the nature of the ancillary ligands [see Eq. (9.44)].

$$\text{(Ir-OH}_2\text{ complex)} \xrightarrow[-H_2O]{H_2} \text{(Ir-}\eta^2\text{-H}_2\text{ complex)} \rightleftharpoons \text{(Ir-H, NH}_3^+\text{ complex)} \quad (10.10)$$

Transfer of a proton from η^2-H_2 to the μ-thiolates in H–ases is also possible, and calculations support such heterolysis (although it is endothermic by 15 kcal/mol).[41] Transfer of a proton to CN is nearly isoenergetic but a high barrier is computed (38 kcal/mol, compared to 17 kcal/mol for transfer to sulfide). The next steps involve movement of protons away from the active site and synchronous or asynchronous electron transfer to the cubane cluster and away from the site via other Fe–S clusters. The electrons in the H–H bond could essentially flow through the Fe–Fe bond and, depending on whether one- or two-electron transfer process takes place, one-electron Fe$\cdots$Fe bonds (2.9–3.1 Å)[42] may be present in the intermediates (one-electron transfer steps are shown in Scheme 2). The flexibility of the M–M separation (2.6-3.2 Å, corresponding to 0, 1, or 2e M–M bonds) could facilitate electron/proton transfer here and in the [NiFe] H–ases.

The reverse reaction, formation of H_2 from $2H^+$ and $2e^-$, involves protonation of the 2Fe center to form a metal hydride. The most basic site for initial protonation in the enzyme active sites may be the electrons in the M–M bonds, which can readily be reversibly protonated to form hydride-bridged species.[43] Electrophilic $[M-H_2]^+$ systems often prefer to form μ-H complexes rather than M–H on deprotonation of η^2-H_2 by external bases:

$$[M]^+ + H_2 \rightleftharpoons [M-H_2]^+ \xrightarrow[-BH^+]{B} M-H \xrightarrow{[M]^+} [M_2(\mu\text{-}H)]^+ \quad (10.11)$$

The Fe–Fe bonds in $[CpFe(CO)(PR_3)(\mu\text{-}CO)]_2$ are as basic as weak amines ($pK_b \sim 6$) and concomitant shift of μ-CO to terminal positions occurs on protonation [Eq. (10.12)].[44] Protonation of the Fe–Fe bond in $[Fe(CO)_2(PR_3)(\mu\text{-}SR')]_2$ occurs in preference to protonation of the sulfur ligands [Eq. (10.13)].[45] These are, however, Fe^I centers, and analogous protonation of Fe^{II}–Fe^{II} is not well known. The basicities of M–M bonds such as in $[CpRu(CO)_2]_2$ are substantially higher than that of the metal sites in related 18e mononuclear complexes and are highly sensitive to the nature of the ancillary ligands.[43] As discussed above, theoretical studies of [NiFe] H-ase mechanisms indicate that μ-H intermediates are energetically favorable. As shown in the reverse sequence of Scheme 2. the hydride could then shift to a terminal position and be protonated to a readily dissociable H_2 ligand, and the cycle would continue.

Such bridging/terminal shifts involving CO as well as H may be more likely in the [Fe] H-ase sites, which are attached to the protein only via the 4Fe–4S cluster, than in the [NiFe] sites, which are more tightly attached via cysteine groups that also bridge the metals. DFT calculations on [(MeS)(CO)(CN)Fe(μ-S)$_2$(μ-CO)-Fe(CO(CN)]^z ($z = 0$ to -2) models show that the μ-CO can easily shift like a gate

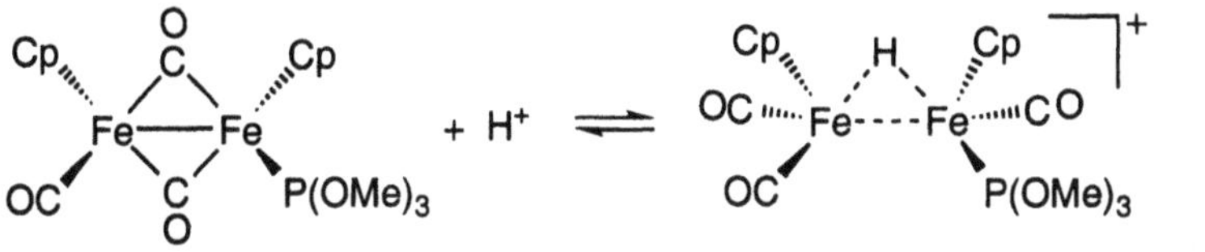

$$\text{Cp}_{\text{''}}\underset{\text{OC}}{\overset{\text{O}}{\underset{\text{C}}{\text{Fe}-\text{Fe}}}}\overset{\text{Cp}}{\underset{\text{P(OMe)}_3}{\text{''}}} + \text{H}^+ \rightleftharpoons \left[\text{OC}_{\text{''}}\overset{\text{Cp}}{\underset{\text{Fe}}{\text{Fe}}}\overset{\text{H}}{\text{''}}\overset{\text{Cp}}{\underset{\text{P(OMe)}_3}{\text{Fe}\text{''}\text{CO}}}\right]^+ \quad (10.12)$$

$$\text{R}_3\text{P}_{\text{''}}\underset{\text{OC}}{\overset{\text{Me}}{\underset{\text{S}}{\text{Fe}-\text{Fe}}}}\overset{\text{PR}_3}{\underset{\text{CO}}{\text{''}\text{CO}}} + \text{H}^+ \rightleftharpoons \left[\text{R}_3\text{P}_{\text{''}}\overset{\text{H}}{\underset{\text{Fe}}{\text{Fe}}}\overset{\text{PR}_3}{\underset{\text{CO}}{\text{''}\text{CO}}}\right]^+ \quad (10.13)$$

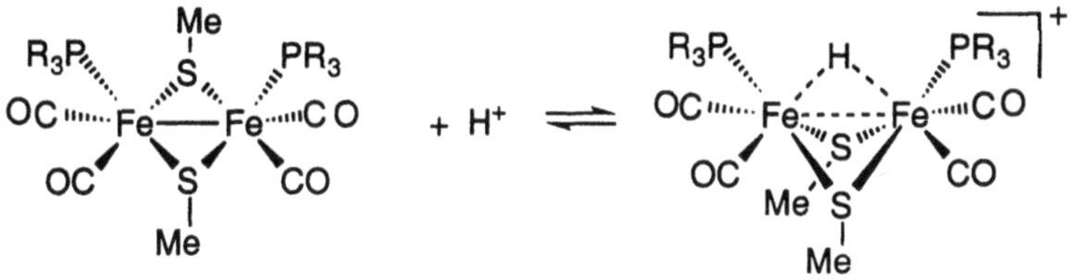

(where the O atom moves little but the carbon swings left or right).[46] Also the μ-S can join via S–S bonds, a variable not even considered above. It should be noted that the transformations among six different isomers at three possible redox levels

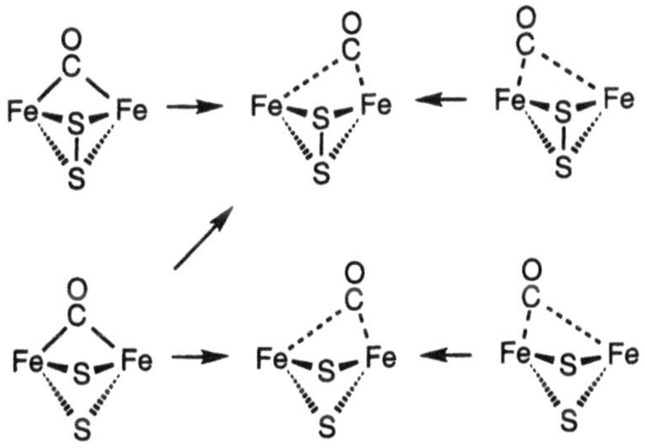

are virtually barrierless. The active site possesses a relatively flat potential energy surface for geometrical changes at Fe, CO, S, and bound H, which is consistent with the extremely rapid rates of H$_2$ production in the enzymes. H$_2$ weakly binds to Fe in the position of the H$_2$O ligand in the protein (5), but it is stabilized by a CO gate shift to the right (6). In its reduced state, 5^{2-} undergoes a mechanistically significant

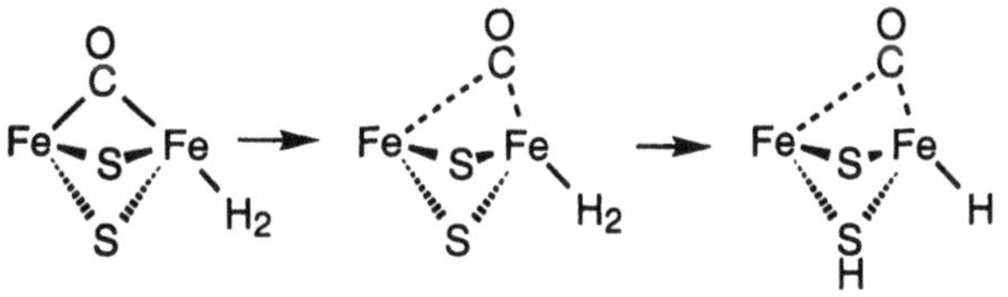

(5)   (6)   (7)

barrierless transfer of one H atom from Fe–H$_2$ to form SH (7^{2-}). An intermediate for proton shuttling between Fe and μ-S involves an α-agostic S–H interaction known experimentally for an Fe–thiolate (see Section 13.2).

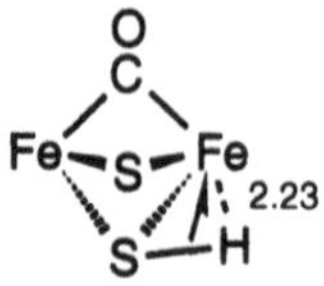

The above concepts serve to bridge the fields of organometallic and biochemistry, and modeling of the functional aspects of H–ases is still in its infancy,[9] e.g., **8**.[25]

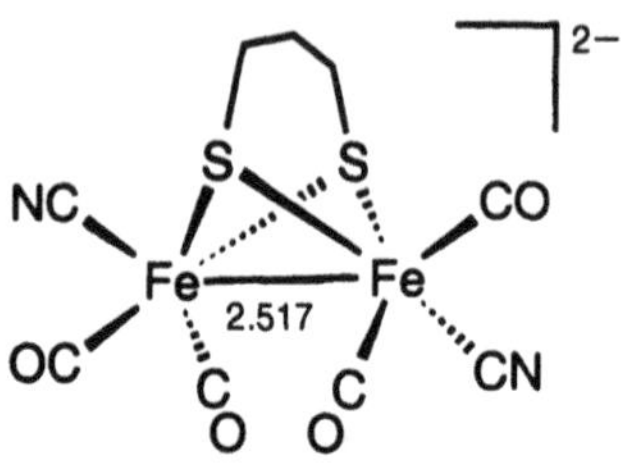

(8)

The bridging propanedithiolate, terminal CO and CN ligands, and the Fe–Fe bond in **8** mimic the site in *D. desulfuricans*, although it lacks the μ-CO and terminal water ligands. More challenging will be the design of the secondary structure critical to electron and proton transfer such as the Fe–S cubane cluster and proximal cysteine ligands. It is of interest that methane monooxygenase (MMO), which activates the C–H bond in CH_4 to produce methanol, also has a dinuclear iron active site, and the mechanism of oxygenases may involve alkane complexes (see Section 10.3).

10.2. NITROGENASES AND NITROGEN FIXATION

Hydrogen conversion is again of prime importance in nitrogen fixation to ammonia by N-ases,[47,48] and can be at least partially understood in terms of inorganometallic chemistry. Massive research efforts[6,7,10,49–52] have attempted to model the structure and function of N-ase for ammonia production. Billions of tons of NH_3 are produced annually worldwide by the Haber process, which actually is efficient and cheap (any major improvement is now recognized to be unlikely, at least from an economic standpoint).

$$N_2 + 3H_2 \xrightarrow[\text{Fe catalyst}]{400-550°C} 2NH_3 \quad (10.14)$$

In the absence of a substrate, N-ase acts like a H-ase and evolves H_2 by proton reduction, which cannot be completely suppressed even at high N_2 pressures. At least one H_2 is produced for every N_2 reduced, seemingly as a waste of reducing equivalents:

$$8e^- + 8H^+ + N_2 \rightarrow 2NH_3 + H_2 \tag{10.15}$$

The isolated Fe–Mo cofactor thought to be the active site of N-ase catalyzes hydrogen evolution at high potentials,[53] and H_2 is in fact an inhibitor of N_2 reduction. H_2 reduction is also proven by the formation of HD from D_2 gas and protons derived from H_2O, which occurs only in the presence of N_2[54]:

$$2H^+ + D_2 + 2e^- \rightarrow 2HD \tag{10.16}$$

This phenomenon implies that the H and D that form HD come from different sources that do not mix their hydrogen atoms and that this reaction is facilitated only when N_2 is bound. This in turn implies that the displacement of H_2 by N_2 at a single active site must be an associative process. A reasonable explanation for this is that N_2 binds to a trihydride species, MH_3 or $MH(H_2)$, with displacement of H_2. Subsequent loss of N_2 by reaction with protons toward NH_3 formation or by dissociation, followed by binding of D_2 would generate MHD_2, an obvious source of HD (see Section 9.3). The Lowe–Thorneley[55] model of the nitrogenase mechanism is consistent with generation of a trihydride species by protons binding to the reduced site prior to N_2 binding. Some H_2 is released during this process, as in labile H_2 complexes that readily exchange N_2 and H_2. Hughes et al. propose a scheme for H_2 evolution, H_2 binding, and reduction at the Mo site of the enzyme wherein a Mo dihydride species eliminates H_2 on reaction with N_2.[56]

However, this model does not explain why, in the comparable experiment performed under HD, no D_2 forms, nor why substrates other than N_2 do not promote HD formation. Also, if H_2 can interact with the active site, why is a substrate of any kind needed to promote HD formation? Displacement of H_2 is not a necessity for binding N_2, but why does HD form only when N_2 is being reduced? One simple answer proposed by Helleren et al. is that HD formation and N_2 binding occur at different places.[54] It is possible that different substrates bind to and are transformed at different parts of the large FeMoco (FeMo cofactor) site of N-ases discussed below. CO inhibits nitrogen fixation in N-ases but *not H_2 evolution*. A single site that binds H_2 and N_2 equivalently should be poisoned by CO for both H_2 and N_2 activity, and evidence increasingly points to multisite processes in the FeMoco cluster. However, a possible model (**10**) for HD formation at the same site as N_2 activation is discussed below.

There are three major types of N-ases: Mo–Fe, V–Fe, and Fe only, and, as for the H-ases, Fe appears to be the key metal. The crystal structures of the more common Mo-containing enzymes have been determined from three sources (resolution in parentheses): *Azotobacter vinelandii* (2 Å), *Clostridium pasteurianum* (2 Å), and *Klebsiella pneumoniae* (1.6 Å). The FeMoco site contains Fe (seven sulfide-bridged Fe atoms), along with a single six-coordinate Mo linked to three Fe atoms by sulfide bridges (Figure 10.4).[57,58] The FeMoco is remarkably similar in all the enzymes structurally characterized, and also present is a separate 8Fe–7S P-cluster that is less likely to be an N_2 binding site. Although the Mo seems to be coordinatively

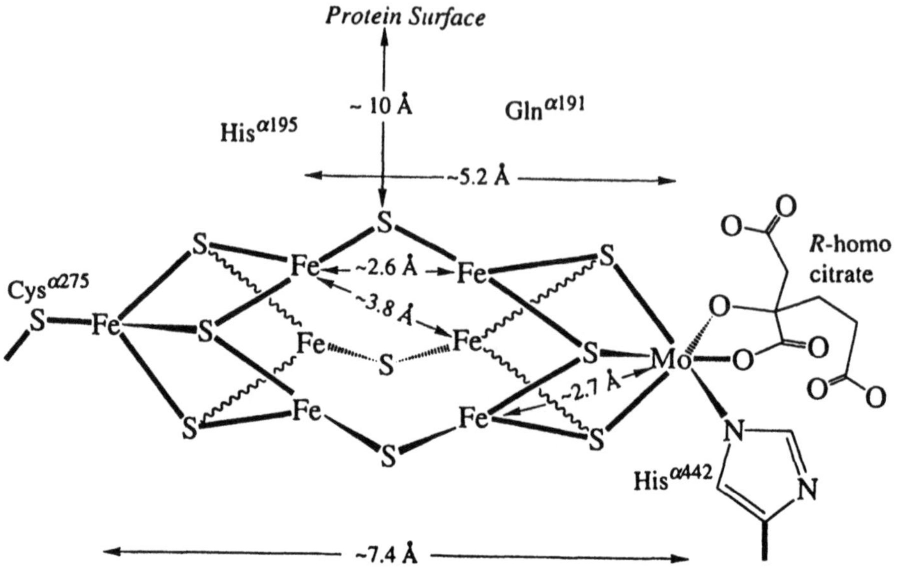

Figure 10.4. Structure of the FeMo cofactor in *Azotobacter vinelandii*. The oxygen donor atoms are from homocitrate and the nitrogen is from histidine imidazole. Reprinted with permission from Coucouvanis, et al. ACS Symp. Ser. **1996**, *653* (*Transition Metal Sulfur Chemistry*), 117. Copyright 1996 American Chemical Society.

saturated, six of the seven Fe atoms *appear* to be three-coordinate (a rare low coordination number for Fe) and thus are suitable sites for substrate binding. However, extensive M–M bonding is present to offset coordinative unsaturation, and Fe–Fe distances range from 2.46(10)–2.74(9) Å.[58] Theoretical studies of the mechanism of N_2 reduction are hampered by the large size and complexity of the problem but provide some insight.[59–61] Most predict extensive M–M bonding and η^1 coordination of N_2 to one or more Fe atoms on the face of the trigonal cavity rather than to Mo. H_2 can also bind and compete with N_2 binding. One model (**9**) only weakly binds N_2, but placing a hydrogen on the μ-S (reducing the irons to a Fe^I, Fe^{II} state) greatly increases the binding and favors a bridging mode.

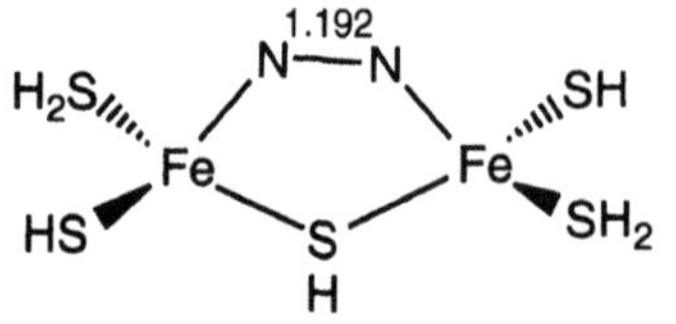

(**9**)

In **9** d_{NN} is elongated over that in free N_2 (1.10 Å). DFT calculations indicate that N_2 may bridge end-on to four Fe atoms with $d_{NN} = 1.29$ Å.[60] Transfer of H from proximal SH functionality on Fe to the μ-N_2 ligand is proposed,[6,59–61] and successive addition of hydrogens to form N_2H_x ($x = 2, 3, 4$) and eventually NH_3 is exothermic.[59-61] The key argument is the critical need for protons to bind to μ-S to activate N_2. These hydrogens are also proposed to form and release one H_2 molecule for every N_2 activated. In addition to the Fe sites, Mo could be involved as a site of hydride formation,[47] but it or other metals such as V may only be "fine-tuners" for electron transfer, and N_2 may be activated at the two Fe atoms in the "waist region" of FeMoco.[52] A core geometry based on a hybrid of the FeMoco structure with a dinuclear diazene complex, [Fe("N_HS_4")]$_2$(μ-N_2H_2), is a proposed N-ase model (**10**), where a diazene between the waist Fe represents partially reduced N_2). This model features η^2-D_2 binding that is claimed to be consistent with the severe constraints imposed on the above discussed N-ase catalyzed "N_2-dependent HD formation" from D_2 and protons.[54] Spectroscopy indicates that CO, the strongest N-ase inhibitor, binds to the same two Fe atoms in N-ase as those depicted in **10** to bind N_2.[62] As for the H-ases, changes in Fe–Fe bonding by electron

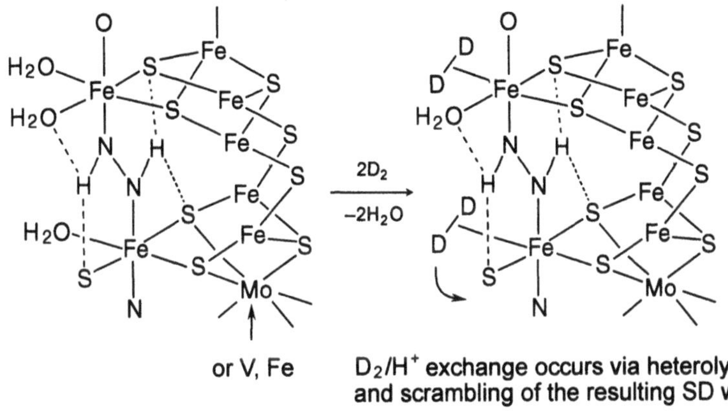

or V, Fe D_2/H^+ exchange occurs via heterolysis on S and scrambling of the resulting SD with NH

(**10**)

addition to the $MoFe_7S_9$ cofactor and/or the Fe_8S_7 "P-cluster" component may be crucial to binding and activation of N_2.[58,61,63] For example, recent synthetic analogues with $MoFe_3S_3$ cuboidal centers such as (catecholate)(py)$MoFe_3S_3$-$(CO)_4(PR_3)_3$, which model the top half of FeMoco, exhibit dramatic core structural changes associated with the ligand environment and total number of valence electrons.[63]

Much effort has been made to model biological N_2 activation synthetically and N≡N bond cleavage.[49-52] The N≡N bond is one of the strongest known; few inorganic complexes can break it, and the nitrogenase process is extremely difficult to duplicate or even fully comprehend. Only a very limited number of reactions of bound N_2 with H_2 are known, e.g., Eq. (10.17), which slowly occurs in toluene over 1–2 weeks for a dinuclear Zr complex capped by macrocyclic ligands with N and P donor atoms.[64,65] Neutron diffraction of **12** shows the formation of a N–H bond (0.93(6) Å) and a μ-H ligand (Zr–H = 1.95(6) and 1.98(6) Å) that lies in an extremely soft potential. DFT calculations indicate that Eq. (10.17) as well as a similar reaction with SiH_4 is exothermic by 13.0 and 19.7 kcal/mol, respectively (reaction with CH_4 is *endothermic* by 10.7 kcal/mol). Although an H_2 complex of **11** is not experimentally or calculationally stable (as for H_2 activation on sulfido ligands below), it would be expected that weak transient binding to Zr or a multicenter intermediate as in **12** occurs. Computations show that such a four-center "metathesis-like" transition state forms with an activation barrier for breaking the H–H bond of 21 kcal/mol and proceeds directly to **13**.[65] Addition of a second H_2 to **11** is calculated

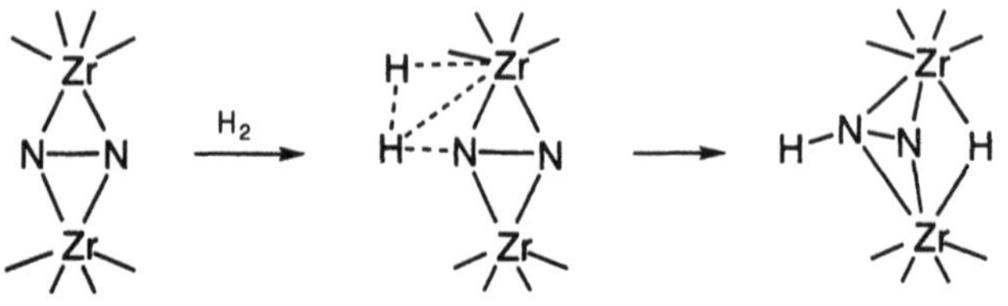

$$(11) \xrightarrow{H_2} (12) \longrightarrow (13) \quad (10.17)$$

(11)   (12)   (13)

to be feasible. This novel direct reaction of H_2 with coordinated N_2 suggests that addition of H_2 to N_2 on metal complexes and biosystems is possible under mild conditions as potential steps in ammonia synthesis.

Protonation of coordinated N_2 by Ru dihydrogen dihydride tautomers[66] and H_2 complexes[51] is also known:

$$trans\text{-}W(N_2)_2(dppe)_2 + 2[CpRuH_2(dtfpe)]BF_4$$
$$\longrightarrow [WF(NNH_2)(dppe)_2]BF_4 + 2CpRuH(dtfpe) + BF_3 \cdot thf + N_2 \quad (10.18)$$

$$cis\text{-}W(N_2)_2(PR_3)_4 + 6[RuCl(\eta^2\text{-}H_2)(dppp)_2]PF_6$$
$$\xrightarrow{55°C} RuCl(H)(dppp)_2 + 2NH_3 + W^{VI} \text{ species} \quad (10.19)$$

Although the production of NH_3 under mild conditions in Eq. (10.19) is encouraging, it is only stoichiometric, and the catalysis of such a reaction remains a major challenge involving a wide variety of scientific disciplines.

10.3. HYDROCARBON ACTIVATION BY OXYGENASE ENZYMES

The strong C–H bond in hydrocarbons, including methane, is also susceptible to biological activation via oxygenase enzymes that oxidize them to alcohols.[50] The mechanistic details are unknown, but intermediates containing σ-CH complexes (see Chapter 12) are possible. The catalytic oxygenation of hydrocarbons by the enzyme family, cytochromes P-450, had widely been considered to proceed through a "rebound mechanism" involving radical species.[67] The active oxidant is believed to be a porphyrinic high-valent iron-oxo species produced by a rate-determining electron transfer step, and a crystal structure of a complex with a camphor substrate hydrogen bonded near the $Fe^{IV}=O$ moiety has recently been determined by Schlichting et al.[68] In the subsequent, product-forming step, an H atom is proposed to be extracted from the substrate C–H bond to form a carbon radical that is transformed to the hydroxylated product in rapid successive electron- and atom-transfer steps.

Evidence for a radical intermediate includes C–H bond selectivities typical for a radical reaction and intramolecular isotope studies using substrates with CHD groups, which show primary isotope effects (k_H/k_D) of 11.3.[69] However, there are also

$$\underset{Fe^{n+2}}{\overset{RH}{\underset{\|}{O}}} \longrightarrow \underset{Fe}{\overset{R\cdot}{\underset{|}{OH}}} \longrightarrow \underset{Fe^n}{ROH} \qquad (10.20)$$

results inconsistent with a long-lived free-radical intermediate in P-450 hydroxylations.[70] Also, hydroxylations by P-450 and MMO (see below) show a regiochemical selectivity very different from that of a *tert*-butoxyl radical.[71] Alternate mechanisms involving σ-alkane complexes as intermediates have been proposed both for P-450[72] and a synthetic "Gif" system containing Fe=O bonds.[73] High-valent transition metal oxides can react with H_2, e.g., the oxidation of H_2 by permanganate ion, which shows second-order kinetics.[74] Thus a σ-alkane complex might be a key intermediate prior to oxygen transfer. As shown theoretically,[75]

$$\underset{Fe^{n+2}}{\overset{O}{\|}} \overset{RH}{\rightleftharpoons} \underset{Fe-}{\overset{O}{\|}}\underset{R}{\overset{H}{|}} \rightleftharpoons \underset{Fe-R}{\overset{O-H}{|}} \longrightarrow \underset{Fe^n}{\overset{R_{\diagdown}O^{\diagup}H}{|}} \longrightarrow \underset{Fe^n}{ROH}+ \qquad (10.21)$$

the coordinated C–H proton should have increased acidity as for bound H_2 and immediately migrate to the basic oxo group, i.e., intramolecular heterolytic cleavage analogous to that for H_2 (see Section 9.4.4). This process may be related to cleavage of H_2 on metal oxides (see Section 4.14) and metal sulfides (see Section 10.4) in heterogeneous catalysis. Reductive elimination of the alkyl and OH groups would then lead to the alcohol product [Eq. (10.21)] in, e.g., P-450.[72] The Gif system functionalizes alkanes to alcohols and ketones by a mechanism proposed to have a σ interaction as the key step, similar to that in Eq. (10.21). A "Sleeping Beauty effect" is described in which a "dormant" Fe-oxo complex is "kissed" by (collides with) the alkane (the prince) to form the reactive σ complex. The highly mobile H^+ on the bound $C^{\delta-}\cdots H^{\delta+}$ can rapidly hop over to the basic oxo ligand. This type of bond activation could explain why alkanes can be oxidized here even in the presence of substances such as H_2S, which are normally far more easily oxidized than saturated hydrocarbons. However, radical processes cannot be completely ruled out, and the mechanisms here and in biological systems are still controversial.

Methane is produced in enormous quantities in nature by methanogenic bacteria, but much of it is immediately consumed in aqueous environments by methanotrophic bacteria, which oxidize CH_4 as their sole source of carbon and energy. Thus for the bonds in H_2 and N_2, the strong C–H bond in methane is activated biologically, in this case converted to methanol by the enzyme MMO[76]:

$$CH_4 + O_2 + NAD(P)H + H^+ \rightarrow CH_3OH + H_2O + NAD(P)^+ \qquad (10.22)$$

Two reducing equivalents from NAD(P)H are used to split the O–O bond of O_2 as the first step. One oxygen atom is then reduced to water while the second is inserted into CH_4 to give methanol. Although P-450 catalyzes similar reactions, only MMO can activate methane. It also can oxidize a large variety of other hydrocarbons up to C_8, including halogenated and heterocyclic species.

The crystal structure of the enzyme contains another workhorse dinuclear iron center, although here it is bridged by oxygen donor ligands such as glutamate in **18**.

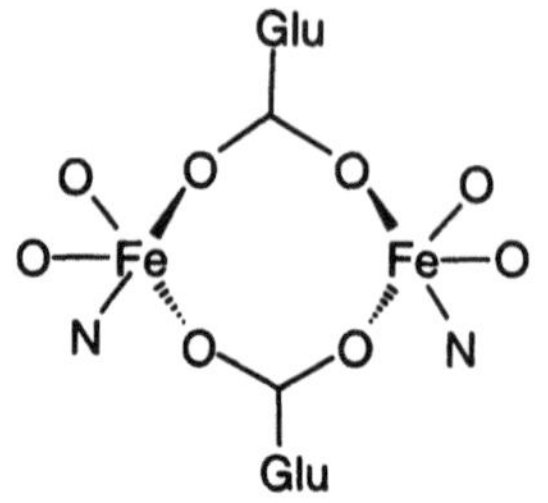

(18)

The reduced form of MMO is oxidized by direct reaction with O_2 and several intermediates are observed. The active species in the methane conversion is denoted as compound **Q** and is proposed to contain Fe^{IV} centers bridged by glutamate and oxo groups, the first of its type in biology.

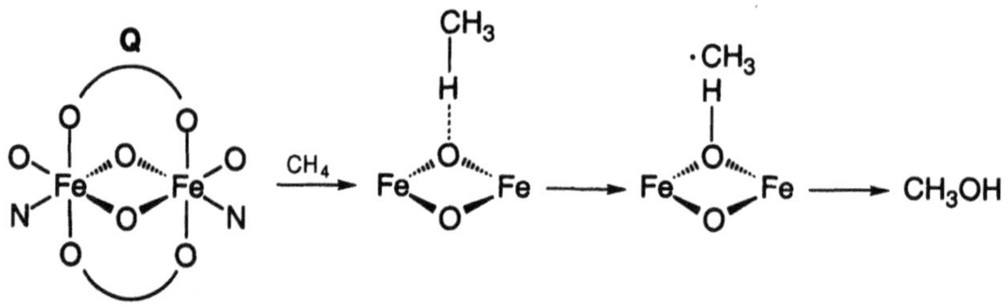

$$\qquad\qquad\qquad\qquad\qquad\qquad\qquad\qquad\qquad\qquad\qquad\qquad\qquad (10.23)$$

EXAFS measurements[77] show an extremely short d_{FeFe} of 2.46 Å, and strong ferromagnetic and antiferromagnetic coupling renders the compound diamagnetic (it is assumed that Fe–Fe bonding as in the H-ases is not present). The mechanism of methane activation is controversial and may occur by abstraction of a hydrogen from CH_4 by one of the μ-oxo to give an OH ligand and a free methyl radical, which combine to form methanol, but this is not clear. Calculations support a transition state such as in Eq. (10.23) consistent with large kinetic isotope effects and experiments based on chiral substrates, and the $\cdot CH_3$ radical then combines with the μ-OH with no barrier.[78] The calculated barrier for the rate-determining step, methane H-abstraction, is 19 kcal/mol. Many mechanisms are possible for C–H cleavage, and it is important to note that a distinctly *nonradical* pathway has recently been identified involving a "cationic" component[80] in a soluble MMO system (*Methylococcus capsulatus*).[79] A heterolytic cleavage type of process was proposed as one of several possible mechanisms on a structure of **Q** with Fe=O bonds by Liu and Lippard [Eq. (10.24)],[76] and calculations suggest that this could occur on

Fe–O–Fe also.[75] An electrophilic cationic system would favor rapid heterolytic cleavage of the C–H bond (see Section 12.4.3), but there is no evidence for interaction of CH_4 with the Fe atoms or a heterolysis of C–H. Carbocation-like

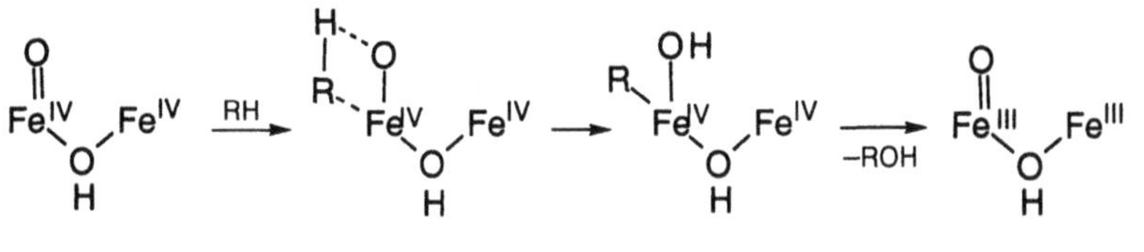

$$Fe^{IV}\underset{H}{\overset{O}{\diagdown}}\overset{O}{\underset{}{\diagup}}Fe^{IV} \xrightarrow{RH} R\cdots\overset{H\cdots O}{\underset{H}{\diagdown}}Fe^{V}\underset{}{\overset{}{\diagdown}}\overset{O}{\underset{}{\diagup}}Fe^{IV} \longrightarrow R\overset{OH}{\underset{H}{\diagdown}}Fe^{V}\underset{}{\overset{}{\diagdown}}\overset{O}{\underset{}{\diagup}}Fe^{IV} \xrightarrow[-ROH]{} Fe^{III}\underset{H}{\overset{O}{\diagdown}}\overset{O}{\underset{}{\diagup}}Fe^{III} \quad (10.24)$$

intermediates as in Eq. (10.8) may be possible on these cationic centers,[79] and C–H activation may be oxo-ligand–based as proposed for base-assisted H_2 splitting on sulfide ligands [Eq. (10.29) below]. Mimics for MMO are being sought, including inorganic Fe complexes in zeolitic matrices (ZSM-5), for selective hydrocarbon oxidations.[81]

10.4. MODELS FOR BIOLOGICAL AND INDUSTRIAL ACTIVATION OF HYDROGEN ON SULFUR LIGANDS AND SULFIDES

Several model complexes for N-ase and H-ase have been investigated in addition to those noted above. Systems with sulfur ligands capable of heterolytically cleaving H_2 are highly relevant, although isolable N_2 or H_2 complexes containing at least one "soft" sulfur coligand are still rare compared with complexes with phosphine, CO, Cp, or other "hard" ligands. The activation of H_2 and other small molecules on transition metal sulfur compounds has been reviewed.[82] In addition to biological systems, hydrodesulfurization (HDS) catalysts vital to the petroleum industry contain metal sulfides, typically MoS_2 and RuS_2, often with Co promoters.[83] HDS is said to be the world's largest man-made chemical reaction, and all crude oil is treated to remove organosulfur compounds:

$$S \text{ (in oil)} + H_2 \xrightarrow{\text{Mo-S}} H_2S \xrightarrow{\text{Claus process}} \text{Sulfur} \quad (10.25)$$

Heterolysis of H_2 to form M–H and M–SH groups can be modeled on a Ni_3S_2 cluster, and a transient $Ni–H_2$ species is calculated to be stable by about 16 kcal/mol and energetically capable of transferring one H to [Eq. (10.25)]

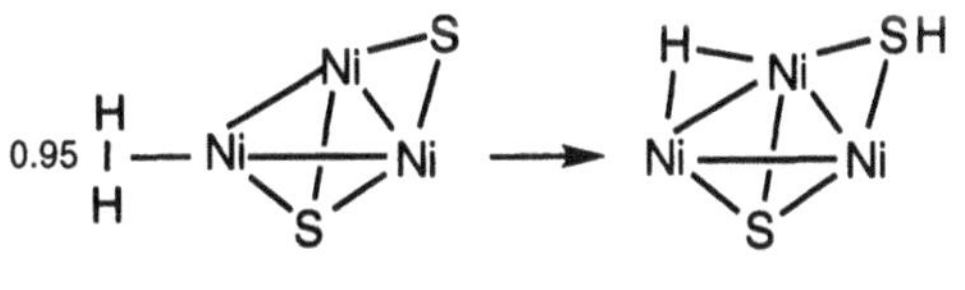

$$0.95\ \overset{H}{\underset{H}{|}} - Ni\overset{\diagup Ni\diagdown S\diagdown}{\underset{\diagdown S\diagup}{\diagup}}Ni \longrightarrow \overset{H-Ni-SH}{\underset{Ni\diagdown S\diagup Ni}{}}$$

H_2 also readily reacts with a few organometallic sulfides to give SH complexes [Eq. (10.26)], which can show exchange behavior [Eq. (10.28)].[85–88] Although the mechanism of Eq. (10.26) is unknown,[85,88] a four-center S_2H_2 or $MoSH_2$

transition state or a reduction coupled with proton transfer as in metalloenzymes can be envisioned. Equation (10.27) represents the first example of H_2 addition to a nonbridging disulfide complex.[86] An undetected H_2 complex may explain NMR evidence for H-atom exchange in Eq. (10.28), including the protons in dissolved H_2

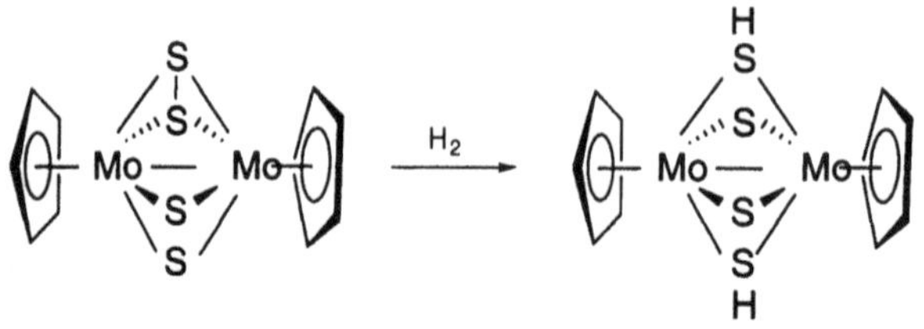

$$(10.26)$$

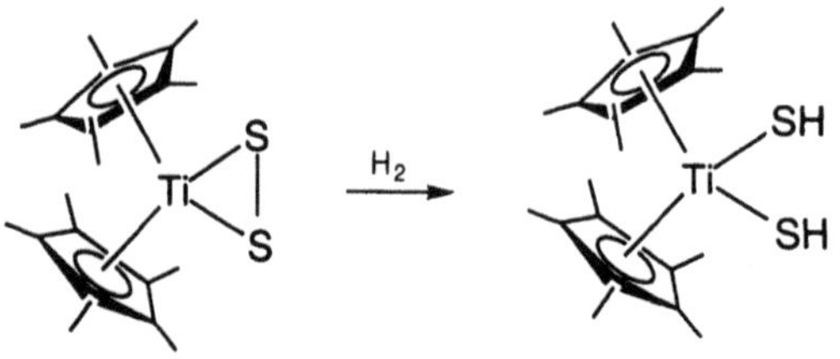

$$(10.27)$$

gas.[86] A related Mo–S system shows reaction of H_2 with saturated cationic sulfide-bridged complexes in the presence of a base, which may be explainable by direct attack of H_2 on S to form a 3c-2e S–H_2 interaction, followed by base-assisted

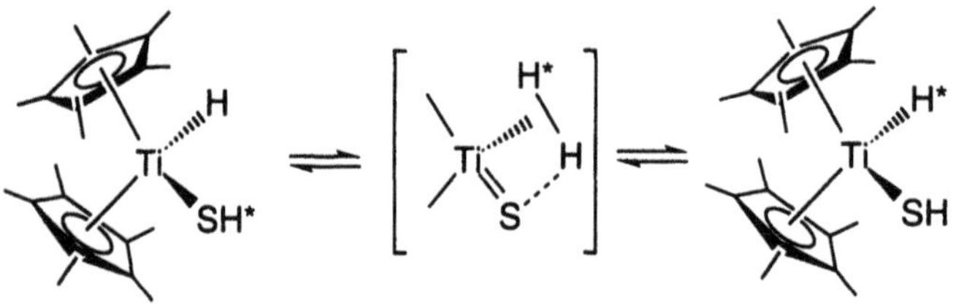

$$(10.28)$$

heterolytic cleavage of H_2.[85,88] Although this type of reaction is quite rare, the possibility that activation of H_2 could be entirely *sulfide ligand-based* in certain biological and industrial catalyst systems must be considered. This may parallel

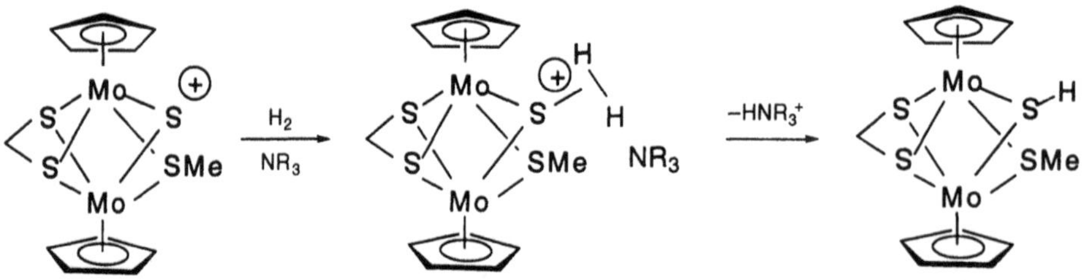

$$(10.29)$$

activation of H_2 on metal-free H-ases (Eq. (10.8) or C–H bond cleavage on cationic MMO oxo sites [Eq. (10.24)]. The richness and versatility of *Mo-based* clusters in undergoing such unique reactions, which can involve internal Mo-S redox processes, could relate to their presence in N-ase and HDS catalysts (W analogues do not display the reactivity in Eqs. (10.26) and (10.29).[89] Much research has been devoted to organometallic models for the mechanism of HDS and HDN (hydrodenitrogenation).[90] The SH ligand could be a key reactive component and ammonia has been produced in 78% yield (after base distillation) by the reaction in Eq. (10.30)[91]:

$$\textit{cis-}W(N_2)_2(PMe_2Ph)_4 + 10[Cp^*_2Ir_2(\mu\text{-SH})_3]Cl \xrightarrow[55°, 24h]{C_2H_4Cl_2-C_6H_6} NH_3 + \cdots \quad (10.30)$$

η^2-H_2 ligands that react intramolecularly with S ligands are known and include thiol species such as that in Eq. (9.46). In order for proton transfer from η^2-H_2 to a coordinated base to occur, the pK_a of the H_2 ligand and the protonated base must be similar. Morris has estimated that coordinated alkanethiol ligands have pK_a values between 5 and 10, which matches well with the acidity of many H_2 ligands.[82] Protonation of an anionic Ru hydride using CD_3OD gives an unstable HD complex with $J_{HD} = 32$ Hz. This reaction can be reversed by displacing the

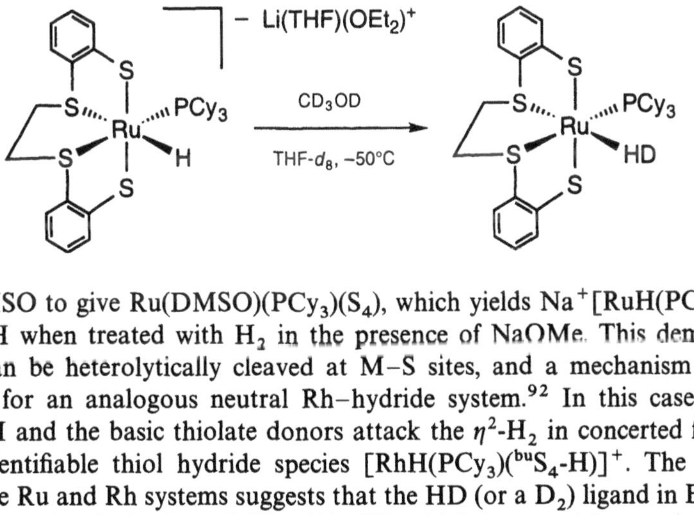

(10.31)

H_2 by DMSO to give $Ru(DMSO)(PCy_3)(S_4)$, which yields $Na^+[RuH(PCy_3)(S_4)]^-$ and MeOH when treated with H_2 in the presence of NaOMe. This demonstrates that H_2 can be heterolytically cleaved at M–S sites, and a mechanism has been elucidated for an analogous neutral Rh–hydride system.[92] In this case the electrophilic M and the basic thiolate donors attack the η^2-H_2 in concerted fashion to give an identifiable thiol hydride species $[RhH(PCy_3)(^{bu}S_4\text{-H})]^+$. The similarity between the Ru and Rh systems suggests that the HD (or a D_2) ligand in Eq. (10.31) can be intramolecularly cleaved [Eq. (10.32)], which is essential to rationalize the D_2/H^+ exchange between D_2 and EtOH that these complexes catalyze. For the Ru

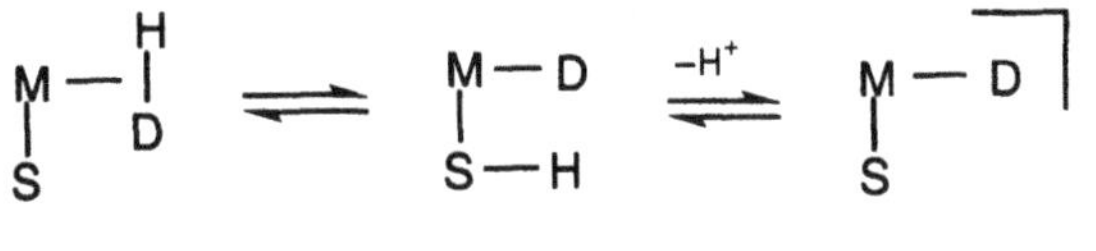

(10.32)

system the thiol hydride could not be detected, while for the Rh system and also $[IrH_2(HS(CH_2)_3SH)(PCy_3)_2]^+$ (which similarly catalyzes D_2/H^+ exchange[93]), the H_2 complex could not be seen but is a transient. A related system, $Ni(NHP^nPr_3(S_3)$, that models the Ni site in [FeNi] H-ases clearly shows that heterolysis of D_2 can also occur at Ni^{II}.[94]

A final question, which also relates to protonation at hydride versus halide ligands (see Section 9.4.2), is what happens when a complex containing both a thiolate (SR^-) and a hydride ligand is reacted with a Brønsted acid? Four products are possible [Eq. (10.33)],[95] including an η^2-thiol ligand structure (15), which has been identified in $Fe(\eta^2\text{-HSMe})(CO)_3(PEt_3)$ formed by protonation of $[Fe(SMe)(CO)_3(PEt_3)]^-$ and is the only example of an agostic complex of an S–H bond.[96] Treatment of trans-$Os(H)(PhS)(dppe)_2$ with $HBF_4 \cdot Et_2O$ leads exclusively to 14, showing that protonation at sulfur is both kinetically and thermodynamically

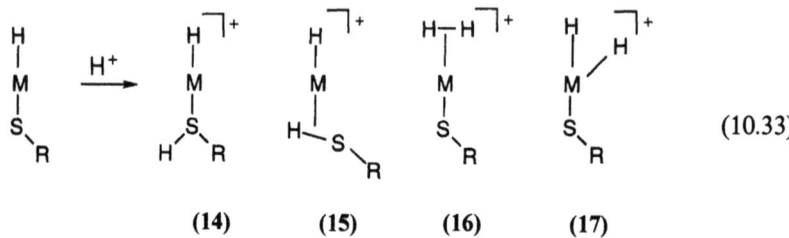

(14) (15) (16) (17)

(10.33)

favored (even for a less electron-rich congener).[95] The resultant thiol ligand is now very labile and can be displaced by weak ligands including H_2, N_2, and O_2, indicating that removal of S ligands, i.e., reactivation of "sulfur poisoned" catalyst sites, is possible by protonation pathways. It is interesting that further protonation of $[Os(H)(PhSH)(dppe)_2]^+$ takes place at the hydride if a large excess of acid is present to give a dicationic $[Os(H_2)(PhSH)(dppe)_2]^{2+}$ containing both H_2 and thiol ligands.

REFERENCES

1. R. H. Crabtree, *The Organometallic Chemistry of the Transition Metals*, Wiley: New York, 1988.
2. Crabtree, R. H. *Inorg. Chim. Acta* **1986**, *125*, L7.
3. Frey, M. *Struct. Bonding (Berlin)* **1998**, *90*, 97; Cammack, R. *Nature* **1995**, *373*, 556; Fontecilla-Camps, J. C.; Ragsdale, S. W. *Adv. Inorg. Chem.* **1999**, *47*, 283; Adams, M. W. W.; Stiefel, E. I. *Curr. Op. Struct. Biol.* **2000**, *4*, 214.
4. Adams, M. W. W.; Stiefel, E. I. *Science* **1998**, *282*, 1842.
5. Cammack, R. *Nature* **1999**, *397*, 214.
6. Sellmann, D.; Sutter, J. *Acc. Chem. Res.* **1997**, *30*, 460.
7. Holm, R. H.; Kennepohl, P.; Solomon, E. I. *Chem. Rev.* **1996**, *96*, 2239.
8. Thauer, R. K.; Klein, A. R.; Hartmann, G. C. *Chem. Rev.* **1996**, *96*, 3031.
9. Darensbourg, M. Y.; Lyon, E. J.; Smee, J. J. *Coord. Chem. Rev.* **2000**, *206–207*, 533.
10. Henderson, R. A. *J. Chem. Soc., Dalton Trans.* **1995**, 503.
11. Albracht, S. P. J. *Biochim. Biophys. Acta* **1994**, *1188*, 167, and references to review articles therein.
12. Montet, Y.; Amara, P.; Volbeda, A.; Vernede, X.; Hatchikian, E. C.; Field, M. J.; Frey, M.; Fontecilla-Camps, J. C. *Nat. Struct. Biol.* **1997**, *4*, 523.
13. Bian, S.; Cowan, J. A. *Coord. Chem. Rev.* **1999**, *190–192*, 1049, and references therein.
14. Wang, H.; Ralston, C. Y.; Patil, D. S.; Jones, R. M.; Gu, W.; Verhagen, M.; Adams, M.; Ge, P.; Riordan, C.; Marganian, C. A.; Mascharak, P.; Kovacs, J.; Miller, C. G.; Collins, T. J.; Brooker, S.; Croucher, P. D.; Wang, K.; Stiefel, E. I.; Cramer, S. P. *J. Am. Chem. Soc.* **2000**, *122*, 10544.

15. Volbeda, A.; Charon, M. H.; Piras, C.; Hatchikian, E. C.; Frey, M.; Fontecilla-Camps, J. C. *Nature* **1995**, *373*, 580.
16. Volbeda, A.; Garcin, E.; Piras, C.; de Lacey, A. L.; Fernandez, V. M.; Hatchikian, E. C.; Frey, M.; Fontecilla-Camps, J. C. *J. Am. Chem. Soc.* **1996**, *118*, 12989.
17. Huyett, J. E.; Carepo, M.,; Pamplona, A.; Moura, I., Moura, J. J. G., Hoffman, B. M. *J. Am. Chem. Soc.* **1997**, *119*, 9291.
18. Bagley, K. A.; Duin, E. C.; Rosebloom, W.; Albracht, S. P. J.; Woodruff, W. H. *Biochemistry* **1995**, *34*, 5527.
19. Bagley, K. A.; Van Garderen, C. J.; Chen, M.; Duin, E. C.; Albracht, S. P. J.; Woodruff, W. H. *Biochemistry* **1994**, *33*, 9229.
20. De Lacey, A. L.; Hatchikian, E. C.; Volbeda, A.; Frey, M.; Fontecilla-Camps, J. C.; Fernandez, V. M. *J. Am. Chem. Soc.* **1997**, *119*, 7181.
21. Happe, R. P.; Rosebloom, W.; Pierek, A. J.; Albracht, S. P. J.; Bagley, K. A. *Nature* **1997**, *385*, 126; Pierek, A. J.; Rosebloom, W.; Happe, R. P.; Bagley, K. A.; Albracht, S. P. J. *J. Biol. Chem.* **1999**, *274*, 3331.
22. Garcin, E.; Vernede, X.; Volbeda, A.; Hatchikian, E. C.; Frey, M.; Fontecilla-Camps, J. C. *Structure* **1999**, *5*, 557.
23. Higuchi, Y.; Yagi, T.; Yasuoka, N. *Structure* **1997**, *5*, 1671; Higuchi, Y.; Ogata, H.; Miki, K.; Yasuoka, N.; Yagi, T. *Structure* **1999**, *7*, 549.
24. Darensbourg, D. J.; Reibenspies, J. H.; Lai, C.-H.; Lee, W.-Z.; Darensbourg, M. Y. *J. Am. Chem. Soc.* **1997**, *119*, 7903; Hsu, H.-F.; Koch, S. A.; Popescu, C. V.; Munck, E. *J. Am. Chem. Soc.* **1997** 8371. Kaasjager, V. E.; Henderson, R. K.; Bouwman, E.; Lutz, M.; Spek, A. L.; Reedijk, J. *Angew. Chem. Int. Ed. Engl.* **1998**, *37*, 1668; Davies, S. C.; Evans, D. J.; Hughes, D. L.; Longhurst, S.; Sanders, J. R. *Chem. Commun.* **1999**, 1935; Liaw, W.-F.; Lee, N.-H.; Chen, C.-H.; Lee, C.-M.; Lee, G.-H.; Peng, S.-M. *J. Am. Chem. Soc.* **2000**, *121*, 488; Cloirec, A. L.; Best, S. P.; Borg, S.; Davies, S. C.; Evans, D. J.; Hughes, D. L.; Pickett, C. *J. Chem. Commun.* **1999**, 2285.
25. Schmidt, M.; Contakes, S. M.; Rauchfuss, T. B. *J. Am. Chem. Soc.* **1999**, *121*, 9736.
26. Hembre, R. T.; McQueen, S.; Day, V. W. *J. Am. Chem. Soc.* **1996**, *118*, 798.
27. Amrhein, P. I.; Drouin, S. D.; Forde, C. E.; Lough, A. J.; Morris, R. H. *J. Chem. Soc. Chem. Commun.* **1996**, 1665.
28. Pavlov, M.; Siegbahn, P. E. M.; Blomberg, M. R. A.; Crabtree, R. H. *J. Am. Chem. Soc.* **1998**, *120*, 548; Siegbahn, P. E. M.; Blomberg, M. R. A. *Chem. Rev.* **2000**, *100*, 421.
29. Niu, S.; Thomson, L. M.; Hall, M. B. *J. Am. Chem. Soc.* **1999**, *121*, 4000.
30. Amara, P.; Volbeda, A.; Fontecilla-Camps, J. C.; Field, M. J. *J. Am. Chem. Soc.* **1999**, *121*, 4468.
31. De Gioia, L.; Fantucci, P.; Guigliarelli, B.; Bertrand, P. *Inorg. Chem.* **1999**, *38*, 2658.
32. Fan, C.; Teixeira, M.; Moura, J.; Moura, I.; Huynh, B.-H.; LeGall, J.; Peck, H. D., Jr.; Hoffman, B. M. *J. Am. Chem. Soc.* **1991**, *113*, 20.
33. Olah, G. A.; Hartz, N.; Rasul, G.; Prakash, G. K. S. *J. Amer. Chem. Soc.* **1995**, *117*, 1336; Olah, G. A. *Angew. Chem., Int. Ed. Engl.* **1995**, *34*, 1393.
34. Cioslowski, J.; Boche, G. *Angew. Chem., Int. Ed. Engl.* **1997**, *36*, 107.
35. Teles, J. H.; Brode, S.; Berkessel, A. *J. Am. Chem. Soc.* **1998**, *120*, 1345.
36. Peters, J. W.; Lanzilotta, W. N.; Lemon, B. J.; Seefeldt, L. C. *Science* **1998**, *282*, 1853.
37. Nicolet, Y.; Piras, C.; Legrand, P.; Hatchikian, C. E.; Fontecilla-Camps, J. C. *Structure* **1999**, *7*, 13.
38. Popescu, C. V.; Munck, E. *J. Am. Chem. Soc.* **1999**, *121*, 7877.
39. Lemon, B. J.; Peters, J. W. *Biochem.* **1999**, *38*, 12969; Lemon, B. J.; Peters, J. W., *J. Am. Chem. Soc.* **2000**, *122*, 3793.
40. Huhmann-Vincent, J.; Scott, B. L.; Kubas, G. J. *Inorg. Chim. Acta* **1999**, *294*, 240.
41. Cao, Z.; Hall, M. B. *J. Am. Chem. Soc.* **2001**, *123*, in press.
42. Connelly, N. G.; Dahl, L. F. *J. Am. Chem. Soc.* **1970**, *92*, 7472; Vergamini, P. J.; Kubas, G. J. *Progr. Inorg. Chem.* **1976**, *21*, 261.
43. Nataro, C.; Angelici, R. J. *Inorg. Chem.* **1998**, *37*, 2975, and references therein.
44. Harris, D. C.; Gray, H. B. *Inorg. Chem* **1975**, *14*, 1215.
45. Fauvet, K.; Mathieu, R.; Poilblanc, R. *Inorg. Chem.* **1976**, *15*, 976; Arabi, M. S.; Mathieu, R.; Poilblanc, R. *J. Organomet. Chem.* **1979**, *177*, 199.
46. Dance, I. *Chem. Comm.* **1999**, 1655; Dance, I., submitted to *J. Am. Chem. Soc.*
47. Burgess, B. K.; Lowe, D. J. *Chem. Rev.* **1996**, *96*, 2983.

48. Eady, R. R. *Chem. Rev.* **1996**, *96*, 3013; Rees, D. C.; Howard, J. B. *Curr. Op. Chem. Biol.* **2000**, *4*, 559.
49. Coucouvanis, D. *Adv. Inorg. Chem.* **1998**, *45*, 1; Coucouvanis, D.; Demadis, K. D.; Malinak, S. M.; Mosier, P. E.; Tyson, M. A.; Laughlin, L. *J. ACS Symp. Ser.* **1996**, *653 (Transition Metal Sulfur Chemistry)*, 117; Smith, B. E. *Adv. Inorg. Chem.* **1999**, *47*, 159; Hidai, M.; Mizobe, Y. *Chem. Rev.* **1995**, *95*, 1115; *Chem. Eng. News*, June 22, 1998, 29. Fryzuk, M. D.; Johnson, S. A. *Coord. Chem. Rev.* **2000**, *200-202*, 379.
50. Shilov, A. E. *Metal Complexes in Biomimetic Chemical Reactions*, CRC Press: Boca Raton, **1997**.
51. Hidai, M. *Coord. Chem. Rev.* **1999**, *185/186*, 99; Nishibayashi, Y.; Iwai, S.; Hidai, M. *Science* **1998**, *279*, 540; Nishibayashi, Y.; Takemoto, S.; Iwai, S; Hidai, M. *Inorg. Chem.* **2000**, *39*, 5949.
52. Sellmann, D.: Utz, J.; Blum, N.; Heinemann, F. W. *Coord. Chem. Rev.* **1999**, *190-192*, 607.
53. Le Gall, T.; Ibrahim, S. K.; Gormal, C. A.; Smith, B. E.; Pickett, C. J. *Chem. Comm.* **1999**, 773.
54. Sellmann, D.; Fursattel, A.; Sutter, J. *Coord. Chem. Rev.* **2000**, *200-202*, 545, and references therein; Helleren, C. A.; Henderson, R. A.; Leigh, G. J. *J. Chem. Soc., Dalton Trans.* **1999**, 1213.
55. Thorneley, R. N. F.; Lowe, D. J. In *Molybdenum Enzymes*, Spiro, T. G., ed., Wiley-Interscience: New York, 1985, p. 161.
56. Hughes, D. L.; Ibrahim, S. K.; Pickett, C. J.; Querne, G.; Lauoenan, A.; Talarmin, J.; Queiros, A.; Fonseca, A. *Polyhedron* **1994**, *13*, 3341.
57. Kim, J.; Rees, D. C. *Science* **1992**, *257*, 1677; Howard, J. B.; Rees, D. C. *Chem. Rev.* **1996**, *96*, 2965; Mayer, S. M.; Lawson, D. M.; Gormal, C. A.; Roe, S. M.; Smith, B. E. *J. Mol. Biol* **1999**, *292*, 871.
58. Peters, J. W.; Stowell, M. H. B.; Soltis, S. M.; Finnegan, M. G.; Johnson, M. K.; Rees, D. C. *Biochemistry* **1997**, *36*, 1181.
59. Siegbahn, P. E. M.; Westerberg, J.; Svensson, M.; Crabtree, R. H. *J. Phys. Chem. B.* **1998**, *102*, 1615; Rod, T. H.; Norskov, J. K. *J. Am. Chem. Soc.* **2000**, *122*, 12751, and references therein.
60. Dance, I. *Chem. Comm.* **1997**, 165; **1998**, 523.
61. Deng, H.; Hoffmann, R. *Angew. Chem. Int. Ed. Engl.* **1993**, *32*, 1062.
62. Lee, H.-I.; Cameron, L. M.; Hales, B. J.; Hoffman, B. M. *J. Am. Chem. Soc.* **1997**, *119*, 10121.
63. Han, J.; Beck, K.; Ockwig, N.; Coucouvanis, D. *J. Am. Chem. Soc.* **1999**, *121*, 10448.
64. Fryzuk, M. D.; Love, J. B.; Rettig, S. J.; Young, V. G. *Science* **1997**, *275*, 1445; Basch, H.; Musaev, D. G.; Morokuma, K.; Fryzuk, M. D.; Love, J. B.; Seidel, W. W.; Albinati, A.; Koetzle, T. F.; Klooster, W. T.; Mason, S. A.; Eckert, J. *J. Am. Chem. Soc.* **1999**, *121*, 523.
65. Basch, H.; Musaev, D. G.; Morokuma, K. *J. Am. Chem. Soc.* **1999**, *121*, 5754; *Organometallics* **2000**, *19*, 3393.
66. Jia, G.; Morris, R. H.; Schweitzer, C. T. *Inorg. Chem.* **1991**, *30*, 594.
67. Groves, J. T.; McClusky, G. A. *J. Am. Chem. Soc.* **1976**, *98*, 859.
68. Meunier, B. *Chem. Rev.* **1992**, *92*, 1411; Schlichting, I; Berendzen, J.; Chu, K.; Stock, A. M.; Maves, S. A.; Benson, D. E.; Sweet, R. M.; Ringe, D.; Petsko, G. A.; Sligar, S. G. *Science* **2000**, *287*, 1615.
69. Groves, J. T.; Nemo, T. E. *J. Am. Chem. Soc.* **1983**, *105*, 6243; Bouy-Debec, D.; Brigaud, O.; Leduc, P.; Battioni, P.; Mansuy, D. *Gazz. Chim. Ital.* **1996**, *126*, 233; Hjelmeland, L. M.; Aronow, L.; Trudell, J. R. *Biochem. Biophys. Res. Commun.* **1977**, *76*, 541; Groves, J. T.; McClusky, G. A.; White, R. E.; Coon, M. J. *Biochem. Biophys. Res. Commun.* **1978**, *81*, 154.
70. Newcomb, M.; Le Tadic-Biadatti, M.; Chestney, D. L.; Roberta, E. S.; Hollenberg, P. F. *J. Am. Chem. Soc.* **1995**, *117*, 12085; Newcomb, M.; Shen, R.; Choi, S.-Y.; Toy, P. H.; Hollenberg, P. F.; Vaz, A. D. N.; Coon, M. J. *J. Am. Chem. Soc.* **2000**, *122*, 2677.
71. Choi, S. Y.; Eaton, P. E.; Hollenberg, P. F.; Liu, K. E.; Lippard, S. J.; Newcomb, M.; Putt, D. A.; Upadhyaya, S. P.; Xiong, Y. *J. Am. Chem. Soc.* **1996**, *118*, 6547.
72. Collman, J. P.; Chien, A. S.; Eberspacher, T. A.; Brauman, J. I. *J. Am. Chem. Soc.* **1998**, *120*, 425.
73. Barton, D. H. R.; Doller, D. *Acc. Chem. Res.* **1992**, *25*, 504; Barton, D. H. R. *Tetrahedron* **1998**, *54*, 5805.
74. Just, G.; Kauko, Y. *Z. Phys. Chem.* **1911**, *76*, 601; Webster, A. H.; Halpern, J. *Trans. Faraday Soc.* **1957**, *53*, 51.
75. Yoshizawa, K.; Shiota, Y.; Kagawa, Y.; Yamabe, T. *J. Phys. Chem. A* **2000**, *104*, 2552; Yoshizawa, K.; Suzuki, A.; Shiota, Y.; Yamabe, T. *Bull. Chem. Soc. Jpn.* **2000**, *73*, 815; Yoshizawa, K. *J. Inorg. Biochem.* **2000**, *78*, 23, and references therein.
76. Feig, A. L.; Lippard, S. J. *Chem. Rev.* **1994**, *94*, 759; Liu, K. E.; Lippard, S. J. *Adv. Inorg. Chem.* **1995**, *42*, 263; Wallar, B. J.; Lipscomb, J. D. *Chem. Rev.* **1996**, *96*, 2625; Du Bois, J.; Mizoguchi, T. J.; Lippard, S. J. *Coord. Chem. Rev.* **2000**, *200-202*, 443; Dunietz, B. D.; Beachy, M. D.; Cao, Y.;

Whittington, D. A.; Lippard, S. J.; Friesner, R. A. *J. Am. Chem. Soc.* **2000**, *122*, 2828; Westerheide, L.; Pascaly, M.; Krebs, B. *Curr. Op. Chem. Biol.* **2000**, *4*, 235.
77. Shu, L.; Nesheim, J. C.; Kauffmann, K.; Munck, E.; Lipscomb, J. D.; Que, L., Jr. *Science* **1997**, *275*, 515.
78. Siegbahn, P. E. M. *Inorg. Chem.* **1999**, *38*, 2880; Basch, H.; Mogi, K; Musaev, D. G.; Morokuma, K. *J. Am. Chem. Soc.* **1999**, *121*, 7249.
79. Choi, S.-Y.; Eaton, P. E.; Kopp, D. A.; Lippard, S. J.; Newcomb, M.; Shen, R. *J. Am. Chem. Soc.* **1999**, *121*, 12198.
80. Ruzicka, F.; Huang, D.-S.; Donnelly, M. I.; Frey, P. A. *Biochemistry* **1990**, *29*, 1696.
81. Panov, G. I.; Uriarte, A. K.; Rodkin, M. A.; Sobolev, V. I. *Catal. Today* **1998**, *41*, 365,
82. Morris, R. H. *NATO ASI Ser., Ser. 3* **1998**, *60* (Transition Metal Sulfides), 57; *ACS Symp. Ser.* **1996**, *653* (Transition Metal Sulfur Chemistry); Bayon, J. C.; Claver, C.; Masdeu-Bulto, A. M. *Coord. Chem. Rev.* **1999**, *193–195*, 73.
83. Startsev, A. N. *Catal. Rev.-Sci. Eng.* **1995**, *37*, 353.
84. Neurock, M.; van Santen, R. A. *J. Am. Chem. Soc.* **1994**, *116*, 4427, and references therein.
85. Rakowski DuBois, M. *Chem. Rev.* **1989**, *89*, 1, and references therein.
86. Sweeney, Z. K.; Polse, J. L.; Andersen, R. A.; Bergman, R. G.; Kubinec, M. G. *J. Am. Chem. Soc.* **1997**, *119*, 4543.
87. Bianchini, C.; Mealli, C.; Meli, A.; Sabat, M. *Inorg. Chem.* **1986**, *25*, 4617.
88. Rakowski DuBois, M.; Jagirdar, B.; Noll, B.; Dietz, S. *ACS Symp. Ser.* **1996**, *653* (Transition Metal Sulfur Chemistry), 269–281.
89. Pan, W.-H.; Harmer, M. A.; Halbert, T. R.; Stiefel, E. I. *J. Am. Chem. Soc.* **1984**, *106*, 459; Shibahara, T. *Coord. Chem. Rev.* **1993**, *123*, 73, and references therein.
90. Angelici, R. J. *Polyhedron* **1997**, *16*, 3073, and references therein.
91. Nishibayashi, Y.; Iwai, S.; Hidai, M. *J. Am. Chem. Soc.* **1998**, *120*, 10559.
92. Sellmann, D.: Rackelmann, G. H.; Heinemann, F. W. *Chem. Eur. J.* **1997**, *3*, 2071; Sellmann, D.; Kappler, J.; Moll, M. *J. Am. Chem. Soc.* **1993**, *115*, 1830; Sellmann, D.; Sutter, J. *ACS Symp. Ser.* **1996**, *653* (Transition Metal Sulfur Chemistry), pp. 101–116.
93. Jessop, P. G.; Morris, R. H. *Inorg. Chem.* **1993**, *32*, 2236.
94. Sellman, D.; Geipel, F.; Moll, M. *Angew. Chem. Int. Ed.* **2000**, *39*, 561.
95. Bartucz, T. Y.; Golombeck, A.; Lough, A. J.; Maltby, P. A.; Morris, R. H.; Ramachandran, R.; Schlaf, M. *Inorg. Chem.* **1998**, *37*, 1555, and references therein.
96. Darensbourg, M. Y.; Liaw, W. F.; Riordan, C. G. *J. Am. Chem. Soc.* **1989**, *111*, 805.

11

Coordination and Activation of Si–H, Ge–H, and Sn–H Bonds

11.1. INTRODUCTION

The large class of chemical compounds known generically as silanes consists of a silicon atom bonded to four other atoms or groups, and the compounds are analogous to hydrocarbons and their derivatives. Hydrosilanes, SiH_nR_{4-n}, which contain at least one Si–H bond, form the largest subgroup, and the parent is silane itself, SiH_4, a congener of methane. Silanes are much more reactive to atmospheric oxygen and moisture than hydrocarbons, and SiH_4 gas explodes in air spontaneously. As in catalytic hydrogenation, OA of a Si–H bond is one of the key steps in the mechanism of transition-metal–catalyzed addition of Si–H bonds to unsaturated organic compounds (hydrosilylation).[1] Similar to H_2, hydrosilanes do not have nonbonding electron pairs or π electrons to ligate to metal centers, but they can bind to M to form stable σ complexes through Si–H bonds to give 3c-2e $M(\eta^2\text{-Si–H})$ bonding. Silanes with more than one Si–H bond usually coordinate through only one Si–H bond, although in one case SiH_4 binds through all four. The Si–H bond is activated toward cleavage remarkably like the H–H bond, and tautomeric equilibria can exist with the OA product.

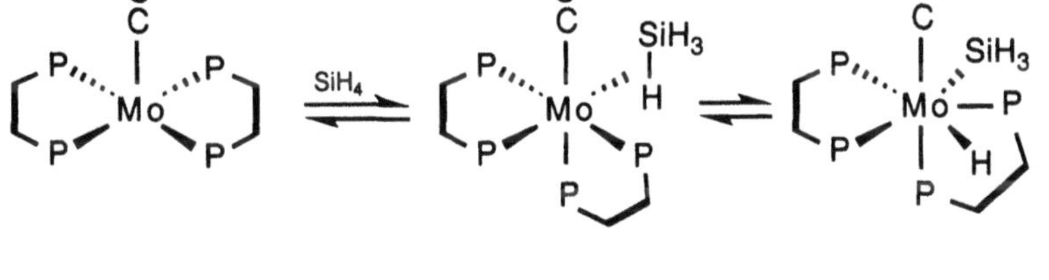

$$ \tag{11.1} $$

The Si–H bond elongates on coordination to the same relative extent as the H–H bond does, and the energetics of binding, cleavage, and exchange with cis hydrides are also startlingly similar. This would not have been expected on the basis of the large difference in steric and electronic factors. $RuH_2(\eta^2\text{-}H_2)\text{-}(\eta^2\text{-SiHPh}_3)(PCy_3)_2$ simultaneously coordinates both H–H and Si–H bonds. The coordination and activation of Ge–H and Sn–H bonds in germanes and stannanes are also directly analogous to Si–H. The general principles of H–H activation thus can most probably be applied to activation of all σ X–H bonds. The η^2-Si–H coordination also represents a model for binding of alkanes and methane to M.

A major difference in comparison to M–H$_2$ complexes is that M–H–Si and M–H–X linkages are asymmetric, i.e., look like hydride-bridged systems with M–H–Si near 90°. Also the presence of substituents R on Si leads to variations in

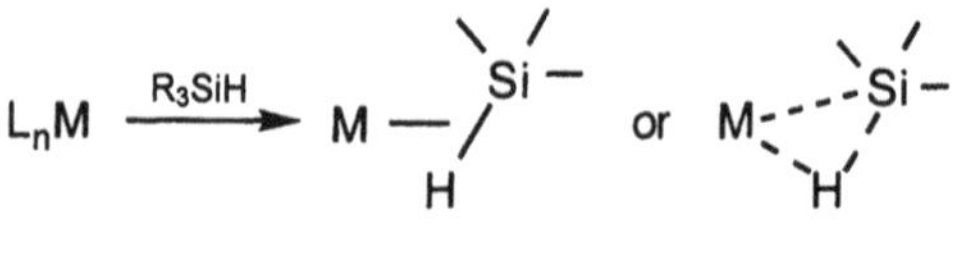

$$L_nM \xrightarrow{R_3SiH} M\underset{H}{\overset{Si-}{-}} \text{ or } M\cdots\underset{H}{\overset{Si-}{\cdots}} \tag{11.2}$$

electronics/sterics, and greatly increases the number of sites of attachment to one or more M. Dinuclear species with μ-silane or μ-silyl groups and chelating systems with agostic-type Si–H interactions are quite common.

Historically, the first silane complexes were prepared by Graham's group in the late 1960s well before H$_2$ complexes.[2–6] However, the bonding situation was not immediately perceived to be nonclassical in the same clearcut manner as M–H$_2$. The properties of the complexes were recognized to be unusual, but their true σ-complex nature was revealed only through a prolonged series of chemical, spectroscopic, and structural investigations by several researchers. Excellent accounts of this are given in a review of Si–H σ coordination by Schubert,[7] a subsequent researcher in the field, as well as by Graham.[6] One of the initial Graham complexes, a Re dimer from photolysis of Re$_2$(CO)$_{10}$ and Ph$_2$SiH$_2$ in benzene, was characterized crystallographically and claimed to be the first compound with a hydrogen bridge between Si and a transition metal.[3] Although the hydrogens in **1** were not located, NMR indicated

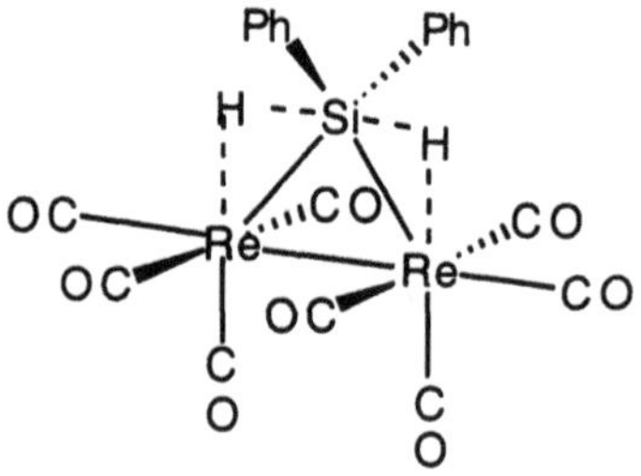

(1)

their presence, e.g., the Me$_2$SiH$_2$ analogue showed a 1:2:1 triplet for the Me and a high field signal for the hydrogens. The next key finding was the photolytic preparation[4] and X-ray structure[8] of CpMn(CO)$_2$(HSiPh$_3$) [Eq. (11.3)]. Ironically this finding was also first reported in *Chemical and Engineering News*[9] (as a highlight of the 1970 Joint ACS/CIC Meeting) in nearly the same manner as the first H$_2$ complex in 1983 (see Chapter 2). Both this article and a later journal paper[5] noted that despite the fact that the (located) hydrogen was "normally" bonded to Mn, the distance between the H and Si atoms (1.76 Å) "must be symptomatic of a weak bonding attraction," i.e., the Si–H bond did not oxidatively add. A resonance hybrid of two canonical forms, **2** and **3**, was suggested to explain this. Form **2** (OA product) was proposed to be the major contributor because of the normal d_{MnH} (1.55 Å) and d_{MnSi} (2.42 Å). IR showed a very broad absorption for the Mn···H···Si unit at 1890

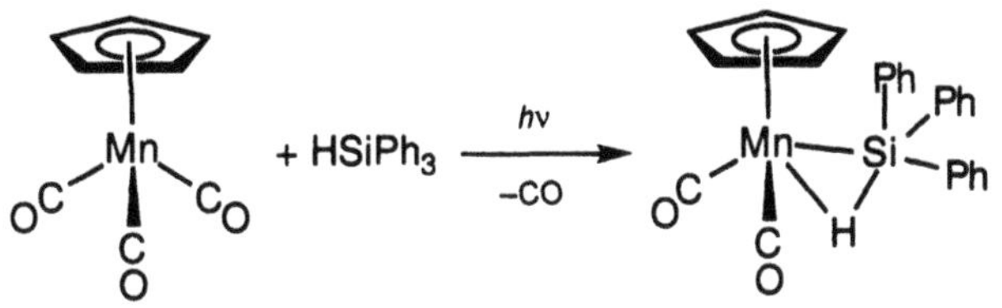

$$\text{CpMn(CO)}_3 + \text{HSiPh}_3 \xrightarrow[-CO]{hv} \text{CpMn(CO)}_2(\eta^2\text{-HSiPh}_3) \quad (11.3)$$

cm^{-1}, which could be characteristic of a bridging hydride. Later on in 1978 Kaesz, who had earlier prepared complexes with coordinated B–H bonds (see Chapter 13),

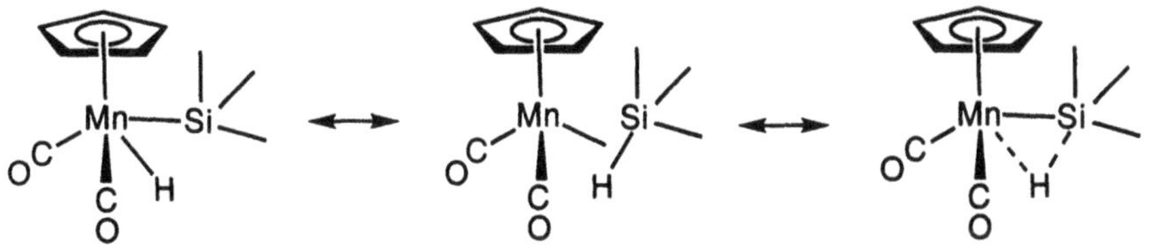

(2)   (3)   (4)

suggested form **4** to describe the situation as "arrested oxidative addition,"[10] which to some degree fits σ-bond complexes in general. However, it was believed at the time that steric congestion was instrumental in disallowing **2**, hence creating the close Si–H contact. As in H$_2$ complexes, steric congestion is not a requisite for σ-bond coordination, so the bonding situation was still not yet well understood, certainly not at a theoretical level.

The following year bent M–H–M bonds such as in W$_2$(CO)$_{10}$(μ-H) were identified as being electron-deficient 3c-2e systems in analogy with boron hydrides and H$_3^+$.[11] In 1982 Colomer et al., then applied this description (and Kaesz's structural representation **4**) to the Mn–Si–H bonding in (CpMe)Mn(CO)$_2$(HSiPh$_3$) and determined a $^2J_{SiH}$ of 65 Hz that clearly showed significant Si–H bonding interaction in analogy with the large J_{HD} in H$_2$ complexes.[12] Furthermore, this interaction was shown to be electronically favored and *not the result of steric hindrance*. The stereochemistry of protonation of [(CpMe)Mn(CO)$_2$(SiPh$_3$)]$^-$ gives only the cis isomer as in **2** or **3** and never the sterically more favorable isomer with hydride trans to Si. Subsequent structure determinations, particularly the neutron structure of (CpMe)Mn(CO)$_2$(SiHFPh$_2$) in 1982 added to the evidence for η^2-Si–H.[13] The key feature in the latter is the short d_{SiH} [1.802(5) Å] that is only 0.3 Å (20%) longer than covalent Si–H bonds in tetrahedral silanes (1.48 ± 0.02 Å, which for reference is twice the d_{HH} in H$_2$). The bonding situation in the Mn complexes was finally clarified by extended Huckel calculations in 1987 that pointed out the similarity of M(η^2-Si–H) and M(η^2-H–H) bonding.[14]

The bonding and activation of hydrosilanes has been expertly reviewed by Schubert[7,15] and more recently by Corey and Braddock-Wilking.[16] Crabtree[17] and Schneider[18] discuss Si–H coordination in the context of other σ complexes, which constitute a growing family of complexes that have some analogy to π-ligand complexes. d_{HH} in $W(H_2)(CO)_3(PR_3)_2$ is ca 20% longer than that in free H_2, similar to the lengthening of d_{SiH} in $(CpMe)Mn(CO)_2(SiHFPh_2)$. However, typical J_{SiH} for the Mn-silane complexes, 40–70 Hz, are far lower than those for free silanes, ca 200 Hz, whereas J_{HD} for $M(\eta^2\text{-HD})$ (30–35 Hz for true H_2 complexes) is much closer to the value for free HD (43 Hz) and correlates linearly with d_{HD}. Thus the situation for silane binding is considerably more complex, primarily because Si has substituents that change both electronic and steric properties, and the Si–H bond is more basic (better σ donor) than H–H (and C–H) bonds. As will be shown, the Si–H bond is also a better π acceptor. Nonetheless, there is a close overall relationship between Si–H, H–H, and other X–H complexes, and as for H_2 complexes a near continuum of degrees of "arrested" bond activation can exist. The effects of variation in M/L sets can be similar to that for H_2 complexes, e.g., $CpMn(CO)_2(HX)$ are σ complexes for both X = H and SiR_3 while $CpRe(CO)_2(H)(X)$ are classical (or have very long d_{XH}) because of the better backdonating Re center.

11.2. SYNTHESIS AND CHARACTERIZATION OF $M(\eta^2\text{-Si–H})$ COMPLEXES

At least a dozen different types of structures possess $M(\eta^2\text{-Si–H})$ interactions and these are summarized in Table 11.1 and Figure 11.1 (see Corey and Braddock-Wilking[16] for detailed listing and literature references if not given). About half of the over 80 various complexes contain Cp or arene ligands and, unlike H_2 complexes, about one-third are dinuclear species, some of which contain α-agostic interactions similar to α-agostic C–H interactions. Several types of agostic interactions are known, including novel multiple Sm···Si–H β-agostic interactions.[19] Structure **XII** where X = P is directly analogous to that for M···H–C interactions of phosphine organo groups. These types of interactions should be differentiated from the Si–H binding in structures **I, V, VIII,** and **IX**, which are true σ complexes (silanes held to the metal *only* by Si–H interactions). Whereas cationic H_2 complexes are exceedingly common, only a few such silane complexes are known and even fewer have been isolated because of their instability to heterolysis of the Si–H bond (see below).

J_{SiH} in η^2-Si–H complexes are observed to be greater than those typically seen for silyl–hydride complexes (<20 Hz). The ^{1}H NMR resonance for the coordinated SiH proton is normally in the metal hydride region (negative chemical shift), whereas free SiH protons generally appear at 4–8 ppm. Reported d_{SiH} vary from 1.43–2.10 Å, compared to the average Si–H distance in free silanes, 1.48 Å, although standard deviations are typically large for these mostly X-ray values (only one neutron structure). Schubert proposed that a d_{SiH} of 2.00 Å represents the upper limit for any significant Si···H interaction,[7] although subsequently characterized $[Fe(CO)_3]_2(\mu\text{-HSiPh}_2)_2$ has a very elongated 2.10 Å d_{SiH} with $J_{SiH} = 23$ Hz at the low end of the range for σ complexes. This represents about a 0.6 Å lengthening over that in the free ligand, which is less than that for the increase in d_{HH} over the entire reaction coordinate for OA of H_2 (0.74 to ~1.6 Å). Also, d_{SiH} in $CpRe(CO)_2(H)(SiPh_3)$ is estimated to be 2.2 Å (X-ray, H not located), which may represent a stretched σ complex on the verge of OA.[22] It is likely that the position

Coordination and Activation of Si-H, Ge-H, and Sn-H Bonds

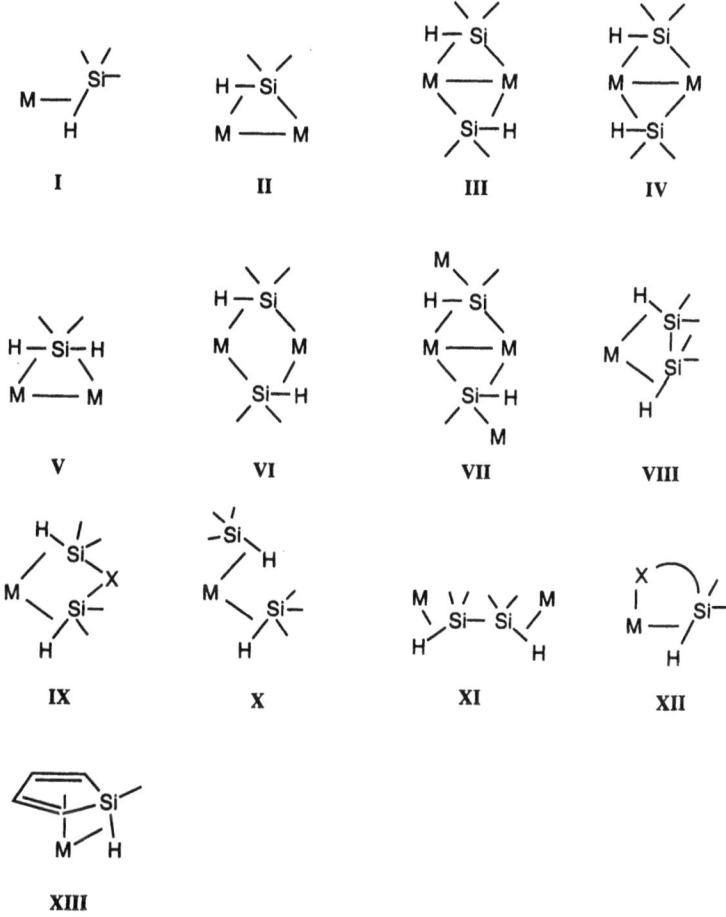

Figure 11.1. Structures of Si-H bond interaction with metal centers.

of H in such elongated M(Si···H) complexes is highly delocalized within a very flat potential energy surface, much like that for stretched M(H···H) complexes.

Although d_{SiH} vary from 1.48–2.2 Å in Table 11.1, the J_{SiH} are always closer to those in OA products and more analogous to J_{HD} in elongated H–H bonds (> 1 Å). Furthermore, J_{SiH} does not correlate with d_{SiH} or even with the extent of OA.[16] In some cases J_{SiH} are nearly as low (22–26 Hz) as those in hydrido–silyl complexes yet d_{SiH} fall in the range of true σ complexes. As will be discussed below, d_{SiH} often varies little with substituents at Si compared to d_{MSi}. However, d_{MSi} is not a reliable indicator of either the presence or absence of a 3c-2e interaction because d_{MSi} can be either longer or shorter than those in classical M–Si bonds.

IR is seldom used to characterize silane complexes, although the stretching mode for the coordinated Si–H bond is observable by both IR and Raman. As for H_2 complexes it is expected that this mode will be highly mixed with M–H types of modes and will be similar to vibrational modes for bridging hydride ligands. The position of the $v_{M···H-Si}$ frequency can be as high as 1890 cm^{-1} but is generally in

Table 11.1. Stable Silane Complexes by Generic Type[a]

Configuration	Complex	Structure type	J_{Si-H}, Hz	d_{SiH}, Å
Group 4				
$M^{II} d^2$	$Cp_2Ti(SiH_2Ph_2)(PMe_3)$	I	28	1.69(5)
	$[Cp_2Ti(\mu\text{-}SiH_nPh_{3-n})]_2$ $n = 2, 3$	VI		1.58(3)[b]
	$[Cp_2Ti]_2(\mu\text{-}H)(\mu\text{-}SiH_nPh_{3-n})$ $n = 2, 3$	II	58[b]	1.56(7)[b]
Group 6				
$M^0 d^6$	$Cr(\eta^6\text{-arene})(CO)_2(SiH_2Ph_2)$	I	71–80	1.61(4)[c]
	$Mo(SiH_nR_{4-n})(CO)(PP)_2$ $n = 2–4$	I	31–61	1.60; 1.77(6)[d]
	$W(CO)_3[SiHCl_2(p\text{-tol})](PP)$	I	35	
	$[W(CO)_3(P^iPr_3)(\mu\text{-SiHPh}_2)]_2$	VI	52	
	$W(CO)_4(HPh_2SiCH_2CH_2PPh_2)$	XII	98	
Group 7				
$M^I d^6$	$Cp'Mn(CO)_2(SiH_nR_{4-n})$ $n = 1, 2$	I	55–70	1.76(4)–1.802(5)[e]
	$Cp'Mn(CO)(SiH_2Ph_2)(P)$	I	38–43	1.78(4)
	$[Cp'Mn(CO)_2]_2(\mu\text{-HR}_2Si\text{-SiHR}_2)$	XI	57	
	$Cp'Mn(CO)(HPh_2SiCH_2CH_2PPh_2)$	XII		1.75(4)
	$CpRe(CO)_2(SiHPh_3)$	I		2.19
	$[Re(CO)_3(P)_2(SiHEt_3)]^+$ [f]	I	66	
$M^{III} d^4$	$Re(N_3N_F)(SiH_2R_2)$[g]	I	38–44	
Group 8				
$M^{II} d^6$	$[CpFe(CO)(SiH_nR_{4-n})(P)]^+$ $n = 1, 2$	I	58–62	1.72(3)
	$[CpRu(SiHCl_3)(P)_2]^+$	I	48	1.70(7)
	$TpRuH(SiHR_3)(PPh_3)$[h]	I	23–53	
	$RuH_2(H_2)(SiHPh_3)(P)_2$	I	41	1.84(2)[i]
	$\{Cp^*Ru[HSi(SiMe_3)C_4Me_4]\}^+$	XIII	22,65[i]	1.73(3), 1.78(3)
	$RuH_2[(SiHMe_2)_2X](P)_2$ $X = O, C_6H_4$	IX	70	1.77(4), 1.81(3)
	$X = (CH_2)_2$	IX	82	
	$X = OSiMe_2O$	IX		

Compound		Type	Values
[Fe(CO)$_3$]$_2$(μ-HSiPh$_2$)$_2$		III	2.10
[(OC)$_4$FeFe(CO)$_3$(HSiPh$_2$)](μ-η^2-HSiPh$_2$)		II	1.66
(Cp*Ru)$_2$(μ-H)(H)(μ-SiHR$_2$)		IV	26–49	1.68,^{j} 1.75(4)^j
[Cp*Ru(H)]$_2$(μ-η^2-η^2-SiH$_2$^{t}Bu$_2$)		V	75
[Cp*Ru(CO)$_2$]$_2$(μ-η^2-η^2-SiH$_2$^{t}Bu$_2$)		V	22	1.75, 1.77(3)

Group 9
$M^{III} d^6$

Compound		Type	Values
CpRh(SiMe$_3$)$_2$(SiHEt$_3$)		I	
[IrH$_2$(SiHEt$_3$)$_2$(P)$_2$]$^+$		X	
[IrH$_2$(HEt$_2$SiSiHEt$_2$)$_2$(P)$_2$]$^+$		VIII	24
M$_2$H(CO)$_2$(μ-PP)$_2$(μ-SiHPh$_2$) M = Rh, Ir		II	
Rh$_2$H$_2$(PP)$_2$(μ-SiH$_2$R)$_2$		III	
Rh$_2$(μ-H)(PP)$_2$(μ-SiH$_2$R)		II	1.66(6),^{k} 1.73(4)^l
Rh$_2$(μ-SiHR)(PP)$_2$(μ-SiH$_2$R)$_2$		III	1.43,^{m} 1.47^m

Group 10
$M^{II} d^8$

Compound		Type	Values
[Pt(P)(μ-SiHR$_2$)]$_2$		III	
[Pt(P)(SiHR{PtH(P$_2$)})]$_2$		VII	31	72,^{n} 78,^{n} 1.674,^{n} 1.72^o

Lanthanide

Compound	Type
Sm{[μ-N(SiHMe$_2$)$_2$]$_2$Sm[N(SiHMe$_2$)$_2$](thf)$_2$}^{p}	XII

^{a} See Corey et al.^{16} for literature citations except as noted. Abbreviations: P, PP = mono- and bidentate phosphine or phosphite.
^{b} $n = 2$.
^{c} arene = C$_6$Me$_6$.
^{d} PP = depe; R = Ph;n = 2 and 3, respectively.
^{e} Neutron diffraction value for SiHFPh$_2$.
^{f} Fang et al.^{49} P = P(OCH$_2$)$_3$CMe.
^{g} Reed et al.^{21}
^{h} Ng et al.^{20}
^{i} X = C$_6$H$_4$.
^{j} R = Et.
^{k} R = Ph; PP = dippe.
^{l} R = Me; PP = dippe.
^{m} R = p-tol; PP = dippe.
^{n} P = PPh$_3$; R$_2$ = H(Ar).
^{o} P = PCy$_3$; R = Me.
^{p} Nagl et al.^{19}

the 1650–1800 cm^{-1} range. The modes for a dinuclear Ru complex with a bridging silane are assigned using deuterium substitution.[23] A broad absorption assignable to the Ru···H–Si stretch at 1790 cm^{-1} shifts to 1290 cm^{-1} for **5** with SiD$_2$^{t}Bu$_2$. The

(5)

ν_{SiH} for free protio silane occurs at 2116 cm^{-1}; hence it represents a significant reduction in the Si–H bond order. A deformational band, δ_{HSiH} shifts from 1050 to 760 cm^{-1} on deuteration. In free SiH$_2$^{t}Bu$_2$ deformations are seen at 928 and 851 cm^{-1}, shifting to 677 and 569 cm^{-1} on deuteration, which are 100–200 cm^{-1} lower than for the coordinated silane. This is ascribed to a less flexible H–Si–H angle on binding, which would raise the deformation energy.

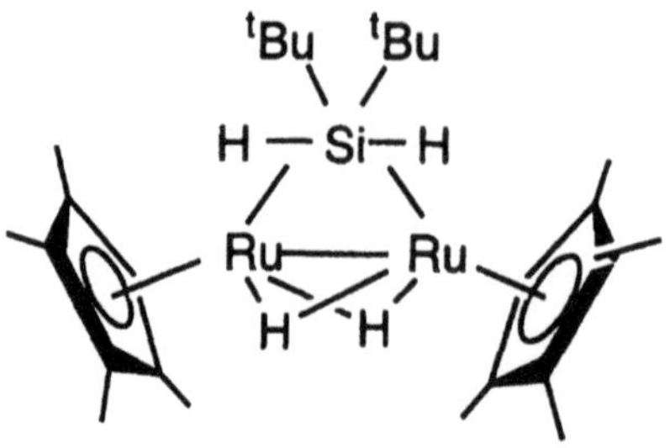

Generally, synthetic procedures for σ-silane complexes are similar to those that produce silyl hydride complexes[16] and involve addition of silane to an unsaturated complex, often generated by photolytic or thermolytic dissociation of CO as for CpMn(CO)$_2$(silane). Agostic, H$_2$, or solvento complexes are also excellent precursors because silanes tend to bind as well or more strongly than H$_2$ or weak donors such as THF. Addition of a primary or secondary silane to solutions of Mo(CO)(R$_2$PC$_2$H$_4$PR$_2$)$_2$ forms yellow silane complexes [Eq. (11.4)].[24] The less

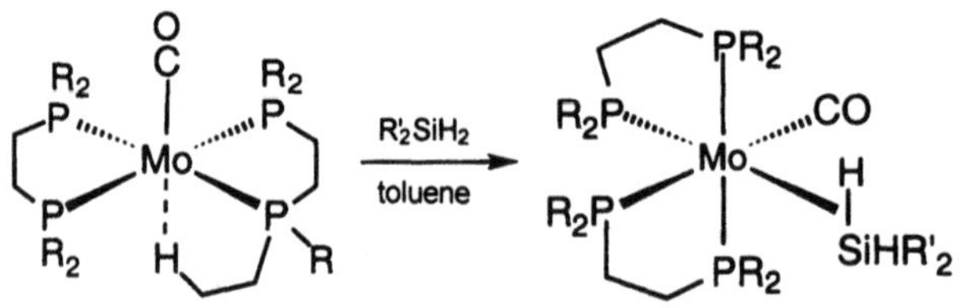

$$R = Et; \quad SiHR'_2 = SiH_2Ph, SiH_2(n\text{-}C_6H_{13}), SiHPh_2$$
$$R = CH_2Ph; \quad SiHR'_2 = SiH_2Ph, SiH_2(n\text{-}C_6H_{13})$$

electron-rich dppe analogue (R = Ph) binds silanes weakly, and only primary silanes SiH$_3$R coordinate, in equilibrium with the agostic precursor. In comparison, H$_2$ complexes can be isolated, possibly because steric hindrance is not a factor (see Section 11.3).

^{31}P{^{1}H} NMR spectra show four signals consistent with an octahedral structure in which the silane is regarded as one ligand with the midpoint of one Si–H σ bond taking up a coordination site cis to CO. If these species were to be formulated as seven-coordinate hydrido–silyl complexes, they would undoubtedly be fluxional

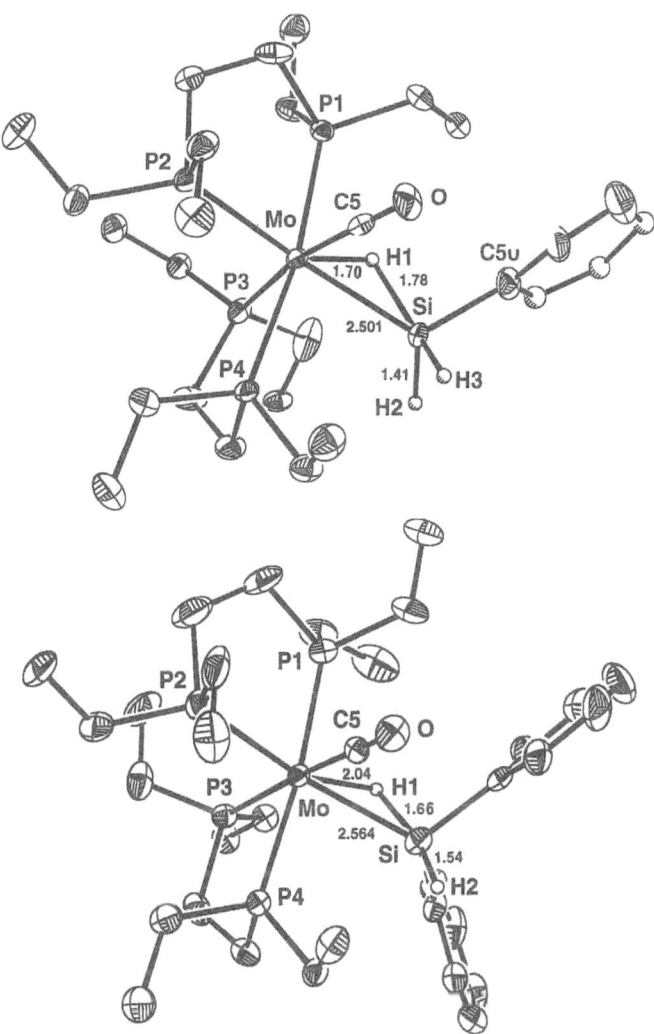

Figure 11.2. ORTEP drawings of Mo(CO)(SiH$_3$Ph)(depe)$_2$ and the SiH$_2$Ph$_2$ analogue.

and show only a single resonance. The X-ray structure of Mo(CO)(SiH$_3$Ph)(depe)$_2$ and the SiH$_2$Ph$_2$ analogue confirm the η^2-Si–H bonding in the solid state (Figure 11.2). The coordinated Si–H(1) bonds are significantly lengthened [1.77(6) and 1.66(6) Å] relative to the terminal Si–H(2) bonds [1.42(6) and 1.54(6) Å] in the SiH$_3$Ph and SiH$_2$Ph$_2$ complexes, respectively. It is noteworthy that both the solution NMR data and the X-ray structures of the Mo complexes indicate that the silane and CO ligands are cis to each other in contrast to the corresponding H$_2$ or agostic complexes where the σ ligand is trans to CO. This could result from both steric and electronic factors (see Section 11.3). The Mo–H–Si angle is 92(3)°, comparable to that in the neutron structure of (CpMe)Mn(CO)$_2$(SiHFPh$_2$), 88.2(2)°.

In rare cases OA occurs [Eq. (11.5)] when silanes are added to precursors that bind η^2-H_2, e.g., an agostic tungsten complex similar to the above Mo system.[25] The OA of silane here is an equilibrium process where the OA product is favored on

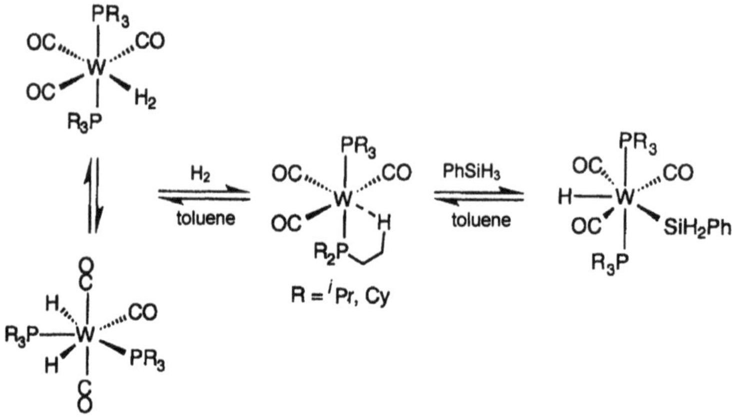

$$R = {}^iPr, Cy \tag{11.5}$$

cooling but cannot be isolated, and the corresponding σ-silane complex is not observed spectroscopically. For H_2 addition, tautomeric equilibrium between η^2-H_2 and dihydride forms occurs, in contrast to the silane systems in Eqs. (11.3) and (11.4), showing that σ coordination versus OA is finely tuned for these electronically and sterically divergent ligands. In Eq. (11.5), the hydrogens in free $PhSiH_3$ resonate in the ^1H NMR spectrum at 4.23 ppm (J_{SiH} = 200.0 Hz), but the Si–H resonance of $W(H)(CO)_3(PCy_3)_2(SiH_2Ph)$ at −45°C shifts downfield to 5.79 ppm, with smaller J_{SiH} (176.8 Hz) as is typically observed in metal-bound silyl.[16] The W–H signal appears at −4.51 ppm (t, J_{HP} = 14.9 Hz; J_{WH} = 28.2 Hz) and is not coupled to Si.

For the analogous reaction of Ph_2SiH_2 with $W(CO)_3(P^iPr_3)_2$, low-temperature ^1H and ^{31}P NMR show a mixture of compounds inconsistent with the presence of either the OA product or a σ complex.[25] At 60°C, a relatively clean reaction with excess Ph_2SiH_2 occurs to form the crystallographically characterized μ-silyl complex

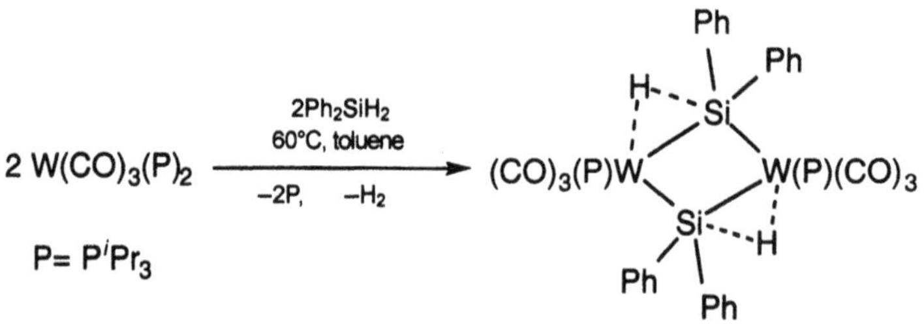

$$2\ W(CO)_3(P)_2 \xrightarrow[-2P,\ -H_2]{2Ph_2SiH_2\ 60°C,\ toluene} (CO)_3(P)W\underset{Ph\ Ph}{\overset{Ph\ Ph}{\cdots Si \cdots}}W(P)(CO)_3 \tag{11.6}$$

$P = P^iPr_3$

in Eq. (11.6). A three-center W···Si···H interaction is suggested on the basis of the ^{29}Si NMR spectrum, which contains only a broad doublet at 146.3 ppm with J_{SiH} = 52 Hz. The large upfield chemical shift of the W–H resonance suggests significant tungsten–hydride character and possibly an elongated Si···H bond. A well-defined species with such a M–H···Si bonding description, $[Fe(CO)_3(\mu\text{-}SiHPh_2)]_2$, has a very long d_{SiH} of 2.10 Å.[26] The value of J_{SiH}, 24.3 Hz, is one of the lowest reported for M···H···Si interaction and is much lower than that in the W complex for which the hydrogens could not be located. Thus a distinction between the M–H···Si and M···H–Si formalisms cannot be drawn, which is a general problem in nonclassical silane coordination that is not seen in agostic C–H interactions, which are usually of the M···H–C type.

Substitution of two cis H_2 ligands conveniently produces the first structurally characterized chelating bis(silane) systems, $RuH_2[(SiHMe_2)_2X](PCy_3)_2$ [Eq. (11.7)].[27] The bis(silane) ligand is relatively strongly bound here [d_{SiH} = 1.73–1.84 Å

$$\begin{array}{c}\text{H}_2\text{'''}\overset{\overset{\text{PCy}_3}{|}}{\underset{\underset{\text{PCy}_3}{|}}{\text{Ru}}}\overset{\text{H}}{\underset{\text{H}}{\text{''}}}\quad\xrightarrow[-2\text{H}_2]{(\text{Me}_2\text{SiH})_2\text{X}}\quad X\begin{pmatrix}\text{Si}-\text{H}\\\text{H}\cdots\text{Ru}\cdots\text{PCy}_3\\\text{H}\cdots\text{PCy}_3\\\text{-Si}-\text{H}\end{pmatrix}\end{array}\qquad(11.7)$$

$X = $ C$_6$H$_4$, $(CH_2)_n$, $Me_2Si\overset{O}{\underset{O}{\diagup}}$

(Table 11.1); cf 1.49 Å in the free disilanes) and cannot be displaced by H_2 or even CO for $X = C_6H_4$. However for $X = O$, the opposite is true despite the fact that J_{SiH} is very low, 22 Hz, seemingly indicative of a stretched Si–H bond. J_{SiH} varies widely from 22–82 Hz and relates to the length of the bridge between the Si: the lowest values are for $X = O$. The degree of activation of the Si–H bond is at least partially related to the chain length: weaker activation occurs for the longer chains, i.e., toward the limit of nonchelating ligands (see Section 11.3.2). Calculated total binding energies are 31–46 kcal/mol for the bis(silane) ligands, hence Si–H is bound more strongly than H_2 in the more stable complexes. The chelate effect and nonbonding Si···H interactions (2.04–2.43 Å, X-ray; 2.25 Å calculated) stabilize the complexes and perhaps promote the unexpected cis disposition of the bulky PCy$_3$. The cis interactions play an important role in the reactivity of the complexes (see Section 11.4.2), similar to cis interactions in M(H$_2$)H complexes that represent incipient σ bond metathesis. Deuterium is easily incorporated into the hydride sites

$$\begin{array}{c}\overset{H_a}{\underset{|}{|}}\text{-Si}\\M-\underset{H_b}{|}\end{array}\rightleftharpoons\begin{array}{c}\overset{H_a}{\underset{}{\diagdown}}\text{Si}\\M\diagup\underset{H_b}{:}\end{array}\rightleftharpoons\begin{array}{c}\overset{H_b}{\underset{|}{|}}\text{-Si}\\M-\underset{H_a}{|}\end{array}$$

using $DSiMe_2(CH_2)_3SiMe_2D$, and exchange between bound hydrogens is facile from NMR studies, especially for $X = O$, the most reactive species.

As for H_2 complexes, silane complexes are known for all group 5–10 metals, and additionally a group 4 complex, $Cp_2Ti(SiH_2Ph_2)(PMe_3)$, is prepared by displacement of labile PMe$_3$ from $Cp_2Ti(PMe_3)_2$.[28] It is close to OA (J_{SiH} is 28 Hz) and the second-row Zr analogue is a silyl hydride.[29] The only group 5 silane complexes, $CpV(CO)_3(HSiEt_{3-n}Cl_n)$ ($n = 0, 2, 3$), are thermally unstable and formed by photolysis of CpV(CO)$_4$ in liquid Xe.[30] The majority of group 6 and 7 complexes have arene- or Cp-type ligands. As for H$_2$ analogues, electron-poor group 6 species such as M(silane)(CO)$_5$ are not stable,[31,32] again because σ-ligand binding is weak (16–19 kcal/mol[32]) and dissociation leads to very unstable M(CO)$_5$. The chelate effect in W(CO)$_4$(HPh$_2$SiCH$_2$CH$_2$PPh$_2$) helps stabilize a γ-agostic silane interaction analogous to the β-agostic C–H interaction in W(CO)$_3$(PR$_3$)$_2$. J_{SiH} is relatively high,

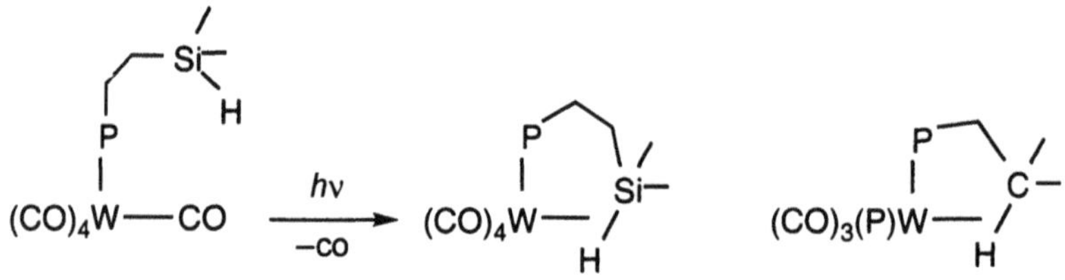

$$(11.8)$$

98 Hz.[33] β-agostic Si–H interactions are found in Zr[34] and Ti[35] metallocenes with similarly high couplings (>90 Hz) and short d_{SiH} (1.41–1.52 Å), which are characteristic of agostic Si–H interactions (other examples and characterizations are given by Corey[16]). As for the C–H bond in CH agostic interactions, the d_{SiH} are not much

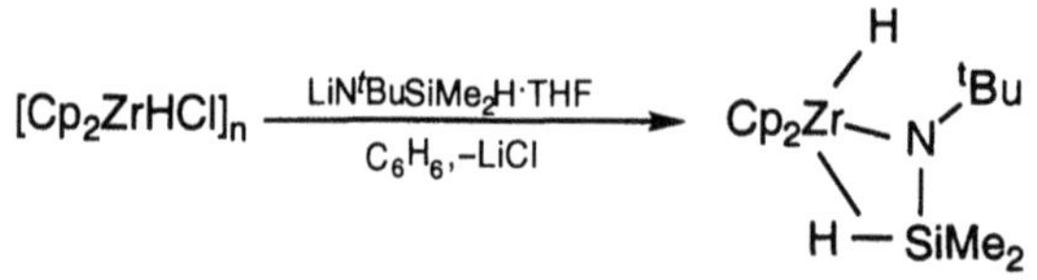

$$(11.9)$$

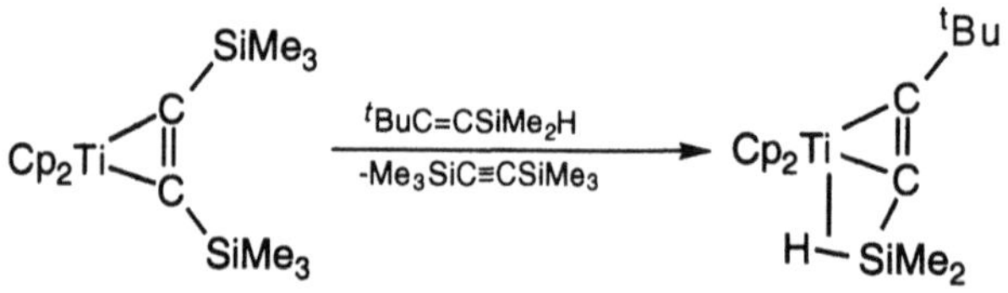

$$(11.10)$$

different than those in the uncoordinated bonds but computations (see Section 11.3.2) indicate that the M–SiH interactions here can be relatively strong. The α atoms in complexes such as in Eqs. (11.9)–(11.10) are nearly always N or C, and most cases involve the –SiHMe$_2$ group as supported by IR data: ν_{SiH} normally at 2100–2200 cm^{-1} for uncoordinated silanes shifts 125–270 cm^{-1} to lower frequency.

The widely investigated and slightly volatile CpMn(CO)$_2$(silane) complexes have been studied by photoelectron spectroscopy (PES), which confirms a d^6 Mn^I state for (MeCp)Mn(CO)$_2$(HSiCl$_3$), consistent with arrested OA.[36–38] By contrast, CpMn(CO)$_2$(HSiCl$_3$) is revealed to have a d^4 Mn^{III} center, i.e., the bonding situation is closer to full OA. Paradoxically, however, J_{SiH} is actually only a little lower (55 Hz) than for the MeCp analogue (64 Hz) and still well within the range for σ complexes. The d_{SiH} is 1.8 Å, which is also in the nonclassical range. The J_{SiH} is rationalized to be a result of the tight angle between the d_{z^2} and d_{yz} orbitals that hold the Si and H atoms in close proximity.[37] Thus there may be a caveat concerning correlating J_{SiH} with degree of OA, at least for complexes with piano-stool geometries. It should be recalled from Chapter 4 that there are anomalies for η^2-H$_2$ versus dihydride binding to such species. Steric factors are also more relevant for silane binding, as exemplified by PES studies of Cp*Mn(CO)$_2$(H$_2$SiPh$_2$) versus the Cp analogue.[38] Unexpectedly the more electron-rich Cp* complex does not give OA, and there is less BD to Si-H σ* than in the Cp complex presumably because the bulky C$_5$Me$_5$ prevents the Si and H atoms from spreading further apart. The electronic factors in silane activation gleaned from PES and other evidence will be discussed in detail below.

Re^{III}(N$_3$N$_F$)(SiH$_2$R$_2$) is a rare example of a relatively high-valent M^{III} silane complex, although it is considered to be a diamagnetic 18e complex.[21] The Re(N$_3$N$_F$)

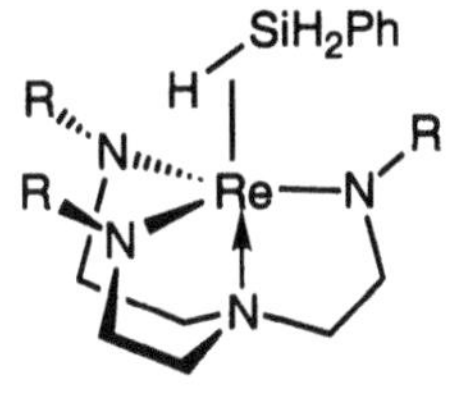

R = C$_6$F$_5$

fragment would appear fairly electron-rich based on ν_{NN} = 2004 cm^{-1} and J_{HD} = 17 Hz for the respective N$_2$ and H$_2$ complexes, indicating that the silane should undergo OA (see Section 11.3). The "high" oxidation state of the metal and the triamidoamine ligand are unusual for a σ complex, however, so such correlations may not be valid.

In contrast to the abundant H$_2$ complexes, group 8 mononuclear complexes are rare and no Os(silane) σ complexes of any type exist. This correlates with the fact that silanes are stronger acceptors and oxidatively add more easily than H$_2$ to better backdonating metals such as Os. Several dinuclear species and a novel silacyclopentadienyl ligand (structure XIII) exist.[39] One of the few silane complexes formed by protonation of M-silyl is shown in Eq. (11.11).[40]:

$$\text{CpFe(CO)(SiEt}_3\text{)(PEt}_3\text{)} \xrightarrow[\text{CD}_2\text{Cl}_2, -78 \text{ C}]{\text{HBAr}_f} [\text{CpFe(CO)(SiHEt}_3\text{)(PEt}_3\text{)}][\text{BAr}_f] \quad (11.11)$$

The cationic complex is stable at room temperature in the presence of excess silane for the nonnucleophilic anion BAr$_f$.

Perhaps the most novel reaction involving σ-bond activation is shown in Eq. (11.12), where Si-H, C-H, Si-C, and H-H bond cleavage/formation all occur in the

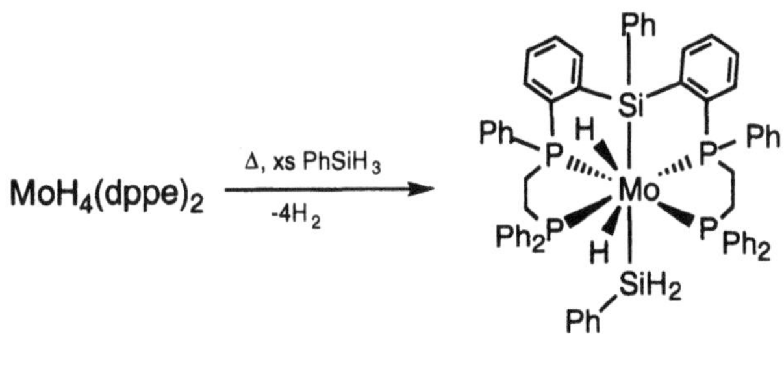

$$\text{MoH}_4(\text{dppe})_2 \xrightarrow[-4\text{H}_2]{\Delta, \text{ xs PhSiH}_3} \text{[product]} \quad (11.12)$$

same system![41] One proposed pathway for this unusual transformation (note one Si now has three Ph groups and eight H are eliminated) involves orthometallation by dppe of 16e MoH$_2$(dppe)$_2$ formed by H$_2$ loss, followed by OA of the silane (with further H$_2$ loss) and the reaction shown in Eq. (11.13). Presumably the process

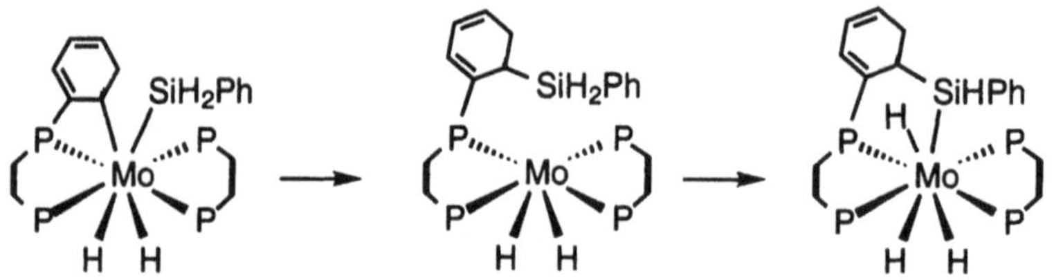

$$\text{(11.13)}$$

repeats to attach Si to the other dppe Ph, and finally a second silane either oxidatively adds [to give the structure illustrated for the product in Eq. (11.13)] or is bound as a σ ligand (hydrogen X-ray positions not located).

Few complexes are derived from silane gas, SiH$_4$, which is spontaneously explosive in air, and SiH bonds tend to undergo additional transformations, even in M(SiH$_3$) species.[16] Matrix-isolated Al(η^2-SiH$_4$) species undergo photoreversible OA to Al(H)(SiH$_3$), but structural details have not been obtained.[42] The first stable organometallic SiH$_4$ complexes are Mo(SiH$_4$)(CO)(PP)$_2$ and are important models for CH$_4$ coordination and activation.[43] The complexes are synthesized in organic solvents by addition of SiH$_4$ gas to Mo(CO)(PP)$_2$ [Eq. (11.14)] and are yellow solids that decompose in air. The silane can be removed by heating although it is bound

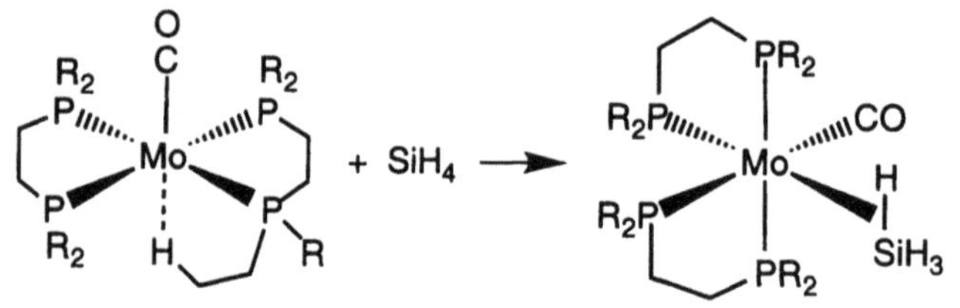

$$\text{(11.14)}$$

somewhat more tightly than η^2-H$_2$. The η^2-coordination is confirmed by $J_{SiH} = 31$, 35, and 50 Hz for R = ^{i}Bu, Et (depe), and Ph, respectively. The more electron-rich alkylphosphines give increased Si–H bond activation (lower J_{SiH}) as for H–H activation. The X-ray structure of the ^{i}Bu congener did not locate the H atoms on the Si, but, as for organosilane complexes, the coordination about Mo is distorted octahedral with cis CO and silane ligands. The geometry of the MoP$_4$CSi core is very similar to that in cis-Mo(η^2-H–SiH$_2$Ph)(CO)(depe)$_2$,[24] where the H atoms on Si were located. The d_{MoSi} in the two compounds (2.556 and 2.501 Å) are also similar. Thus the structure is consistent with η^2-SiH$_4$ coordination, as also observed in solution by ^{1}H and ^{31}P NMR. The involvement of only one SiH$_4$ hydrogen rather than two or three in the bonding is confirmed by IR, which shows three bands for the terminal Si–H groups (2047, 1995, 1972 cm^{-1}) and one for the Mo–H–Si stretch at 1732 cm^{-1} in solid Mo(η^2-H–SiH$_3$)(CO)(depe)$_2$. This pattern is similar to that for an analogous GeH$_4$ complex (see Figure 11.5) and suggests that CH$_4$ probably also coordinates in η^2-H–C fashion. However, the difference in electronegativity between Si and C and the consequent much higher degree of BD to Si–H may invalidate direct comparison.

The Mo–SiH$_4$ complex for the more electron-rich depe complex is in tautomeric equilibrium with the OA product [Eq. (11.15)]. Both forms are present in 1:1

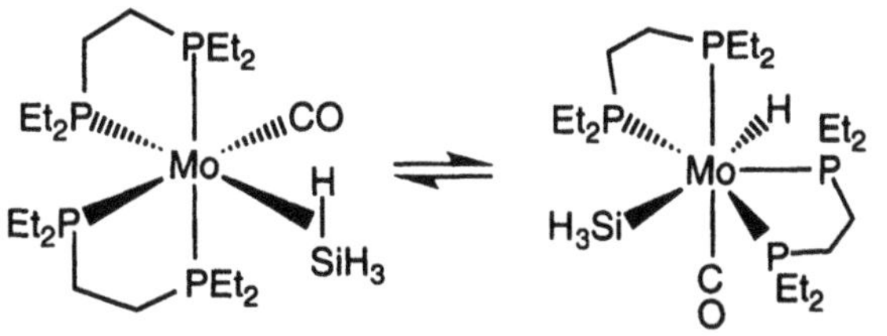

$$ \text{(11.15)} $$

ratio in solution and rapidly interchange, just as in M–H$_2$ systems. Although an unambiguous structure of the silyl–hydride tautomer is not obtainable from spectroscopic evidence, it undoubtedly is similar to that for a crystallographically characterized germyl–hydride analogue, which has a very similar ^{31}P NMR pattern (see Section 11.5). MoH$_2$(CO)(depe)$_2$ also has a similar pentagonal bipyramid structure. Equation (11.15) is the first case of equilibrium X–H bond breaking/reforming on metals other than that for H–H, and Ozin's observation of reversible Al(η^2-SiH$_4$) $\rightleftharpoons$ Al(H)(SiH$_3$) interconversion also indicates that thermodynamically there is little energy difference between such tautomers.

A novel μ-SiH$_4$ species shows coordination of all four Si–H bonds,[44] which are significantly elongated [1.69–1.73(3) Å]. The d_{RuSi} is among the shortest seen

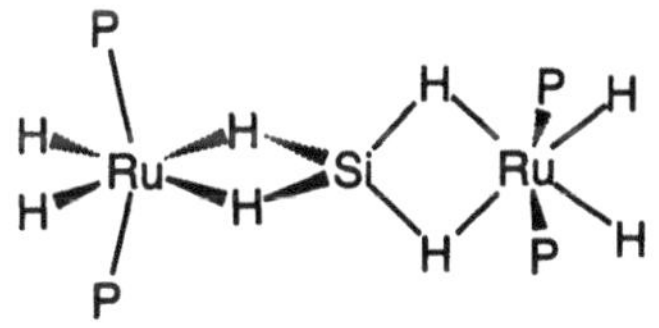

[2.1875(4) Å], even shorter than that in the silylene [Cp*(PMe$_3$)$_2$Ru=SiMe$_3$]$^+$ [2.238(2) Å]. These are shown to be σ-bond interactions based on NMR studies ($J_{SiH} = 36$ Hz) and DFT/B3LYP calculations, which reveal novel bonding involving σ donation from two Si–H bonds to one Ru and BD from the second Ru to σ^* of the other two Si–H bonds. The short d_{RuSi} cannot be considered to represent

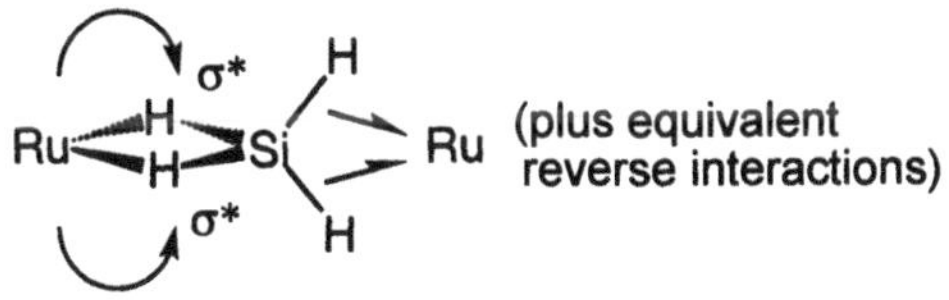

classical double bonds but result from the multiple σ interactions, which are conceivable in a CH$_4$ analogue. A high total binding energy of 61 kcal/mol between the SiH$_4$ and RuH$_2$(PH$_3$)$_2$ fragments is calculated, although this would be much lower in a similar CH$_4$ complex. This complex is not formed from SiH$_4$ but from a redistribution reaction at Si on addition of two equivalents of SiH$_2$PhMe to RuH$_2$(H$_2$)$_2$(PR$_3$)$_2$ (R = Cy, Pr) in pentane. NMR shows a mixture of Ru-containing products and silanes form en route. Si–C cleavage produces SiH$_4$ which is immediately trapped by the Ru fragments.

11.3. Si–H BONDING TO METALS COMPARED TO M–H$_2$ BONDING

Calculations support a similar bonding picture for silane complexes as to that for H$_2$ complexes. Early extended Huckel calculations in 1987 by Saillard on binding of SiH$_4$ to the CpMn(CO)$_2$ fragment showed that the Si–H overlap population is reduced to 0.24 by the σ(SiH) → M donation and M → σ*(SiH) BD, compared with 0.74 for uncoordinated SiH$_4$.[14] Thus the Si–H bond is weakened but not fully broken, and the aforementioned PES studies show that the electronic structure of the interaction of MeCpMn(CO)$_2$ with SiHPh$_3$ and GeHPh$_3$ (see Section 11.5) is consistent with an early stage of X–H bond addition to M, and donation of X–H electrons to M predominates over BD. As for H$_2$ coordination, there is a full range of M(SiH) interactions that span the extremes of complete OA to hardly any Si–H bond stretching. In most cases, however, the d_{MSi} remains relatively short regardless of the extent of activation, and there is no general explanation for this. On the other hand, the distance of the hydrogen from either the metal or Si can vary greatly; hence the extent of bond activation is more problematic than for H$_2$ activation. A trajectory proposed[7] for the reaction of silanes with the CpMn(CO)$_2$ fragment is similar to Crabtree's model for C–H bond cleavage deduced from X-ray structures of agostic complexes (illustrated in Chapter 11). This also relates to H$_2$ binding and

cleavage although the Si–H bond remains tilted throughout the OA process, but pivots so as to decrease d_{MSi} and to a lesser extent d_{MH} while the Si–H bond weakens and breaks. In the characterized silane complexes, the reaction coordinate seems arrested mostly in the middle to late part of this process where the d_{MSi} does not vary as much, although the fluctuation in experimental d_{MH} seems out of line.

The Si–H bond is more basic than either H–H or C–H and according to Crabtree, is probably a better σ donor to M,[17] Also, the Si–H bond is weaker than H–H and C–H, and hence is a better π acceptor because the energy of the SiH σ* orbital is lower, and is a better match with d-orbital energies. Si–H bonds undergo OA much more rapidly than C–H: 4.4 ps compared to 230 ns.[45] There is experimental evidence, described below, that silanes appear to be *both better σ donors and π acceptors* in comparison with H$_2$. Substituents at both M *and at Si* have large effects on the degree of activation toward OA and also on the reverse process, reductive elimination, i.e., silane dissociation.[5,13,37,45] Silanes dissociate more slowly for complexes with electron-rich M or more electronegative substituents at Si; e.g., Cp(CO)$_2$Re(HSiPh$_3$) with elongated d_{SiH} undergoes silane elimination much more

Table 11.2. $J_{SiH}{}^a$ for cis-Mo(η^2-H-SiHR$'_2$)(CO)(R$_2$PC$_2$H$_4$PR$_2$)$_2$ and (MeCp)Mn(CO)(L)(η^2-H-SiR$'_3$)^a

Complex	R or L	SiHR$'_2$	J_{SiH}, Hz^b	J_{SiH}, Hz^c
Mo	Et	SiH$_3$	35	164
	Et	SiH$_2$Ph	39	164, 170
	Et	SiH$_2$(n-C$_6$H$_{13}$)	42	155, 168
	Et	SiHPh$_2$	50	172
	CH$_2$Ph	SiH$_2$Ph	41	164, 165
	CH$_2$Ph	SiH$_2$(n-C$_6$H$_{13}$)	42	160, 166
	Ph	SiH$_3$	50	181
	Ph	SiH$_2$Ph	57	187, 194
	Ph	SiH$_2$(n-C$_6$H$_{13}$)	61	180, 181
Mn	CO	SiHPh$_2$	63.5	205
	CNBu^n	SiHPh$_2$	57.5	194
	PPh$_3$	SiHPh$_2$	43	191
	PMe$_3$	SiHPh$_2$	38	188
	PMe$_3$	SiCl$_3$	20	
	CO	SiCl$_3$	54.8	

^{a}Determined by ^{1}H{^{31}P} or ^{29}Si NMR for Mo and Mn, respectively.
$^b J_{SiH}$ for η^2-bound Si-H bonds.
$^c J_{SiH}$ for uncoordinated Si-H bonds.

slowly than the corresponding Mn complex, analogous to the stronger binding of H$_2$ to third-row M.[22] Also, HSiCl$_3$ is eliminated 10^5 times more slowly than HSiPh$_3$ from Cp(CO)$_2$Mn(silane) at 100°C,[5] which might imply that d_{SiH} is shorter in the HSiPh$_3$ complex than in the HSiCl$_3$ complex.[13] Electron-withdrawing R groups on HSiR$_3$ such as Cl increase Si-H activation and decrease J_{SiH}, presumably because they lower the SiH σ^* orbital energy, which favors increased BD. Withdrawing groups are also suggested to increase the s contribution in the Si orbital directed toward M, thereby increasing J_{SiH}, which opposes and diminishes the latter effect somewhat.[7] Electron-donating substituents such as alkyls decrease the Si-H interaction and raise J_{SiH}, fostering σ coordination over OA. This is well illustrated in the products of silane reaction with the (MeCp)Mn(CO)(L) and Mo(CO)(PP)$_2$ fragments (Table 11.2). For the Mo system the activation of the Si-H bond is controlled by the electronic properties of both the R groups on the phosphine and the R' groups on Si in opposing fashion, and a moderate J_{SiH} of 50 Hz is observed either for R = Et and R' = Ph or for R = Ph and R' = H. The X-ray structures of Mo(CO)(SiH$_2$Ph$_2$)(depe)$_2$ and its SiH$_3$Ph analogue have d_{SiH} of 1.66 and 1.77 Å; hence replacement of just one Ph by less donating H lengthens d_{SiH}, which correlates with a reduction in J_{SiH} from 50 to 39 Hz. The lowest J_{SiH}, 35 Hz, is seen for the most electron-rich R = Et and the least electron-rich R' = H (SiH$_4$), and the Si-H bond is so activated that it undergoes equilibrium OA [Eq. (11.15)]. The Mn complexes show similar behavior, and, as the donor strength of L increases down Table 11.2, J_{SiH} decreases. Therefore, although J_{SiH} can behave irregularly across a wide variety of complexes, it correlates well with electronic factors within a given system.

Paradoxically, however, theoretical studies on Cp$_2$MXH(SiCl$_n$H$_{3-n}$) (M = Nb; Ta; X = SiH$_3$, Cl, H, CH$_3$; n = 0-3) at both the MP2 and B3LYP levels show that

d_{SiH} is *shorter* for complexes that have more Cl substituents at Si.[46] This may be related to the ease with which Si becomes hypervalent with an adjacent electronegative ligand. Similarly, for a given M in $Cp(CO)_2M(SiCl_nH_{4-n})$ (M = group 7), calculations show that substitution of H atoms by Cl gives moderate decreases in d_{SiH}, in opposition to the experimental higher degree of activation of the Si–H toward OA and attendant decreased J_{SiH}.[47] The experimental d_{SiH} do not change much upon substitution (Table 11.1), e.g., X-ray d_{SiH} of $Cp(CO)_2Mn(HSiPh_3)$ and $MeCp(CO)_2Mn(HSiCl_3)$ are 1.76(4) Å and 1.79(4) Å, respectively [cf calcd d_{SiH} for $Cp(CO)_2Mn(HSiCl_3)$, 1.806[47]].[7] However, this trend is at least in line with J_{SiH}, which decreases on sustitution by Cl (Table 11.2), indicating a more activated (longer) Si–H bond. In contrast to d_{SiH}, d_{MSi} shortens considerably on placing electron-withdrawing Cl on Si, both calculationally [from 2.297 to 2.216 Å on changing n from 1 to 3 in $Cp(CO)_2Mn(SiCl_nH_{4-n})$] and experimentally [from 2.424(2) Å for $Cp(CO)_2Mn(HSiPh_3)$ to 2.25 Å for $MeCp(CO)_2Mn(HSiCl_3)$]. Calculationally d_{MH} remains almost constant upon the substitutions by Cl but decreases experimentally [1.55(4) to 1.47(3) Å for the above]. Thus the changes in d_{SiH} on changing substituents at Si in $Cp(CO)_2M(HSiR_3)$ systems are much lower when compared to the changes for d_{MSi} and d_{MH}, particularly from calculations. The calculated silane dissociation energies increase for increased n in $Cp(CO)_2M(SiCl_nH_{4-n})$ (and also down the group 7 row),[47] in agreement with experiment. These findings suggest that the increased silane binding strength results from increased M–Si interaction, and contrary to what might be expected, the Si⋯H interaction is not weakened in parallel, unlike for $M(\eta^2-H_2)$ activation. The behavior of $M(\eta^2-Si-H)$ may differ because of the tendency of the silicon center to become hypervalent.

Lastly, there are exceptions to the above: For interaction of silanes with strongly *electrophilic* fragments such as $Cr(CO)_5$, the binding energy *decreases* with more electron-withdrawing substituents on Si.[31] This reversal, which will be discussed below, may reflect the relatively greater importance of SiH → M σ donation compared to BD for electrophilic systems. Clearly the M–Si–H interaction can be finely tuned in several seemingly conflicting ways to give a series of complexes arrested along different points on the reaction coordinate to OA.

11.3.1. Comparisons of H–H versus Si–H σ-Bond Activation and OA Processes

The number of metal fragments that coordinate both silanes and H_2 as σ ligands, especially near the point of OA, is fewer than might be expected at this time. Although the similarities in binding and activation are remarkable, there are some differences and it is instructive to examine them in detail. A comparison is given in Table 11.3, which also includes electronic marker ligands N_2 and SO_2 that indicate relative electrophilicities of the 16e fragments [higher v_{NN} and v_{SO} correspond to more electron-poor centers (see Chapter 4)].[25] The most electrophilic complexes are at the top of Table 11.3, and the Re system is unstable toward heterolysis of both H_2 and silanes. Whereas H_2 binds or undergoes OA to most unsaturated metal fragments, silanes do not bind to certain fragments even at low temperatures.

Table 11.3. Comparison of Silane and Small-Molecule Binding/Activation on Various Metal Fragments[a]

Metal fragment	H_2, J_{HD}[b]	PhSiH$_3$, J_{SiH}[c]	Ph$_2$SiH$_2$, J_{SiH}[c]	N_2, ν_{NN}[d]	SO_2, ν_{SO}[e]
[Re(CO)$_4$(PCy$_3$)]$^+$	33.8[f]		61.6[f]	NR	
[CpFe(CO)(PEt$_3$)]$^+$	31.6		58		~1282[g]
TpRuH(PPh$_3$)	29.7	27.4[h]	23.5[h]		
CpMn(CO)$_2$	33		63.5[i]	2173	1270
[Mn(CO)(dppe)$_2$]$^+$	33	NR	NR	2167	Weak
[Mn(CO)$_3$(P)$_2$]$^+$	34.5	Weak		NR	
Cr(CO)$_3$(PR$_3$)$_2$	35	NR		2128	
W(CO)$_3$(PR$_3$)$_2$	34, equil OA	OA	Eq. (11.6)	2120	1237
Mo(CO)(dppe)$_2$	34	57, weak	Weak	2090	1209
Mo(CO)(dBzpe)$_2$	30	41		2062	1180
Mo(CO)(depe)$_2$	OA	39	50	2050	1164
Re^{III}(N$_3$N$_F$)	17	38		2004	

[a]Abbreviations: P = P(OCH$_2$)$_3$CMe; dBzpe = dibenzylphosphinoethane.
[b]Hz. R = i-Pr.
[c]Hz. For coordinated Si–H bond.
[d]cm^{-1}. R = Cy.
[e]cm^{-1}; asymmetric stretch. R = Cy.
[f]At −40°C or lower (unstable towards heterolytic cleavage). The silane in the silane complex is Et$_3$SiH.
[g]Value for [CpFe(dppe)(SO$_2$)]$^+$.
[h]Average of J_{SiH} for the bound η^2-SiH and the coupling between Si and the hydrogen not involved in the σ bond (Ng et al.[20]).
[i]Value for MeCp derivative.

For example, W(CO)$_3$(PR$_3$)$_2$ oxidatively adds silanes yet binds H$_2$,[25] which unexpectedly is opposite to the reactivity on more electron-rich Mo(CO)(depe)$_2$. The possible reasons for this will be discussed here. For H$_2$ binding on a wide variety of

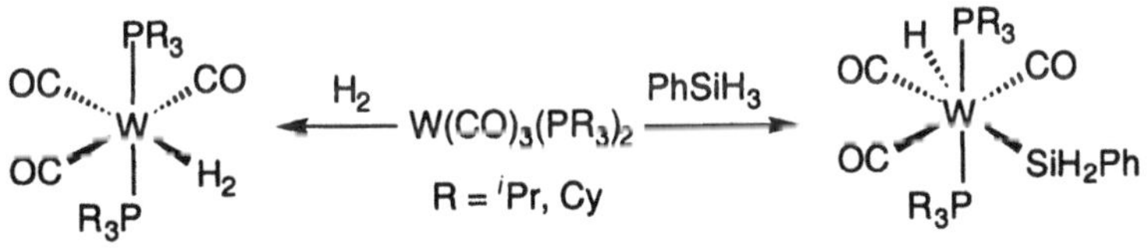

$$\text{[H}_2\text{ complex]} \xleftarrow{H_2} W(CO)_3(PR_3)_2 \xrightarrow{PhSiH_3} \text{[PhSiH}_2\text{ complex]} \quad R = {}^iPr, Cy \quad (11.16)$$

M/L fragments, a near continuum of d_{HH} (0.85–1.5 Å) exists that blurs the point at which the H–H bond can be considered to be fully broken (1.6 Å is oft-quoted). In Chapter 4, use of Bader's AIM formalism showed that complexes with $d_{HH} > 1.3$ Å are best viewed as dihydrides with weak H···H interactions, while a d_{HH} of 0.82 Å corresponds to a true H$_2$ complex, i.e., a clear bond path connects the H.[48] A d_{HH} near 1.0 Å represents advanced activation with some residual H–H bonding, and similarly stretched Si–H bonds (e.g., 2.10 Å) do exist.

BD from M to the X–H σ* orbital generally controls d_{XH} (X = H, C, Si, etc.) along the path toward OA and helps stabilize σ complexes,[5] but electrophilic cations such as [Re(CO)$_4$(PR$_3$)]$^+$ still bind H$_2$ and silanes at −40°C by increased σ donation to M, offsetting decreased BD.[49,50] cis-{Re(CO)$_3$[P(OCH$_2$)$_3$CMe]$_2$-(SiHEt$_3$)}$^+$ can even be isolated (J_{SiH} = 66). The influence of cis ligands (e.g. PPh$_3$

versus more electron donating PCy$_3$) on J_{HD} and J_{SiH} (hence d_{XH}) is minimal compared to the dominating effect of the trans ligand. Et$_3$SiH is bound somewhat more strongly than H$_2$, consistent with the notion that the Si–H bond is a better σ donor than the H–H bond (ignoring much weaker BD effects). This situation goes

	X = H	X = Si
R = Ph	J_{HD} = 33.8	J_{SiH} = 60.9
R = Cy	J_{HD} = 33.9	J_{SiH} = 61.6

beyond simple trade-off effects for silane coordination to electrophilic fragments such as Cr(CO)$_5$. Here increasing the electron richness of the Si center (e.g., going from SiHCl$_3$ to SiHR$_3$) leads to increased silane → M σ donation and more stable complexes, in contrast to the situation on electron-rich centers where BD predominates.[31] The increased σ donation overrides the diminished role of BD. There is a nice parallel to H$_2$ binding except that substituent effects provide additional (and potentially counterintuitive) electronic fine-tuning for silane binding.

Increased σ donation by itself cannot cause X–H bond rupture: BD is responsible for this task. Sorting out these two components is important in understanding σ-bond activation, and the BD component, E_{BD}, of the total M–H$_2$ orbital interaction energy can be separated from σ-donation E_D and is comparable to or exceeds the latter even when d_{HH} are short (see Chapter 4). Thus except on highly electrophilic fragments, H$_2$ complexes might best be pictured as triangulo species, with partial bonds joining the three atoms rather than the usual three-center representation. Because silanes are better π acceptors, the oft-drawn triangulo M⋯Si⋯H

representation should apply even more, as supported by calculations (see Section 11.3.2).

Tautomeric equilibria between σ complexes and OA products can exist for H–H [Eq. (11.5)], Si–H [Eq. (11.1)], and also Ge–H bond activation. Remarkably, $\Delta G^{\ddagger}$ for H$_2$ cleavage (16.0 kcal/mol at 298 K) is very similar to that for Si–H rupture (15.3 kcal/mol at 333 K); hence the reaction coordinates for OA are similar. However, there are subtle differences in H–H and Si–H activation, as seen in Eq. (11.16). Although the *energetics of OA of silanes versus H$_2$ are comparable, whether the H–H or Si–H bond breaks more easily depends on the M/L system.* It is instructive to further examine Table 11.3 where J_{HD} and J_{SiH} are indicators of the activation of the σ bond, and v_{NN} and v_{SO} measure the metal's BD ability. M–H$_2$ binding is generally stable toward OA if v_{NN} > 2060 cm^{-1}, and the H–H bond is usually less than 0.9 Å when trans to CO. For group 6 systems, the H–H bond breaks suddenly when the electron richness of M is increased too much, e.g., changing R from Ph to Et in Eq. (11.17), rather than elongating it as for later metals.

It is surprising that organosilanes do *not* oxidatively add to the Mo(CO)(depe)$_2$ fragment that cleaves H$_2$ because silanes are better π acceptors than H$_2$ and rearrange position to be cis to CO, ostensibly to avoid competing with CO for BD.

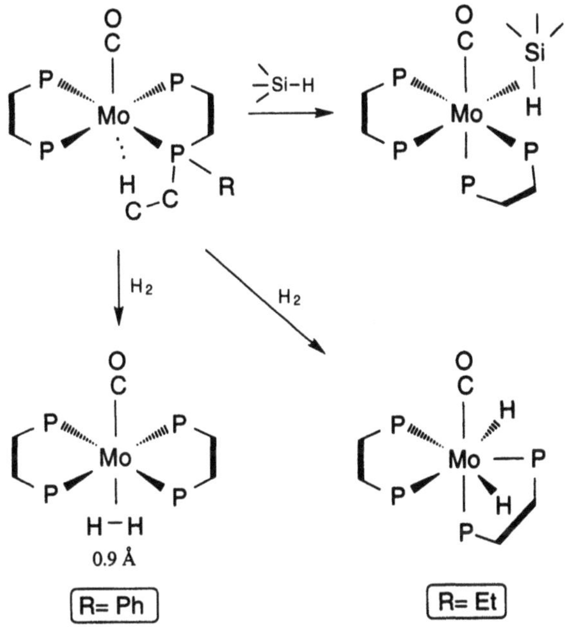

(11.17)

Calculations[51] on a Mo(CO)(PH$_3$)$_4$(H$\cdots$SiH$_3$) model also favor coordination of SiH$_4$ cis to CO for this electronic reason. The reversal of results compared to that for H$_2$ and PhSiH$_3$ addition to W(CO)$_3$(PR$_3$)$_2$ demonstrates the fine balance of electronic and steric forces here that may include structural rearrangement barriers for six versus seven coordination as discussed in Section 4.7.4. For silane and σ ligands other than H$_2$, steric influences are magnified and perhaps favor the cis (CO)(silane) orientation as well as disfavor OA to a seven-coordinate product with a silyl ligand that is quite large compared to a hydride. Theoretical and experimental studies indicate that overall bond energetics (e.g., energy of X–H bond versus M–X bond in OA product) can also be crucial, as discussed in Chapter 7.[52-56] Calculations by Maseras and Lledos[53] show little energy difference between OsCl(CO)(PH$_3$)$_2$(H)(η^2-HSiH$_3$) and OsCl(CO)(PH$_3$)$_2$(SiH$_3$)(η^2-H$_2$) although experimentally there is preference for the latter [7 in Eq. (11.18)][54]:

$$\text{OsHCl(CO)(P}^i\text{Pr}_3)_2 \xrightarrow{\text{HSiEt}_3} \text{OsH}_2\text{(SiEt}_3\text{)Cl(CO)(P}^i\text{Pr}_3)_2$$
$$(6)$$
$$\rightleftharpoons \text{Os}(\eta^2\text{-H}_2)\text{(SiEt}_3\text{)Cl(CO)(P}^i\text{Pr}_3)_2$$
$$(7)$$

(11.18)

Conversely, highly dynamic hydrido(η^2-silane) structures are found for [IrH$_2$(η^2-HSiEt$_3$)$_2$(PPh$_3$)$_2$]$^+$ [55] and TpRuH(η^2-HSiR$_3$)(PPh$_3$)[20] which will be further discussed in Section 11.4. The structure of the latter complex demonstrates that, as for η^2-H_2 analogues (see Section 4.9), the Tp ligand appears to stabilize the σ complex against OA relative to Cp-type analogues (Cp*RuH$_2$(SiR$_3$)(P^{i}Pr$_3$) is a silyl(hydride)[57])

σ Complex	OA complex
TpRuH(H$_2$)(PPh$_3$)	CpRuH$_3$(PR$_3$)
TpRuH(η^2-HSiR$_3$)(PPh$_3$)	Cp*RuH$_2$(SiR$_3$)(P^{i}Pr$_3$)

In RuH$_2$(H$_2$)(SiHPh$_3$)(PCy$_3$)$_2$, which contains both H$_2$ and silane ligands, the (η^2-H$_2$)(η^2-HSiR$_3$) isomer (**8**) is only about 2 kcal/mol more stable experimentally and calculationally (as a PH$_3$ model complex with R = H) than the (η^2-H$_2$)$_2$(H)(SiR$_3$) isomer (**9**).[58] However, differences in phosphine orientation and

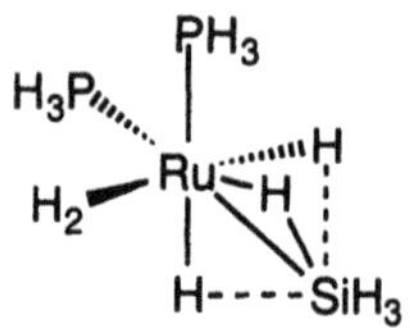

(**8**) 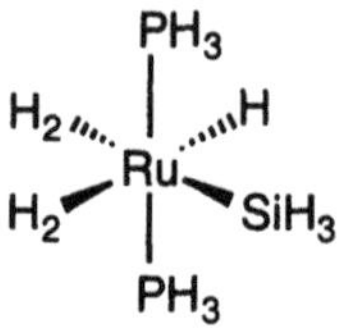   (**9**)

stabilizing nonbonded Si···H cis interactions exist here and contribute both electronically and sterically to the energetics as for η^2-H$_2$···hydride interactions (see Section 4.11). The cis interactions (~2.1 Å, calcd) offset the unfavorable rare cis arrangement of the bulky PCy$_3$ found experimentally. Addition of H$_2$ to RuH$_2$(CO)(P^{t}Bu$_2$Me)$_2$ gives the η^2-H$_2$ adduct, but HSiMe$_3$ undergoes OA.[56] The relative stabilities of the two possible products may depend on a subtle balance of thermodynamics alone or in combination with BD arguments.

According to Table 11.3, silanes seem to undergo OA more easily than H$_2$ on more electrophilic M, perhaps because of the superior BD ability of Si–H. For more electron-rich M, differences in π-acceptor strength of σ ligands should not be as critical in bond cleavage, and H$_2$ could even undergo OA more easily than silanes for steric or other reasons. Based on both IR evidence (Table 11.3) and calculations in Chapter 4, the W(CO)$_3$P$_2$ center is more electrophilic than Mo(CO)P$_4$ and indeed cleaves silanes more easily than H$_2$. Although direct comparisons of silane and H$_2$ activation are valid only for silanes with similar substituents, some inferences can be drawn. One of the more electrophilic fragments in Table 11.3, CpMn(CO)$_2$, readily forms a stable Ph$_2$SiH$_2$ complex and oxidatively adds HSiCl$_3$,[7,15] but the H$_2$ complex is less stable and has J_{HD} = 33 Hz; i.e., relatively unactivated. More electrophilic CpV(CO)$_3$[59] and [CpFe(CO)(PEt$_3$)]$^+$[40] fragments give unstable H$_2$ adducts but easily bind silanes and cleave HSiEtCl$_2$.[59] Reactions of bare metal cations such as Fe$^+$ with SiH$_4$ give Si–H cleavage products rather than [M(L)]$^+$ adducts observed by similar guided ion-beam techniques.[60] Agostic electrostatic

Si–H⋯M (M = Li$^+$, Mg^{2+}) interactions are observed by X-ray diffraction in main-group species such as Mg$_2$[Me$_2$Si(H)NBu]$_4$,[61] but H$_2$ interactions with alkali metal cations are very weak (see Chapter 4). For organometallic systems, steric

	Si–H	H–H	C–H
Fe$^+$	OA	[Fe(H$_2$)]$^+$	[Fe(CH$_4$)]$^+$
Li$^+$	Si–H⋯Li$^+$	Very weak H$_2$⋯Li$^+$	

congestion appears to inhibit silane binding for smaller first-row metals with bulky phosphines: Cr(CO)$_3$(PCy$_3$)$_2$ and its Mn$^+$ congener coordinate H$_2$ weakly but do not interact with excess PhSiH$_3$ unlike the W system. In contrast, the less congested [Mn(CO)$_3$(P)$_2$]$^+$ where P is a tied-back phosphite (see Section 7.2) does coordinate PhSiH$_3$. First-row [Mn(CO)(dppe)$_2$]$^+$ binds H$_2$ almost as well as the isoelectronic

Silanes do not bind to:	Cr(CO)$_3$(PCy$_3$)$_2$	[Mn(CO)$_3$(PCy$_3$)$_2$]$^+$	Mn(CO)(dppe)$_2$]$^+$
Silanes bind to:	W(CO)$_3$(PCy$_3$)$_2$	[Mn(CO)$_3$(phosphite)$_2$]$^+$ [Re(CO)$_4$(PR$_3$)]$^+$	Mo(CO)(dppe)$_2$

second row Mo congener, but does not coordinate silanes.[62] The sterically uncrowded, very electrophilic [Re(CO)$_4$(PR$_3$)]$^+$ fragment binds silanes better than H$_2$ below −40°C, although both σ ligands are unstable toward heterolytic cleavage at room temperature.[49,50]

11.3.2. Theoretical Studies of Si–H versus H–H σ-Bond Activation

The model complexes Mo(CO)(L)(PH$_3$)$_4$ (L = H$_2$, SiH$_4$) offer a good calculational comparison of Si–H and H–H bond activation, and the cis silane isomer is favored by 9.9 kcal over the trans isomer where the silane competes with CO for BD.[51] The optimized cis structure shows a d_{SiH} of 1.813 Å close to the experimental value of 1.77 Å in cis-Mo(CO)(depe)$_2$(SiH$_2$Ph$_2$). Analysis of the electron-density Laplacian around the Mo(H⋯Si) triangle (Figure 11.3) shows significant differences compared to the situation for true (unstretched) H$_2$ coordination (see Chapter 4). A covalent-type Mo–Si bond is indicated by one local concentration for both Mo and Si, while the Mo–H bond is essentially a dative H$^-$ → Mo interaction similar to P → Mo. The bond path of H⋯Si is curved inward with the turning point sloped toward Si, implying that the H–Si covalent bond is significantly weakened. In true M–H$_2$ coordination, the H–H covalent bonding is retained for the most part, i.e., there is higher electron density between the two H and less M → H$_2$ BD than H$_2$ → M σ donation. However, for the silane bonding here, the situation is more like stretched H$_2$ complexes because silanes are stronger π acceptors than H$_2$. There are high degrees of density concentrations in both Mo–H and Mo–Si bonds, although there is still appreciable concentration between H and Si. Thus silane complexes lie closer to the OA product than H$_2$ complexes, as depicted below, and true nonclassical Si–H interactions occur primarily for highly electrophilic centers, d^0 systems

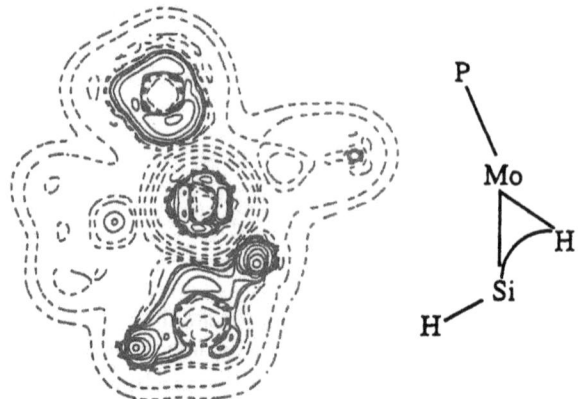

Figure 11.3. Laplacian of the valence electron density of Mo(CO)(PH$_3$)$_4$(H···SiH$_3$) on the plane defined by the Mo–H···Si triangle, with atom labels and bond paths on the right. Reprinted with permission from Fan et al.[51] Copyright 1996 American Chemical Society.

where BD is weak, or β-agostic systems. Fan et al. conclude that the hydrogen in M(H)(SiR$_3$) is more hydridic than in MH$_2$ and that its attraction to Si to form stretched σ complexes is facilitated by the ease with which Si becomes hypervalent

with an adjacent electronegative ligand.[51] However, the similarity in the overall reaction coordinate for OA of H–H and Si–H bonds suggested by Table 11.3 and Chapter 7 indicates that the bonding situation is not vastly different, but mainly that *M-silane interactions are arrested further along the reaction coordinate and steric factors are more important.*

DFT methods are also used to analyze a bis(silane)ruthenium complex, RuH$_2$-[(SiHMe$_2$)$_2$X](PR$_3$)$_2$ [X = *o*-phenylene, see Eq. (11.7)].[63] Although the Si–H units are nearly trans, a significant distortion *away* from a direct trans geometry avoids the competition for BD to the strong acceptor silane ligands [a trans bis(H$_2$) complex is unknown]. The distortion is enhanced by the chelation, which forces a small Si–Ru–Si angle (88°). The calculated d_{SiH} is 1.842 Å, within the 1.65–2.0 Å range of stretched silane complexes, and agrees well with the experimental value. Prominent direct covalent R–Si and Ru–H bonds are observed as indicated by local density concentrations lying along the Ru–Si and Ru–H bonds in the density

Laplacian. Compared to Figure 11.3, significantly curved and weakened H···Si "bonds" are present, hence a higher degree of BD. Natural-bonding orbital analysis shows significant occupation of the H–Si σ* orbital (an occupation number of about

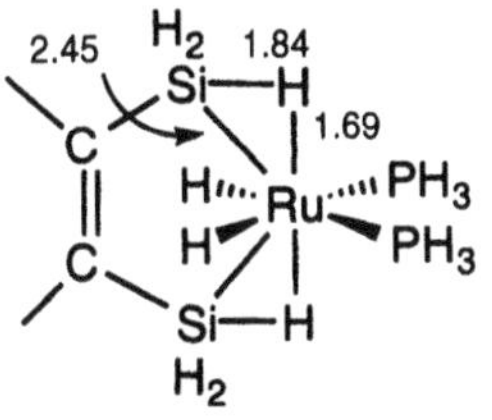

(10)

1.5). Calculations on an unchelated model without the phenylene bridge give shorter d_{SiH} of about 1.75 Å; hence the chelation gives more activated Si···H bonds in accordance with the decrease in J_{SiH} as the Si–Ru–Si bite angle shrinks {smaller X in RuH$_2$[(SiHMe$_2$)$_2$X](P)$_2$}.

Further *ab initio* investigations and electron-density analysis[64] of silane coordination to early-metal metallocenes reveal that the agostic Si–H interaction in the Ti complex in Eq. (11.10) is quite strong despite the relatively short X-ray d_{SiH} of 1.42 Å and high J_{SiH}. The theoretical d_{SiH} is 1.64–1.71 Å; hence the X-ray d_{SiH} is probably too short (not surprising in light of the general unreliability of such data). The complex is better formulated as d^2 rather than d^0, and strong BD again occurs here. The powerful synergism between experimental and theoretical analysis of M σ-bond interactions is further demonstrated in the metallocene system, where X-ray data showed the unsymmetrical structure (11) with the hydride closer to one Si than the other (1.83 versus 2.48 Å for M = Ta; X = H).[65]

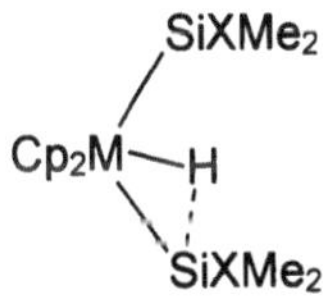

(11) 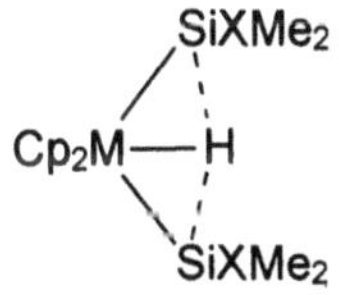   (12)

However, solution ^{1}H NMR shows that the two silyl groups are equivalent. This would indicate that the hydride is positioned on the metal symmetrically with respect to the silyl groups, as shown in structure **12**. In agreement with this, calculations show that the symmetrical structure (**12**) is more likely for both M = Ta and Nb,[46] and subsequent neutron diffraction studies of these *interligand hypervalent interactions* (*IHI*) support this (M = Ta; X = H by Berry and M = Nb; X = Cl by Nikonov).[66] Computationally, the extent of H···Si interaction is increased either by replacing one silyl by halide or Me, or by using halide-substituted silyl ligands. This is shown experimentally as well by changing X to Cl in **12** and Nikonov proposes that donation of Nb–H electron density to the Si–Cl antibonding orbital gives rise to the IHI. An extreme case is Cp$_2$MHCl(SiCl$_n$H$_{3-n}$) for $n = 1$, where d_{SiH} are as low as 1.88 Å and a curved H–Si bond path is seen in Laplacian plots now indicative of nonclassical $M(\eta^2\text{-Si-H})$ coordination.[46] This pathway toward Si–H bond formation (reversal of OA) is favored by increasing the electrophilicity of M by halide substitution and also by increasing the electronegativity of the Si substituents, which enhances attraction between the hydride and Si. This "reversed" Si substituent effect is similar to that seen in silane coordination to electron-deficient centers such as $Cr(CO)_5$ discussed above. Changing M also can be crucial: Nikonov finds that reaction of $CpM(=NAr)(PMe_3)_2$ with $HSiMe_2Cl$ for M = Ta yields $CpTa(=NAr)(H)(SiMe_2Cl)(PMe_3)$ which possesses IHI as in 11 but for M = Nb gives $CpNb[\eta^3\text{-}N(Ar)SiMe_2\text{-}H]Cl(PMe_3)$ with a β-agostic silylamine Si–H⋯M interaction.[66]

In summary, there are at least five primary variables in $L_nM(\eta^2\text{-X-H})$ systems influencing the electronics and hence activation toward OA: the nature of M and X, the substituents at both M and X, and also L, especially when trans to XH. Additionally, steric factors and overall energetics of OA processes could play important roles in determining the point at which a σ bond breaks. Silanes bind less easily than the diminutive H_2 ligand to smaller first-row metals with bulky phosphines. This is counterbalanced somewhat by fact that $H-SiR_3$ ligands are both better σ donors and π acceptors than H_2 and can approach M in tilted fashion with the H atom in front and the bulky SiR_3 out of the way. The binding and activation of σ bonds other than H_2 will always be much more complex electronically and sterically and even lead to counterintuitive behavior. Thus, on certain fragments such as $Mo(CO)(depe)_2$, H_2 undergoes OA more easily than silanes, but on others such as $W(CO)_3(PR_3)_2$, the reverse is true. Although the reaction coordinates for H–H and Si–H bond cleavage are quite similar in most aspects, the stronger π-acceptor strength of the Si–H bond often leads to σ complexes arrested further along the reaction coordinate toward OA than for H_2 binding. As for H–H bond coordination, Si–H bond interactions with metals encompass the full range from weak, reversible binding where the bond length is barely perturbed to systems where the Si–H bond is nearly completely ruptured.

11.4. REACTIONS AND DYNAMICS OF σ-SILANE COMPLEXES

The reactions of hydrosilanes have been well reviewed[16] and only major points concerning σ-silane complexes will be addressed here. Dynamics and reactivity have not been as extensively studied as those for H_2 complexes, but are remarkably similar, e.g., hydrosilylation is well known and directly analogous to homogeneous hydrogenation. It is generally assumed that the mechanisms are the same.[1] Alkene insertion into a M–Si bond is quite common also, and dienes can be photocatalytically hydrosilylated in the presence of $Cr(CO)_6$ [Eq. (11.19)].[67] The mechanism is

$$\diagup\!\!\!\diagdown + H-SiEt_3 \xrightarrow[h\nu]{Cr(CO)_6} \diagdown\!\!\!\!\diagup\!\!\!\!\diagdown\!-SiEt_3 \quad\quad (11.19)$$

proposed to involve photolytically promoted displacement of three meridonal CO by the diene and silane, followed by OA of η^2-silane and transfer of hydride and silyl groups to diene. As for η^2-H$_2$, direct reaction of η^2-silane with a substrate is not well established but is conceivable in Eq. (11.19) and in other reactions, especially those involving heterolytic cleavage.

11.4.1. Heterolytic Cleavage of M(η^2-Si–H) and Related Reactions

Although not as many examples of heterolysis of silane Si–H bonds are known as compared to splitting of H–H bonds, the Si is highly activated toward nucleophilic attack when coordinated to electrophilic cationic centers because of depletion of the electron density from the Si–H bond.[55] The Si–H bond in free silanes is

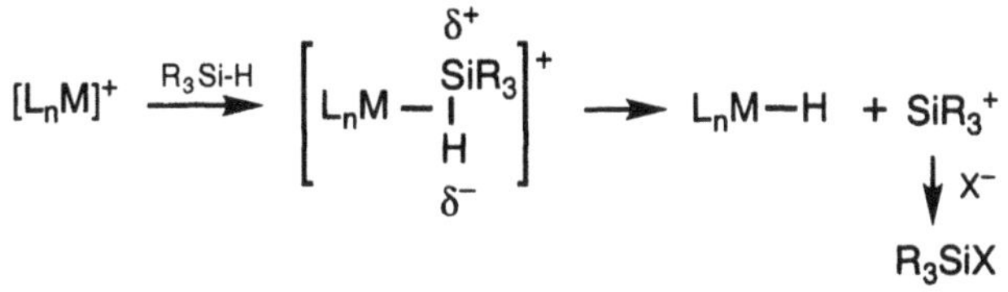

$$[L_nM]^+ \xrightarrow{R_3Si-H} \left[L_nM - \overset{\delta^+}{\underset{\underset{\delta^-}{H}}{SiR_3}} \right]^+ \longrightarrow L_nM-H + SiR_3^+ \downarrow X^-$$
$$R_3SiX$$

X= OH from H$_2$O, F from anion

already polarized in the sense $Si^{\delta+}-H^{\delta-}$ and coordination to an electrophilic M increases the positive charge on Si. This gives rise to its elimination as a silylium cation, R_3Si^+, a powerful electrophile that can abstract OH$^-$ from trace water to give R_3SiOH and also extract fluoride from counteranions such as SbF_6 [Eq. (11.20)] and even BAr$_f$.[68] Cationic silane complexes are normally stable only in solution at low temperature and rarely isolable. Thermally unstable bis(silane) adducts such as [IrH$_2$(η^2-HSiEt$_3$)$_2$(PPh$_3$)$_2$]SbF$_6$ (13) are detectable by NMR at low temperatures[55] and show intriguing reactivity, structures, and dynamics, as will be shown later.

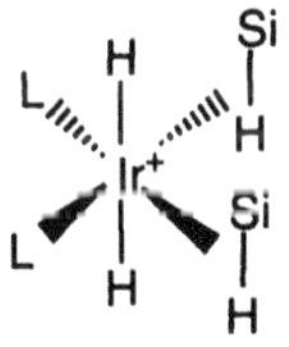

(13)

At room temperature heterolytic cleavage of the Si–H bonds in 13 occurs in CH$_2$Cl$_2$:

$$2[\text{IrH}_2(\text{THF})_2(\text{PPh}_3)_2]\text{SbF}_6 + \text{Et}_3\text{SiH} \rightarrow [\text{IrH}_2(\eta^2\text{-HSiEt}_3)_2(\text{PPh}_3)_2]\text{SbF}_6$$
$$\rightarrow [\text{Ir}_2(\mu\text{-H})_3\text{H}_2(\text{PPh}_3)_4]\text{SbF}_6 + \text{Et}_3\text{SiF}$$
$$+ \text{SbF}_5 + 4\text{THF} \qquad (11.20)$$

If this reaction is carried out in the presence of alcohols, however, homogeneous catalysis of silane alcoholysis occurs with high efficiency and selectivity:

$$R'_3\text{SiH} + \text{ROH} \rightarrow R'_3\text{SiOR} + H_2 \qquad (11.21)$$

It is important to note that kinetic and mechanistic studies are consistent with direct nucleophilic attack by ROH on a η^2-Si–H bond rather than initial OA.[69] Cationic [CpFe(CO)(PR$_3$)(HSiEt$_3$)]$^+$ is observed by NMR at room temperature but only in the presence of excess silane (sacrificial removal of trace H$_2$O as Et$_3$SiOH) and cannot be isolated as a solid.[40] [CpFe(CO)(PR$_3$)]$^+$ as well as [Mn(CO)$_3$(P)$_2$]$^+$ (P = P(OCH$_2$)$_3$CMe) also catalyze silane alcoholysis [Eq. (11.21)] in the presence of the BAr$_f^-$ counterion.[69,70] In the Fe system, Brookhart proposes that the H$^+$

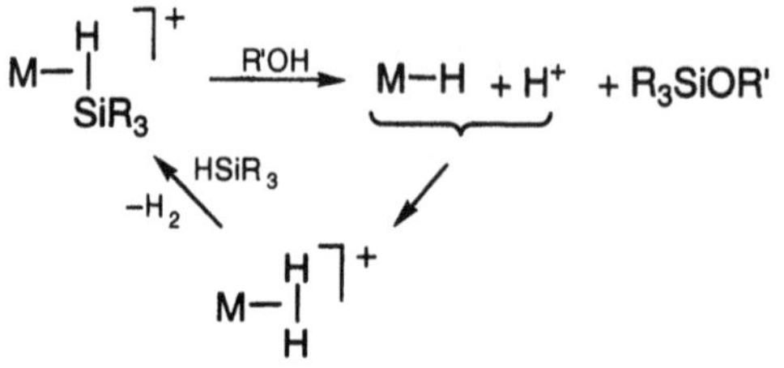

released from reaction of alcohol with R$_3$Si$^+$ protonates M–H to form [M–H$_2$]$^+$ {the known [CpFe(CO)(PR$_3$)(H$_2$)]$^+$} and silane displaces H$_2$ to give back the starting silane complex for further alcoholysis.[40,69]

Heterolysis of Et$_3$SiH in highly electron-poor [cis-Re(CO)$_4$(PR$_3$)(η^2-HSiEt$_3$)]-[A] (R = Ph, Cy) [Eq. (11.22)] occurs much like that in the above systems and is analogous to H$_2$ heterolysis in giving a hydride-bridged dimer (see Scheme 3 in Chapter 9).[50] Injection of one equivalent of Et$_3$SiH into a solution of

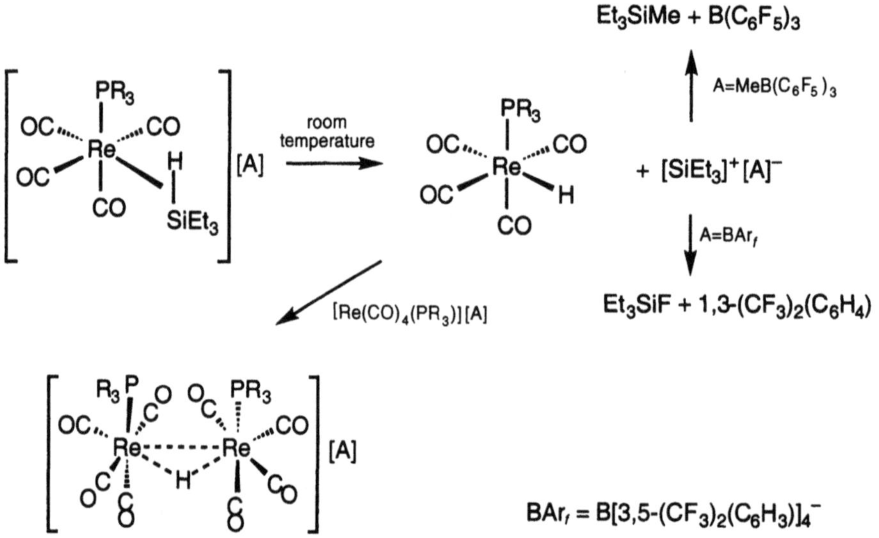

(11.22)

[Re(CO)$_4$(PPh$_3$)(ClCD$_2$Cl)][BAr$_f$] in CD$_2$Cl$_2$ at −40°C forms [Re(CO)$_4$(PPh$_3$)-(HSiEt$_3$)][BAr$_f$]. On warming to 25°C, complete conversion to the μ-H dimer occurs in minutes. However, a silicon cation equivalent "Et$_3$Si$^+$" is eliminated instead of H$^+$, which then reacts with both trace water (to give Et$_3$SiOSiEt$_3$) and the anion to form Et_3SiF and $1,3-(CF_3)_2(C_6H_4)$ [Eq. (11.22)]. This suggests that the $Re(HSiEt_3)$ complex is unstable at room temperature even with the noninteracting anion BAr_f^-, and that elimination of Et_3Si^+ probably proceeds with attack on the BAr_f anion. The Et_3Si^+ eliminated may exist transiently as a highly reactive solvento complex.[68] Silanes and H_2 parallel each other in reactivity with organometallic compounds,[71] and silanes react with coordinated sulfide and N_2 ligands much like H_2 (see Section 10.4).

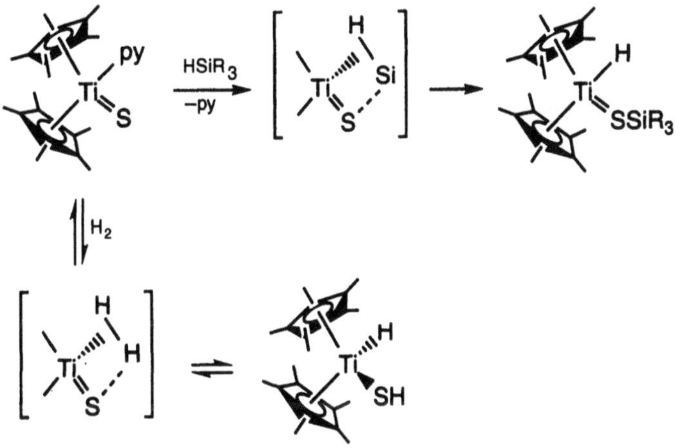

(11.23)

Reaction of silanes with the Ti=S complex to form a S–Si bond is reversible as for H_2 addition.[72] A small k_H/k_D of 1.3 for addition of $(H/D)SiMe_3$ is consistent with hydride transfer from the silane via a four-center transition state. Another remarkable reaction involves the macrocyclic Zr system of Fryzuk and coworkers [see also Eq. (10.17)]. The crystal structure of the product shows Si attached to

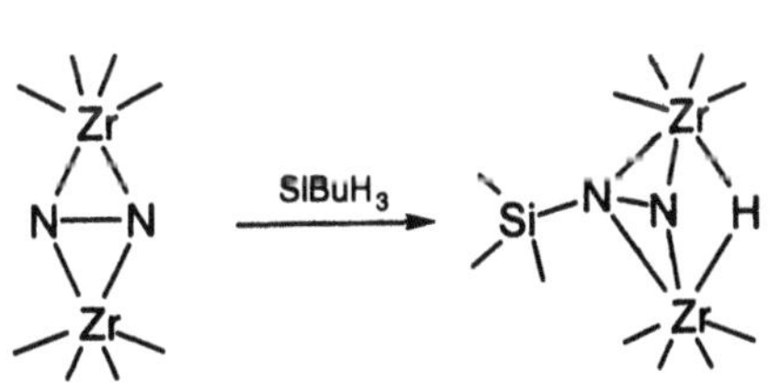

(11.24)

nitrogen [Si–N = 1.735(4)] with d_{NN} = 1.530(4), elongated from 1.43(1) in the reactant, and DFT calculations show that Eq. (11.24) is exothermic by 19.7 kcal/mol for SiH_4 reaction.[73] Thus a rare novel example of heterolytic cleavage of the Si–H bond occurs where the proton takes up a bridging-hydride position.

11.4.2. Dynamics of Silane Ligands and Exchange with cis Hydrides

The bis(silane) system $[IrH_2(\eta^2-HSiEt_3)_2(PPh_3)_2]SbF_6$ offers several structural possibilities. The disilane $Et_2HSiSiHEt_2$ gives a chelate structure (**17**) but is no more stable than the bis(silane) complex, and lower temperatures (180 K) are required to freeze out fluxionality. Equilibration of **14** with silyl(H_2) species **15** and/or **16** should be possible (and may facilitate intramolecular exchange), but only **14** is seen in low-temperature NMR. This is significant thermochemically because **15** and **16** with

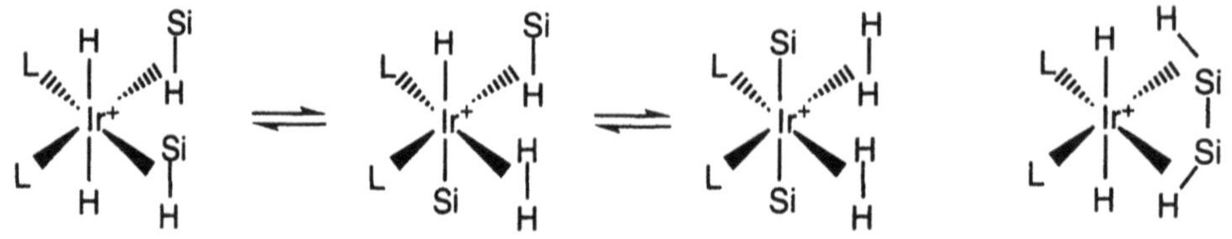

(14)   (15)   (16)   (17)

Ir–Si and Ir–H$_2$ bonds might have been expected because Si–H bonds are weaker than H–H. Even though Si–H bonds in organosilanes are stronger than originally believed (as high as 95 kcal/mol[74] in HSiMe$_3$, which is actually stronger than the C–H bond in HCMe$_3$, 90 kcal/mol),[55] they are still weaker than the H–H bond (104 kcal/mol). Thus either the silane is a better ligand than H$_2$ or the Ir–H bonds are much stronger than Ir–Si bonds.

For [IrH$_2$(η^2-HSiEt$_3$)$_2$(PPh$_3$)$_2$]$^+$ separate NMR signals for the silane and hydride protons are seen at 193 K consistent with **14**, and they coalesce into a single broad resonance as the temperature is raised to 250 K.[55] Thus intramolecular exchange of the Si–H proton with hydride readily occurs but does not appear to be as facile as for H$_2$ complexes since complexes with H$_2$ cis to hydride rarely give decoalesced signals, even at the lowest attainable temperature. The chelating disilane system RuH$_2$[(HSiR$_2$)$_2$X](PCy$_3$)$_2$ shows barriers of 11–16 kcal/mol for exchange and nonbonded separations between Si and hydride ligands of 2.04–2.43 Å (see Section 11.2). Such cis interactions are calculated to be present even for d_{SiH} of 2.1–2.3 Å in ReH$_x$(SiR$_3$) species that do not contain σ ligands.[75] Bridging hydrides also exchange with η^2-SiH protons in the dinuclear Ru complex in Section 11.2 where decoalescence is seen at −60°C.[23] However, much greater fluxionality is observed in the pyrazolylborates, TpRuH(η^2-HSiR$_3$)(PPh$_3$), and ^{1}H NMR shows that the hydride and η^2-SiH ligands are chemically equivalent down to −100°C as for the hydride and H$_2$ in TpRuH(H$_2$)(PPh$_3$).[20] Calculations on a model with SiH$_4$, PH$_3$, and Tp ligands confirm a RuH(η^2-HSiH$_3$) structure (d_{SiH} = 1.823 Å) that is about 2 kcal/mol more stable than the Ru(η^2-H$_2$)(SiH$_3$) isomer. The dynamic solution processes in Eq. (11.25) that interconvert the various species with barriers as low as 0.5 kcal/mol are reminiscent of those for MH(H$_2$) systems that proceed via M(H$_3$) transients (see Section 4.12) and rotation of η^2-H$_2$. Curiously, the hydrogens on the η^2-HSi and hydride ligands experimentally do not interchange with the terminal Si–H hydrogens on the NMR timescale for SiH$_2$R$_2$ and SiH$_3$R complexes.

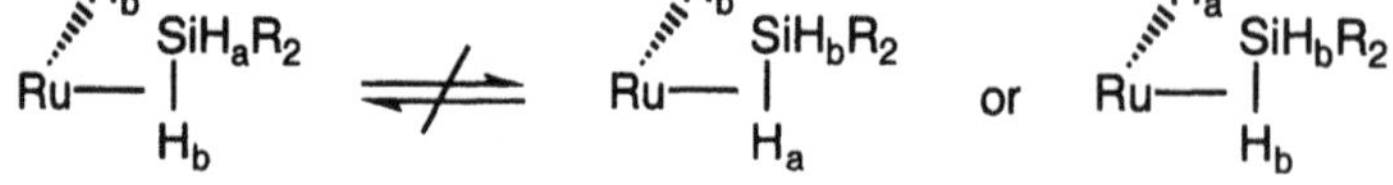

This is also the case at 50°C for the diasteromeric complexes with chiral Fe

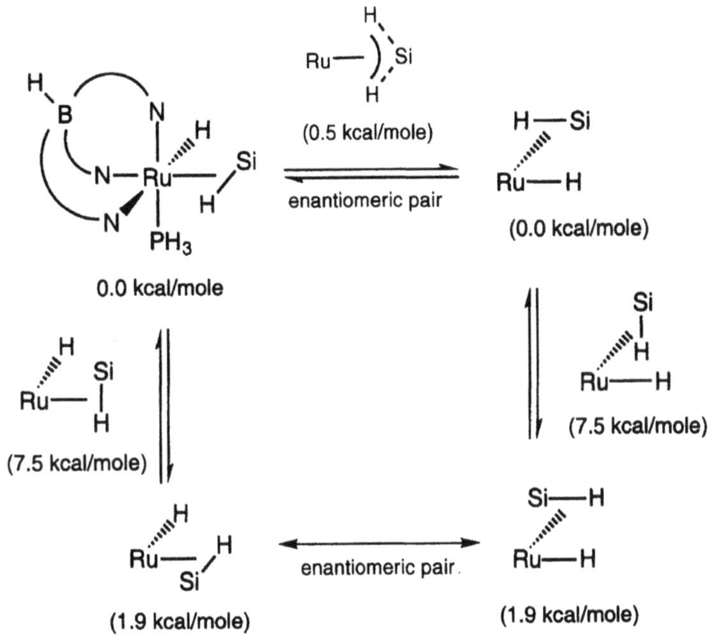

(11.25)

centers [CpFe(CO)(η^2-HSiHMePh)(PEt$_3$)]$^+$, although diastereomeric interconversion occurs.[40] The silane is reversibly displaced by H$_2$ (15 atm) and other ligands, although forcing conditions are necessary:

$$\text{TpRuH}(\eta^2\text{-HSiR}_3)(\text{PPh}_3) + \text{L} \xrightarrow{60-80°C} \text{TpRuH(L)(PPh}_3) + \text{HSiR}_3 \quad (11.26)$$

$$R_3 = Et_3, Ph_3, (EtO)_3, HPh_2, H_2Ph \quad L = H_2, CH_3CN, PPh_3$$

This is consistent with the somewhat higher bonding energies of silanes compared to H$_2$ ligands.

11.5. GERMANE AND STANNANE σ COMPLEXES

Germanes and stannanes give very similar σ complexes to those for silanes, although many fewer have been prepared. M(η^2-HGeR$_3$) systems are particularly close and exhibit solution equilibria with MH(GeR$_3$) OA products. The CpMn(CO)$_2$

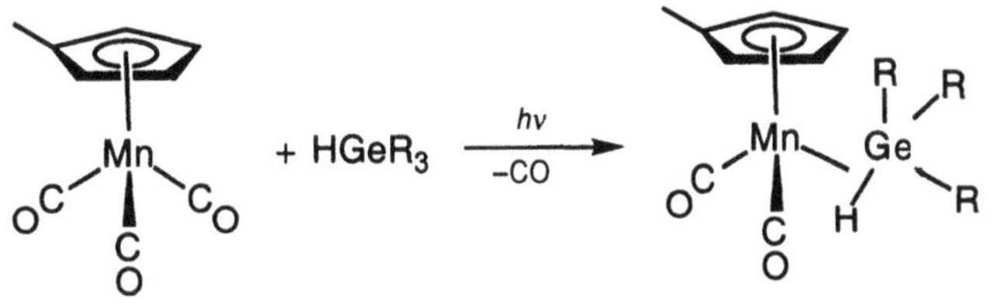

(11.27)

fragment again provided the first examples [Eq. (11.27)], wherein the bound germanes are GeHPh$_3$ and GeHMePh(naphthyl).[76] Although X-ray and NMR evidence were not obtained, the properties and reactivities are similar to silane analogues. Both strong and weak bases deprotonate the complexes, so the germane proton is fairly acidic. Reaction with PPh$_3$ eliminates germane at the same rate as that of the SiHPh$_3$ complex. As for M(η^2-Si–H), PES confirms interaction of the Ge–H σ and σ* orbitals with M.[77] The electronic structure of the interaction is consistent with an early stage of Ge–H bond addition to M, and donation of Ge–H electrons to M predominates over BD. In comparison to the silane analogue, MeCpMn(CO)$_2$(SiHPh$_3$), the Cp and M ionization energies are at nearly the same position, indicative of similar charges on the Mn centers. The Si–H and Ge–H σ interactions with Mn are nearly the same in magnitude. However, the extent of BD appears to be slightly higher in M(η^2-Ge–H), consistent with the situation in Mo complexes discussed below and the expectation that the Ge–H σ* level is lower in energy than that of the corresponding Si–H σ* orbital.

From NMR data a dimer containing Cp$_2$Ti centers isolated as an intermediate in the dehydrocoupling reaction of Ph$_2$GeH$_2$ with Cp$_2$TiMe$_2$ is proposed to have either one or two bridging Ph$_2$GeH units.[78] Related dinuclear germyl-bridged Fe complexes provide the first structural evidence for M···H–Ge interaction, although they are agostic as in structure VI in Figure 11.1[79] The d_{GeH} is 2.03(8) Å, which is

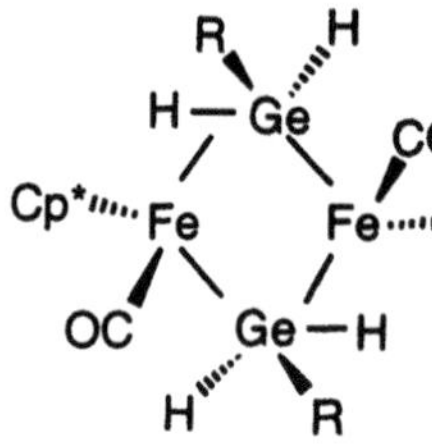

significantly longer than the normal length of 1.529 Å,[80] and the d_{FeH} is 1.64(16) Å. The IR stretch for the coordinated Ge–H bond occurs at 1849–1898 cm^{-1}.

Organogermanes and GeH$_4$ form complexes with Mo(CO)(PP)$_2$ fragments analogous to the organosilane and SiH$_4$ complexes described in Section 11.2. Synthesis proceeds according to Eqs. (11.4) and (11.14), and X-ray crystallography of Mo(CO)(GeH$_2$Ph$_2$)(depe)$_2$ confirms the η^2-coordination of Ge–H in the solid state as in Figure 11.2 for the silane analogue.[81] The germane, as for the silane, is

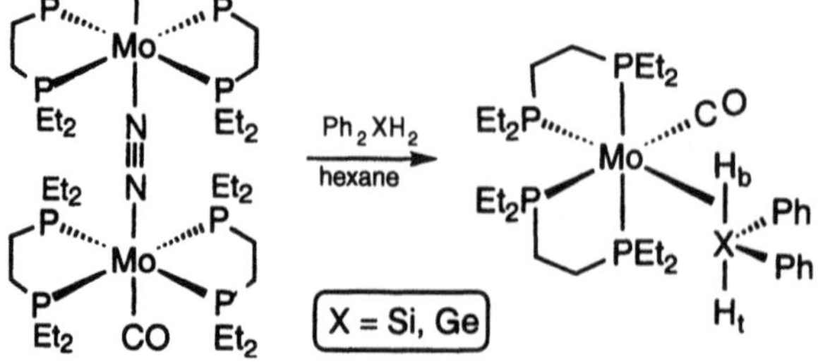

(11.28)

Table 11.4. Interatomic Distances and Angles for
Mo(CO)(η^2-Ph$_2$XH$_2$)(depe)$_2$ (X = Si, Ge)

	Distances, Å	
	X = Si	X = Ge
Mo–X	2.564(3)	2.6367(7)
Mo–H$_b$	2.04(2)	2.58(7)
X–H$_b$	1.54(6)	1.52(7)
X–H$_t$	1.66(6)	1.78(7)
	Angles, Å	
	X = Si	X = Ge
X–Mo–H$_b$	40.3(12)	40(2)
Mo–X–H$_b$	52.6(11)	68(2)
H$_b$–X–H$_t$	161(2)	171(3)
C$_{Ph}$–X–C$_{Ph}$	103.0(4)	99.2(2)

cis to the CO most likely for the electronic and steric reasons discussed in Section 11.3.1. Distances and angles for the GeH$_2$Ph$_2$ versus SiH$_2$Ph$_2$ complexes are compared in Table 11.4. The coordinated Ge–H$_b$ bond is lengthened [1.78(7) Å] relative to the terminal Ge–H$_t$ bond [1.52(7) Å] to a greater extent than the silane analogue. This reflects *higher activation toward OA for a germane compared to the silane congener*, as will be further shown. Unexpectedly, the Mo–H$_b$ distance is very long for the germane [(2.58(7) Å] and indicates a very weak Mo–H interaction [this distance is 1.70 Å in Mo(CO)(SiPhH$_3$)(depe)$_2$, where the silane is on the brink of OA]. In this regard, the geometry about X is distorted trigonal bipyramidal and appears to approach that for an octahedral fragment where the Ph–X–Ph angle is getting close to 90° and H$_b$–X–H$_t$ approaches 180° (more octahedral for X = Ge). It would appear that sterically the Ge–H$_b$ bond could still tilt or bend so that H$_b$ gets closer to Mo despite any geometric preference around Ge.

As for silanes, decreasing the number of electron-donating substituents on Ge increases the Ge–H activation and favors OA over formation of the σ complex. The OA product MoH(GePhH$_2$)(CO)(depe)$_2$ then results on addition of GePhH$_3$, which has only one substituent [Eq. (11.29)]. As expected, decreasing phosphine basicity (dppe instead of depe) on M gives back the σ complex, which is in solution

[Mo(CO)(depe)$_2$]$_2$(N$_2$) + GePhH$_3$ ⟶ (structure shown) (11.29)

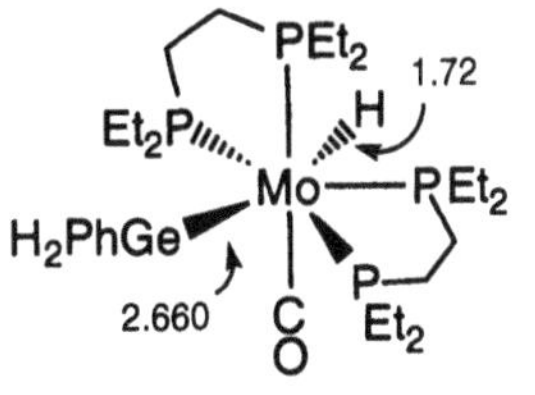

equilibium with the germyl–hydride [Eq. (11.30)]. Solutions of Mo(CO)-(GePh$_2$H$_2$)(depe)$_2$ also show very similar equilibria, but there is evidence for a third

$$\text{Mo(CO)(dppe)}_2 + \text{GePhH}_3 \longrightarrow \underset{(18)}{\text{[complex]}} \rightleftharpoons \underset{(19)}{\text{[complex]}}$$

^{31}P NMR (–80 °C): four singlets at 47–76 ppm   four singlets at 41–85 ppm

(11.30)

isomer showing only two ^{31}P signals indicative of two phosphine environments, e.g., the depe ligands in the same plane. This isomer could represent an initial OA product and is normally not seen in either the silane or H$_2$ systems because of rearrangement to the pentagonal bipyramid geometry as in 19.

For silanes, tautomeric equilibrium as in Eq. (11.30) is observed *only* for unsubstituted SiH$_4$, the most electrophilic Si center studied, and only on the most electron-rich Mo–depe center. By comparison GeH$_4$ forms a σ complex for the less basic dppe system only, undergoing complete OA for the depe system, i.e., germanes undergo OA more easily than silanes. Although X-ray data are not available, IR of Mo(η^2-GeH$_4$)(dppe)$_2$(CO) is similar to that for the SiH$_4$ derivative (Figure 11.4) and shows v_{MoHGe} at 1756 cm^{-1}. For MoH(GeH$_3$)(depe)$_2$(CO), v_{MoH} is at 1728 cm^{-1}, and v_{CO} shifts down from 1795 to 1756 cm^{-1}. ^{31}P NMR shows four signals in the 34–85 ppm range as for the analogous silyl hydride complex and the organogermyl hydrides. Although calculations have not been performed, BD appears to be better for germanes than silanes on the basis of ease of OA, giving the following comparison: Strength of BD: Ge–H > Si–H > H–H ≫ C–H.

Stannane complexes can be prepared by photolysis at low temperatures, analogous to silane complexes on the MeCpMn(CO)$_2$ fragment.[82] In Eq. (11.31) the use of a nonpolar solvent such as pentane is crucial because the product thermally

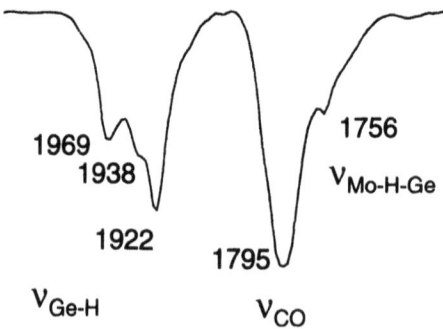

Figure 11.4. Nujol mull IR spectrum of Mo(η^2-HGeH$_3$)(CO)(dppe)$_2$.

reacts with excess HSnPh$_3$ to give the bis(stannyl) complex in ether or THF [Eq. (11.32)].

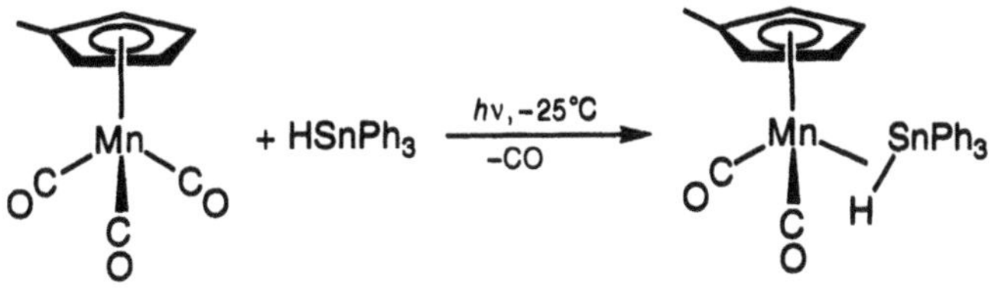

$$\text{MeCpMn(CO)}_3 + \text{HSnPh}_3 \xrightarrow[-\text{CO}]{h\nu, -25°\text{C}} \text{MeCpMn(CO)}_2(\text{HSnPh}_3) \quad (11.31)$$

$$\text{MeCpMn(CO)}_2(\text{HSnPh}_3) + \text{HSnPh}_3 \xrightarrow[-\text{H}_2]{-25°\text{C}} \text{MeCpMn(CO)}_2(\text{SnPh}_3)_2 \quad (11.32)$$

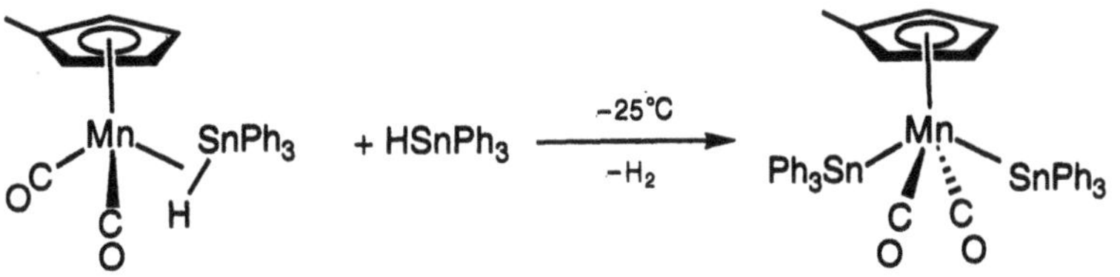

Presumably loss of a second CO, a second OA, H$_2$ loss, and CO scavenging occur. The analogous bis(silyl) complexes can only be prepared photochemically and with SiR$_3$ = SiCl$_3$ or SiMeCl$_2$. The crystal structure of MeCpMn(CO)$_2$(HSnPh$_3$) is very similar to the silane complex and the d_{SnH} of 2.16(4) Å is only ca 0.45 Å or 20% longer than in free methyl-substituted stannanes.[83] This elongation is similar to that for silane coordination to this fragment, but there is evidence that the OA of the Sn—H bond is further along the reaction coordinate than for the Si—H bond. The d_{MnSn} is 2.636(1) Å compared to 2.424(2) Å for d_{MnSi}, a difference less than expected (0.27 Å) based on relative Sn—C versus Si—C bond distances (Table 11.5). Also the $J_{^{117}SnMnH}$ is 252 Hz ($J_{^{119}SnMnH} = 270$ Hz), which is closer to the lower limit values for J_{SnCH} in free stannanes (50–70 Hz) or in oxidatively added M(H)(SnR$_3$) than for the silane case. The $J_{^{117}SnOsH}$ is 129.5 Hz in cis-Os(CO)$_4$(H)(SnCl$_3$), where there should be little interaction between hydride and SnCl$_3$.[84] The upper limit values for J_{SnH} in HSnR$_3$ are nearly an order of magnitude larger than those for $^1J_{SiH}$ in HSiR$_3$ (Table 11.5).

Only a few other examples of stannane σ complexes are known. Group 6 complexes include half-sandwich complexes (arene)Cr(CO)$_2$(HSnPh$_3$) synthesized by photochemical reaction of (arene)Cr(CO)$_3$ with HSnPh$_3$.[85,86] Two independent

Table 11.5. Comparison of Parameters for MeCpMn(CO)$_2$(HEPh$_3$) (E = Sn, Si)

Bond (X = Mn, C)	d_{EH}, Å	d_{EX}, Å	$^2J_{EXH}$, Hz	$^1J_{EH}$, Hz
Mn(Sn–H)	2.16(4)	2.636(1)	252	
Mn(Si–H)	1.76(4)	2.424(2)	65	
C–Sn–H in HSnR$_3$	1.71	2.153^a	50–70	1500–1800
C–Si–H in HSiR$_3$	1.48	1.88^a	3–20	200

^{a}X = C, average values for coordinated HEPh$_3$ in MeCpMn(CO)$_2$(HEPh$_3$).

X-ray structures of the mesitylene complex show that the Sn–H bond is coordinated in an η^2 fashion as for the Mn complexes (d_{SnH} = 2.02(4) Å[85]; 1.95 Å[86] and $J_{119SnCrH}$ is 327.6 Hz, which along with the d_{SnH} values correspond to a less activated bond than in MeCpMn(CO)$_2$(HSnPh$_3$). Bis(stannyl) complexes are formed as in Eq. (11.32)[86]. Other complexes (**20–22**) contain phosphine or dppe ligands and HSnR$_3$ (R = Ph, Me) and are prepared either by thermal reaction of W(CO)$_4$(PR$_3$)(THF) or M(CO)$_3$(dppe)(L) (L = THF, acetone) with the stannane or by photolysis of M(CO)$_4$(dppe).[85] The dppe complexes (**21 and 22**) decompose in benzene at 25°C to give hexaphenyldistannane.

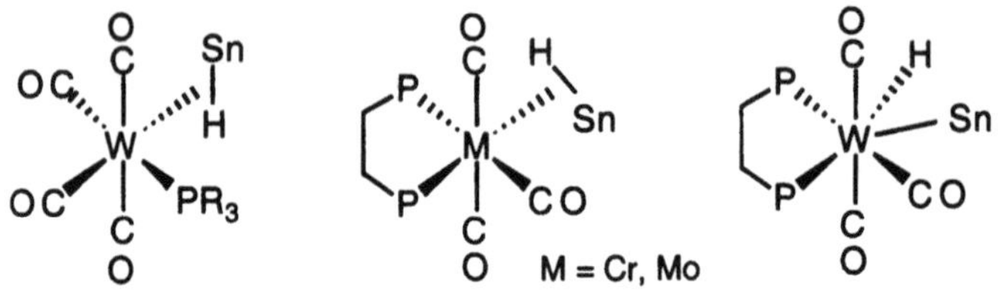

(20)   (21)   (22)

The $J_{119SnMH}$ couplings are in the range of 250–315 Hz, but are considerably lower (70–90 Hz) in W(CO)$_3$(dppe)(H)(SnR$_3$), which suggests complete or nearly complete OA of the H–Sn bond. This once again is probably a consequence of the greater BD from the third-row metal. Although the analogous H$_2$ and silane complexes are unknown, they would be expected to be σ complexes even for W, which indicates that Sn–H bonds, like Ge–H bonds, undergo OA more easily than H–H or Si–H. A large series of five-coordinate complexes of the type RhX(H)(SnPh$_3$)(PPh$_3$)(L) possess more distant Sn$\cdots$H interactions (J_{SnRhH} = 29–170 Hz) that may be analogous to elongated H$\cdots$H complexes.[87] Although J_{SnRhH} for Rh(NCBPh$_3$)(H)(SnPh$_3$)(PPh$_3$)$_2$ is quite low, 29 Hz, and would appear to indicate complete OA, d_{SnH} is 2.31(5) Å, which is only about 0.15 Å longer than that in MeCpMn(CO)$_2$(HSnPh$_3$). Steric factors may be important because electronically RhX(H)(SnPh$_3$)(PPh$_3$)$_2$ would be expected to be a hydrido-stannyl complex (analogous RhXH$_2$(PPh$_3$)$_2$ systems clearly are dihydrides).

REFERENCES

1. Chalk, A. J; Harrod, J. F. *J. Am. Chem. Soc.* **1965**, *87*, 1133; **1967**, *89*, 1640; Speier, J. L. *Adv. Organomet. Chem.* **1979**, *17*, 407; Marciniec, B., Ed., *Comprehensive Handbook on Hydrosilylation*, Pergamon Press, Oxford, 1992.
2. Jetz, W.; Graham, W. A. G. *J. Am. Chem. Soc.* **1969**, *91*, 3375.
3. Hoyano, J. K.; Graham, W. A. G. *J. Am. Chem. Soc.* **1969**, *91*, 4568.
4. Jetz, W.; Graham, W. A. G. *Inorg. Chem.* **1971**, *10*, 4.
5. Hart-Davis, A. J.; Graham, W. A. G. *J. Am. Chem. Soc.* **1971**, *93*, 4388.
6. Graham, W. A. G. *J. Organomet Chem.* **1986**, *300*, 81.
7. Schubert, U. *Adv. Organomet. Chem.* **1990**, *30*, 151.
8. Hutcheon, W. L., Ph.D. thesis, University of Alberta, Edmonton, Alberta, Canada, 1971.

9. Graham, W. A. G.; Bennett, M., *J. Chem. Eng. News* **1970**, *48*(24), 75.
10. Andrews, M. A.; Kirtley, S. W.; Kaesz, H. D. *Adv. Chem. Ser.* **1978**, *167*, 229.
11. Bau, R.; Teller, R. G.; Kirtley, S. W.; Koetzle, T. F. *Acc. Chem. Res.* **1979**, *12*, 176.
12. Colomer, E.; Corriu, R. J. P.; Marzin, C.; Vioux, A. *Inorg. Chem.* **1982**, *21*, 368.
13. Schubert, U.; Ackermann, K.; Worle, B. *J. Am. Chem. Soc.* **1982**, *104*, 7378; Schubert, U.; Scholz, G.; Muller, J.; Ackermann, K.; Worle, B.; Stansfield, R. F. D. *J. Organomet Chem.* **1986**, *306*, 303.
14. Rabaa, H.; Saillard, J.-Y.; Schubert, U. *J. Organomet Chem.* **1987**, *330*, 397.
15. Schubert, U. In *Advances in Organosilicon Chemistry*, B. Marciniec and J. Chojnowski, eds, Gordon and Breach: Yverdon-lesBains, Switzerland, 1994.
16. Corey, J. Y.; Braddock-Wilking, *J. Chem. Rev.* **1999**, *99*, 175.
17. Crabtree, R. H. *Angew. Chem. Int. Ed. Engl.* **1993**, *32*, 789.
18. Schneider, J. J. *Angew. Chem. Int. Ed. Engl.* **1996**, *35*, 1068.
19. Nagl, I.; Scherer, W.; Tafipolsky, M.; Anwander, R. *Eur. J. Inorg. Chem.* **1999**, 1405.
20. Ng, S. M.; Lau, C. P.; Fan, M.-F.; Lin, Z. *Organometallics* **1999**, *18*, 2484.
21. Reid, S. M.; Neuner, B.; Schrock, R. R.; Davis, W. M. *Organometallics* **1998**, *17*, 4077.
22. Smith, R. A.; Bennett, M. J. *Acta Crystallogr., Sect. B* **1977**, *33*, 1113; Dong, D. F.; Hoyano, J. K.; Graham, A. G. *Can. J. Chem.* **1981**, *59*, 1455.
23. Takao, T.; Yoshida, S.; Suzuki, H.; Tanaka, M. *Organometallics* **1995**, 14.
24. Luo, X.-L.; Kubas, G. J.; Bryan, J. C.; Burns, C. J.; Unkefer, C. J. *J. Am. Chem. Soc.* **1994**, *116*, 10312.
25. Butts, M. D.; Kubas, G. J.; Luo, X.-L.; Bryan J. C. *Inorg. Chem.* **1997**, *36*, 3341.
26. Simons, R. F.; Tessier, C. A. *Organometallics* **1996**, *15*, 2604.
27. Delpech, F.; Sabo-Etienne, S.; Chaudret, B.; Daran, J.-C. *J. Am. Chem. Soc.* **1997**, *119*, 3167; Delpech, F.; Sabo-Etienne, S.; Daran, J.-C.; Chaudret, B.; Hussein, K.; Marsden, C. J.; Barthelet, J.-C. *J. Am. Chem. Soc.* **1999**, *121*, 6668.
28. Spaltenstein, E.; Palma, P.; Kreutzer, K. A.; Willoughby, C. A.; Davis, W. M.; Buchwald, S. L. *J. Am. Chem. Soc.* **1994**, *116*, 10308.
29. Kreutzer, K. A.; Fisher, R. A.; Davis, W. M.; Spaltenstein, E.; Buchwald, S. L. *Organometallics* **1991**, *10*, 4031.
30. George, M. W.; Haward, M. T.; Hamley, P. A.; Hughes, C.; Johnson, F. P. A.; Popov, V. K.; Poliakoff, M. *J. Am. Chem. Soc.* **1993**, *115*, 2286.
31. Zhang, S.; Dobson, G. R.; Brown, T. L. *J. Am. Chem. Soc.* **1991**, *113*, 6908.
32. Burkey, T. J. *J. Am. Chem. Soc.* **1990**, *112*, 8329.
33. Schubert, U.; Gilges, H. *Organometallics* **1996**, *15*, 2373.
34. Procopio, L. J.; Carroll, P. J.; Berry, D. H. *J. Am. Chem. Soc.* **1994**, *116*, 177.
35. Ohff, A.; Kosse, P.; Baumann, W.; Tillack, A.; Kempe, R.; Gorls, H.; Burlakov, V. V.; Rosenthal, U. *J. Am. Chem. Soc.* **1995**, *117*, 10399.
36. Lichtenberger, D. L.; Rai-Chaudhuri, A.; Hogan, R. H. In *Inorganometallic Chemistry*, Fehlner, T. P., ed., Plenum Press, New York, 1992.
37. Lichtenberger, D. L.; Rai-Chaudhuri, A. *J. Am. Chem. Soc.* **1989**, *111*, 3583.
38. Lichtenberger, D. L.; Rai-Chaudhuri, A. *Organometallics* **1990**, *9*, 1686.
39. Freeman, W. P.; Tilley, T. D.; Rheingold, A. L. *J. Am. Chem. Soc.* **1994**, *111*, 3583.
40. Scharrer, E.; Chang, S.; Brookhart, M. *Organometallics* **1995**, *14*, 5686.
41. Zhou, D.-Y.; Minato, M.; Ito, T.; Yamasaki, M. *Chem. Lett.* **1997**, 1017; Zhou, D.-Y.; Zhang, L.-B.; Minato, M.; Ito, T.; Osakada, K. *Chem. Lett.* **1998**, 187.
42. Lefcourt, M. A.; Ozin, G. A. *J. Phys. Chem.* **1991**, *95*, 2616, 2623.
43. Luo, X.-L.; Kubas, G. J.; Burns, C. J.; Bryan, J. C.; Unkefer, C. J. *J. Am. Chem. Soc.* **1995**, *117*, 1159.
44. Atheaux, I.; Donnadieu, B.; Rodriguez, V.; Sabo-Etienne, S.; Chaudret, B,; Hussein, K.; Barthelet, J.-C. *J. Am. Chem. Soc.* **2000**, *122*, 5664.
45. Yang, H.; Asplund, M. C.; Kotz, K. T.; Wilkens, M. J.; Frei, H.; Harris, C. B. *J. Am. Chem. Soc.* **1998**, *120*, 10154.
46. Fan, M.-F.; Lin, Z. *Organometallics* **1998**, *17*, 1092.
47. Choi, S.-H.; Feng, J.; Lin, Z. *Organometallics* **2000**, *19*, 2051.
48. Maseras, F.; Lledos, A.; Costas, M.; Poblet, J. M. *Organometallics* **1996**, *15*, 2947.
49. Huhmann-Vincent, J.; Scott, B. L.; Kubas, G. J. *J. Am. Chem. Soc.* **1998**, *120*, 6808; Fang, X.; Scott, B. L.; John, K. D.; Kubas, G. J. *Organometallics* **2000**, *19*, 4141.
50. Huhmann-Vincent, J.; Scott, B. L.; Kubas, G. J. *Inorg. Chim. Acta* **1999**, *294*, 240.

51. Fan, M.-F.; Jia, G.; Lin, Z. *J. Am. Chem. Soc.* **1996**, *118*, 9915.
52. Musaev, D.; Morokuma, K. *J. Am. Chem. Soc.* **1995**, *117*, 799.
53. Maseras, F.; Lledos, A. *Organometallics* **1996**, *15*, 1218.
54. Esteruelas, M. A.; Oro, L. A.; Valero, C. *Organometallics* **1991**, *10*, 462.
55. Luo, X.-L.; Crabtree, R. H. *J. Am. Chem. Soc.* **1989**, *111*, 2527.
56. Poulton, J. T.; Sigala, M. P.; Eisenstein, O.; Caulton, K. G. *Inorg. Chem.* **1993**, *32*, 5490.
57. Campion, B. K.; Heyn, R. H.; Tilley, T. D.; *J. Chem. Soc., Chem. Commun.* **1992**, 1201.
58. Hussein, K.; Marsden, C. J.; Barthelat, J.-C.; Rodriguez, V.; Conjero, S.; Sabo-Etienne, S.; Donnadieu, B.; Chaudret, B. *Chem. Comm.* **1999**, 1315.
59. George, M. W.; Haward, M. T.; Hamley, P. A.; Hughes, C.; Johnson, F. P. A.; Popov, V. K.; Poliakoff, M. *J. Am. Chem. Soc.* **1993**, *115*, 2286.
60. Kickel, B. L.; Armentrout, P. B. *J. Am. Chem. Soc.* **1995**, *117*, 764.
61. Goldfuss, B.; Schleyer, P. v. R.; Handschuh, S.; Hampel, F.; Bauer, W. *Organometallics* **1997**, *16*, 5999, and references therein; Veith, M.; Zimmer, M.; Kosse, P. *Chem. Ber.* **1994**, *127*, 2099.
62. King, W. A.; Scott, B. L.; Eckert, J.; Kubas, G. J. *Inorg. Chem.* **1999**, *38*, 1069.
63. Fan, M.-F.; Lin, Z. *Organometallics* **1999**, *18*, 286.
64. Fan, M.-F.; Lin, Z. *Organometallics* **1997**, *16*, 494.
65. Jiang, Q.; Carroll, P. J.; Berry, D. H. *Organometallics* **1991**, *10*, 3648.
66. Tanaka, I.; Ohhara, T.; Niimura, N.; Ohashi, Y.; Jiang, Q.; Berry, D. H.; Bau, R. *J. Chem. Res., Synop.* **1999**, *14/15*, 180; Bakhmutov, V. I.; Howard, J. A. K.; Keen, D. A.; Kuzmina, L. G.; Leech, M. A.; Nikunov, G. I.; Vorontsov, E. V.; Wilson, C. C. *J. Chem. Doc. Dalton Trans.* **2000**, 1631; Nikonov, G. I.; Mountford, P.; Green, J. C.; Cooke, P. A.; Leech, M. A.; Blake, A. J.; Howard, J. A. K.; Lemenovski, D. A. *Eur. J. Inorg. Chem.* **2000**, 1917.
67. Abdelqader, W.; Chmielewski, D.; Grevels, F.-W.; Ozkar, S.; Peynircioglu, N. B. *Organometallics* **1996**, *15*, 604.
68. Reed, C. A. *Acc. Chem. Res.* **1998**, *31*, 325.
69. Chang, S.; Scharrer, E.; Brookhart, M. *J. Mol. Catal. A: Chem.* **1998**, *130*, 107.
70. Fang, X.; Huhmann-Vincent, J.; Scott, B. F.; Kubas,G. J., *J. Organometal. Chem.* **2000**, *609*, 95.
71. Tilley, T. D. In *The Chemistry of Organic Silicon Compounds*, Patai, S.; Rappoport, Z., eds.; Wiley: New York, 1989.
72. Sweeney, Z. K.; Polse, J. L.; Andersen, R. A.; Bergman, R. G.; Kubinec, M. G. *J. Am. Chem. Soc.* **1997**, *119*, 4543.
73. Fryzuk, M. D.; Love, J. B.; Rettig, S. J.; Young, V. G. *Science* **1997**, *275*, 1445; Basch, H.; Musaev, D. C.; Morokuma, K.; Fryzuk, M. D.; Love, J. B.; Seidel, W. W.; Albinati, A.; Koetzle, T. F.; Klooster, W. T.; Mason, S. A.; Eckert, J. *J. Am. Chem. Soc.*, **1999**, *120*, 523.
74. Ding, L.; Marshall, P. *J. Am. Chem. Soc.* **1992**, *114*, 5754.
75. Lin, Z.; Hall, M. B. *Inorg. Chem.* **1991**, *30*, 2569.
76. Carre, F.; Colomer, E.; Corriu, R. J. P.; Vioux, A. *Organometallics* **1984**, *3*, 1272.
77. Lichtenberger, D. L.; Rai-Chaudhuri, A. *J. Chem. Soc. Dalton Trans.* **1990**, 2161.
78. Aitken, C.; Harrod, J. F.; Malek, A.; Samuel, E. *J. Organomet Chem.* **1988**, *349*, 285.
79. El-Maradny, A.; Tobita, H.; Ogino, H. *Organometallics* **1996**, *15*, 4954; Tobita, H.; Ogino, H., private communication.
80. Emsley, J. *The Elements*, Clarendon Press: Oxford, 1989.
81. Huhmann-Vincent, J. L.; Scott, B. J.; Kubas, G. J., unpublished results.
82. Schubert, U,; Kunz, E. K.; Harkers, B.; Willnecker, J.; Meyer, J. *J. Am. Chem. Soc.* **1989**, *111*, 2572.
83. Beagley, B.; McAloon, K.; Freeman, J. M. *Acta Crystallogr., Part B* **1974**, *30*, 444.
84. Moss, J. R.; Graham, W. A. G. *J. Organomet Chem.* **1969**, *18*, P24.
85. Piana, H.; Kirchgaessner, U.; Schubert, U. *Chem. Ber.* **1991**, *124*, 743.
86. Khaleel, A.; Klabunde, K. *J. Inorg. Chem.* **1996**, *35*, 3223.
87. Carleton, L. *Inorg. Chem.* **2000**, *39*, 4510; Carleton, L.; Weber, R.; Levendis, D. C. *Inorg. Chem.* **1998**, *37*, 1264.

12

C–H Bond Coordination and Activation

12.1. INTRODUCTION

The conversion of alkanes, particularly methane, to liquid fuels or other useful products is one of the "Holy Grails" of chemistry. Alkanes are the most abundant hydrocarbon resource and an important potential feedstock for the chemical industry, but very few selective methods are available for conversion into more valuable products. Even where alkanes are used, their transformations are inefficient, e.g., steam reforming of methane to synthesis gas (CO and H_2), which is often then converted into methanol for use as a fuel or as a chemical feedstock. The direct *selective* oxidation of CH_4 is a much more efficient pathway for methanol synthesis but has two significant problems. In regard to selectivity, the initial product of alkane oxidation is often more reactive toward the oxidant than the alkane itself. For example, the C–H bond in CH_3OH is 11 kcal/mol weaker than that in CH_4 (93 and 104 kcal/mol, respectively), ultimately leading to overoxidation of methanol to the thermodynamic sinks, CO_2 and H_2O. Thus the main challenge in alkane transformations is selectivity, not reactivity. Surface-catalyzed methods as well as many radical reactions often lead to overoxidation. Although organometallic complexes in homogeneous solutions do undergo alkane reaction with great selectivity, most of these systems are not catalytic, which is the second major hurdle. The recent observation of alkane σ complexes near ambient conditions represents hope for future design of catalytic systems.

In 1965 Chatt discovered OA of an arene C–H bond to a metal complex and in 1976 predicted that "in 25 years methane will be the most popular ligand in coordination chemistry."[1] Since 1982 when Bergman and Graham independently reported the first examples of aliphatic C–H bond OA by soluble organometallic complexes,[2] a large number of organotransition metal complexes have been shown to undergo this reaction with alkanes.[3–11] This surprising facility of unsaturated M to cleave such strong bonds has industrial potential as a means of selectively functionalizing hydrocarbons. Great numbers of synthetic, kinetic, mechanistic, and theoretical studies of aliphatic C–H activation in homogeneous and heterogeneous systems have been carried out, most of which cannot be presented in detail here since they would fill several books such as that by Shilov and Shul'pin.[5] However, data on alkane σ complexes are extremely limited because of their transient nature along the reaction coordinate to OA to alkyl hydrides [Eq. (12.1)].

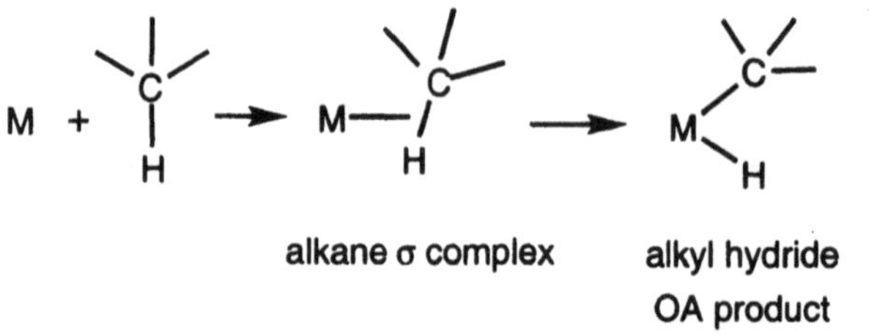

$$M + \underset{H}{\overset{|}{C}} \longrightarrow M\!-\!\underset{H}{\overset{|}{C}} \longrightarrow M\underset{H}{\overset{C}{\diagdown}} \tag{12.1}$$

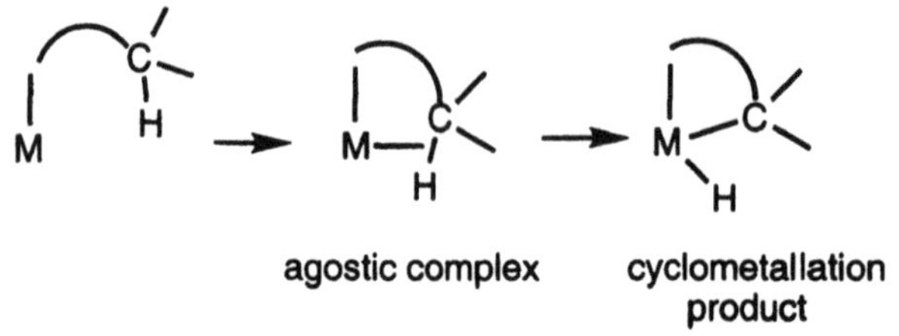

alkane σ complex   alkyl hydride
                    OA product

$$\underset{M}{\overset{\frown}{}}\overset{C}{\underset{H}{}} \longrightarrow \underset{M-C}{\overset{\frown}{}}\overset{}{\underset{H}{}} \longrightarrow \underset{M\!-\!C}{\overset{\frown}{}}\overset{}{\underset{H}{}} \tag{12.2}$$

agostic complex   cyclometallation
                   product

Methane complexes with M ions, $[M(CH_4)_n]^+$ (see Section 4.13), and matrix-isolated CH_4 complexes are known but are unstable. Only recently has it been possible to observe alkane binding in solution at low temperature by NMR. As will also be shown, crystallographic evidence exists for heptane interaction with an iron–porphyrin system, although the alkane here is perhaps at least partially confined sterically, i.e., entropically stabilized. *Intramolecular* (agostic) coordination of C–H bonds to transition metals [Eq. (12.2)] predated the discovery of H_2 complexes, although this interaction is clearly entropically stabilized and must be distinguished from the far less stable binding of an external C–H bond as in an alkane complex. As for silane coordination, the precise nature of the σ-type interaction in agostic species only became apparent after M–H_2 bonding was understood, and the close analogy was noted by Brookhart, Green, and Wong in their comprehensive review of these interactions.[12] The term "agostic" was coined and popularized by these authors to denote a variety of intramolecular C–H interactions and is derived from a Greek word which translates "to clasp, to draw toward, to hold to oneself." Such interactions, which have been reviewed more recently by Crabtree,[3] were proposed and/or studied by several researchers well before this terminology was adopted, as will be shown below. Agostic interactions are normally weak, but strong bonding can occur wherein M ⋯ H–C intramolecular bonding lies along the reaction coordinate for OA of the C–H bond in cyclometallation or orthometallation [Eq. (12.2)]. The term "agostic" should not be used when describing external ligand binding solely through a σ bond, which is best referred to as a "σ complex."

Alkane coordination and C–H bond cleavage have been extensively studied theoretically; which is important because of the instability of alkane complexes—they are either are too weak to be observed or rapidly proceed to products of OA. In the latter case, computed binding energies of the transient σ complexes seem sufficient to allow observation, but an "arrested" complex analogous to an elongated

H–H or Si–H σ complex has not been isolated (stretched agostic C–H interactions are known). As will be shown in Section 12.4, clear evidence exists for intermediate alkane σ complexes in OA/RE of C–H bonds. Equilibria between $M(\eta^2\text{-RH})$ and

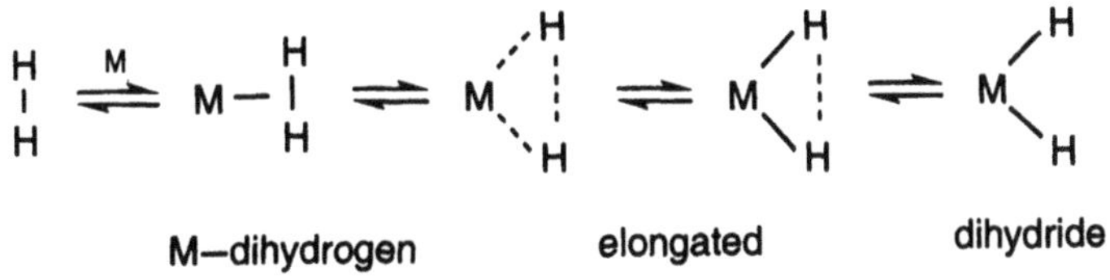

M–dihydrogen   elongated   dihydride

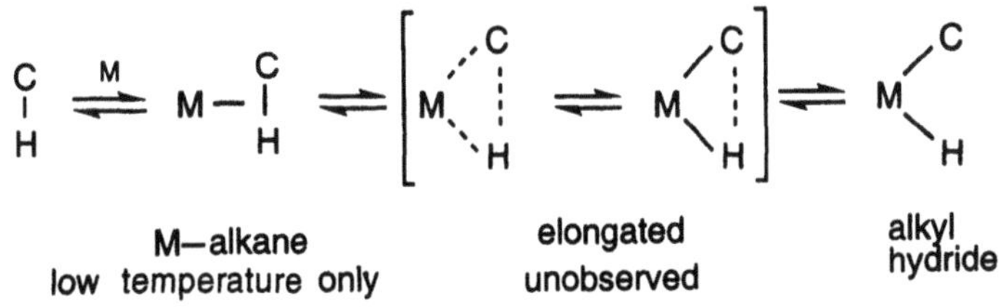

M–alkane
low temperature only   elongated
unobserved   alkyl
hydride

M(R)(H) tautomers has not been seen perhaps because the weak M → C–H backbonding prevents a good balance of σ donation and backdonation (BD).

12.2. AGOSTIC C–H COORDINATION AND CYCLOMETALLATION

12.2.1. Structures and Strengths of M ··· H–C Interaction

In 1965 Ibers[13] and Mason[14] independently provided the first evidence for close approach of C–H bonds to Ru and Pd metal centers as in 1. The d_{MH} were long (e.g., 2.59 Å in 1) but less than the sum of the van der Waals radii (3.1 Å). However, a

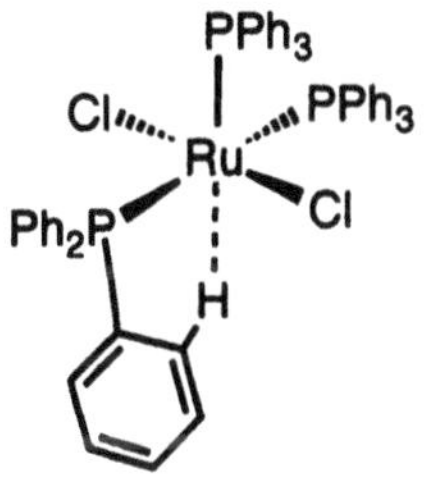

(1)

bonding interaction was not claimed in 1. Several years later Maitlis and coworkers found a shorter d_{MH} of 2.23 Å, where the C–H bond was drawn toward M to form a chelate structure unforced by outside steric factors.[15] In addition, the presence of NMR coupling between the agostic H atom and the P atoms led to a proposal of direct Pd ⋯ H–C interaction. However, the bonding involves the filled Pd d_{z^2} orbital and is therefore of a hydrogen-bonding nature (3c–4e, see below).

The most important advances were made by Trofimenko[16] and especially crystallographically by Cotton[17,18] in pyrazolylborate systems. Unusual low-field chemical shifts for methylene groups and low values for v_{CH} in complexes such as Ni[Et$_2$B(pz)$_2$]$_2$ were seen. The crystal structure of Mo[Et$_2$B(pz)$_2$](CO)$_2$-(CH$_2$CPhCH$_2$) showed close approach of a C–H bond and a 3c-2e bond was proposed by Cotton in analogy to a similar B–H interaction in Mo[H$_2$B(pz)$_2$]-(CO)$_2$(η_3-C$_7$H$_7$).[18] Very low values of v_{CH}, e.g., 2704 and 2664 cm^{-1}, were observed

and NMR studies revealed the strength of the interaction to be 17–20 kcal/mol, which importantly is competitive with M-olefin binding.

A large variety of agostic M ⋯ H–X interactions are known, some of which are shown in Figure 12.1 for X = C (see Chapters 11 and 13 for other systems). The α-agostic interactions are quite common in unsaturated alkyl and alkylidene complexes. The structure with C=C π coordination and C–H σ coordination to M nicely illustrates the close parallel between the two bonding types. Virtually any C–H or other bond in a ligand is capable of forming an agostic interaction, and AIM theory shows that these are σ-bonded interactions and *not* special types of hydrogen

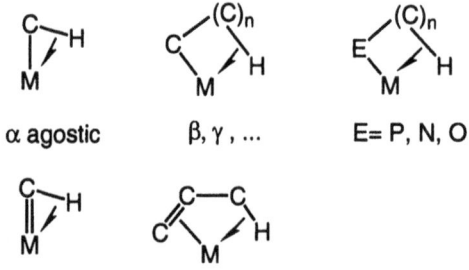

Figure 12.1. Types of agostic interactions (half-arrow convention).

bonds.[19] Interactions are even identifiable on metal *surfaces* by techniques such as reflection–absorption IR and Auger electron spectroscopies.[20] Below 200 K, cycloheptane binds weakly to Pt(111) and exhibits a broad v_{CH} at 2630 cm^{-1} consistent with agostic M $\cdots$ H–C (sharper bands for free cycloheptane occur at 2845–2907 cm^{-1}). In contrast, cyclic alkenes bind strongly and irreversibly via their π systems.

Multiple agostic interactions are common, in which case each interaction may represent fractional donation of electron pairs and is correspondingly weaker (longer M–H and M–C distances). In some cases *hydrogen bonding* between M and the proton may predominate, as found by Brammer in [R$_3$N–H$^+$ $\cdots$ Co(CO)$_4^-$], where Co $\cdots$ H is 2.613(2) Å by neutron diffraction.[21] In this case and others (see Section 13.2) it is clear that the 18e Co(CO)$_4^-$ ion cannot accept an electron pair from the NH bond, and the interaction is a 3c-4e bond as in a conventional hydrogen bond except that M is now the weak-base bond acceptor.

$$N-H \cdots M \approx N-H \cdots O$$

In contrast to a η^2 agostic interaction, the arrangement is linear to maximize dipole–dipole interaction. For M $\cdots$ H–C interactions, most are agostic, but for 16e square planar d^8-ML$_4$ the situation is ambiguous because M has available both a lone pair and an empty orbital. A comparison with those for unambiguously agostic d^6-ML$_5$ systems shows that M $\cdots$ H–N systems typically have M–H–N angles greater than 160° and are clearly hydrogen-bonded.[9] However d^8 M $\cdots$ H–C species, which have more bent M–H–C, may have only weak H bonding.

The first neutron diffraction study of an agostic interaction was performed on [Fe(η-C$_8$H$_{13}$){P(OMe)$_3$}$_3$]$^+$.[22] The d_{FeH} and d_{FeC} in **2** are 1.874(3) and 2.384(4) Å

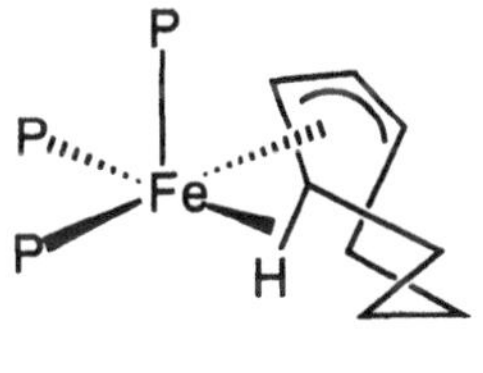

(2)

and the d_{CH} is 1.164(3) Å. A recent survey of agostic complexes showed that there are eight d_{CH} (four sp^2 and four sp^3 C–H groups) determined in six neutron structures and these range from 1.09–1.19 Å,[23] only a 10% elongation over normal d_{CH}. However, an example of a stretched agostic C–H bond has recently been characterized in a "pincer" ligand.[24]

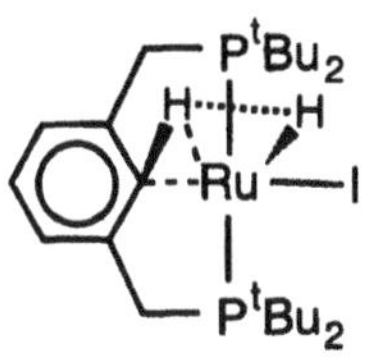

(3) 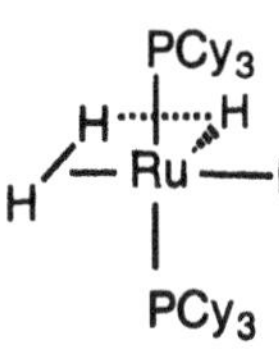   (4)

The d_{CH} in **3** is 1.32(4) Å, a 22% lengthening over d_{CH} in benzene (1.08 Å), which would be comparable to stretching d_{HH} in free H_2 to 0.9 Å. A comparable H_2 complex (**4**) shows that d_{HH} actually elongates much further, to 1.03(7) Å, in keeping with the higher BD to H–H than to C–H bonds. However, J_{CH} in **3** is quite low, 58 Hz, which is half that in normal C–H bonds, 120–130 Hz, indicating that the CH bond order may be lower than that suggested by the d_{CH}. However J_{CH} may not correlate to d_{CH} and/or bond order in the same manner as shown for J_{HD} in H_2 complexes. There is also an attractive interaction between the agostic C–H hydrogen and the neighboring hydride in **3** [1.68(7) Å], which may be considered to be analogous to cis interactions in H_2–hydride complexes (**4** also shows such an interaction [1.66(6) Å]). This attraction can also be considered another form of *dihydrogen bonding* (see Section 9.4.4.1). The analogue of **3** with Cl instead of I shows a slightly more activated agostic C–H (J_{CH} = 52 Hz) in keeping with the stronger π-donor character of Cl (see Section 4.10). The pincer ligands are important for catalytic alkane dehydrogenation on Ir^{III} complexes and C–C versus C–H activation on cationic Rh centers (see also Sections 12.2.2 and 13.3.2).[25] Calculations by both Hall and Martin on Rh–Ir pincer ligand systems as in **3** but with a Me substituent on the arene proximal to the metal show that both agostic C_{Me}–H and C_{ring}–C_{Me} interactions can form and are likely intermediates in the observed insertion of the metal into these bonds.

The d_{MH} in agostic interactions (typically 1.75–2.8 Å) are generally 10–20% longer than for a terminal M–H bond. The d_{MC} usually range from 1.9–3.5 Å, but as for M–H_2, putting boundaries on these and other parameters is difficult because of the continuum nature of metal σ-bond interactions. Table 12.1 summarizes structural (including X-ray data) and spectroscopic characteristics for agostic interactions that can be considered to range from "strong" interactions to marginal interactions bordering on van der Waals contacts. Diffraction data show that all agostic bonds are bent and H is nearly always closer to M than C. The C–H ⋯ M angle α ranges from 63–137° and can be correlated with C–H and M ⋯ (H–C) distances. The angle approaches 90° as the d_{CH} increases in accord with the kinetic

$$M\text{----}H\text{--------}$$
$$\alpha \searrow \diagdown C$$

pathway for the reaction M + C–H → C–M–H derived by Crabtree[8,26] from structural data for agostic systems using the method of Burgi and Dunitz (Figure 12.2).[27] The C–H bond initially approaches M end-on as for incipient M–H_2 interaction, and then rotates to bring the C closer to M as the C–H bond weakens

Table 12.1. Characteristic Parameters for Agostic Interactions

d_{MH}	1.75–2.8 Å
d_{MC}	1.9–3.5 Å
d_{CH}	1.07–1.19 Å
J_{CH}	60–90 Hz
v_{CH}	2250–2800 cm^{-1}

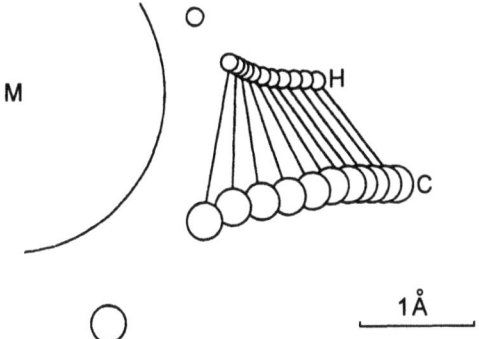

Figure 12.2. Trajectory for the reaction M + C–H → C–M–H derived from a series of structures of agostic complexes. The isolated circles are the final positions of C and H in the alkyl hydride OA product. Reprinted with permission from Crabtree et al.[26] Copyright 1985 American Chemical Society.

and elongates. This experimental trajectory for C–H OA mirrors that in theoretical studies of CH_4 addition to model complexes such as 16e $RhCl(PH_3)_3$[28] and 14e $IrX(PH_3)_2$ (X = H, Cl)[29] where a two-stage mechanism occurs for the latter. In the early part of the reaction (electrophilic stage), which closely parallels agostic interactions, σ donation to M is important. Thus a more strongly bound CH_4 adduct

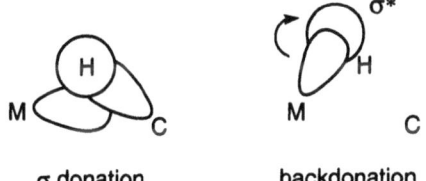

σ donation   backdonation

occurs for the Ir complex where X is the more electronegative Cl ($\Delta H_{binding}$ = −15.6 versus −6.8 kcal/mol for X = H) because M is a better acceptor for σ donation from the C–H bond. At this stage along the reaction coordinate, BD is not critical and is weak compared to that in $M-H_2$ or $M-(\eta^2\text{-SiH})$. Although later on (nucleophilic stage), BD is necessary to cleave C–H, agostic M ⋯ H–C models the first stage of OA of C–H very well.

The strength of the agostic interaction and d_{MC} and d_{MH} vary greatly as for $M-H_2$ complexes, and a better yardstick is the distance from M to the bonding electron pair, d_{bp}.[26]

$$d_{bp} = [d_{MH}^2 + 0.0784 d_{CH}^2 - 0.28(d_{MH}^2 + d_{CH}^2 - d_{MC}^2)]^{1/2} \qquad (12.3)$$

Because C and H have similar electronegativity, the bonding pair is thought to be

Table 12.2. Structural Parameters (Å) for Agostic Interactions (from Stahl et al.[10])

Complex	Type[a]	d_{MH}	d_{MC}	d_{CH}	d_{bp}	r_{bp}	α, deg
$Pt(PPh_2{}^tBu)_2$	γ	2.77	3.45	0.97	2.9	1.6	126
$PdBr(C_4Me_4)(PPh_3)$	δ	2.31	3.19	1.08	2.54	1.26	137
$TiEtCl_3(dmpe)$	β	2.29	2.52	1.02	2.30	0.98	88
$[Fe(\eta^3\text{-}C_8H_{13})\{P(OMe)_3\}_3]^+$	β	1.874	2.384	1.164	1.96	0.79	101
$TaCl_3(CH^tBu)L$	α	2.119	1.898	1.131	2.00	0.68	63
$Fe_4(CH)(H)(CO)_{12}$	α	1.80	1.926	1.18	1.76	0.59	77

[a]First three are X-ray structures; last three are neutron structures.

located on the C–H vector at the point where the covalent radii of C and H meet. The values calculated for d_{bp} using Eq. (12.3) are derived from diffraction data and range from 1.76 to 2.9 Å for a large variety of complexes (selected data are shown in Table 12.2).[26] In order to compare interactions on metals of different covalent radii, the covalent radius of M, r_M, is subtracted:

$$r_{bp} = d_{bp} - r_M \qquad (12.4)$$

where r_{bp} is effectively the covalent radius of the C–H bonding electrons and has a surprisingly broad range. For the lowest values near 0.5 Å, the d_{MC} and d_{MH} are similar to those observed for single bonds in, e.g., a metal alkyl or hydride. For r_{bp} greater than about 1.9 Å, these distances resemble van der Waals contacts. Just as for M–H_2, there is a near continuum of structural variation reflecting the degree of activation of the σ bond.

Experimental determinations of the energy of the agostic interaction are rare, but estimates[30] are 7–15 kcal/mol from calculations for CH_4 adducts of $IrX(PH_3)_2$ as well as thermodynamic data for alkane complexes (see Chapter 7). As for H_2 complexes, d^6 systems are the most abundant, but d^0, lanthanide, and actinide agostic complexes are also common. Examples of d^0 systems are $TiCl_3(\eta^2\text{-}CH_3)(dmpe)$ and its ethyl analogue, which have α-agostic and β-agostic interactions[31] and the α-agostic alkylidene complex $Cp_2ClTa=C-H(Bu^t)$.[32] Lanthanide complexes include tetrameric $Ln_4(OCH_2{}^tBu)_{12}$ (Ln = La, Nd) and Evan's Nd complex with seven agostic interactions arising from C–H groups on three different ligand systems.[33]

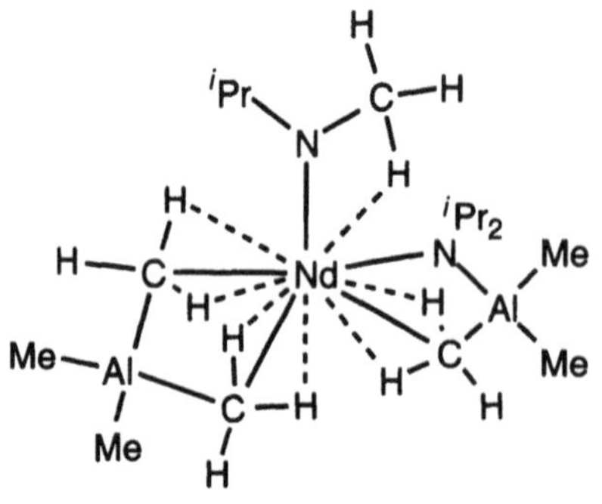

The d_{NdH} are 2.54–2.73 Å and are at the high end of the range in keeping with the expected weakness of polyagostic interactions. A recent neutron structure of Mo(Me)$_2$[N(2,6-^{i}Pr-C$_6$H$_3$)]$_2$ not included in Braga's survey also has multiple agostic interactions.[34] Four α-agostic interactions from trans Me are present (d_{MoH} = 2.59 Å) and d_{MoC} is very short [2.112(2) Å]. The agostic d_{CH} are not very elongated [1.068(7) Å versus 1.049(8) Å in the noninteracting H that is 2.75 Å from Mo]. Although the long d_{MH} are not good evidence for agostic interactions, the methyl groups markedly distort from tetrahedral geometry, showing the hydrogens to be drawn toward M, e.g., abnormally small Mo–C–H angles (104°).

NMR is also useful for detecting and analyzing M⋯H–C interactions. The signal for agostic H is usually upfield of tetramethylsilane as in a hydride ligand but can be difficult to locate because of high fluxionality. In Eq. (12.5), each proton of a

$$\text{(12.5)}$$

methyl group will be in the agostic site one-third of the time because of free rotation. For bis(P^{i}Pr$_3$) complexes, there are 36 β hydrogens that can interact with the open site on M! Thus J_{CH} for the agostic CH is averaged and cannot be

distinguished from the normal value, and the upfield chemical shift is also greatly reduced. IPR can aid in situations where fewer protons are involved.[35] IR is also useful, and ν_{CH} occurs at 2250–2800 cm^{-1}, although the band can be weak and broad as in Figure 12.3.[36]

In comparison to Si–H coordination, agostic C–H interactions (1) correspond to an earlier stage of OA. With few exceptions, the relative lengthening of d_{CH} in 1 (1.07–1.19 Å; unbound 1.05–1.10 Å) is usually less than that for η^2-Si–H [20% on the CpMn(CO)$_2$ fragment]. J_{CH}, which range from 52–90 Hz, are normally closer to the values for normal C–H bonds (120–130 Hz), while J_{SiH} in M(Si–H) complexes are closer to those in oxidatively added M(H)(SiR$_3$) complexes. There are exceptions to both generalities, which emphasize that the degree of σ-bond activation follows a continuum. [Cp*Co(Et){P(OMe)$_3$}]$^+$ (5a) well illustrates this and the sensitivity to changes in substituents at carbon.[12] Placement of a Me at the β carbon (5b) dramatically decreases J_{CH} to 38 Hz so that it is now closer to the lower limit of <10 Hz for an alkyl hydride complex. Just as for elongated H$_2$ complexes, the value of J_{PH} rises considerably on methyl substitution because d_{MH} is shortened. This *increased activation by placing a better electron-donating substituent at carbon* is in direct contrast to the situation in Si–H coordination, which reflects the much weaker

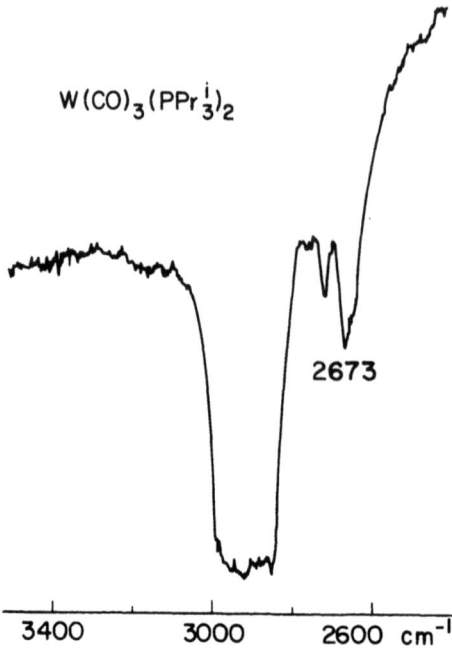

Figure 12.3. Nujol mull IR spectrum of $W(CO)_3(P^iPr_3)_2$ showing ν_{CH} at 2673 cm^{-1} for the agostic interaction.

BD in C–H complexes. The C–H bond simply becomes a better σ donor to M when the carbon in **5b** becomes more electron-rich and BD effects are inconsequential.

J_{CH} = 61 Hz J_{CH} = 38 Hz
J_{PH} < 4 Hz J_{PH} = 32 Hz
(**5a**)   (**5b**)

12.2.2. Cyclometallation and OA of Agostic C–H Bonds

OA of the agostic C–H bond is very common and referred to as cyclometallation [Eq. (12.2)].[37,38] Cyclometallated species are important in organic synthesis, catalysis, asymmetric synthesis, and photochemistry. Although the agostic species is normally an unobserved transient in Eq. (12.2), equilibrium between agostic and cyclometallated species is observable in W(CO)(dppe)$_2$[39] and a few other systems [Eq. (9.47)]. Addition of H$_2$ to W(CO)(dppe)$_2$ in Eq. (12.6) [and also the agostic

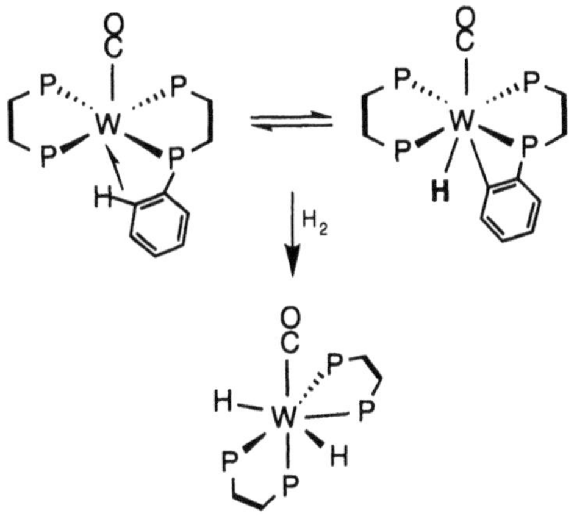

(12.6)

complex Mo(CO)(^{i}Bu$_2$PC$_2$H$_4$P^{i}Bu$_2$)$_2$[40]] gives complete OA to the dihydride, which suggests that the H–H bond undergoes OA more readily than agostic C–H bonds. For metallations of *aryl* C–H bonds, the term "orthometallation" is used interchangeably with cyclometallation and has been likened to classical electrophilic arene substitution. One accepted mechanism for metal-assisted inter- or intramolecular cleavage of arene C–H involves initial formation of an η^2-arene complex. Another mechanism centers on electrophilic attack at the ipso-C to form a metal arenium (Wheland) complex followed by proton elimination [Eq. 12.7]. However,

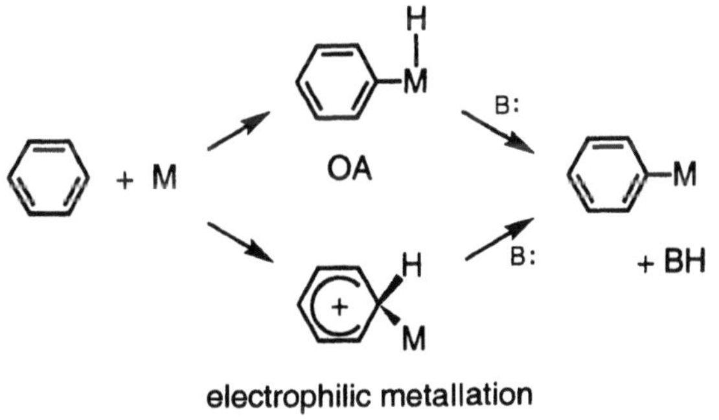

(12.7)

electrophilic metallation

such an intermediate has not been isolated, and the kinetics of electrophilic metallation of arenes do not always correspond to that for aromatic electrophilic substitution. In intramolecular processes, aromatic C–H activation might occur via an agostic intermediate or transition state,[41] but examples of such complexes are rare.[42] This interaction may increase the acidity of the agostic proton as observed in aliphatic agostic complexes and lead to facile deprotonation of the arene to give metallated product without actual insertion of M into the C–H bond, i.e., a process similar to electrophilic metallation. The agostic intermediate for aromatic C–H cyclometallation in a Rh system is isolable and can be easily deprotonated.[42]

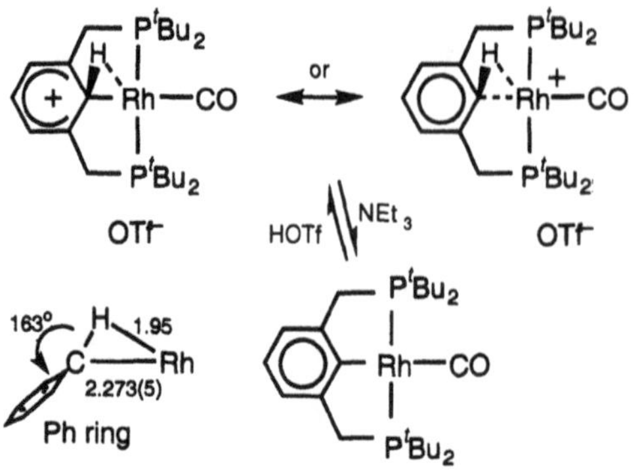

Equation (12.8) shows the two possible tautomeric forms for the X-ray characterized cationic agostic complex, which exhibits rare air stability and $J_{CH} = 123$ Hz for the agostic C–H. The H atom strongly interacts with Rh (calculated $d_{bp} = 0.73$) and is bent away from the aromatic plane by about 17°. The 0.93-Å d_{CH} is typical of unactivated hydrocarbons. Although the Ph ring is not distorted from aromaticity, a slight bending of the ipso-C–H bond from the plane suggests that contribution of the arenium form cannot be excluded. Previously seen only for aliphatic protons, the acidic agostic proton can be removed with weak bases such as NEt_3 to give the metallated complex, which can be reprotonated [Eq. (12.8)]. NMR, deuterium-labeling, and DFT analyses show very little, if any, contribution of a metal arenium structure. Thus, deprotonation of a species with a strong agostic interaction as on the right side of Eq. (12.8) is more likely than electrophilic aromatic metallation involving arenium intermediates. This system is directly analogous to base-assisted heterolytic cleavage of H_2, and is important in that it shows that the proton in M(C–H) attains partial positive charge, i.e., $M(C^{\delta-}-H^{\delta+})$ rather than the reverse. Transfer of such an "acidic" proton from an aromatic agostic C–H to a hydride ligand to give an unobserved metallated bis(H_2) complex, followed by substitution of H_2 by THF are proposed to be equilibrium processes [Eq. (12.9)].[43]

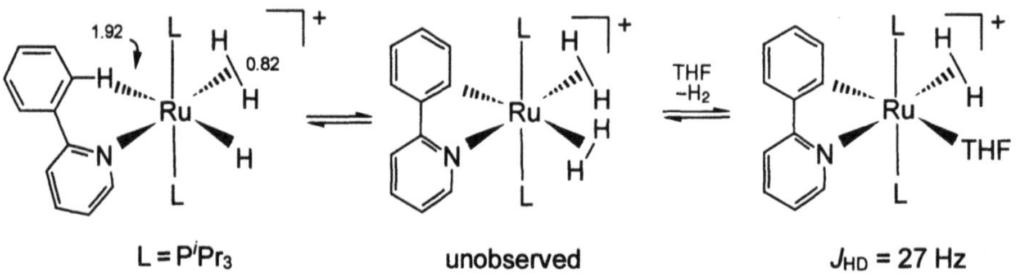

Isotopic exchange with D_2 to give exclusive deuteration of the ortho positions of the phenylpyridine supports the equilibrium between agostic and metallated forms.

Bridging agostic bonds exist and can also undergo equilibrium C–H cleavage [Eq. (12.10)]. The factors that favor **6** over **7** are similar to that for H_2 coordination

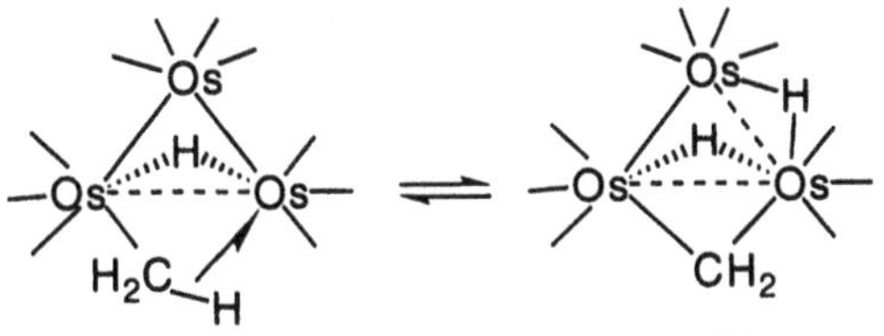

$$\text{(6)} \rightleftharpoons \text{(7)} \quad (12.10)$$

versus dihydride formation and are also reflected in the preference for the β-agostic interaction (8) over the olefin–hydride structures in 9 and 10. Most agostic complexes involving β carbons require an electrophilic M that is either cationic, contains more than one π-acceptor ligand, or is an early transition metal in a high oxidation state.[12] Thus 9 with neutral Fe has the ethylene–hydride structure[44] whereas 8 with cationic Co is agostic[45] [cf $MoH_2(CO)(depe)_2$ versus the cationic σ complex, $[Mn(H_2)(CO)(depe)_2]^+$]. There is a fine balance: the Rh congener (10) is not agostic even though cationic presumably because increased BD from the second-row M cleaves the C–H bond.[46]

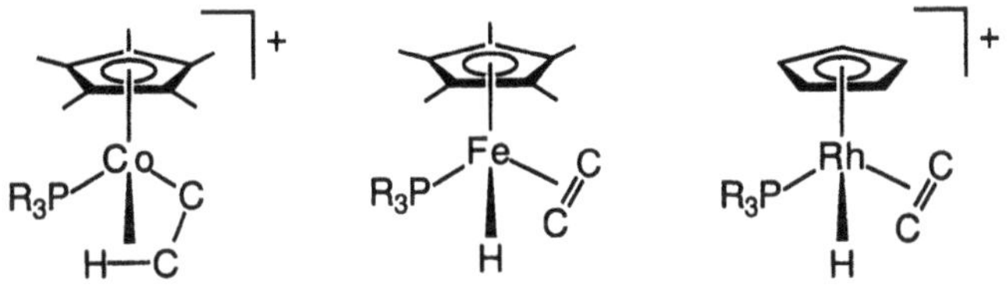

(8)   (9)   (10)

The above structures are relevant to organometallic reactions such as alkene insertion [Eq. 12.11] and Ziegler–Natta alkene polymerization catalysis [Eq. 12.12]. Calculations indicate a β-agostic transition state in Eq. (12.11) for $RhH(CO)_2(PH_3)(C_2H_4)$.[47] In Eq. (12.12) the α-agostic binding rotates the Me toward the alkene to facilitate insertion.[48]

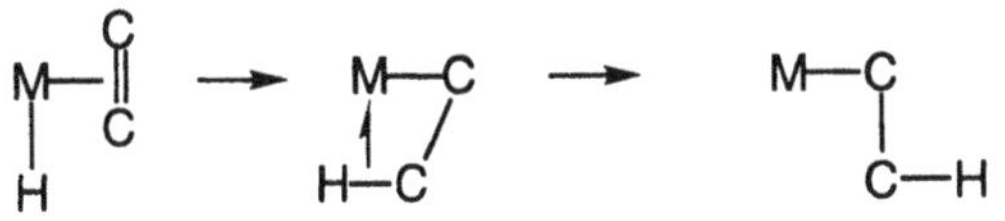

$$\text{(12.11)}$$

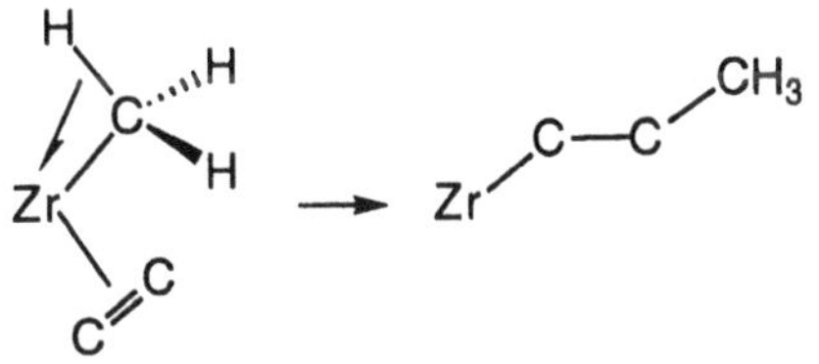

$$\text{(12.12)}$$

An important paradoxical question posed by Crabtree is why complexes containing C–H bonds in their ligand systems break the C–H bonds of alkanes (see below) rather than cyclometallate their own C–H bonds, a normally facile process.[26] Intermolecular C–H cleavage often occurs in systems containing PR_3 and Cp* ligands, i.e., cyclometallation must be kinetically disfavored, most likely because of sterics. The internal C–H bonds of these ligands cannot easily follow the trajectory of Figure 12.2 without bond strain (although as the R groups of PR_3 become larger, there is more freedom to metallate).[26] Steric congestion around M greatly accelerates cyclometallation[38] and would inherently increase for alkane C–H activation because of the addition of another ligand to the coordination sphere. Thus the less sterically crowded a complex is, the more it should favor external over internal C–H cleavage. The PPh_3-containing systems that activate alkanes have low coordination numbers (ML_2), and more congested (C_5Me_5)ML requires smaller phosphines such as PMe_3 to avoid metallation. Alkane activation is relatively rare not for kinetic reasons but because it is limited to uncrowded systems that do not readily cyclometallate.

12.2.3. Agostic Interactions in Phosphine Complexes

Agostic interactions involving phosphines are quite significant because they "reserve" binding sites for other small molecules such as H_2, and phosphine systems are crucial in many homogeneous catalysts. The first H_2 complexes were prepared from agostic $M(CO)_3(PCy_3)_2$ (M = Mo, W), which are valuable precursors for a large series of complexes [Eq. (12.13)].[36] The agostic interactions are relatively

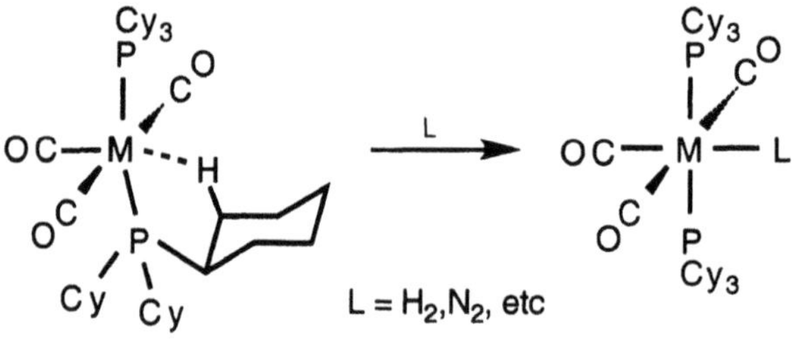

$$L = H_2, N_2, \text{etc} \qquad (12.13)$$

strong and help offset the electronic unsaturation. The P–M–P angle of 160° and other angular distortions in the M–P–C–C–H ring clearly shows that the C–H bond is drawn toward M (Figure 12.4). The energy of the agostic interaction is estimated to be 10–15 kcal/mol based on the energetics of heptane binding to $W(CO)_5$ discussed in Chapter 7, which presents additional information on agostic interactions of phosphines. Five-coordinate zero-valent group 6 complexes are unstable, especially photolytically generated $M(CO)_5$, which have extremely short lifetimes without sixth ligands and even coordinate rare gases. It is important to note that ligand displacement reactions for $W(CO)_3(PCy_3)_2(L)$ (and conceivably other PR_3 complexes) proceed via a 16e agostic intermediate as in Eq. (12.13)[49] Several other 16e agostic PR_3 complexes and their X-ray parameters are summarized in Table 12.3. An novel cationic Ru^{II} complex contains both a strong ortho-phenyl

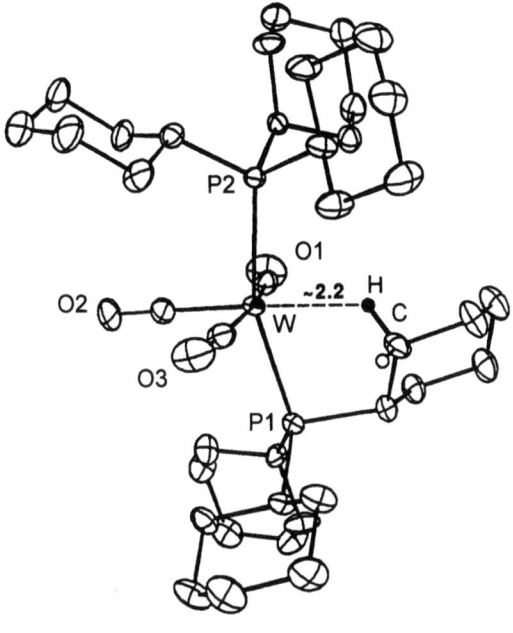

W(CO)$_3$(PCy$_3$)$_2$
X-RAY STRUCTURE, 25 °C

Figure 12.4. ORTEP drawing of W(CO)$_3$(PCy$_3$)$_2$

Table 12.3. Comparison of Metal-Agostic X-Ray Distances in Formally 16e Phosphine Complexes

Complex	d_{MH}, Å[a]	d_{MC}, Å	References
Cr(CO)$_3$(PCy$_3$)$_2$	2.240(1)	2.884(1)	50
Mo(CO)(dppe)$_2$ (12)	2.98(11)	3.50(2)	51
Mo(CO)(d^{i}Bupe)$_2$ (13)	2.20	3.007(4)	52
W(CO)$_3$(PCy$_3$)$_2$	2.27	2.945(6)	36
[Mn(CO)(dppe)$_2$]$^+$ (11)	2.89(6), 2.98(6)	3.456(4), 3.589(4)	53
[Mn(CO)$_3$(PCy$_3$)$_2$]$^+$	2.01(9)	2.75(3)[b]	54
[Re(CO)$_3$(PCy$_3$)$_2$]$^+$		2.89(5)	55
[RuH(dppp)$_2$]$^+$ (14)	2.75	3.282(8)	56
[RuX(CO)(P^{t}Bu$_2$Me)$_2$]$^{+c}$		3.049(1)	57
[Ir(H)$_2$(PCy$_2$Ph)$_3$]$^+$		2.923(10)	58

[a]Idealized distances if no esd's are given.
[b]Average value for disordered carbons.
[c]X = CHC(SiMe$_3$)Ph; d_{MC} given for phosphine CH.

C–H agostic interaction from a pendant alkyl (d_{RuC} = 2.588 Å) and a weaker interaction from a phosphine methyl group (d_{RuC} = 3.049 Å).[57]

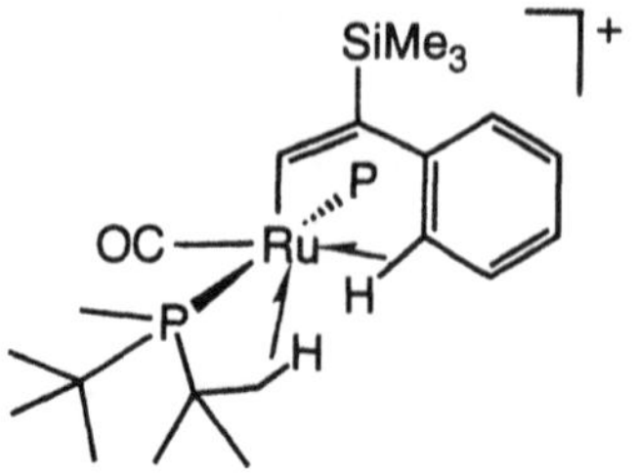

The [Mn(CO)(dppe)$_2$]$^+$ cation (**11**) contains multiple distant agostic interactions with Mn involving phenyl C–H on the phosphines (M $\cdots$ H near 3.0 Å and M $\cdots$ C near 3.5 Å). These weak interactions are more suggestive of van der Waals contacts with a small contribution from agostic binding. Also, only a maximum of 2e from the agostic interactions need to be donated to these 16e fragments, so polyagostic interactions represent donation of less than 2e from each C–H. Such species can be viewed as representing very early positions along the reaction coordinate for OA of σ bonds, stabilized primarily by favorable entropic factors. The

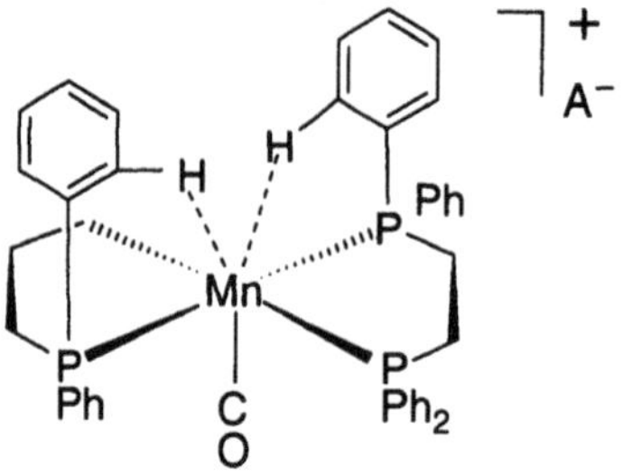

(**11**)

Mn systems contrast with the neutral Mo(CO)(PP)$_2$ analogues, which contain only one interaction from either a phenyl C–H (**12**) or an isobutyl γ-C–H (**13**).

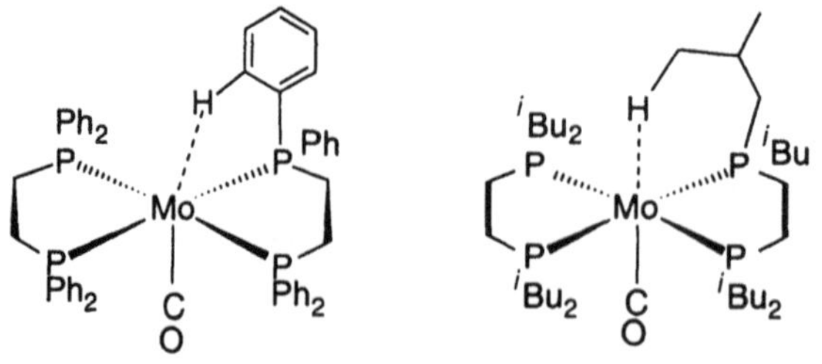

dppe (**12**)   d^{i}Bupe (**13**)

Steric factors are critical (see below) and are no doubt responsible for the long M $\cdots$ H–C distances in the rigid chelate systems, which do not allow easy approach of the agostic C–H (Mo $\cdots$ H in **12** is quite long, 2.98 Å). The d_{MoH} in **13** is much shorter (2.20 Å) because the γ-C–H can more easily approach M because of the larger six-membered ring (cyclometallation is disfavored by the strong trans effect of the CO). The d_{MC} also reflect the relative strengths of the agostic interactions: 3.007(4) Å in **13**, which is much shorter than in **12** [3.50(2) Å] and **11**. The weak multiple agostic interactions in **11** most likely result from reduced Mn–P–C–C–H ring size as compared to the larger-radius Mo system, preventing full interaction of C–H. The d_{MnH} are much shorter for the single agostic interaction in [Mn(CO)$_3$(PCy$_3$)$_2$]$^+$, where the Mn–P–C–C–H ring is more flexible.

The agostic bonding in [RuH(dppp)$_2$]$^+$ is noteworthy in that the overall geometry (**14**) differs from that of the corresponding H$_2$ complex, which has the usual trans geometry (**15**).[56] The octahedral geometry in **14** also contrasts to that in

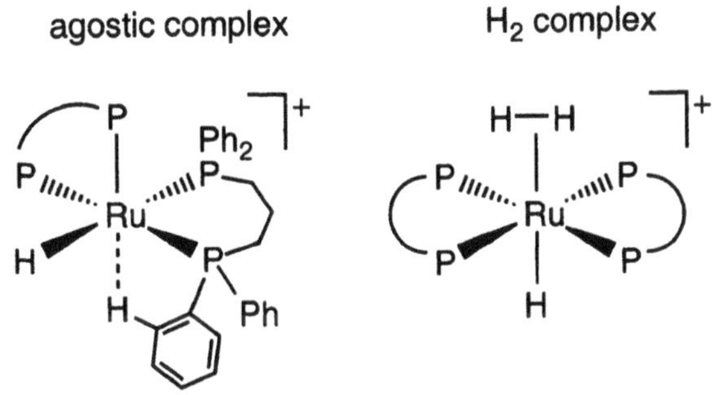

agostic complex   H$_2$ complex

(14)   (15)

the dcpe and R,R'-Me-DuPHOS analogues, which have a true square pyramidal geometry with the hydride in the apical site and no agostic bonding.[59] In yet another variation, the dppe analogue obtained by reaction of [RuH(NH$_2$NMe$_2$)$_3$-(COD)]PF$_6$ with dppe in acetone under Ar (identical synthesis to that for **14**) is markedly different in color (pale beige) from **14**, which has the characteristic deep color of agostic phosphine complexes (ranging from brown-black here to deep purple). The X-ray structure of [RuH(acetone)(dppe)$_2$]PF$_6$ shows that it is a *solvento* complex with acetone occupying the sixth site trans to hydride. Once again color is an important clue to the structure, which is remarkably diverse in these unsaturated or "lightly bound" bis(diphosphine) systems (cis agostic, trans agostic, trans L, or nonagostic sq pyr). Minor steric and electronic differences greatly influence the geometry, e.g., larger bite angle of dppp (89–91°) versus dppe (82-84°) allows easier approach of C–H. As will be further shown below, the presence of a hydride (or a π donor such as Cl) ligand stabilizes 16e complexes, so agostic interactions are less common for MX(diphosphine)$_2$ (X = H, Cl) than M(CO)(diphosphine)$_2$ (**11–13**).

Formally 14e complexes with two open sites possess polyagostic interactions that are stronger than those in less unsaturated 16e systems.[58,60,61] A fourfold interaction (two M–P-*ortho*-phenyl-H and two M-*t*-butyl-H) exists in M(PPh^{t}Bu$_2$)$_2$ (M = Pt, Pd; estimated M $\cdots$ H = 2.5–2.6 Å),[60] and [Ru(Ph)(CO)(P^{t}Bu$_2$Me)$_2$]$^+$ has two Ru $\cdots$ H–C interactions originating from one ^{t}Bu on each phosphine.[61] The preference for donation from ^{t}Bu over Me substituents is attributed to better steric approach of the former (giving a five-membered ring) than the latter (four-mem-

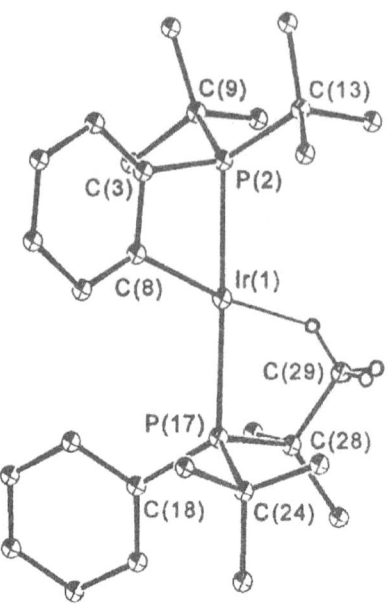

Figure 12.5. ORTEP drawing of [IrH(η^2-C$_6$H$_4$P^{t}Bu$_2$)(P^{t}Bu$_2$Ph)]$^+$. Reprinted with permission from Cooper et al.[58] Copyright 1999 American Chemical Society.

bered). The d_{RuC}, 2.87 Å and 2.88 Å, are quite short compared to most of the 16e complexes in Table 12.3. Related 14e [Ir(H)$_2$(P^{t}Bu$_2$Ph)$_2$]$^+$ also shows one strong interaction from *each* phosphine ^{t}Bu, but an ortho-metallated analogue with a very acute angle (84°) Ir–P(2)–C(3) gives only one total despite the presence of two vacant sites and three *seemingly* available pendant ^{t}Bu groups (Figure 12.5).[58] The constraints imposed by orthometallation on the geometry about P inhibit additional agostic bonding to the remaining open site. 16e [Ir(H)$_2$(PCy$_2$Ph)$_3$]$^+$ gives one strong agostic interaction, but the less bulky P^{i}Pr$_2$Ph analogue shows only very long d_{MC} (e.g., 3.46 Å) as for the Mn–dppe system, apparently simply because the C–H is further away from M in its unstrained position. These systems demonstrate the importance of steric effects on agostic bonding. Attendant computational studies also support the precept that, assuming favorable electronics, the natural proximity of C–H bonds to an unsaturated M (or their ability to approach M with little strain) is *the* critical factor in the strength and/or number of interactions.[58]

The d^6 16e Rh^{III} system (16–18)[62–64] and related Ru^{II} complexes[61] also show the delicate electronic and steric influences on coordination geometries and weak interactions, including interligand hydrogen bonding. In the alkylphosphine complexes, 16 has a tbp geometry and 17 prefers distortion to a sq pyr because of the poorer σ-donor character of the two Cl.[62] However, neither has close contacts with phosphine hydrogens as in W(CO)$_3$(P^{i}Pr$_3$)$_2$. In 18 where Ph replaces hydride and PPh$_3$ replace P^{i}Pr$_3$, ortho hydrogens on the Ph are proximal to not only Rh [2.911(2) Å] but also Cl [2.72(1) Å], i.e., both agostic interactions and hydrogen bonding to Cl may be occurring.[63,64] DFT calculations show, however, that the shortest Rh $\cdots$ H contact in the optimized model structure is 3.691 Å and the shortest

d_{CIH} is 2.828 Å.[64] This suggests that Rh···H–C interactions may not be present (as in **17**) or at best are weaker than Cl···H–C hydrogen bonds and may only be

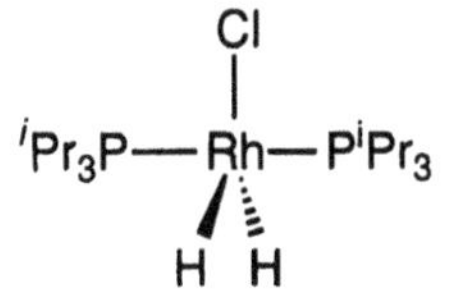

Cl   H   Ph

(structures **16**, 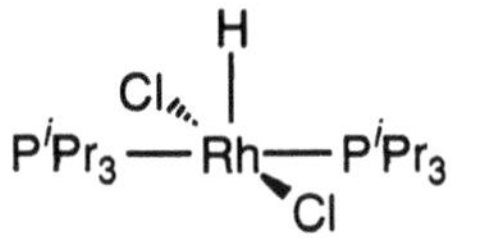 **17**, 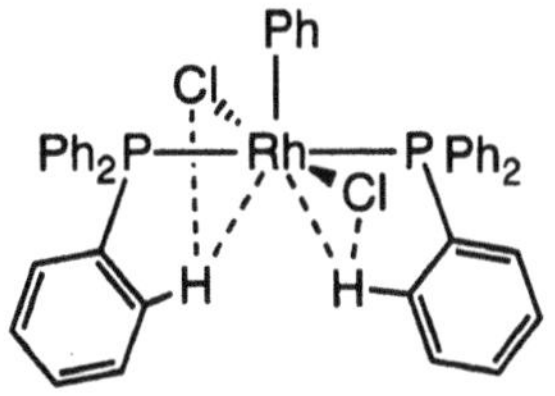 **18**)

present because of the steric repulsion within the bulky phosphines, i.e., the proximity effect as in the above Ir system. P^iPr_3 has a larger cone angle than PPh_3, so the absence of agostic interactions in **17** and all of these complexes is probably electronic in origin, i.e., π-donating Cl relieves electronic unsaturation (see Section 4.10). Similarly, $RuCl_2(CO)(P^iPr_3)_2$ and $Ru(Ph)Cl(CO)(P^tBu_2Me)_2$ show no agostic interaction, but removal of Cl from the latter to give 14e $[Ru(Ph)(CO)(P^tBu_2Me)_2]^+$ produces two interactions.[61] Another factor is the high trans-labilizing influence of the hydride ligand. Structure **17** and related hydrides such as $RuHCl(CO)(P^iPr_3)_2$, $RuH_2(CO)(P^tBu_2Me)_2$, and $[RuH(CO)_2(^tBu_2PC_2H_4P^tBu_2)]^+$ with hydride trans to the vacant site have nonagostic, five-coordinate structures.[61,65,66] Such behavior is also seen for the diphosphine hydride systems above, and the Ru complexes may not even add strong ligands such as CO trans to the hydride, although $RuH_2(CO)(P^tBu_2Me)_2$ binds H_2 weakly and reversibly.[66] These 16e species may be stabilized by strong σ donation from the hydrides, but 16e Ru complexes can be stable without hydrides or π donors: $[Cp*Ru(PMe^iPr_2)_2]^+$ shuns agostic interaction even though C–H bonds are available.[67]

12.3. ALKANE COORDINATION

12.3.1. Experimental Evidence

Along with the noble gases and perfluorocarbons, alkanes are the poorest ligands because of their strong nonpolar C–H bonds, and a stable complex has not yet been prepared. Nonetheless alkane complexes have been detected spectroscopically and are well-established intermediates in C–H cleavage as described in excellent reviews of alkane binding and activation.[3–9] Hall and Perutz's article focuses on alkane coordination while Shilov and Shul'pin discuss primarily activation of C–H bonds. Methane present in large reserves of natural gas and methane hydrate in marine environments is of particular interest.[6] Although enthalpically CH_4 is among the weakest polyatomic ligands known, CH_4 has one of the lowest absolute entropies ($S° = 44.5$ cal/mol·deg), and its binding is thus favored *entropically* over coordination of virtually any other molecule except H_2 (see Section 7.2.1). This could be a significant factor in competitive binding with other weak ligands (including other alkanes) on vacant M sites, especially at higher temperatures where

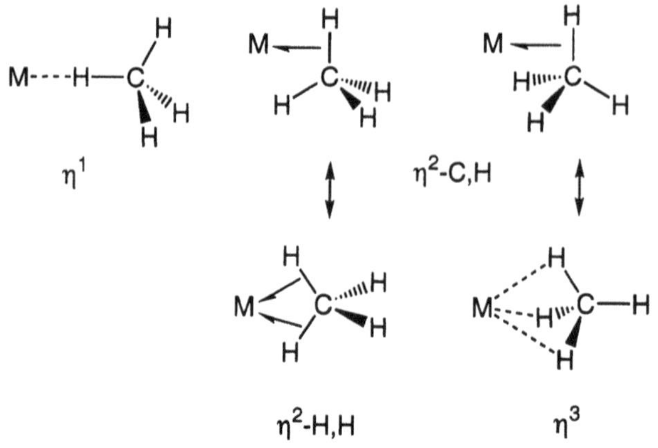

Figure 12.6. Possible coordination modes for methane and related alkane binding.

$T\Delta S$ is larger. CH_4 is also intriguing because there are several possible coordination modes in analogy with isoelectronic metal borohydride complexes (Figure 12.6). A beautiful model complex for η^3 coordination is the Lewis acid–base adduct of CH_3BeCp^* with Cp_2^*Yb synthesized by Burns and Andersen.[68] The d_{YbC}, 2.766(4) Å, is similar to that in $Cp_2^*Yb(\mu\text{-}C_2H_4)Pt(PPh_3)_2$ but shorter than in a MeC≡CMe

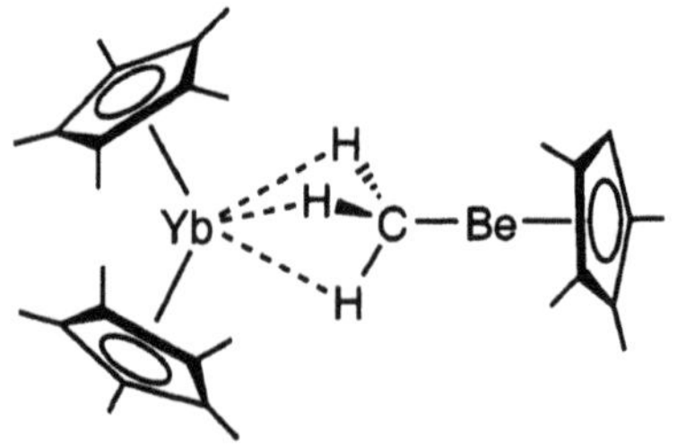

complex, 2.85 Å. The average d_{YbH} is 2.59 ± 0.08 Å, virtually identical to that in the ethylene complex. As will be shown below, however, the η^2-C, H form is a more favorable geometry for CH_4 binding experimentally and theoretically (there are actually two η^2-C,H geometries, differing only in orientation of the unbound hydrogens.

The first examples of alkane complexes were discovered by matrix isolation studies of photochemically generated $Cr(CO)_5$ or Co atoms.[69] The classic experiments performed by Perutz and Turner showed that CH_4 binds to $Cr(CO)_5$ in CH_4 matrices and later Billups detected $Co(CH_4)$. Because IR shifts resulting from coordination of a sixth ligand are difficult to distinguish from solvent effects, UV/vis spectroscopy best demonstrates CH_4 complexation in the Cr species. The absorption frequency (489 cm^{-1}) is similar to that for Xe coordination and much higher than those for H_2 and N_2 binding (370 and 364 cm^{-1}).[70] Along with IR evidence (see also Table 4.3), these spectra demonstrate that CH_4 is a much weaker acceptor than H_2 or N_2. However, the coordination mode of the alkane complex cannot be determined by such spectroscopic studies.

Solution studies of transient alkane binding at room temperature were first performed in 1973 by flash photolysis of $Cr(CO)_6$ in cyclohexane, which gave a UV/vis band at 503 cm^{-1} similar to that seen in CH_4 matrices[71,72]:

$$Cr(CO)_6 \xrightarrow{h\nu, C_6H_{12}} Cr(CO)_5(C_6H_{12}) + CO \qquad (12.14)$$

The same experiments in perfluorocyclohexane showed no binding of this much weaker ligand, only generation of highly unstable "naked" $Cr(CO)_5$ with a thousand-fold lower lifetime.[72] These and subsequent experiments provide kinetic information to quantify the reactivity and bond strengths of metal–alkane species such as $M(CO)_5$(alkane), which have been measured in both solution and gas phases (see Chapter 7). Ultrafast spectroscopy shows that addition of alkane to $Cr(CO)_5$ in solution is extremely fast, within the first picosecond after UV irradiation of $Cr(CO)_6$, but decomposition of $Cr(CO)_5$(cyclohexane) occurs within 50 μs.[73] The binding enthalpies range from only ca 5 kcal/mol for CH_4 to about 12 kcal/mol for cyclohexane, i.e., energies much less than that for H_2 binding and too low for isolation. Larger alkanes and cycloalkanes coordinate more strongly here and to Pd atoms in the gas phase[74] (which do not bind CH_4), as will be discussed further below. Both cis- and trans-$W(CO)_4(L)$(heptane) form at different rates from photolysis of $W(CO)_5(L)$ in heptane, and the cis isomer is shorter-lived (L = PR_3 or phosphite).[75]

Because alkanes bind primarily by σ donation to M, one approach to forming a stronger bond might be to increase the electrophilicity of M as much as possible, e.g., using even better withdrawing ligands such as fluorophosphines like $(C_2F_5)_2$-$PC_2H_4P(C_2F_5)_2$, dfepe. However, photochemically generated $Cr(CO)_3$(dfepe) in matrices do not bind CH_4 any better than $Cr(CO)_5$.[76] As an aside, a fluorine on a C_2F_5 group could also competitively bind to the sixth site on **20**, although this

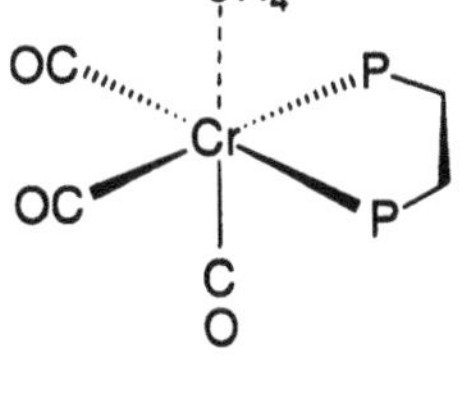

(19) 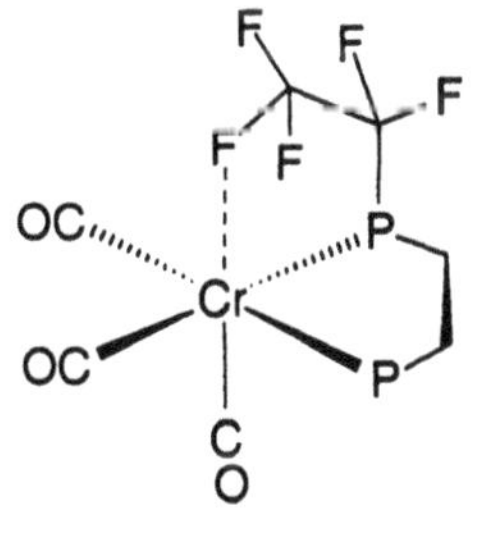   (20)

agostic-type species was 50 times as reactive as the CH_4 complex (**19**). Recent theoretical and experimental studies show that C–F bonds preferentially coordinate over C–H bonds on the same C (see below). In another test of maximizing the Lewis acidity of M, CH_4 binds to highly electrophilic naked metals in $[M(CH_4)_n]^+$ even more strongly than H_2, on the order of 20 kcal/mol as discussed in Chapter 4. Polarization effects are important, and alkanes are more polarizable. However, similar collision-induced mass spectral studies show only weak (<7 kcal/mol)

Table 12.4. Second-Order Rate Constants (mol^{-1} dm^{-1} s^{-1}) for the Reaction of Metal–Heptane Complexes with CO in Heptane at 298 K

CpV(CO)$_3$(heptane) 1×10^8	(Benzene)Cr(CO)$_2$(heptane) 2×10^6	CpMn(CO)$_2$(heptane) 8×10^5
CpNb(CO)$_3$(heptane) 7×10^6		
CpTa(CO)$_3$(heptane) 5×10^6		CpRe(CO)$_2$(heptane) 2×10^3

binding of CH_4 to $[Mn(CO)_5]^+$, where σ donation should be near maximum for an organometallic complex.[77] Thus BD must still be important in alkane binding (certainly in OA), and attempts to increase σ donation diminishes BD. Related studies of organometallic ions show the first solvation shell of CpCo$^+$ can accommodate either two H_2 or two CH_4 ligands, and C–H activation can involve the Cp, e.g., H from bound CH_4 can transfer to Cp.[78]

$$(\eta^5\text{-}C_5H_5)Co(CH_4)_2^+ \rightarrow (\eta^4\text{-}C_5H_6)Co(CH_3)(CH_4)^+$$

Conversely, CH_4 also coordinates to *electron-rich* non-Cp fragments such as MH(dmpe)$_2$ (M = Mn, Re) in matrices.[15] Binding of alkanes to d^8 fragments such as Fe(CO)$_4$ and M(CO)$_n$P$_{4-n}$ (M = Fe, Ru) as well as half-sandwich complexes such as CpMn(CO)$_2$ and (arene)Cr(CO)$_2$ has also been observed in matrices and by flash photolysis.[79-82] The Mn–heptane bond dissociation energy is about 9 kcal/mol from photoacoustic calorimetry (see Section 7.1.2),[80] comparable to the 10-kcal/mol value for Cr(CO)$_5$(heptane). The UV/vis shifts (3620, 3330 cm^{-1}) on coordination of CH_4 to Ru(CO)$_2$(dmpe) are the largest yet observed for any matrix-isolated species (cf 1690 cm^{-1} for Cr(CO)$_5$).[81] Significantly, CpRe(CO)$_2$(heptane) generated photolytically in heptane has the longest lifetime of any known alkane complex.[82] At 25 °C the decay rate in the absence of added ligand is only 40 s^{-1}. The rate constants for reaction with CO in heptane are more than four orders of magnitude slower than that for CpV(CO)$_3$(heptane) (Table 12.4). This indicates that the Re–heptane bond may be among the strongest of any alkane complex of the group 5–7 metals. Thus, although CpRe(CO)$_2$ cleaves H_2 to form a dihydride, it is apparently optimally suited for *alkane* coordination without OA, and a more strongly backbonding third-row M helps stabilize C–H coordination. Table 12.4 shows that the stability rises both across and down the groups. Although this might predict that later transition metals would give even more stable complexes, CpCo(CO) formed by flash photolysis does not react with cyclohexane[83] and the Cp* Rh and Ir analogues give OA of alkanes.[10,84] Fe(CO)$_4$(cyclohexane) is less stable and OA occurs for Os–arene systems.[85]

The above findings provided inspiration for the *first direct observation by NMR of a transition metal alkane complex*, CpRe(CO)$_2$(cyclopentane), reported by Geftakis and Ball in 1998.[86] The key to this discovery is an apparatus allowing an alkane solution of CpRe(CO)$_3$ to be continuously irradiated with UV light brought into the NMR tube via a fiber optic cable so as to increase the concentration of photolytically generated species in Eq. (12.15). Photolysis of CpRe(CO)$_3$ in cyclopentane at $-80°C$ gives a high-field ^1H resonance at $\delta - 2.32$, which integrates with a ratio of 2:5 with respect to a shifted Cp peak. The peak at $\delta - 2.32$ is assigned to one

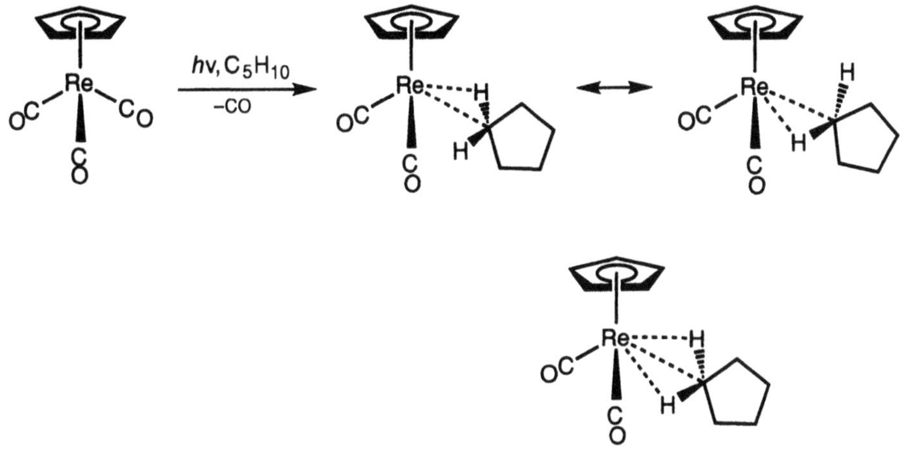

$$\text{(12.15)}$$

methylene of the bound cyclopentane in CpRe(CO)$_2$(cyclopentane) and is a first-order quintet with $^3J_{HH} = 6.6$ Hz, consistent with four similar couplings to protons on adjacent carbons. This coupling is similar to that in free cyclopentane, which suggests little geometric perturbation. Both protons attached to the same C are involved in binding to Re, with the most likely explanation for their equivalence being rapid exchange of the two η^2-C–H bonds interacting with M. Alternatively, an η^2-H,H interaction is possible, but theoretical studies on W(CO)$_5$(propane) suggest that the lowest energy form for methylene binding is the η^2-C–H structure (see below).[87] Silanes coordinate in η^2-Si–H fashion as shown in Chapter 11. Pentane also forms a complex with CpRe(CO)$_2$ that shows three high-field resonances ascribed to the three possible binding sites on this ligand.

A ^{13}C-labeled sample of the cyclopentane complex shows a ^{13}C signal at $\delta - 31.2$ and J_{CH} to be 112.9 Hz compared with 129.4 Hz in free cyclopentane, consistent with a weakly bound alkane. If CpRe(CO)$_2$(cyclopentane) does have the η^2-C–H structure, the observed J_{CH} value represents the average value from one coordinated and one unbound C–H bond, allowing estimation of J_{CH} in the bound C–H to be about 96 Hz, which is similar to that in agostic interactions. Use of cyclopentane-d_{10} in the experiments gives an isotopically shifted Cp resonance but no signal at $\delta - 2.32$. A 1:1 mixture of C_5H_{10}/C_5D_{10} results in two close Cp resonances that show by integration that the C_5H_{10} ligand is preferentially bound in the ratio 1.33:1. Thus, the EIE is smaller and "normal" compared to the much larger "inverse" effect in Cp*Rh(alkane) complexes and also M–H$_2$ complexes in general (see Section 7.5.2). Because isotope effects can be complex and involve multiple vibrational modes, the origin of the different effects is not yet clear.

The first crystallographic evidence for alkane interaction with a M was reported by Reed's group in 1997.[88] Heptane interaction with a double A-frame porphyrin

(DAP) complex is shown, although the heptane molecule could also be considered to be held in place by a host–guest effect. There is a short Fe–C contact in **21** and a much longer one with a second Fe(DAP) center, although crystallographic

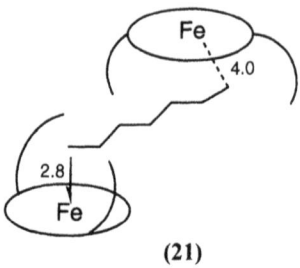

(21)

disorder is present (d_{FeC} are modeled to be 2.5 and 4.8 Å for one orientation and 2.8 Å and 4.0 Å for the other). The short contacts are typical of agostic distances. Since hydrogens were not located, DFT calculations of model Fe(porphyrin) complexes of alkanes ranging from methane to butane were used to guide placement. An unsymmetrical η^2-H,H geometry is found where d_{FeC} are 2.68–2.70 Å and d_{FeH} are 2.01–2.13 Å and 2.62–2.65 Å for the proximal and distal C–H, respectively. Binding energies are 10.5–16.7 kcal/mol, generally increasing with size, and are somewhat higher than those calculated for alkane binding to $W(CO)_5$ discussed below.

Evidence for alkane and arene binding comes from many sources, including high solvent dependence of the thermodynamics of reversible H_2 dissociation from $IrXH_2(H_2)(PR_3)_2$.[89] Apparently alkanes and toluene can compete with H_2 for binding to the five-coordinate fragments for X = I. The surprising rapid loss of H_2 on dissolving $Cr(CO)_3(PCy_3)_2(H_2)$ in H_2-saturated toluene indicates that toluene may be aiding dissociation of H_2 by mass action effects.[90] The solubility of H_2 is low

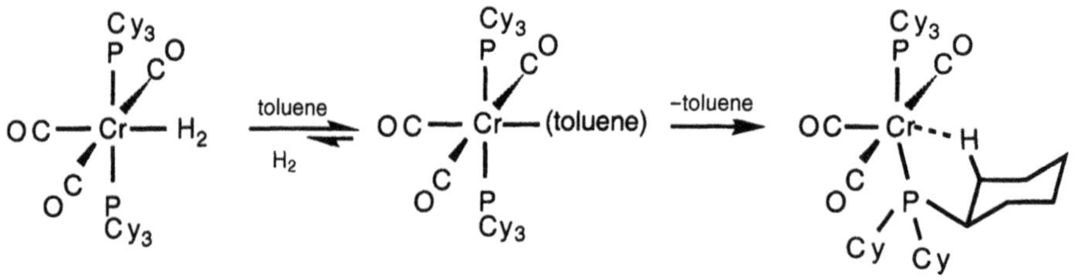

(12.16)

and equilibrium would lie toward an unstable toluene complex that rapidly loses toluene to form the entropically favored agostic complex.

12.3.2. Theoretical Studies of Alkane Activation and C–H versus C–F Bond Coordination

The first notable theoretical investigation of alkane binding was performed by Saillard and Hoffmann, who compared H_2 and CH_4 interaction with $Cr(CO)_5$, $Rh(CO)_4^+$, and $CpRh(CO)$ fragments by extended Huckel methods.[91] The HOMO

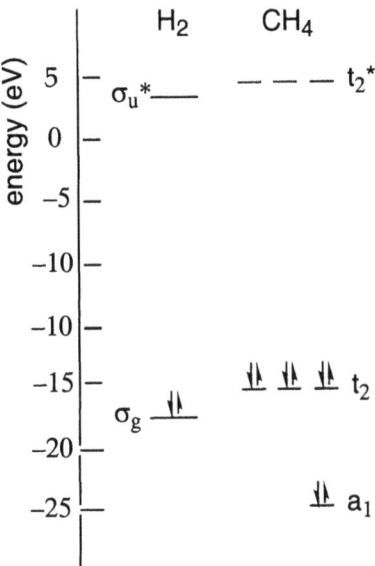

Figure 12.7. Comparison of the frontier orbital energies of H_2 and CH_4. Reprinted with permission from Saillard and Huffmann.[91] Copyright 1984 American Chemical Society.

of CH_4 is a t_2 C–H bonding orbital (12.54 eV versus 15.45 eV for H–H in H_2) while the LUMO is the antibonding counterpart slightly higher in energy than H_2 σ^* (Figure 12.7). For M–H_2, η^2 bonding is clearly preferable to take advantage of BD, but for CH_4 this is disfavored both sterically and electronically. Thus the C–H bond approaches M initially η^1 and then in tilted fashion as in Figure 12.2 to allow some BD. Si–H bonds also bind asymmetrically but the BD is far stronger because of better overlap with lower energy Si–H σ^*. The calculated energy for CH_4 binding η^1 to $Cr(CO)_5$ is only 6 kcal/mol, which is actually several kcal/mol higher than for H_2 coordination.[91] However, in general η^2-H_2 (and η^2-SiH_4) binding is worth about 20 kcal/mol, whereas η^2-CH_4 is half this even on more favorable fragments such as CpRh(CO).

Ziegler employed DFT methods to study H_2 and CH_4 binding to CpML (M = Rh, Ir; L = CO, PH_3) and $M(CO)_4$ (M = Ru, Os) fragments as intermediates in OA.[92] The bonding of H_2 in the transition state shows elongated H–H bonds (1.0–1.1 Å) despite the presence of electron-withdrawing CO ligands (see Chapter 4). CH_4 initially binds η^1 then goes to η^2 (with substantially elongated C–H), before undergoing OA [Eq. (12.17)]. The binding energy of CH_4 adducts is strongest for CpIrL (12–14 kcal/mol), which rapidly oxidatively adds alkanes. However, interaction with $M(CO)_4$ is extremely weak, less than 1 kcal/mol.

Other calculations[93–97] focus mainly on adducts of CH_4 and other small molecules with CpML and unsaturated Pd/Pt fragments because of their extensive use in experimental studies of alkane activation (see below). The binding and activation is similar to Eq. (12.17), and η^2-C,H coordination of CH_4 is shown by Hall for CpRh(CO)(CH_4) and [CpIr(PH_3)(CH_3)(CH_4)]$^+$ and by Smith for CpM(NO)(CH_2)(CH_4) (M = Mo,W).[93,97] The energy profiles are in agreement with experimental energies for, e.g., OA of alkanes by [CpRh(CO)] in the gas phase.[98] The calculated barrier for OA of CH_4 is much lower than for lone-pair donors such as H_2O and NH_3, but is not zero for SiH_4 and H_2 where the σ adducts do not even

(12.17)

exist as true intermediates.[94] The effect of metal variation on adduct stability and ease of OA is not clearcut and varies with computational methodology. Calculations encompass group 7–9 metals, and potential energy profiles are summarized in Figure 12.8 for CH_4 addition to 16e CpM(CO).[95] The stabilization energies (14–18 kcal/mol) of the η^2-C,H "precursor" σ complexes relative to reactants are remarkably similar for all M, although the potential well shapes and transition state energies vary markedly. The well depth is sufficient to allow isolation of the σ complex, but OA to alkyl hydrides is too rapid. Third-row metals (Ir, Os) facilitate OA while the barrier to the reverse process, reductive elimination, is lower for electrophilic centers (Pd^+, Pt^+) and the second-row metal (Rh). Ancillary ligand effects also play a role, as will be discussed below.

Calculations show that η^2-H,H coordination of CH_4 to 14e $RhCl(PH_3)_2$ is favored and stabilized by BD. This and 16e $RhCl(CO)(PH_3)_2$ fragments give higher

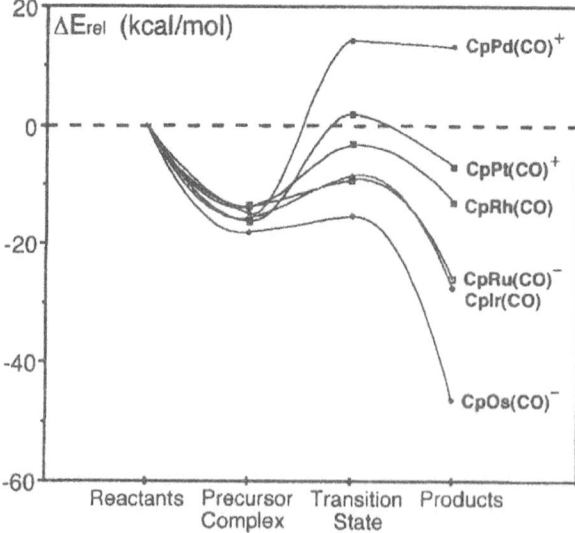

Figure 12.8. Potential energy profiles of the reaction of CpM(CO) with CH_4. Reprinted with permission from Su and Chu.[95] Copyright 1997 American Chemical Society.

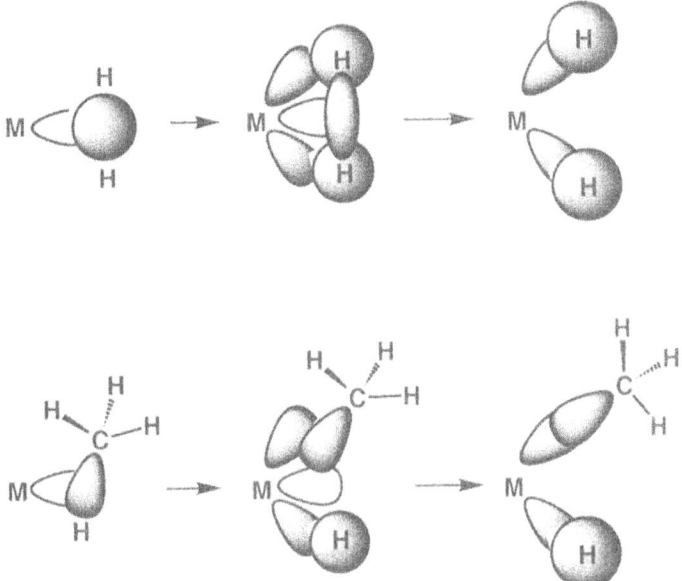

Figure 12.9. Comparison of OA of CH_4 versus H_2.

barriers, with those for $CH_4 > H_2$.[99–101] OA of CH_4 to the latter complex is actually endothermic by 20 kcal/mol, whereas it is 2 kcal/mol exothermic for H_2.[101] Thus despite the similarity of H–H and C–H bond strengths, the former is cleaved much more easily, partly because of the stronger Rh–H bond (50 kcal/mol) compared to the Rh–Me bond (38 kcal/mol) in, e.g., $RhClH(X)(CO)(PH_3)_2$ for X = H versus Me.[101] Also the directionality of the nascent Rh–X bonds is more favorable for the spherically symmetric H orbitals, so the Rh–H bond can start to form at the same time as H–H weakens, and most of the energy lost in bond-breaking is regained in the formation of new bonds (Figure 12.9). However, the CH_3 group cannot form bonds in several directions as easily, and C–H has to break before the new Rh–CH_3 bond forms, leading to a higher barrier. This is a general feature of H–H versus C–H activation, first noted by Siegbahn in 1983.[102] Addition of SiH_4 to $RhCl(PH_3)_2$ is also more facile than CH_4 because $RhCl(PH_3)_2(SiH_4)$ is only a transition state and not a potential minimum as for the CH_4 complex.[100]

Ab initio calculations address several questions about structures and bonding energies of alkane versus fluoroalkane interactions with 16e fragments, specifically $W(CO)_5$. Particularly perplexing is the geometry of CH_4 and –CH_3 and –CH_2 group coordination and whether C–F bonds are better ligands than C–H bonds. In general, the computations of Zaric and Hall show that η^2-C–H structures (**22**) are the most stable for CH_4, ethane, and propane complexes, while the various fluoromethanes CH_nF_{4-n} are coordinated through η^2-C–F bonds with longer d_{WC} (**23**).[87] This is the first computation to show that in partially fluorinated alkanes the C–F bond coordinates rather than the C–H bond and it is supported by Su and Chu for reaction of CH_3F with 14e systems $MX(PH_3)_2$ (M = Rh, Ir; X = H, Me, Cl).[103] Here OA occurs via an η^2-C–F intermediate with $d_{MC} = 3.4$ Å and binding energies up to 13 kcal/mol for IrCl(PH$_3$)$_2$(CH$_3$F). Activation of C–halogen bonds will be discussed further in Section 13.3.1.

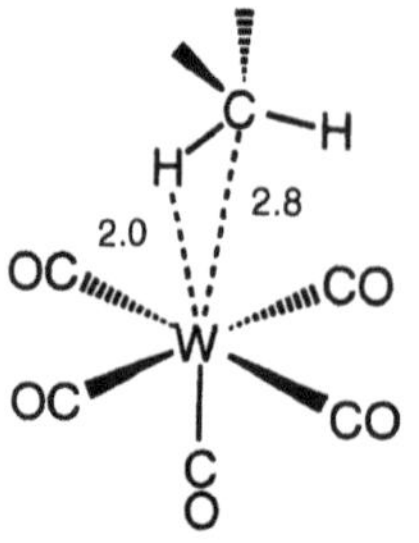

(22) 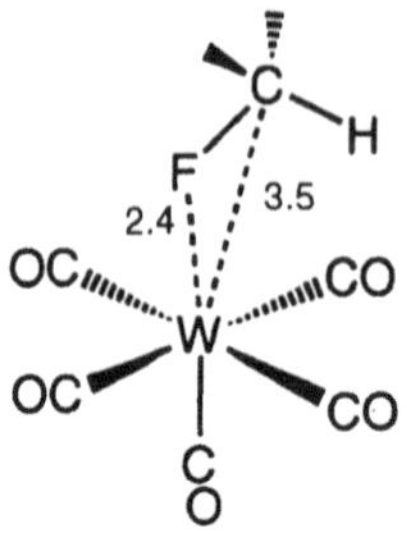   (23)

For all W(CO)$_5$(alkane) complexes, the η^2-C–H structure is most stable. All other possible structures with different H positions on the same C are so close in energy and the barriers are so small that complexes will be quite fluxional, as for agostic interactions, although the exchange of C atoms has a significantly higher barrier. Optimized geometrical parameters and calculated versus experimental bonding energies (see Chapter 7) are given in Table 12.5 for W(CO)$_5$(η^2-C–H-alkane) and W(CO)$_5$(CH$_n$F$_{4-n}$) complexes. For propane, coordination of the secondary C is more stable than the primary carbons by about 1 kcal/mol. The calculations support the experimental trend that bonding energies increase with increasing alkane size (see Section 12.4.2), much of which is attributed to increasing the substitution on C. Thus, propane bound at a primary C–H is only 0.26 kcal/mol more stable than the ethane complex, but propane coordinated via the secondary C–H is 1.63 kcal/mol more stable.

Regarding fluoroalkane bonding via fluorine, bonding energies decrease with increasing F substitution because of decreasing charge on F. Thus CF$_4$ bonds very weakly and complexation to W(CO)$_5$ is not observed experimentally in the gas phase.[104] Because of this, it had been believed that CH$_3$F coordinates via a C–H bond rather than through fluorine. The calculations of Zaric and Hall inspired

Table 12.5. Geometrical Parameters (Å) and Bonding Energies (kcal/mol) for W(CO)$_5$(L)

L	d_{WC}	d_{WH}, d_{WF}[a]	d_{CH}, d_{CF}[a]	calcd ΔH	exptl ΔH
CH$_4$	2.804	2.096	1.147	6.39	<5
CH$_3$CH$_3$	2.814	2.047	1.153	8.53	7.4 ± 2
(CH$_3$)$_2$CH$_2$	2.815	2.012	1.157		
CH$_3$CH$_2$CH$_3$	2.805	2.047	1.153	10.16	8.1 ± 2
CH$_3$F	3.509	2.349	1.508	11.28	11.2 ± 2
CH$_2$F$_2$	3.614	2.405	1.478	7.73	>5
CHF$_3$	3.508	2.400	1.456	4.73	<5
CF$_4$	3.592	2.469	1.432	1.85	<5

[a]For fluorine-bound structure for fluoroalkanes.

experimental resolution of this problem, and Dobson et al. demonstrated by flash photolysis that 1-fluorohexane binds to $W(CO)_5$ via F.[105] From both computations and experiments, the trends in the relative bonding energies (i.e., the increase with alkane size, secondary C–H > primary, $CH_3F > CH_4$, and $CH_3F > CF_4$) indicate that both $R_3CH \cdots W(CO)_5$ and $R_3CF \cdots W(CO)_5$ interactions contain a major electrostatic component and BD from W is too weak to be a factor.

Methane is the most weakly binding alkane yet is the most important for conversion chemistry. A calculational technique (reduced variational space analysis) used by Cundari indicates that three energetic components control binding of CH_4: Coulomb and exchange repulsion, polarization of CH_4, and $CH_4 \to M$ charge transfer.[106] The latter two favor binding, and trends in bond strengths as a function of M, L, and charge arise mainly from changes in the amount of charge transfer. Not surprisingly, cationic systems modeled by fragments such as $[Cl_2Ta = X]^+$ (X = O, NH, CH_2) clearly favor CH_4 coordination over neutral counterparts. Current efforts

$$L_nM^+ \leftarrow \cdots \overset{H}{\underset{C}{|}} \gg L_nM \leftarrow \cdots \overset{H}{\underset{C}{|}}$$

to activate CH_4 focus on highly electrophilic M such as $[Pt^{II}L_4]^+$, which will be discussed below. As shown in Chapter 4, it is significant that *the binding energies of CH_4 to naked metal cations $[M]^+$ are greater than those for H_2*. The high electrophilicity of cations promotes electron donation from the σ bond to M, and $M-H_2$ complexes can presumably be stabilized by strong σ donation with less BD equally as well as strong BD and less σ donation.

Another important consideration is *heterolytic cleavage* of the C–H bond, which does not change the oxidation state of M, versus *homolytic cleavage* (OA), which formally increases it by 2.

Homolytic: ML_n + R–H → $M(R)(H)L_n$

Heterolytic: MXL_n + R–H → $M(R)L_n$ + HX (X = anionic ligand)

Calculations by Sakaki[97] address this problem for both benzene and methane and show that $M(\eta^2-O_2CH)_2$ (M = Pd, Pt) favors facile heterolysis of RH to $M(\eta^2-O_2CH)(R)(\eta^1-HCOOH)$ while $M(PH_3)_2$ gives OA. The heterolysis is more exothermic because the O–H bond formed is much stronger than the M–H bond. Heterolytic cleavage may be the critical step in CH_4 conversions on electrophilic metal centers (see Section 12.4.3). The C–H bond in benzene is more easily activated than that in CH_4 because of relative bond energies and the bonding interaction of benzene π and π* orbitals with M d orbitals.

The ligand set has a big effect on alkane binding, and DFT calculations[100] show that the binding energy of CH_4 to 16e $Os(CO)_4$ is only about 0.5 kcal/mol (in contrast H_2 undergoes OA, see Chapter 4). However, studies of two isomers of 14e $OsCl_2(PH_3)_2$ show adduct formation with both H_2 and CH_4.[107] H_2 addition to the square pyramidal isomer of $OsCl_2(PH_3)_2$ gives an elongated H_2 adduct (**24**) while OA occurs for the tetrahedral isomer to give a dihydride (**25**) similar to known[108] $OsH_2Cl_2(P^iPr_3)_2$. For CH_4 addition, $\eta^2\sigma$ complexes form for both isomers, but the bonding energy is substantially higher for **27** (11.4 kcal/mol) than **26** (4 kcal/mol). It is interesting that H_2 and CH_4 can both bind simultaneously as σ ligands to $OsCl_2(PH_3)_2$ in either cis or trans fashion. The most stable CH_4 binding is for

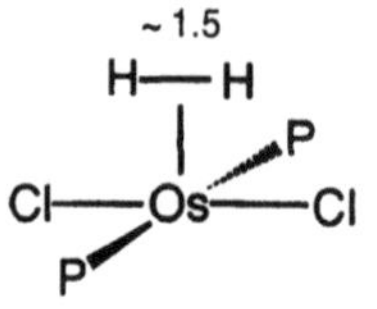

structures with weakly bonded cis H_2 ligands (d_{HH} = 0.79–0.86 Å) and vice versa. However, the most stable CH_4 complexes are not necessarily the most active in C–H activation, i.e., OA to a methyl–hydride complex requires the overall reaction to be quite exothermic to give a low activation energy. Thus a low barrier to OA is possible without a strong alkane complex intermediate and conversely a stable σ complex may exist with a high barrier to OA, as found theoretically for CH_4 activation by $RhX(L)_2$ (X = H, Cl).[109] This gives hope for isolation of an alkane complex.

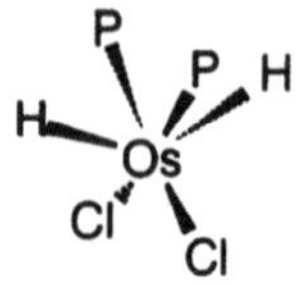

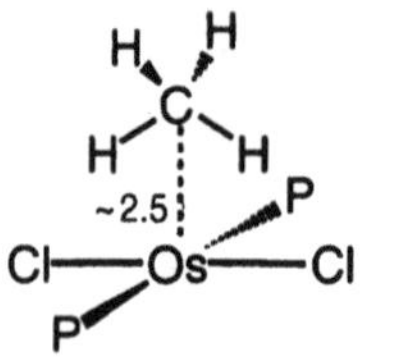

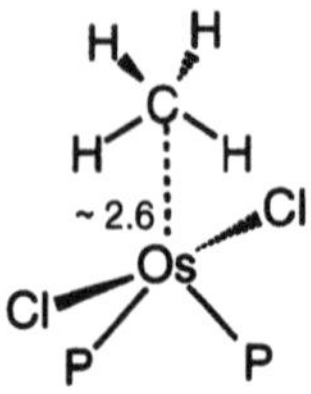

DFT calculations on OA of CH_4 to $IrX(PH_3)_2$ (X = H, Cl) show η^2-H,H coordination and binding energies near 16 kcal/mol for the σ-complex intermediate with X = Cl but only ca 7 kcal/mol for X = H.[29] This is analogous to the much stronger binding of H_2 when trans to the π-donor Cl versus trans to H. OA of CH_4 to the Rh analogue shows a lower barrier for Cl than H.[99] The computations indicate that significant BD is required for C–H rupture, and although the C–H bond is a much weaker acceptor than H–H, apparently sufficient BD can occur for more donating ligand sets. Unlike for M–H_2 coordination, the relative amounts of σ donation and BD energies have not been partitioned. Analysis of 14e T-shaped $M(L')(L)_2$ (M = Ru, Os; L' = ligand trans to incoming CH_4) shows that a better electron-donor ligand L' such as PH_3 and Cl, and a heavier metal (i.e., third-row) facilitate rapid OA of CH_4.[110] Conversely, a complex with a stronger π-accepting L' such as CO and a lighter metal favors reductive coupling of C–H. Not surprisingly these trans effects are again similar to that for H_2 activation and, for the former case, correlate with greater BD to C–H σ^*, which may not be as strong as for M–H_2 but is enough to break C–H.

Steric effects also play a role in C–H activation and explain why there can be a surprising *inverse* correlation between the initial C–H bond strength and ease of cleavage, e.g., the C–H bond strengths in CH_4, ethylene, and acetylene increase yet

CH$_4$ is the most difficult to activate. The required η^2 orientation of C–H is most easily attained by acetylene with less sterically encumbered sp bond character, followed by sp^2 and sp^3 bonds.[111] Similarly, cycloalkanes with a higher degree of sp^2

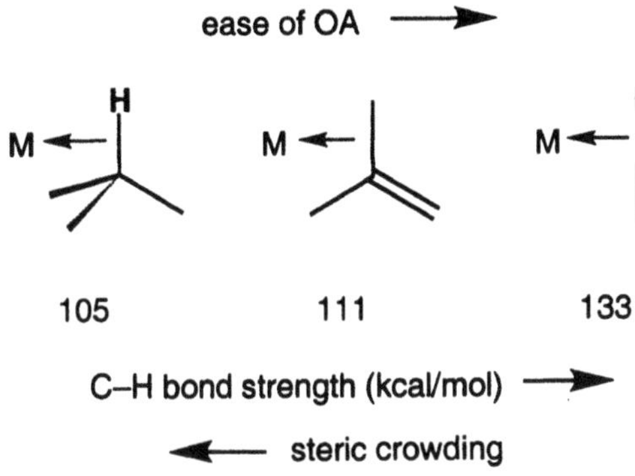

ease of OA ⟶

M⟵ (sp³, 105)   M⟵ (sp², 111)   M⟵ (sp, 133)

C–H bond strength (kcal/mol) ⟶

⟵ steric crowding

character are easier to activate, although electronic reasons are probably more important. The decrease in σ(C–H) → σ*(C–H) triplet excitation energy on going from propane to cyclopropane can also rationalize the more facile OA of the latter.

12.4. EVIDENCE FOR σ COMPLEXES IN THE OXIDATIVE ADDITION AND REDUCTIVE ELIMINATION OF C–H BONDS

12.4.1. Isotopic Exchange in Intermediates in RE and OA of Alkanes and Other Transformations

Although alkanes are the "noble gases of organic chemistry" in terms of reactivity, it is well known that transition metal systems oxidatively add alkanes to form stable alkylhydride complexes.[2–11,112–114] C–H bond insertion by a coordinatively unsaturated fragment has been examined using fast spectroscopic techniques in solid, liquefied, or supercritical rare gases[10,115–117] and hydrocarbon solution.[118,119] There is now substantial evidence for the intermediacy of alkane σ complexes in the activation of alkanes via OA [Eq. (12.18)] exactly as for transient H$_2$ complexes in OA of H$_2$.

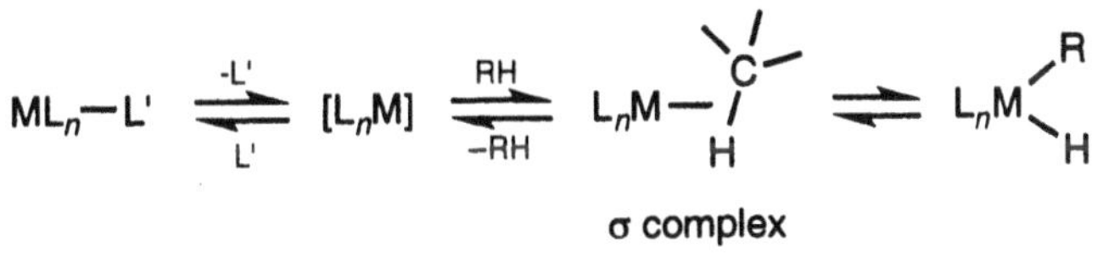

$$ML_n-L' \underset{L'}{\overset{-L'}{\rightleftharpoons}} [L_nM] \underset{-RH}{\overset{RH}{\rightleftharpoons}} L_nM\!-\!\overset{\diagdown\!\!C\!\diagup}{\underset{H}{|}} \rightleftharpoons L_nM\overset{R}{\underset{H}{\diagup\!\!\diagdown}} \quad (12.18)$$

σ complex

The coordinated C–H dissociates and recoordinates many times before it cleaves, as seen for H$_2$ complexes (see Section 7.4), but on a much faster timescale (picoseconds rather than milliseconds). Crabtree's reaction trajectory in Figure 12.2 can be used to model the approach of the free alkane. Initially, and as for H–H approach, the H end of C–H is directed toward M with the C end pointed away and is considered a transient structure, albeit fleeting. Significant C–H cleavage then occurs in a triangular transition state structure in which the C end approaches M more closely. Unlike for H_2 activation, much of the current understanding of C–H activation by complexes in solution comes from mechanistic investigations of alkane *reductive elimination* (RE) from alkyl hydride complexes, the microscopic reverse of the OA in Eq. (12.18). Hence, as emphasized by Jones[120] and other researchers,[121–128] there is a very effective synergism in these studies: the mechanism of OA in Eq. (12.18) must involve the same intermediates and transition states as in the RE reaction.

Several trails of evidence are left by transient alkane σ complexes, including a common observation that hydrogen exchange takes place between the hydride and alkyl ligands of isotopically labeled alkyl hydride prior to RE.[113,121–128] Such intramolecular rearrangement mediated by a σ complex with an appreciable lifetime as in Eq. (12.18) is observed for instance in Cp*Rh(PMe$_3$)(^{13}CH$_2$CH$_2$D) by Periana and Bergman [Eq. (12.19)].[122] Migration and insertion of [Cp*Rh(PMe$_3$)] occurs

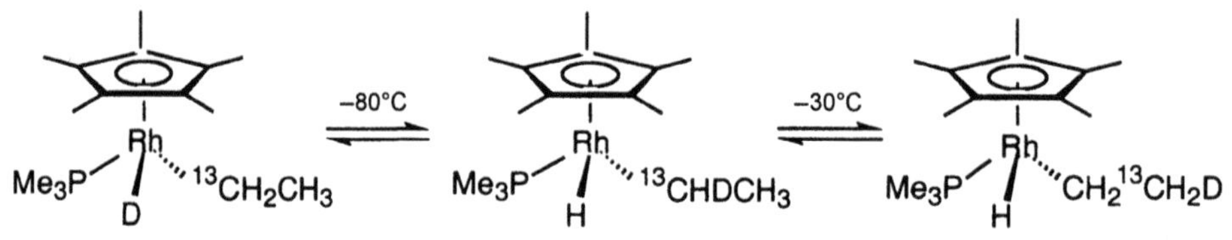

$$(12.19)$$

most rapidly into the α position of the ethyl group, followed by migration to the β position. The deuterium label remains attached to the ^{13}C throughout this process, which is competitive with RE of ^{13}C, D–ethane. The intramolecular nature of the rearrangements here and in other systems[123] is confirmed by mass spectroscopy of the RE products from crossover experiments. Exchange of H for D in a dimethyl cyclopropyl group is linked to interconversion of diasteromers in Eq. (12.20).

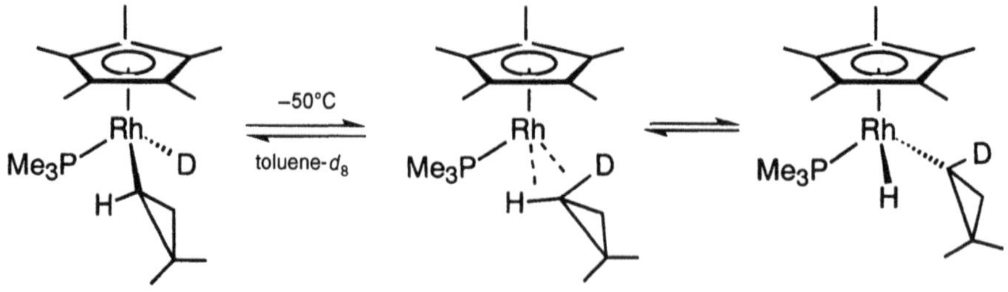

$$(12.20)$$

With Ir rather than Rh the diastereomers can be isolated and the kinetics of exchange studied independently, providing further evidence for intramolecular H/D exchange with σ complexes as intermediates.[127] Other similar exchanging systems include Cp$_2$W(Me)(H)[123] and its Cp*[124] and cationic Re[125] congeners. DFT calculations on the former support this mechanism and, for reasons of symmetry, an

η^2-H,H structure for bound CH_4 is found at the midpoint of exchange (~19 kcal/mol above the ground state).[128] In all experimental cases, OA of added CH_4 does not occur, ruling out scrambling by an elimination/insertion mechanism and confirming the intramolecular character. Other studies include RE of CH_4 from an analogue (28) with a Cn ligand in the presence of isotopically labeled CH_4 or

[Structure of complex (28): a cationic Rh complex with a tridentate N,N,N ligand (Cn), R_3P, H, and CH_3 ligands]

(28)

benzene.[113] Here the alkyl hydride complexes are unusually stable, possibly because of the positive charge and the hard donor Cn ligand. Thermolysis of 28 at 50°C is required to induce unimolecular elimination of CH_4 or CD_4.

Gross and Girolami found the first example of a methyl–hydride complex in which the hydrogens of the alkyl and hydride ligands exchange at a rate sufficient to be dynamic on the NMR timescale.[129] In Cp*Os(dmpm)$(CH_3)(H)^+$ the exchange is fast enough at $-100°C$ to give NMR signal-broadening, but can be halted at $-120°C$ to give distinct Me and H resonances. This suggests that an alkane complex Cp*Os(dmpm)$(CH_4)^+$ is formed reversibly from the Me/H complex over 100 times per second at $-100°C$. Spin saturation transfer experiments provide definitive evidence for exchange, wherein $\Delta H^\ddagger$ is 7.1 ± 0.9 kcal/mol ($\Delta G^\ddagger = 8.1$ kcal/mol at $-100°C$). DFT calculations by Martin on a Cp model system depict an energy surface for RE in excellent accord with experiment (Scheme 1).[130] The most stable point on the surface is the Me/H complex (1), which lies about 12 kcal/mol below the products of elimination, CpOs(dmpm)$^+$ + CH_4. A stable η^1-CH_4 complex (3) lies only some 6 kcal/mol above 1 and has $d_{OsH} = 1.92$ Å, which is much shorter than d_{OsC} (2.70 Å). These minima are connected via a transition state (2) that most closely resembles 1. The theoretical barrier for formation of the CH_4 complex from 1, 9.2 kcal/mol, agrees well with the experimental enthalpy of activation for the exchange process. If the rate-limiting step for H-atom scrambling is determined by the [Os]H*$(CH_3) \rightarrow$ [Os](CH_3H^*) conversion, then there must exist a lower-lying path from the CH_4 complex that exchanges the H bound to Os with one of the H atoms of the Me. This path is shown in Scheme 1, where the CH_4 complex (3) goes through a transition state with η^2-H,H structure (4) to the analogous CH_4 complex (3′) with a barrier calculated to be only 2 kcal/mol. There is only about a 1 kcal/mol difference in the relative energies for Cp versus Cp* complexes in Scheme 1. The calculated k_H/k_D for the [Os]H*$(CH_3) \rightarrow$ [Os](CH_3H^*) step only is 1.6, showing that formation of the CH_4 complex is the bottleneck in the H exchange. For full alkane RE from alkyl hydride complexes, *inverse kinetic isotope effects* ($k_H/k_D < 1$) are frequently observed that support the intermediacy of alkane complexes in these systems, as discussed in Section 7.5.2. Systems showing intramolecular isotope exchange have experimental $k_H/k_D = 0.2$–0.8 for alkane RE.

Most of the complexes discussed above contain electron-donating Cp, Cp*, or phosphine ligands, but the mechanism of alkane RE from an *octahedral* alkylhydrido Rh complex containing the Tp' ligand and a π-acid ligand has been studied.[120] The

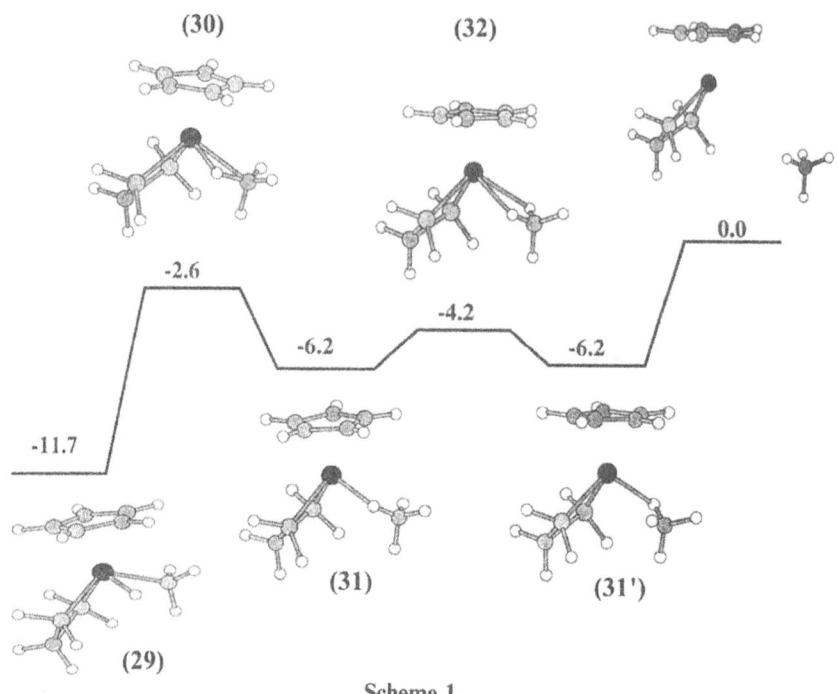

Scheme 1

electronic and steric demands of Tp' and Cp* ligands are usually assumed to be analogous. However, RE of benzene from Tp'Rh(L)(Ph)H in the presence of CNR is associative in CNR, and Tp' goes from tridentate to bidentate, while a Cp* analogue shows concerted benzene loss and no Cp* hapticity changes.[131] Two sets of experiments provide indirect evidence for involvement of alkane complexes in OA/RE reactions of Tp'Rh(L)(R)H (L = CNCH$_2$CMe$_3$). First, the methyl–deuteride complex Tp'Rh(L)(CH$_3$)D rearranges to Tp'Rh(L)(CH$_2$D)H prior to loss of CH$_3$D and *no other isotopomer*, certifying that the exchange is entirely intramolecular [Eq. (12.21)]. Similarly, from IR and NMR evidence, Tp'Rh(L)(CD$_3$)H

$$\text{Rh}\genfrac{}{}{0pt}{}{\diagup \text{D}}{\diagdown \text{CH}_3} \rightleftharpoons \text{Rh}(\eta^2\text{-CH}_3\text{D}) \rightleftharpoons \text{Rh}\genfrac{}{}{0pt}{}{\diagup \text{H}}{\diagdown \text{CH}_2\text{D}} \quad (12.21)$$

$$\text{Rh}\genfrac{}{}{0pt}{}{\diagup \text{H}}{\diagdown \text{CD}_3} \rightleftharpoons \text{Rh}(\eta^2\text{-CD}_3\text{H}) \rightleftharpoons \text{Rh}\genfrac{}{}{0pt}{}{\diagup \text{D}}{\diagdown \text{CD}_2\text{H}} \quad (12.22)$$

rearranges to Tp'Rh(L)(CD$_2$H)D before eliminating CD$_3$H. The rates of hydrogen exchange between the hydride and Me ligands of these species are competitive with RE of CH$_4$, and k_H/k_D is inverse (0.62) for RE in the fully protio versus deutero system. Secondly, the rate of CH$_4$ elimination from Tp'Rh(L)(CH$_3$)H in benzene/perfluorobenzene mixtures depends on benzene concentration, hence there is an associative component to RE of CH$_4$. All these results strongly implicate intermediacy of a σ complex prior to RE of CH$_4$.

For earlier metals, reaction of anionic [L$_3$WH]K with primary alkyl halides provides evidence of alkane complexation (L = ^{t}Bu$_3$SiN=).[132] Exposure to CH$_3$I gives L$_2$(LH)WCH$_3$, but use of CD$_3$I produces two isotopomers, L$_2$(LH)WCD$_3$ and L$_2$(LD)WCHD$_2$, consistent with intermediacy of the σ complex L$_3$W(CHD$_3$). ^{2}H NMR analysis of the CHD$_2$/CD$_3$ product ratio gives k_H/k_D = 9.6(6), a plausible value for partitioning from an L$_3$W(CHD$_3$) intermediate. There is a propensity for early metal imido complexes to selectively activate sp^2 C-H bonds over sp^3 bonds. Cyclopropylmethyl bromide reacts with [L$_3$WH]K to produce the trans methylcyclopropyl derivative (L)$_2$(LH)W[*trans*-(*c*-C$_3$H$_4$)Me] and a small amount of ring-opened product (L)$_2$(LH)W(*trans*-CH=CHCH$_2$CH$_3$), while 5-hexenyl bromide affords 1-*trans*-CH=CH(CH$_2$)$_3$CH$_3$. Although the results are also consistent with a solvent cage comprised of [L$_3$W] and RH, calculations strongly support alkane adducts L$_3$W(RH). The model (HN=)$_3$W(CH$_4$) shows ΔH_{bind} = -15.6 kcal/mol and ΔG_{bind} = -8.4 kcal/mol at 25°C.[133]

12.4.2. Direct Spectroscopic Evidence for Alkane Complexes as Intermediates in OA of Alkanes

As for studies of unstable H$_2$ complexes such as M(CO)$_5$(H$_2$), liquefied rare gas solvents are also highly desirable for the above types of studies. These solvents are inert and transparent to IR, and can be used over a broad temperature range, especially at very low temperatures if needed to stabilize reactive intermediates and slow down reactions. Although noble gases can bind very weakly to unsaturated fragments (see Chapter 7), unlike hydrocarbon or fluorocarbon solvents, they do not have bonds that can be broken and cause interference. Bergman in particular has made exquisite use of this medium to obtain evidence for the intermediacy of alkane complexes in C-H activation using ultrafast-timescale techniques such as TRIR.[2,134] Photolysis of Cp*Rh(CO)$_2$ in liquid Xe or Kr gives the highly reactive Cp*Rh(CO)(Xe) and Cp*Rh(CO)(Kr) fragments characterized by their ν_{CO} modes. In the presence of a small amount of cyclohexane, these species are rapidly transformed into the alkyl hydride, Cp*Rh(CO)(Cy)H [Eqs. (12.23)–(12.25)]:

$$\text{Cp*Rh(CO)}_2 \xrightarrow[\text{-CO}]{h\nu,\text{ liq Kr}} \text{Cp*Rh(CO)(Kr)} \qquad (12.23)$$

$$\text{Cp*Rh(CO)(Kr)} + \text{CyH} \xrightarrow{K_{eq}} \text{Cp*Rh(CO)(CyH)} + \text{Kr} \qquad (12.24)$$

$$\text{Cp*Rh(CO)(CyH)} \xrightarrow{K_2} \text{Cp*Rh(CO)(Cy)(H)} \qquad (12.25)$$

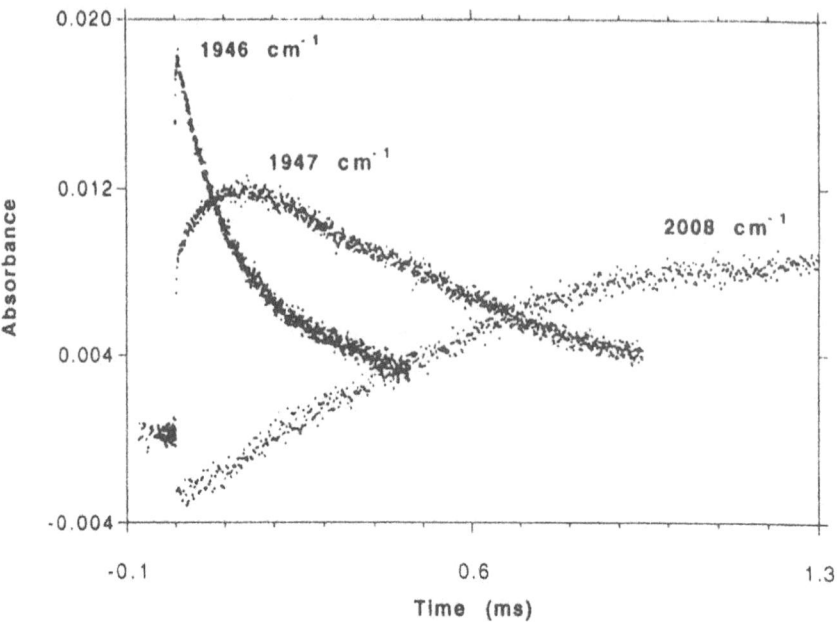

Figure 12.10. The time-dependent absorbances observed when Cp*Rh(CO)$_2$ is photolyzed in liquid Kr at 165 K in the presence of 3.8 mM neopentane-d_{12}. The transient at 1946 cm^{-1} shows biexponential behavior with a fast decay followed by a slower one. A second transient absorption which grows in and then decays is observed at 1947 cm^{-1} and assigned to the σ complex. The signal growth at 2008 cm^{-1} is due to the Cp*Rh(CO)(C$_5$D$_{11}$)(D) product. Reprinted with permission from Bengali et al.[84] Copyright 1994 American Chemical Society.

The kinetics of formation in liquid Kr are inconsistent with a dissociative mechanism but are rationalized by an equilibrium between Kr and cyclohexane complexes prior to OA. ν_{CO} for Cp*Rh(CO)(CyH) is unobservable, probably because of overlap with that of Cp*Rh(CO)(Kr).[135] Use of cyclohexane-d_{12} to examine isotopic substitution effects shows an inverse EIE (K_H/K_D = ca 0.1) for the interconversion of Kr and alkane adducts, i.e., C$_6$D$_{12}$ binds more strongly to M than C$_6$H$_{12}$.[116] Although initially contrary to expectations based on simple zero-point energy arguments, this is explainable by the increase in the number of vibrational modes for bound alkane versus free alkane (see Section 7.5.2). As observed for the formation of H$_2$ complexes, the inverse effect is perfectly consistent with formation of an alkane σ complex. The first-row analogue Cp*Co(CO)$_2$ does not display the behavior in Eqs. (12.23)–(12.25), possibly because calculations show a triplet ground state for Cp*Co(CO).[83]

Use of neopentane-d_{12} in Eqs. (12.23)–(12.25) leads to the crucial observation of two distinct transient absorption bands for ν_{CO}: 1946 cm^{-1} for Cp*Rh(CO)(Kr) and a very close band at 1947 cm^{-1} (Figure 12.10) believed to originate from the σ complex, Cp*Rh(CO)[C(CD$_3$)$_4$].[117] Fortuitously, only one signal could be resolved for the protio species, again demonstrating the value of deuterium substitution. Although structural inferences such as the number of C–D bond interactions cannot be made from one ν_{CO}, the similarity in position to the Kr complex is consistent with

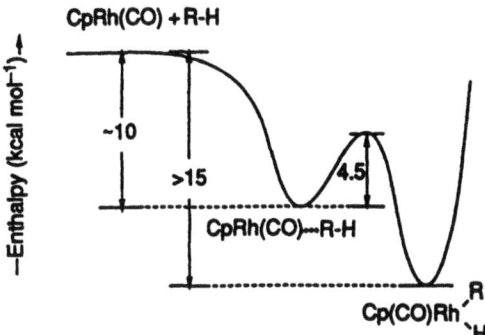

Figure 12.11. Reaction profile for the gas-phase reaction of CpRh(CO) with alkanes to give alkyl hydrides. Estimated energy differences are based on both solution and gas-phase determinations.

weak L → M donation and poor M → L BD, i.e., the electronic interactions in M(C–H) are not unlike those in M(Kr). The equilibrium between the Kr and alkane complexes and the direct conversion of the neopentane-d_{12} complex to the neopentyl deuteride (2008 cm^{-1}), could both be monitored by IR. The spectral results and complete reaction profiles obtained for each step elegantly demonstrate the intermediacy of alkane complexes in OA of alkanes.

Similar TRIR evidence for alkane coordination to CpRh(CO) in the *gas phase* also comes from Bergman and Moore's group, although no signal for the alkane complex is seen.[98] Reaction with various alkanes occurs at every collision, implying a minimal barrier to C–H activation consistent with formation of alkane complexes before C–H activation. Conversion of the alkane complexes to alkyl hydride complexes is proposed to be exothermic, with an activation energy of about 4.5 kcal/mol (Figure 12.11). Studies of CH$_4$ activation on related CpM(PMe$_3$) (M = Rh, Ir) in matrices by FTIR and UV/vis also offer indirect evidence for intermediacy of CH$_4$ complexes.[136]

Faster-timescale studies (microsecond to femtosecond scale) probe the mechanism of C–H activation on, e.g., CpRh(CO) in *alkane solution at room temperature*.[137] A similar Tp*Rh(CO) fragment is generated photolytically from Tp*Rh(CO)$_2$ in less than 100 fs and reacts with alkanes to give observable intermediates. The 16e fragment is quickly solvated by alkane in a barrierless reaction and then cools vibrationally in 20 ps. The replacement of one CO by alkane increases the electron density at Rh, which weakens the bonds between Rh and the σ-donating Tp* pyrazole ligands and breaks a Rh–N bond (η^3- to η^2-Tp*). The loss of the electrons formerly donated to Rh by the now detached arm of the Tp* ligand reduces the electron density at Rh, shifts ν_{CO} to higher frequency (1990 cm^{-1}), and increases the electronic unsaturation of the alkane complex, driving it toward OA. The nanosecond timescale of the reaction indicates a significant thermal barrier, and for cyclohexane reactions, the height of the free-energy barrier is about 8.3 kcal/mol for going from σ complex to alkyl hydride [third step in Eq. (12.26)]. This compares to a 7.2 kcal/mol barrier calculated for the Cp* system by extrapolating the results of 183 K liquid Kr-cyclohexane experiments to room temperature. The η^2-Tp* complex is formally Ru^{III}, which reduces its electron density and provides the driving force for rechelation of the dangling pyrazole to form the final product, Tp*Rh(CO)(R)(H). Thus the steps before and after activation very effectively mold the structural and electronic environment around Rh to facilitate OA. DFT

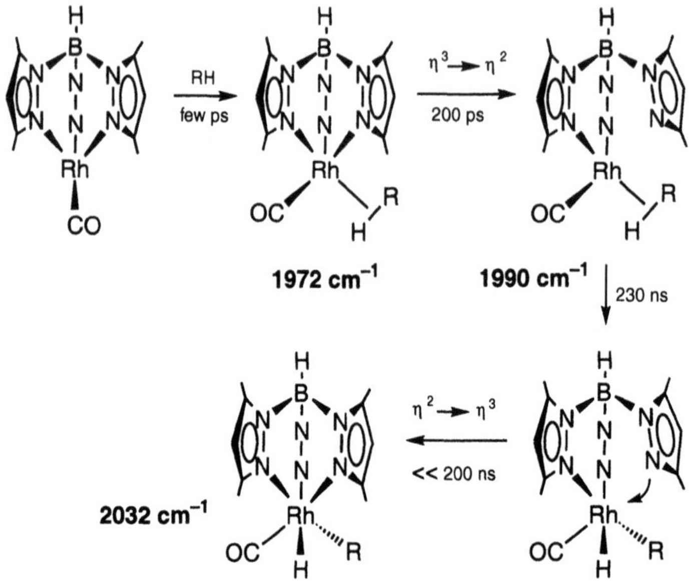

$$\text{(12.26)}$$

calculations by Zaric and Hall not only confirmed the above but also predicted the (η^3-Tp*)Rh(CO)(RH) and (η^2-Tp*)Rh(CO)(RH) intermediates.[138] The former shows very weakly bound η^1-CH$_4$ (1.7 kcal/mol) while the latter has more strongly bound η^2-C,H methane (9.4 kcal/mol).

Unlike the studies in rare gas media at low temperatures, the above neat alkane studies do not allow a comparison of the binding affinities of different alkanes because the binding step is too fleeting. More recent studies of alkane reaction with Cp*Rh(CO) in liquid Kr or Xe on the slower microsecond timescale support a strong influence of alkane size and structure on the binding affinity of the alkane to Rh.[119] In the initially formed alkane σ complex, larger alkanes interact more strongly with Rh than do smaller ones, and CH$_4$ does not appear to bind at all in competition with Kr. Ethane binds an order of magnitude less strongly than the longer-chain alkanes, and cycloalkanes coordinate better than comparably sized linear alkanes. Table 12.6 presents ΔG_{bind} relative to Kr along with polarizabilities and IPs of the alkanes. The above trends extend to photogenerated W(CO)$_5$ in the gas phase (Table 7.2) where the much less electron-rich W does not induce OA of the C–H bonds. The binding energies increase as alkane size increases, and again CH$_4$ does not bind. This trend is attributed to improved orbital interaction between the LUMO on W and the C–H molecular orbital on the alkane. As alkane size increases, so does the energy of the C–H σ MO as determined by photoelectron spectroscopy, and as the energy gap shrinks between this orbital and the LUMO, M–alkane interaction increases (Table 12.6). BD from M is weak and of little

Table 12.6. Rare Gas and Alkane Ionization Potentials and Polarizabilities and Free Energies for the Kr–Alkane Complex Equilibria at $-90°C$

Substrate	First IP, eV	Polarizability, Å	ΔG_{183K}, kcal/mol
Kr	14.00	2.74	
Xe	12.13	4.46	
Methane	12.75	2.61	>0
Ethane	11.56	4.5	−0.9
Propane	10.9	6.36	−1.8
Hexane	9.97	11.9	−2.1
Octane	9.71	15.6	−2.3
Cyclopentane	10.33	9.11	−2.4
Cyclohexane	9.82	10.96	−2.6
Cycloheptane	9.96	12.1	−3.5
Cyclooctane	9.75	13.9	−3.5

consequence in M(CH) interactions, so substituent effects at C are less relevant than in silane binding.

M–alkane interactions are thus fundamentally "soft" in the parlance of hard/soft acid–base principles. The low dipole moments, low basicities, and high IP associated with alkanes and the larger noble gases point to the importance of an induced distortion of electron clouds in the bonding of these weak ligands to soft M. Binding affinities should then correlate with increasing polarizability of the ligand, as shown in Table 12.6. Significantly, Kr and CH_4 have comparable polarizabilities, as do Xe, ethane, and propane. Thus CH_4 is ineffective in competing with the equally polarizable Kr that is present in 10- to 1000-fold excess, and similarly ethane and propane compete poorly with excess Xe in this solvent. However, the larger, more polarizable alkanes interact more strongly with Rh than do rare gases, and the linear alkane binding affinities correlate well with polarizability. Despite comparable polarizabilities, *cyclic alkanes bond more strongly than their linear counterparts* (see Section 12.3.2), and a large discontinuity occurs on going from cyclohexane to cycloheptane.

Rates of C–H OA follow the opposite trend, and reaction of the Rh–ethane complex to form Cp*Rh(CO)(ethyl)(H) is most rapid, while reaction of the cycloheptane and cyclooctane complexes is anomalously slow. It is likely that solvent effects in the noble gas media play a significant role in this and the above trends. It is important to treat C–H activation at late M as a two-step process, with potentially different factors influencing the binding step and the OA step.

12.4.3. Shilov Chemistry and Related Electrophilic Systems for Homogeneous Alkane Oxidation

Alkanes are extremely poor acids, but on analogy with H_2 binding to metal complexes, coordination to highly electrophilic M greatly enhances the acidity of the C–H bond and promotes heterolytic cleavage. Soft electrophiles such as Pt^{II} and Hg^{II} are ideal because they bind CH_4 and other alkanes transiently even in aqueous solution. Thus superacid catalysis in a conventional acid medium is possible, a very important potential route to alkane oxidation. Such metal centers can stabilize the product of cleavage in the form of an alkyl ligand for conversion to, e.g., alcohols.

$$[M^{II}]^+ \xrightarrow{CH_4} \left[M^{II}\underset{H}{-}CH_3\right]^+ \xrightarrow{B^-} M^{II}-CH_3 + BH$$
$$X \downarrow \text{oxidant}$$
$$M = Pt, Hg \quad X = Cl^-, OH^- \quad M^{II} + CH_3X \quad (12.27)$$

In oxygenase enzymes and related metal–oxo systems, intramolecular heterolytic cleavage of C–H may be a critical step facilitated by the high mobility of H^+, which can rapidly hop to a more basic site (see Section 10.3). This may also be true in the organometallic systems that will be discussed here. The reactive species responsible for alkane activation is often incompatible with oxidants (e.g., O_2, H_2O_2) required for catalytic oxidation. An important exception is the reaction of alkanes with chloroplatinum salts in aqueous solution, which is also one of the earliest examples of C–H activation. In 1969 Shilov and coworkers observed the incorporation of deuterium into alkanes in solutions of $K_2[PtCl_4]$ in D_2O/acetic acid-d_1 and subsequently reported that addition of $H_2[PtCl_6]$ to the reaction mixture generated oxidized alkane products RCl and ROH [Eq. (12.28)]:[5]

$$CH_4 + [PtCl_6]^{2-} + H_2O\ (Cl^-) \xrightarrow[H_2O, 120°C]{[PtCl_4]^{2-}} CH_3OH\ (CH_3Cl) + [PtCl_4]^{2-} + 2HCl$$
(12.28)

This system is remarkably robust, unlike most organometallic reagents which are air- and moisture-sensitive. The Pt complexes are soluble in water, and the reactions are unaffected by atmospheric O_2 or N_2 (Pt^{II} does not stably bind the latter or π acceptors such as SO_2 or H_2). It is also surprising that alkane binding is at all competitive with H_2O. H_2 can be a better ligand than H_2O (see Chapter 7), but alkane binding is much weaker, so a soft electrophile is important (water is a hard base). In regard to selectivity of bond activation, the initial product of alkane oxidation is often more reactive toward the oxidant than the alkane itself. On Pt systems, however, CH_4 is six times more reactive than the methanol product. The doubly solvated complex $PtCl_2(H_2O)_2$ has been proposed as the active catalyst, and the observed selectivity patterns reflect those of other organometallic activation systems, namely, primary > secondary > tertiary C–H. This and related chemistry has been well reviewed[5,6,10] and only aspects dealing with evidence for σ complexes from more recent studies will be discussed. Several new oxidation systems utilize electrophilic late metals such as Pt^{II} and Hg^{II} in strong acid media (e.g., CF_3CO_2H, H_2SO_4) that exhibit similarities to Shilov chemistry and represent promising approaches to alkane functionalization.[10]

In 1983 Shilov et al. proposed a mechanism for Pt-catalyzed alkane oxidation consisting of three basic transformations [Eq. (12.29)]: (a) activation of the alkane by Pt^{II} to generate an alkyl Pt^{II} intermediate, (b) two-electron oxidation of the alkyl Pt^{II} intermediate to generate an alkyl Pt^{IV} species, and (c) reductive elimination of

RX (X = Cl or OH) to give back the Pt^{II} catalyst.[5] Many features of the individual steps have only recently been identified, particularly through the work of Bercaw and Labinger, and each of the three steps can potentially proceed by at least two different pathways not discussed in detail here.[10] However, the first step determines both the

$$\text{Pt}^{II} + RH \rightleftharpoons \text{Pt}^{II}\text{-R} + H^+ \quad \rightarrow \text{Pt}^{IV} \rightarrow \text{Pt}^{II} \rightarrow \text{Pt}^{IV}(R)(Cl) \xrightarrow{H_2O} RCl/ROH + H^+ + Cl^-$$

(12.29)

rate and selectivity of the oxidation and bears close scrutiny to understand mechanistic details. It is the most interesting in terms of σ-complex participation and is the most difficult to study. The reaction stoichiometry involves electrophilic displacement of a proton on the alkane by Pt^{II}, which essentially represents heterolytic cleavage of C–H. Once again two different mechanisms have been considered: (1) deprotonation of an intermediate Pt^{II}-alkane σ complex by a base as in Eq. (12.27), or (2) OA of the C–H bond at Pt^{II} producing an alkyl(hydrido)Pt^{IV} complex that is subsequently deprotonated. Considerable evidence has been obtained supporting both paths.

Because reaction conditions (>100°C) for C–H activation do not allow detection of intermediates, most of the mechanistic information for this reaction has been obtained by studying the microscopic reverse of this reaction, where protonolysis of alkyl-Pt^{II} complexes is the key step. Working backward in the mechanism in Eq. (12.29), alkyl(hydrido)Pt^{IV} complexes have been established in a series of model complexes with stabilizing N or P donor ligands or Tp ligands.[10] Three-coordinate (η^2-Tp′)Pt^{II}(CH$_3$) reacts with a solvent alkane molecule with cleavage of C–H bonds to afford stable dialkyl(hydrido) complexes (η^3-Tp′)Pt^{IV}(CH$_3$)(R)(H).[114] Most Pt^{IV}(R)(H) complexes are generated by addition of HX acids to the corresponding Pt^{II}(R) complexes at low temperatures. Significantly, certain classes of these alkyl-hydride complexes in deuterated acidic media give multiple H/D exchange into the alkyl positions prior to alkane elimination,[10] supporting the mechanism in Scheme 2 for the protonolysis of alkyl-Pt^{II} complexes. In the presence of D$^+$, deuterium incorporation into the alkyl group can be explained by the reversibility of the steps leading up to alkane elimination, as in the systems in Section 12.4.1. The Pt–tmeda system of Bercaw and Labinger[126,139] is an excellent model in this regard and also shows multiple exchange with hydrocarbon solvents as in Scheme 3. Prolonged heating (85°C) of [Pt(CH$_3$)(NC$_5$F$_5$)(tmeda)][BAr$_f$] with benzene-d_6 gives the methane isotopomers CH$_4$, CH$_3$D, CH$_2$D$_2$, CHD$_3$, as observed by ^{1}NMR.[140,141] The possibility that exchange occurs by reversible loss and readdition of CH$_4$ is ruled out by carrying out the experiment under ^{13}CH$_4$: No methane containing both ^{13}C and D is observed. Variation of the ligand trans to CH$_3$ affects the exchange,

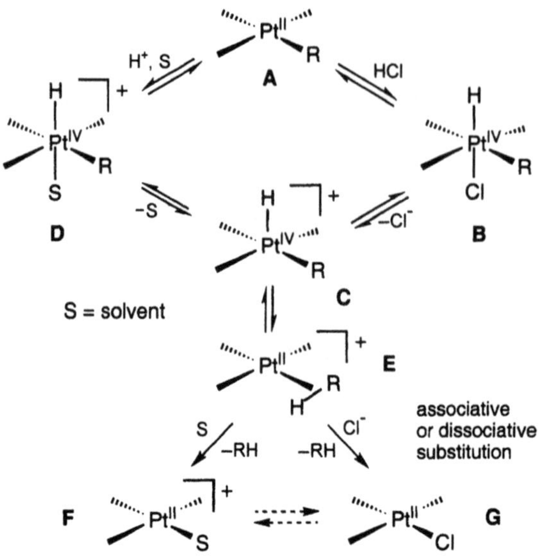

Scheme 2

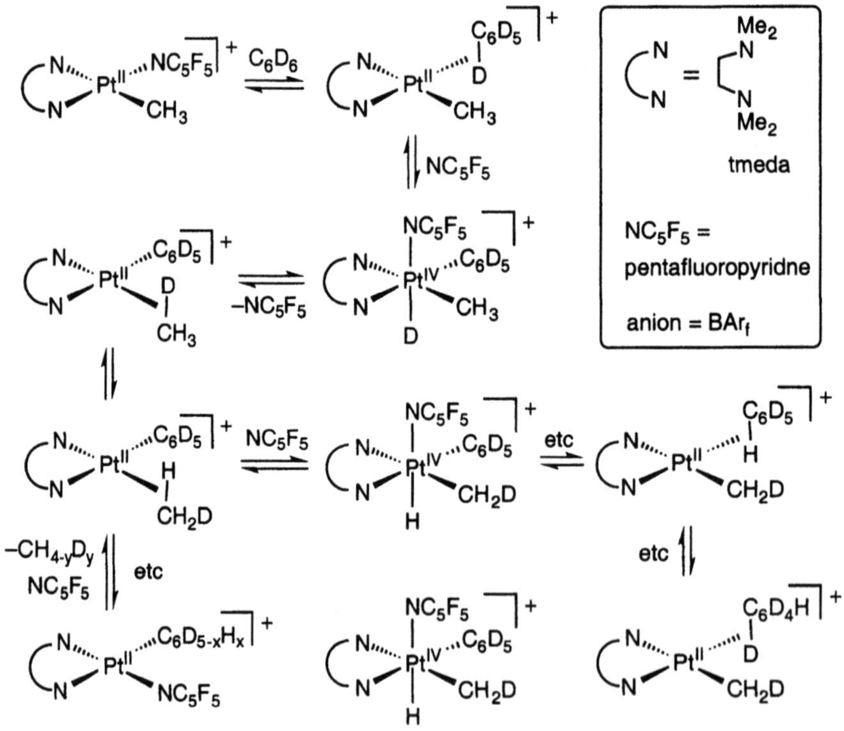

Scheme 3

which is inhibited by soft ligands such as PR_3 and facilitated by hard ligands such as Cl, which would stabilize the σ complex critical to exchange.[141] This parallels H_2 ligand stability: η^2-H_2 is rarely found trans to PR_3 but is stable trans to halides.

The above exchange results require an intermediate with significant lifetime such as a σ complex, and logical extension to the Shilov system suggests that $Pt^{IV}(R)(H)$ and alkane σ adducts are transients in alkane activation. It is important, however, that formation of the $Pt^{II}(R)$ intermediate (**A**) in Scheme 2 does not seem to arise from direct deprotonation of σ complex **E**, but rather from the $Pt^{IV}(R)(H)$ tautomer **C** in equilibrium with it after it is converted to **B** or **D**. Such a $Pt^{II}(RH) \rightleftharpoons Pt^{IV}(R)(H)$ equilibrium is analogous to well-known $M(H_2) \rightleftharpoons M(H)_2$ equilibria. H/D isotopic exchange with benzene-d_6 very similar to that in Scheme 3 is seen in closely related complexes such as **33**.[142] Although the C,H-η^2-benzene

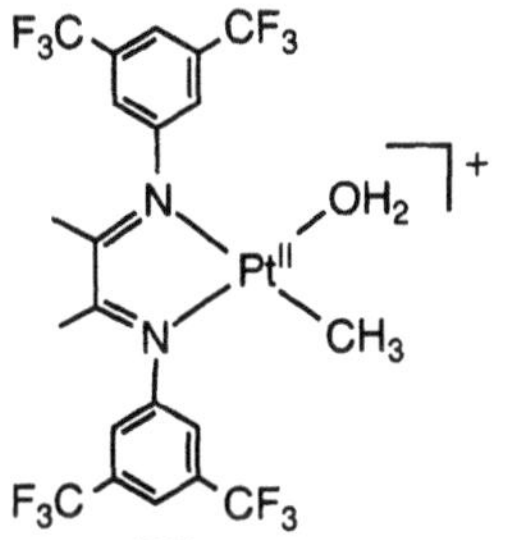

(33)

adduct in Scheme 3 is not seen, protonation of (N–N)Pt(CH_3)(C_6H_5) with triflic acid at $-70°C$ allows NMR observation of a C,C-η^2-benzene species that could lead to it. Exposure of **33** to 20–25 atm of $^{13}CH_4$ at 45°C in trifluoroethanol gives exchange of $^{13}CH_3$ for CH_3, and exchange reactions with CD_4 implicated methane σ complexes as intermediates. Significantly, recent studies indicate that methane may enter the coordination sphere of **33** by *associative substitution* of H_2O, i.e., without need for three-coordinate 14e $L_2Pt(CH_3)^+$ intermediates.[142] CH_4 elimination from an illuminatively constructed Pt^{IV} complex that contains two inequivalent CH_3 (Scheme 4) provides further detailed mechanistic information.[143] The flexible tri-

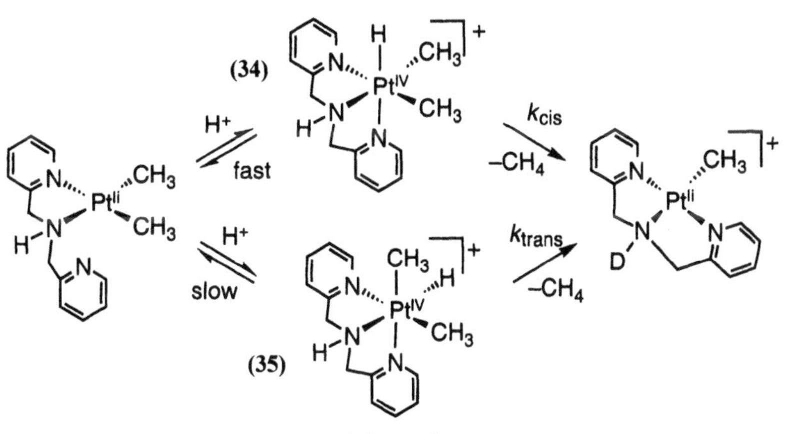

Scheme 4

dentate N-donor ligand can coordinate η^2 or η^3 in either octahedral or square planar geometries. CH_4 elimination occurs at significantly different rates for the two different isomers **34** and **35** generated on protonation ($k_{cis} = 1.1 \times 10^{-4}$ s^{-1}; $k_{trans} = 4.8 \times 10^{-6}$ s^{-1} at 25°C). Apparently the amine nitrogen cannot dissociate easily but the pyridyl residues can dechelate to form a five-coordinate species, shown in Scheme 5 for isomer **34** with deuterium labeling. The nitrogen arm trans to

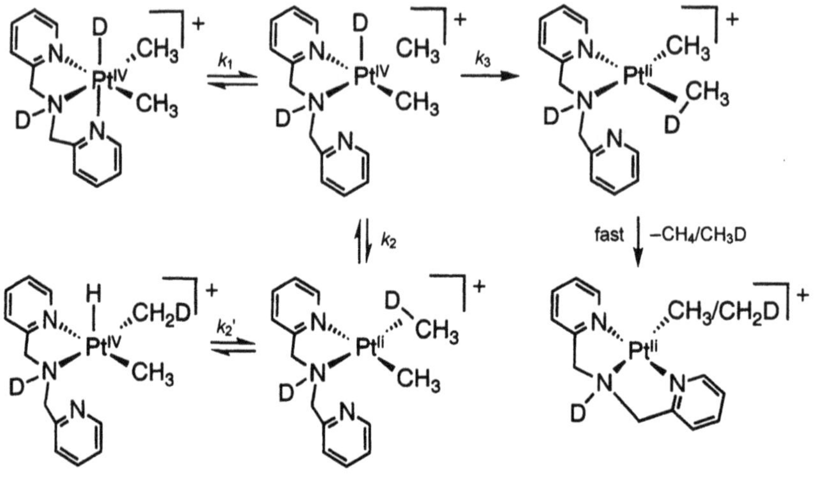

Scheme 5

deuteride readily dissociates because of the strong trans influence of the latter, which is greater than that for CH_3. This rationalizes why isomer **34** eliminates CH_4 some 20 times faster than isomer **31**, which does not have a pyridyl trans to hydride. Scheme 5 also offers a straightforward explanation for selective deuterium incorporation trans to amine. The formation of a CH_4 ligand trans to amine must be reversible since no fast and irreversible step leads to CH_4 elimination. Conversely, a similar bound CH_4 formed cis to amine can be substituted very rapidly by the uncoordinated pyridylmethyl residue. The similar orders of magnitude for the rates of scrambling and elimination confirm the hypothesis that a common mechanism operates.

Theoretical studies indicate the mechanism of C–H activation by Pt^{II} or other metals in Shilov and related systems could include σ-bond metathesis pathways, although OA paths are generally favored for Pt^{II} systems {and clearly for [CpIrL(CH$_3$)]$^+$ + RH $\rightarrow$ [CpIrL(R)]$^+$ + CH$_4$}.[142,144] Siegbahn finds that hydrogen could transfer from a Pt–CH$_4$ complex first to a cis Cl and then to H$_2$O solvent. This avoids formation of M^{IV} oxidation states, which are unstable or nonexistent for Pd and Hg systems, which also catalyze Shilov chemistry (see below) and is more like the mechanism Shilov himself proposes. DFT calculations by Hush show that the binding energy of CH_4 to Pt depends markedly on the coligands: 7.3 kcal/mol for cis-PtCl$_2$(NH$_3$)(CH$_4$) and 21.6 kcal/mol for trans-PtCl$_2$(H$_2$O)(CH$_4$), a surprisingly high value.[144] The overall reaction barriers are calculated to be 24 kcal/mol by Siegbahn and 34 kcal/mol for cis-PtCl$_2$(NH$_3$)$_2$ (Hush), close to the measured barrier of about 28 kcal/mol.

Related systems have been intensely investigated, and recent mechanistic studies by Sen and coworkers indicate that Zeise's salt, $[PtCl_3(C_2H_4)]^-$, is an intermediate in the oxidation of both ethane and ethanol.[145] This is significant because direct oxidation of ethane to ethylene glycol is a very attractive alternative to the oxidation of ethylene. However, a formidable general challenge for use of Shilov systems is replacing Pt^{IV} with an inexpensive oxidant such as oxygen. Considerable attention has been directed toward the oxidation of alkanes by electrophilic metal ions in strong acid media using other metals in their highest stable oxidation state such as Hg^{2+}. Hg^{II} and subsequent Pt^{II} mediated systems discovered by Periana and coworkers represent a major advance for the direct, selective oxidation of CH_4.[146,147] For example, the reaction in Eq. (12.30) can be carried out at 180°C in fuming sulfuric acid (!) and is catalytic in Hg^{II}, generating methyl bisulfate (CH_3OSO_3H) in 43% yield (85% selectivity at 50% conversion):

$$CH_4 + 2H_2SO_4 \xrightarrow[180°C]{Hg^{II}} CH_3OSO_3H + 2H_2O + SO_2 \qquad (12.30)$$

The high selectivity suggests that CH_4 is about 100 times as reactive as CH_3OSO_3H. The exact mechanism of C–H activation by Hg remains debatable. Species such as CH_3HgOSO_3H that reductively eliminate CH_3OSO_3H are likely intermediates. Many of the features of this reaction are consistent with simple electrophilic substitution, i.e., deprotonation of a transient Hg^{II}–CH_4 σ complex. However, an electron transfer pathway in which Hg^{II} is acting as a one-electron outer-sphere oxidant may also account for the observations.[145] Recall here that Hg^{2+} is one of the very few metal systems that oxidizes H_2 to H^+ via direct electron transfer (Eq. 2.2).

Inorganic salts of Tl, Au, Pd, and Pt also promote oxidation of CH_4 in strong oxidizing acids such as H_2SO_4. However, only stoichiometric oxidations are observed with Au and Tl salts because reoxidation of the reduced forms of these salts is not possible without suitable oxidants. Catalysis is observed with Pd or Pt salts, with Pt being the more efficient, as shown in Scheme 6 for a bipyrimidine complex of Pt^{II}.[147] At 200°C in fuming sulfuric acid, methyl bisulfate can be obtained in 70% one-pass yield based on CH_4 (90% conversion, 81% selectivity) when X is HSO_4. Remarkably, the organic ligand is stable to hot H_2SO_4, probably because the uncoordinated nitrogen atoms are protonated, reducing oxidative degradation. The high temperatures may actually favor CH_4 binding *relative* to that of other potential weak ligands because of low absolute entropy of CH_4 (see Section 7.2.1). The mechanism appears to be closely related to Eq. (12.29), and H/D exchange and other experiments show that the C–H activation step must be faster than either the oxidation step or the functionalization step. The process is believed to be electrophilic and to occur via a 14e complex (Scheme 6), which presumably promotes very rapid heterolytic cleavage of the C–H bond. However, associative substitution[142] of X by CH_4 may occur here as in 33, and H_2 is known to displace Cl [Eq. (3.8)].

Intramolecular transfer of the proton to either the cis X or the anion X is possible, as for heterolytic cleavage of H_2 (see Section 9.4.4). The M here and in related systems may be considered to be a "superelectrophile" isolobal with H^+, mimicking superacid-induced carbocation chemistry (i.e., CH_5^+). Evidence for Scheme 6 includes the observation that although H/D exchange occurs with CH_4

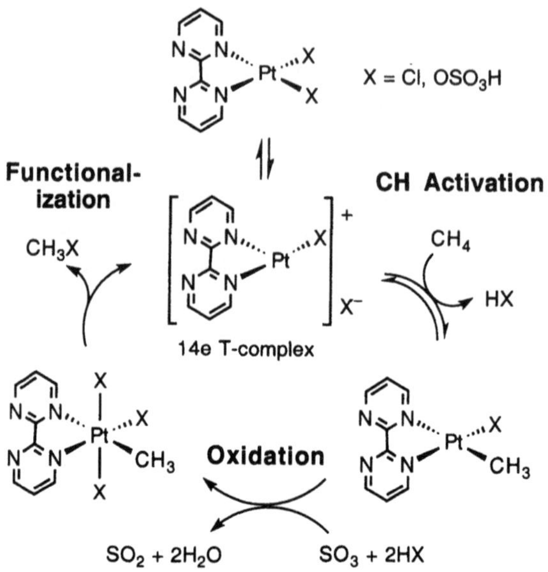

Scheme 6

gas dissolved in D_2SO and treated with (bpym)$PtCl_2$, no exchange is seen within the C–H bonds of methyl bisulfate. Thus the electron-withdrawing bisulfate group inhibits electrophilic reaction between the metal complex and the C–H bonds of the

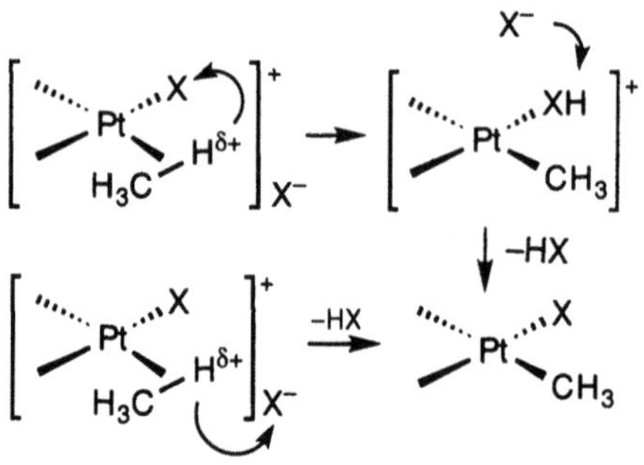

(12.31)

methyl bisulfate produced in the reaction, relative to those of CH_4. This process may be related to that in biological activation of alkanes by, e.g., P-450 and MMO, where a proton from a transient σ alkane complex is transferred to an oxo ligand (see Section 10.3).

Multiple D incorporation into the methanes produced upon reaction of a synthesized sample of (bpym)Pt(CH_3)(Cl) with D_2SO_4 at 25°C supports the four-coordinate methyl complex in Scheme 6 as a key intermediate. A single C–H activation can lead to multiple H/D exchange, again most likely via a CH_4 σ complex before CH_4 loss or oxidation of $Pt-CH_3$ to methyl bisulfate occurs, similar to the above findings of Bercaw and others. DFT calculations on model systems support a σ complex intermediate that undergoes OA.[148] A model with bisulfate ligands and with NH_3 replacing bpym suggests that an H_2SO_4 solvated complex of the form $[Pt(NH_3)_2(OSO_3H)(H_2SO_4)]^+$ is a good candidate for the active catalyst. An alternative scheme in which the catalyst is a Pt^{IV} complex, $[PtCl_2(NH_3)_2(OSO_3H)]^+$, also appears to be feasible on the basis of thermodynamics. Regardless of mechanism, the experimental results are very encouraging since catalytic alkane oxidation in 40–70% yields is unprecedented. However, in order for such systems to be useful for conversion of CH_4 to CH_3OH, a cheap oxidant such as O_2 in a catalytic cycle is needed. There are good possibilities for recycle here, such as the oxidation by O_2 of the SO_2 product in Eq. (12.30) to SO_3 followed by hydrolysis to sulfuric acid starting material.

Other electrophilic systems investigated include Pd^{II}-acetate mediated oxidation of CH_4 to $CF_3CO_2CH_3$ in trifluoroacetic acid by Sen, including a catalytic version of this reaction using H_2O_2 as oxidant.[149] Direct carboxylation of alkanes including CH_4 can be catalyzed by Rh and Pd complexes:[150]

$$RH + CO + \tfrac{1}{2}O_2 \xrightarrow{M^{n+}} RCOOH \quad (12.32)$$

These systems not only effect the catalytic functionalization of C–H bonds, but also utilize SO_2 as the oxidant. However, the complexity of such reactions prevents elucidation of detailed mechanisms. While OA of the C–H bond may be possible in the Pt^{II}-mediated reactions, this is not a viable pathway for many of the other systems (e.g., Hg^{II}). It is likely that a single mechanism does not operate for all alkane conversions, although σ complexes are involved in some, especially for platinum.

REFERENCES

1. Chatt, J.; Davidson, J. M. *J. Chem. Soc.* **1965**, *843*; for quotation see Chapter 2 in reference 50 in Chapter 10.
2. Janowicz, A. H.; Bergman, R. G. *J. Am. Chem. Soc.* **1982**, *104*, 352; Hoyano, J. K.; Graham, W. A. G. *J. Am. Chem. Soc.* **1982**, *104*, 3723.
3. Crabtree, R. H. *Angew. Chem. Int. Ed. Engl.* **1993**, *32*, 789.
4. Hall, C.; Perutz, R. N. *Chem. Rev.* **1996**, *96*, 3125.
5. Shilov, A. E.; Shul'pin, G. B. *Chem. Rev.* **1997**, *97*, 2879; *Catalysis by Metal Complexes*, Vl. 21. *Activation and Catalytic Reactions of Saturated Hydrocarbons in the Presence of Metal Complexes*, A. E. Shilov and G. B. Shul'pin (eds.), Kluwer: Dordrecht, Boston, and London, 2000.
6. Crabtree, R. H. *Chem. Rev.* **1995**, *95*, 987.
7. Schneider, J. J. *Angew. Chem. Int. Ed. Engl.* **1996**, *35*, 1068.
8. Crabtree, R. H. *Chem. Rev.* **1985**, *85*, 245.
9. Yao, W.; Eisenstein, O.; Crabtree, R. H. *Inorg. Chim. Acta.* **1997**, *254*, 105.
10. Stahl, S. S.; Labinger, J. A.; Bercaw, J. E. *Angew. Chem. Int. Ed. Engl.* **1998**, *37*, 2181, and references therein.
11. Arndtsen, B. A.; Bergman, R. G.; Mobley, T. A.; Peterson, T. H. *Acc. Chem. Res.* **1995**, *28*, 154; Dyker, G. *Angew. Chem. Int. Ed. Engl.* **1999**, *38*, 1698; Sen, A. *Acc. Chem. Res.* **1998**, *31*, 550; Jones, W. D.; Feher. F. J. *Acc. Chem. Res.* **1989**, *22*, 91; Jones, W. D. *Science* **2000**, *287*, 1942.

12. Brookhart, M.; Green, M. L. H.; Wong, L.-L. *Progr. Inorg. Chem.* **1988**, *36*, 1.
13. LaPlaca, S. J.; Ibers, J. A. *Inorg. Chem.* **1965**, *4*, 778.
14. Bailey, N. A.; Jenkins, J. M.; Mason, R.; Shaw, B. L. *J. Chem. Soc., Chem. Commun.* **1965**, *237*.
15. Roe, D. M.; Bailey, P. M.; Moseley, K.; Maitlis, P. M. *J. Chem. Soc., Chem. Commun.* **1972**, *1273*.
16. Trofimenko, J. *J. Am. Chem. Soc.* **1967**, *89*, 6288; Trofimenko, J. *Progr. Inorg. Chem.* **1986**, *34*, 115.
17. Cotton, F. A.; Jeremic, M.; Shaver, A. *Inorg. Chim. Acta.* **1972**, *6*, 543; Cotton, F. A.; Stanislowski, A. G. *J. Am. Chem. Soc.* **1974**, *96*, 5074.
18. Cotton, F. A.; Stanislowski, A. G. *J. Am. Chem. Soc.* **1974**, *96*, 5074; Cotton, F. A.; Day, V. W. *J. Chem. Soc., Chem. Commun.* **1974**, 415.
19. Popelier, P. L. A.; Logothetis, G. *J. Organometal. Chem.* **1998**, *555*, 101.
20. Manner, W. L.; Hostetler, M. J.; Girolami, G. S.; Nuzzo, R. G. *J. Phys. Chem. B* **1999**, *103*, 6752.
21. Brammer, L.; McCann, M. C.; Bullock, R. M.; McMullan, R. K.; Sherwood, P. *Organometallics* **1992**, *11*, 2339.
22. Brown, R. K.; Williams, J. M.; Schultz, A. J.; Stucky, G. D.; Ittel, S. D.; Harlow, R. L. *J. Am. Chem. Soc.* **1980**, *102*, 981.
23. Braga, D.; Grepioni, F.; Biradha, K.; Desiraju, G. R. *J. Chem. Soc., Dalton Trans.* **1996**, *3925*.
24. Gusev, D. G.; Madott, M.; Dolgushin, F. M.; Lyssenko, K. A.; Antipin, M. Y. *Organometallics* **2000**, *19*, 1734, and references therein
25. Lee, D. W.; Kaska, W. C.; Jensen, C. M. *Organometallics* **1998**, *17*, 1, and references therein; Niu, S.; Hall, M. B. *J. Am. Chem. Soc.* **1999**, *121*, 3992; Rybtchinski, B.; Milstein, D. *J. Am. Chem. Soc.* **1999**, *121*, 4528, and references therein; Rybtchinski, B.; Milstein, D. *Angew. Chem. Int. Ed. Engl.* **1999**, *38*, 870; Cao, Z.; Hall, M. *Organometallics* **2000**, *19*, 3338; Sunderman, A.; Uzan, O.; Milstein, D.; Martin, J. M. L. *J. Am. Chem. Soc.* **2000**, *122*, 7095.
26. Crabtree, R. H.; Holt, E. M.; Lavin, M. E.; Morehouse, S. M. *Inorg. Chem.* **1985**, *24*, 1986.
27. Burgi, H. B.; Dunitz, J. *Acc. Chem. Res.* **1983**, *16*, 153.
28. Koga, N.; Morokuma, K. *Chem. Rev.* **1991**, *91*, 823.
29. Cundari, T. R. *J. Am. Chem. Soc.* **1994**, *116*, 340.
30. Zhang, K.; Gonzalez, A. A.; Hoff, C. D. *J. Am. Chem. Soc.* **1989**, *111*, 3627; Zhang, K.; Gonzalez, A. A.; Mukerjee, S. L.; Chou, S.-J.; Hoff, C. D.; Kubat-Martin, K. A.; Kubas, G. J. *J. Am. Chem. Soc.* **1991**, *113*, 9170.
31. Dawoodi, Z.; Green, M. L. H.; Mtetwa, V. S. B.; Prout, K. *J. Chem. Soc., Chem. Commun.* **1982**, *1410*; Dawoodi, Z.; Green, M. L. H.; Mtetwa, V. S. B.; Prout, K.; Schultz, A. J.; Williams, J. M.; Koetzle, T. F. *J. Chem. Soc., Dalton Trans.* **1986**, *1629*.
32. Schrock, R. R. *Acc. Chem. Res.* **1979**, *12*, 98.
33. Barnhart, D. M.; Clark, D. L.; Gordon, J. C.; Huffman, J. C.; Watkin, J. G.; Zwick, B. D. *J. Am. Chem. Soc.* **1993**, *115*, 8461; Evans, W. J.; Anwander, R.; Ziller, J. W.; Khan, S. I. *Inorg. Chem.* **1995**, *34*, 5927.
34. Cole, J. M.; Gibson, V. C.; Howard, J. A. K.; McIntyre, G. J.; Walker, G. L. P. *Chem. Comm.* **1998**, *1829*.
35. Calvert, R. B.; Shapley, J. R. *J. Am. Chem. Soc.* **1978**, *100*, 7726.
36. Wasserman. H. J.; Kubas, G. J.; Ryan, R. R.; *J. Am. Chem. Soc.* **1986**, *108*, 2294.
37. Bruce, M. I. *Angew. Chem. Int. Ed. Engl.* **1977**, *16*, 73; Ryabov, A. D. *Chem. Rev.* **1990**, *90*, 403.
38. Cheney, A. J.; Shaw, B. L. *J. Chem. Soc., Dalton Trans.* **1972**, *754*.
39. Ishida, T.; Mizobe, Y.; Hidai, M. *Chem. Lett.* **1989**, *2077*.
40. Luo, X.-L.; Kubas, G. J.; Burns, C. J.; Butcher, R. J.; Bryan, J. C. *Inorg. Chem.* **1995**, *34*, 6538.
41. Lavin, M. E.; Holt, E. M.; Crabtree, R. H. *Organometallics* **1989**, *8*, 99.
42. Vigalok, A.; Uzan, O.; Shimon, L. J. W.; Ben-David, Y.; Martin, J. M. L.; Milstein, D. *J. Am. Chem. Soc.* **1998**, *120*, 12539, and references therein.
43. Toner, A. J.; Grundemann, S.; Clot, E.; Limbach, H.-H.; Donnadieu, B.; Sabo-Etienne, S.; Chaudret, B. *J. Am. Chem. Soc.* **2000**, *122*, 6777.
44. Green, M. L. H.; Wong, L.-L. *J. Chem. Soc., Dalton* **1987**, *411*.
45. Crackness, R. B.; Orpen, A. G.; Spencer, J. L. *J. Chem. Soc., Chem. Commun.* **1984**, *326*.
46. Werner, H.; Feser, R. *Angew. Chem. Int. Ed. Engl.* **1979**, *18*, 157.
47. Koga, N.; Kitaura, K.; Obara, S.; Morokuma, K. *J. Am. Chem. Soc.* **1985**, *107*, 7109.
48. Krauledat, H.; Brintzinger, H. H. *Angew. Chem. Int. Ed. Engl.* **1990**, *29*, 1412; Piers, W. E.; Bercaw, J. E. *J. Am. Chem. Soc.* **1990**, *112*, 9406; Clawson, L.; Soto, J.; Buchwald, S. L.; Steigerwald, M. L.; Grubbs, R. H. *J. Am. Chem. Soc.* **1985**, *107*, 3377.

49. Gonzalez, A. A.; Zhang, K.; Nolan, S. P.; de la Vega, R. L.; Mukerjee, S. L.; Hoff, C. D.; Kubas, G. J. *Organometallics* **1988**, *7*, 2429.
50. Zhang, K.; Gonzalez, A.A.; Mukerjee, S. L.; Chou, S.-J.; Hoff, C. D.; Kubat-Martin, K. A.; Barnhart, D.; Kubas, G. J. *J. Am. Chem. Soc.* **1991**, *113*, 9170.
51. Sato, M.; Tatsumi, T.; Kodama, T.; Hidai, M.; Uchida, T.; Uchida, Y. *J. Am. Chem. Soc.* **1978**, *100*, 447.
52. Luo, X. L.; Kubas, G. J.; Burns, C. J.; Butcher, R. J.; Bryan, J. C. *Inorg. Chem.* **1995**, *34*.
53. King, W. A.; Luo, X-L.; Scott, B. L.; Kubas, G. J.; Zilm, K. W. *J. Am. Chem. Soc*, **1996**, *118*, 6782; King, W. A.; Scott, B. L.; Eckert, J.; Kubas, G. J. *Inorg. Chem.* **1999**, *38*, 1069.
54. Toupadakis, A. I.; Kubas, G. J.; King, W. A.; Scott, B. L.; Huhmann-Vincent, J., *Organometallics* **1998**, *17*, 5315.
55. Heinekey, D. M.; Schomber, B. M.; Radzewich, C. E. *J. Am. Chem. Soc.* **1994**, *116*, 4515.
56. Saburi, M.; Fujii, T.; Shibusawa, T.; Masui, D.; Ishii, Y. *Chem. Lett.* **1999**, *1343*.
57. Huang, D.; Folting, K.; Caulton, K. G. *J. Am. Chem. Soc.* **1999**, *121*, 10318.
58. Cooper, A. C.; Clot, E.; Huffman, J. C.; Streib, W. E.; Maseras, F.; Eisenstein, O.; Caulton, K. G. *J. Am. Chem. Soc.* **1999**, *121*, 97; Ujaque, G.; Cooper, A. C.; Maseras, F.; Eisenstein, O.; Caulton, K. G. *J. Am., Chem. Soc.* **1998**, *120*, 361.
59. Winter, R. F.; Hornung, F. M. *Inorg. Chem.* **1997**, *36*, 6197; Schlaf, M.; Lough, A. J.; Morris, R. H. *Organometallics* **1997**, *16*, 1253.
60. Otsuka, S.; Yoshida, T.; Matsumoto, M.; Nakatsu, K. *J. Am. Chem. Soc.* **1976**, *98*, 588; Baratta, W.; Herdtweck, E.; Rigo, P. *Angew. Chem. Int. Ed. Engl.* **1999**, *38*, 1629.
61. Huang, D.; Streib, W. E.; Bollinger, J. C.; Caulton, K. G.; Winter, R. F.; Scheiring, T. *J. Am. Chem. Soc.* **1999**, *121*, 8087.
62. Harlow, R. L.; Thorn, D. L.; Baker, R. T.; Jones, N. L. *Inorg. Chem.* **1992**, *31*, 993.
63. Fawcett, J.; Holloway, J. H.; Saunders, G. C. *Inorg. Chim. Acta* **1992**, *202*, 111.
64. Cini, R.; Cavaglioni, A. *Inorg. Chem.* **1999**, *38*, 3751.
65. Gottschalk-Gaudig, T.; Folting, K.; Caulton, K. G. *Inorg. Chem.* **1999**, *38*, 5241.
66. Heyn, R. H.; Macgregor, S. A.; Nadasdi, T. T.; Ogasawara, M.; Esenstein, O.; Caulton, K. G. *Inorg. Chim. Acta* **1997**, *259*, 5.
67. Jimenez-Tenorio, M.; Mereiter, K.; Puerta, M. C.; Valerga, P. *J. Am. Chem. Soc.* **2000**, *122*, 11230.
68. Burns, C. J.; Andersen, R. A. *J. Am. Chem. Soc.* **1987**, *109*, 5853.
69. Perutz, R. N.; Turner, J. J. *J. Am. Chem. Soc.* **1975**, *97*, 4791; Turner, J. J.; Burdett, J. K.; Perutz, R. N.; Poliakoff, M. *Pure Appl. Chem.* **1977**, *49*, 271; Billups, W. E.; Chang, S.-C.; Hauge, R. H.; Margrave, J. L. *J Am. Chem. Soc.* **1993**, *115*, 2039.
70. Upmacis, R. K.; Gadd, G. E.; Poliakoff, M.; Simpson, M. B.; Turner, J. J.; Whyman, R.; Simpson, A. F. *J. Chem. Soc., Chem. Commun.* **1985**, *27*; Burdett, J. K.; Downs, A. J.; Gaskill, G. P.; Graham, M. A.; Turner, J. J.; Turner, R. F. *Inorg. Chem.* **1978**, *17*, 523.
71. Kelly, J. M.; Hermann, H.; Von Gustorf, E. K. *J. Chem. Soc., Chem. Commun.* **1973**, 105
72. Bonneau, R.; Kelly, J. M. *J. Am. Chem. Soc.* **1980**, *102*, 1220; Kelly, J. M.; Long, C.; Bonneau, R. *J. Phys. Chem.* **1983**, *87*, 3344.
73. Simon, J. D.; Xie, X. *J. Phys. Chem.* **1986**, *90*, 7651; Joly, A. G.; Nelson, K. A. *J. Phys. Chem.* **1989**, *93*, 2876.
74. Carroll, J. J.; Weisshaar, J. C. *J. Am. Chem. Soc.* **1993**, *115*, 800.
75. Dobson, G. R.; Hodges, P. M.; Healy, M. A.; Poliakoff, M.; Turner, J. J.; Firth, S.; Asali, K. *J. Am. Chem. Soc.* **1987**, *109*, 4218.
76. Brookhart, M.; Chandler, W.; Kessler, R. J.; Liu, Y.; Pienta, N. J.; Santini, C. C.; Hall, C.; Perutz, R. N.; Timney, J. A. *J. Am. Chem. Soc.* **1992**, *114*, 3802.
77. Hop, C. E. C. A.; McMahon, T. B. *J. Am. Chem. Soc.* **1991**, *113*, 355.
78. Carpenter, C. J.; van Koppen, P. A. M.; Bowers, M. T.; Perry, J. K. *J. Am. Chem. Soc.* **2000**, *122*, 392.
79. Poliakoff, M.; Weitz, E. *Acc. Chem. Res.* **1987**, *20*, 408; Rest, A. J.; Sodeau, J. R.; Taylor, D. J. *J. Chem. Soc., Dalton Trans.* **1978**, 651. Creaven, B. S.; George, M. W.; Ginzburg, A. G.; Hughes, C.; Kelly, J. M.; Long, C.; McGrath, I. M.; Pryce, M. T. *Organometallics* **1993**, *12*, 3127; Nayak, S. K.; Burkey, T. J. *J. Am. Chem. Soc.* **1993**, *115*, 6391; Mawby, R. J.; Perutz, R. N.; Whittlesey, M. K. *Organometallics* **1995**, *14*, 3268.
80. Klassen, J. K.; Selke, M.; Sorensen, A. A.; Yang, G. K. *J. Am. Chem. Soc.* **1990**, *112*, 1267.
81. Whittlesey, M. K.; Perutz, R. N.; Virrels, I. G.; George, M. W. *Organometallics* **1997**, *16*, 268.

82. Sun, X.-Z.; Grills, D. C.; Nikiforov, S. M.; Poliakoff, M.; George, M. W. *J. Am. Chem. Soc.* **1997**, *119*, 7521.
83. Bengali, A. A.; Bergman, R. G.; Moore, C. B. *J. Am. Chem. Soc.* **1995**, *117*, 3879.
84. Bengali, A. A.; Schultz, R. H.; Moore, C. B.; Bergman, R. G. *J. Am. Chem. Soc.* **1994**, *116*, 9585.
85. Grevels, F.-W. *NATO ASI Series C* **1992**, *376*, 141; Brough, S.-A.; Hall, C.; McCamley, A.; Perutz, R. N.; Stahl, S.; Wecker, U.; Werner, H. *J. Organomet. Chem.* **1995**, *504*, 33.
86. Geftakis, S.; Ball, G. E. *J. Am. Chem. Soc.* **1998**, *120*, 9953.
87. Zaric, S.; Hall, M. B. *J. Phys. Chem. A*, **1997**, *101*, 4646.
88. Evans, D. R.; Drovetskaya, T.; Bau, R.; Reed, C. A.; Boyd, P. D. W. *J. Am. Chem. Soc.* **1997**, *119*, 3633.
89. Lee, D. W.; Jensen, C. M. *Inorg. Chim. Acta.* **1997**, *259*, 359.
90. Kubas, G. J.; Nelson, J. E.; Bryan, J. C.; Eckert, J.; Wisniewski, L.; Zilm, K. *Inorg. Chem.* **1994**, *33*, 2954; Eckert, J.; Kubas, G. J.; White, R. P. *Inorg. Chem.* **1992**, *31*, 1550.
91. Saillard, J.-Y.; Hoffmann, R. *J. Am. Chem. Soc.* **1984**, *106*, 2006.
92. Ziegler, T.; Tschinke, V.; Fan, L.; Becke, A. D. *J. Am. Chem. Soc.* **1989**, *111*, 9177.
93. Song, J.; Hall, M. B. *Organometallics* **1993**, *12*, 3118; Poli, R. J. Smith, K. M. *Organometallics* **2000**, *19*, 2858.
94. Musaev, D. G.; Morokuma, K. *J. Am. Chem. Soc.* **1995**, *117*, 799.
95. Su, M.-D.; Chu, S.-Y. *Organometallics* **1997**, *16*, 1621; *J. Am. Chem. Soc.* **1997**, *119*, 5573; *Chem. Eur. J.* **1999**, *5*, 198.
96. Siegbahn, P. E. M. *J. Am. Chem. Soc.* **1996**, *118*, 1487.
97. Niu, S.; Strout, D. L.; Zaric, S.; Bayse, C. A.; Hall, M. B. *ACS Symp. Ser.* **1999**, *721* (*Transition State Modeling for Catalysis*), 138; Niu, S.; Hall, M. B. *Chem. Rev.* **2000**, *100*, 353; Dedieu, A. *Chem. Rev.* **2000**, *100*, 543; Biswas, B.; Sujimoto, M.; Sakaki, S. *Organometallics* **2000**, *19*, 3895.
98. Wasserman, E. P.; Moore, C. B.; Bergman, R. G. *Science* **1992**, *255*, 315.
99. Koga, N.; Morokuma, K. *J. Phys. Chem.* **1990**, *94*, 5454.
100. Koga, N.; Morokuma, K. *J. Am. Chem. Soc.* **1993**, *115*, 6883.
101. Musaev, D. G.; Morokuma, K. *J. Organomet. Chem.* **1995**, *504*, 93.
102. Blomberg, M. R. A.; Brandemark, U. B.; Petterson, L. G. M.; Siegbahn, P. E. M. *Int. J. Quan. Chem.* **1983**, *23*, 855. Blomberg, M. R. A.; Brandemark, U. B.; Siegbahn, P. E. M. *J. Am. Chem. Soc.* **1983**, *105*, 5557.
103. Su, M.-D.; Chu, S.-Y. *J. Am. Chem. Soc.* **1997**, *119*, 10178.
104. Brown, C. E.; Ishikawa, Y.; Hackett, P. A.; Rayner, D. M. *J. Am. Chem. Soc.* **1990**, *112*, 2530.
105. Dobson, G. R.; Smit, J. P.; Ladogana, S.; Walton, W. B. *Organometallics* **1997**, *16*, 2858.
106. Cundari, T. R.; Klinckman, T. R. *Inorg. Chem.* **1998**, *37*, 5399.
107. Avdeev, V. I.; Zhidomirov, G. M. *Catal. Today* **1998**, *42*, 247.
108. Aracama, M.; Esteruelas, M. A.; Lahoz, F. J.; Lopez, J. A.; Meyer, U.; Oro, L. A.; Werner, H. *Inorg. Chem.* **1991**, *30*, 288.
109. Siegbahn, P. E. M.; Svensson, M. *J. Am. Chem. Soc.* **1994**, *116*, 10124.
110. Su, M.-D.; Chu, S.-Y. *J. Phys. Chem. A* **1998**, *102*, 10159.
111. Siegbahn, P. E. M.; Blomberg, M. R. A. *Theoretical Aspects of Homogeneous Catalysis*, van Leeuwen, P. W. N. M.; Morokuma, K.; van Lenthe, J. H., eds., Kluwer Academic Publishers, Dordrecht, 1995.
112. Janowicz, A. H.; Bergman, R. G. *J. Am. Chem. Soc.* **1983**, *105*, 3929; Hoyano, J. K.; McMaster, A. D.; Graham, W. A. G. *J. Am. Chem. Soc.* **1983**, *105*, 7190; Jones, W. D.; Feher, F. J. *Organometallics* **1983**, *2*, 562; Bergman, R. G.; Seidler, P. F.; Wenzel, T. T. *J. Am. Chem. Soc.* **1985**, *107*, 4358; Baker, M. V.; Field, L. S. *J. Am. Chem. Soc.* **1987**, *109*, 2825; Ghosh, C. K.; Graham, W. A. G. *J. Am. Chem. Soc.* **1987**, *109*, 4726; Hackett, M.; Whitesides, G. M. *J. Am. Chem. Soc.* **1988**, *110*, 1449; Harper, T. G. P.; Shinomoto, R. S.; Deming, M. A.; Flood, T. C. *J. Am. Chem. Soc.* **1988**, *110*, 7915; Belt, S. T.; Grevels, F. W.; Klotzbcher, W. E.; McCamley, A.; Perutz, R. N. *J. Am. Chem. Soc.* **1989**, *111*, 8373; Kiel, W. A.; Ball, R. G.; Graham, W. A. G. *J. Organomet. Chem.* **1993**, *259*, 481; Jones, W. D.; Hessell, E. T. *J. Am. Chem. Soc.* **1993**, *115*, 554.
113. Wang, C.; Ziller, J. W.; Flood, T. C. *J. Am. Chem. Soc.* **1995**, *117*, 1647.
114. Wick, D. D.; Goldberg, K. I. *J. Am. Chem. Soc.* **1997**, *117*, 10235.
115. Haddleton, D. M.; McCamley, A.; Perutz, R. N. *J. Am. Chem. Soc.* **1988**, *110*, 1810.
116. Schultz, R. H.; Bengali, A. A.; Tauber, M. J.; Weiller, B. H.; Wasserman, E. P.; Kyle, K. R.; Moore, C. B.; Bergman, R. G. *J. Am. Chem. Soc.* **1994**, *116*, 7369; Van Zee, R. J.; Li, S.; Weltner, W. Jr. *J. Am. Chem. Soc.* **1993**, *115*, 2976.

117. Bengali, A. A.; Schultz, R. H.; Moore, C. B.; Bergman, R. G. *J. Am. Chem. Soc.* **1994**, *116*, 9585.
118. Lian, T.; Bromberg, S. E.; Yang, H.; Proulx, G.; Bergman, R. G.; Harris, C. B. *J. Am. Chem. Soc.* **1996**, *118*, 3769; Osman, R.; Pattison, D. I.; Perutz, R. N.; Bianchini, C.; Peruzzini, M. J. C. S., *Chem. Comm.* **1994**, 513; Lees, A. J.; Purwoko, A. A. *Coord. Chem. Rev.* **1994**, *132*, 155.
119. McNamara, B. K.; Yeston, J. S.; Bergman, R. G.; Moore, C. B. *J. Am. Chem. Soc.* **1999**, *121*, 6437.
120. Wick, D. D.; Reynolds, K. A.; Jones, W. D. *J. Am. Chem. Soc.* **1999**, *121*, 3974.
121. Buchanan, J. M.; Stryker, J. M.; Bergman, R. G. *J. Am. Chem. Soc.* **1986**, *108*, 1537.
122. Periana, R. A.; Bergman, R. G. *J. Am. Chem. Soc.* **1986**, *108*, 7332.
123. Bullock, R. M.; Headford, C. E. L.; Hennessy, K. M.; Kegley, S. E.; Norton, J. R. *J. Am. Chem. Soc.* **1989**, *111*, 3897.
124. Parkin, G.; Bercaw, J. E. *Organometallics* **1989**, *8*, 1172.
125. Gould, G. L.; Heinekey, D. M. *J. Am. Chem. Soc.* **1989**, *111*, 5502.
126. Stahl, S. S.; Labinger, J. A.; Bercaw, J. E. *J. Am. Chem. Soc.* **1996**, *118*, 5961.
127. Mobley, T. A.; Schade, C.; Bergman, R. G. *J. Am. Chem. Soc.* **1995**, *117*, 7822.
128. Green, J. C.; Jardine, C. N. *J. Chem. Soc. Dalton Trans.* **1998**, 1057.
129. Gross, C. L.; Girolami, G. S. *J. Am. Chem. Soc.* **1998**, *120*, 6605.
130. Martin, R. L. *J. Am. Chem. Soc.* **1999**, *121*, 9459.
131. Jones, W. D.; Hessell, E. T. *J. Am. Chem. Soc.* **1992**, *114*, 6087; Jones, W. D.; Feher, F. J. *J. Am. Chem. Soc.* **1984**, *106*, 1650.
132. Schafer, D. F., II; Wolczanski, P. T. *J. Am. Chem. Soc.* **1998**, *120*, 4881.
133. Cundari, T. R. *Organometallics* **1993**, *12*, 1998.
134. Bengali, A. A.; Arndtsen, B. A.; Burger, P. M.; Schultz, R. H.; Weiller, B. H.; Kyle, K. R.; Moore, C. B.; Bergman, R. G. *Pure Appl. Chem.* **1995**, *67*, 281.
135. Weiller, B. H.; Wasserman, E. P.; Bergman, R. G.; Moore, C. B.; Pimentel, G. C. *J. Am. Chem. Soc.* **1989**, *111*, 8288.
136. Partridge, M. G.; McCamley, A.; Perutz, R. N. *J. Chem. Soc., Dalton Trans.* **1994**, 3519.
137. Bromberg, S. E.; Yang, H.; Asplund, M. C.; Lian, T., McNamara, B. K.; Kotz, K. T.; Yeston, J. S.; Wilkens, M.; Frei, H.; Bergman, R. G.; Harris, C. B. *Science* **1998**, *278*, 260; Asbury, J. B.; Ghosh, H. N.; Yeston, J. S.; Bergman, R. G.; Lian, T. *Organometallics* **1998**, *17*, 3417; Asbury, J. B.; Hang, K.; Yeston, J. S.; Cordaro, J. G.; Bergman, R. G.; Lian, T. *J. Am. Chem. Soc.* **2000**, *122*, 12870.
138. Zaric, S.; Hall, M. B. *J. Phys. Chem. A*, **1998**, *102*, 1963.
139. Stahl, S. S.; Labinger, J. A.; Bercaw, J. E. *J. Am. Chem. Soc.* **1995**, *117*, 9371.
140. Holtcamp, M. W.; Labinger, J. A.; Bercaw, J. E. *J. Am. Chem. Soc.* **1997**, *119*, 848; Holtcamp, M. W.; Labinger, J. A.; Bercaw, J. E. *Inorg. Chim. Acta* **1997**, *265*, 117.
141. Holtcamp, M. W.; Henling, L. M.; Day, M. W.; Labinger, J. A.; Bercaw, J. E. *Inorg. Chim. Acta* **1998**, *270*, 467.
142. Johansson, L.; Ryan, O. B.; Tilset, M. *J. Am. Chem. Soc.* **1999**, *121*, 1974; Heiberg, H.; Johansson, L.; Gropen, O.; Ryan, O. B.; Swang, O.; Tilset, M. *J. Am. Chem. Soc.* **2000**, *122*, 10831; Johansson, L.; Tilset, M.; Labinger, J. A.; Bercaw, J. E. *J. Am. Chem. Soc.* **2000**, *122*, 10846. Johansson, L.; Tilset, M. *J. Am. Chem. Soc.* **2001**, *123*, 739.
143. Fekl, U.; Zahl, A.; van Eldik, R. *Organometallics* **1999**, *18*, 4156.
144. Siegbahn, P. E. M.; Crabtree, R. H. *J. Am. Chem. Soc.* **1996**, *118*, 4442; Milet, A.; Dedieu, A.; Kapteijn, G.; van Koten, G. *Inorg. Chem.* **1997**, *36*, 3223; Mylvaganam, K.; Bacskay, G. B.; Hush, N. S. *J. Am. Chem. Soc.* **2000**, *118*, 2041; Niu, S.; Hall, M. B. *J. Am. Chem. Soc.* **1998**, *120*, 6169; Bartlett, K. L.; Goldberg, K. I.; Borden, W. T. *J. Am. Chem. Soc.* **2000**, *122*, 1456.
145. Sen, A.; Benvenuto, M. A.; Lin, M.; Hutson, A. C.; Basickes, N. *J. Am. Chem. Soc.* **1994**, *116*, 998.
146. Periana, R. A.; Taube, D. J.; Evitt, E. R.; Loffler, D. G.; Wentrcek, P. R.; Voss, G.; Masuda, T. *Science* **1993**, *259*, 340.
147. Periana, R. A.; Taube, D. J.; Gamble, S.; Taube, H.; Satoh, T.; Fujii, H. *Science* **1998**, *280*, 560.
148. Mylvaganam, K.; Bacskay, G. B.; Hush, N. S. *J. Am. Chem. Soc.* **1999**, *121*, 4633; Heilberg, H.; Swang, O.; Ryan, O. B.; Gropen, O. *J. Phys. Chem. A* **1999**, *103*, 10004.
149. Sen, A. *Acc. Chem. Res.* **1988**, *21*, 421; Kao, L.-C.; Hutson, A. C.; Sen, A. *J. Am. Chem. Soc.* **1991**, *113*, 700.
150. Lin, M.; Hogan, T. E.; Sen, A. *J. Am. Chem. Soc.* **1996**, *118*, 4574; Fujiwara, Y.; Takaki, K.; Taniguchi, Y. *Synlett* **1996**, 591.

13

Coordination and Activation of B–H and Other X–H and X–Y Bonds

13.1. B–H BONDS

13.1.1. Metal Borohydride and Agostic-Like Systems

Complexes containing interactions of B–H bonds with transition metal centers were among the earliest electron-deficient compounds, and more recent systems have been shown to be genuine σ complexes. This is most appropriate because main-group boranes with B–H–B bridges were the first well-recognized examples of nonclassical chemical bonding. The 3c-2e bonding in covalent metal borohydride complexes, $M(BH_4)_n$, which feature M–H–B bridges, was indeed compared to the electron-deficient bonding in diborane.[1] Borohydride complexes of all the transition metals and many of the lanthanides and actinides are known and have a variety of structures. In 1960, Noth and Hartwimmer proposed a double M–H–B bridged structure in $Cp_2Ti(H_2BH_2)$ on the basis of IR data.[2]

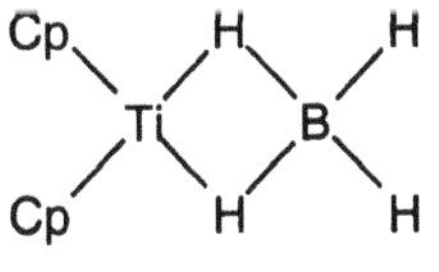

(1)

In analogy with methane complexes, this geometry should be referred to as an η^2-H, H complex as in Chapter 12 but is usually written (as will be done here) as η^2-BH_4. In 1965 the X-ray structure of a novel trinuclear polyborane complex determined by Kaesz and Dahl provided the first crystallographic confirmation of such bridging structures for transition metal systems.[3] $Mn_3(CO)_{10}(H)(B_2H_6)$ has six Mn–H–B interactions and also a Mn–H–Mn bridge. The Mn–B distances in **2** were equal, 2.30(2) Å, and ^1H NMR showed a weak broad resonance at −19 ppm for the Mn–H–B hydrogens, which are bent and highly dynamic here and in borohydride complexes. The latter presumably also exchanges with the μ-H ligand, as later shown experimentally and calculationally in other systems.[1,4] Although the bonding in **2** was noted to be electron-deficient, the nominal dinegative charge on the B_2H_6 and the close to linear Mn–H–B angles makes the true 3c-2e nature of the interactions ambiguous (see below).

(2)

The initial crystal structures of borohydride complexes of transition metals were obtained in 1967 for $Cu(BH_4)(PPh_3)_2$,[5] and $Zr(BH_4)_4$.[6] The X-ray data for the latter show beautifully symmetric *triple* hydrogen (η^3) bridges from four borons that tetrahedrally surround M (Figure 13.1). Metalloborane cluster complexes of the type $[Fe(CO)_3]_{5-n}B_nH_{n+4}$ and $HFe(CO)_9(BH_4)$ (**3**) along with [Cp*Ru] moieties connected to B_nH_{n+4} clusters ($n = 2$–4) provide an elegant "borane analogy."[7] The

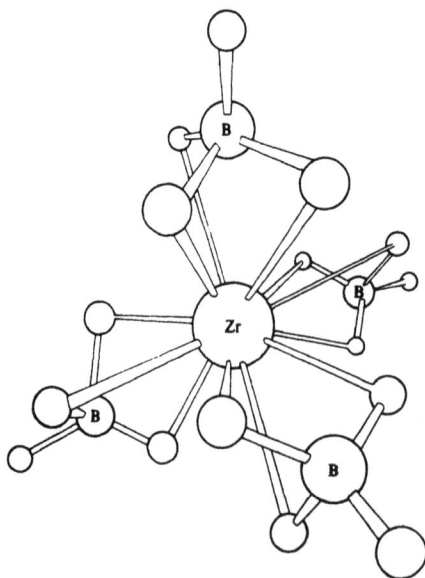

Figure 13.1. Molecular structure of $Zr(BH_4)_4$ at $-160°C$, showing one of the possible cube-octahedral or icosahedral arrangements of borohydride groups about Zr (Bird and Churchill[6]). Reproduced by permission of the Royal Society of Chemistry.

tridentate BH_4^- capping a triangle of Fe atoms in **3** is isolobal with B_4H_8 and the d_{FeH} are short, as in hydrides [e.g., 1.56–1.66(4) Å in a bidentate $Fe_2(CO)_6(B_3H_7)$ derivative]. Borohydride structures can be considered to model CH_4 coordination

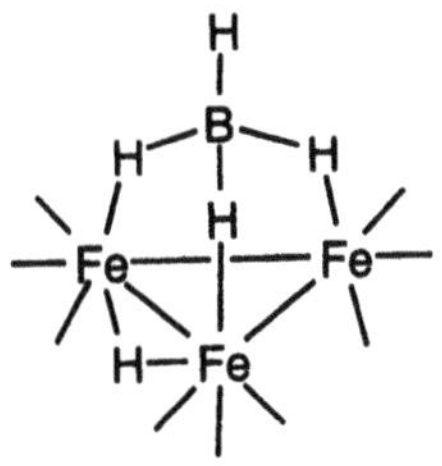

(3)

because BH_4^- and CH_4 are isoelectronic, although the negative charge on the former contributes a strong ionic component to the bonding, which detracts from the comparison. Thus the borohydride and related anionic ligand systems fit more into the same category as metal bridged hydrides, M–H–M, which like M–H–B systems have bent three-center bonds but are not true σ complexes in the manner of $M-H_2$ complexes. As will be shown later, legitimate borane σ complexes are known.

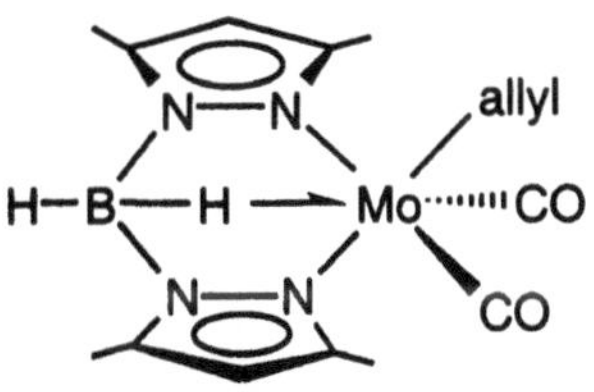

(4) 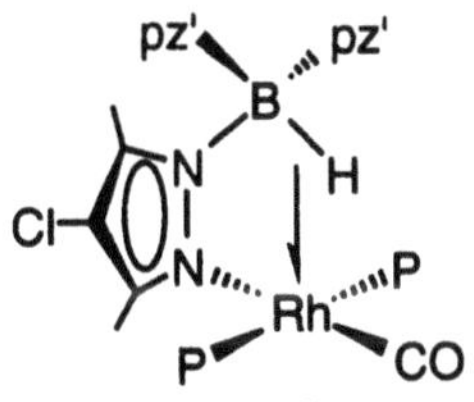   (5)

Polypyrazolyborate (scorpionate) complexes[8] came onto the scene several years later and several show agostic Mo···H–B interactions.[9] The structure of the Bp ligand in **4**, determined in the 1970s, show that d_{MoH} is 2.14 Å, and the coordinated d_{BH} is 1.26 Å compared to 1.05 Å for the terminal B–H.[8] This stretching is analogous to that in other σ complexes. Malbosc et al. showed that the Tp ligand in **5** adopts a $\kappa^2_{N,B-H}$ coordination mode with two dangling pyrazole rings (pz′) and a M–B–H interaction. d_{RhH} is long here, 2.35(3) Å [Rh–H–B = 126(2)°], but is much shorter (~1.83 Å) in Herberhold's analogous $Tp*Rh[P(C_7H_7)_3]$, which shows a reduced $J_{BH} = 71.6$ Hz in the ^{11}B spectrum (cf. 105 Hz in KTp). Chaudret's $(Bp^{(CF3)2})RuH(H_2)(PR_3)_2$ shows interaction of the metal with both H–H and B–H (as in **5** except H replaces a pz′).[9]

Regarding elongation of the normal d_{BH} observed in **4** and in other σ-bond interactions, there are several exceptions, including $Cu(BH_4)(PMePh_2)_3$,[10] for which neutron diffraction data provided the first accurate characterization of an unsupported M–H–B bridge bond. Surprisingly, the long distance in **6** is not the one associated with the Cu–H–B bridge, and no satisfactory explanation could be given for the unusual distribution of d_{BH} except to invoke possible disorder of the BH_4 group (a potential problem in such systems). This result along with other crystallographic inconsistencies indicates that, unlike d_{HH}, structurally determined d_{BH} may may not correlate well with the degree of activation. There is a major problem in that X-ray determined d_{BH} lack accuracy, few neutron studies have been performed, and

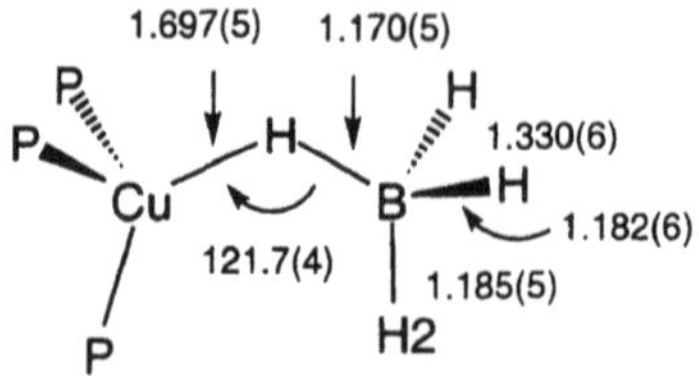

(6)

NMR methods have not been established comparable to those for determining d_{HH}. The d_{CuH} is shorter than in other Cu–borohydride complexes (1.87 Å average) and even in Cu–hydride (1.73 Å). It is interesting that the 2.72 Å distance from Cu to H2 in **6** is close enough to be described as an incipient bonding interaction ready to fully coordinate when a phosphine dissociates.

Other complexes with related M···H–B interactions include Fe[η^3-HC(SMe)S–B(H)H$_2$](CO)(PMe$_3$)$_2$,[11] CpMo(CO)$_2${P(BH$_3$)Ph[N(SiMe$_3$)$_2$]} (**7**)[12] and metallocyclopentadienyl[13] and metallocarborane[14–17] complexes such as **8** and **9** with agostic geometries similar to a Si analogue (Structure **XIII** in Table 11.1). d_{MH} in **7–9** are short as in hydrides [as low as 1.47(3) Å for d_{FeH}]. Extended Hückel

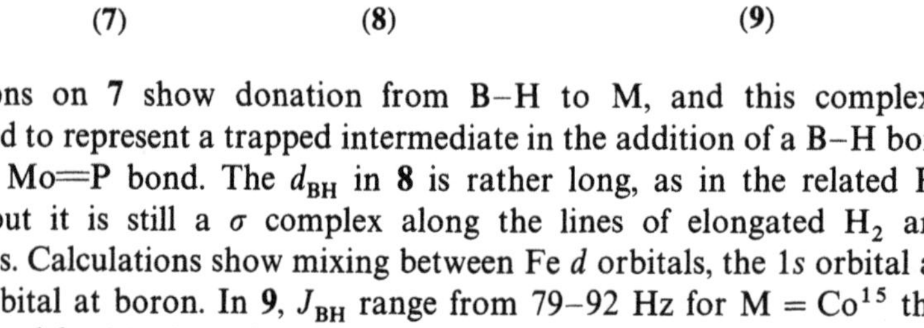

(7)   (8)   (9)

calculations on **7** show donation from B–H to M, and this complex can be considered to represent a trapped intermediate in the addition of a B–H bond across a formal Mo=P bond. The d_{BH} in **8** is rather long, as in the related Ru–Si–H system, but it is still a σ complex along the lines of elongated H$_2$ and silane complexes. Calculations show mixing between Fe d orbitals, the 1s orbital at H, and the 2p orbital at boron. In **9**, J_{BH} range from 79–92 Hz for M = Co[15] that are at the high end for M···H–B interactions, but still well below the benchmark value of ca 130 Hz for terminal B–H groups.[16] There is a good comparison here with the reduction in J_{CH} for related agostic M···H–C interactions (see Section 12.2.1).[9]

Most of the above metal–borane systems can indeed be considered to be agostic σ bonds, but the X-ray structure of Ti(BH$_4$)$_3$(PMe$_3$)$_2$ by Jensen and Girolami was reported to be different in having an η^2-BH$_4$ and two η^1-BH$_4$ ligands that coordinate via one B–H bond in a "side-on" manner.[18] However, *ab initio* calculations disputed this and showed that an η^2–η^2–η^3– structure is more energetically favorable.[19]

Subsequent X-ray studies of analogues do show this structure [disorder occurred in Ti(BH$_4$)$_3$(PMe$_3$)$_2$],[20] a good example of the power of computational chemistry.

A recent borohydride complex of Re is interesting synthetically and dynamically. Complex **10** can be prepared from either an H$_2$ complex or a dihydride,

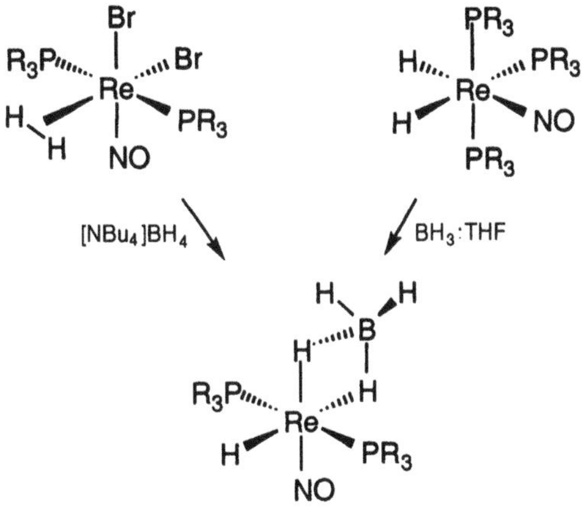

$$\text{(13.1)}$$

(10)

where one equivalent of BH$_3$ removes a phosphine as BH$_3 \cdot$PR$_3$ and a second adds to M to form the η^2-borohydride. In this complex, the BH$_4$ is not fluxional and does not exchange with the hydride as in several other systems (disorder between BH$_4$ and NO precludes meaningful metrical data). Conversely, theoretical and experimental analysis of OsH$_3$(BH$_4$)(PR$_3$)$_2$ shows a wealth of dynamic phenomena.[4] Three different intramolecular hydrogen exchange processes occur in order of increasing barrier: hydride–hydride (fast), hydride–BH$_4$, and BH$_4$ bridging–terminal H (slow). The last is not directly observed by NMR, but the other calculated processes are consistent with variable-temperature NMR studies.

13.1.2. σ Complexes of Neutral Boranes

In the above complexes, uncertainties about ionic contributions to the bonding and/or intramolecular character of the interaction raise questions as whether they are true σ complexes.[21] To more clearly define the bonding and properties of a σ B–H complex, an indisputably *neutral* ligand coordinated only by a M–B–H linkage is needed. In other words, a covalent M–H–B linkage must be identified as *the* bond between two species each capable of independent existence as in LM$_n$ + H$_2$ ↔ LM$_n$–H$_2$ systems. Neutral boranes such as BH$_3$ are Lewis acids but their base adducts such as BH$_3 \cdot$PR$_3$ and B$_2$H$_4 \cdot$2PR$_3$ are potential σ ligands more analogous to silanes than the negatively charged borohydrides. The first complexes of the type L$_n$M(B$_2$H$_4 \cdot$2PR$_3$) were isolated in 1984 simply by adding B$_2$H$_4 \cdot$2PMe$_3$ to ZnCl$_2$ or CuX (X = halide).[21,22] The X-ray structures of the Zn complex[22] and subsequent group 6 pentacarbonyl analogues[23–25] such as Cr(CO)$_5$(B$_2$H$_4 \cdot$2PMe$_3$) and W(CO)$_5$(BH$_3 \cdot$PMe$_3$) clearly show bidentate or monodentate η^2-B–H coordination.

In **11**, two M can coordinate to one borane ligand (for the Cr ligand system, the two monodentate PMe$_3$ are replaced by a bidentate Me$_2$PCH$_2$PMe$_2$).[24] Structure **12** is the first example of intermolecular coordination of a single B–H bond in

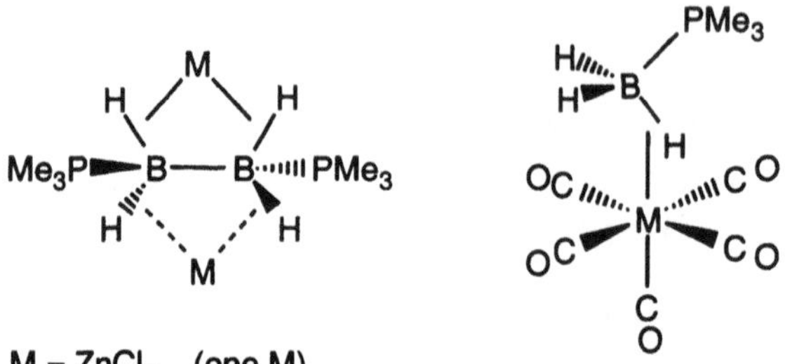

M = ZnCl$_2$ (one M)
Cr(CO)$_4$ (one or two M)
(**11**)

M = Cr, W
(**12**)

a neutral borane to a transition metal, and all of these species can be regarded as models for alkane coordination. The CO complexes, which can form either bidentate or monodentate B$_2$H$_4$·2PMe$_3$ species, are prepared by photolysis [Eq. (13.2)] in direct analogy with H$_2$ complexes. The stability of these complexes varies greatly: the ZnCl$_2$ bidentate complexes are stable to 150°C while M(CO)$_5$(BH$_3$·PMe$_3$) readily dissociates in solution or vacuum. The Mo analogues in Eq. (13.2) are the least stable and cannot be isolated,

$$M(CO)_6 + BH_3 \cdot L \xrightarrow[-CO]{h\nu} M(CO)_5(BH_3 \cdot L) \quad M = Cr, W; L = PR_3, NMe_3 \quad (13.2)$$

whereas in comparison to M(CO)$_5$(H$_2$), the M–H$_2$ binding energy is 3 kcal/mol higher for Mo than for Cr (see Chapter 7). This instability may be attributable to the greater lability of Mo among the group 6 metals. The stabilities are nonetheless generally greater than those for the H$_2$ complexes and much higher than for alkane complexes. ^{1}H NMR shows only one signal for the B*H* atoms in the range −1 to −3.8 ppm *depending on L*, indicating rapid exchange even at −80°C. Coupling to ^{11}B is seen, however, and J_{BH} is 80–87 Hz, about 10 Hz less than for free BH$_3$·L. Because of the fast scrambling of the BH hydrogens, these values represent an average, and the actual J_{BH} for bound B–H is estimated to be about 60 Hz. This is some 65% of that in the free ligand, and comparable to the 50–80% reduction in J_{HD} in unstretched HD complexes and the 74% reduction in J_{CH} for cyclopentane coordination in CpRe(CO)$_2$(C$_5$H$_{10}$). This contrasts to J_{SiH} in silane complexes, which are always closer to those in OA products, so the behavior for the B–H bond is more like that in H$_2$ and alkane complexes. In terms of reactivity, H–D isotopic exchange with solvent occurs in borane complexes as for H$_2$ and other σ complexes: Heating a benzene-d_6 solution of BH$_3$·L and ReH$_7$(PPh$_3$)$_2$ forms deuterated BD$_n$H$_{3-n}$·L for L = PMe$_3$ and NMe$_3$ via a transient Re–borane complex.[26] σ-Bond metathesis via a borane adduct has been proposed in the reaction of Cp*W(CO)$_3$(CH$_3$) with BH$_3$·PMe$_3$ to give the boryl, Cp*W(CO)$_3$(BH$_2$PMe$_3$), and CH$_4$.[27] Reaction of BH$_3$·PPhH$_2$ with Pt(PEt$_3$)$_2$ gives OA of the *P–H* bond.

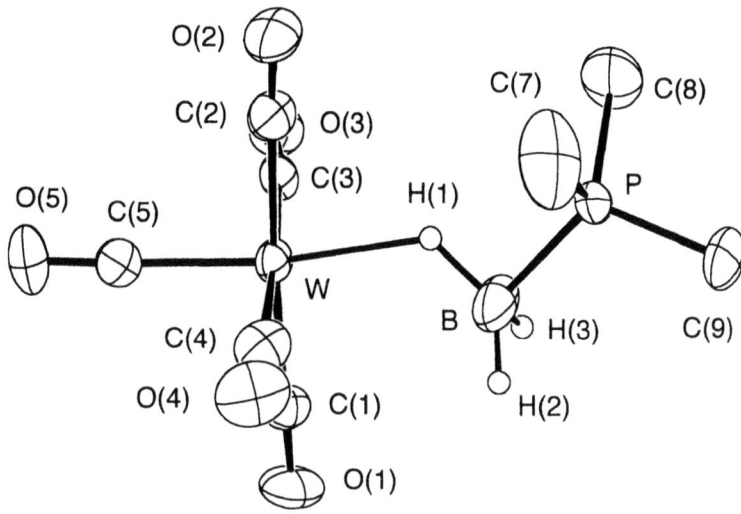

Figure 13.2. ORTEP drawing of W(CO)$_5$(BH$_3$·PMe$_3$). Reprinted with permission from Kawano and Shimoi.[26] Copyright 1999 American Chemical Society.

Regarding vibrational spectroscopic comparison, the coordinated B–H stretch cannot be observed in M(CO)$_5$BH$_3$·L, possibly because of overlap with the ν_{CO}, which are highly split into six bands that occur from 1840 to 2075 cm^{-1}. These values are significantly lower than those in M(CO)$_5$(H$_2$), which typically show only two ν_{CO} at 2090–2100 and 1970–1980 cm^{-1}. This indicates that the borane here is much more of a donor ligand and less of a π-acceptor ligand than H$_2$ or silanes.

It is of interest to compare the crystal structures of borane complexes (Figure 13.2) with those for other octahedral σ complexes, particularly silane complexes. The d_{BH} (1.1–1.3 Å) are not well determined (large standard deviations) and are not very meaningful. The M–H–B angles in M(CO)$_5$(BH$_3$·PMe$_3$) are near 130° (again with large esd's), and as high as 167° in the more sterically crowded W(CO)$_5$(B$_2$H$_4$·2PMe$_3$)$_2$[4] (Table 13.1). These are much larger than the angles

Table 13.1. Comparison of Structural Parameters in Borane (X = B) and Silane (X = Si) Complexes

Complex	M–H–X, °	M–X, Å	X–H$_b$, Å[a]	X–H$_t$, Å[b]
W(CO)$_5$(BH$_3$·PMe$_3$)	128(7)	2.86(2)	1.14(10)	1.14(28), 0.92(12)
Cu(BH$_4$)(PMePh$_2$)$_3$[c]	121.7(4)	2.518(3)	1.170(5)	
Ti(BH$_4$)$_3$(PMe$_3$)$_2$	112(5)	2.27(1)	0.95(6)	0.82(6), 0.99(6), 1.06(6)
Cp$_2$Ti(HBcat)$_2$	101(2)	2.335(5)	1.25(3)	
Mo(CO)(depe)$_2$(SiH$_3$Ph)	92(3)	2.501(2)	1.77(6)	1.41(6), 1.42(6)

[a] Coordinated XH.
[b] Terminal XH.
[c] Neutron structure.

observed in M–H–Si σ complexes and the Ti(HBcat) σ complexes discussed below. Thus the bonding for these neutral borane ligands, while still bent, is closer to end-on than side-on as in the borohydride complexes, where M–H–B can be as high as 162°. The hydrogen indeed appears to be nearly trans to CO in Figure 13.2, and the d_{MH} (<2 Å with large esd) appear to be in the range for hydrides as for **7–9**. Because these borane ligands contain four-coordinate boron when free, they are more like BH_4 compounds even though they are neutral. There is certainly more hydridic character than in the base-free catecholboranes discussed below.

It is clear that d_{BH} vary unpredictably, and the M–H–B angles perhaps also lack sufficient accuracy. However, d_{MB} is meaningful and is much longer, by about 0.6 Å, in $W(CO)_5(BH_3 \cdot PMe_3)$ than in $Cp_2W(BR_2)$, which has direct W–B bonds.[28] This would suggest that the electron density in the B–H bond being donated to M resides closer to H than B (see below for a discussion of bonding).

Transition metal–catalyzed hydroboration of carbon–carbon multiple bonds using organoborane derivatives and coupling with alkanes to form linear alkylboranes has provided valuable applications in organic synthesis.[29–31] Titanocene catecholborane complexes recently synthesized and characterized by Hartwig also appear to be the best examples of genuine η^2-BH σ complexes (Figure 13.3).[32–34]

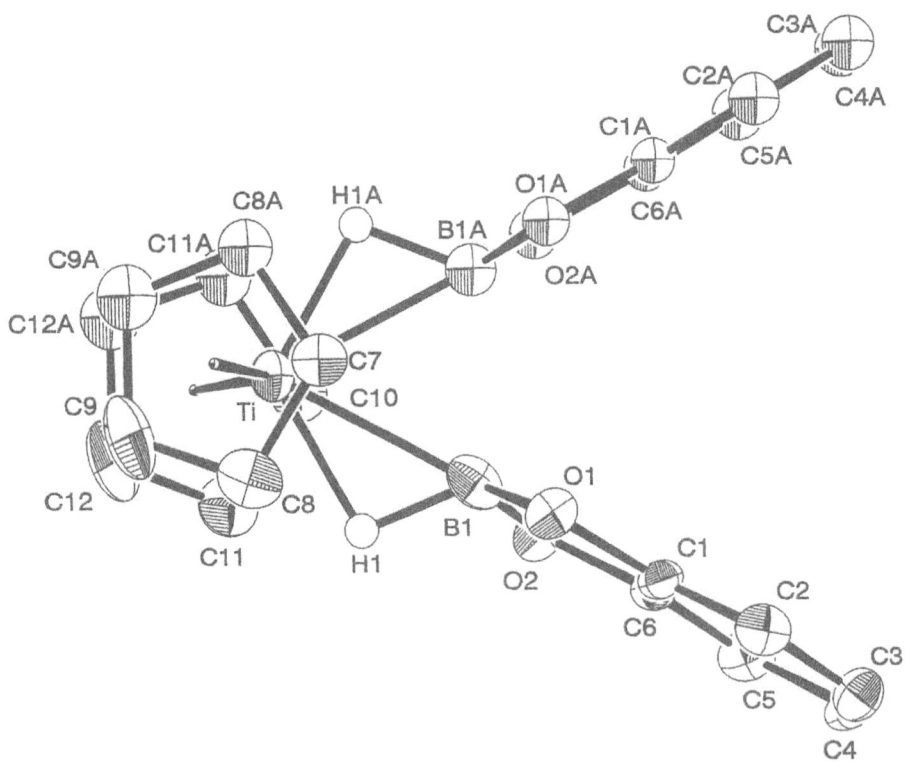

Figure 13.3. ORTEP drawing of $Cp_2Ti(HBcat)_2$ (R = R' = R" = H). Reprinted with permission from Hartwig *et al.*[32] Copyright 1996 American Chemical Society.

The ligand is neutral and the asymmetric side-on bonding geometry with Ti–H–B near 100° is more like that in η^2-silanes than in the borohydrides or $W(CO)_5$-$(BH_3 \cdot PMe_3)$ (Figure 13.2, Table 13.1). The complexes are initially synthesized as bis(borane) complexes [Eq. (13.3)] from which one borane can be displaced as in Eq. (13.4) to form, e.g., an analogous σ complex (15) with an η^2-silane instead of an η^2-borane.

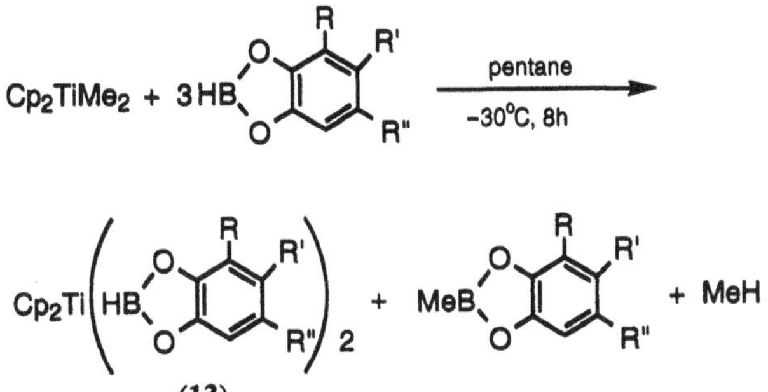

$$Cp_2TiMe_2 + 3\,HB\text{(cat)} \xrightarrow[-30°C,\,8h]{\text{pentane}} Cp_2Ti(HB\text{cat})_2 + MeB\text{cat} + MeH \quad (13.3)$$

(13)

Both borane ligands in **13** are rapidly displaced by CO to give $Cp_2Ti(CO)_2$. The d_{MB} [2.335(5) Å] in **13** (Figure 13.3) are longer than in metallocene boryl complexes, indicative of a bond order of less than one, but are significantly shorter than in the $M(CO)_5(BH_3 \cdot PMe_3)$ species. The d_{BH} is elongated more in **14** [1.35(5) Å] than in

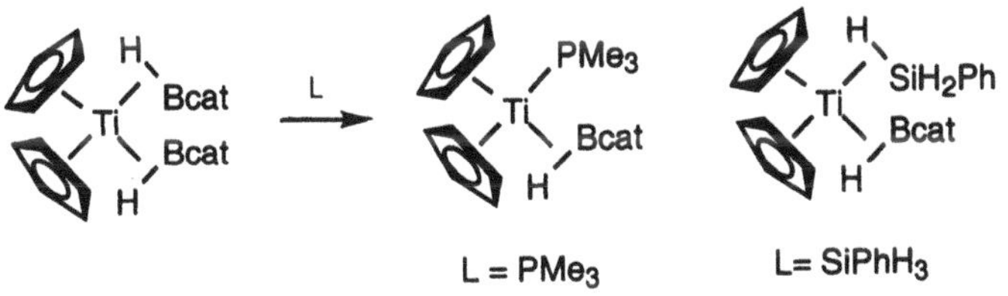

$$\text{(13)} \xrightarrow{L} \text{(14)} \quad L = PMe_3 \qquad \text{(15)} \quad L = SiPhH_3 \quad (13.4)$$

13 [1.25(3) Å], indicating higher activation toward OA because of the electron-donating phosphine, although d_{BH} correlations are not reliable. The d_{TiH} of 1.61(5) Å in **14** and 1.74(4) Å in **13** are relatively short, indicating substantial interaction. The most striking feature of the structure of **13** is the unusual geometry about boron, which has bonding interactions to four other atoms. The sum of the three angles at B that do not include H is 360°, indicating that the Ti, two oxygens, and B all lie in the same plane, i.e., the boron is highly distorted from tetrahedral geometry. The IR of the various analogues of **13** show ν_{BH} at 1600–1722 cm^{-1}, reduced from the 2660 cm^{-1} value in free catecholborane, in analogy with such reduction in other σ complexes.

In Eq. (13.4), the silane reaction is first order with a rate constant of $4.2 \times 10^{-4}\,s^{-1}$, and facile exchange of the BH and SiH hydrogens in the silane–borane product occurs even at $-30°C$. This lability and exchange dynamics are typical of true σ complexes and favor reactivity such as the hydroboration process in Eq. (13.5).[29] The alkyne borane is unstable at $-30°C$ and reacts with diphenylacetylene to give the vinyl boronate ester. Structure **13** is also a highly active catalyst for the hydroboration of vinylarenes. Several mechanisms for related Rh^I-catalyzed olefin hydroboration are possible from *ab initio* studies, involving either OA of the B–H bond or less favorable σ-bond metathesis pathways.[30]

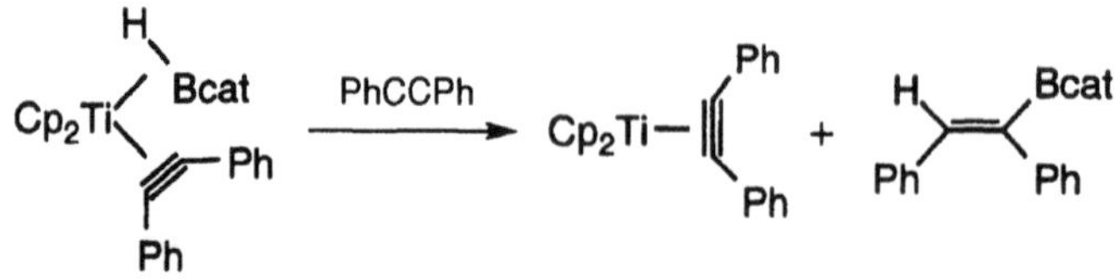

$$Cp_2Ti\begin{pmatrix}H\\Bcat\\-Ph\\Ph\end{pmatrix} \xrightarrow{PhCCPh} Cp_2Ti-\|\begin{pmatrix}Ph\\Ph\end{pmatrix} + \begin{pmatrix}H\\Ph\end{pmatrix}=\begin{pmatrix}Bcat\\Ph\end{pmatrix} \quad (13.5)$$

Theoretical studies of model complexes give insight into the nature of the M(η^2-BH) bonding in **13**, which *differs substantially from that in other σ complexes*.[4,32] In an *ab initio* study of hydride exchange processes in OsH$_3$(BH$_4$)(PR$_3$)$_2$, an intermediate with a η^2-BH$_3$ ligand is found to be stable calculationally, although such a species is not known experimentally.[4] The M–H–B angle of 85° in **16** is

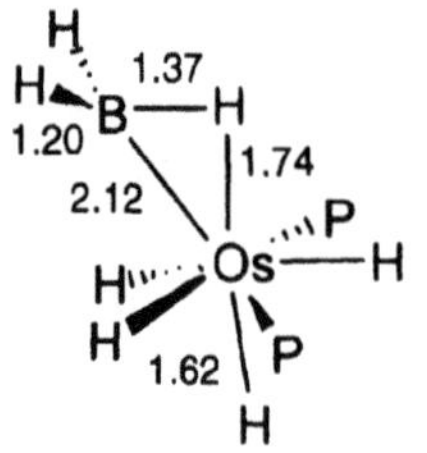

(16)

slightly more acute than that in silane complexes. Donation from the B–H bond to M occurs, but unlike for most other σ complexes, *BD from M does not go to the σ* B–H orbital but rather to a boron p_π orbital that is nonbonding with the hydrogen atom.* It is important to note that this interaction, no matter how strong, would not be expected to promote breaking of the B–H bond in an OA process. The geometry is also unusual in that it looks like a BH$_2$ plus an H bridging the B and Os atoms, i.e., the two H atoms bend away from M instead of keeping BH$_3$ trigonal planar.

Calculations on a Cp$_2$Ti[HB(OH)$_2$]$_2$ model for **13** verify the above bonding situation and also indicate a 3c-2e bond involving the B–Ti–B triangle.[32] There is

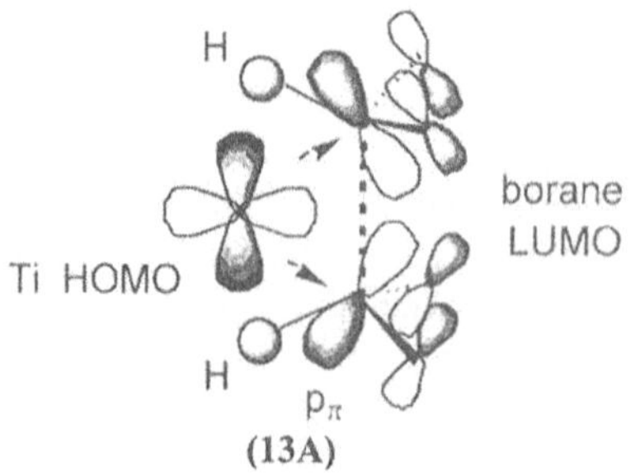

(13A) 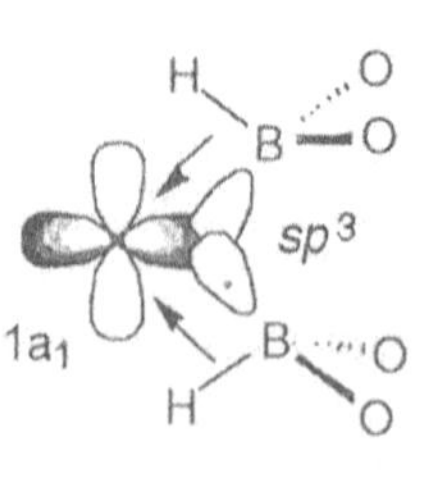   (13B)

overlap between the borons, which are close together ($d_{BB} = 2.11$ Å, which is about 0.25 Å longer than that in **2**).

Hartwig et al. showed that BD from Ti again goes into boron p orbitals (shown in **13A**), which are lower in energy than the σ^* orbitals of X–H in general. Lin's representation (**13B**) shows that the major interaction involves a filled Ti $1a_1$ orbital and the in-phase combination of the two empty sp^3-hybridized orbitals from the HB(OH)$_2$ units. Both cases give a closer energy match with the d orbitals and hence a stronger BD interaction than in other σ complexes. Equilibrium studies show that HBcat binding is thermodynamically favored over silane binding here, partly because the borane is more Lewis acidic and enhances BD (HBcat forms adducts with amines whereas SiHPh$_3$ does not). Furthermore, invoking this BD explains the unusual geometry at B also seen in the Os complex: Maximum overlap exists when the Ti, B, and O atoms are coplanar. The complete series of complexes Cp$_2$Ti(HBcat)(L) may be viewed as being additionally stabilized by some degree of intramolecular interaction with boron such as the B---B interaction above and B···H and B···C interactions in the silane (**17**) and alkyne (**18**) species, which undoubtedly facilitate hydrogen exchange and hydroboration.

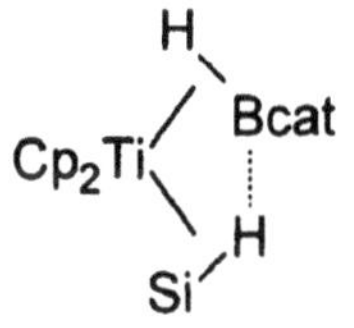

(17) 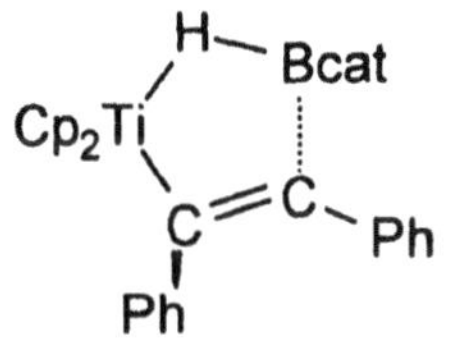   (18)

In terms of the influence of the electronic nature of L only, the stabilities follow the usual trends: donors such as PR$_3$ stabilize the complexes and CO destabilizes. Finally, as for silanes the stability of the σ complexes depends on the substituents on B, in this case the substituents R, R′, and R″ on the catechol aromatic ring. The rates for borane dissociation decrease with increasingly electron-withdrawing R groups, just as such groups on R$_3$SiH give more activated M(Si–H) bonds and more stable complexes.

As for M–silane and other M σ-bond interactions, there are several degrees of activation of the B–H bond (forms **19–21**).[31] Complete OA of boranes to

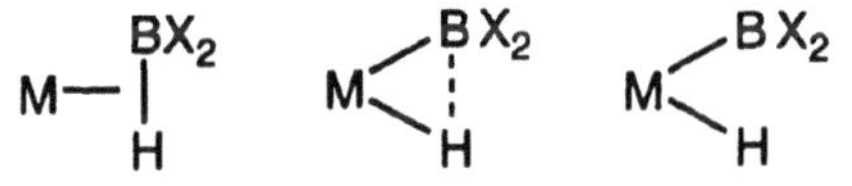

(19)   (20)   (21)

RhCl(PPh$_3$)$_3$ is known (**21**),[35] and calculations modeling Eq. (13.6) show that homolytic cleavage of B–H occurs without formation of an intermediate σ complex, i.e., without barrier.[30] Although the electronics for the bond-breaking process have not yet been well defined, the stereochemistry of OA to IrX(CO)(dppe) differs from that of H$_2$ and silane OA.[31] The kinetic product of Eq. (13.7) has boryl trans to halide whereas for H_2 and silane OA, the initial product has CO and hydride trans, suggesting that the mechanism for B–H addition differs substantially.

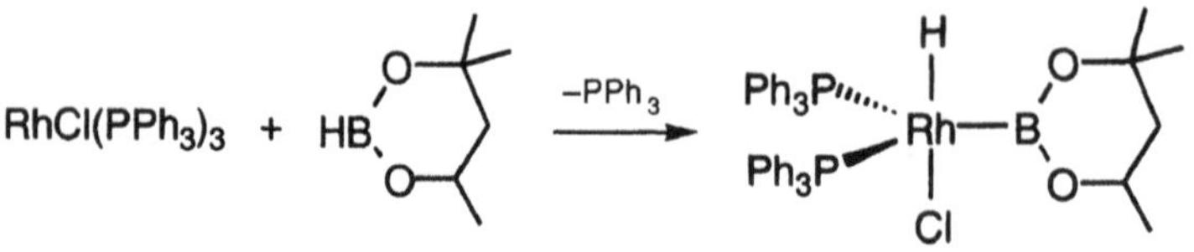

$$RhCl(PPh_3)_3 + HB\begin{pmatrix}O\\O\end{pmatrix} \xrightarrow{-PPh_3} Ph_3P\cdots Rh\begin{pmatrix}H\\Cl\end{pmatrix}B\begin{pmatrix}O\\O\end{pmatrix} \quad (13.6)$$

Examples of complexes that most closely resemble form **20** include exo-$Cp_2TaH_2(Bcat)$ and exo-$Cp_2NbH(Bcat)_2$, where d_{BH} is 1.12(13) Å.[31] The d_{MB} (~2.3 Å) is similar to that in endo-$Cp_2TaH_2(Bcat)$, which has nonbonding d_{BH} of

$$Br\cdots Ir\begin{pmatrix}P\\P\end{pmatrix}OC \xrightarrow{HBcat} H\cdots Ir\begin{pmatrix}Bcat\\Br\end{pmatrix}\begin{pmatrix}P\\P\end{pmatrix}OC \quad (13.7)$$

1.85 and 2.06 Å. Large isotopic perturbations in the NMR chemical shifts of the hydrides adjacent to B indicate significant B$\cdots$H interactions. The hydride and boryl environments are spectroscopically distinct, and the complexes are stereochemically rigid on the NMR timescale over a broad temperature range. Also, borane elimination does not readily occur, unlike for the Ti complexes. The W species $Cp_2WH(Bcat)$ on the other hand is an example of form **21**, where there is no structural or spectroscopic evidence for a B–H interaction and d_{WB} is considerably shorter.

$CpM(CO)_2(HBcat)$ and other group 7 derivatives with pinacolborane (HBpin) and HBR_2 ligands can be prepared by three different routes [Eqs. (13.8)–(13.10)][36]:

$$(MeCp)Mn(CO)_3 + HBcat \xrightarrow[\text{THF, 30 min}]{h\nu} (MeCp)Mn(CO)_2(HBcat) + CO$$

$$(13.8)$$

$$K[(MeCp)Mn(CO)_2H] + ClBcat \xrightarrow[\text{pentane, 10 min}]{\text{room temperature}} (MeCp)Mn(CO)_2(HBcat) + KCl$$

$$(13.9)$$

$$cis\text{-}Cp^*Re(CO)_2(Bpin)_2 + MeOH \xrightarrow[\text{benzene, 2 h}]{\text{room temperature}} Cp^*Re(CO)_2(HBpin) + MeOBpin$$

$$(13.10)$$

The structure of $(MeCp)Mn(CO)_2(HBcat)$ shows $d_{MnB} = 2.08$ Å, $d_{MnH} = 1.57$ Å, $d_{BH} = 1.29$ Å, and Mn–H–B = 93°, and the borane can be displaced by silanes, stannanes, and alkynes. The nonclassically bound H atoms in the borane complexes show extended self-decoupling in the NMR, which has also been observed in [BH$_4$]$^-$ complexes of transition metals.[37]

In summary, coordination and activation of B–H bonds closely parallels other M σ-bond interactions. The bonding is clearly 3c-2e, giving bent M–H–B geometries. Hydroboration reactions and σ-bond metathesis pathways similar to those discussed in Chapter 4 for other X–H bonds and are well summarized by Smith.[31] However, there are notable differences, and the variables of charge and coordination number for boron come into play and cloud the comparisons. Structural and spectroscopic gauges such as d_{BH} and J_{BH} are not as well defined or available as for H–H and Si–H activation. The most significant difference appears to be the nature of M → borane BD for certain systems where a p_π orbital on boron is lower in energy than B–H σ* and receives BD. Theoretical studies of OA of (HO)$_2$B–XH$_3$ to Pt(PH$_3$)$_2$ (X = C, Si, etc.) show that the σ B–X complex in the transition state is highly stabilized by the charge-transfer interaction between a metal d orbital and a B(OH)$_2$ p orbital as in **13A** and is the main reason for the high reactivity of (HO)$_2$B–XH$_3$ in the OA reaction.[38] OA occurs with moderate activation energy for X = C and with either a very small or no barrier for X = Ge, Si, and Sn. Further studies are needed to resolve the mechanism of OA of B–X, which differs both calculationally and experimentally from that of other OA processes.

13.2. X–H σ INTERACTIONS WHERE X IS ALSO A LONE-PAIR DONOR (N, P, S)

There is evidence that σ-bond interactions with M can exist for X–H groups where X is a classical lone-pair donor atom such as N, P, or S. In these cases the interaction is more akin to α-agostic M–C–H or M–P–C interactions (e.g., structure (**10**) in Section 12.2.2). Oxidative addition could occur here as an extreme case of cyclometallation.

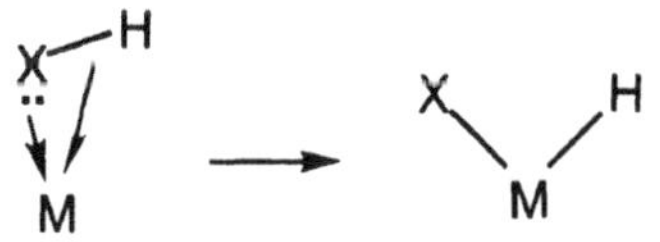

Protonation of a μ-phosphanido complex produces a species containing an apparent agostic P–H interaction.[39] X-ray data do not reveal the H positions but

(13.11)

NMR shows four separate signals for the P atoms. Most important, an ^{1}H signal for a hydridic type proton appears at δ – 0.16 as a doublet of doublets due to a reduced but still relatively high J_{PH} of 151 Hz compared to the normal values of 322 and 324 Hz for the terminal P–H bonds. NOE experiments reveal a strong interaction between the agostic P and H atoms and an internuclear distance of 1.9 Å.

As discussed in Chapter 12, interactions of N–H bonds with M are also known but are essentially 3c-4e and typically η^1 in nature as in a conventional hydrogen bond.[40-42] The prototypical example is $[R_3N-H^+\cdots Co(CO)_4^-]$, where the M–H–N angle is 180°,[41] but neutral N–H groups also interact as in **22**. Most involve 16e

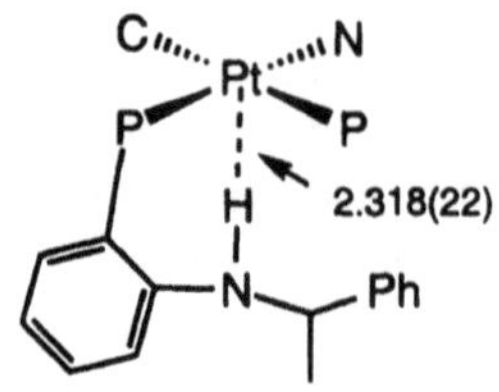

(22)

d^8-ML$_4$Pt^{II} or Pd^{II} centers and contain nearly linear M–H–N interactions (>140°), although an example is known with two interactions with Pd–H–N angles of 105° and 112°.[42] The proton in **22** can undergo H/D exchange with benzene-d_6 and other deuterated solvents as in many σ complexes.[43] Activation of N–H bonds toward OA does not occur here but is known for d^8 Ir complexes as in Eq. (9.22). Either N–H or C–H activation can occur in Eq. (13.12) depending on phosphine size, indicating

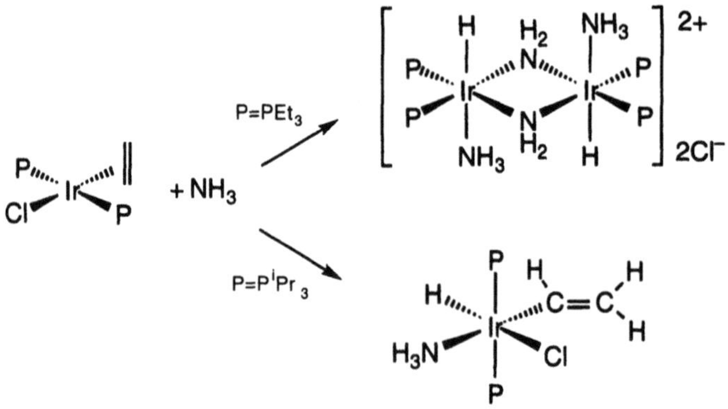

(13.12)

the fine balance of this system.[44] Such N–H OA, which could involve a η^2-NH intermediate, is rare compared to C–H OA in general and is apparently more reliant on steric effects.

Evidence for M···H–S interaction is less clear, but protonation of [Fe(SMe)-(CO)$_3$(PEt$_3$)]$^-$ at −80°C gives a species with an abnormal high-field NMR resonance for the S–H proton near δ − 8.[45] This resonance decays irreversibly on warming, and an isomeric mixture of hydride–thiolate complexes results. Low-temperature (−57°C) IR spectra of the initial protonation product in Eq. (13.13) shows a v_{CO} band at 1870 cm^{-1} consistent with an Fe^0 rather than an Fe^{II} complex. On warming, the band converts to others that match those observed for the hydride–

thiolate final product mixture. Thus the kinetic product is proposed to represent arrested RS–H OA. As proposed in calculations, such α-agostic structures could represent intermediates in proton transfer between Fe and μ-S ligands in hydro-

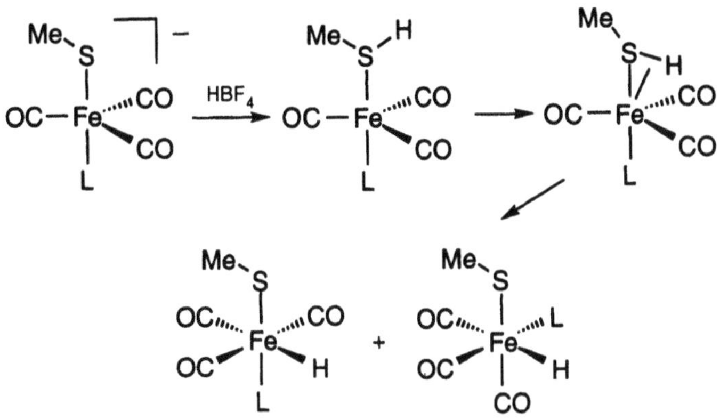

(13.13)

genases (see Section 10.1.4). Similarly, OA of O–H bonds and proton transfer between oxo ligands and M could involve analogous structures.

13.3. INTERACTIONS OF σ BONDS X–Y WHERE X AND Y ARE NOT HYDROGEN

13.3.1. C–Halogen Bonds

As discussed in Chapter 12, fluorocarbons and other molecules with C–halogen bonds coordinate and oxidatively add to M via the C–X bond rather than C–H. Although this should be expected for the heavier halocarbons such as CH_2Cl_2, which form isolable complexes through halogen lone pairs,[46] it is not so obvious for C–F bonds that are energetically at least similar to C–H bonds. However, both experimental and calculational evidence show that C–F bonds preferentially coordinate and cleave in H–C–F systems. Many examples of binding and activation of C–F bonds are known.[47] In regard to coordination of C–F, agostic-like interactions and low-temperature and gas-phase intermolecular binding [e.g., $W(CO)_5(CH_3F)$ with 11–12 kcal/mol energies[48]] are known, as for C–H interactions. Theoretical studies indicate that the bonding has a $M(\eta^2$-F–C) geometry, where d_{MC} is quite a bit longer (3.4–3.5 Å) than that for the analogous $M(\eta^2$-H–C), 2.8–2.9 Å (see Chapter 12).[49] The degree of purely nonclassical versus lone-pair bonding is not clear from computations since the M(X–C) system can be regarded as containing both lone-pair and α-agostic interaction (3c-4e). The interaction of aromatic C–F bonds with Cp_2Zr^{IV} types of species has been reported or postulated, and several of these complexes have been characterized by X-ray crystallography.[50]

As for OA of C–H bonds, calculations indicate that π-donating ligands such as Cl and third-row M facilitate OA of C–F bonds via a mechanism not unlike that for C–H cleavage in cyclometallation reactions. Theoretical studies of OA of CX_4 (X = F, Cl, Br, I) on 14e $IrCl(PH_3)_2$ suggest this pathway for X = F, Cl but that a radical mechanism proceeding by single-electron transfer is competitive for X = Br, I.[51]

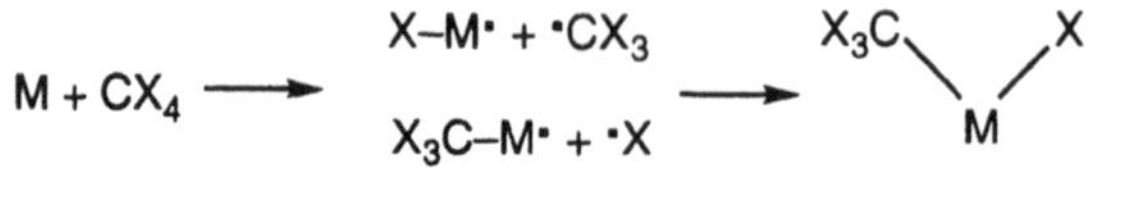

$$M + CX_4 \longrightarrow \begin{matrix} X\text{–}M^\bullet + {}^\bullet CX_3 \\ X_3C\text{–}M^\bullet + {}^\bullet X \end{matrix} \longrightarrow \begin{matrix} X_3C \diagdown \diagup X \\ M \end{matrix} \qquad (13.14)$$

13.3.2. C–C, C–Si, C–N, and C–B Bond Interactions with Metal Centers

Many examples of C–C bond cleavage and coupling reactions on M are known[52] and may proceed through a C–C σ complex. DFT calculations on $[Pt(CH_3)(CH_4)L_2]^+$ predict that C–C bond OA and RE are more difficult than C–H, although activation energies are similar when L = NH_3 or PH_3.[53] Conversely, computations on pincer complexes show that direct M (Rh, Ir) insertion into C–C is not only thermodynamically favored but is kinetically competitive with OA of C–H.[54] Formation of an approximately four-coordinate η^1-arene complex makes the C–C bond quite accessible to M, and a strong M-arene bond in the transition state significantly reduces the barrier for C–C activation.

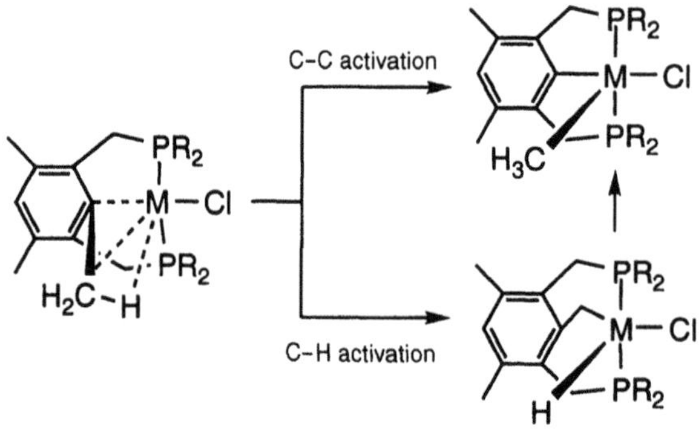

However, until recently no complexes clearly showed interactions of single C–C bonds with M because of steric interference of the substituents on tetrahedral carbon and the presence of more accessible C–H bonds that can give competing agostic interactions. The first example of well-characterized agostic C–C interactions is present in a 14e titanium system (**23**).[55] The saturated carbon C7 is only 2.293(7) Å from Ti (and even less for C3, C6, and C8). Remarkably these carbons are closer to Ti than any of the bound olefinic carbons C4 and C5 (2.30–2.34 Å) and Cp carbons, clearly indicative of several C–C bond interactions. The Ti–C2 distance of 2.579(7) Å also indicates that this carbon is involved in the agostic C–C interactions, which are verified by DFT calculations to occur for C2–C7 and even more strongly for the C2–C3, C6–C7, and C7–C8 bonds. Bond indexes of 0.168 for the Ti–C7 bond are comparable to the 0.182–0.209 values for the Ti bonds to the C4, C5, and the Cp carbons. The computed stabilizing energies for the donor–acceptor interactions of the four C–C bonds with Ti totaled a surprisingly high 57 kcal/mol.

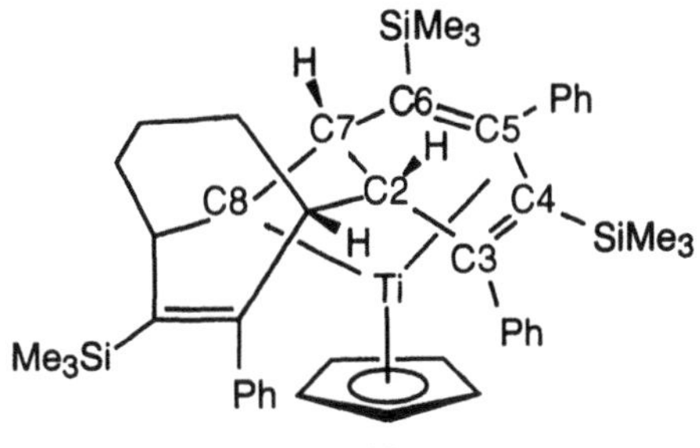

(23)

Although this is not a quantitative value for the overall binding energy, it is obvious that substantial interaction is occurring. The ^{13}C–^{13}C coupling constants for the agostic C–C bonds range from 18–30 Hz and inversely correlate with the calculated stabilization energies for each bond (10–20 kcal/mol). The strongest interaction is for the C6–C7 bond, which is significantly lengthened to 1.596(8) Å. Somewhat weaker interactions of Ti with the sterically less accessible C–H groups also occur in **23**, including two "backside" interactions involving C2–H and C7–H. The

presence of the M–C interactions here is obviously relevant to C–C bond activation and C–C coupling reactions.

There is striking precedence for the above C–C interactions in an agostic C–Si interaction also with a Ti center.[56,57] The d_{TiC} of 2.52 Å in **24** is similar to that for

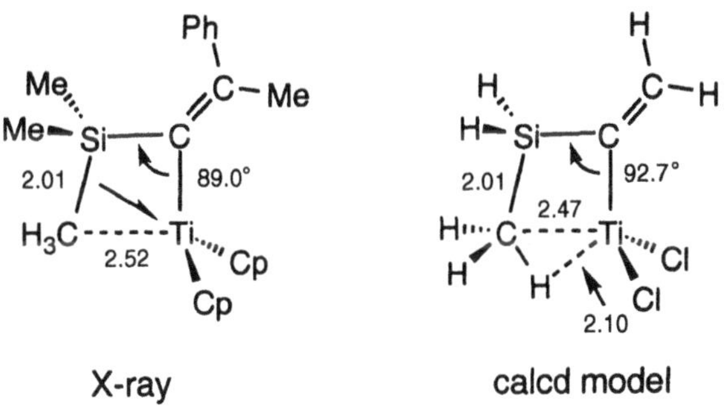

X-ray

(24)

calcd model

(25)

Ti–C2 in **23**, and the distortion of the alkenyl group (Ti–C–Si = 89°) is evidence for a β-agostic C–Si interaction, which was not initially recognized as such until subsequent *ab initio* calculations clarified the situation. Restricted Hartree–Fock energy gradient computations on a model (25) give an optimized structure similar to 23 with a slightly shorter d_{TiC}, and the alkenyl distortion is shown to be a direct consequence of the agostic interaction.[57] A Ti···H–C interaction (unlocated by X-ray) is also predicted to assist in the distortion, and the Ti–C overlap population is 0.288. Similarly distorted ligands such as $CH(SiMe_3)_2$ in the neutron structures of an electron-deficient lanthanide complex, $Cp^*La[CH(SiMe_3)_2]_2$, and a related yttrium complex, point to agostic Ln···Si–C_{Me} stabilization (Ln···C = 2.97 Å; Ln···Si = 3.28–3.42 Å).[58] Significant elongation (0.037 Å) of two Si–C bonds (one from each Si) is present both experimentally and calculationally. These interactions predominate over much weaker C–H···Ln interactions in stabilizing the lanthanide center, as previously found in the Ti system.[57]

The type of C–C interaction in 23 was essentially predicted 10 years in advance by Morokuma when he calculationally replaced the Si in 25 by C to test whether a C–C interaction could occur here.[57] A d_{TiC} of 2.36 Å and an elongated d_{CC} of 1.592 Å resulted [cf 1.596(8) Å for the C6–C7 in 23], along with an agostic C–H as in 25. The power of computational chemistry in both characterizing and predicting σ-bond interactions is again graphically demonstrated here. Regarding nonagostic intermolecular interactions, a theoretical study of OA of H–CH_3, CH_3–CH_3, H–SiR_3, and SiR_3–CH_3 to $Pt(PH_3)_2$ shows that OA of C–H and Si–H bonds occurs through a transition state with a planar structure as expected (X–H bond in the P–Pt–P plane).[59] However, OA of CH_3–CH_3 and SiH_3–CH_3 unexpectedly (from electronic factors) takes place through a nonplanar transition state structure where the dihedral angle between PtP_2 and PtXC planes is about 70° and 80° for X = Si and C,

respectively. This indicates that steric effects, i.e., avoidance of repulsions with the phosphines, takes precedence. Precursor complex A in which the CH_3 group approaches Pt in a similar way to the precursor complex of CH_4 is favored by 2.2 kcal/mol over B where the C–C is almost perpendicular to P–Pt–P. For OA of SiR_3–CH_3, the B form can be favored depending on R.

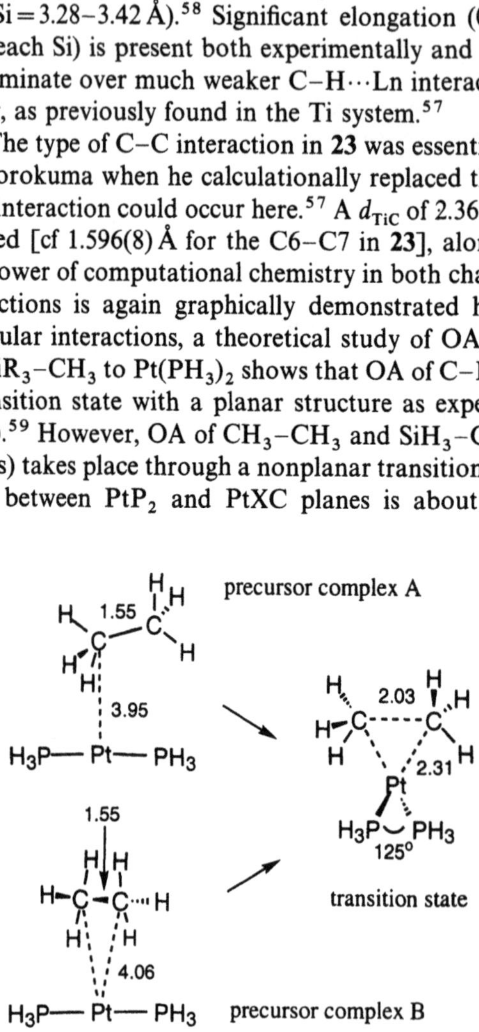

Another distorted system offers evidence for agostic C–B interactions with close M–C contacts.[60] Both the B–C–B (149°) and H–Zr–H (126°) angles in 26 are significantly greater than expected, drawing the C closer to Zr. The 2.419(4) Å separation is only slightly longer than an actual Zr–C bond [2.300(7) Å] in an

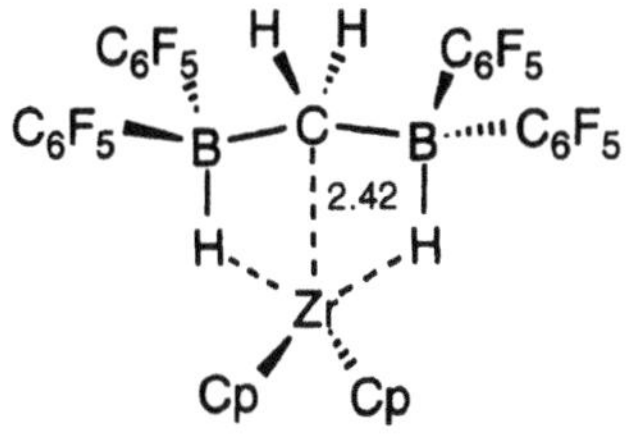

(26)

analogous complex. The agostic C atom has an unusual trigonal bipyramidal geometry, which is reflected in its upfield shift (δ 0.5) in the ^{13}C{^{1}H} NMR, although no direct evidence for the agostic interaction is found in the solution NMR. Because **26** can be formally viewed as a complex of $[Cp_2Zr]^{2+}$ with the dianion $[CH_2\{HB(C_6F_5)_2\}_2]^{2-}$ there is ambiguity in the ionic versus covalent nature of the bonding as in borohydride complexes in general.

Another Zr system shows evidence for an α-agostic C–N interaction, where the approach of the N-bound C to Zr is only 0.14 Å longer than one of the Zr-bonded C

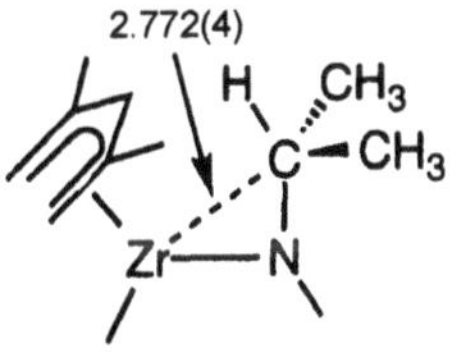

atoms.[55] Although a C–H interaction is possible here, the ^{13}C–H coupling constant is not low, indicating that interaction involves the N–C(^{i}Pr) bond.

13.3.3. Activation of C–P Bonds in Phosphines

Direct observation of C–P bond interaction with a M is quite rare despite the many known electronically unsaturated phosphine complexes (see Section 12.2.3). A known case is a trinuclear Ru cluster with a μ-phosphido interaction (**27**).[61] The Ru2···C$_{Ph}$ distance in **27** is 2.518(4) Å and the Ru(2)···P distance is 2.779(1) Å [cf.

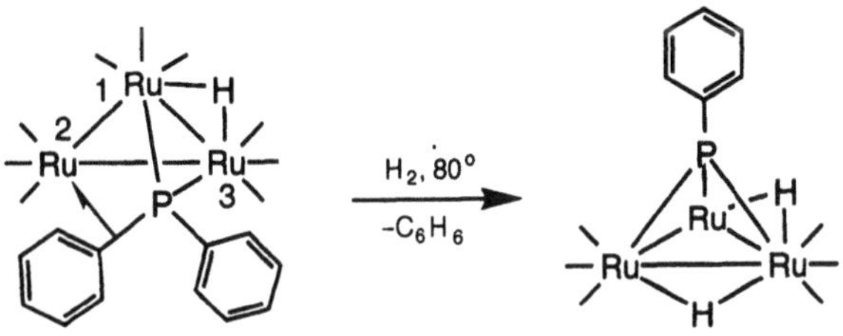

(13.15)

(27)   (28)

2.336(1) for Ru3–P]. The agostic C–P length [1.825(5) Å] does not vary significantly from that for the nonagostic C–P [(1.832(5) Å, cf. 1.87(2) Å in $[RuH(dpmb)_2]^+$. The ^{31}P NMR is not unusual, and **27** readily adds Lewis bases such as PR_3 to the electron-deficient Ru2 site. Most significantly, reaction with H_2 leads to cleavage of P–C_{Ph} to give benzene and a μ_3-phosphinidene complex (**28**) with two μ-H [Eq. (13.15)]. In the absence of H_2, heating **27** to 80°C for 4 h in heptane also produces a phosphinidene species $Ru_4(CO)_{13}(\mu_3$-PPh). Thus the agostic C–P interaction represents arrested bond rupture as for other σ complexes.

C–P bond cleavage of tertiary phosphines has indeed often been encountered.[62] The first examples in 1972 shown by Nyholm involved cleavage of PPh_3 on $Os_3(CO)_{12}$ in refluxing xylene to give nine novel products resulting from C–P and C–H cleavage and also C–C formation.[63] Products contained μ_2-PPh_2 (**29**), μ_3-Ph, and in some cases hydride and biphenylated ligands (**30**), all derived from decomposition of $Os_3(CO)_{10}(PPh_3)_2$ and CO loss. Intermediates with agostic C–P as in Eq. (13.15) are no doubt involved. Many more reactions comprising net addition of C–P

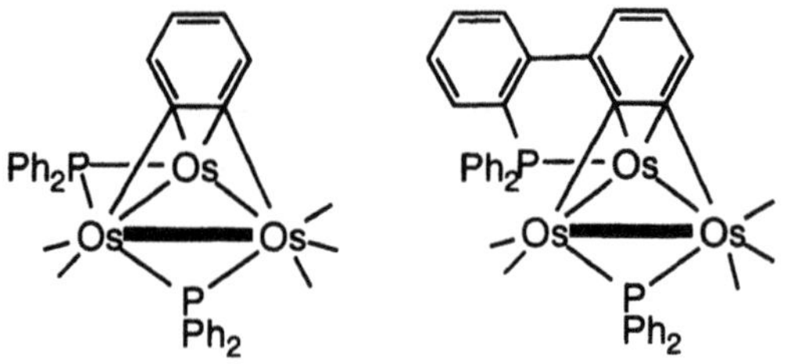

(**29**)   (**30**)

bonds to organometallic complexes have since been observed, even at room temperature.[62] This can be a concern in "heterogenized" or polymer-supported catalysis, wherein phosphines are anchored via C–P bonds to a support. Also such bond-breaking would be unacceptable in asymmetric synthesis and homogeneous catalysis of organic coupling reactions. Palladium-catalyzed cross-coupling reactions, for instance, can give problems involving interchange between P-bound aryl

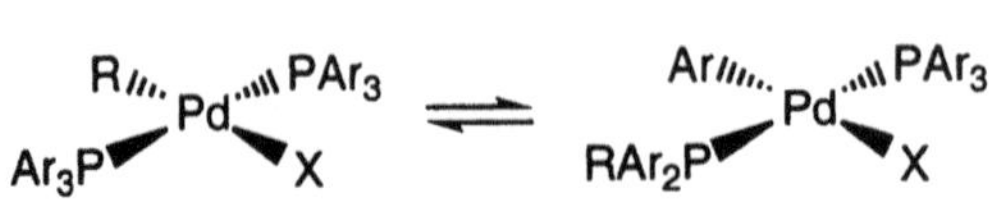

groups and Pd-bound aryl or alkyl groups in $RPdL_2X$ intermediates (X = Br, I) Mechanistic studies indicate that reductive elimination to form a phosphonium salt may occur, followed by OA of a different C–P bond.[64]

13.4. SUMMARY AND A GLANCE TO THE FUTURE

In summary for this chapter as well as this book, the above complexes along with the dihydrogen, silane, and alkane complexes suggest that the two electrons in any σ bond X–Y can and undoubtedly will be found to interact with electronically unsaturated M centers. The fact that Xe is now known to donate its closed-shell electrons to an electrophilic Au^{2+} center to form a structurally characterized complex, $[AuXe_4][Sb_2F_{11}]_2$,[65] solidifies the fundamental notion intimated by the original Kubas $M-H_2$ complex that *any* type of electron pair (Lewis base) can be attracted to a M binding site (Lewis acid) to form a Lewis acid–base complex. Clearly the atoms in X–Y do not have to be hydrogen, although the bonding picture can become more complex, and *p* and possibly *d* orbitals on XY can become involved. This becomes increasingly so the further away a σ complex gets from being an "ideal" nonclassical 3c-2e system such as $H^+-(H_2)$, i.e., as X and Y becomes atoms other than H and as the M center becomes less electrophilic (less like H^+). The electronic picture for X–Y cleavage is relatively straightforward for H–H (increased M → σ* BD leads to weakening of the M–H bond), but appears to involve *p* orbitals for B–H. Further theoretical studies may show that this or other orbital involvement is true for Si–H and other σ-bond activations as well.

Regardless of electronic considerations, there can be little dispute that the first step in σ-bond cleavage is formation of a transient or stable σ complex and that ever increasingly weaker interactions, including various types of hydrogen-bonded systems, will be subjects of future research. The latter will include and aid a variety of biomimetic efforts, particularly modeling the fascinating organometallic iron active sites in hydrogenases and the possibly related site in nitrogenase for hydrogen and ammonia production. Iron, the fourth most abundant element in the Earth's crust, will undoubtedly receive increasing focus in organometallic chemistry in this regard. Alkane binding and conversion discussed in the previous chapter is already well underway. As predicted by Chatt some 25 years ago and spurred on by the discovery of H_2 coordination, methane (along with the noble gases and other "unreactive" species) may indeed soon be among the most popular ligands in chemistry.

REFERENCES

1. James, B. D.; Wallbridge, M. G. H. *Progr. Inorg. Chem.* **1970**, *11*, 99; Marks, T. J.; Kolb, J. R. *Chem. Rev.* **1977**, *77*, 263.
2. Noth, H.; Hartwimmer, R. *Chem. Ber.* **1960**, *93*, 2238.
3. Kaesz, H. D.; Fellman, W.; Wilkes, G. R.; Dahl, L. F. *J. Am. Chem. Soc.* **1965**, *87*, 2753.
4. Demachy, I.; Esteruelas, M. A.; Jean, Y.; Lledos, A.; Maseras, F.; Oro, L. A.; Valero, C.; Volatron, F. *J. Am. Chem. Soc.* **1996**, *118*, 8388.
5. Lippard, S. J.; Melmed, K. M. *J. Am. Chem. Soc.* **1967**, *89*, 3929.
6. Bird, P. H.; Churchill, M. R. *J. Chem. Soc., Chem. Commun*, **1967**, 403.
7. Haller, K. J.; Andersen, E. L.; Fehlner, T. P. *Inorg. Chem.* **1981**, *20*, 309; Vites, J. C.; Eigenbrot, C.; Fehlner, T. P. *J. Am. Chem. Soc.* **1984**, *106*, 4633; Kawano, Y.; Matsumoto, H.; Shimoi, M. *Chem. Lett.* **1999**, 489.
8. Trofimenko, S., *Scorpionates: The Coordination Chemistry of Polypyrazolyborate Ligands*; Imperial College Press: London, 1999; Trofimenko, S., *Chem. Rev.* **1993**, *93*, 943.
9. Kosky, C. A.; Ganis, P.; Avitabile, G. *Acta Cryst.* **1971**, *27b*, 1859; Cotton, F. A.; Calderon, J. C.; Jeremic, M.; Shaver, A. *J. Chem. Soc., Chem. Commun.* **1972**, 777; Corrochano, A. E.; Jalon, F. A.; Otero, A.; Kubicki, M. M.; Richard, P. *Organometallics* **1997**, *16*, 145; Takahashi, T.; Akita, M.; Hikichi, S.; Moro-oka, Y. *Organometallics* **1998**, *17*, 4884; Malbosc, F.; Kalck, P.; Daran, J.-C.; Etienne, M. *J. Chem. Soc., Dalton Trans.* **1999**, 271; Herberhold, M.; Eibl, S.; Milius, W.; Wrackmeyer, B. *Z. Anorg. Allg. Chem.* **2000**, *626*, 552; Rodriguez, V.; Atheaux, I.; Donnadieu, B.; Sabo-Etienne, S.; Chaudret, B. *Organometallics* **2000**, *19*, 2916.

10. Takusagawa, F.; Fumagalli, A.; Koetzle, T. F.; Shore, S. G.; Schmitkons, T.; Fratini, A. V.; Morse, K. W.; Wei, C.-Yu.; Bau, R. *J. Am. Chem. Soc.* **1981**, *103*, 5165.
11. Khasnis, D. V.; Toupet, L.; Dixneuf, P. H. *J. Chem. Soc., Chem. Commun.* **1987**, 230.
12. McNamara, W. F.; Duesler, E. N.; Paine, R. T.; Ortiz, J. V.; Kolle, P.; Noth, H. *Organometallics* **1986**, *5*, 380.
13. Herberich, G. E.; Carstensen, T.; Kofer, D. P. J.; Klaff, N.; Boese, R.; Hyla-Krispin, I.; Gleiter, R.; Stephan, M.; Meth, H.; Zenneck, U. *Organometallics* **1994**, *13*, 619.
14. Behnken, P. E.; Márder, T. B.; Baker, R. T.; Knobler, C. B.; Thompson, M. R.; Hawthorne, M. F. *J. Am. Chem. Soc.* **1985**, *107*, 932.
15. Hendershot, S. L.; Jeffery, J. C.; Jelliss, P. A.; Mullica, D. F.; Sappenfield, E. L.; Stone, F. G. A. *Inorg. Chem.* **1996**, *35*, 6561, and references therein.
16. Brew, S. A.; Stone, F. G. A. *J. Adv. Organomet. Chem.* **1993**, *35*, 135.
17. Grimes, R. N., *Metal Interactions with Boron Clusters*, Plenum Press: New York, 1982.
18. Jensen, J. A.; Girolami, G. S. *J. Chem. Soc., Chem. Commun.* **1986**, 1160; Jensen, J. A.; Wilson, S. R.; Girolami, G. S. *J. Am. Chem. Soc.* **1988**, *110*, 4977.
19. Volatron, F.; Duran, M.; Lledos, A.; Jean, Y. *Inorg. Chem.* **1993**, *32*, 951.
20. Girolami, G., private communication.
21. Parry, R. W.; Kodama, G. *Coord. Chem. Rev.* **1993**, *128*, 245.
22. Snow, S. A.; Shimoi, M.; Ostler, C. D.; Thompson, B. K.; Kodama, G.; Parry, R. W. *Inorg. Chem.* **1984**, *23*, 511.
23. Shimoi, M.; Katoh, K.; Ogino, H. *J. Chem. Soc., Chem. Commun.* **1990**, 811; Katoh, K.; Shimoi, M.; Ogino, H. *Inorg. Chem.* **1992**, *31*, 670.
24. Hata, M.; Kawano, Y.; Shimoi, M. *Inorg. Chem.* **1998**, 37, 4482.
25. Shimoi, M.; Nagai, S.; Ichikawa, M.; Kawano, Y.; Katoh, K.; Uruichi, M. Ogino, H. *J. Am. Chem. Soc.* **1999**, *121*, 11704.
26. Kawano, Y.; Shimoi, M. *Chem. Lett.* **1999**, 869.
27. Piers, W. E. *Angew. Chem. Int. Ed. Engl.* **2000**, *39*, 1923; Dorn, H.; Jaska, C. A.; Singh, R. A.; Lough, A. J.; Manners, I. *Chem. Comm.* **2000**, 1041.
28. Hartwig, J. F.; DeGala, S. R. *J. Am. Chem. Soc.* **1994**, *116*, 3661.
29. He, X.; Hartwig, J. F. *J. Am. Chem. Soc.* **1996**, *118*, 1696; Hartwig, J. F.; Muhoro, C. N. *Organometallics* **2000**, *19*, 30, and references therein; Chen, H.; Schlecht, S.; Semple, T. C.; Hartwig, J. F. *Science* **2000**, *287*, 1995, and references therein.
30. Musaev, D. G.; Mebel, A. M.; Morokuma, K. *J. Am. Chem. Soc.* **1994**, *116*, 10693, and references therein.
31. Smith, M. R., III, *Progr. Inorg. Chem.* **1999**, *48*, 505.
32. Hartwig, J. F.; Muhoro, C. N.; He, X.; Eisenstein, O.; Bosque, R.; Maseras, F. *J. Am. Chem. Soc.* **1996**, *118*, 10936; Lam, W. H.; Lin, Z. *Organometallics* **2000**, *19*, 2625.
33. Muhoro, C. N; Hartwig, J. F. *Angew. Chem. Int. Ed. Engl.* **1997**, *36*, 1510.
34. Muhoro, C. N; He, X.; Hartwig, J. F. *J. Am. Chem. Soc.* **1999**, *121*, 5033.
35. Churchill, M. R.; Hackbarth, J. J.; Davison, A.; Traficante, D. D.; Wreford, S. S. *Am. Chem. Soc.* **1974**, *96*, 4041.
36. Schlecht, S.; Hartwig, J. F., *J. Am. Chem. Soc.* **2000**, *122*, 9435.
37. Marks, T. J.; Shimp, L. A., *J. Am. Chem. Soc.* **1972**, *94*, 1542.
38. Sakaki, S.; Kai, S.; Sugimoto, *Organometallics* **1999**, *18*, 4825.
39. Albinati, A.; Lianza, F.; Pasquali, M.; Sommovigo, M.; Leoni, P.; Pregosin, P. S.; Ruegger, H. *Inorg. Chem.* **1991**, *30*, 4690; Leoni, P.; Pasquali, M.; Sommovigo, M.; Laschi, F.; Zanello, P.; Albinati, A.; Lianza, F.; Pregosin, P. S.; Ruegger, H. *Organometallics* **1993**, *12*, 1702.
40. Yao, W.; Eisenstein, O.; Crabtree, R. H. *Inorg. Chim. Acta.* **1997**, *254*, 105, and references therein.
41. Brammer, L.; McCann, M. C.; Bullock, R. M.; McMullan, R. K.; Sherwood, P. *Organometallics* **1992**, *11*, 2339.
42. Chen, T.-R.; Wu, Y.-Y.; Chen, C.-T.; Chen, J.-D.; Keng, T.-C.; Wang, J.-C. *Inorg. Chem. Commun.* **1999**, *2*, 207, and references therein.
43. Hedden, D.; Roundhill, D. M.; Fultz, W. C. Rheingold, A. L. *Organometallics* **1986**, *5*, 336.
44. Schultz, M.; Milstein, D. *J. Chem. Soc. Chem. Commun.* **1993**, 318.
45. Darensbourg, M. Y.; Liaw, W.-F.; Riordan, C. G. *J. Am. Chem. Soc.* **1989**, 111, 8051.

46. Kulawiec, R. J.; Crabtree, R. H. *Coord. Chem. Rev.* **1990**, *99*, 89; Butts, M. D.; Scott, B. L.; Kubas, G. J. *J. Am. Chem. Soc.* **1996**, *118*, 11831.
47. Kiplinger, J. L.; Richmond, T. G.; Osterberg, C. E. *Chem. Rev.* **1994**, *94*, 373; Murphy, E. F.; Murugavel, R.; Roesky, H. W. *Chem. Rev.* **1997**, *97*, 3425; Plenio, H. *Chem. Rev.* **1997**, *97*, 3363.
48. Brown, C. E.; Ishikawa, Y.; Hackett, P. A.; Rayner, D. M. *J. Am. Chem. Soc.* **1990**, *112*, 2530.
49. Zaric, S.; Hall, M. B. *J. Phys. Chem. A*, **1997**, *101*, 4646; Su, M.-D.; Chu, S.-Y. *Organometallics* **1997**, *16*, 1621; *J. Am. Chem. Soc.* **1997**, *119*, 5573; *Chem. Eur. J.* **1999**, *5*, 198.
50. Horton, A. D.; Orpen, A. G. *Organometallics* **1991**, *10*, 3910; Yang, X.; Stern, C. L.; Marks, T. J. *J. Am. Chem. Soc.* **1994**, *116*, 10015; Karl, J.; Erker, G.; Frohlich, R. *J. Am. Chem. Soc.* **1997**, *119*, 11165; Dahlmann, M.; Erker, G.; Nissinen, M.; Fröhlich, R. *J. Am. Chem. Soc.* **1999**, *121*, 2820; Watson, P. L.; Tulip, T. H.; Williams, I. *Organometallics* **1990**, *9*, 1999; Lancaster, S. J.; Thornton-Pett, M.; Dawson, D. M.; Bochmann, M. *Organometallics* **1998**, *17*, 3829; Edelbach, B. L.; Rahman, A. K. F.; Lachicotte, R. J.; Jones, W. D. *Organometallics* **1999**, *18*, 3170.
51. Su, M.-D.; Chu, S.-Y. *J. Am. Chem. Soc.* **1999**, *121*, 1045.
52. Rybtchinski, B.; Milstein, D. *Angew. Chem. Int. Ed. Engl.* **1999**, *38*, 870.
53. Hill, G. S.; Puddephatt, R. J. *Organometallics* **1998**, *17*, 1478.
54. Cao, Z.; Hall, M. *Organometallics* **2000**, *19*, 3338; Sundermann, A.; Uzan, O.; Milstein, D.; Martin, J. M. L. *J. Am. Chem. Soc.* **2000**, *122*, 7095.
55. Tomaszewski, R.; Hyla-Kryspin, I.; Mayne, C. L.; Arif, A. M.; Gleiter, R.; Ernst, R. D. *J. Am. Chem. Soc.* **1998**, *120*, 2959; Ernst, R. D. *Comm. Inorg. Chem.* **1999**, *21*, 285.
56. Eisch, J. J.; Piotrowski, A. M.; Brownstein, S. K.; Gabe, E. J.; Lee, F. L. *J. Am. Chem. Soc.* **1985**, *107*, 7219.
57. Koga, N.; Morokuma, K. *J. Am. Chem. Soc.* **1988**, *110*, 108.
58. Klooster, W.; Brammer, L.; Schaverien, C. J.; Budzelaar, P. H. M. *J. Am. Chem. Soc.* **1999**, *121*, 1381.
59. Sakaki, S.; Mizoe, N.; Musashi, Y.; Biswas, B.; Sugimoto, M. *J. Phys. Chem. A* **1998**, *102*, 8027.
60. Spence, R. E. von H.; Parks, D. J.; Piers, W. E.; MacDonald, M.-A.; Zaworotko, M. J.; Rettig, S. J. *Angew. Chem. Int. Ed. Engl.* **1995**, *34*, 1230.
61. MacLaughlin, S. A.; Carty, A. F.; Taylor, N. J. *Can. J. Chem.* **1982**, *60*, 87.
62. Garrou, P. E. *Chem. Rev.* **1985**, *85*, 171.
63. Bradford, C. W.; Nyholm, R. S. *J. Chem. Soc., Chem. Commun.* **1972**, 87; *J. Chem. Soc., Dalton Trans.* **1973**, 529.
64. Goodson, F. E.; Wallow, T. I.; Novak, B. M. *J. Am. Chem. Soc.* **1997**, *119*, 12441.
65. Seidel, S.; Seppelt, K. *Science* **2000**, *290*, 117.

Abbreviations Commonly Used in the Text

1D, 2D	one-dimensional, two-dimensional
3c-2e	three-center, two-electron
Å	angstrom unit, 10^{-10} m
AIM	atoms-in-molecules
Ar	aryl or arene
asym	asymmetric or antisymmetric
B	rotational constant
B3LYP	Becke gradient-corrected DFT
BAr$_f$	$[B\{C_6H_3(3,5\text{-}CF_3)_2\}_4]^-$ anion
Bcat	B(catechol) unit
BD	backdonation or backbonding from M to ligand
BDE	bond dissociation energy
bipy	2,2'-dipyridine or bipyridine
Bp	dihydrobis(pyrazo-1-yl)borate, $H_2B(pz)_2$
Bp$^{(CF_3)_2}$	dihydrobis(3,5-ditrifluoromethylpyrazo-1-yl)borate
bpym	bipyrimidine
bq	benzoquinolinate
Bu	butyl (^{i}Bu, iso-; ^{t}Bu, tertiary-butyl)
Bz	benzyl
CASSCF	complete-active-space self-consistent-field
CDA	charge decomposition analysis
cht	cycloheptatriene
CI	configuration interaction
CID	collision-induced dissociation
CISD	configuration interaction with all-single and -double excitations
Cn	1,4,7-triazacyclononane
CNDO	complete neglect of differential overlap
COD or cod	cyclooctadiene
COT or cot	cyclooctatetraene
Cp	cyclopentadienyl
Cp*	pentamethylcyclopentadienyl

CpMe or MeCp	methylcyclopentadienyl
C_Q	deuterium quadrupole coupling constant
cus	coordinatively unsaturated site
Cy	cyclohexyl
cyttp	$PhP(CH_2CH_2CH_2PCy_2)_2$
dcype	1,2-bis(dicyclohexylphosphino)ethane
D_e	bond energy
depe	1,2-bis(diethylphosphino)ethane
dfepe	$(C_2F_5)_2PC_2H_4P(C_2F_5)_2$
DFT	density functional theory
d_{HH}, d_{MH}, etc.	interatomic distance
diphos or PP	any chelating diphosphine, but usually dppe
diphpyH	2-(Ph)-6-(o-C_6H_4)(C_5H_3N)
dippe	1,2-bis(di-isopropylphosphino)ethane
DMF or dmf	N,N'-dimethylformamide
dmpe	1,2-bis(dimethylphosphino)ethane
dmpm	1,2-bis(dimethylphosphino)methane
DMSO	dimethylsulfoxide
DPB	diporphyrinatobiphenylene tetraanion
dppb	1,2-bis(diphenylphosphino)butane
dppe	1,2-bis(diphenylphosphino)ethane
dppf	1,1'-bis(diphenylphosphino)ferrocene
dppm	1,2-bis(diphenylphospino)methane
dppp	1,2-bis(diphenylphosphino)propane
dtfpe	1,2-bis[di-(p-trifluoromethylphenyl)phosphino]ethane
DVR	discrete variable representation
E_{BD}	backdonation energy
E_D	donation energy
EELS	electron energy loss spectroscopy
EH	extended Huckel
EIE	equilibrium isotope effect
en	ethylenediamine
ENDOR	electron double resonance
Et	ethyl
ETS	extended transition state
EXAFS	extended X-ray absorption fine structure
F	force constant
FeMoco	Fe–Mo cofactor of nitrogenase
FWHM	full width at half maximum
H_2bim	2,2'-bi-imidazole
H-ase	hydrogenase (enzyme)
HDS	hydrodesulfurization
HOMO	highest occupied molecular orbital
Hz	hertz
INS	inelastic neutron scattering
IPR	isotopic perturbation of resonance
KIE	kinetic isotope effect

Abbreviations Commonly Used in the Text

L	ligand
Ln	any lanthanide element
LUMO	lowest occupied molecular orbital
M	central metal in compound
MCSCF	multiconfigurational self-consistent field
Me	methyl
MM2	molecular mechanics calculation
MMO	methane monooxygenase (enzyme)
MO	molecular orbital
MP2, MP3, etc.	Moller–Plesset
MRCI	multireference configuration interaction
$N_3N_F^{3-}$	$[(C_6F_5NCH_2CH_2)_3N]^{3-}$
N-ase	nitrogenase (enzyme)
NBD or nbd	norbornadiene
NHE	normal hydrogen electrode
OA	oxidative addition
OAc	acetate anion
OEP	octaethylporphyrinato dianion
OTf	triflate anion
PAC	photoacoustic calorimetry
P–C–P	$HC[C_2H_4P^tBu_2]_2$
PE	photoelectron spectroscopy
PES	potential energy surface
Ph	phenyl
PHIP	parahydrogen induced polarization
PP or P–P	diphosphine ligand
pp_3	$P(CH_2CH_2PPh_2)_2$
Pr	propyl (^iPr, isopropyl)
py	pyridine
pz	pyrazine or pyrazolyl
QEC	quantum-mechanical exchange coupling
QNS	quasielastic neutron scattering
R	alkyl (preferably) or aryl group
RE	reductive elimination
RHF	restricted Hartree–Fock
S	solvent
S_4	1,2-bis[(2-mercaptophenyl)thioethane]
SCF	self-consistent field
sq pyr	square pyramid
sym	symmetric
T	tritium
T_1	NMR relaxation time
T_1^{min}	minimum value of the NMR relaxation time
TBP or tbp	trigonal bipyramid
tetraphos	$PPh_2(CH_2CH_2PPh_2)_2CH_2CH_2PPh_2$
THF or thf	tetrahydrofuran
THT or tht	tetrahydrothiophene

tmeda	$Me_2NC_2H_4NMe_2$
Tp	hydrotris(pyrazo-1-yl)borate, $HB(pz)_3$
Tp* or Tp^{Me2}	hydrotris(3,5-dimethylpyrazo-1-yl)borate
Tp'	derivative of Tp (scorpionate)
triflic acid	CF_3SO_3H
triflate (ion)	$CF_3SO_3^-$
triphos	$MeC(CH_2PPh_2)_2$
TRIR	time-resolved infrared
TS	transition state
V_2, V_4	energy barrier for twofold or fourfold rotation
X (used alone)	halide (typically), pseudohalide, or ligand anion
XH, XY, etc.	X, Y, or Z is any covalently-bonded element
ZPE	zero-point energy

Index

Acidity of H$_2$ ligand, hydride, etc.: *see* Dihydrogen ligand or complex, acidity of, etc.
Actinide-dinitrogen complex, 218
   importance of M−N$_2$ backdonation in stabilizing, 218
Activation energy: *see* Energy barrier
Agostic interactions
   favored entropically over external ligand binding, 220–221
   B−H bonds in
      metallocyclopentadienyl, metallocarborane, and other borane-like complexes, 420
      pyrazolylborate (Bp and Tp) complexes, 419
   of C−B bonds, 434–435
   of C−C bonds, 432–433
   C−H bonds, 3–4, 59, 366–383
      acidity of, 376
      in bis(pyrazolylborate) complexes, 368
      bond energy of: *see* Bond energy
      bridging, 376–377
      cleavage: *see* Cleavage, heterolytic, of C−H; Cleavage, homolytic, of C−H
      comparison to agostic Si−H bonds, 337, 338
      d^0 complexes of, 372
      deprotonation of, 376
      derivation of the term "agostic," 366
      discovery and development of new examples of, 367–369
      distance from M to CH bonding electron pair, d_{bp}, 371–372, 376
      effect of substituents at carbon on binding and activation, 373
      elongated, 369
      fluxionality in, 373
      as intermediate in ligand displacement reactions, 240–241
      lanthanide complexes of, 372
      multiple, 86, 368, 372–373
      on metal surfaces, 368
      multiple interactions, 86, 268, 372–373
      reversible displacement by H$_2$ and other ligands, 23–24, 42, 80, 86, 222, 232–233, 264, 378–379

Agostic interactions (*cont.*)
   of C−H bonds (*cont.*)
      review of, 366
      structural parameters for, table of, 372
      types of, 368
      use in modeling the trajectory for the reaction M + C−H → C−M−H, 371, 395–396
   of C−H bonds in phosphines, 378–383
      ancillary ligand effects in stabilizing or destabilizing, 381–383
      characteristic deep color of complexes with, 22, 381
      comparison of bond distances in, table of, 380
      multiple interactions, 379–383
      reservation of binding sites for other small molecules, 23, 378
      reversible displacement by H$_2$ and other ligands: *see* Agostic interactions, of C−H bonds, reversible displacement by H$_2$ and other ligands
      steric factors, critical effects on bonding of, 381–383
      use in preparation of first H$_2$ complexes, 23–24, 378
      in W(CO)$_3$(PR$_3$)$_2$, 23–24, 378–380
      X-ray structure of W(CO)$_3$(PCy$_3$)$_2$, 378–379
   C−N bonds, 435
   C−P bonds, 435–436
   C−Si bonds, 433–434
   Ge−H bonds, 358
   N−H and P−H bonds, 429–431
   S−H bonds, 312, 322, 429–431
   Si−H bonds, 330–331
      in bridging silyl complexes, 336–337
      multiple interactions, 330
      synthesis of complexes with, 338
      to early-metal metallocenes, 351–352
      to main-group metal cations, 348–349
      γ-agostic interaction, 338
   σ bonds, 4, 14
Alkane
   activation on Shilov and related systems: *see* Shilov and related systems

Alkane (cont.)
   cleavage of C—H bond: see Cleavage, heterolytic; Cleavage, homolytic; Oxidative addition, of alkanes and C—H bonds
   dehydrogenation, 370
   exchange reactions on Cp_2M, 126
   oxidation and conversion to liquid fuels, 5, 14, 365, 404–411
      importance of selectivity, 365, 404
Alkane ligand or complex, 223–226, 383–388; see also Methane coordination
   acidity of, 403–404, 409–410
   in alkane activation on Shilov and related systems: see Shilov and related systems
   alkane versus fluoroalkane binding, 391–393
   bond energy: see Bond energy
   coordination modes for, 384
   to $CpRe(CO)_2$ fragment
      $CpRe(CO)_2$(heptane) as longest-lived at room temperature, 223, 386
      observation of $CpRe(CO)_2$(cyclopentane) by NMR, 386–387
   direct spectroscopic evidence for, in TRIR studies of alkane OA, 399–403
      correlation of binding affinities of alkanes and rare gases to $Cp*Rh(CO)$ with their effect of alkane size and structure on binding affinity, 402–403
      ionization potentials and polarizabilities, 403
      in photolysis of $Cp*Rh(CO)_2$ in liquid rare gas solvents, 399–401
      in photolysis of $CpRh(CO)_2$ in gas phase, 401
      in photolysis of $Tp*Rh(CO)_2$ in alkane solvents at room temperature, 401–402
   disfavored by the presence of agostic interactions, 220–221
   equilibrium isotope effect for alkane binding, 238
   first direct observation by NMR, 386–387
   to group 6 $M(CO)_5$, $CpRh(CO)$, and similar fragments, 211–212, 223–226, 384–387, 396–403
      effect of alkane size and structure on binding affinity, 402–403
      kinetic studies and rate constants for reaction of H_2 and other ligands with, 223–226, 385–386
   on highly electrophilic metal centers such as Pt^{II} cations, 403–411
   isotope effects and isotopic exchange in, 238–241, 395–399, 405–411
   oxidative addition and reductive elimination of alkanes, 317–319; see also Oxidative addition; Reductive elimination as possible intermediate in the function of oxygenases

Alkane ligand or complex (cont.)
   reaction of heptane complexes of metal carbonyls with CO, dependence of rate constants on M, 386
   as soft ligand in terms of hard/soft acid-base principles, 403–404
   thermodynamics of the bonding and cleavage of C—H bonds: see Thermodynamics of the bonding and cleavage of σ ligands, for C—H bonds
   as transient and/or intermediate species in OA/RE of C—H bonds, 366–367, 385, 395–411
   evidence for in isotopic (H/D) involving σ-alkane and CH_4 ligands, 396–399, 405–411
   trajectory for the approach of alkane to M as modeled by agostic interactions of C—H bonds, 371, 395–396
   X-ray structure of heptane interaction with iron in porphyrin complex, 387–388
Alkene: see also Ethylene; Catalytic hydrogenation
   displacement of coordinated H2 by, 288
   equilibrium isotope effect for binding of, 238
   insertion reactions, 377
   insertion into M—Si bond, 352
   photocatalytic hydrosilylation of dienes, 352
   Ziegler-Natta polymerization of, 113, 377
AlkyneL see also Catalytic hydrogenation
   addition to $[MX(H_2)P_4]^+$, 288–289
   alkyne-borane complex, reaction with diphenylacetylene to give vinyl boronate ester, 426
   dimerization of 1-alkynes catalyzed by $[RuH(H_2)(pp_3)]^+$, 289
   displacement of coordinated borane by, 428
   displacement of coordinated H2 by, 288
   insertion reactions, 377
Allyl ligand, analogy to H_3^-, 12, 122
Ammonia production, 6, 312, 437
   by protonation of N_2 ligand by hydride or H_2 complexes, 316
   by reaction of N_2 complex with SH complex, 321
Anion
   binding to M, 42
   low-coordinating, to stabilize cationic σ complexes, 42, 46
Aquo ligand
   comparison of bonding electronics to that for H_2 ligand, 10
   competition with alkane ligands, 215, 404
   displacement by H_2 or competition with H_2 ligand, 10, 40, 45, 213–216
      relevance to biological activation of H_2, 215, 297, 308

Aquo ligand (cont.)
 displacement by H_2 or competition with H_2 ligand (cont.)
  thermodynamics of the displacement of H_2 ligand by H_2O, 215–216
 displacement by methane by associative substitution process, 407–408
 X-ray structure of $W(CO)_3(PR_3)_2(H_2O) \cdot thf$, 213–214
 hydrogen bonding of H_2O with thf solvate and CO ligand in, 213–214
Argon coordination to $M(CO)_5$ (M = group 6), 23, 76

Backdonation (BD)
 comparisons among σ and classical ligands, 3, 10–12, 360
 effect of trans ligand on: see Trans effect
 influence on oxidative addition of σ ligands, 11, 15
 M to B–H
  to boron p_π orbital rather than B–H σ^*, 426–427, 429
  compared to BD to H–H and Si–H, 423
 M to C–H σ^*,
  influence of metal and metal row on, 377
  weaker than for other σ ligands, but still necessary for CH activation, 76, 86, 211, 367, 371, 374, 386, 390, 393–394
 M to CO, 76, 78
 M to germane Ge–H σ^*, comparison to BD in other σ complexes, 358–360
 M to H_2 σ^*, 60–61, 80–86: see also Rotation and libration of H_2 ligand, energy barrier for
  analogy to M-olefin BD, 60
  balanced with σ donation, 60, 84, 253
  competition for, 70, 73, 97–98
  correlation with reactivity and structural trends, 271
  for dihydrogen ligand, first illustration of, 265
  effect of orientation of H_2, 73, 172–187
  effect of skeletal ligand distortions on, 65, 184–185
  effect of varying cis-halide, 184
  experimental evidence for, 15, 75–76, 172–173
  for Fe versus Ru complexes, 45, 118–119, 180
  influence of metal and metal row on, 178–182
  influence on H–H activation towards OA, 10, 11, 64, 72, 73, 109, 167
  measurement of and separation from σ donation, 67, 77–79, 178–184
  in $Pd(H_2)$, 67
  reduction of on electrophilic and positively-charged centers, offset by increased σ donation, 11, 61, 66–67, 78, 84–86, 105, 167, 253, 255, 275, 304

Backdonation (cont.)
 M to $H_2\sigma^*$ (cont.)
  relative strength of and comparison to BD to other ligands, 76, 78, 167, 213, 218–219, 423
  stabilization of $M-\eta^2-H_2$ binding, 3, 11, 19, 60–61
 M to N_2, importance in stabilizing N_2 binding, 218
 M to silane Si–H σ^*, 12, 213, 339
  comparison to $M{\rightarrow}H_2$ BD, 213, 219, 330, 342, 348, 349
  reduction of on electrophilic and positively-charged centers, offset by increased σ donation, 344, 345
 M to stannane Sn–H σ^*, comparison to BD in other σ complexes, 362
Badger's rule, 252
Barrier: see Energy barrier
B–H bond coordination and activation, 417–429; see also Borane σ-complex; Borohydride and borohydride-like ligands; Agostic interactions, of B–H bonds
B–H bond distance,
 in coordinated borohydrides and boranes, 419, 423–429
 elongated, 427–428
 lack of correlation with degree of activation of B–H, 419–420, 424
 in terminal (uncoordinated) B–H bond, 419
Biomimetic methods, 6, 437
Bis(pyrazolylborate) complexes (Bp), agostic interactions in, 368, 419
Bis-H_2 complexes: see Dihydrogen ligand or complex, bis-H_2 complexes
Bond cleavage: see Cleavage
Bond energy
 of agostic C–H interaction, 208, 210, 232–233, 372, 378
 C–H in methane, and comparison to that in ethylene and acetylene, 5, 356, 365, 394–395
 C–H in methanol, 365
 comparison of H_2 and CH_4 binding energies in $[M(L)_n]^+$, 129, 210–211
 comparison of H_2 and N_2 binding energies, 77, 212–213, 216–222
 gas phase measurements of H_2 and alkane binding energies, 210–213
 H–H, 5, 6, 356
 of H_2, alkane, silane, and other ligands in thermally unstable complexes, 210–213, 385–386, 389
 table of, 212
 in $M(CO)_5(L)$ in gas versus solution phases, 211–212

Bond energy (*cont.*)
  in hydrogen bonds, 281
  M−alkane, 220−221, 385−386, 388: *see also*
    Bond energy, of H_2, alkane, silane, and
    other ligands in thermally unstable
    complexes
  M−CH_4, 4, 129, 210−212, 385, 389, 391−394
    in $[M(CH_4)_n]^+$, 129−131
    to Pt in Shilov system, 408−409
    versus fluoromethanes, 391−393
  M−fluoroalkane, 391−393
  M−H, in classical hydrides, 208, 237, 391
  M−H_2
    calculated in $[Cp^*Ru(dppm)(H_2)]^+$ and
      $[OsCl(dppe)_2(H_2)]^+$, 98−100
    calculated in a Ti-allyl-H_2 complex, 113
    calculated in $M(H_2)(CO)_3(PH_3)_2$ and other
      group 6 complexes, 69−70, 73−75, 78
    comparison of calculated and observed, table
      of, 73
    in $[M(H_2)_n]^+$, 129−131
    in $M(CO)(H_2)(dppe)_2$ (M = Mo, Mn^+, Fe^{2+}),
      76, 85−86
    in $M(H_2)(CO)_3(PR_3)_2$ (M = group 6), 105,
      179, 207−208, 216, 228−229, 234−235
    in $OsCl_nH_{6-n}(PR_3)_2$, 209−210
    relative to that for other M−L, 213−222
    separated into donation and BD components,
      78−79
    in stable H_2 complexes, 207−210, 213−222
    in thermally unstable H_2 complexes: *see*
      Bond energy, in thermally unstable H_2,
      alkane, silane, and other complexes
    versus bond energy in MH_2 tautomer, 73−75
    versus M-D_2, 235−238
  M−silane, 210, 212−213, 228, 337
  Rh-CH_3, 391
  Si−H, 356
  for σ-ligand coordination, relative to other
    ligands, 14, 211−222
Bond-no-bond resonance, 13
Borane σ-complex, 421−429
  bonding in, compared to H_2 and silane
    complexes, 423
  comparison of crystal structures to those in
    borohydride and silane complexes, 423-424
  coordination of neutral boranes such as
    $BH_3 \cdot PH_3$, 421−424
  $Cp_2Ti(HBcat)_2$ and related complexes, 424−429
  degrees of activation of the B−H bond, 427−428
  displacement of borane by silane and other
    ligands, 425, 428−429
  electronics of bonding in, different from that in
    other σ complexes 426−427, 429

Borane σ-complex (*cont.*)
  as hydroboration catalysts, 426
  influence of ancillary ligands and substituents at
    B on stability, 427
  synthesis of, 425, 428
  synthesis and stability of compared to other σ
    complexes, 422
  as true η^2-BH σ complexes, 424−425
  unusual geometry about boron, 425, 427
Borohydride and borohydride-like ligands, 417−421; *see also* Bis(pyrazolylborate) ligand;
    Tris(pyrazolylborate) ligand
  comparison of crystal structures to those in
    borane and silane σ complexes, 423−424
  coordination geometry of, 417−418
  discovery of, in $Cp_2Ti(BH_4)$ and development of
    the field, 417−419
  metalloborane and metallocarborane clusters,
    418−420
  as model for methane coordination, 419
  neutron diffraction of $Cu(BH_4)(PMePh_2)_3$
    showing first accurate characterization
    of an unsupported M−H−B bridge, 419−420
  structure of $Mn_3(CO)_{10}(H)(B_2H_6)$, 417−418
  structure of $Zr(BH_4)_4$, 418
  synthesis of, from H_2 complex, 421
Buckminsterfullerene, 24

Carbocation
  bonding and reactivity compared to metal
    σ-bond coordination, 2, 13, 14, 27, 60, 410
  as possible intermediate in mechanism of
    hydrogenases and oxygenases, 306−307, 319
Carbon dioxide
  binding to Rh pincer complex, 220, 286-287
  hydrogenation to formic acid, 127, 281
  insertion into an H_2 complex, 286−287
  noninteraction with group 6 $M(CO)_3(PR_3)_2$, 43,
    220
  supercritical, as reaction solvent, 49
Carbon monoxide ligand
  in active site of hydrogenases, 298, 302-312
  bonding and electronic effect at M, 15, 76, 78
  energy of dissociation from $[M(CO)_6]^+$, 79
  hydrogen bonding to H_2O in
    $W(CO)_3(PR_3)_2(H_2O) \cdot$ thf, 213−214
  as moderator of σ-bond activation and BD when
    trans to σ ligand, 33, 75, 81−82, 147, 167,
    193, 381, 394
  in hydrogenases, 298, 304, 308
  nonclassical complexes of, 78−79
Catalytic conversion of methane to methyl
  bisulfate by Pt and Hg bipyrimidine
  complexes, 409−410
Catalytic hydroboration, 424, 425−426, 429

Index

Catalytic hydrogenation, 1, 5, 7, 284–291
  of alkenes and arenes, 285, 289–290
  of alkynes to alkenes, 40, 119
  of alkynes, 286–289
  of benzylideneacetone, 289
  direct transfer of H from η^2-H_2, 6, 13, 15, 20, 261–262, 284–287
    to norbornadiene ligand, 286
    in styrene hydrogenation, 13, 285
    in σ-bond metathesis type reactions, 125–128, 262
    to vinylidene ligand, 287–286
  heterogeneous, 17–18
  homogeneous, 17–20
    by cobalt carbonyl complexes, 18–20
    by copper salts, 17–19
    by $RhCl(PPh_3)_3$, 261
  hydroformylation, *see* Hydroformylation
  involving heterolytic cleavage of η^2-H_2, mechanism of, 287
  ionic hydrogenation, 286, 290–291
  of nitriles and nitro compounds, 289
  using zeolites, 291
Catalytic hydrosilylation and silylation, 289–290, 352
C–B bond interactions with metal centers, 434–435
C–C bond distances in agostic C–C interactions, 432–433
C–C bond interactions with metal centers
  in 14e titanium system, 432–433
    bond index for and bonding in, 432–433
  in Rh and Ir pincer complexes, 432
C–F bond, coordination of versus C–H coordination, 385, 391–393, 431–432
C–H bond coordination and activation, 365–411; *see also* Alkane σ-complex; Agostic interactions, of C–H bonds; Methane coordination
  acidity of, 403–404, 409–410
  activation in oxygenases, 316–319
  activation in σ-bond metathesis, 126
  cleavage of and oxidative addition of: *see* Cleavage, heterolytic; Cleavage, homolytic; Oxidative addition, of alkanes and C–H bonds
  coordination of, versus C–F bond, 385
  oxidative addition of: *see* Oxidative addition, of C–H bond
  polarization towards $M(C^{\sigma-}-H^{\delta+})$ when coordinated, 14, 317, 376
  steric effects in C–H bond activation, 394–395
  trajectory for the reaction M + C–H → C–M–H as modeled by X-ray diffraction of a series of structures of agostic complexes, 371, 395–396

C–H bond distance,
  in agostic CH bonds, 369–373, 376
    elongated, 369
  in σ-bond metathesis intermediates, 126–127
CH_3^+, isolobal with transition metal center, 2, 60
CH_5^+, 14, 132
  viewed as H_2 complex of CH_3^+, 60
C-halogen bond interaction with metal center, 431–432
Cis interaction between agostic C–H and hydride, 370
Cis interaction between H atoms in cis-H_2/H ligands (cis effect), 12, 41, 112, 115–121, 195–198, 370
  calculations for, 110, 116–120
  evidence from quasielastic neutron scattering, 195–198
  in $FeH_2(H_2)(PR_3)_3$, 165–166, 180–181, 195
  giving rise to asymmetric η^2-H_2 binding, 164
  involving bridging hydride, 194
  lowering barrier to H_2 rotation, 116, 181–183
  in $[OsH_5(PH_3)_3]^+$, 110
  in $RuH_2(H_2)_2(PR_3)_2$, 165, 183, 195
Cis interaction between Si in silane ligand and hydride ligand, 119–120, 337, 348, 356
Claus process, 319
Cleavage, heterolytic
  C–H bond
    in agostic bond, 282, 376
    calculations on, 393
    on Fe=O complexes and oxygenase enzymes, 317–319
    on highly electrophilic complexes, 403–405, 409–410
  H–H bond, 17–18, 115, 270–284
    on Cr_2O_3, 133–134
    in hydrogenases, 298, 304, 308–312
    intermolecular, 11–15, 18, 271, 278–279
    intermolecular, by external bases, 278–279
    intramolecular, 10–15, 18, 271, 280–284, 289
    intramolecular, by pendant ligands, 127, 280–283, 310
    on metal sulfide complexes, 319–322
    protonation of ethers, 278
  of Si–H bond, 13, 219, 353–355
  of X–H bond, 11–12
Cleavage, homolytic, of, *see also* Oxidative addition; Equilibrium
  B–H bond, 427–428
  C–H bond
    agostic: *see* Cyclometallation
    of arene, 375–376
    calculations on, 393
    ease of, correlated with C–H bond strength and *sp*, *sp*2, and *sp*3 bond character, 394–395

Cleavage, homolytic, of (cont.)
    C—H bond (cont.)
        external versus internal, 377–378
    of C—P bond, in tertiary phosphines, 436
    of H—H bond, 10–12, 17–20, 261–265
        as rate-determining step in OA of H_2, 226
        intermolecular, rare examples of, 261, 265
    of Si—H bond, 229–230
Cleavage, of XY bonds, 5
Cleavage/formation of Si—H, C—H, H—H, and Si—C bonds in the same reaction, 339–340
C—N bond interactions with metal centers, 435
Cold fusion, 27
Collision-induced dissociation in a guided ion-beam mass spectrometer, 129–131, 210, 348, 385–386

Complexes; see also Tables on pp. 36–37, 39, 41, 49, 73, 76, 80, 82, 133, 148, 149, 212, 224, 226, 227, 250, 274, 332–333, 343, 345, 380

Cationic and anionic complexes (Cp and Tp derivatives are in separate groups):

$[AuXe_4][Sb_2F_{11}]_2$, 4, 437
$[Cp*Co(Et)\{P(OMe)_3\}]^+$, 374, 377
$[CpFe(CO)(H_2)(PR_3)]^+$, 345, 348, 354
$[CpFe(CO)(silane)(PR_3)]^+$, 339, 345, 348, 354
$[Cp'Fe(H_2)(L)(L')]^+$, 46, 111, 160–161
$[Cp*Fe(N_2)(PP)]^+$, 106
$[Cp*FeH(PR)_3(C_2H_4)]^+$, 377
$[Cp*FeH_2(PP)]^+$, 106
$[CpIrH_3(PR_3)]^+$, 120–121, 158, 198–199, 201
$[CpMoH_4(H_2)(PR_3)]^+$, 188, 189
$[Cp'_2Nb(H_2)L]^+$, 173, 201–202
$[CpOsH_2(L)_2]^+$, 113
$[Cp'Os(H_2)(L)(L')]^+$, 46, 111, 160–161
$[Cp*Os(CO)_2(H_2)]^+$, 263
$[Cp*OsH(CH_3)(dmpm)]^+$, 397–398
$[Cp*Os^{IV}H_2(H_2)(PPh_3)]^+$, 38, 42, 120, 121
$[Cp*Re(CO)(NO)(H_2)]^+$, 45, 274, 278
$[CpRhH(PR)_3(C_2H_4)]^+$, 377
$[CpRuH_2(PP)]^+$, 111, 272, 316
$[CpRu(H_2)(CO)(PCy_3)]^+$, 183
$[CpRu(H_2)(PP)]^+$, 111, 247, 272, 274, 276, 316
$[CpRu(H_2)(PPh_3)_2]^+$, 111
$[Cp'Ru(H_2)(L)(L')]^+$, 46, 111, 160–161
$[Cp*Ru(CO)_2(H_2)]^+$, 279
$[Cp*Ru(D_2)(dppm)]^+$, 163
$[Cp*Ru(H_2)(dppm)]^+$, 157–158, 236
$[Cp*Ru(H_2)(PP)]^+$, 98–103
$[Cp*Ru(PMeiPr_2)_2]^+$, 383
$[Cp*Ru^{IV}H_2(H_2)(PPh_3)]^+$, 38, 42, 120, 121
$[Cp_2Ta(CO)(H_2)]^+$, 161
$[Cp'_2Ta(H_2)L]^+$, 173, 201–202

Complexes, cationic and anionic (cont.)
$[Cp_2WH(H_2)]^+$, 238
$[CpWH_2(CO)_3]^+$, 87
$[CpWH_4(H_2)(PR_3)]^+$, 189
$[FeH_2]^{+,0,-}$, 130131
$[Fe(CN)(PP)_2(H_2)]^+$, 304
$[Fe(CO)(PP)_2(H_2)]^{2+}$, 76, 82, 85, 167, 217, 304
$[Fe(\eta^3-C_8H_{13})\{P(OMe)_3\}_3]^+$, 369, 372
$[FeH(H_2)(PP)_2]^+$, 30, 45, 81–82, 116, 180–181, 183, 187, 195, 218, 223, 225, 279
$[FeH(H_2)(pp_3)]^+$, 40, 42, 45, 118–119, 180–181, 183, 285, 289
$[FeH(H_2)(PR_3)_4]^+$, 118, 122, 164, 188, 223, 224, 269, 288
$[FeH(H_2)(tetraphos)]^+$, 266
$[FeH(N_2)(PP)_2]^+$, 218
$[Ir(C-N-C)(H_2)(PMe_3)_4]^+$, 282
$[IrH(H_2)(PPh_3)_2(bq)]^+$, 29, 39, 161, 215, 269, 272, 274, 274
$[IrH(OH)(PMe_3)_4]^+$, 117, 281
$[IrH(\eta_2-C_6H_4PBu^t_2)(PBu^t_2Ph)]^+$, 382
$[IrH_2(H_2)(PMe_2Ph)_3]^+$, 289
$[IrH_2(H_2)(P^tBu_2Ph)_2]^+$, 33, 83, 192
$[IrH_2(H_2)_2(PCy_3)_2]^+$, 37, 193–194
$[IrH_2(PBu^t_2Ph)_2]^+$, 382
$[IrH_2(PCy_2Ph)_2]^+$, 380, 382
$[IrH_2(silane)_2(PR_3)_2]^+$, 233, 353, 355–356
$[IrH_2(triphos)(C_2H_4)]^+$, 37
$[IrH_2(triphos)(H_2)]^+$, 48, 284
$[M(CN)(H_2)(PP)_2]^+$ (M = Fe, Ru), 283, 304
$[M(H_2)(L)(PP)_2]^{2+}$ (M = group 8), 82
$[M(H_2)_n]^+$, 50
$[MCl(H_2)(depe)_2]^+$ (M = Ru, Os), 45
$[MH(H_2)(depe)_2]^+$ (M = Fe, Ru, Os), 45
$[MH(H_2)(PP)_2]^{2+}$ (M = group 8), 159–160
$[MH(H_2)(PR_3)_4]^+$ (M = group 8), 36
$[MH(H_2)P_4]^+$, 109
$[Mn(CO)(PP)_2(H_2)]^+$, 76, 82, 85–86, 159, 167, 217, 218, 349, 377
$[Mn(CO)(PP)_2(N_2)]^+$, 216–218
$[Mn(CO)(PP)_2]^+$, 86, 106, 217–219, 345, 349, 379–380
$[Mn(CO)\{P(OR)_3\}_4(H_2)]^+$, 41
$[Mn(CO)_3(phosphite)_2(silane)]^+$, 218, 349, 354
$[Mn(CO)_3(PR_3)_2(H_2)]^+$, 77, 82, 216–218, 345, 349
$[Mn(CO)_3(PR_3)_2]^+$, 217–218, 345, 349, 380–381
$[Ni(SiO_2)(H_2)_n]^+$, 50
$[Ni(H_2)(zeolite)]^{n+}$, 50
$[Os(H_2)(en)_2(acetate)]^+$, 8, 147, 167
$[Os(H_2)(L)(PP)_2]^{2+}$, 81, 82, 85, 275, 322
$[Os(H_2)(L)N_4]^{+/2+}$ (N_4 = 4NH_3 or 2en), 36, 47, 115, 275
$[Os(H_2)(MeCN)(PP)_2]^{+2}$, 275, 283
$[Os(H_2)(MeCN)_3(PR_3)_2]^{+2}$, 36
$[Os(H_2)(NH_3)_5]^+$, 42

Index

Complexes, cationic and anionic (cont.)
[Os(H$_2$)(NH$_3$)$_5$]$^{2+}$, 33, 36, 47
[Os(H$_2$)(NH$_3$)$_5$]$^{3+}$, 291
[Os(H$_2$)$_2$(MeCN)$_2$(PPr$^i_3$)$_2$]$^{+2}$, 37
[Os(NH$_3$)$_4$(L)(H$_2$)]$^{z+}$, 82, 92, 153
[Os(N−S)(H$_2$)(CO)(PPh$_3$)$_2$]$^+$, 282
[OsCl(H$_2$)(PP)$_2$]$^+$, 82, 98–103, 147, 157, 236
[OsD(D$_2$)(PP)$_2$]$^+$, 163
[OsD(D$_2$)(pp$_3$)]$^+$, 163
[OsH(H$_2$)(CO)(P^{i}Pr$_3$)$_2$]$^+$, 45, 290
[OsH(H$_2$)(PP)$_2$]$^+$, 107, 157, 217
[OsH(H$_2$)(pp$_3$)]$^+$, 269, 289
[OsH$_3$(H$_2$)$_2$(PPr$^i_3$)$_2$]$^+$, 37
[OsH$_3$(PPh$_3$)$_4$]$^+$, 107
[OsH$_5$(PR$_3$)$_3$]$^+$, 109–110, 143, 144, 159
[Pd$_2$(PHR$_2$)$_2$(μ-PR$_2$)(μ-PHR$_2$)]$^+$, 429
[Pt(CH$_3$)(H$_2$O)(diimine)]$^+$, 407
[Pt(CH$_3$)(NC$_5$F$_5$)(tmeda)]$^+$, 405–406
[PtCl$_2$(NH$_3$)$_2$(OSO$_3$H)]$^+$, 411
[PtCl$_3$(C$_2$H$_4$)]$^-$, 409
[PtCl$_6$]^2\$^{m}i and [PtCl$_4$]$^{2-}$, 404
[PtH(CH$_2$Cl$_2$)(PR$_3$)$_2$]$^+$, 218
[PtH(Et$_2$O)(PR$_3$)$_2$]$^+$, 218
[PtH(H$_2$)(PR$_3$)$_2$]$^+$, 39, 217, 277–278
[Pt^{IV}H(CH$_3$)$_2$(NNN)]$^+$, 407–408
[Re(CN^{t}Bu)$_3$(PCy$_3$)$_2$(H$_2$)]Cl, 45
[Re(CO)(PR$_3$)$_4$(D$_2$)]$^+$, 163
[Re(CO)(PR$_3$)$_4$(H$_2$)]$^+$, 193
[Re(CO)$_2$(PR$_3$)$_3$(H$_2$)]$^+$, 229–230
[Re(CO)$_2$(triphos)(H$_2$)]$^+$, 217
[Re(CO)$_3$(phosphite)$_2$(silane)]$^+$, 345–346
[Re(CO)$_3$(PR$_3$)$_2$(H$_2$)]$^+$, 36, 42, 76, 82, 217, 267, 275
[Re(CO)$_3$(PR$_3$)$_2$(N$_2$)]$^+$, 217
[Re(CO)$_3$(PR$_3$)$_2$]$^+$, 217, 380
[Re(CO)$_4$(PR$_3$)(CH$_2$Cl$_2$)]$^+$, 218, 284, 354
[Re(CO)$_4$(PR$_3$)(Et$_2$O)]$^+$, 218
[Re(CO)$_4$(PR$_3$)(H$_2$)]$^+$, 82, 217, 275, 284, 307, 308, 345–346, 349
[Re(CO)$_4$(PR$_3$)(silane)]$^+$, 219, 345–346, 349, 354
[Re(CO)$_n${P(OEt)$_3$}$_{5-n}$(H$_2$)]$_+$, 82, 86
[Re$_2$(μ-H)(CO)$_8$(PR$_3$)$_2$]$^+$, 284, 354
[ReH$_2$(H$_2$)(CO)(PR$_3$)$_3$]$^+$, 114, 156, 188, 192–193, 229–230, 237, 262–263
[ReH$_4$(CO)(PR$_3$)$_3$]$^+$, 114, 192–193, 262–263, 274, 278
[ReH$_4$(PR$_3$)$_4$]$^+$, 114
[ReH$_9$]$^{2-}$, 94–95
[Rh(Cn)H(CH$_3$)(PR$_3$)]$^+$, 397
[RhD$_2$(D$_2$)(triphos)]$^+$, 163
[Ru(CO)(PP)$_2$(H$_2$)]$^{2+}$, 85
[Ru(H$_2$)(o-C$_6$H$_4$py)(P^{i}Pr$_3$)$_2$(THF)]$^+$, 377
[Ru(H$_2$O)$_5$(H$_2$)]$^{2+}$, 10, 36, 45, 82
[Ru(NH$_3$)$_5$(N$_2$)]$^{2+}$, 6
[Ru(Ph)(CO)(P^{t}Bu$_2$Me)$_2$]$^+$, 381, 383

Complexes, cationic and anionic (cont.)
[RuCl(H$_2$)(PP)$_2$]$^+$, 39, 82, 167, 217, 220, 269, 287, 316
[RuCl(PP)$_2$]$^+$, 217, 220
[RuD(D$_2$)(PP)$_2$]$^+$, 163
[RuD(D$_2$)(pp$_3$)]$^+$, 163
[RuH(CO)$_2$(^{t}Bu$_2$PC$_2$H$_4$P^{t}Bu$_2$)]$^+$, 383
[RuH(H$_2$)(CO)$_2$(PR$_3$)$_2$]$^+$, 188, 189
[RuH(H$_2$)(o-C$_6$H$_5$py)(P^{i}Pr$_3$)$_2$]$^+$, 377
[RuH(H$_2$)(PP)$_2$]$^+$, 45, 86, 103–104, 107, 147, 283, 381
[RuH(H$_2$)(pp$_3$)]$^+$, 40, 118, 180–181, 183, 192, 289
[RuH(H$_2$)(PR$_3$)$_4$]$^+$, 188, 224, 288
[RuH(PP)$_2$]$^+$, 380, 381, 436
[RuH$_3$(dppf)$_2$]$^+$, 107
[RuH$_5$(P^{i}Pr$_3$)$_2$]$^-$, 47
[RuX(CO)(P^{t}Bu$_2$Me)$_2$]$^+$, 380
[TpIrH(H$_2$)(PR$_3$)]$^+$, 155–156, 188
[TpOs(H$_2$)(L)$_2$]$^+$, 113
[TpRu(H$_2$)(PR$_3$)$_2$]$^+$, 276
[TpRu(L)(H$_2$)(PR$_3$)]$^+$, 276, 277
[W(=NSi^{t}Bu$_3$)$_3$(alkane)]$^-$, 399
[WH(=NSi^{t}Bu$_3$)$_3$]$^-$, 399

Neutral complexes (Cp and Tp derivatives are in separate groups):

CeNiInH$_{1.0}$, 39
Co(CO)$_2$(NO)(H$_2$), 49
Co$_2$(CO)$_8$, 18, 231
CoNa-A zeolite(H$_2$), 50
CoX(H$_2$)(CO)$_3$ (X = H, Me), 49
Cr(arene)(CO)$_2$(H$_2$), 49, 212
Cr(arene)(CO)$_2$(L) (L = alkane, silane, stannane, N$_2$), 212, 361–362, 386
Cr(CO)$_3$(PCy$_3$)$_2$(N$_2$), 221, 345
Cr(CO)$_3$(PCy$_3$)$_2$, 217, 219, 221, 234, 345, 349, 380, 388
Cr(CO)$_3$(PR$_3$)$_2$(H$_2$), 40, 42–44, 75, 82, 92, 106, 145, 173, 221, 234, 345, 388
Cr(CO)$_3$(PP)(stannane), 362
Cp*CrH(CO)$_3$, 230–231
CpIr(CO)(H$_2$), 49
Cp*La[CH(SiMe$_3$)$_2$]$_2$, 434
Cp$_2$LuMe, 126
Cp*MH$_5$(PMe$_3$) (M = Mo, W), 112
CpMH(CO)$_2$(H$_2$) (M = Mo, W), 49
Cp'Mn(CO)$_2$(HBcat), 428–429
CpMn(CO)$_2$(alkane), 212, 224, 386
Cp'Mn(CO)$_2$(germane), 357–358
Cp'Mn(CO)$_2$(silane), 28, 50, 212, 328–335, 339–339, 342–345, 348, 361
Cp'Mn(CO)$_2$(stannane), 360–362
CpMn(CO)$_2$(C$_2$H$_4$), 50
CpMn(CO)$_2$(H$_2$), 49, 80, 111, 226, 345
CpMn(CO)$_2$(N$_2$), 50, 224, 226, 345

Complexes, neutral (cont.)
CpMo(CO)$_2${P(BH$_3$)Ph[N(SiMe$_3$)$_2$]}, 420
Cp'M(CO)$_3$(H$_2$) (M = V, Nb), 49
Cp$_2$Mo$_2$(μ-S)$_2$(μ-SH)$_2$, 320
CpNb(CO)$_3$(alkane), 386
CpNb(CO)$_3$(H$_2$), 111–112, 238, 263
Cp$_2$NbH(Bcat)$_2$, 428
Cp$_2$NbH(H$_2$)·BH$_3$, 110
Cp$_2$NbH(silyl)$_2$, 351–352
Cp'$_2$NbH$_3$, 110, 158, 198
CpNbH$_2$(CO)$_3$, 238, 263
CpRe(CO)$_2$(alkane), 223–224, 386–387, 422
CpRe(CO)$_2$(silane), 342–343
Cp*Re(CO)$_2$(HBpin), 428
CpReH(silyl)(CO)$_2$, 330
CpReH$_2$(CO)$_2$, 80, 87, 111
Cp*Rh(Et)(PMe$_3$), 396
Cp'Rh(CO) fragment, 225–226, 386, 390, 399–402
Cp'Rh(CO)(alkane), 226, 389–390, 399–402
Cp'Rh(R)(H)(CO), 399–402
Cp*Ru(cod)(H$_2$), 36
Cp*RuH$_2$(silyl)(PR$_3$), 348
Cp*RuH3(PR$_3$), 198
CpRuH$_3$(L), 158, 348
Cp*$_2$Ru2(μ-H)$_2$(μ-silane), 334
Cp$_2$Ta(silyl)$_2$, 351–352
Cp$_2$Ta(=C−H(Bu^t)Cl, 372
Cp$_2$TaH$_2$(Bcat), 428
CpTa(CO)$_3$(alkane), 386
Cp$_2$Ti system with agostic C−Si, 433–434
Cp$_2$Ti(H$_2$BH$_2$), 417
Cp$_2$Ti(HBcat)(C$_2$Ph$_2$), 426
Cp$_2$Ti(HBcat)(PMe$_3$), 425
Cp$_2$Ti(HBcat)(silane), 425, 427
Cp$_2$Ti(HBcat)$_2$, 423–427
Cp$_2$Ti(silane)(PR$_3$), 337
CpTi system with agostic C−C, 433
Cp*$_2$Ti(S)(py), 355
Cp*$_2$Ti(SH)$_2$, 320
CpV(CO)$_3$ fragment, 223–224, 348
CpV(CO)$_3$(H$_2$), 212, 223–224
CpV(CO)$_3$(L) (L = alkane, silane, N$_2$), 212, 224, 337, 348, 386
Cp$_2$WH(Me), 396
Cp*$_2$Yb(CH$_3$BeCp*), 384
Cp*$_2$ZrH$_2$, 189
Cp$_2$Zr system with agostic C−B, 434–435
Cr(CO)$_3$(dfepe) fragment, 385
Cr(CO)$_4$(B$_2$H$_4$·2PMe$_3$), 421–423
Cr(CO)4(H$_2$)$_2$, 48–49, 77, 173, 188, 267
Cr(CO)$_5$ fragment, 22, 65, 211–212, 217, 223, 384–385
Cr(CO)$_5$(H$_2$), 36, 48–49, 73–75, 77, 105–106, 179, 211–212, 223, 238, 253, 267, 422, 423

Complexes, neutral (cont.)
Cr(CO)$_5$(L) (L = alkane, silane, Xe, N$_2$), 211–212, 223, 225, 384–386
Cr$_2$O$_3$(surface-bound H$_2$), 50, 133–134
CrH$_2$[P(OMe)$_3$]$_5$, 72
CuCl(H$_2$), 49
Cu$_x$(H$_2$), 50, 131–132
Cu(BH$_4$)(PMePh$_2$)$_3$, 419–420, 423
Cu(BH$_4$)(PPh$_3$)$_2$, 418
Fe(CO)(NO)$_2$(H$_2$), 49, 105
Fe(CO)$_5$, 260
Fe(η^2-H$_2$)(CO)$_4$, 88
Fe(C$_4$H$_4$)(CO)$_2$(H$_2$), 49
Fe$_4$(CH)(H)(CO)$_{12}$, 372
Fe(DAP)(heptane), 387–388
Fe(η^2-HSMe)(CO)$_3$(PEt$_3$), 322, 431
Fe[η^3-HC(SMe)S−B(H)H$_2$](CO)(PMe$_3$)$_2$, 420
Fe$_2$(CO)$_6$(B$_3$H$_7$), 419
FeH(CO)$_9$(BH$_4$), 418
FeH$_2$(CO)$_4$, 17, 61, 87–88
FeH$_2$(H$_2$)(PR$_3$)$_3$, 45, 116, 159, 165, 180–181, 183, 191, 195
FeH$_2$(pp$_3$), 42
Ir(H$_2$)(H)(diphpyH)(PR$_3$)$_2$, 41
IrCl(CO)(PPh$_3$)$_2$ (see Vaska's complex)
IrH$_2$(PPh$_3$)$_2$(bq), 46
IrH$_2$X(H$_2$)(PR$_3$)$_2$ (X = Cl, Br, I), 44, 63, 82, 149, 164, 181, 184, 185, 187, 190–191, 195–196, 229, 235, 259, 289, 388
IrH$_2$X(PR$_3$)$_2$ (X = Cl, Br, I), 190
IrH$_5$(PR$_3$)$_2$, 108–109
IrHBr(Bcat)(CO)(PP), 428
IrHX$_2$(H$_2$)(PR$_3$)$_2$ (X = Cl, Br), 63, 81–82, 89, 147, 166, 167, 185–186, 208–209, 224, 289
La$_4$(OCH$_2^t$Bu)$_{12}$, 372
[M(L)$_n$]$^+$ (L = H$_2$, CH$_4$, C$_2$H$_4$, H$_2$O, CO), 128–131, 210–211, 348–349, 366, 368, 385–386, 393
M(CO)$_4$(H$_2$)(L) (M = group 6), 48, 285
M(CO)$_5$ fragment, 211–212
M(CO)$_5$(Xe) (M = group 6), 4
M(CO)$_{5-n}$(olefin)$_n$(H$_2$) (M = group 6), 49
M$_2$H$_x$(H$_2$)$_y$ (M = Fe, Co, Ni), 50
Mn$_3$(CO)$_{10}$(H)(B$_2$H$_6$), 417–418
MnX(CO)$_4$(H$_2$) (X = Cl, Br), 49
Mo(arene)(CO)$_3$(H$_2$), 49
Mo(CO)(H$_2$)(PP)$_2$, 73, 76, 80, 81, 84, 86, 103–106, 145–146, 167, 183–185, 255, 345, 347
Mo(CO)(N$_2$)(PP)$_2$, 76, 84, 217, 345
Mo(CO)(PP)$_2$(germane), 358–360
Mo(CO)(PP)$_2$(silane), 210, 230, 327, 334–335, 340–341, 343, 345, 347, 358–360, 423
Mo(CO)(PP)$_2$, 80, 158, 210, 217, 219, 345, 347, 349, 375, 380–381
Mo(CO)$_2$(PPh$_3$)$_2$(SO$_2$)L, 21–22

Complexes, neutral (cont.)
Mo(CO)$_3$(PCy$_3$)$_2$(H$_2$), 25, 40, 41–44, 82, 106, 173, 177–180, 183, 221, 255
Mo(CO)$_3$(PCy$_3$)$_2$(N$_2$), 221, 378
Mo(CO)$_3$(PP)(stannane), 362
Mo(CO)$_3$(PCy$_3$)$_2$, 22–23, 25, 40, 42–44, 219, 221, 378–380
Mo(CO)$_4$(H$_2$)$_2$, 48–49, 77
Mo(CO)$_4$(PCy$_3$)$_2$, 24
Mo(CO)$_5$(H$_2$), 48–49, 77, 105, 179, 211–212, 253, 422, 423
Mo(CO)$_5$(L) (L = alkane, silane, N$_2$), 211–212
Mo[CN(p-Me$_2$NC$_6$H$_4$)](H$_2$)(PP)$_2$, 84, 105–106
Mo[CN(p-Me$_2$NC$_6$H$_4$)](N$_2$)(PP)$_2$, 84, 105–106
Mo[Et$_2$B(pz)$_2$](CO)$_2$(CH$_2$CPhCH$_2$), 368
Mo[H$_2$B(pz)$_2$](CO)$_2$(η^3-C$_7$H$_7$), 368
MoH(germyl)(CO)(PP)$_2$, 359–360
MoH(silyl)(CO)(PP)$_2$, 230, 327, 341
MoH$_2$(CO)(depe)$_2$, 73, 80, 86, 103–104, 106, 152, 264, 341, 347, 377
Mo(Me)$_2$[N(2,6-Pri-C$_6$H$_3$)]$_2$), 373
MoH$_2$(PR$_3$)$_5$, 72, 80, 87, 106
MoH$_4$(PP)$_2$, 171
MoH$_4$(PR$_3$)$_4$, 339–340
M$_x$(H$_2$)$_y$ (M = Fe, Co, Ni), 50
Nd$_4$(OCH$_2^t$Bu)$_{12}$, 372
Ni(510)–(H$_2$), 50–51, 68
Ni(CO)$_3$(H$_2$), 49, 105
Ni(H$_2$), 68
Ni[Et$_2$B(pz)$_2$]$_2$, 368
Os$_3$(CO)$_{10}$(PPh$_3$)$_2$, 436
Os$_3$(CO)$_{10}$(μ-H$_2$(μ-CH$_2$), 377
Os$_3$(CO)$_{12}$, 436
OsCl(H$_2$)(NH=C(Ph)C$_6$H$_4$)(P^{i}Pr$_3$)$_2$, 162
OsCl(silyl)(H$_2$)(CO)(PR$_3$)$_2$, 290, 347
OsCl$_2$(D$_2$)(CO)(PR$_3$)$_2$, 163–164, 173
OsDCl(H$_2$)(CO)(PR$_3$)$_2$, 163
OsH(η^2-H$_2$BH$_2$)(CO)(P^{i}Pr$_3$)$_2$, 48
OsH$_2$(CO)$_4$, 88, 94–96
OsH$_2$(H$_2$)(CO)(PR$_3$)$_2$, 48, 155, 191, 224
OsH$_2$(X)(Y)(H$_2$)(PPr$^i_3$)$_2$ (X, Y = Cl, Br, I), 41
OsH$_2$I$_2$(PR$_3$)$_2$(H$_2$), 209
OsH$_3$(BH$_4$)(PR$_3$)$_2$, 421
OsH$_3$(py–S)(PPr$^i_3$)$_2$, 201
OsH$_3$X(PPr$^i_3$)$_2$(H$_2$), 209
OsH$_3$X(PPr$^i_3$)$_2$, 189–190, 201, 202, 209
OsH$_4$(PR$_3$)$_3$, 79, 110, 143, 147
OsH$_6$(P^{i}Pr$_3$)$_2$, 143
OsHCl(H$_2$)(CO)(PR$_3$)$_2$, 81, 209, 229, 288, 290
Pd(PPh^{t}Bu$_2$)$_2$, 381
Pd$_2$(PHR$_2$)$_2$(μ-PR$_2$)$_2$, 429
PdBr(C$_4$Me$_4$)(PPh$_3$), 372
Pd$_n$(H$_2$)$_x$, 50, 67–68
PdXR(PR$_3$)$_2$, 436
Pt(CH$_3$)(C$_6$H$_5$)(diimine), 407

Complexes, neutral (cont.)
Pt(PPh$_2^t$Bu)$_2$ or Pt(PPh^{t}Bu$_2$)$_2$, 372, 381
PtCl(CH$_3$)(bpym), 410
PtCl$_2$(bpym), 409–411
Re(CN^{t}Bu)$_3$(PCy$_3$)$_2$Cl, 45
ReBr$_2$(H$_2$)(NO)(PCy$_3$)$_2$, 150, 421
ReCl(H$_2$)(PMePh$_2$)$_4$, 42, 47, 82, 164
ReH(BH$_4$)(NO)(PCy$_3$)$_2$, 421
ReH$_7$(PR$_3$)$_2$, 45, 107–108, 143, 422
Re^{III}(N$_3$NF)(H$_2$), 38, 106
Re^{III}(N$_3$NF)(silane), 339, 345
Rh(P–C–P)(C$_2$H$_4$), 220
Rh(P–C–P)(CO), 376
Rh(P–C–P)(CO$_2$), 220, 286–287
Rh(P–C–P)(H$_2$), 220, 286–287
Rh(P–C–P)(N$_2$), 220
Rh(Ph)Cl$_2$(PPh$_3$)$_2$, 382–383
RhCl(P^{i}Pr$_3$)$_2$, 22
RhCl(PPh$_3$)$_3$ fragment, 225
RhCl(PPh$_3$)$_3$, 19, 63, 64, 261, 427
RhH$_2$Cl(P^{i}Pr$_3$)$_2$, 382–383
RhHCl$_2$(P^{i}Pr$_3$)$_2$, 382–383
RhHX(SnPh$_3$)(L)(PPh$_3$), 362
Ru("S4")(H$_2$)(PCy$_3$), 321
Ru(CO)$_4$ fragment, 66
Ru(OEP)(H$_2$), 49, 265
Ru(Ph)Cl(CO)(PBu$^t_2$Me)$_2$, 383
Ru(PR$_3$)$_2$(H$_2$)(μ-Cl)$_2$(μ-H)RuH(PR$_3$)$_2$, 194
Ru(PR$_3$)$_2$(H$_2$)(μ-Cl)$_3$RuCl(PR$_3$)$_2$, 222–223
Ru(PR$_3$)$_2$(H$_2$)(μ-H)$_3$RuH(PR$_3$)$_2$, 194
Ru$_2$(μ-DPB)(H$_2$)(Im)$_2$, 164–165
Ru$_2$H$_4$(PR$_3$)$_4$(μ-SiH$_4$), 341–342
Ru$_3$(CO)$_9$(μ_3-PPh$_2$)(μ_2-H), 436
RuCl$_2$(H$_2$)(P-N)(PR$_3$), 41
RuCl$_2$(PPh$_3$)$_3$, 367
RuCl$_2$CO)(P^{i}Pr$_3$)$_2$, 383
RuH$_2$(CO)(P^{t}Bu$_2$Me)$_2$, 83, 383
RuI$_2$(CO)$_4$, 66, 87–88
RuH$_2$(H$_2$)(CO)(PR$_3$)$_2$, 191, 348
RuH$_2$(H$_2$)(cyttp), 45
RuH$_2$(H$_2$)(PR$_3$)$_3$, 29–30, 38, 45, 109, 191, 288, 289
RuH$_2$(H$_2$)$_2$(PR$_3$)$_2$ (R = Cy, ^{i}Pr), 37, 47, 119, 165–166, 183, 194, 195–198, 260, 288, 289, 337, 342
RuH$_2$(silane)$_2$(PR$_3$)$_2$, 337
RuH$_2$(η^2-H$_2$)(η^2-SiHPh$_3$)(PCy$_3$)$_2$, 37–38, 119
RuH$_2$(NO)(PR$_3$)$_2$, 201
RuH$_3$(silyl)(CO)(PR$_3$)$_2$, 348
RuHCl(CO)(P^{i}Pr$_3$)$_2$, 383
RuHCl(H$_2$)(CO)(PR$_3$)$_2$, 161, 209
RuHX(H$_2$)(PR$_3$)$_2$ (X = Cl, I, SR), 33, 82, 83, 120, 289, 369
TaCl$_3$(CHBu^t)L, 372
TcCl(dppe)$_2$, 217, 220
TcCl(H$_2$)(dppe)$_2$, 80, 81, 167, 220

Complexes, neutral (*cont.*)
TcCl(N_2)(dppe)$_2$, 220
TcH$_3$(dppe)$_2$, 83
Ti(BH$_4$)$_3$(PMe$_3$)$_2$, 420–421, 423
TiCl$_3$(η^2-CH$_3$)(dmpe), 372
TiEtCl$_3$(dmpe), 372
(Bp$^{(CF3)2}$)RuH(H$_2$)(PR$_3$)$_2$, 419
(η^2-Bp′)Mo(CO)$_2$(allyl), 419
(η^2-Tp′)Pt^{II}(CH$_3$), 405
(η^3-Tp′)Pt^{IV}(CH$_3$)(R)(H), 405
($\kappa^2_{N, B-H}$-Tp′)Rh(CO)(PR$_3$)$_2$, 419
Tp*Rh(CO) fragment, 401–402
Tp*Rh(CO)(alkane), 401–402
Tp*Rh(R)(H)(CO), 401–402
Tp*Rh[P(C$_7$H$_7$)$_3$], 419
Tp*RhH$_2$(H$_2$), 33, 45, 62, 113, 181, 288
Tp*RuH(H$_2$)$_2$, 37, 47, 194, 247
Tp′RhH(R)(L), 398
TpRuH(H$_2$)(PR$_3$), 113, 345, 348, 356
TpRuH(silane)(PR$_3$), 345, 348, 356–357
[U(NN′$_3$)]$_2$(μ_2-N$_2$), 218
V(CO)$_5$(H$_2$), 49
W(CO)(dppe)$_2$, 375
W(CO)$_3$(PCy$_3$)$_2$(C$_2$H$_4$), 70, 76
W(CO)$_3$(PCy$_3$)$_2$(H$_2$), 25, 44, 88, 149–150, 175, 177–180, 208, 221, 222, 235–236, 248–254
W(CO)$_3$(PCy$_3$)$_2$(N$_2$), 221, 222, 235, 345
W(CO)$_3$(P^{i}Pr$_3$)$_2$(D$_2$), 163
W(CO)$_3$(P^{i}Pr$_3$)$_2$(H$_2$), 7, 26–28, 40, 44, 145, 151–152, 160, 177–178, 216, 374
W(CO)$_3$(P^{i}Pr$_3$)$_2$(H$_2$O), 213–216, 308
W(CO)$_3$(P^{i}Pr$_3$)$_2$(HD), 26
W(CO)$_3$(PR$_3$)$_2$(H$_2$), 36, 40–44, 81–82, 101–103, 105–106, 116, 167, 173, 175–180, 228–229, 235, 245–247, 262, 267, 330, 336, 345
W(CO)$_3$(PR$_3$)$_2$(L), 76, 208, 214, 217, 345, 378–379
W(CO)$_3$(PR$_3$)$_2$, 23–24, 43, 70, 158, 208, 214–217, 219–222, 228–230, 240, 336, 345, 349, 378–380, 382
W(CO)$_3$[P(C$_6$D$_{11}$)$_3$]$_2$(H$_2$), 26
W(CO)$_4$(H$_2$)$_2$, 48–49, 77
W(CO)$_4$(P^{i}Pr$_3$)$_2$, 24
W(CO)$_4$(PR$_3$)(stannane), 362
W(CO)$_5$(BH$_3$·PMe$_3$), 421–445 **page range OK?**
W(CO)$_5$(H$_2$), 48–49, 73–78, 105, 179, 225, 253, 422, 423
W(CO)$_5$(L) (L = alkane, silane, N$_2$), 211–212, 225, 385, 431
W(CO)$_5$(L), 76, 211–212, 385
W(N$_2$)$_2$(PR$_3$)$_4$, 316, 321
W[P(OMe)$_3$]$_5$, 72
W(CO)$_6$, 260
W$_2$(CO)$_6$(PR$_3$)$_2$(μ-silyl)$_2$, 336
WH(silyl)(CO)$_3$(PR$_3$)$_2$, 336, 345

Complexes, neutral (*cont.*)
WH(stannyl)(CO)$_3$(PP), 362
WH$_2$(CO)(PP)$_2$, 80–81, 106, 375
WH$_2$(CO)$_3$(PR$_3$)$_2$, 72–73, 228–229, 336
WH$_2$(PR$_3$)$_5$, 72
WH$_2$I$_2$(PR$_3$)$_4$, 237
WH$_4$(PR$_3$)$_4$, 171
ZnCl$_2$(B$_2$H$_4$·2PMe$_3$), 421–422
Zr(BH$_4$)$_4$, 418
Zr$_2$[P2N2]$_2$(μ-η^2-N$_2$) (P2N2 = PhP(CH$_2$SiMe$_2$NSiMe$_2$CH$_2$)$_2$PPh), 316, 355

CoNa–A zeolite interaction with H$_2$, 38, 50, 177
Coordinatively unsaturated complex or site, 4, 14, 15, 33, 42–43; *see also* Sixteen-electron complexes; Fourteen-electron complexes
  on Cr$_2$O$_3$, H$_2$ binding to, 133–134
  deep color characteristic of, 43
  exchange between hydrogens in [IrH$_2$(H$_2$)(PtBu$_2$Ph)$_2$]$^+$, 192
  operationally unsaturated precursors, 42
  stabilized by halide or other π-donor ligands, 33
Coupling reactions of organic molecules, 436
C–P bond distance, in agostic C–P bonds, 435–436
Crystal engineering, 282
C–Si bond distances in agostic C–Si interactions, 433–434
C–Si bond interactions with metal centers, 433–434
Cu(I) acetate catalyzed hydrogenation, 17–19
Cyanide ligand
  in active site of hydrogenases, 302–312
  electronics of bonding to M, 77
  protonated; migration of protons between H$_2$ and CNH ligands and relevance in hydrogenase, 283, 304
Cyclometallation, 375–378
  for bridging agostic CH, 377
  comparison to OA of H–H, 375
  versus external C–H bond cleavage, 377–378
Cyclopentadienyl ligand
  in alkane complexes, 384, 386–390, 396–403
  in borohydride and borane complexes, 417, 420, 422–429
  containing ligand system with C–C bond interactions with metal center, 432–433
  electronics of bonding of, 111–114
  in germane and stannane σ complexes, 357–362
  in H$_2$ ligands, 33, 111–114, 223–227, 230–232, 238–240, 263, 278
  in silane σ complexes, 328–334, 337–339, 342–345, 348, 351, 354–356
Cytochrome P-450, 316–318, 410

Dehydrofluorination reaction, 283
Dehydrohalogenation reactions, 282–283, 289

Delocalized bonding
  of H_2 ligands: *see* Dihydrogen ligand or complex, elongated of silane ligands, 9
Deprotonation of H_2 complexes: *see* Reactions of H_2 complexes, deprotonation
Deuterium NMR: *see* NMR, deuterium
Deuterium, *see also* Exchange processes, isotopic
  incorporation of D into ancillary ligands and alcohols, 269–270
  substitution of D for H in IR spectroscopy, 26, 245, 252, 334, 400
Dewar-Chatt-Duncanson model for $M-H_2$
  bonding compared to olefin binding, 3, 10, 60–61, 69, 219, 253
Diborane, 3c-2e bonding in, compared to metal-borohydride complexes, 417
Dichloromethane ligand, 213, 217–220, 431
Dihydrogen bonding: *see* Hydrogen bonding, unconventional
Dihydrogen gas
  acidity of, 272
  HH stretch, *see* Vibrational modes and analysis for H_2 complexes, HH stretch
  low absolute entropy of, 221–222
  reactions of: *see* Reactions of H_2 gas; Catalytic hydrogenation
Dihydrogen ligand or complex
  acidity of, 13, 47, 270–278
    high acidity in electrophilic and cationic complexes, 274–277
    highly acidic complexes synthesized from H_2 gas, 278
    kinetic acidity of, 13, 272–278
    pK_a of H_2 ligands, 269, 272–277
    pK_a of H_2 ligands, table of, 274
    spectacular increase over that of free H_2, 272
    thermodynamic acidity of, 272–278
    trends with ancillary ligand and metal variation, 276–277
  activation of toward oxidative addition: *see* Oxidative addition
  ambiguity with dihydride, 144
  bis-H_2 complexes, 35, 38, 47, 83, 108, 119–120, 165, 183 193–198
  bonding of, *see also* Theoretical calculations for H_2 and hydride systems; Dewar-Chatt-Duncanson model
    amphoteric nature (Lewis acid and base), 7, 12, 19, 61, 78; *see also* versatility of binding of
    asymmetric side-on, 117–118, 164
    bond index and bond order, 90–93, 154–155
    comparison to H_2O and other lone-pair donors, 10

Dihydrogen ligand or complex (*cont.*)
  bonding of (*cont.*)
    comparison to N_2, CO, and other π-acceptor ligands, 76–78, 216–222
    comparison to other σ-ligands, 10–12
    to Cp versus Tp complexes, 111–114, 182
    to d^0 complexes, 112–113, 120–121
    effects of M/L/charge, 80, 84–88, 104–107, 167
    favored on electrophilic and positively-charged metal centers, 11, 60–61, 76, 78, 84–86, 105, 128–131, 167, 255
    influence of ligand trans to H_2, 81–84, 105, 147
    Mulliken overlap population, 94–96
    to $RhCl(PH_3)_3$, 63–65
    side-on, first illustration of, 265
  bridging, 37–38, 143, 164–165, 230–231, 265
  cis-interaction in: *see* Cis interaction between hydrogens
  clarity of, in understanding bonding of, 28–29
  coordination of
    to binary metal hydrides, 134–135
    to early transition metals, 113
    equilibrium isotope effect for: *see* Equilibrium isotope effect, KH/KD
    favored by low absolute entropy of H_2, 221–222
    kinetic isotope effect for, 238–239
    to $M(CO)_3(PR_3)_2$ (M = group 6), 24
    to main group Lewis acids/bases, 38, 51, 132–133
    to metal ions, 38, 50, 128–131
    to metal oxides, 132–134
    to metal-atom and metal-atom clusters, 50
    to surfaces of metals and compounds, 38, 50–51, 131–135
  development of new examples of, 29–30
  diagnostics for: *see* "evidence for H_2 coordination" subentries under: NMR coupling constant, JHD; NMR spectroscopy, solid state; NMR relaxation time (T_1); Inelastic neutron scattering, Vibrational modes and analysis for H_2 complexes
  direct transfer of H from, 158, 316; *see also* Catalytic hydrogenation, direct transfer of H from η^2-H_2
  discovery of, 20–29
  displacement of: *see* Reactions of H_2 complexes, displacement of H_2
  electrochemistry of, 41–42
  elongated (stretched), 8, 38, 88, 104, 108, 112, 143, 161, 247, 254–257
    delocalized bonding in, 9, 15, 97–104, 157–158, 255–257

Dihydrogen ligand or complex (*cont.*)
  elongated (stretched) (*cont.*)
    H—H bond distance in, 8, 88, 97–98, 108, 143, 149, 161, 247, 370
    high strength of bonding for as in a classical hydride, 213
    in IrCl$_2$H(H$_2$)(PR$_3$)$_2$, 63, 82
    influence of ligand trans to H$_2$, 81–83
    nuclear motion quantum calculations in, 98–102, 236, 255–256
    in Os-amine-H$_2$ complexes, 47, 83, 97–99, 108, 167
    theoretical analysis of bonding in, 97–104, 255–257
    versus equilibrium between η^2-H$_2$ and dihydride tautomers, 264
    vibrational analysis of, 254–257
  end-on (η^1-H$_2$ coordination), 19, 50–51, 63–66, 70
  equilibrium with dihydride tautomer: *see* Equilibrium, between H$_2$ and hydride tautomers
  favored in d^6 octahedral complexes, 33, 61
  favored in electrophilic and cationic complexes, 33, 34, 61, 131, 216–218
  high acidity of H$_2$ ligand, 274–277
  first spectroscopic evidence for, 30
  "gray zone" between H$_2$ and dihydride ligands, 38–39
    table of complexes in, 39
  hidden nature of, 29
  with lowest molecular weight and highest percentage of H$_2$, 49–50
  M—H$_2$ bond energy: *see* Bond energy, M—H$_2$
  in models for hydrogenase enzymes: *see* Hydrogenase enzymes
  nineteen-electron complexes as intermediates, 33
  NMR studies of, 143–167; *see also* "NMR" entries
  number of complexes containing and papers on, 30, 33, 34
  oxidation state of M in, 38
  π-acceptor behavior of, 75–79, 86; *see also* Backdonation, M to H$_2$ σ*
  in polynuclear complexes, 35
  precursor for catalytic and other reactions, 288–291
  properties of, 33–51
    air-sensitivity of, 39–40
    air-stable complexes, 39–40
    dissociation pressure of, 40
    lability of, 40, 43–45, 47, 225, 288–289
    reversible binding of, 40, 43–44, 165
    solubilities of, 43–44

Dihydrogen ligand or complex (*cont.*)
  reactions of: *see* Reactions of H$_2$ complexes
  as a reservoir of coordinated unsaturation, 288
  rotation of: *see* Rotation of H$_2$
  sixteen-electron complexes of, 15, 35, 120, 267
  stability of, 47; *see also* Dihydrogen ligand, properties of
    in CpMn(CO)$_2$(H$_2$), 49
    for Fe versus Ru complexes, 40, 45
    in solution versus solid states, 40, 44
    towards H$_2$ loss, 40–41, 43–44
    towards H$_2$ loss, importance of backbonding, 180
    towards H$_2$ loss, table of, 41
    in various solvents, 40, 43–44
    with respect to metal row, 40, 80–81, 179–180
  structural studies of, 143–167; *see also* X-ray diffraction; Neutron diffraction
  synthesis of stable complexes of, 33–48
    from H$_2$ addition to precursor, 33, 40–45
    by H$_2$ addition to precursor, table of, 41
    by protonation of hydride: *see* Hydride complex, protonation of
    by reduction and other methods, 45–48
  table of complexes possibly containing H$_2$ and/or having d_{HH} <1.6 Å, 39
  table of generic types of stable H$_2$ complexes, 36–37
  true H$_2$ complex (d_{HH} <0.9 Å), 39, 41, 81, 88, 93, 105–107, 143, 157, 167, 173, 213, 237, 287
  unrecognized or thought to be hydrides, 29–30
  unrecognized or thought to be hydrides, table of, 29
  unstable, transient, or low-temperature stable, 38, 48–49, 112, 223–226, 348
    in hydride site exchange in polyhydrides, 171, 199–202
    in H$_2$ reaction with sulfide complexes, 319–321
    in H$_2$ reaction with metalloenzymes: *see* Hydrogenase enzymes; Nitrogenase enzymes
    as intermediate in water-gas shift reaction, 260
    as intermediate in exchange between a hydride and a deutero acid, 279
    as intermediate in OA and RE of H$_2$ and dihydride-dihydrogen interconversions, 261, 264–265
    observation of, 48–49, 223–226; *see also* Photochemical reactions, preparation of H$_2$ complexes using
    table of, 49

Dihydrogen ligand or complex (cont.)
  versatility of binding of, 12, 61, 78, 213, 218
  versus dihydride coordination, 69–75, 79–88, 104–115
    in Cp versus Cp* complexes, 111
    in Cp versus Tp complexes, 111–114, 263
    influence of ancillary ligand stereochemistry or rearrangement energetics, 103–104, 107, 263
    influence of cis versus trans ligands, 86, 105
    influence of metal row and periodic trends, 109–111, 263
    influence of metal charge, 87
  vibrational spectra and analysis: see Vibrational modes and analysis
  weakly bound, 38, 43–44, 48–51, 105–107, 127, 274
Dinitrogen ligand
  direct reaction of H_2 with, 316
  discovery of, compared to H_2 ligand, 67, 27
  displacement by H_2 (and vice-versa), 48, 216–218, 220–222, 235, 297
    dependence on absolute entropies of H_2 and N_2, 221–222
  electronics of bonding, 77, 180
    charge decomposition analysis of, 77, 218
  importance of M to N_2 backdonation in stabilizing N_2 binding, 218
  in actinide complex, 218
  in models for nitrogenase enzymes, 314–316
  M–N_2 bond strength of, compared to that for H_2 and other ligands, 77, 212–213, 216–222, 228–229
  NN bond activation in, 180, 314–316, 355
  as predictor of dihydrogen versus dihydride coordination or gauge of M–H_2 BD, 105–107, 182–183, 263
  protonation of, by hydride and H_2 complexes, 316
  rate of reaction of N_2 compared to that for H_2, 228–229
  reaction with silanes, 355
  reversible binding to group 6 complexes, 22–23, 43–44, 76
  reversible binding to Rh pincer complex, 220
  weakness of binding to electrophilic centers, 216–217
Dioxygen ligand, rare coexistence with H_2 on the same metal center, 220
Diphosphine ligand, effect of bite angle or chelate effect on H_2 versus dihydride coordination, 107, 109, 111
Dynamic processes for σ ligands, 5
Dynamic processes, for dihydrogen complexes, 172–202

Electrochemical potential
  as predictor of H_2 versus dihydride coordination or gauge of M–H_2 BD, 106–107, 182–183, 263
  use in determination of pK_a of dihydrogen ligands, 275–276
Electrochemistry of H_2 complexes, 41–42
Electron energy loss spectroscopy (EELS) of surface-bound H_2, 50–51, 247
Electron transfer reactions of H_2: see Reactions of H_2
Electrophilic arene substitution, 375–376
Elongated H_2 complex: see Dihydrogen ligand or complex, elongated
Energy barrier for
  addition of H_2, SiH_4, and CH_4 to CpRh(CO) in gas phase, 231–232
  binding and cleavage (OA) of alkanes and CH_4, 389–394, 397, 401, 409
  binding and cleavage (OA) of H_2, 228–233
  dihydrogen-hydride ligand exchange, 13
  hydride site exchange in polyhydrides, 189–190
  rotation of H_2: see Rotation of H_2 ligand
  silane-hydride ligand exchange, 356
  stretching the H–H bond, 9
Enthalpies of ligand bonding: see Bond energy
Entropy
  agostic interactions favored over external ligand binding by, 44, 383–384
  importance of in σ ligand and other weak ligand binding, 209–210, 216, 218, 220–222, 383–384, 409
  low absolute entropy of H_2, CH_4, and other ligands, 221–222, 383–384, 409
Equilibrium
  between acidic η^2-H_2 complex and hydride complex with protonated ancillary ligand, 282
  between agostic and cyclometallated tautomers, 375–377
  between η^2-germane and germyl-hydride tautomers, 346, 360
  between η_2-H_2 and dihydride tautomers, 7 18, 15, 72, 88-89, 97, 103, 111, 114, 226–229, 262–264
    enthalpy and K_{eq} for $W(CO)_3(PR_3)_2(H_2)$-$WH_2(CO)_3(PR_3)_2$, 216
    isotope effect for, 237–238
    NMR and IR spectral data for $W(CO)_3(P^iPr_3)_2(H_2)$-$WH_2(CO)_3(P^iPr_3)_2$, 151–152
    nonexistence of for first-row metals, 264
    pK_a of tautomers, 274–275
    specific deprotonation of H_2 tautomer, 272, 277

Equilibrium (cont.)
  between η_2-H_2 and dihydride tautomers (cont.)
    steric factors in, 226–227
    thermodynamics and kinetics for, and table of thermodynamic parameters, 226–229
    versus stretching of the H–H bond, 264
  between η^2-silane and silyl-hydride tautomers, 9, 229–230, 341, 343, 346
Equilibrium isotope effect, KH/KD
  for binding of alkanes and alkenes, 238, 387, 400
  for formation of metal hydride versus deuteride, 234, 237–238
  for H_2 versus D_2 coordination, 233–238
  for occupation of deuterium in hydride versus η^2-H_2 binding sites, 237–238
  origin of, 235–238
  for protonation of metal hydride to η^2-$_{H2}$, 234
  for silane oxidative addition, 238
Ethylene ligand or complex
  binding to $M(CO)_3(PR_3)_2$ and unreactivity towards catalytic hydrogenation, 24, 43
  coordination geometry in trans bis-ethylene complexes, 70
  electronics of bonding and similarity to H_2 ligand, 76, 78
  equilibrium isotope effect for ethylene binding, 238
  hydrogenation to give H_2 complex, 14
  M–C_2H_4 bond strength compared to H_2 and other ligands, 217, 219220
  as model for H_2 coordination, 62
  noncorrelation of C=C bond distance with ν_{CC}, 253
  orientation in $W(CO)_3(PR_3)_2(C_2H_4)$, 70
Exchange processes
  between a hydride and a deutero acid, with intermediate H_2 complex, 278–279
  between dihydrogen and hydride ligands, 12, 15, 103, 120–125, 150–151, 187–202, 264–270
    associative versus dissociative, 188, 192
    involving bridging hydride, 194
    in $IrXH_2(H_2)(PR_3)_2$, 190–191
    migratory type versus replacement type, 189–190, 192
    observed by quasielastic neutron scattering, 195–198
    in solid state, low barrier for, 191
  between H in agostic CH and H in unbound ligand CH, 368
  between labile ligands L on $CpMn(CO)_2L$, 50
  between SH and hydride ligands, 320
  between silane and borane ligands, 425–426, 427
  between silane and hydride ligands, 337, 356–357

Exchange processes (cont.)
  between σ-ligands and ligands bound cis to them, 12
  in borohydride complexes, 417–418, 421, 422
  coupling with rotation of H_2 ligand, 181–182, 184, 191
  energy barriers: see Energy barrier
  fluxionality in agostic CH interactions, 373
  for H + H_2 → H_2 + H, 122
  for H_2 + D_2 → 2HD or MH_2 + D_2 → MHD + HD, 134, 189
  hydrogen exchange between NH and CH bonds, 268–269
  isotopic (H/D) involving agostic CH, 376–377
  isotopic (H/D) involving coordinated NH bonds with solvents, 430
  isotopic (H/D) involving σ-alkane ligands,
    in Cp*Rh(PMe$_3$) system, 396
    in Cp*Os(dmpm) system, 397–398
    in Tp'RhL system, 398–399
    in W(=NSi^{t}Bu$_3$) system, 399
    in cationic Pt(II) systems, 405–411
  isotopic (H/D) involving σ-borane ligands, 422
  isotopic (H/D/^{13}C) involving σ-alkane and CH_4 ligands, 396–399, 405–411
  isotopic (H/D/T) involving H_2 or hydride ligands, 121, 150, 155–158, 189, 265–270, 279, 281
    for Cr carbonyls, 267
    in hydrogenase enzymes, 313, 315
    incorporation of deuterium from D_2 or deuterated solvents into ancillary ligands, 269–270, 279
    intramolecular versus intermolecular, 266
    mechanism of, 268, 279
    scrambling with H_2O and alcohols, 268–270, 279, 281, 297–298, 321
    in nitrogenase enzymes, 297–298
    in solid-gas reactions, 267, 269
    in $W(CO)_3(PR_3)_2(H_2)$, 267–268
  isotopic (H/D/T) involving silane and hydride ligands, 337
[FeH_2]$^{+, 0, -}$, comparison of H_2 activation with charge, 130–131
Fluoroalkane bonding, 211–212, 391–393
Fluoroalkane, lack of binding to $Cr(CO)_5$, 385
Fluorophosphines, 385
Four-center interaction, 123–128: see also Sigma-bond metathesis
Four-center transition state: see Sigma-bond metathesis
Fourteen-electron (14e) complexes, 371, 381–383, 390–391, 393–394, 409–410, 432; see also Coordinatively unsaturated complexes or sites
Fuel cells, 6

Ge—H bond distance in coordinated germanes, 358–359
Germane (GeH$_4$) ligand or complex
  bonding and cleavage on Mo(CO)(PP)$_2$, 103, 358–360
  IR spectrum of Mo(CO)(dppe)$_2$(GeH$_4$), 360
Germane (R$_{4-n}$GeHn) ligand or complex, 357–360
  asymmetric η^2-Ge—H bonding of, 358–359
  bonding and activation of compared to silane ligands, 359–360
  bonding of, σ-donation versus backdonation, 358
  comparison to silane binding/activation, 358–360
  coordination to CpMn(CO)$_2$ type fragments, 357–360
  coordination to Mo(CO)(PP)$_2$, 103, 358–360
  dinuclear complexes, 358
  discovery and development of new examples of, 358–360
  effect of substituents on Si on bonding and activation of, 359–360
  photoelectron spectroscopic studies of, 358
  properties and characterization of, 358–360
  synthesis of, 358–360
  versus germyl-hydride coordination, influence of ancillary ligand, 359–360
  influence of substituents on Ge, 359
Gif system, 317

H$_2$O complex or ligand: see Aquo ligand
H$_3$, H$_4$, or H$_n$ ligand: see Trihydrogen ligand; Polyhydrogen ligand
H$_3^+$ versus H$_3^-$ as a ligand, 121–122
H$_3^+$, 121, 132, 329, 437
Haber process, 5, 312
Halide ligand
  effect of variation of on barrier to rotation of H$_2$ in Ir—H$_2$ complexes, 184
  protonation of, 277–278, 282
    by intramolecular heterolysis of H$_2$ ligand, 282–283, 289
  trans effect of: see Trans effect
  stabilization of coordinative unsaturation by π-donation, 33, 44, 115, 383
Hard/soft acid-base principles and alkane as soft ligand, 403–404
HD gas
  J_{HD} for: see NMR coupling constant J_{HD}
Heterolytic bond cleavage: see Cleavage, heterolytic
H—H bond; see also Dihydrogen ligand; H—H bond distance
  at what point is it broken, 9, 89–97
  cleavage of: see Cleavage; Oxidative addition of H$_2$

H—H bond (cont.)
  energy of: see Bond energy, H—H
  orientation of, on H$_2$ complexes, 117, 165, 166, 173–174, 178, 180–181, 184, 186–187
  polarization of when coordinated, 13, 270, 277
H—H bond distance
  comparison of
    experimental with calculated, 73, 81–83, 97, 89–90, 108, 118, 164
    in the series M(CO)(H$_2$)(dppe)$_2$ (M = Mo, Mn$^+$, Fe^{2+}), 167
    in solid versus solution states, 146, 148, 149, 153, 164
    in TcCl(H$_2$)(dppe)$_2$ versus Mo(CO)(H$_2$)(dppe)$_2$, 167
  continuum behavior of, 8, 15, 97, 143, 271, 345
  correlation with electrophilicity of metal and other parameters, 271
  determined by solid state NMR: see NMR spectroscopy, solid state
  determined by two-dimensional parameterized model for H$_2$ rotation, 186–187
  determined in
    [OsH$_5$(PPhMe$_2$)$_3$]$^+$, saga of, 144
    "gray zone" between H$_2$ and dihydride ligands (d_{HH} ~1.5 Å), 38–39, 110, 143–144
    intermediates in H$_2$/hydride and hydride site exchange processes, 171, 191–192, 200
    Ir complexes, 63
    Pd(H$_2$), 67–68
    polyhydrides with short H···H contacts, 143
    rare-earth hydrides, 39
    true H$_2$ complexes, see Dihydrogen ligand, true H$_2$ complexes
    unconventional H···H hydrogen bonds, 279, 281
    W(CO)$_3$(PR$_3$)$_2$(H$_2$), 28, 70, 88
  effect of trans versus cis ligand on, 81–83, 167
  effects of H$_2$ rotation on measurement of, 144–147, 165
  elongated: see Dihydrogen ligand or complex, elongated
  elongation of relative to other coordinated σ bonds, 329–331, 369–370
  estimated from NMR: see NMR coupling constant J_{HD}; NMR relaxation time
  isotope and temperature dependence of, 103, 157–158
  moderated by CO trans to H$_2$, 75, 81–82, 89, 147, 167
  tables of, 73, 82, 89, 148, 149
  unequal in η^2-H$_2$, 117–118, 164
Homolytic bond cleavage: see Cleavage, homolytic

Hydride ligand or complex
  "compressed dihydride" (d_{HH} = 1.2–1.6 Å), 88–89, 143; *see also* Dihydrogen ligand, elongated
  acidity of and kinetic acidity of, 13
  bond index and bond order of, 90–93
  bridging
    3c-2e bonding in, 329, 419
    formation in hydrogenases, 303–306, 309–311
    formation on deprotonation of H_2 complex, 272, 278, 284, 310
    formation on heterolysis of silane complex, 354
    in Mn-borohydride complex, 417–418
    viewed as protonated M−M bond, 15
    viewed as rapidly interconverting tautomers of a metal coordinated to a M−H bond, 60
  coordination of H_2 to binary metal hydrides, 130–131
  discovery of first, 17
  exchange coupling between hydrides: *see* Quantum-mechanical exchange coupling
  why $FeH_2(CO)_4$ is not an H_2 complex, 87–88
  fluxionality and hydride interchange in via transient H_2 ligand, 171, 189–190
  geometry of $WH_2(CO)_3(PR_3)_2$, 72
  hydrogen bonding of hydride: see Hydrogen bonding
  intermetallic rare earth, with short d_{HH}, 38–39
  isotope effects for formation of, 234, 237–238
  later recognized to contain H_2 ligand, table of, 29
  with long-range H····H interactions" (d_{HH} 1.6 Å), 94–96
  polyhydride-dihydrogen complexes, 34, 107–110, 225–227, 237, 262–263
    acidities of the H and H_2 ligands in, 274
    with CO versus halide ligand, 114–115
    facilitation of isotopic exchange reactions in, 269–270
    as precursors in catalytic hydrogenation, 288–291
    selective deprotonation of H_2 ligand in, 274, 277
  protonation of N_2 ligand by, 316
  protonation to give H_2 complex, 13, 15, 30, 34, 46–47, 113, 262, 270–273, 279, 283
    in Cp and Tp complexes, 111–113, 278
    kinetic versus thermodynamic site of, 272–273
    kinetics and rate of, 223, 272–273
    mechanism of, 46, 223, 277–279
    in sulfido complexes, 321–322
  trans influence of: *see* Trans effect
  vibrational modes for, 24, 237

Hydrocarbon oxidation by oxygenases and mimics for, 316–319
Hydrodesulfurization and hydrodenitrogenation processes, 5, 280, 319
  models for, 319–321
Hydroformylation reaction, 18–20, 127, 231, 265
Hydrogen bonding
  between Cl and H in agostic CH, 382–383
  between Cl and H_2 ligands, 166–167, 185–186, 283
  between hydride ligand and HOR, 279
  between hydride ligand and HY (Y = acid anion, OR), as first step in protonation, 223, 279
  between non-H_2 ligands in H_2 complexes, 167
  in $[R_3N-H^+\cdots Co(CO)_4^-]$ and other M···H−N interactions, 369
  unconventional (dihydrogen bonding), 117, 281–282, 370
    H···H distances in, 279, 281, 370
  in $W(CO)_3(PR_3)_2(H_2O)\cdot thf$, 213–214
    role in thermodynamics of formation from H_2 complex, 216
Hydrogen storage, 6
Hydrogen, production of, 6
Hydrogenase enzymes, 6, 15, 297–312
  evolution, general properties, and function of, 297–300
  [Fe] hydrogenases, 307–312
    displacement of aquo ligand by H_2 or CO, 308–310
    proposed mechanism for H_2 activation, 308–312
    X-ray structure of active site, 307–308
  [Ni-Fe] hydrogenases, 300–306
    proposed mechanism for H_2 activation, 304–306
    states of active site and plausible catalytic cycle, 300–301
    structural components in mechanism of, 299–300
    X-ray structure of active site, 301–303
  metal-free, 306–307
  organometallic models for, 303–304, 310–312
  theoretical calculations for, 304–305, 309, 311
Hydrogenation: *see* Catalytic hydrogenation
Hydrohalic acid ligands formed by protonation of halide ligands, 277–278
Hydrosilane ligand: *see* Organosilane ligand
Hydrosilylation, 352
Hypervalent behavior of Si, 344, 351
  interligand hypervalent interaction in silyl-hydride complexes, 351–352

Inelastic neutron scattering
  determination of barrier to H_2 rotation, 15, 165–166, 172–187

Index

Inelastic neutron scattering (*cont.*)
  determination of barrier to H_2 rotation (*cont.*)
    model for hindered H_2 rotation and experimental measurement of, 174–178
  determination of vibrational modes of $M-H_2$, 245–250
  as evidence for H_2 coordination, 172
Infrared spectroscopy: *see* Vibrational modes and analysis
Interligand hypervalent interaction in silyl-hydride complexes, 351–352
Intramolecular σ-bond interactions: *see* Agostic interaction
Ionic hydrogenation, 290–291
Ionization potentials of alkanes and rare gases, correlation with binding affinities to Cp*Rh(CO), 403
Iron, as the active site of metalloenzymes in
  hydrogenases, 297–312
  nitrogenases, 313–316
  oxidases, 316–319
Isolobal species or relationship, 2, 14, 60, 121
Isotope effects in σ ligand coordination and OA, 233–241; *see also* Equilibrium isotope effect; Kinetics of the bonding and cleavage of σ ligands, isotope effects; Exchange processes, isotopic
Isotopic exchange: *see* Exchange processes, isotopic
Isotopic Perturbation of Resonance (IPR), 155-157, 266, 373

Kinetic isotope effect, *see* Kinetics of the bonding and cleavage of σ ligands, isotope effects
Kinetics of the bonding and cleavage of σ ligands, 222–240
  for coordination and OA of H_2 to $W(CO)_3(PR_3)_2$, 228–229
  for displacement of Cl ligand by H_2, 222–223
  isotope effects for
    H_2 binding and loss from $W(CO)_3(PCy_3)_2(H_2)$, 238
    H_2 and alkane OA and RE, 238–241, 397
  mechanism of cytochrome P-450, 317
  mechanism of hydrogenases, 301
  protonation of M-H, 223
  protonation of metal hydrides to give H_2 complexes, 241
  reaction of $SiHMe_3$ with $Cp*_2TiS(py)$, 355
  for protonation of metal hydrides to give H_2 complexes, 223, 241
  rate constants for reaction of metal-alkane complexes with CO, 386
  for reaction of H_2 with 17e radical, 230–231
  for reaction of alkane σ-complexes ligands with H_2 and other ligands, 223–226

Kinetics of the bonding and cleavage of σ ligands (*cont.*)
  for $W(CO)_3(PR_3)_2(H_2)$, 222
  stop-flow data for substitution of H_2 in $W(CO)_3(PCy_3)_2(H_2)$ by pyridine, 228
Krypton ligand, 399–403
Krypton, liquid, as solvent, for preparing and characterizing alkane complexes, 399–403

Lanthanide H_2 complex, 33
Libration of $M-H_2$: *see* Rotation of H_2

Mass action effects, 44, 213, 215, 388
Matrix isolation of, 48–51
  H_2 complexes, table of, 49, 50
  methane complexes, 384–386
  $Pd(H_2)$, 50, 68
M-B bond distance in coordinated borohydrides and boranes, 417, 423–429
M-C bond distance in
  agostic C-B complexes, 435
  agostic C-H complexes, 369–373, 379–383
  agostic C-N complexes, 435
  agostic C-P complexes, 435–436
  agostic C-Si complexes, 433
  alkane versus fluoroalkane binding and CH_4 versus fluoromethanes, 392, 431
  heptane interaction with Fe in porphyrin complex, 388
Metal clusters: *see* Dihydrogen ligand, coordination, *to* metal-atom and metal-atom clusters; Theoretical calculations, for M_2 or M_3 clusters + H_2
Metal ion complexes, $[M(L)_n]^+$ (L = H_2, CH_4, C_2H_4, H_2O, CO), 128–131, 210–211, 366
  comparison of H_2 and CH_4 binding energies in, 129, 210–211, 385–386, 393
  reactions of silane with, 348–349, 368
Metal oxides, σ-bond binding and activation on, 15, 132–134
Metal surfaces: *see* Dihydrogen ligand, coordination, to surfaces of metals and compounds; Electron energy loss spectroscopy (EELS) of surface-bound H_2; Agostic interactions, of C-H bonds, on M surfaces
Metal-metal bond
  hydrogenation of, 230–231
  in hydrogenases, 302–312
  protonation of, 278, 303
Methane
  abundance of, 383
  activation on Shilov and related systems: *see* Shilov and related systems
  biological activation by MMO, 317–319, 410

Metal-metal bond (cont.)
  conversion to liquid fuels and selective oxidation of, 4, 5, 365, 404–411
    catalytic conversion to methyl bisulfate by Pt and Hg bipyrimidine complexes, 409-410
    first example of activation by an organometallic complex, 126
    heterolytic cleavage of C–H bond: see Cleavage, heterolytic
    low absolute entropy of, favoring competitive binding, 222, 383–384, 409
  Methane coordination; see also Alkane σ complex
    acidity of CH_4 σ-complex, 403–404, 409–410
    in alkane activation on Shilov and related systems: see Shilov and related systems
    bond energy for: see Bond energy
    coordination modes for, 384
    displacement of aquo ligand by CH_4 by associative substitution process, 407–408
    to electrophilic metal centers such as Pt^{II} cations, 393, 403–411
    to $M(CO)_5$ (M = group 6), 23, 76, 211–212, 384–385
    to metal-ions, 128–131, 366, 385–386, 393
    in methane exchange reactions on Cp_2M, 126
    to organometallic metal ions such as $[Mn(CO)_5]^+$, 385–386
    possibly favored by low absolute entropy of methane, 222, 383–384
    prediction that CH_4 would now be the most popular ligand, 365, 437
    in σ-bond metathesis, 125–126
    theoretical calculations: see Theoretical calculations, for C–H bonding and activation, for CH_4
    as transient and/or intermediate species in OA/RE of CH_4, 397–399, 401–411
    versus fluoromethanes, 391–393
Methane hydrate, 383
Methane mono-oxygenase: see Oxygenase enzymes
Methanol
  C–H bond strength in compared to that in CH_4, 365
  production from methane, 365, 404, 411
  reactivity of C–H bond in compared to methane, 404
M–Ge bond distance, in M–germane complexes, 359
M–H bond distance in
  agostic C–H complexes, 367, 369–373, 379–383
  alkane versus fluoroalkane binding and CH_4 versus fluoromethanes, 392
  M–borane and borohydride complexes, 419–428
    comparison to those in silane complexes, 423

M–M bond distance in (cont.)
  M–germane complexes, 359
  in M–H_2 complexes, 73, 89, 147, 159, 164
    comparison of calculated and observed, tables of, 73, 89
    NMR relaxation time, as estimate of, 159
  M–silane complexes, 329, 344, 359, 423
    comparison to those in borane complexes, 423
M–H_3 intermediate: see trihydrogen ligand
MMO: see methane mono-oxygenase
Mossbauer spectroscopy, as probe of BD versus σ donation, 77
M–P bond distance, in agostic C–P complexes, 435–436
M–Si bond distance, in M–silane complexes, 331, 342, 344, 359

Natural gas, 5, 6, 383; see also Methane
Neutron diffraction (single crystal)
  agostic C–H interactions, 369, 372–373
  $(CpMe)Mn(CO)_2(SiHFPh_2)$, 329, 335
  $Cp_2MH(SiR_3)$ (M = Nb, Ta) showing interligand hypervalent interactions, 351–352
  $Cu(BH_4)(PMePh_2)_3$, 419–420
  H_2 complexes, 89, 145–149
    foreshortening of observed d_{HH} by effects of H_2 rotation, 144–147
  $Mo(CO)(dppe)_2(H_2)$, 145–146
  $[OsH_5(PPhMe_2)_3]^+$, 143–144
  polyhydrides, 89, 143
  $W(CO)_3(P^iPr_3)_2(H_2)$, 26–28, 145, 148
N–H bond, interaction with metal centers, 369, 429–430
Ni^+/silica interaction with H_2, 38
Nineteen-electron (19e) complexes, as intermediates in reactions of 17e radicals with H_2, 230–231
Nitrogenase enzymes, 6, 15, 312–316
  general properties, and function of, 312–314
  inorganic models for, 315–316
    splitting of H_2 on, 315
  proposed mechanism for N_2 activation, 314–316
  theoretical calculations for, 314–316
  X-ray structure of active site, 313–314
NMR coupling constant, $J(^{11}BH)$
  in coordinated B–H bonds, 419–422, 429
  reduction of value of, 9, 420
  in terminal B–H groups and K[Tp], 419, 420
NMR coupling constant, J_{CC}, in complex with agostic C–C bonds, 433
NMR coupling constant, J_{CH}
  in agostic CH, 370, 373, 376
    averaging of, 373
  in $CpRe(CO)_2(cyclopentane)$, 387
  reduction of value of, 9

NMR coupling constant, J_{HD}
  averaging of, in fluxional H_2/hydride complexes, 155–156
  calculation of, 154–155
  as estimate of H–H bond distance, 148–157, 161
  as evidence of H_2 coordination, 26, 38, 152
  for HD gas, 26, 153
  influence of ligand trans to HD versus cis, 81–83, 193
  influence on by ancillary ligands, 81–83, 86
  lowest value observed for complex with CO trans to H_2, 86–87
  proportional to H–H bond order, 154–155
  reduction of value of, 9, 26
  tables of, 82, 148, 149
  temperature and field dependence of, 98, 103, 157–158
NMR coupling constant, J_{HH}, extraordinarily large values of in quantum-mechanical exchange coupling, 198–202
NMR coupling constant, J_{HT} and J_{DT}, of partially tritiated complexes, 155–158
NMR coupling constant, J_{PH}
  low values or nonobservance of in H_2 complexes, 150
  in P–H bond interacting with metal center, 430
  in terminal P–H bond, 430
NMR coupling constant, J_{SiH}
  in agostic Si–H bonds, 338
  in bridging silyl complex, 336
  in chelating bis(silane) complex, 337
  influence of ligand trans to silane versus cis, 345–346
  lack of correlation with Si–H distance, 331, 338
  reduction of value of in silane complexes, 9, 329–331
  similarity to J_{HD} in elongated H_2 complexes, 26, 329, 331
  tables of, 343, 345
NMR coupling constant, J_{SnH}, 361–362
NMR deuterium (^{2}H), solution and solid state, 162–164
  deuterium quadrupole coupling constant ($\underline{C}_Q$), 162–163
  relaxation time (T_1) for ^{2}H NMR, 159–163
  solid state, 163–164
NMR proton resonance for
  B–H binding to metals, 417, 422
  for C–H binding to metals in
    agostic CH, 373
    CpRe(CO)$_2$(cyclopentane), 387
  H_2 ligand
    averaging of, for exchanging hydrogens
    broadness of, 26, 150, 159, 161–162

NMR proton resonanance for (cont.)
  H_2 ligand (cont.)
    spectra for W(CO)$_3$(PiPr$_3$)$_2$(H$_2$) and HD isotopomer, 151–152
  P–H bond interaction with a metal center, 429–430
  S–H bond interaction with a metal center, 429–430
  silane and silyl ligands, 330, 336, 356
NMR relaxation time (T_1 and T_1^{min})
  contribution from
    dipole-dipole relaxation in H_2 ligands, 159
    metals with high magnetogyric ratios, 159
    nearby H on ligands, 159
  as estimate of H–H bond distance, 30, 143, 148, 149, 158–163, 279
    correction for H_2 rotation, 159–160
  as estimate of M–H bond distance, 159
  evidence for H_2 coordination, 30, 110, 112, 159, 287
    as indicated by T_1 of dissolved H_2, 161
  field strength dependence of, 159
  for ^{2}H NMR, 159–163
  for hydride-hydride interactions, 159
  variation with temperature, 159
NMR spectroscopy, solid state, 148–150
  ^{2}H NMR, 163-164
  determination of d_{HH} as evidence for H_2 coordination, 30, 38
  determination of d_{HH}, 8, 144, 148–150
  determination of dH_H, compared to calculated, 72, 75
  evidence for mechanism of H_2/H exchange in IrXH$_2$(H$_2$)(PR$_3$)$_2$ using, 190–191
  Pake doublet in, 39, 148–150
  studies of rare-earth hydrides
NMR, exchange coupling between hydrides: see Quantum-mechanical Exchange Coupling
NMR, extended self-decoupling in borane and borohydride complexes, 429
NMR, tritium, 155–156
  J_{HT} and J_{DT} of partially tritiated complexes, 155–158
Nonclassical carbonium ion: see Carbocation

Olefin coordination, compared to M–H$_2$ bonding, 2, 3, 10, 60–61, 69, 253
Olefin metathesis, 13, 122
Olefin polymerization, 113
Organosilane complex or ligand: see Silane (R3SiH) ligand
Orthometallation: see Cyclometallation
Oxidative addition (OA) of; see also Cleavage, homolytic
  alkanes and C–H bonds
    of agostic CH: see Cyclometallation

Oxidative addition (OA) of (*cont.*)
  alkanes and C−H bonds (*cont.*)
    barrier to, 229
    for CH_4, and comparison to H_2, 389−394
    discovery and development of, 365−367
    ease of, correlated with C−H bond strength and *sp*, *sp*2, and *sp*3 bond character, 394−395
    evidence for σ complexes as intermediates, 395−399
    external versus internal, 377−378
    isotope effects and isotopic exchange in, 238−241, 395−399
    isotope effects for, 239−241
    mechanism and reaction coordinate for, 367
    reviews of and books on, 365
    trajectory for derived from a series of structures of agostic complexes, 371
  boranes, 427−429
  C−P bonds, 436
  C−X bonds (X = halogen), 432
  dihydrogen
    barrier to, 226, 228−233
    comparison of H_2 activation on $[FeH_2]^{+, 0, -}$ with charge, 130−131
    effects of stereochemistry of ancillary ligands, 107
    energy profile for addition to CpML, 112
    factors that disfavor, 47, 61, 86
    influence of backdonation (BD) on: *see* BD, influence on OA
    isotope effects for, 234, 237−239
    late transition state for, 96−97, 254
    mechanism and reaction coordinate for, 8, 88−97, 105−107, 172, 254
    to $Mo(CO)(PP)_2$ (M = Mo, W), 80−81, 86
    overview of, 79−81
    rearrangement of coordination sphere upon, 103−104, 262−263
    theoretical studies of: *see* Theoretical calculations for H_2 and hydride systems
  germanes, 359−360
    compared to silanes, 359
  N−H bonds, 430
  S−H bonds, 430−431
  silanes, 83
    effect of metal row and ancillary ligand, 337, 338
    mechanism and reaction coordinate for, 342, 344, 350, 352
    more favorable than for H_2 OA, 219, 336, 339
    to $Mo(CO)(PP)_2$
    rate of, compared to alkanes, 342
    to $W(CO)_3(PR_3)_2$, 345

Oxidative addition (OA) of (*cont.*)
  stannanes, 362
  σ ligands
    "arrested" along reaction coordinate, 5, 7, 8, 10, 28, 98, 103, 105, 329−330, 338, 436
    effects of bond energetics on, 231−233
    equilibrium processes: *see* Equilibrium
Oxo ligand (Fe=O), proton transfer to, in oxygenases, 317−319
Oxo reaction: *see* Hydroformylation
Oxygenase enzymes and MMO, 14−15, 316−319
  general properties and function of, 316−318
  inorganic models for, 317−316, 320−321
  proposed mechanisms for, 317−319
  theoretical calculations for, 318−319
  X-ray structure of active site of MMO, 318

Palladium-catalyzed cross-coupling reactions, 43
Parahydrogen Induced Polarization (PHIP) effect, 158, 261
Paramagnetic σ-complexes, virtually unknown existence of, 33
P−H bond distance in P−H bond interacting with metal center, 430
P−H bond, interaction with metal centers, 429−430
PH_3, use of to model PR_3 computationally, 62−63, 108−109, 118, 164, 184, 280−281
Photoacoustic calorimetry, for bond energy measurements, 208, 211−212, 386
Photochemical reactions
  of $CpM(CO)_2$ (M = Rh, Ir), 225−226
  of $CpMn(CO)_3$, 49, 223−226, 328−329, 360−361
  in hydrosilylation reactions, 352−353
  of $M(CO)_6$ (M = group 6), 48−49, 211−213, 246, 223, 384−385
  of $M(CO)_6$, in gas phase, 211−213
  of $Mo(CO)_4(dppe)$, 362
  preparation of alkane complexes using, 384−387; *see also* Alkane σ complexes
  preparation of H_2 complexes using, 48−50, 223−226
    in gas phase, 225−226
    in polyethylene disks, 48−49
    in rare-gas media, 48−49, 76, 246, 263
    in supercritical CO_2, 49
  preparation of silane complexes using, 328−329, 337−338
  preparation of stannane complexes using, 360−362
  to give reductive elimination (RE) of H_2, 260, 264
  of $W(CO)_4(P^iPr_3)_2$, 24
Photoelectron spectroscopic studies of silane complexes, 338−339
Piano-stool complexes, 111−114; *see also* Cyclopentadienyl complexes

Photoelectron spectroscopic studies of silane complexes (cont.)
  effect of bite angle of diphosphines on σ coordination in, 107
  unpredictability of σ bond coordination to, 87, 338–339
Pincer (P–C–P) ligand complexes, 220, 286–287, 369–370, 376, 432
Platinum complexes for alkane activation: see Shilov and related systems for alkane and methane activation
Polarizabilities of alkanes and rare gases, correlation with binding affinities to Cp*Rh(CO), 403
Polyhydrogen ligand, 121–123, 267
Polymer-supported catalysis, 436
Polywater, 27
Proton
  high mobility of, 14, 15, 270–271, 317, 404
  transfer or exchange, 14, 270–284; see also Cleavage, heterolytic; Exchange processes; Reactions of H_2 complexes, deprotonation
Protonation
  double, 47, 225
  of hydride complex: see Hydride complex, protonation of
  of hydride-halide complex, site selectivity of, 46
Pseudohalide ligand, π-donation from stabilizes coordinative unsaturation, 33

Quantum-mechanical behavior of H_2 and hydride ligands, 158 172–175, 178, 195, 99–202
  at room temperature, 195
Quantum-mechanical exchange coupling (QEC), 158, 161, 190, 198–202
  extraordinarily large values of J_{HH} in, 198–202
  involving only an H_2 ligand having blocked rotation, 201–202
  η^2-H_2 as intermediate in, 199–202
Quasielastic neutron scattering, 182, 194–198
  as evidence for cis interaction between hydrogens, 195–198
  observation of H_2/hydride site exchange by, 195–198

Radicals: see Seventeen-electron complexes; Nineteen-electron complexes
Raman spectroscopy: see Vibrational modes and analysis
Rare gas ligands, 4; see also Ar, Xe, etc.
Reactions of H_2 complexes, 269–291; see also Catalytic hydrogenation; Exchange process; Oxidative addition of H_2; Cleavage, heterolytic, of H–H; Cleavage, homolytic, of H–H

Reactions of H_2 complexes (cont.)
  with CO_2 to give hydrido-formate complex, 286–287
  decomposition
    in halogenated solvents, 40
    of $M(CO)_5(H_2)$, 49
  deprotonation of H_2 ligand, 47, 268, 270–284
    by ancillary ligand, 280–282, 310
    by ancillary ligand, in mechanism of hydrogenases, 305–306, 309–310
    by anion, 268, 283–284
    calculated energies of, 276–277
    by H_2O, 268, 277
    by N_2 ligand to give ammonia, 316
    selective deprotonation of H_2 over H ligand in $MH(H_2)$ complexes, 274, 277
    by strong bases, 272
    direct transfer of H from η^2-H_2: see Catalytic hydrogenation, direct transfer of H; Dihydrogen ligand, direct transfer of H from
  displacement of H_2 by, 40
    alkenes and alkynes, 288
    anion, 42, 46–47, 259–260
    H_2O: see Aquo ligand, displacement by H_2 or competition with H_2 ligand
    hydrocarbon solvent, 44, 213, 259–260, 388
    N_2 (and vice-versa), 39, 49, 216–218, 220–222
    N_2 (and vice-versa), dependence on absolute entropies of H_2 and N_2, 221–222
    nitriles, 40, 224–225
    O_2, 39
    pyridine, 228
    solvent, 40, 46–47, 259–260, 285, 288
    THF, 40
  dissociation of H_2, 40, 224–225
    color changes upon, 40
    to create open site for ligand binding in catalysis, 288–289
    rates and kinetics of, 224–225
    stability towards, 43–44
  photochemical, 41
  protonation of coordinated N_2 to give ammonia, 316
  reduction of acetone to isopropanol, 42
  reviews of, 259
  in solid state, 260–261
  water-gas shift reaction, H_2 complex as intermediate in, 260
Reactions of H_2 gas
  displacement of bound chloride, 45, 222–223
  displacement of bound N_2, see Reactions of H_2 complexes, displacement of H_2 by N_2
  displacement of labile ligands, 45, 223–226, 357

Reactions of H_2 gas (*cont.*)
  electron transfer, 18
  with electrophilic precursor to form acidic H_2 complex, 278, 284
  homogeneous activation on metal complexes, 17–20
  kinetic isotope effect for, 238–241
  across M−L multiple bonds, 127–128
  with metal sulfide complexes, 319–321
  with metalloenzymes: *see* Hydrogenase enzymes; Nitrogenase enzymes
  with methyl radical, 128
  with RhCl(PPh$_3$)$_2$, 225–226
  with 17e Cp*Cr(CO)$_3$ radical, 230–231
  with 16e metal carbonyl fragments, 223–226; *see also* Photochemical reactions, preparation of H_2 complexes using
  with a trinuclear Ru cluster with a μ-phosphido ligand, 436
  with Vaska complex, 19, 34
Reductive elimination (RE) of, 4
  alkanes and C−H bonds,
    evidence for σ complexes as intermediates, 395–399
    isotope effects and isotopic exchange in, 238–241, 395–399
    in Shilov systems, 405–411
  dihydrogen
    isotope effects for, 238–241
    as microscopic reverse of H_2 cleavage, 264
    photochemical promotion of, 260
  functionalized alkane (RX) in Shilov systems, 405–411
  hydrohalic and triflic acids from H_2 complexes, 277–284
  silanes, 342–343
Rotation and libration of H_2 ligand, 15, 159–163, 172–187, 194–198
  blocked (restricted), 160–162, 173, 201–202
  coupling with exchange dynamics and cis-effect, 181–182, 184, 191, 195–198
  effect on measurement of d_{HH}, *see* Neutron diffraction (single crystal), foreshortening observed d_{HH} by effects of H_2 rotation
  energy barrier for, 15, 112, 165–166, 172–187
    correlation with *n*NN and electrochemical parameters, 182–183
    determination of by INS, 174–178
    as direct proof of M→H_2 backdonation, 172–173
    effect of crystallographic disorder on, 185
    effect of distortions in M−ligand geometries on, 184–185
    lowering by cis interaction between H_2 and H ligands, 172, 180–181

Rotation and libration of H_2 ligand (*cont.*)
  energy barrier for (*cont.*)
    lowering by symmetric ligand sets, 181, 184
    for M(CO)$_3$(PR$_3$)$_2$(H$_2$) and other group 6 complexes, 173, 179
    origins of, effect of ancillary ligand, 182–184
    origins of, effect of the metal center, 179–182
    origins of, intramolecular interactions and crystal packing effects, 184–186
    origins of, metal-hydrogen binding, 178–179
  four-fold component of, 160, 181
  physical description of, 159–160
  in quantum-mechanical exchange coupling and hindered rotational phenomena, 198–202
  rotational tunneling, 174–178, 195–198, 202
  two-dimensional parametrized model for, 186–187
Rotation of D_2 ligand and energy barrier for, 163–164
Rotational tunneling and rotational tunneling spectroscopy of H_2 ligand, 174–178, 195–198

Seventeen-electron (17e) complexes, reaction with H_2, 230–231
Shilov and related systems for alkane and methane activation, 14, 215, 403–411
  ancillary ligand effects, 407–409
  discovery and proposed mechanism of original Shilov Pt-chloro system, 404–405
  isotopic exchange in, 404–411
  Pt and Hg bipyrimidine complexes, 409–411
    catalytic conversion of methane to methyl bisulfate in strong acids, 409–410
  Pt-diimine type complexes, 407–408
  Pt-tmeda complexes, 405–408
  reviews of, 404
  theoretical calculations in, 408–409
  σ bond metathesis versus OA pathways, 408–409
Sigma (σ) complexes and σ-bond interactions; *see also entries such as* Dihydrogen ligand and complexes
  as arrested intermediates in OA: *see* Oxidative addition, arrested
  coordination of H−H, Si−H, and C−H bonds to Mo(CO)(PP)$_2$, 43
  general similarity of bonding and activation of all X−H bonds, 327–328, 373
  oxidation state of M in, 38
  stabilized by highly electrophilic M centers; *see also* Dihydrogen ligand, bonding of, to electrophilic metal centers
  with two different types of σ-ligand on same M, 35, 120, 327, 348, 394, 419, 425

Index

Sigma (σ) complexes and σ-bond interactions (cont.)
  X–H interactions where X (N, P, S) is also a lone-pair donor, 429–431
  X–Y interactions where X and Y are not hydrogen, 431–436
Sigma (σ)-bond metathesis, 12–13, 116, 123–128, 189
  in intramolecular heterolytic cleavage of H_2 and silane ligands, 280, 355
  involving borane ligands, 422, 426, 429
  mechanism of, 13, 116, 123–128, 280
  via six-membered ring, 127, 281
Si–H bond
  agostic interactions of: see Agostic interactions, of Si–H bonds
  basicity and relative donor/backdonor properties of, compared to H–H and C–H, 342
  cleavage of: see Cleavage
  coordination and activation, 327–357; see also Silane ligand; Agostic interactions, of Si–H bonds
  elongated, 97, 336, 345, 350
  polarization towards $M(Si^{\delta+}-H^{\delta-})$ when coordinated, 13–14, 353
Si–H bond distance
  in agostic Si–H bonds, 338
  in chelating bis(silane), 337
  comparison of experimental with calculated d_{SiH}, 344, 350
  elongation of compared to H–H bond, 330–331
  in free silanes, 329
  high variability of, 342
  in $Mo(CO)(depe)_2(SiH_2Ph_2)$ versus SiH_3Ph analogue, 335
  in neutron structure of $(CpMe)Mn(CO)_2(SiHFPh_2)$, 329–330
  table of, 332–333
Silane (SiH_4)
  properties of, 327, 340
  reaction with metal ions, M^+, 130
Silane (SiH_4) ligand, see also Silane $(R_{4-n}SiH_n)$ ligand for general characteristics of
  bonding and cleavage on $Mo(CO)(PP)_2$, 103–104, 229–230, 327, 340–341
  bridging, 341–342
  in matrix-isolated $Al(\eta^2-SiH_4)$ species, 340, 341
  model for CH_4 ligand, 9, 328, 340–341
Silane $(R_{4-n}SiH_n)$ ligand, 327–357
  asymmetric η^2-Si–H bonding of, 328, 342
  bis-silane complexes, 233, 337, 353, 355–356
  bond energies for M-silane: see Bond energies
  bonding and activation of compared to H_2 and other ligands, 219, 327, 329, 336, 346–347
  bonding of, 327, 330, 342, 346, 349–351

Silane $(R_{4-n}SiH_n)$ ligand (cont.)
  bonding of (cont.)
    to electrophilic fragments, 344, 349
    influence of ligand trans to silane, 345–346
    σ-donation versus backdonation, 346–347, 349
  cationic complexes, and rareness of, 330, 345, 353–355
  chelating bis(silane), 337, 350–351, 355–356
  comparison of silane and H_2 reaction with $W(CO)_3(PR_3)_2$, 345
  comparison of silane, H_2, and small-molecule binding/activation to 16e fragments, 344–346
  coordination to $CpMn(CO)_2$ type fragments, 328–334, 337–339, 342–345, 348, 351, 354–356
  coordination to $Mo(CO)(PP)_2$, 103–104, 210, 228, 334–335, 345, 347
  dinuclear complexes, 328, 330–334, 336
  discovery and development of new examples of, 328–329
  dissociation of, rates and kinetics of, 225
  effect of metal row and ancillary ligands on strength of bonding, 219
  effect of substituents on Si on bonding and activation of, 15, 219, 225, 328, 330, 342–346, 352
  elongated, 330–331, 337, 342, 345, 350
  with H_2 ligand on same complex, 327, 348
  high-valent M^{III} complex, 339
  hydrosilylation, 352
  M–silane bond strength compared to H_2 and other ligands, 212, 217, 219, 233, 327
  photoelectron spectroscopic studies of, 338–339
  properties and characterization of, 330–342
  reactions of, 352–355
    with coordinated sulfide and N_2 ligands, 355
    dissociation and RE of, 342–343
    heterolytic cleavage: see Cleavage, heterolytic, of Si–H
    photocatalytic hydrosilylatoin of dienes, 352
    reviews of, 352
    in silane alcoholysis, 353–354
  reviews of, 329–330
  steric effects on binding, 219
  synthesis by, 334–342
    addition of silane to precursor complex, 334–337
    displacement of borane ligand, 425, 428–429
    protonation of silyl ligand, 339
    substitution of H_2 ligand, 337
    substitution of phosphine ligand, 337
  table of stable complexes by generic type, 332–333
  true σ-silane complexes, 330–331, 349–350
  types of structures of $M(\eta^2-Si-H)$ interactions and table of, 330–331

Silane ($R_{4-n}SiH_n$) ligand (cont.)
  unstable or weakly bound, 337–338, 345, 353
  versus silyl-hydride coordination, influence of
    ancillary ligand, 342–343, 345, 352
    ancillary ligand stereochemistry or
      rearrangement energetics, 347, 352
    cis versus trans ligands, 345–346
    metal row and periodic trends, 330, 349
    substituents on Si, 15, 219, 225, 328, 330, 342–346
    vibrational spectra and analysis: see Vibrational modes and analysis
Silanols, catalytic formation of, from silanes and alcohols, 353–354
Silylium cation, R_3Si^+, 14, 353–355
Sixteen-electron (16e) complexes, 2224, 44, 80, 83, 105, 369, 370, 379–383, 389–394; see also Coordinatively unsaturated complexes or sites
  comparison of silane, H_2, and small-molecule binding/activation to 16e fragments, 344–346
  H_2 binding to: see Dihydrogen ligand, 16e complexes of
  J_{HD} and d_{HH} for H_2 coordination to various 16e fragments, table of, 82
  phosphine complexes with agostic interactions, and table of X-ray distances in 379–383
  stabilized by halide or other π-donor ligands, 33, 44, 115, 383
  stabilized by σ-donation from hydride ligands, 83, 383
Sleeping Beauty effect, 14
Sn–H bond distance in coordinated stannanes, 361–362
Solar energy conversion, 6
Stannane ($R_{4-n}Sn_nH_n$) ligand, 360–362
  asymmetric η^2-Sn–H bonding of, 361–362
  bonding of, σ-donation versus backdonation, 362
  comparison of X-ray and NMR parameters of MeCpMn(CO)$_2$(SnHPh$_3$) to silane analogue, 361–362
  coordination to CpMn(CO)$_2$ type fragments, 360–362
  elongated, 362
  properties and characterization of, 361–362
  reactions and decomposition of, 361–362
  synthesis of, 360–362, 428–429
Steric effects in
  agostic interactions of C–H bonds in phosphines, critical nature of, 381–383
  C–C bond coordination, 432
  C–H bond activation, 394–395
  cyclometallation versus external C–H bond cleavage, 377–378

Steric effects in (cont.)
  H_2 complexes see also Theoretical calculations for H_2 and hydride systems, molecular mechanics calculations
    [CpRu(H$_2$)(PP)]$^+$, 111
    equilibria between η^2-H_2 and dihydride tautomers, 226–227, 264
    influence of electronic properties, 104
    influence on barrier to rotation of H_2, 185
    minor importance of, 29, 79–80
    MoL(H$_2$)(CO)(dppe)$_2$ (L = CO, CNR), 84
    in silane and other σ complexes, 103–104, 219, 329, 334, 339, 347, 362
Stretched H_2 complex: see Dihydrogen ligand or complex, elongated
Sulfido ligands
  as models for biological and industrial activation of H_2, 319–322
  heterolytic cleavage of H_2 on
    in hydrogenase, 298, 304, 308–312
    in model for nitrogenase, 315
    in sulfide complexes, 319–322
    in sulfide ligand alone, 320
Sulfur dioxide ligand
  electronics of bonding, 21, 78
  group 6 complexes, 21–23, 76
  M–SO$_2$ bond strength compared to H_2 and other ligands, 217–220
  versatility of binding to metals, 12, 21
Sulfur dioxide, use in methane oxidation, 410–411
Summary of book and a glance to the future, 436–437
Supercritical gases, as solvents for study of σ complexes, 49, 395
Superelectrophile, 3, 14

Tautomeric equilibrium: see Equilibrium
Temperature-dependent equilibrium measurements of M$^+$ ion reactions with H_2 and CH$_4$, 129–131, 210
Theoretical calculations for actinide-dinitrogen complex, 218
Theoretical calculations for B–H bonding and activation
  for agostic B–H interactions, 420
  for Cp$_2$Ti[HB(OH)$_2$]$_2$ and OsH$_3$(BH$_4$)(PH$_3$)$_2$ models, showing difference in bonding compared to other σ complexes, 426–427
  for Ti(BH$_4$)$_3$(PMe$_3$)$_2$, versus experimental structure, 420–421
Theoretical calculations for C–H bonding and activation
  for agostic C–H interactions, 370, 376, 377
  for alkane versus fluoroalkane binding and CH$_4$ versus fluoromethanes on W(CO)$_5$, 391–393

Theoretical calculations for C—H bonding and activation (cont.)
  for CH$_4$ binding and activation in, 65–66, 77, 388–395
    addition of CH$_4$, SiH$_4$, and H$_2$ to RhCl(PH$_3$)$_2$, 390–391
    addition of CH$_4$, SiH$_4$, and H$_2$ to CpML (M = Rh, Ir), 231–232, 388–390
    Cr(CO)$_5$ + CH$_4$ versus H$_2$, 65, 77, 388–389
    Cp*Os(dmpm) system, 397–398
    M(CO)$_4$ + CH$_4$ versus H$_2$ (M = Ru, Os), 388–389, 393
    OsCl$_2$(PH$_3$)$_4$ + CH$_4$ and H$_2$, 393–394
    [Rh(CO)$_4$]$^+$ + CH$_4$ versus H$_2$, 65, 388–389
    reduced variable space analysis of, 393
  effect of ligand set on alkane binding, 393-394
  in Shilov systems, 408–409
  in σ-bond metathesis, 125–127
Theoretical calculations for coordination and OA/RE of C—C and C—Si bonds, 432–434
Theoretical calculations for H$_2$ and hydride systems
  addition of H$_2$, SiH$_4$, and CH$_4$ to CpML (M = Rh, Ir), 231–232, 388–390
  alkane activation, 388–395
  C—F versus C—H bond coordination, 391–393, 431–432
  cis and trans-IrCl$_2$H(H$_2$)(PH$_3$)$_2$, 81–82, 89–95
  cis-H$_2$/H interactions in group 8 complexes, 116–120
  cis-OsH$_2$(CO)$_4$, 94–96
  comparison of frontier orbital energies of H$_2$ and CH$_4$, 388–389
  computational methods, 61–63
    atoms-in-molecules (AIM) formalism, 89, 201, 345
    Carr-Parrinello ab initio molecular dynamics
    charge decomposition analysis and extended transition state analysis of BD, 77–79, 218
    comparison of DFT versus MP2, 63
    density functional theory (DFT) as leading methodology, 62
    discrete variable representation analysis, 255–257
    orbitally ranked symmetry analysis, 109
  [CpMH$_4$(H$_2$)(PH$_3$)]$^+$ (M = Mo, W), 112
  CpNbH(H$_2$)·BH$_3$, 110
  [CpRuH$_2$(PH$_3$)$_2$]$^+$ and other Cp complexes, 111–112
  Cr(CO)$_5$ + H$_2$ versus CH$_4$, 65, 77, 388–389
  deuterium quadrupole coupling constant, $\underline{C}_Q$, 163
  deprotonation energies, 276–277
  d_{HH}, compared with neutron diffraction data, 89–90, 108, 164

Theoretical calculations for H$_2$ and hydride systems (cont.)
  elongated H$_2$ complexes
    [CpRu(H$_2$)(H$_2$PCH$_2$PH$_2$)]$^+$ and [OsCl(H$_2$)(H$_2$PC$_2$H$_4$PH$_2$)]$^+$, 98–103
    [Os(NH$_3$)$_4$(Lz)(H$_2$)]$^{z\$^pL^2}$, 83, 97–99, 108, 213
    [OsH$_5$(PH$_3$)$_3$]$^+$, 109–110
    MH$_7$(PH$_3$)$_2$ (M = Re, Tc), 107–108
  equilibria between η^2-H$_2$ and dihydride tautomers, 72
  [FeH(H$_2$)(PH$_3$)$_4$]$^+$ versus [FeH(H$_2$)(PMe$_3$)$_4$]$^+$, 118, 122
  H—H interactions in rare-earth hydrides, 38–39
  η^2-H$_2$ versus dihydride coordination, 79
  MH$_2$(CO)$_4$ (M = group 8), 87–88
  H$_2$ addition to W(CO)$_3$(PH$_3$)$_2$ and W(PH$_3$)$_5$, 69–75, 98
  [Fe(H$_2$)(PH$_3$)$_5$]$^+$ versus MoH$_2$(PH$_3$)$_5$, 87
  H$_2$ rotation and energy barriers, 178–187
  H$_2$/H and other hydride ligand site exchange processes, 188, 191–194, 199–202
  H$_3^+$ and analogues M(H$_2$) (M = Li$^+$, Be^{2+}, BeO), 132
  heterolytic cleavage of H$_2$ on metal sulfide complexes, 319
  importance of electron correlation, 107–108
  interplay between experiment and theory, 5, 14, 30, 62, 80, 109–110
  [IrH$_2$(H$_2$)(P^{t}Bu$_2$Ph)$_2$]$^+$, 83
  IrH$_5$(PH$_3$)$_2$, 108–109
  J_{HD}, 154–155
  [M(L)$_n$]$^+$ and [FeH$_2$]$^{+, 0, -}$, 128–131
  M(CO)$_3$(PH$_3$)$_2$(H$_2$), 69–75, 78, 89–91, 101–102
  M(CO)$_4$ + H$_2$ versus CH$_4$ (M = Ru, Os), 388–389
  M(CO)$_5$L and M(CO)$_4$(H$_2$)$_L$ (M = group 6), 77–78, 83
  M(CO)$_n$(PH$_3$)$_{5-n}$(H$_2$), 72–75
  M$_2$ or M$_3$ clusters + H$_2$ (M = Pd, Pt), 68
  metalloenzymes, 304–305, 309, 311, 314–316, 318
  molecular mechanics calculations, 103–105, 165, 185
  Mo(CO)(H$_2$)(PH$_3$)$_4$, 72–73, 78
  nuclear motion quantum calculations: see H$_2$ ligand, elongated
  OA of CH$_4$, 389–395
    CH$_4$ versus SiH$_4$, 389–391
    effect of metal and ancillary ligand, 394
    potential energy profiles for CpM(CO) + CH$_4$, 390
    to IrX(PH$_3$)$_2$, 394
  for OA of H$_2$, 63–66
    CpRh(CO), 231–232
    Ir system, 64

Theoretical calculations for H_2 and hydride
systems (cont.)
  for OA of H_2 (cont.)
    $M(PH_3)_4$ (M = Fe, Ru, Rh^+), 66
    Pt, 67–68
    $Ru(CO)_4$, 66
    $W(PR_3)_5$, 72
    Wilkinson's catalyst, 63–65
  $OsCl_2(PH_3)_4 + H_2$ and CH_4, 393–394
  $OsH_4(PH_3)_3$, 89–95
  $[Os(HCO_2)(H_2NCH_2CH_2NH_2)_2(H_2)]^+$, 89–95
  $Pd(H_2)$, 67–68
  quantum-mechanical exchange coupling, 199–202
  $ReBr(H_2)(NO)(PR_3)_2$, 97–98
  $ReH_6^{2-}\$MI$ and $[ReH_8(PR_3)]^-$, 94–95
  relativistic effects in, 79
  replacement of PR_3 by PH_3 or use of PR_3, 62–63, 108–109, 118, 164, 184, 191, 280–281
  $[Rh(CO)_4]^+ + H_2$ versus CH_4, 65, 388–389
  $RuH_2(H_2)_2(PH_3)_2$, three isomers of, 119, 195–197
  σ-bond metathesis, 123–128
  $TpRhH_2(H_2)$ versus $CpRhH_2(H_2)$, 114, 182
  water gas shift reaction mechanism, 260
Theoretical calculations for OA of B−X bonds, 429
Theoretical calculations for OA of C−X bonds, 432
Theoretical calculations for silane complexes
  for bridging SiH_4 ligand, 341–342
  for chelating bis(silane), 337, 350–351
  containing H_2 ligand or either silane or H_2 ligand, 347–348
  for $Cp_2MH(SiR_3)$ (M = Nb, Ta) showing interligand hypervalent interactions, 351–352
  for $Cp_2MXH(SiCl_nH_{3-n})$ (M = Nb, Ta), 343–344
  interplay between experiment and theory, 351
  on a $Mo(CO)(PH_3)_4(H\cdots SiH_3)$ model, 347, 349–350
    plot of Laplacian of valence electron density, 350
  showing similarity of bonding to that for H_2 complexes, 329, 342
  SiH_4 versus CH_4, 389–391
Thermodynamics of the bonding and cleavage of
  σ ligands, 207–240; see also Bond energy
  addition of H_2, SiH_4, CH_4 and other small molecules to CpRh(CO) in gas phase, 231–232
  C−H bonds, 229, 231–233
    versus H−H bonds, 232–233
    calculations of, 389–395
  displacement of H_2 ligand by H_2O, 215–216
  entropic effects, importance of: see Entropy

Thermodynamics of the bonding and cleavage of
  σ ligands (cont.)
  H_2 versus D_2 coordination: see Equilibrium isotope effect
  $M(CO)_3(PR_3)_2(L)$ (L = H_2, N_2; M = group 6), 208, 221–222, 234–235
    dependence on absolute entropies of H_2 and N_2, 221–222
    profile for reaction of H_2 with $W(CO)_3(PR_3)_2$, 228–229
  $MHCl(H_2)(CO)(P^iPr_3)_2$ (M = Ru, Os), 209
  reaction profile for $[Cp^*Cr(CO)_3]_2 + H_2$, 230–231
  Si−H bonds, 228–233
Three-center, four-electron (3c-4e) bonding in $[R_3N-H^+\cdots Co(CO)_4^-]$ and other M···H−N interactions, 369, 430
M(X−C) interactions (X = halogen), 431–432
Three-center, two-electron (3c-2e) bonding in, 60, 437
  agostic M···H−C bonds, 368
  bent M−H−M bonds, 329
  metal-borane complexes, 426
  metal-borohydride complexes, 417
    comparison to that in diborane, 417
  metal-H_2 bonding; see Dihydrogen ligand, bonding of
  metal-silane complexes, 327, 331, 342, 346, 349–351
  sulfide-H_2 interaction, 320
Time-resolved IR (TRIR), for study of alkane activation, 395–403
Toluene, binding to M as possible transient species, 44, 388
Trans effect, 81–84; see also Dihydrogen ligand, bonding of, effect of trans ligand
  CO ligand: see Carbon monoxide ligand, as moderator of σ-bond activation
  halide ligand, 114–115, 167, 184, 209, 220
  hydride ligand, 33, 81–82, 103, 147, 220, 383, 408
Triazacyclononane ligand, 35, 275, 397
Triflic acid, elimination from H_2 complex, 283
Trihydrogen ligand, as transient in H/H_2 and isotopic exchange, 12–13, 15, 121–123, 150–151, 188–189, 192–193, 267, 356
Tris(pyrazolylborate) complexes (Tp)
  agostic interaction of B−H bond in, 419
  bonding of alkanes to, 398, 401–401
  bonding of H_2 to and comparison to Cp complexes, 35, 113–114 182
  bonding of silanes to and comparison to Cp complexes, 347–348, 356–357
Tritium NMR, 155–156
Twenty-electron (20e) complexes, as intermediates, 267

Vaska complex, reaction with H_2, 19, 34, 264
Vibrational modes and analysis for H_2 complexes
see also Electron energy loss spectroscopy; Inelastic neutron scattering
  anharmonicity of M−H_2 potential function, 255−257, 236−237
  clue to discovery of and evidence for H_2 complexes, 24−26, 245−248
  coupling of HH and MH modes, 96−97, 180, 247, 252−257
  in elongated complexes, 102, 254−257
    redefinition of normal modes in, 102, 256−257
  for HD and D_2 ligand, 245−249, 252, 254
  fundamental modes for, 245−252
  H_2 torsion, 245−252; *see also* Rotation of H_2 ligand
  HH stretch, 245−257
    broadness of, 246−247
    effect of M or L on, 246−248, 250, 253−255
    force constant for, 251−254
    in free H_2, 70, 246
    importance of backbonding in influencing the frequency of, 180
    lack of correlation with d_{HH}, 247−248, 254
    in $M(CO)_5(H_2)$ (M = group 6), 246−247, 251, 253
    use of perdeuterated phosphines to locate, 26, 246, 248
    in $W(CO)_3(PR_3)_2(H_2)$, 26, 70, 245−248, 252−255
    indicator of bonding character, 253−255
    influence on equilibrium isotope effect for H_2 versus D_2 bonding, 235−238
    in low-temperature stable complexes, 246−247, 251
  MH stretch, 24, 245−257
    effect of M or L on, 255
    force constant for, 252−254
  MH_2 deformations, 245−252
  normal coordinate analysis of $W(CO)_3(PCy_3)_2(H_2)$, 249−254
  Raman spectroscopic observation of, 245, 247, 249, 250, 255, 257
  tables of frequencies, 179, 250, 252
  in $W(CO)_3(PR_3)_2(H_2)$, 245−252
  in $W(CO)_3(PR_3)_2(HD)$, 245−247, 249, 252
Vibrational modes, BH stretch in borane complexes, 423, 425
Vibrational modes, CH stretch
  in agostic CH, 368, 370, 373
    in IR spectrum of $W(CO)_3(P^iPr_3)_2$, 374
    on surface-bound, 369
Vibrational modes, CO stretch
  coupling with ν_{MC} in hexacarbonyl complexes, 253

Vibrational modes, CO stretch (*cont.*)
  gauge of a metal's backbonding ability, 11, 76, 86
  in hydrogenases, 302−304
  time-resolved infrared spectroscopic measurement of in transient species, 211, 223, 395−403
  in $W(CO)_5(BH_3 \cdot PH_3)$ compared to those in $W(CO)_5(H_2)$, 423
Vibrational modes for germane complexes, GeH stretch, 358, 360
Vibrational modes, NN stretch
  gauge of a metal's backbonding ability, 11, 84, 180, 182−183, 344−346
  measure of NN bond activation, 180
  predictor of H_2 versus dihydride coordination, 84, 105−107
Vibrational modes for silane complexes, 257, 331, 334
  in bridging silane, 334
  deformational modes, 334
  SiH stretch, 257, 331, 334
    in agostic interactions, of Si−H bonds, 338
    mixing with M−H modes, 331
Vibrational modes, SO stretch
  as gauge of a metal's backbonding ability, 11, 344−346

Water complex or ligand: *see* Aquo ligand
Water gas shift reaction, 6, 260
  mechanism of catalysis by $W(CO)_6$ and $Fe(CO)_5$, and H_2 complex as intermediate in, 260
Water, photoreduction of, 6
Werner-type complexes, 2, 8
Wheland complex, 375
Wilkinson's catalyst, 19, 63, 261; *see also* $RhCl(PPh_3)_3$

Xenon ligand, 4, 212−213, 384, 399−403, 437
  stable binding in $[AuXe_4][Sb_2F_{11}]_2$, 4, 437
Xenon, liquid, as solvent for preparing and characterizing, 4
  alkane complexes, 395, 399−403
  H_2 complexes, 48−49, 76, 111−112, 246, 251, 263, 267
  silane complexes, 338
X-ray diffraction, studies of
  adduct of CH_3BeCp^* with Cp^*_2Yb as model for methane coordination, 384
  agostic C−B bonds, 435
  agostic C−C bonds, 432−433
  agostic C−H bonds, 370−372
    bond distances for 16e phosphine complexes, table of, 380

X-ray diffraction, studies of (*cont.*)
  agostic C−H bonds (*cont.*)
    [IrH(η²-C₆H₄PBuᵗ₂)(PBuᵗ₂Ph)]⁺, 382
    trajectory for the reaction M + C−H →
      C−M−H derived from a series of
      structures of agostic complexes, 371
    W(CO)₃(PCy₃)₂, 378–379
  agostic C−N bonds, 435
  agostic C−Si bonds, 433–434
  agostic N−H bonds, 430
  borohydride and borane complexes, 417–425
    Cp₂Ti(HBcat)₂, 423–425
    inaccuracy of determination of B−H
      distances, 420
    Ti(BH₄)₃(PMe₃)₂, versus theoretical
      predictions for, 420–421
    W(CO)₅(BH₃·PH₃), 423
    Zr(BH₄)₄, 418
  disorder problem in H₂ complexes, 25–26
  germane complexes, 358–359
  H₂ complexes, 84–85, 145–149, 164

X-ray diffraction, studies of (*cont.*)
  H₂ complexes (*cont.*)
    M(CO)₃(PiPr₃)₂(H₂) (M = Cr, W), 25–27,
      145, 148
    heptane interaction with iron in porphyrin
      complex, 387–388
    location of hydride positions and determination
      of d_{HH}, difficulty of, 144–145
    metalloenzymes, 301–303, 313–314, 318
    silane complexes, 332–335, 343
      Mo(CO)(PP)₂(silane) complexes, 335
    stannane complexes, 361–362

Zeolite matrices for Fe complexes for hydrocarbon
  oxidation, 319
Zeolite-catalyzed hydrogenation, 291
Zeolite-H₂ interactions, 38; *see also* CoNa−A
  zeolite interaction with H₂
Zero-point energy: *see* Equilibrium isotope effect
Ziegler-Natta polymerization, 113, 377
Zinc alkyl hydride complexes, 20–21

GPSR Compliance

The European Union's (EU) General Product Safety Regulation (GPSR) is a set of rules that requires consumer products to be safe and our obligations to ensure this.

If you have any concerns about our products, you can contact us on

ProductSafety@springernature.com

In case Publisher is established outside the EU, the EU authorized representative is:

Springer Nature Customer Service Center GmbH
Europaplatz 3
69115 Heidelberg, Germany